I. Barin O. Knacke O. Kubaschewski

Thermochemical properties of inorganic substances

Supplement

1977
Springer-Verlag Berlin Heidelberg GmbH

IHSAN BARIN
Privatdozent Dr.-Ing.

OTTMAR KNACKE
o. Professor Dr. rer. nat.

OSWALD KUBASCHEWSKI
Professor Dr. habil., Dres. h. c.

Rhein.-Westf. Technische Hochschule Aachen
Lehrstuhl für Metallurgie der Kernbrennstoffe und
Theoretische Hüttenkunde

ISBN 978-3-662-02295-5 ISBN 978-3-662-02293-1 (eBook)
DOI 10.1007/978-3-662-02293-1

Library of Congress Cataloging in Publication Data Barin, Ihsan. Thermochemical properties of inorganic substances. Preface in English, French, and German. Supplement to the work by I. Barin and O. Knacke. Bibliography: p. Includes index. 1. Thermochemistry--Tables, etc. 2. Chemistry, Inorganic--Tables. I. Knacke, Ottmar, 1920– joint author. II. Kubaschewski, Oswald, joint author. III. Title. QD511.8.B372
546 76–57688

Originally published by Springer-Verlag Berlin Heidelberg New York in 1977
Softcover reprint of the hardcover 1st edition 1977

Preface

For the practical application of thermochemistry to the development and control of technical processes, the data for as many substances as possible are needed in conjunction with rapid and simple methods of calculating equilibrium constants, heat balances and the EMF of galvanic cells. For these three types of calculation the following three thermodynamic functions are suitable: The Planck function, the enthalpy and the Gibbs free energy, which are here defined and tabulated as unambigous functions of temperature for pure substances.

The first edition of the tables was published in 1973 under the title "Thermochemical Properties of Inorganic Substances". The present supplementary volume contains the data and functions for a further 800 inorganic substances. In addition, the data for about 250 substances from the first volume have been up-dated. These usually small corrections produce better consistency with the data from more recent publications.

The comments of users and reviewers of the first volume have largely been concerned with the difference between the present thermodynamic functions and the system used in the JANAF tables, the somewhat unconventional handling of heat balances adopted here, the notation of cell reactions, the description of non-stoichiometric phases and the accuracy of the tabulated data. To answer these questions and criticims the theoretical concepts and the practical use of the tables are dealt with in more detail in the introduction, following the recommendation of some reviewers.

Aachen, autumn 1976

I. Barin O. Knacke O. Kubaschewski

Préface

L'application pratique de la thermochimie au développement et au contrôle de procédés techniques suppose la disponibilité de données pour un ensemble de substances aussi étendu que possible et de méthodes simples et rapides pour le calcul de constantes d'équilibre, de bilans thermiques et de forces électromotrices. A ces buts correspondent trois fonctions thermodynamiques: la fonction de Planck, l'enthalpie et l'enthalpie libre (énergie libre de Gibbs). Ces fonctions thermodynamiques, dont les valeurs numériques sont consignées dans les tables, sont définies ici comme des fonctions univoques de la température pour des corps purs.

La première édition des tables a paru en 1973 sous le titre "Thermochemical Properties of Inorganic Substances". Le présent supplément présente des données pour 800 substances inorganiques additionnelles. Il comporte également des mises à jour pour environ 250 substances traitées dans le premier volume. Les révisions sont en général mineures et servent surtout à réaliser une meilleure cohérence des données.

Les remarques des utilisateurs et des commentateurs de la première édition ont porté surtout sur la différence entre les fonctions thermodynamiques retenues ici et celles utilisées dans les Tables JANAF, sur le calcul non conventionnel des bilans thermiques, sur la formulation des réactions galvaniques, sur la description des composés non-stoechiométriques et sur la précision des données publiées. En réponse aux questions, critiques, et suggestions, les concepts théoriques et l'utilisation pratique des tables sont décrits avec plus de détails dans l'introduction.

Aix-la-Chapelle, automne 1976

I. Barin O. Knacke O. Kubaschewski

Vorwort

Für die praktische Anwendung der Thermochemie bei der Entwicklung und Kontrolle technischer Prozesse benötigt man eine möglichst umfassende Stoffübersicht und eine einfache und schnelle Berechnung von Gleichgewichtskonstanten, Wärmebilanzen und der EMK galvanischer Zellen. Diesen drei Aufgaben sind folgende drei thermodynamische Funktionen angemessen: die Plancksche Funktion, die Enthalpie und die freie Enthalpie, die hier als eindeutige Temperaturfunktionen für reine Stoffe definiert und tabelliert werden.

Die erste Ausgabe dieser Tabellen erschien 1973 unter dem Titel "Thermochemical Properties of Inorganic Substances". Der vorliegende Ergänzungsband enthält die Daten und Funktionswerte von weiteren rd. 800 anorganischen Stoffen. Außerdem sind die Daten von rd. 250 Stoffen aus dem ersten Band auf den neuesten Stand gebracht worden. Diese im allgemeinen geringfügigen Korrekturen dienen hauptsächlich der Konsistenz der Daten aufgrund neuerer Veröffentlichungen.

Die Kommentare von Benutzern und Rezensenten des ersten Bandes betrafen vor allem den Unterschied zwischen den hier gewählten thermodynamischen Funktionen und dem System der JANAF-Tables, die hier empfohlene unkonventionelle Art der Wärmebilanz, die Schreibweise von Zellreaktionen, Beschreibung nichtstöchiometrischer Phasen und die Genauigkeit der mitgeteilten Daten. Zur Beantwortung der vorgebrachten Fragen und Kritik werden, der Empfehlung einiger Rezensenten folgend, in der Einleitung zu diesem Band das theoretische Konzept und die Anwendung der Tabellen etwas ausführlicher erläutert.

Aachen, Herbst 1976

I. Barin   O. Knacke   O. Kubaschewski

Contents

Introduction

1. Equilibrium constants

The thermodynamic equilibrium of a macroscopically homogeneous or heterogeneous system is phenomenologically defined by the time-independent compositions of the various phases and, respectively, the time-independent concentrations of all components. In addition to stable equilibria, metastable states of equilibrium occur, the transitions of which into the stable states are subject to kinetic checks. There are systems which exhibit, at a given temperature, fast or slow reaction rates, and equilibrium may be attained either immediately or after months or even years. In dynamic equilibrium, which may be stable or metastable, reversible chemical fluctuations, that is virtual or real differential chemical reactions in both directions occur - described by chemical equations between the components, B_i, of the reaction with the mole numbers ν_i,

$$\nu_1 B_1 + \nu_2 B_2 + \ldots = \nu_j B_j + \ldots \tag{1}$$

B_i may represent the components in a mixed phase like Zn in brass or like Zn, or O, or ZnO in the near-stoichiometric zinc oxide of the chemical composition $Zn_{1+y}O$ with $y \approx 0$. The statistical fluctuations in composition at equilibrium do not affect the macroscopic concentrations of the reacting components in their phases. Reversible chemical fluctuations are compatible with the fundamental principles of equilibrium, namely maximum entropy, minimum Gibbs energy, etc.

The equilibrium concentrations of homogeneous and heterogeneous systems are interdependent and functions of temperature and pressure. For most technically important systems, pressure may be considered as isotropic. Many problems concerning equilibrium are properly treated by means of the equilibrium constant which derives from the Law of Mass Action for ideal mixed phases. In the limiting case, that is where it pertains to perfect gases, ideal or ideal diluted non-electrolytic solutions, the mass action law assumes the following simple forms:

$$K(T) = \prod \left(p_i/p_i^o\right)^{\nu_i} , \tag{2a}$$

$$K(T) = \prod x_i^{\nu_i} . \tag{2b}$$

In these formulations, the equilibrium constant, $K(T)$, is a dimensionless quantity that depends on temperature and the adopted standard state of the components, e.g. the pure substances under a pressure $p_i^o = 1$ atm. In equation (2a), p_i represents the partial pressures defined by the product $p_i = x_i P$, where x_i are mole fractions in the gas, P is total pressure, and ν_i the molar numbers of equation (1). In equation (2b) x_i represents the mole fractions in the condensed solutions.

The right hand sides of equations (2a and b) have been generalised for real systems by introducing the dimensionless quantities fugacity and activity while retaining the mathematical form of the mass action law. In fact fugacity could be considered as a particular case of activity. Then the mass action law assumes the general form:

$$K(T) = \prod a_i^{\nu_i}, \tag{3}$$

$$\log K(T) = \sum \nu_i \beta_i(T) . \tag{4}$$

The activity of pure substances under 1 atm pressure equals unity by definition. For technical applications, one may accept the validity of the ideal gas law in the first approximation and hence for gaseous components put $a_i = p_i/p_i^o$ where $p_i^o = 1$ atm. The activity of the components in condensed phases is a function that depends little on pressure and must be determined by experiment. The compilation of activities in solid and liquid solutions is no concern of the present tables. For approximate calculations, if experimental data are lacking, one may put $a_i \approx x_i$ or use estimation methods described elsewhere.

Equations (3) and (4) are only valid for equilibrium reactions of type (1), i.e. for given molar numbers. According to equation (4), the equilibrium constant, K(T), is obtained from the dimensionless β-functions, values of which are tabulated here for the pure substances under 1 atm pressure. The relationship between the β-function (or the Planck function which is proportional to it) and the enthalpy and entropy of pure substances is discussed in Section 4. β-functions for substances in the reference state of infinitely dilute solution are not given in these tables. Non-stoichiometric compounds are discussed in Section 6.

2. Heat balances

The heat balance of a technical process may be obtained either as a physical balance, by measuring the heat exchange with the environment due to heat conduction and radiation, or as a chemical heat balance related to the material balance and the temperatures of the incoming and outgoing materials. In technical processes material flow and temperatures are continuously recorded. Since material flow is inseparably connected with transport of internal energy, the heat requirement or loss may be calculated from the material balance and the measured temperatures by means of the enthalpies tabulated here. According to the First Law of Thermodynamics, the correlation between the material balance or the corresponding irreversible overall reaction and heat balance for an isobaric process at 1 atm pressure is given by the following two equations:

$$\nu_1 B_1(T_1) + \nu_2 B_2(T_2) + \ldots \rightarrow \nu_j B_j(T_j) + \ldots , \tag{5}$$

$$Q = \sum \nu_i H_i^o(T_i) + \sum \Delta H_{mix} . \tag{6}$$

In equation (5), the chemical symbols B_i denote phases. This is in contrast to the symbols B_i in equation (1), which denote differential chemical fluctuations of the different atoms, atom combinations or species of an equilibrium system in which the macroscopic concentrations of the phases involved are conserved. The chemical equations (1) and (5) are therefore quite different in nature. The different thermodynamic significance of the chemical formulae and reaction equations reflects the historical development of inorganic chemistry and chemical thermodynamics. The overall reaction (5) must be stoichiometric and in addition the temperatures T_i of the incoming and outgoing substances must be noted and considered in equation (6), where $H_i^o(T_i)$ is the tabulated enthalpy of the pure substances and ΔH_{mix} the heat of mixing, which is approximately independent of temperature. $Q > 0$ indicates heat consumption and $Q < 0$ heat evolution by the technical process. In the first approximation, the heats of mixing, ΔH_{mix}, may often be neglected. Equations (5) and (6) are valid for a complete process, single stages or differential sections. A detailed investigation and discussion of the different contributions to the balance, e.g. the influence of the preheat of the individual reactants or of an incomplete reaction, is best based on equations (5) and (6).

The deviations from the equation PV = const for real gases normally lie within the error limits of a chemical or metallurgical heat balance so that equations (5) and (6) also suffice for non-isobaric processes. The contributions of kinetic and potential mechanical energy to the heat balance of such processes may, in general, also be disregarded and only be considered for detailed analysis or special processes.

For electrochemical processes, the evaluations of the heat balance according to equations (5) and (6) are to be supplemented as follows:

$$\nu_1 B_1(T_1) + \ldots \nu_A e_A \rightarrow \nu_j B_j(T_j) + \ldots \nu_K e_K , \quad (7)$$

where $|\nu_A| = \nu_K$,

$$Q = \sum \nu_i H_i^o(T_i) + \sum \Delta H_{mix} + \nu_K F(\varphi_K - \varphi_A) . \quad (8)$$

Here, e_A and e_K are the electrons that enter and leave the anode and cathode, respectively, and φ_A and φ_K the inner electrical potentials of the anode and cathode, $\varphi_A - \varphi_K > 0$ is the terminal voltage, which is different from the equilibrium voltage. F is the Faraday = 23066 cal $\cdot$ Volt^{-1} mol^{-1}. We recommend the retention and differentiation of the electrons, $\nu_A e_A$ and $\nu_K e_K$ in equation (7), and the differentiation of the potentials, φ_A and φ_K in equation (8), both for fundamental considerations (see Section 3) and for the sake of expediency, since this achieves a correct correlation of electrical energy consumption and material conversion - correct in sign and amount.

The conventional calculation of a heat balance with the arbitrary division of the energy flow into "heat contents" and heats of reaction for a fictitious constant temperature in the reactor (e.g. room temperature) is cumbersome and does not reflect the scientific concept of the internal energy or the enthalpy of matter.

3. EMF of galvanic cells

The application of galvanic cells, in particular cells using solid electrolytes, to the development and control of metallurgical and chemical processes, is rapidly expanding. Experience shows that difficulties may arise in the calculation of the sign and magnitude of the cell potential difference from thermochemical data, in particular with cells the potential difference of which may change sign, such as during the course of measurement of the oxygen content of liquid steel employing a Cr/Cr_2O_3 reference electrode, or with the conversion of potential differences measured with different reference electrodes. The cause of these difficulties may be sought in the conventional but unsatisfactory representation of the cell reactions in which the participating electrons are finally cancelled. The expediency of the usual representation of electrochemical equilibrium reactions with open galvanic cells may be doubted. Retaining and differentiating the thermodynamically non-identical electrons e_α and e_β of the leads α and β, the following general form of a cell reaction results and from this the thermodynamic equation for the calculation of the cell potential difference from the chemical potentials, μ, and the chemical activities, a, respectively:[+)]

$$\nu_1 B_1 + \nu_2 B_2 + \ldots \nu_\alpha e_\alpha = \nu_j B_j + \ldots \nu_\beta e_\beta \,, \qquad (9)$$

$$|\nu_\alpha| = \nu_\beta \,,$$

$$\nu_\beta F(\varphi_\beta - \varphi_\alpha) = \sum \nu_i \mu_i = \Delta\mu^o + RT \ln \prod a_i^{\nu_i} . \qquad (10)$$

Equations (9) and (7) are fundamentally different in character as shown by the explanations of equations (5) and (1). The denoting of anode and cathode should be reserved for irreversible processes. In an open galvanic cell the sign of the voltage can change with the concentrations in the system. By means of equations (9) and (10) the sign and amount of the voltage $(\varphi_\beta - \varphi_\alpha)$ will be correctly obtained; $F = 23066$ cal · Volt^{-1} mol^{-1}. The electrons e_α and e_β usually have a different inner electrical potential and hence a different electrochemical potential and must therefore not be cancelled out in the equation for the electrochemical reaction of the cell. Since the leads consist of the same material, the chemical potentials of the lead electrons e_α and e_β are the same and cancel in equation (10). Therefore the summation of chemical potentials in equation (10) comprises electrically neutral or ionised components of reaction only. Cell reactions of the type expressed by equation (9) may be linearly combined like other chemical reactions thereby providing a correct conversion of potential differences measured using different reference electrodes.

As an example, the calculation of the cell potential difference of a repeatedly investigated formation cell is given:

[+)] Knacke, O., Krahe, J., Weber, R.: Z. Physik. Chem. Neue Folge 67 (1969) 194

$$\alpha / Ag_{sol} / AgCl_{liq} / C(Cl_2) / \beta$$

$$Ag^+ + e_\alpha = Ag \qquad \text{Electrode reaction } \alpha ,$$

$$Cl^- = \frac{1}{2} Cl_2 + e_\beta \qquad \text{Electrode reaction } \beta$$

$$AgCl_{liq} + e_\alpha = Ag_{sol} + \frac{1}{2} Cl_2 + e_\beta \qquad \text{Cell reaction ,}$$

$$F(\varphi_\beta - \varphi_\alpha) = \mu^o_{Ag} + \frac{1}{2}\mu^o_{Cl_2} - \mu^o_{AgCl} + \frac{RT}{2} \ln p_{Cl_2}$$

$$= \Delta G^o_r + \frac{RT}{2} \ln p_{Cl_2} .$$

Here the relations $\mu^o_i = G^o_i$ for elements and $\mu^o_i \approx \overset{o}{G}_i$ for near-stoichiometric compounds respectively are used, and therefore $\sum \nu_i \mu^o_i = \Delta G^o_r$ according to conventional notation in several fields of applied thermochemistry. With the tabulated values of the Gibbs free energies of the pure substances, one obtains at 1000 K ($R = 1.9869$ cal · mol^{-1}K^{-1})

$$\Delta G^o_r = 19.347 \text{ kcal} \cdot \text{mol}^{-1},$$

$$\varphi_\beta - \varphi_\alpha = 0.839 + 0.099 \lg p_{Cl_2} \text{ Volt .}$$

4. Tabulated functions

The following functions of the pure substances at 1 atm pressure are tabulated:

molar heat, CP, in cal · mol^{-1}K^{-1},

enthalpy, H, in kcal · mol^{-1},

entropy, S, in cal · mol^{-1}K^{-1},

Gibbs free energy, G, in kcal · mol^{-1},

β function, BT, as a dimensionless quantity,

vapour pressure, P, in Torr (mm Hg).

The symbols, as given here, are the same as printed by the computer.

The integration constant of the entropy of pure crystalline substances is conventionally defined according to the Third Law of Thermodynamics, zero at Absolute Zero. This assumes that attainment of equilibrium is unhindered and that the definition of the zero point entropy according to statistical mechanics is adopted. Thermochemical isotope effects, which are only noticeable with light elements and compounds, are not taken into account. The natural isotopic mixture of the elements is assumed.

For the definition of the integration constant of the enthalpy, there is no scientific argument corresponding to the one for the entropy. In accordance with common convention, the enthalpies of the pure elements in their stable state at 25°C and 1 atm pressure are therefore taken to be zero. The sole exception is phosphorus, for which the reference state chosen is not the thermodynamically most stable state but white phosphorus, which can be prepared in a reproducible manner. It is incidentally permissible to choose different zero points for entropy and enthalpy because only the differences ΔS and ΔH enter the calculation of equilibria separately. By fixing the zero points of entropy and enthalpy, the Gibbs free energies and Planck functions of pure substances also become unambiguous functions of temperature.

From the above definitions, the following unambiguously defined temperature functions result.+)

The molar enthalpy, $H_i^o(T)$, of a pure substance i at 1 atmosphere pressure and temperature T is defined by

$$H_i^o(T) = \Delta H_i^f + \int_{298}^{T} dH_i^o \,, \tag{11}$$

where ΔH_i^f is the enthalpy of formation of the substance from the stable states of the elements at 298.15 K. The integration limit of 298 denotes for simplicity 298.15 K. The enthalpy of formation of the elements in their stable states is zero by definition.

The temperature dependence of the enthalpy is expressed conventionally by the molar heat, $c_i(T)$, at constant pressure,

$$c_i(T) = dH_i^o/dT \,. \tag{12}$$

In the case of isobaric transformations at temperature T_t, the enthalpy of transformation ΔH_i^t replaces the molar heat,which becomes infinite. Equation (11) therefore becomes

$$H_i^o(T) = \Delta H_i^f + \int_{298}^{T} c_i(T)\,dT + \sum \Delta H_i^t \,, \tag{13}$$

where the summation includes all transformations between 298.15 and T[K].

The relationship between entropy and enthalpy of pure substances at 1 atm pressure is given by

$$dS_i^o = dH_i^o/T \,, \qquad \Delta S_i^t = \Delta H_i^t/T_t \,, \tag{14}$$

$$S_i^o(T) = S_i^{298} + \int_{298}^{T} dH_i^o/T \,. \tag{15}$$

By using molar heats these equations become

+) Barin, I., Knacke, O.: Met. Trans. 5 (1974) 1769,
Hirschwald, W., Knacke, O., and Reinitzer, P.: Erzmetall 10 (1957) 127,
Gerlach, J., Kleist, H.G., and Knacke, O.: Erzmetall 8 (1955) 377

$$dS_i^o = (c_i(T)/T)\,dT\ , \qquad \Delta S_i^t = \Delta H_i^t/T_t\ , \tag{16}$$

$$S_i^o(T) = S_i^{298} + \int_{298}^{T} (c_i(T)/T)\,dT + \sum \Delta H_i^t/T_t\ . \tag{17}$$

The molar Gibbs free energy $G_i^o(T)$ is related by definition to enthalpy and entropy, thus

$$G_i^o(T) = H_i^o(T) - T \cdot S_i^o(T)\ . \tag{18}$$

The Gibbs free energy changes continuously at the transformation points whereas the enthalpy and entropy changes are discontinuous.

The equilibrium constant $K(T)$ and the standard Gibbs free energy, ΔG_r^o, of any reaction r are related by definition,

$$\log K(T) = -\frac{\sum \nu_i G_i^o}{RT \ln 10} = -\frac{\Delta G_r^o(T)}{RT \ln 10}\ . \tag{19}$$

The mathematical form of equation (19) suggests using the Planck function of a pure substance, $-G_i^o/T$, and a factor $R \ln 10 = 4.575\ \mathrm{cal \cdot mol^{-1} K^{-1}}$ for the most simple calculation of equilibrium constants. From this follows the definition of the β-function and equation (21):

$$\beta_i(T) = -G_i^o(T)/RT \ln 10\ , \tag{20}$$

$$\log K(T) = \sum \nu_i \beta_i(T)\ . \tag{21}$$

Whereas enthalpy and Gibbs free energy are approximately linear functions of the temperature T and therefore permit linear interpolations, the β-function is in general approximately a linear function of $1/T$ and calls for appropriate interpolation.

The relationship between the present β_i functions and the values, $\log K_{p,i}$, of the JANAF tables for a compound, i, that consists of several elements, ε, is as follows:

$$\nu_1 \varepsilon_1 + \nu_2 \varepsilon_2 + \ldots \longrightarrow B_i\ , \tag{22a}$$

$$\log K_{p,i} = \beta_i - \sum |\nu_\varepsilon| \beta_\varepsilon\ .$$

The main difference between the present tables and the JANAF tables lies in the choice by the latter of an infinite number of reference states for the enthalpy of pure substances, and consequently also for the Gibbs free energy and the function $K_{p,i}$; these reference states are the elements at each temperature, whereas it follows from theory that the enthalpy of a pure substance is numerically determined by a single integration constant. The present system has in a way, with regard to enthalpy, its parallel in geography, namely the measurement of altitude above sea level.

The isothermal-isobaric enthalpy differences, $\Delta H_i(T)$, of a compound given in the JANAF Tables are in the same relation to the enthalpy values $H_i(T)$ of the compounds and enthalpies $H_\varepsilon(T)$ of the elements tabulated here as is log $K_{p,i}$ to β in equation (22), i.e.

$$\Delta H_i(T) = H_i(T) - \sum |\nu_\varepsilon| H_\varepsilon(T) \ . \tag{22b}$$

Which of these functions one uses is not a question of correctness but of expediency. The numerical evaluation of heat balances of real, non-isothermal processes is easier to carry out with the aid of the enthalpy values $H_i(T)$ given here than by using the common method involving enthalpy difference values.

It is seen above that the 15°C calorie has been chosen as the calorific unit in preference to the international joule (1 cal = 4.184 J). The primary reason for this choice is obviously that it was the unit used in the first volume. It is also the unit employed in all critical compilations mentioned below and, above all, it is the unit that the majority of users in technological practice are still accustomed to. As students are now frequently taught chemical thermodynamics in terms of SI units, a change-over will gradually take place, but it is not generally appreciated by the academic purists how slow this process is.

5. Vapour pressures

In the last column of a number of tables, the total vapour pressures P (in Torr) are recorded in the range from 0.001 to 760 Torr; in the case of polymerizing vapours P is the sum of the partial pressures of the simple and polymer species. For substances with a single, or only one predominant species in the vapour, the thermochemical functions for the vapour are listed from the normal boiling point upward. For vapours forming polymers, the tabulation of the properties of the condensed stable phase ends at the natural boiling point where the total pressure of all species in the vapour reaches 760 Torr. The composition of the vapour at this pressure is recorded at the top of the table. The heat of vaporization at the natural boiling point is given in the last line of the table. With polymerizing vapours, the thermochemical data of the various species from room temperature upward are printed separately. Although there is consistency in the tables between a vapour pressure of 760 Torr and the boiling point, the experimental vapour pressure and boiling point values from which the tabulated data are derived often show considerable scatter.

6. Non-stoichiometric phases

For the numerous phases exhibiting near-stoichiometry, such as $Zn_{1+y}O$ with $y \approx 0$, the standard chemical potential of the simple stoichiometric combination ZnO in the reference state of the pure compound equals numerically the molar Gibbs free energy of the phase,

$$\mu^o_{ZnO} = \partial G/\partial n_{Zn} + \partial G/\partial n_O \approx G_{Zn_{1+y}O} . \tag{23}$$

This simple relationship greatly facilitates the thermodynamic description of "pure substances". However, it must be borne in mind that the essential thermodynamic quantity for the calculation of equilibrium constants is the chemical potential in the reference state or the corresponding β function, in any event a partial quantity.

There exist quite a number of carbides, nitrides, sulphides, oxides, etc., in particular of transition metals, that exhibit wide homogeneity ranges within which the stoichiometric composition is not stable. Then the simple relationship (23) fails, and difficulties result. A typical and technically important example is wüstite, the metal deficient oxide $Fe_{1-y}O$ with $0.046 < y < 0.168$. Although the stoichiometric combination FeO is not stable, numerous double oxides occur involving stoichiometric FeO. In this case it appears advantageous to tabulate the measured chemical potential $\mu_{FeO} = \mu_{Fe} + \mu_O$ for the experimentally accessible phase of composition $Fe_{0.95}O$ at temperatures above 572°C. Hence the composition $Fe_{0.95}O$ (i.e. $y^o = 0.05$) of wüstite is chosen as reference state,

$$\mu^o_{FeO} = \left(\partial G_{Fe_{1-y}O} / \partial n_{FeO} \right)_{y^o} ; \quad y^o = 0.05. \tag{24}$$

The standard chemical potential, μ^o, of a non-stoichiometric phase is interrelated with a partial enthalpy, $\overline{H}^o$, and a partial entropy, $\overline{S}^o$, which differ from the integral enthalpy, H, and integral entropy, S, of that phase.

The chemical potential, μ, and the activity, a, are functions of temperature and composition. When y is chosen as the concentration variable, the activity is defined by

$$RT \ln a(T,y) = \mu(T,y) - \mu^o(T,y^o). \tag{25}$$

In equilibrium calculations non-stoichiometric phases should be treated like solutions. For the three-phase equilibrium wüstite/iron/oxygen, for instance,

$$FeO(\text{wüstite}) = Fe + \frac{1}{2} O_2, \tag{26}$$

$$\Delta\mu^o = -RT \ln\left[\sqrt{p_{O_2}} / a_{FeO}(T,y) \right]. \tag{27}$$

Whilst the equilibrium of a reaction is determined by the chemical potentials, the chemical heat balance is determined by the integral enthalpies of the reacting phases. In the case of wüstite, the following simple relationship holds between the integral enthalpy of $Fe_{1-y}O$, the partial enthalpy of FeO in the reference state and the integral enthalpy of metallic Fe,

$$H_{Fe_{1-y}O} = \overline{H}^o_{FeO} - y\, H^o_{Fe}. \tag{28}$$

Wüstite is one of the best-investigated non-stoichiometric compounds for which the partial thermodynamic values have often been measured. However, the present treatment can in principle be applied to other non-stoichiometric phases.

7. Notes on the contents of the tables

The constants A, B, C and D in the captions of the tables pertain to the expression for the molar heat CP in $cal \cdot mol^{-1}K^{-1}$

$$CP = A + B \cdot 10^{-3}T + C \cdot 10^{5}T^{-2} + D \cdot 10^{-6}T^{2} \tag{29}$$

In some cases the last term in equation (29) has been substituted by the term $D \cdot 10^{8}T^{-3}$. In all such cases this is indicated at the bottom of the respective table. Also indicated are the temperature range for the interpolation formulae and the literature references, given respectively in the last two columns. The values of this function at different temperatures are given in the appropriate column, CP, of the tables. In the temperature column, 298 always indicates the temperature of 298.15 K.

The enthalpy of formation and the standard entropy at 298.15 K are given under the appropriate columns of H and S with the number of digits warranted by the accuracy given in the cited literature references.

Other values calculated from these data and the molar heat are given to three decimal figures. In the case of phase transformations, the transformation temperature and heat are given with the same accuracy as in the literature cited. For convenience, the values of the functions are given at the transformation temperature for both phases in equilibrium with one another.

In certain cases, where there is a deviation from the usual behaviour of the substances, the tables also indicate available additional information concerning transformation points, thermal decomposition, range of composition, and composition of the vapour phase. In this connection, the following abbreviations are used: allotropic transformation temperature: TPT, melting point: MPT, normal boiling point: BPT, normal sublimation point: SPT, thermal decomposition temperature at P = 1 atm: DPT, and heat of transformation in $kcal\ mol^{-1}$: L.

The absence of data for high temperature phases means that these are not available or that such phases do not exist because of thermal decomposition for example. This is indicated in the tables, e.g. CaO_2 and $CaCO_3$.

For some substances, e.g. As, the sublimation point is lower than the melting point, and to avoid any confusion both temperatures are indicated at the head of the relevant table.

Where temperatures of fusion are known to a higher accuracy than 1 K, this is indicated separately at the head of the table. In these cases the functions were calculated for the precise transformation temperature, although the transformation temperature in the temperature column is only given to the nearest 1 K, as also for the temperature 298.15 K.

8. Origin of basic data

The thermochemical parameters usually listed in critical tables are the heat of formation and the entropy at 25°C. Heat capacities, heats of transformation, fusion and evaporation are also available from critical compilations. The basic data of the present tables are taken from different monographs, tables and original literature, the references for which are always cited. The main sources are the monograph of Kubaschewski, Evans and Alcock, the JANAF Tables, which give full thermochemical descriptions of a selected number of substances, the 'Selected Values' of the National Bureau of Standards (U.S.A.), which give an extensive coverage of the relevant values at room temperature, the corresponding work from the U.S.S.R., which lists in addition heats of transformations, fusion and evaporation, a number of reports from the National Physical Laboratory (U.K.) pertaining to thermochemical properties of individual substances and the comprehensive assessment of the thermodynamics of the sulphides, selenides and tellurides of metals by Mills. The present data for the pure metals have largely been taken from the most recent edition of the 'Selected Values of the Thermodynamic Properties of Metals and Alloys' by Hultgren and his colleagues.

Representatives of CODATA have agreed on certain 'key values' for substances that enter many calculations of chemical equilibria such as CO_2, CO, H_2O, HCl, S_2 etc. These have been accepted here with some exceptions.

Data from all these sources have been compared and checked for consistency. In addition, a substantial number of recent publications on calorimetric or Gibbs free energy measurements have been consulted and perused. Even so, quite a number of gaps remained and have been filled by educated guesses denoted by an asterisk * in the following tables. Here, it may be pointed out that in many cases the molar heat equations are either estimates or, at least, estimates combined with the CP value at 298 K obtained from low temperature calorimetric measurements. The estimation methods used here are those described in the monograph of Kubaschewski et al. A fuller description of the modes of evaluation, as in the JANAF Tables, is not presented here, simply in the interests of economy of space, but may be obtained from the present authors on request.

9. Accuracies

Not all the data have the same accuracy. The basic data in the tables are given with the same number of digits as appear in the literature reference cited. In practical applications, depending on the particular problem and the state of the experimental work, different accuracies are required. In a few cases, it is sufficient to have an accuracy of, say, ± 10 kcal · mol^{-1} in the Gibbs free energy in order to predict whether an envisaged process is likely to be feasible or not. For the development and operation of a practical process, however, much more accurate data are necessary in order to calculate with sufficient precision, e.g. the composition of the product or the optimum operating temperatures and pressures.

To indicate accuracies, an old suggestion made by F.D. Richardson to use the letters A, B, C, D is followed here. These letters appear in the Substance Index on p. LIII and signify the following:

A: key values or equally well-established values;
B: "good values", ± 0.1 for β_i; no immediate need for redetermination;
C: "moderately good values", ± 0.5 for β_i; redetermination desirable;
D: "better than nothing".

The error in the β-functions is different in different temperature ranges. In many cases it is smallest in the temperature range in which experimental investigations have been carried out i.e. where the substance exists as one of the reactants. The accuracy is related to this temperature range.

The aim of a critical assessment of thermodynamic data is both to obtain the best accuracy for the individual data as well as over-all consistency, i.e. the selected values must correctly represent the behaviour of the substance in different chemical equilibria and phase diagrams. Consistent data must produce the correct equilibrium constants, transition temperatures, triple points etc., which can in most cases be measured more precisely than can be calculated from thermodynamic values obtained by other methods. Therefore, it is justified to fit the data from calorimetric, emf, or vapour pressure measurements, within their experimental accuracy, to transition temperatures, triple points etc. with a resultant greater number of digits.

The definition of the enthalpy of formation as an enthalpy difference implies e.g. that for double oxides, enthalpies of formation from the elements have large absolute values, whereas the enthalpy of formation from the constituent oxides is rather small and accurately known. An error in the standard Gibbs free energy of reaction $d\Delta G_r^o$ is related to the relative error in the equilibrium constant, dK/K, according to:

$$\frac{dK}{K} = -\frac{d\Delta G_r^o}{RT}\,. \tag{30}$$

Complete consistency of all the data is not attainable at the present time but it remains an essential requirement for future basic thermochemical research and assessment of data.

10. Arrangement of substances

The sequence of the elements and compounds in the tables is based both on alphabetic order and the Periodic Chart. Every element is listed in the alphabetic order of its chemical symbol. Every element is then followed by a list of its compounds in the sequence of the Periodic Chart starting with the hydrides and then the halides (Group VIIa) and continuing with the chalcogenides (Group VIa), the nitrogen group (Va) and so on via the b groups to group Ia.

It is believed that the reader will quickly get used to the present arrangement which is the same in the Index of Substances on p. LIII and enables rapid reference to be made. The Index of Substances contains the substances appearing in both the first and the present supplementary volume. For each substance, the volume and page number are given. When the data for a substance appearing in the first volume have been revised, the page numbers in both volumes are listed. The page numbers for certain substances appear in brackets. This applies mostly to substances which can be prepared under particular conditions but are unstable at higher temperatures, and therefore their thermochemical data are very uncertain.

Finally, in the Substance Index, the error limits are given as far as possible in the manner described in Section 9.

11. Acknowledgments

The authors are very grateful to Dr. H. Ünal both for his untiring help in the checking of the data and for the computer calculation of the tables.

The authors also wish to thank those who have discussed certain aspects and evaluations in the first volume or who have written to point out errors and those who have supplied them with information concerning the second volume.

Finally the authors thank the Deutsche Forschungsgemeinschaft for financial help and the Data Processing Centre of the Technische Hochschule Aachen for their willing assistance.

12. Symbols and abbreviations used in the tables

A, B, C, D: temperature coefficients in the equations for the molar heat CP

CP: molar heat in cal · $mol^{-1}K^{-1}$

$$CP = A + B\,10^{-3}T + C\,10^{5}T^{-2} + D\,10^{-6}T^{2}$$

BPT: normal boiling point in K

BT: β function, $\beta(T) = -10^{3}\,G(T)/4.575\,T$

DEC.: decomposition

DPT: thermal decomposition temperature in K

G: Gibbs energy in $kcal \cdot mol^{-1}$

GAS: "ideal gas"

H: enthalpy in $kcal \cdot mol^{-1}$

L: heat of transformation in $kcal \cdot mol^{-1}$

LIQ: liquid phase

MPT: melting point in K

P: vapour pressure in Torr

RANGE: temperature range, in K, for the interpolation formulae for CP

REF.: literature references

S: entropy in $cal \cdot mol^{-1}K^{-1}$

SOL: solid phase

SOL-A: α phase (A : α, B : β, C : γ, D : δ, E : ε, A2 : α_2)

SPT: normal sublimation temperature in K

T: temperature in K

TPT: allotropic transformation temperature in K

In the Substance Index the letters A, B, C, D represent error limits as described in Section 9.

Introduction

1. Constantes d'équilibre

L'équilibre thermodynamique d'un système macroscopique homogène ou hétérogène est défini phénoménologiquement par la constance au cours du temps de la concentration des divers constituants ou de la composition des diverses phases. A côté d'états d'équilibre stable, peuvent exister des états metastables, dont l'évolution vers l'équilibre stable est influencée par des facteurs cinétiques. Selon le système et le température, les réactions sont très rapides ou très lentes et l'équilibre est établi quasi instantanément ou seulement après des jours ou des années. L'équilibre dynamique stable ou métastable est caractérisé par des fluctuations chimiques réversibles, c'est-à-dire des transformations chimiques différentielles virtuelles ou réelles dans l'un et dans l'autre sens, qui sont décrites par des réactions chimiques où figurent des constituants B_i et leurs coefficients stoechiométriques ν_i:

$$\nu_1 B_1 + \nu_2 B_2 + \dots = \nu_j B_j + \dots \tag{1}$$

Les réactifs et les produits B_i peuvent être soit les constituants d'une solution solide p.ex. Zn dans le laiton, soit des constituants d'un corps pur comme Zn, ou O, ou ZnO dans l'oxyde de zinc pur approximativement stoechiométrique correspondant à la composition chimique $Zn_{1+y}O$ avec $y \approx 0$. Les fluctuations statistiques ne mettent pas en cause les concentrations macroscopiques des constituants dans leurs phases respectives. Ces fluctuations réversibles sont compatibles avec les principes fondamentaux de l'équilibre, entropie maximale, enthalpie libre minimale etc.

Tant pour les systèmes homogènes qu'hétérogènes les concentrations à l'équilibre sont interdépendantes et fonction de la température et de la pression. Pour la majorité des systèmes d'intérêt technique, on peut considérer que la pression est isotrope. De nombreux problèmes d'équilibre peuvent être traités efficacement par la méthode des constantes d'équilibre, développée à partir de la loi d'action de masse appliquée aux solutions idéales. Dans les cas limites des gaz parfaits, des solutions idéales et des solutions de non-électrolytes idéalement diluées, la loi d'action de masse prend la forme simple:

$$K(T) = \prod \left(p_i / p_i^o \right)^{\nu_i} , \tag{2a}$$

$$K(T) = \prod x_i^{\nu_i} . \tag{2b}$$

La constante d'équilibre K(T) est une grandeur sans dimension, dont la valeur numérique dépend de la température et de l'état de référence adopté pour chaque partenaire de la réaction, p.ex. les corps purs pour une pression $p_i^o = 1$ atm. Dans la relation (2a), p_i représente la pression partielle, définie par le rapport $p_i/P = x_i$ où P est la pression totale et x_i le titre molaire; ν_i représente le coefficient stoechiométrique comme dans la relation (1). Dans la relation (2b), x_i représente le titre molaire dans les solutions en phase condensée.

Dans le cas des systèmes réels, les relations (2a) et (2b) ont été généralisées par l'introduction des concepts de fugacité et d'activité, qui sont également des grandeurs sans dimension. En fait, la fugacité pourrait être traitée comme un cas particulier de l'activité. La loi d'action de masse s'écrit alors de façon générale:

$$K(T) = \prod a_i^{\nu_i}, \qquad (3)$$

$$\log K(T) = \sum \nu_i \beta_i(T) . \qquad (4)$$

L'activité d'un corps pur pour p = 1 atm. est par définition égale à 1. Dans les applications techniques, la loi des gaz parfaits peut en général être utilisée avec une bonne approximation; par conséquent, l'activité des constituants gazeux est définie par $a_i = p_i/p_i^o$ avec $p_i^o = 1$ atm. L'activité des constituants des phases condensées est une grandeur déterminée expérimentalement, qui dépend peu de la pression: $a_i = a_i(T, P, x_i, \ldots) \approx a_i(T, x_i, \ldots)$. La compilation des valeurs numériques des activités dans les solutions en phase condensée ne fait pas l'objet des tables présentées ici. Pour des calculs approximatifs, on peut, à défaut de valeurs expérimentales, poser $a_i \simeq x_i$ ou utiliser des méthodes d'estimation décrites ailleurs.

Les relations (3) et (4) ne sont valables que pour des réactions d'équilibre du type (1), c'est-à-dire avec des coefficients stoechiométriques définis. La constante d'équilibre K(T) est calculée selon la relation (4) à partir des fonctions β, dont les valeurs numériques sont consignées dans les tables pour les corps purs à la pression de référence $p^o = 1$ atm. La relation de la fonction β (ou de la fonction de Planck qui lui est proportionnelle) avec l'enthalpie et l'entropie est discutée dans le paragraphe 4. Lorsque l'état de référence est la solution infiniment diluée, la fonction β n'est pas donnée dans les tables. Les composés non-stoechiométriques sont traités dans le paragraphe 6.

2. Bilans thermiques

Le bilan thermique d'un processus technique peut être déterminé comme bilan physique basé sur la détermination des échanges thermiques avec l'environnement par conduction et par rayonnement ou comme bilan chimique basé sur le bilan de matière et la connaissance des températures des substances à l'entrée et à la sortie des installations.

Les flux de matière et les températures sont déterminés de manière continue dans les procédés techniques. Comme le transport d'énergie interne ne peut être séparé du flux de matière, il est possible de calculer aussi bien l'exo- que l'endothermicité du processus considéré à l'aide des enthalpies données dans les tables. Selon le premier principe de la thermodynamique, les équations (5) et (6) relient, pour un processus isobare, à pression atmosphérique, le bilan de matière ou la réaction irréversible globale et le bilan thermique:

$$\nu_1 B_1(T_1) + \nu_2 B_2(T_2) + \ldots \rightarrow \nu_j B_j(T_j) + \ldots , \qquad (5)$$

$$Q = \sum \nu_i H_i^o(T_i) + \sum \Delta H_{mix} . \qquad (6)$$

Dans la relation (5), les symboles chimiques B_i représentent des phases macroscopiques, alors que dans la relation (1) ces symboles représentent les fluctuations chimiques différentielles des atomes, combinaisons d'atomes ou espèces dans un système à l'équilibre dans lequel les concentrations macrosopiques des phases en équilibre sont constantes.

Les équations chimiques (1) et (5) ont donc une signification complètement différente. La signification thermodynamique distincte des formules chimiques et des équations de réaction reflète le développement historique de la chimie inorganique et de la thermodynamique chimique. La réaction globale (5) doit être stoechiométrique; de plus les températures T_i des substances à l'entrée et à la sortie doivent être déterminées puisqu'il y a lieu d'en tenir compte dans le relation (6). Dans cette dernière relation, $Q > 0$ et $Q < 0$ correspondent respectivement à une endo - et une exo-thermicité dans le processus technique envisagé; $H_i^o(T_i)$ représente les enthalpies des corps purs et ΔH_{mix} les enthalpies de mélange qui sont approximativement indépendantes de la température. Pour un calcul de première approximation, les enthalpies de mélange peuvent souvent être négligées. Les relations (5) et (6) peuvent être appliquées aussi bien à un procédé complet qu'à des étages individuels ou à des tranches différentielles. Tant du point de vue mathématique que chimique, un examen détaillé et critique des diverses contributions au bilan thermique, p.ex. l'incidence du préchauffage de certaines substances ou du déroulement incomplet d'une réaction, se fait le mieux à partir des relations (5) et (6).

Les écarts à la relation PV = constante pour les gaz réels sont en général inférieurs aux incertitudes dans les bilans thermiques d'un processus chimique ou métallurgique, ce qui implique que les relations (5) et (6) sont des approximations suffisantes pour des processus non-isobares. La contribution d'apports d'énergie cinétique ou potentielle peut en général être négligée, sauf si un examen détaillé du bilan énergétique ou de processus particuliers est considéré.

Pour des procédés électrochimiques, le bilan thermique est complété comme suit

$$\nu_1 B_1(T_1) + \ldots \nu_A e_A \rightarrow \nu_j B_j(T_j) + \ldots \nu_K e_K , \qquad (7)$$

où $|\nu_A| = \nu_K$,

$$Q = \sum \nu_i H_i^o(T_i) + \sum \Delta H_{mix} + \nu_K F(\varphi_K - \varphi_A) . \qquad (8)$$

e_A et e_K représentent les électrons captés par l'anode ou cédés par la cathode, φ_A et φ_K sont les potentiels électriques internes de l'anode et de la cathode, exprimés en Volt. Il en résulte que $\varphi_A - \varphi_K > 0$ est la tension aux bornes, qui diffère de la tension d'équilibre. $F = 23066 \text{ cal} \cdot \text{Volt}^{-1} \text{mol}^{-1}$ représente le facteur de conversion. Pour des raisons tant fondamentales (voir paragraphe 3) que pratiques, nous recommendons de faire la distinction entre les électrons $\nu_A e_A$ et $\nu_K e_K$ et de les maintenir explicitement dans le relation (7). De même, nous préconisons de faire la distinction entre les tensions φ_A et φ_K dans la relation (8); il en résulte un calcul correct du signe et des quantités d'énergie électrique et de matière. Formellement les équations (7) et (8) sont également valables pour des transformations électrothermiques, chimiques et métallurgiques.

Le calcul conventionel des bilans thermiques avec décomposition des flux d'énergie en "quantités de chaleurs" et en enthalpies de réaction à une température constante fictive dans le réacteur, p. ex. la température ambiante, est malcommode et ne correspond pas à la représentation que l'on se fait de l'énergie interne et de l'enthalpie de la matière.

3. Force électromotrice de cellules galvaniques

La mise en oeuvre de cellules galvaniques, en particulier celles utilisant des électrolytes solides s'étend rapidement dans le recherche et le contrôle de processus métallurgiques et chimiques. Des difficultés surgissent souvent dans le calcul du signe et de la valeur numérique des tensions galvaniques à partir des données thermochimiques. C'est le cas notamment pour des cellules galvaniques dont la tension s'inverse au cours de la détermination de l'oxygène dans l'acier fondu à l'aide d'une électrode de référence (Cr, Cr_2O_3) ou lors de la comparaison de la force électro-motrice mesurée à l'aide d'électrodes de référence différentes.

La cause des difficultés réside dans la représentation habituelle mais peu appropriée des réactions aux demi-éléments, dans laquelle les électrons impliqués sont simplifiés de part et d'autre. De même, la représentation courante des réactions électrochimiques d'équilibre dans les cellules galvaniques ouvertes est discutable. Les électrons e_α et e_β qui interviennent au niveau des électrodes de contact ne sont pas équivalents du point de vue thermodynamique. Il convient pour cela de les faire figurer explicitement dans la description de la réaction galvanique. Sa formulation générale et celle de la relation thermodynamique qui donne la force électromotrice à partir des potentiels chimiques μ ou des activités a, est alors:[+)]

+) Knacke, O., Krahe, J., Weber, R.: Z. Physik. Chem. Neue Folge 67 (1969) 194

$$\nu_1 B_1 + \nu_2 B_2 + \ldots \nu_\alpha e_\alpha = \nu_j B_j + \ldots \nu_\beta e_\beta , \qquad (9)$$

$$|\nu_\alpha| = \nu_\beta ,$$

$$\nu_\beta F(\varphi_\beta - \varphi_\alpha) = \sum \nu_i \mu_i = \Delta\mu^o + RT \ln \prod a_i^{\nu_i} . \qquad (10)$$

Les relations (9) et (7) sont aussi fondamentalement différentes que les relations (5) et (1). Les dénominations anode et cathode devraient être réservées à des processus irréversibles. Dans une chaîne galvanique ouverte, le signe des tensions peut changer avec les concentrations dans le système. Les relations (9) et (10) donnent la tension $(\varphi_\beta - \varphi_\alpha)$, exprimée en volt, correcte en signe et en valeur numérique. Les électrons e_α et e_β se trouvent en général à des potentiels électriques internes différents et ont dès lors des potentiels électrochimiques différents. Pour une cellule ouverte, ils ne peuvent pas être éliminés dans l'équation de la réaction galvanique. Comme les électrodes de contact α et β sont constituées par le même matériau, le potentiel chimique des électrons e_α et e_β se compense dans l'équation (10) et seuls les potentiels chimiques des partenaires de réaction neutres ou ionsés sont à additionner. Des réactions galvaniques du type (9) peuvent donc être combinées linéairement, comme il est fait pour d'autres réactions chimiques; il en résulte aussi une conversion correcte en signe et valuer numérique par rapport à différentes électrodes de référence. Une réaction étudiée à maintes reprises, est traitée à titre d'exemple:

$$\alpha / Ag_{sol} / AgCl_{liq} / C(Cl_2) / \beta$$

$Ag^+ + e_\alpha = Ag$ Réaction d'électrode α ,

$Cl^- = \frac{1}{2} Cl_2 + e_\beta$ Réaction d'électrode β

$AgCl_{liq} + e_\alpha = Ag_{sol} + \frac{1}{2} Cl_2 + e_\beta$ Réaction galvanique ,

$$F(\varphi_\beta - \varphi_\alpha) = \mu^o_{Ag} + \frac{1}{2}\mu^o_{Cl_2} - \mu^o_{AgCl} + \frac{RT}{2} \ln p_{Cl_2}$$

$$= \Delta G^o_r + \frac{RT}{2} \ln p_{Cl_2} .$$

Les relations $\mu_i^o = G_i^o$ ou $\mu_i^o \approx G_i^o$ ont été utilisées ici pour le potentiel standard des éléments et des composés approximativement stoechiométriques, d'où résulte $\sum \nu_i \mu_i^o = \Delta G^o_r$, ce qui correspond à la représentation habituelle dans les applications de la thermochimie.

Les valeurs numériques des enthalpies libres pour les corps purs à 1000 K donnent:

$$\Delta G^o_r = 19,347 \text{ kcal} \cdot \text{mol}^{-1}$$

$$\varphi_\beta - \varphi_\alpha = 0,839 + 0,099 \lg p_{Cl_2} \text{ Volt} .$$

$(R = 1,9869 \text{ cal} \cdot \text{mol}^{-1} K^{-1})$.

4. Les fonctions consignées dans les tables de données numériques

Les tables donnent les valeurs numériques des fonctions suivantes pour les corps purs à 1 atm.:

Chaleur spécifique molaire, CP, en cal · $mol^{-1}K^{-1}$;

Enthalpie, H, en kcal · mol^{-1};

Entropie, S, en cal · $mol^{-1}K^{-1}$;

Enthalpie libre, G, en kcal · mol^{-1};

Fonction β, BT, sans dimension;

Pression de vapeur, P, en torr.

Les symboles sont ceux imprimés par l'ordinateur.

Pour les substances pures et cristallines, la constante d'intégration de l'entropie est par convention définie par le 3e principe de la thermodynamique, et est posée égale à zéro au zéro absolu de la température. Ceci suppose une mise à l'équilibre idéale et l'adoption de la définition du zéro de l'entropie acceptée lors du calcul de l'entropie dite spectroscopique par les méthodes de la mécanique statistique. L'incidence des effets isotopiques sur les grandeurs thermochimiques, dont il convient de tenir compte pour les éléments et composés légers, est négligée ici. La composition isotopiques est supposée être naturelle. Le choix de la constante d'intégration de l'enthalpie, contrairement à celle de l'entropie, n'est pas imposé. Puisque les chaleurs de formation sont généralement données à 25°C, l'enthalpie des éléments purs dans leur forme stable à 25°C, pour une pression d'une atmosphère, peut être posée égale à zéro. Pour le phosphore toutefois, l'état de référence retenu n'est pas la forme thermodynamiquement la plus stable, mais le phosphore blanc, qui peut être préparé de façon très reproductible. Il est en effet permis de choisir des points zéro différents pour l'entropie et l'enthalpie car le calcul de constantes d'équilibre fait intervenir ΔS et ΔH séparément. Ayant défini les niveaux de référence de l'entropie et de l'enthalpie, l'enthalpie libre et la fonction de Planck sont des fonctions univoques de la température pour des corps purs.+)

Les conventions décrites conduisent aux fonctions thermodynamiques suivantes:

L'enthalpie molaire $H_i^o(T)$ d'une substance pure i à la pression d'une atmosphère et à la température T est définie comme suit:

$$H_i^o(T) = \Delta H_i^f + \int_{298}^{T} dH_i^o. \qquad (11)$$

+) Barin, I., et Knacke, O.: Met. Trans. 5 (1974) 1769,
Hirschwald, W., Knacke, O., et Reinitzer, P.: Erzmetall 10 (1957) 127,
Gerlach, J., Kleist, H.G., et Knacke, O.: Erzmetall 8 (1955) 377

ΔH_i^f est l'enthalpie de formation du composé à partir des éléments dans leur état stable à 298,15 K. La limite inférieure d'intégration 298 représente l'abréviation utilisée ici pour 298,15 K. Par définition, l'enthalpie de formation des éléments dans leur état stable est nulle. La variation de l'enthalpie avec la température est, par convention, décrite par la chaleur spécifique molaire à pression constante, $c_i(T)$:

$$c_i(T) = dH_i^o/dT \, . \tag{12}$$

Lors d'une transformation isobare à la température T_t, la chaleur molaire de transformation, ΔH_i^t est utilisée au lieu de la chaleur spécifique molaire, qui deviendrait infinie. La relation (11) donne ainsi

$$H_i^o(T) = \Delta H_i^f + \int_{298}^{T} c_i(T)\,dT + \sum \Delta H_i^t \, , \tag{13}$$

la sommation portant sur toutes les enthalpies des transformations qui ont lieu entre 298,15 K et T[K].

La relation entre l'entropie et l'enthalpie est

$$dS_i^o = dH_i^o/T \, , \qquad \Delta S_i^t = \Delta H_i^t/T_t \, , \tag{14}$$

$$S_i^o(T) = S_i^{298} + \int_{298}^{T} dH_i^o/T \, . \tag{15}$$

En utilisant les chaleurs spécifiques molaires, ces relations deviennent

$$dS_i^o = (c_i(T)/T)\,dT \, , \qquad \Delta S_i^t = \Delta H_i^t/T_t \, , \tag{16}$$

$$S_i^o(T) = S_i^{298} + \int_{298}^{T} (c_i(T)/T)\,dT + \sum \Delta H_i^t/T_t \, . \tag{17}$$

L'enthalpie libre molaire $G_i^o(T)$ est, par définition, fonction de l'enthalpie et de l'entropie, suivant la relation

$$G_i^o(T) = H_i^o(T) - T \cdot S_i^o(T) \, . \tag{18}$$

Contrairement à l'enthalpie et l'entropie, qui varient de façon discontinue au point de transition, l'enthalpie libre y varie de façon continue.

La constante d'équilibre K(T) et l'enthalpie libre standard de réaction ΔG_r^o d'une réaction r sont reliées par la relation

$$\log K(T) = -\frac{\sum \nu_i G_i^o}{RT \ln 10} = -\frac{\Delta G_r^o(T)}{RT \ln 10} \, . \tag{19}$$

Pour faciliter le calcul des constantes d'équilibre, la forme de la relation (19) suggère l'utilisation de la fonction de Planck pour les substances pures, $-G_i^o/T$, et du facteur $R \ln 10 = 4{,}575\ \mathrm{cal \cdot mol^{-1} K^{-1}}$. Il en résulte, de par la définition de la fonction β_i pour les substances pures,

$$\beta_i(T) = -G_i^o(T)/RT \ln 10 , \tag{20}$$

$$\log K(T) = \sum \nu_i \beta_i(T) . \tag{21}$$

L'enthalpie et l'enthalpie libre varient de façon approximativement linéaire avec la température et peuvent donc être interpolées linéairement; la fonction β varie de façon approximativement linéaire avec la température réciproque et doit être interpolée en conséquence.

Il convient de noter la relation qui existe entre les fonctions β_i adoptées ici et les valeurs de $\log K_{p,i}$ utilisées dans les Tables JANAF. Pour un composé i, formé à partir de plusieurs éléments ε, on a:

$$\nu_1\varepsilon_1 + \nu_2\varepsilon_2 + \ldots \rightarrow B_i , \tag{22a}$$

$$\log K_{p,i} = \beta_i - \sum |\nu_\varepsilon| \beta_\varepsilon .$$

La différence principale entre les présentes tables et les Tables JANAF réside dans le choix par ces dernières d'un nombre illimité d'états de référence pour l'enthalpie des corps purs, et en conséquence également pour l'enthalpie libre et pour les constantes d'équilibre $K_{p,i}$; ces états de référence sont les éléments à chaque température, alors qu'il découle de la théorie que l'enthalpie des corps purs est déterminée numériquement par une seule constante d'intégration. La normalisation des enthalpies adoptée ici a son analogie en géographie, où l'altitude est donnée par rapport au niveau de la mer.

La relation entre les différences d'enthalpie à température et pression constante ΔH_i des Tables JANAF pour un composé et les enthalpies $H_i(T)$ des composés et les enthalpies $H_\varepsilon(T)$ des éléments adoptées ici est analogue à la relation qui existe entre $\log K_{p,i}$ et la fonction β_i, à savoir

$$\Delta H_i(T) = H_i(T) - \sum |\nu_\varepsilon| H_\varepsilon(T) . \tag{22b}$$

Le choix des fonctions n'est évidemment pas déterminé par leur validité mais par leur commodité. Le calcul de bilans thermiques pour des processus réels, non isothermes est en effet plus simple avec les enthalpies $H_i(T)$ définies ici que dans le système habituel.

La calorie à 15°C est retenue comme unité d'énergie de préférence au joule international (1 cal = 4,184 J). La calorie reste utilisée dans la majorité des compilations thermochimiques et les utilisateurs de données thermochimiques y sont accoutumés. Les unités SI sont toutefois de plus en plus utilisées dans l'enseignement universitaire, ce qui entraîne une évolution dans les préférences.

5. Tensions de vapeur

La dernière colonne de quelques tableaux donne les pressions totales P comprises entre 0.001 et 760 torr. Il s'agit de la somme des pressions partielles des monomères et des polymères dans le cas de vapeurs de composition complexe. Pour les substances dont la phase gazeuse est constitué d'une seule espèce ou d'une espèce prédominante, les données thermochimiques pour cette dernière sont données à partir de la température d'ébullition. Dans le cas de phases gazeuses complexes, le tableau pour la phase condensée se termine à la température d'ébullition normale où la pression totale atteint 760 torr. La composition de la phase gazeuse et la chaleur de vaporisation à la température d'ébullition sont indiquées respectivement dans le haut du tableau et dans sa dernière ligne. Les propriétés des diverses espèces constituant la phase gazeuse sont données à la suite du tableau pour la phase condensée. Il est bon de noter que la température pour laquelle la pression atteint 760 torr résulte souvent d'un choix pondéré entre diverses déterminations de la tension de vapeur et de la température d'ébullition, caractérisées par une dispersion parfois importante.

6. Composés non-stoechiométriques

Pour de nombreux composés virtuellement stoechiométriques comme $Zn_{1+y}O$ avec $y \approx 0$, le potentiel chimique standard de la combinaison stoechiométrique ZnO dans l'état de référence du composé pur est pratiquement égal à l'enthalpie libre molaire pour la phase

$$\mu^{o}_{ZnO} = \partial G/\partial n_{Zn} + \partial G/\partial n_{O} \approx G_{Zn_{1+y}O}. \qquad (23)$$

Cette représentation simple facilite la description thermochimique des "corps purs". Il faut toutefois se rappeler que la grandeur thermochimique déterminante pour le calcul des constantes d'équilibre est ou bien le potentiel chimique dans un état de référence donné, ou bien la fonction β correspondante, mais il s'agit toujours d'une grandeur partielle.

Certains carbures, nitrures, oxydes ou autres composés et en particulier ceux des métaux de transition existent dans un large intervalle de stoechiométrie à l'intérieur duquel il n'existe pas de composition stoechiométrique simple. La relation (23) cesse alors d'être valable et il en résulte des difficultés. Un exemple typique et techniquement important est celui de la wüstite, l'oxyde de fer déficitaire en métal $Fe_{1-y}O$ pour lequel $0,046 < y < 0,168$. La combinaison stoechiométrique FeO n'existe pas comme composé stable, mais cette même combinaison stoechiométrique FeO est connue dans de nombreux oxydes mixtes. Nous proposons d'utiliser alors comme état de référence le potentiel chimique $\mu_{FeO} = \mu_{Fe} + \mu_O$ mesuré pour le composition $Fe_{0,95}O$ $(y^o = 0,05)$, accessible expérimentalement à des températures supérieures à 573°C. La composition $Fe_{0,95}O$ ou $y^o = 0,05$ de la wüstite est donc choisie comme état de référence:

$$\mu^o_{FeO} = \left(\partial G_{Fe_{1-y}O} / \partial n_{FeO}\right)_{y^o} ; \quad y^o = 0,05 . \tag{24}$$

Au potentiel standard d'une combinaison stoechiométrique simple dans un intervalle de stoechiometrie correspondent une enthalpie partielle $\overline{H}^o$, et une entropie partielle $\overline{S}^o$, qui diffèrent de l'enthalpie intégrale H et de l'entropie intégrale S du composé.

Le potentiel standard μ ou l'activité a sont des fonctions de la température et de la composition, qui est déterminée dans une phase binaire par la connaissance d'une seule concentration ou de toute autre grandeur équivalente. L'activité est alors défine par:

$$RT \ln a(T,y) = \mu(T,y) - \mu^o(T,y^o) . \tag{25}$$

Des composés non-stoechiométriques sont à traiter comme des solutions pour le calcul des constantes d'équilibre. L'équilibre faisant intervenir les trois phases Wüstite/Fer/Oxygène peut par exemple être décrit par la réaction suivante:

$$FeO(\text{wüstite}) = Fe + \frac{1}{2} O_2 , \tag{26}$$

$$\Delta\mu^o = - RT \ln\left[\sqrt{p_{O_2}} / a_{FeO}(T,y)\right] . \tag{27}$$

Alors que les équilibres de réaction sont déterminés par les potentiels chimiques, les bilans thermiques dépendent des enthalpies intégrales des phases qui réagissent. Pour la wüstite, l'enthalpie intégrale de $Fe_{1-y}O$, l'enthalpie partielle de FeO dans son état de référence et l'enthalpie intégrale du fer sont reliées par l'approximation:

$$H_{Fe_{1-y}O} = \overline{H}^o_{FeO} - y\, H^o_{Fe} . \tag{28}$$

La wüstite, dont les propriétés thermochimiques partielles ont été mesurées maintes fois, est un des composés non-stoechiométriques les mieux connus. La description thermochimique proposée pour cette substance peut en principe être utilisée pour d'autres composés non-stoechiométriques.

7. Présentation des tables et commentaires

Les constantes A, B, C et D qui figurent dans la partie supérieure des tables sont les coefficients de la température dans les équations donnant la chaleur spécifique molaire CP en $cal \cdot mol^{-1} K^{-1}$

$$CP = A + B \cdot 10^{-3} T + C \cdot 10^{5} T^{-2} + D \cdot 10^{-6} T^{2} \tag{29}$$

Les intervalles de validité des formules d'interpolation et les références bibliographiques sont cités dans les dernières colonnes.

Dans quelques cas des notes aux bas des tableaux indiquent que les équations d'interpolation pour les chaleurs spécifiques comportent un terme $D \cdot 10^{8} T^{-3}$ au lieu du dernier terme de la relation (29). Les valeurs numériques des fonctions sont données pour différentes températures dans les colonnes intitulées CP dans les tableaux. Dans la colonne T, 298 est l'abréviation de 298,15 K. La chaleur de formation et l'entropie standard à 298,15 K figurent dans les colonnes H et S avec la même précision que dans les sources bibliographiques. Les grandeurs thermodynamiques calculées à partir de ces données et des chaleurs spécifiques sont par contre imprimées avec trois décimales. Les températures et les chaleurs de transformation sont elles aussi reprises avec la même précision que dans les sources citées. Pour faciliter l'emploi des tables, les valeurs numériques des diverses fonctions sont imprimées pour les deux phases en équilibre aux points de transformation.

Pour les substances dont le comportement s'écarte de la normale, des indications additionnelles figurent dans la partie supérieure des tableaux. Elles concernent les points de transformation, la décomposition thermique, le domaine d'homogénéité et la composition de la phase gazeuse. Les abréviations utilisées sont: température de transition allotropique, TPT; température de fusion, MPT; température d'ébullition, BPT; température de sublimation, SPT; température de décomposition, DPT; chaleur molaire de transformation L en $kcal \cdot mol^{-1}$. L'omission des valeurs numériques pour certaines fonctions à des températures élevées implique l'absence de données bibliographiques ou la non existence des phases correspondantes, par exemple par suite d'une décomposition thermique, ce que des remarques dans les tableaux rappellent également (voir p. ex. CaO_2, $CdCO_3$). Pour de nombreuses substances, par exemple As, la température de sublimation est inférieure à la température de fusion. Dans de tels cas les deux températures sont indiquées pour éviter des ambiguités.

Quelques températures de fusion sont connues à mieux qu'un degré près. Ces températures figurent dans le haut des tableaux, tandis que dans les colonnes, la température de fusion est arrondie comme il est fait pour 298,15 K. Les valeurs numériques des fonctions restent toutefois celles calculées avec la valeur exacte de la température de fusion (voir p. ex. Pb).

8. Origine des données

Les données thermochimiques habituellement citées dans la plupart des compilations critiques sont l'enthalpie de formation, l'entropie standard des corps purs à 25°C, les chaleurs spécifiques molaires et les enthalpies de transformation; dans quelques monographies, les pressions de vapeur sont également reprises. Les données de base utilisées dans les tables présentées ici sont extraites de diverses monographies ou tables de données numériques et en partie de la littérature originale; un renvoi bibliographique signale l'origine des données. Les sources principles de données sont: la monographie de Kubaschewski, Evans et Alcock; les Tables JANAF qui donnent pour de nombreuses substances sélectionnées une description thermodynamique complète; les "Selected Values" du National Bureau of Standards (U.S.A.), qui présentent un ensemble étendu de données à température ambiante; l'ouvrage soviètique correspondant, qui rapporte également des enthalpies de transformation; quelques rapports du National Physical Laboratory (G.B.) et la compilation des propriétés thermochimiques pour des sulfures, séléniures et tellurures établie par Mills. Les données pour les métaux purs proviennent presque toutes de la dernière de l'ouvrage "Selected Values of Thermodynamic Properties of Metals and Alloys" de Hultgren et co-auteurs.

CODATA a recommandé des "valeurs clés" pour quelques substances qui interviennent dans de nombreux équilibres, par exemple CO_2, CO, H_2O, HCl, S_2 etc. Ces valeurs sont retenues ici à quelques exceptions près.

Les données de base des différentes sources ont été comparées entr'elles et vérifiées quant à leur cohérence. Dans certains cas des valeurs quelque peu différentes de celles publiées dans les compilations citées ont été retenues, tandis que de nombreuses données manquantes ont été estimées; ces valeurs sont identifiées par un astérisque * . L'estimation des entropies standard et des chaleurs spécifiques est généralement basée sur les méthodes décrites dans l'ouvrage de Kubaschewski et al. mentionné ci-dessus. Par manque de place, la description détaillée des données expérimentales telle que réalisée dans les Tables JANAF, n'a pu être faite. Sur demande, les auteurs fourniront volontiers les informations.

9. Incertitudes

L'incertitude des données est fort variable. Les tables reprennent les valeurs de base avec le nombre de décimales citées dans la référence dont elles proviennent. Pour l'utilisation pratique l'incertitude acceptable dépend de la nature du problème à résoudre et du degré de développement de la recherche dans un domaine déterminé. Dans certains cas, une enthalpie libre incertaine à $\pm$ 10 kcal · mol^{-1} peut permettre de décider si un procédé proposé doit être discuté d'avantage ou abandonné. Une précision nettement meilleure est toutefois requise pour la mise au point et l'exploitation de procédés, si l'on désire calculer avec une précision suffisante des grandeurs telles que la composition des produits ou les températures et les pressions optimales.

Selon la suggestion de F.D. Richardson, les incertitudes sont caractérisées dans l'index de la p. LIII par les lettres A, B, C ou D, dont les significations sont:

A: valeur clé ou valeur de précision équivalente;
B: "bonne valeur", β_i connu à ± 0,1; de nouvelles déterminations ne sont pas nécessaires;
C: "valeur moyennement bonne", β_i connu à ± 0,5; de nouvelles déterminations sont à souhaiter;
D: "mieux que rien".

L'incertitude sur les valeurs numériques de la fonction β_i varie avec la température. Elle est habituellement la plus faible dans l'intervalle de température où des mesures sont effectuées et où la substance intervient effectivement comme partenaire de réaction. L'incertitude citée se rapporte à cet intervalle de température.

Le but d'une évaluation critique de données thermochimiques est, à côté de l'obtention de la meilleure précision pour les données individuelles, la recherche d'une bonne cohérence, c'est-à-dire une interdépendance correcte des grandeurs thermodynamiques pour une même substance dans les diverses réactions chimiques et les diagrammes de phase où elle intervient. Des données cohérentes doivent conduire à des valeurs correctes pour les constantes d'équilibre, les températures de transformation, les points triples etc. Souvent la mesure directe d'une telle grandeur peut se faire avec une précision supérieure à celle qui caractérise la valeur calculée à partir de diverses grandeurs thermochimiques déterminées par d'autres méthodes. Pour cette raison, les données peuvent être rendues plus significatives par ajustement, endéans leur limites d'erreur, des résultats des mesures calorimétriques, de f.e.m., de tensions de vapeur etc., aux valeurs expérimentales des températures de transition, des points triples etc.

L'établissement d'enthalpies de formation est basé sur des déterminations de différences d'enthalpie. Il en résulte que les enthalpies de formation des oxydes doubles à partir des éléments ont des valeurs absolues très élevées, alors que les enthalpies de formation à partir des oxydes binaires constituants sont souvent faibles et connues avec précision. Une incertitude $d\Delta G_r^o$ sur l'enthalpie de réaction conduit à une incertitude relative pour la constante d'équilibre donnée par

$$\frac{dK}{K} = -\frac{d\Delta G_r^o}{RT}. \qquad (30)$$

Une cohérence complète de toutes les données ne peut être réalisée à l'heure actuelle mais reste une exigence pour toute recherche future dans le domaine de la thermochimie et pour toute compilation critique.

10. Ordre de présentation des substances

Les éléments et les composés sont présentés dans les Tables en tenant compte de l'ordre alphabétique et de la position dans le tableau périodique. Les éléments sont présentés dans l'ordre alphabétique de leur symbole. Chaque élément est suivis de ses composés en tenant compte de la structure du tableau périodique, d'abord les hydrures, ensuite les halogénures (Groupe VIIa), puis les chalcogénures (Groupe VIa), les azotides (Groupe Va) et ainsi de suite, via les Groupes b jusqu'au Groupe Ia.

L'index (p. LIII) permet une orientation rapide et mentionne aussi bien les substances traitées dans le présent volume que celles traitées dans le premier volume. Pour chaque substance, le volume et la page sont donnés. Lorsque le deuxième tome donne des valeurs corrigées, renvoi est fait aux deux volumes. Pour quelques substances, les pages sont marquées entre parenthèses, ce qui signifie le plus souvent que ces substances peuvent être préparées sous des conditions particulières mais qu'elles sont instables à haute température et que les données thermochimiques sont très incertaines. Finalement, l'index cite les incertitudes d'après le code décrit dans le paragraphe 9.

11. Remerciements

Les auteurs expriment leur gratitude au Dr. H. Ünal pour son aide permanente dans la vérification de la cohérence des données et le calcul par ordinateur des tables.

Les auteurs remercient tous ceux qui leur ont envoyé des remarques et des critiques, qui leur ont signalé des erreurs dans le premier volume ou qui leur ont communiqué des informations utilisées dans la préparation du second volume.

Ils expriment leur reconnaissance envers le Deutsche Forschungsgemeinschaft pour son soutien financier et envers le Centre de Calcul de la Technische Hochschule Aachen pour sa collaboration serviable.

12. Symboles et abréviations utilisés dans les tables

A, B, C, D: coefficients de la température dans les équations donnant la chaleur spécifique molaire CP

CP: chaleur spécifique molaire en $cal \cdot mol^{-1} K^{-1}$

$$CP = A + B\,10^{-3}T + C\,10^{5}T^{-2} + D\,10^{-6}T^{2}$$

BPT: température d'ébullition en K

BT: fonction β, $\beta(T) = -10^{3}\,G(T)/4{,}575\,T$

DEC.: décomposition

DPT: température de décomposition en K

G: enthalpie libre (de Gibbs) en $kcal \cdot mol^{-1}$

GAS: gaz supposé parfait

H: enthalpie en $kcal \cdot mol^{-1}$

L: chaleur molaire de transformation en $kcal \cdot mol^{-1}$

LIQ: phase liquide

MPT: température de fusion en K

P: pression de vapeur en torr

RANGE: intervalle de validité, en K, des formules d'interpolation donnant CP

REF.: référence bibliographique

S: entropie en $cal \cdot mol^{-1}K^{-1}$

SOL: phase solide

SOL-A: Phase α (A : α, B : β, C : γ, D : δ, E : ε, A2 : α_2)

SPT: température de sublimation en K

T: température en K

TPT: température de transition allotropique en K

Dans l'index des substances, les lettres A, B, C, D représentent les incertitudes décrites dans le paragraphe 9.

Einführung

1. Gleichgewichtskonstanten

Das thermodynamische Gleichgewicht eines makroskopischen homogenen oder heterogenen Systems ist durch die zeitlich konstante chemische Zusammensetzung der verschiedenen Phasen bzw. die zeitlich konstanten Konzentrationen aller Komponenten phänomenologisch definiert. Neben stabilen Gleichgewichtszuständen gibt es metastabile Zustände, deren Übergang in das stabile Gleichgewicht reaktionskinetisch gehemmt ist. Es gibt Systeme mit großer und geringer Reaktionsgeschwindigkeit, deren Gleichgewicht sich bei gegebener Temperatur fast momentan oder erst in Jahren einstellt. Im dynamischen Gleichgewicht, das stabil oder metastabil sein kann, laufen reversibel chemische Fluktuationen ab, d.h. virtuelle oder wirkliche differentielle chemische Umsetzungen in Hin- und Rückrichtung, die durch Reaktionsgleichungen zwischen den reagierenden Komponenten B_i mit den Molzahlen ν_i beschrieben werden,

$$\nu_1 B_1 + \nu_2 B_2 + \ldots = \nu_j B_j + \ldots \tag{1}$$

Die Reaktionspartner B_i sind Komponenten einer Mischphase wie Zn im Messing oder Bestandteile eines reinen Stoffes wie Zn oder O oder ZnO im annhähernd stöchiometrischen reinen Zinkoxid von der chemischen Zusammensetzung $Zn_{1+y}O$ mit $y \approx 0$. Die statistischen chemischen Schwankungen im Gleichgewicht tangieren nicht die makroskopischen Konzentrationen der Reaktionspartner in ihren Phasen. Reversible chemische Fluktuationen sind mit den Extremalprinzipien des Gleichgewichts, also maximaler Entropie, minimaler freier Enthalpie usw. verträglich.

Die Gleichgewichtskonzentrationen homogener und heterogener Systeme sind interdependent und von der Temperatur und dem Druck abhängig. Der Druck darf für die meisten technisch signifikanten Systeme als isotrop betrachtet werden. Zahlreiche Gleichgewichtsprobleme lassen sich zweckmäßig nach der Methode der Gleichgewichtskonstanten behandeln, die sich aus dem Massenwirkungsgesetz (MWG) für ideale Mischphasen entwickelt hat. Das MWG hat im Grenzfall (a) perfekter Gase sowie (b) idealer und ideal verdünnter nichtelektrolytischer Lösungen folgende einfache Form:

$$K(T) = \prod \left(p_i/p_i^o\right)^{\nu_i}, \tag{2a}$$

$$K(T) = \prod x_i^{\nu_i}. \tag{2b}$$

Die Gleichgewichtskonstante $K(T)$ ist eine dimensionslose Größe, die von der Temperatur und dem vereinbarten Referenzzustand der Reaktionspartner, z.B. den reinen

Stoffen unter $p_i^o = 1$ atm Druck, abhängt. In Gl. (2a) sind p_i die Partialdrücke, die als Produkte $p_i = x_i \cdot P$ aus den Molenbrücken x_i und dem Gesamtdruck P definiert sind, ν_i die Molzahlen aus Gl. (1). In Gl. (2b) sind x_i die Molenbrüche in kondensierten Mischphasen.

Unter Beibehaltung der mathematischen Form des MWG sind die rechten Seiten von Gl. (2) durch die Definition der dimensionslosen Größen Fugazität und Aktivität generalisiert und damit für reale Systeme allgemein anwendbar geworden. Man könnte auf die Bezeichnung Fugazität verzichten und nur von Aktivitäten sprechen. Dann lautet das Massenwirkungsgesetz in allgemein gültiger Form:

$$K(T) = \prod a_i^{\nu_i}, \tag{3}$$

$$\log K(T) = \sum \nu_i \beta_i(T). \tag{4}$$

Zunächst wird Gl. (3), dann Gl. (4) erläutert. Die Aktivität reiner Stoffe unter 1 atm Druck ist definitionsgemäß gleich 1. Bei technischen Anwendungen kann im allgemeinen näherungsweise mit dem idealen Gasgesetz gerechnet und deshalb die Aktivität gasförmiger Komponenten $a_i = p_i/p_i^o$ mit $p_i^o = 1$ atm gesetzt werden. Die Aktivität von Komponenten kondensierter Phasen ist eine experimentell zu ermittelnde Funktion mit nur geringer Druckabhängigkeit, $a_i = a_i(T, P, x_1, \ldots) \approx a_i(T, x_1, \ldots)$. Die Kompilation von Aktivitätswerten in kondensierten Mischphasen ist nicht Inhalt dieser Tabellen. Zu einer ersten Orientierung kann man beim Fehlen experimenteller Werte $a_i \approx x_i$ setzen oder gewisse Schätzmethoden verwenden.

Gl. (3) und (4) gelten nur in Verbindung mit einer bestimmten Gleichgewichtsreaktion Gl. (1), d.h. für bestimmte Mohlzahlen ν_i. Nach Gl. (4) berechnet sich die Gleichgewichtskonstante K(T) aus den dimensionslosen β-Funktionen der reinen Stoffe bei 1 atm Druck, die hier tabelliert werden. Die β-Funktionen sind der Planckschen Funktion proportionale, eindeutige Temperaturfunktionen. Der Zusammenhang der β-Funktion bzw. Planckschen Funktion reiner Stoffe mit der Enthalpie und Entropie wird in Abschnitt 4 dargestellt. β-Funktionen für den Referenzzustand der unendlich verdünnten Lösung sind in diesen Tabellen nicht enthalten. Nichtstöchiometrische Verbindungen werden in Abschnitt 6 behandelt.

2. Wärmebilanzen

Die Wärmebilanz eines technischen Prozesses kann nach zwei grundsätzlich verschiedenen Methoden ermittelt werden: a) als physikalische Wärmebilanz durch Messung des Wärmeaustausches mit der Umgebung infolge Wärmeleitung und -strahlung, b) als chemische Wärmebilanz aufgrund der Stoffbilanz und der Temperaturen der zu- und abfließenden Stoffe. Bei technischen Prozessen werden Stoffströme und Temperaturen laufend gemessen, sind also bekannt. Da mit dem Stoffstrom der Transport von innerer Energie untrennbar verbunden ist, kann aus der Stoffbilanz und den gemessenen Temperaturen der Wärmebedarf bzw. -verlust des Prozesses mit Hilfe der hier tabellierten Enthalpiewerte berechnet werden. Nach dem ersten Hauptsatz der Thermodynamik gilt für einen isobaren Prozeß bei 1 atm Druck folgende Verknüpfung von Stoffbilanz bzw. irreversibler Bruttoreaktion und Wärmebilanz:

$$\nu_1 B_1(T_1) + \nu_2 B_2(T_2) + \ldots \rightarrow \nu_j B_j(T_j) + \ldots , \quad (5)$$

$$Q = \sum \nu_i H_i^o(T_i) + \sum \Delta H_{mix} . \quad (6)$$

In Gl. (5) bezeichnen die chemischen Symbole B_i makroskopische Phasen, während die Symbole B_i in Gl. (1) die differentiellen chemischen Fluktuationen der verschiedenen Atome, Atomkombinationen oder Spezies eines Gleichgewichtssystems unter Erhaltung der makroskopischen Gleichgewichtskonzentrationen der betroffenen Phasen symbolisieren. Die chemischen Gleichungen (1) und (5) haben daher einen völlig verschiedenen Charakter. Die unterschiedliche thermodynamische Bedeutung der chemischen Formeln und Reaktionsgleichungen spiegelt die geschichtliche Entwicklung der anorganischen Chemie und der chemischen Thermodynamik wieder. Die Bruttoreaktion Gl. (5) muß stöchiometrisch sein, und die gemessenen Temperaturen T_i der ein- und austretenden Stoffe müssen notiert und in Gl. (6) berücksichtigt werden. In Gl. (6) ist $Q > 0$ der Wärmebedarf, $Q > 0$ die Wärmeabgabe eines technischen Prozesses, $H_i^o(T_i)$ sind die tabellierten Enthalpien der reinen Stoffe, ΔH_{mix} die Mischungsenthalpien von Mischphasen, die annähernd unabhängig von der Temperatur sind. Für eine erste Näherung können die Mischungsenthalpien oft vernachlässigt werden. Gl. (5) und (6) gelten für einen vollständigen Prozeß sowie für einzelne Stufen oder differentielle Abschnitte eines Prozesses. Eine detaillierte Untersuchung und Diskussion der verschiedenen Beiträge zur Bilanz, z.B. der Einfluß der Vorwärmung einzelner Stoffe oder eines unvollständigen Reaktionsablaufs, geschieht mathematisch und chemisch am zweckmäßigsten auf der Grundlage von Gl. (5) und (6).

Die Abweichungen realer Gase von der Gleichung $PV = \text{const}$ fallen im allgemeinen in die Fehlergrenze einer chemischen oder metallurgischen Wärmebilanz, so daß dann Gl. (5) und (6) auch für nicht-isobare Prozesse ausreichend sind. Der Beitrag kinetischer und potentieller mechanischer Energie zur Wärmebilanz ist im allgemeinen ebenfalls zu vernachlässigen und nur bei detaillierter Analyse des Energiehaushalts oder speziellen Prozessen zu berücksichtigen.

Für elektrochemische Prozesse wird die Wärmebilanz wie folgt ergänzt:

$$\nu_1 B_1(T_1) + \ldots \nu_A e_A \rightarrow \nu_j B_j(T_j) + \ldots \nu_K e_K , \quad (7)$$

$$|\nu_A| = \nu_K ,$$

$$Q = \sum \nu_i H_i^o(T_i) + \sum \Delta H_{mix} + \nu_K F(\varphi_K - \varphi_A) . \quad (8)$$

Hier sind e_A und e_K die Elektronen, die an Anode und Kathode ein- und austreten, φ_A und φ_K die inneren elektrischen Potentiale von Anode und Kathode in Volt, $\varphi_A - \varphi_K > 0$ die von der Gleichgewichtsspannung verschiedene Klemmspannung. $F = 23066\ \text{cal} \cdot \text{Volt}^{-1} \cdot \text{mol}^{-1}$. Wir empfehlen die Beibehaltung und Unterscheidung der Elektronen $\nu_A e_A$ und $\nu_K e_K$ in Gl.(7) und die Unterscheidung der Potentiale φ_A und φ_K in Gl.(8) sowohl aus grundsätzlichen Erwägungen (s. Abschnitt 3) als auch der Zweckmäßigkeit wegen, weil hierdurch zwangsläufig eine nach Vorzeichen und Betrag richtige Zuordnung von elektrischer Energie und Stoffumsatz erzielt wird. Gl.(7) und (8) gelten formal auch für elektrothermische chemische oder metallurgische Umsetzungen.

Die konventionelle Berechnung einer Wärmebilanz mit der willkürlichen Aufspaltung der Energieströme in "fühlbare Wärmen" und Reaktionsenthalpien bei einer fiktiven konstanten Reaktionstemperatur im Reaktor, z.B. Raumtemperatur, ist umständlich und entspricht nicht der wissenschaftlichen Vorstellung von der inneren Energie bzw. der Enthalpie der Materie.

3. EMK galvanischer Zellen

Die Anwendung galvanischer Zellen, insbesondere von Zellen mit festem Elektrolyten, breitet sich in der Forschung und in der Prozeßkontrolle metallurgischer und chemischer Verfahren schnell aus. Man kann nicht leugnen, daß manchmal Schwierigkeiten bei der Berechnung von Vorzeichen und Betrag der Zellspannung aus thermochemischen Daten auftreten, z.B. bei Zellen, deren Spannung das Vorzeichen ändern kann, wie bei der Sauerstoffbestimmung im flüssigen Stahl mit einer Cr, Cr_2O_3-Referenzelektrode und bei der Umrechnung von Spannungen, die mit verschiedenen Referenzelektroden gemessen worden sind. Die Ursache dieser Schwierigkeit liegt in der üblichen, aber unzweckmäßigen Schreibweise der Zellreaktion, bei der die beteiligten Elektronen schließlich gestrichen werden. Die Zweckmäßigkeit der üblichen Schreibweise von elektrochemischen Gleichgewichtsreaktionen offener galvanischer Zellen muß bezweifelt werden. Mit Beibehaltung und Unterscheidung der thermodynamisch nicht identischen Elektronen e_α und e_β der beiden Ableitelektroden α und β ergibt sich folgende generelle Form einer Zellreaktion und hieraus die thermodynamische Gleichung zur Berechnung der Zellspannung aus den chemischen Potentialen μ bzw. Aktivitäten a:+)

$$\nu_1 B_1 + \nu_2 B_2 + \dots \nu_\alpha e_\alpha = \nu_j B_j + \dots \nu_\beta e_\beta , \qquad (9)$$

$$|\nu_\alpha| = \nu_\beta ,$$

$$\nu_\beta F(\varphi_\beta - \varphi_\alpha) = \sum \nu_i \mu_i = \Delta\mu^o + RT \ln \prod a_i^{\nu_i} . \qquad (10)$$

+) Knacke, O., Krahe, J., Weber, R.: Z. Physik. Chem. Neue Folge 67 (1969) 194

Gl. (9) und (7) sind ihrem Charakter nach grundsätzlich verschieden, wie dies bereits bezüglich Gl. (5) und (1) erläutert worden ist. Die Bezeichnungen Anode und Kathode sollten für irreversible Prozesse vorbehalten bleiben. Bei einer offenen galvanischen Kette kann sich das Vorzeichen der Spannung mit den Konzentrationen des Systems ändern. Durch Gl. (9) und (10) werden Betrag und Vorzeichen der Spannung $(\varphi_\beta - \varphi_\alpha)$ Volt richtig geliefert; $F = 23066\ \mathrm{cal \cdot Volt^{-1} mol^{-1}}$. Die Elektronen e_α und e_β befinden sich im allgemeinen auf unterschiedlichen inneren elektrischen Potentialen, haben deshalb unterschiedliche elektrochemische Potentiale und dürfen folglich in der Gleichgewichtsreaktion der offenen Zelle nicht gegeneinander aufgehoben werden. Da die Ableitelektroden α und β aus demselben Material bestehen, hebt sich das chemische Potential der Elektronen e_α und e_β in Gl. (10) auf, und es sind nur die chemischen Potentiale von elektrisch neutralen oder ionisierten Reaktionspartnern zu addieren. Zellreaktionen in der hier vorgeschlagenen Schreibweise Gl. (9) können wie andere chemische Reaktionen linear kombiniert werden, woraus auch eine nach Vorzeichen und Betrag richtige Umrechnung von Spannungen gegenüber unterschiedlichen Referenzelektroden folgt.

Als Beispiel sei die Zellspannung einer mehrfach experimentell untersuchten Bildungszelle berechnet:

$$\alpha/Ag_{sol}/AgCl_{liq}/C(Cl_2)/\beta$$

$Ag^+ + e_\alpha = Ag$	Elektrodenreaktion α
$Cl^- = \frac{1}{2} Cl_2 + e_\beta$	Elektrodenreaktion β
$AgCl_{liq} + e_\alpha = Ag_{sol} + \frac{1}{2} Cl_2 + e_\beta$	Zellreaktion

$$F(\varphi_\beta - \varphi_\alpha) = \mu^o_{Ag} + \frac{1}{2}\mu^o_{Cl_2} - \mu^o_{AgCl} + \frac{RT}{2} \ln p_{Cl_2}$$

$$= \Delta G^o_r + \frac{RT}{2} \ln p_{Cl_2} \ .$$

Hier ist die Beziehung $\mu^o_i = G^o_i$ bzw. $\mu^o_i \approx G^o_i$ für das Standardpotential der Elemente bzw. annähernd stöchiometrische Verbindungen und damit $\sum \nu_i \mu^o_i = \Delta G^o_r$ benutzt worden, was der üblichen Schreibweise in verschiedenen Gebieten der angewandten Thermochemie entspricht. Mit den tabellierten Werten der molaren freien Enthalpie der reinen Stoffe berechnet sich für 1000 K ($R = 1,9869\ \mathrm{cal \cdot mol^{-1}K^{-1}}$):

$$\Delta G^o_r = 19,347\ \mathrm{kcal \cdot mol^{-1}},$$

$$\varphi_\beta - \varphi_\alpha = 0,839 + 0,099\ \lg p_{Cl_2}\ \mathrm{Volt}\ .$$

4. Die tabellierten Funktionen

Folgende Funktionen der reinen Stoffe bei 1 atm Druck werden tabelliert:

Molwärme CP in $cal \cdot mol^{-1}K^{-1}$,

Enthalpie H in $kcal \cdot mol^{-1}$,

Entropie S in $cal \cdot mol^{-1}K^{-1}$,

freie Enthalpie G in $kcal \cdot mol^{-1}$,

die dimensionslose β-Funktion BT,

der Dampfdruck P in Torr.

Die Symbole sind vorstehend in der vom Computer ausgedruckten Form angegeben.

Die Integrationskonstante der Entropie kristalliner reiner Stoffe wird konventionell nach dem 3. Hauptsatz der Thermodynamik festgelegt, also ihre Entropie am absoluten Nullpunkt null gesetzt. Hierbei ist ungehemmte Gleichgewichtseinstellung vorausgesetzt, also die statistische bzw. spektroskopische Definition des Nullpunktes der Entropie gemeint. Thermochemische Isotopieeffekte, die sich nur bei leichten Elementen und Verbindungen bemerkbar machen, bleiben unberücksichtigt. Es wird die natürliche Isotopenmischung der Elemente vorausgesetzt. Für die Integrationskonstante der Enthalpie gibt es keine entsprechenden zwingenden Argumente. Daher wird mit Rücksicht auf die übliche Angabe von Bildungswärmen bei 25°C folgendes Null-Niveau festgelegt: Die Enthalpie der reinen Elemente in ihrer stabilen Form bei 25°C und 1 atm Druck wird null gesetzt. Lediglich im Falle von Phosphor ist nicht die thermodynamisch stabile Modifikation als Referenzzustand gewählt, sondern der gut reproduzierbar herstellbare weiße Phosphor. Es ist statthaft, die Nullpunkte von Entropie und Enthalpie in unterschiedlicher Weise zu vereinbaren, da in die Berechnung von Gleichgewichten stets nur Differenzen ΔS und ΔH getrennt eingehen. Mit der vorgeschlagenen Festlegung der Nullpunkte von Entropie und Enthalpie sind auch die freien Enthalpien und die Planckschen Funktionen reiner Stoffe als eindeutige Temperaturfunktionen festgelegt. Aus Vorstehendem ergeben sich die nachstehend definierten eindeutigen Temperaturfunktionen.+)

Die molare Enthalpie $H_i^o(T)$ einer reinen Substanz i bei 1 atm Druck und der Temperatur T wird definiert durch

$$H_i^o(T) = \Delta H_i^f + \int_{298}^{T} dH_i^o , \qquad (11)$$

wobei ΔH_i^f die Bildungsenthalpie der Verbindung aus den stabilen Zuständen der

---

+) Barin, I., Knacke, O.: Met. Trans. 5 (1974) 1769,
Hirschwald, W., Knacke, O., und Reinitzer, P.: Erzmetall 10 (1957) 127,
Gerlach, J., Kleist, H.G., und Knacke, O.: Erzmetall 8 (1955) 377.

Elemente bei 298,15 K bezeichnet. Die untere Integrationsgrenze 298 ist die hier übliche Abkürzung für 298,15 K. Die Bildungsenthalpie der Elemente im stabilen Zustand ist definitionsgemäß Null. Die Temperaturabhängigkeit der Enthalpie wird konventionell durch die Molwärme bei konstantem Druck $c_i(T)$ beschrieben,

$$c_i(T) = dH_i^o/dT \,. \tag{12}$$

Bei einer isobaren Umwandlung bei der Temperatur T_t tritt an die Stelle der Molwärme, die hier formal unendlich groß würde, die molare Umwandlungsenthalpie ΔH_i^t. Damit folgt aus Gl. (11)

$$H_i^o(T) = \Delta H_i^f + \int_{298}^{T} c_i(T)\,dT + \sum \Delta H_i^t \,, \tag{13}$$

worin alle Umwandlungsenthalpien zwischen 298,15 K und T[K] zu summieren sind.

Der Zusammenhang zwischen Entropie und Enthalpie ist bei reinen Stoffen von 1 atm Druck gegeben durch

$$dS_i^o = dH_i^o/T \,, \qquad \Delta S_i^t = \Delta H_i^t/T_t \,, \tag{14}$$

$$S_i^o(T) = S_i^{298} + \int_{298}^{T} dH_i^o/T \,. \tag{15}$$

Bei Verwendung der Molwärme lauten diese Gleichungen

$$dS_i^o = (c_i(T)/T)\,dT \,, \qquad \Delta S_i^t = \Delta H_i^t/T_t \,, \tag{16}$$

$$S_i^o(T) = S_i^{298} + \int_{298}^{T} (c_i(T)/T)\,dT + \sum \Delta H_i^t/T_t \,. \tag{17}$$

Die molare freie Enthalpie $G_i^o(T)$ ergibt sich definitionsgemäß aus der Enthalpie und Entropie,

$$G_i^o(T) = H_i^o(T) - T \cdot S_i^o(T) \,. \tag{18}$$

Die freie Enthalpie ändert sich am Umwandlungspunkt stetig im Gegensatz zur sprunghaften Änderung von Enthalpie und Entropie.

Die Gleichgewichtskonstante K(T) und die freie Standard-Reaktionsenthalpie ΔG_r^o einer Reaktion r sind definitionsgemäß durch folgende Gleichung verknüpft,

$$\log K(T) = -\frac{\sum \nu_i G_i^o}{RT \ln 10} = -\frac{\Delta G_r^o(T)}{RT \ln 10} \,. \tag{19}$$

worin R die allgemeine Gaskonstante bezeichnet. Die Form der Gl. (19) legt es nahe, zur möglichst einfachen Berechnung der Gleichgewichtskonstante die Plancksche Funktion der reinen Stoffe $-G_i^o/T$ zu verwenden und den Faktor $R \ln 10 = 4{,}575$ cal · mol^{-1}K^{-1} einzubeziehen. Dementsprechend ergibt sich durch Definition der β-Funktionen der reinen Stoffe:

$$\beta_i(T) = -G_i^o(T)/RT \ln 10 \,, \tag{20}$$

$$\log K(T) = \sum \nu_i \beta_i(T) \,. \tag{21}$$

Während Enthalpie und freie Enthalpie sich annähernd linear mit der Temperatur ändern und deshalb auch linear interpoliert werden dürfen, ändert sich die β-Funktion im allgemeinen annähernd linear mit 1/T, was eine entsprechende Interpolation erfordert.

Es sei auf den Zusammenhang zwischen den hier gewählten β_i-Funktionen und den Werten $\log K_{p,i}$ der JANAF-Tables hingewiesen. Für eine Verbindung i, die aus mehreren Elementen ε besteht, gilt

$$\nu_1\varepsilon_1 + \nu_2\varepsilon_2 + \ldots \rightarrow B_i \,, \tag{22a}$$

$$\log K_{p,i} = \beta_i - \sum |\nu_\varepsilon| \beta_\varepsilon \,.$$

Der hauptsächliche Unterschied zwischen diesen Tabellen und den JANAF-Tables liegt darin, daß in den JANAF-Tables unendlich viele Referenzzustände für die Enthalpie der reinen Stoffe und folglich auch für die freie Enthalpie und die Funktionen $K_{p,i}$ gewählt werden, nämlich die Elemente bei variabler Temperatur. Von der Theorie her ist aber klar, daß die Enthalpie reiner Stoffe durch eine Integrationskonstante numerisch bestimmt wird. Die von uns gewählte Normierung der Enthalpie entspricht in gewissem Sinne der geographischen Höhenangabe über Normal-Null. Die in den JANAF-Tables aufgeführten isotherm-isobaren Enthalpiedifferenzen ΔH_i einer Verbindung stehen mit den hier tabellierten Enthalpien $H_i(T)$ von Verbindungen und $H_\varepsilon(T)$ ihrer Elemente in dem entsprechenden Zusammenhang wie $\log K_p$ mit den β-Funktionen; d.h.

$$\Delta H_i(T) = H_i(T) - \sum |\nu_\varepsilon| H_\varepsilon(T) \,. \tag{22b}$$

Es ist keine Frage der Richtigkeit, sondern der Zweckmäßigkeit, welche Funktionen man verwendet. Die Aufstellung von Wärmebilanzen realer, nicht-isothermer Prozesse ist mit Hilfe der hier definierten Enthalpien $H_i(T)$ rechnerisch einfacher als nach dem üblichen Differenzensystem.

Die 15°C Kalorie ist hier als Energieeinheit beibehalten und dem internationalen Joule vorgezogen worden (1 cal = 4,184 J). Die 15°C Kalorie wird noch in fast allen kritischen thermochemischen Kompilationen verwendet, und sie ist auch den meisten Benutzern thermochemischer Daten gewohnt. Da im akademischen Unterricht zunehmend die SI-Einheiten verwendet werden, bereitet sich eine Umstellung vor.

5. Dampfdrücke

In der letzten Spalte einiger Tabellen sind die totalen Dampfdrücke P von 0,001 bis 760 Torr angegeben; dies ist bei polymerisierenden Dämpfen die Summe der Partialdrücke der einfachen und polymeren Spezies. Bei Stoffen mit nur einer oder einer überwiegenden Spezies in der Dampfphase sind die thermochemischen Daten des Dampfes ab Siedetemperatur ausgedruckt. Bei polymerisierendem Dampf endet die Tabelle der kondensierten stabilen Phasen am natürlichen Siedepunkt, d.h. beim Totaldruck von 760 Torr. Die Zusammensetzung der Dampfphase am Siedepunkt ist am Kopf der Tabelle, die Verdampfungswärme in der letzten Zeile der Tabelle angegeben. Bei polymerisierenden Dämpfen werden im Anschluß an die kondensierten Phasen die thermodynamischen Daten der verschiedenen dampfförmigen Spezies ab Raumtemperatur ausgedruckt. Die konsistente Angabe eines Druckes von 760 Torr am Siedepunkt sollte nicht darüber hinwegtäuschen, daß die der Auswertung zugrundeliegenden Dampfdruckmessungen und Siedepunktbestimmungen oft erhebliche Streuungen zeigen.

6. Nicht-stöchiometrische Phasen

Bei den zahlreichen nahezu stöchiometrischen Verbindungen wie $Zn_{1+y}O$ mit $y \approx 0$ ist das chemische Standardpotential der einfachen stöchiometrischen Kombination ZnO im Referenzzustand der reinen Verbindung praktisch gleich der molaren freien Enthalpie der Phase,

$$\mu^o_{ZnO} = \partial G/\partial n_{Zn} + \partial G/\partial n_O \approx G_{Zn_{1+y}O}. \tag{23}$$

Diese einfache Beziehung erleichtert erheblich die thermochemische Beschreibung reiner Stoffe. Man muß aber im Auge behalten, daß nach der Theorie die entscheidende thermodynamische Größe zur Berechnung der Gleichgewichtskonstanten das chemische Potential eines Reaktionspartners in einem bestimmten Referenzzustand ist bzw. die entsprechende β-Funktion, jedenfalls eine partielle Größe.

Einige Karbide, Nitride, Oxide usw., insbesondere der Übergangsmetalle haben einen beträchtlichen Homogenitätsbereich, innerhalb dessen keine einfache stöchiometrische Zusammensetzung auftritt. Dann versagt die einfache Relation Gl. (23). Damit ergeben sich vorläufig Schwierigkeiten, die besonders beim Wüstit, dem Metallunterschuß-Oxid $Fe_{1-y}O$ mit $0,046 < y < 0,168$, zutage getreten sind. Während die stöchiometrische Kombination FeO als stabile Phase nicht vorkommt, gibt es zahlreiche Doppeloxide mit stöchiometrischem FeO.

Wir schlagen in diesem Falle vor, das gemessene chemische Potential $\mu_{FeO} = \mu_{Fe} + \mu_O$ in der für Temperaturen über 572°C experimentell zugänglichen Phase der Zusammensetzung $Fe_{0,95}O$ zu tabellieren. Damit wird die Zusammensetzung $Fe_{0,95}O$ oder $y^o = 0,05$ des Wüstits als Standardzustand gewählt,

$$\mu^{o}_{FeO} = \left(\partial G_{Fe_{1-y}O} / \partial n_{FeO} \right)_{y^{o}} ; \quad y^{o} = 0,05. \tag{24}$$

Zu dem Standardpotential einer einfachen stöchiometrischen Kombination in einer nichtstöchiometrischen Phase gehören eine entsprechende partielle Enthalpie $\overline{H}^{o}$ und partielle Entropie $\overline{S}^{o}$, die von der integralen Enthalpie H und integralen Entropie S der Phase verschieden sind.

Das chemische Potential μ bzw. die Aktivität a sind Funktionen der Temperatur und der Zusammensetzung, die bei binären Phasen durch nur eine Konzentrationsangabe oder äquivalente Angabe gekennzeichnet wird. Die Aktivität ist dann definiert durch

$$RT \ln a(T,y) = \mu(T,y) - \mu^{o}(T,y^{o}). \tag{25}$$

Nicht-stöchiometrische Phasen sind bei Gleichgewichtsberechnungen wie Mischphasen zu behandeln. Das 3-Phasen-Gleichgewicht Wüstit/Eisen/Sauerstoff kann z.B. durch folgende Gleichgewichtsreaktion beschrieben werden:

$$FeO\,(\text{Wüstit}) = Fe + \frac{1}{2} O_2, \tag{26}$$

$$\Delta \mu^{o} = - RT \ln \left[\sqrt{p_{O_2}} / a_{FeO}(T,y) \right]. \tag{27}$$

Während die Reaktionsgleichgewichte durch die chemischen Potentiale bestimmt werden, hängen die chemischen Wärmebilanzen von den integralen Enthalpien der reagierenden Phasen ab. Für den Wüstit gilt folgende einfache Näherungsbeziehung zwischen der integralen Enthalpie von $Fe_{1-y}O$, der partiellen Enthalpie des FeO im Referenzzustand und der integralen Enthalpie des Eisens,

$$H_{Fe_{1-y}O} = \overline{H}^{o}_{FeO} - y\, H^{o}_{Fe}. \tag{28}$$

Bislang ist der Wüstit eine der am besten untersuchten nichtstöchiometrischen Verbindungen, deren partielle thermochemische Daten wiederholt gemessen worden sind. Die für den Wüstit vorgeschlagene thermochemische Beschreibung läßt sich grundsätzlich auch auf andere nicht-stöchiometrische Phasen anwenden.

7. Anmerkungen zu den Tabellen

Die Konstanten A, B, C und D über der Tabelle sind die Temperaturkoeffizienten der Gleichungen für die Molwärme CP in $cal \cdot mol^{-1} K^{-1}$

$$CP = A + B \cdot 10^{-3} T + C \cdot 10^{5} T^{-2} + D \cdot 10^{-6} T^{2}. \tag{29}$$

In der Interpolationsgleichung für die Molwärmen wird in einigen Fällen statt des letzten Terms in Gl.(29) ein Glied $D \cdot 10^8 T^{-3}$ verwendet, was unter der Tabelle vermerkt wird. Der Gültigkeitsbereich der Interpolationsformeln und die Literaturstellen sind in den letzten Spalten angegeben. Die Funktionswerte sind für verschiedene Temperaturen in der Spalte CP der Tabelle ausgedruckt. In der Temperaturspalte T steht 298 abkürzend für 298,15 K.

Die Bildungswärme und Normalentropie bei 298,15 K sind in den Spalten H und S aufgeführt, und zwar mit derselben Genauigkeit wie in der zitierten Literatur. Hingegen werden die aus diesen Daten und der Molwärme berechneten Werte mit drei Dezimalen aufgeführt. Umwandlungstemperaturen und -wärmen werden wiederum mit derselben Genauigkeit wie in der zitierten Literatur notiert. Zur einfacheren Anwendung werden an Umwandlungspunkten die Funktionswerte beider miteinander im Gleichgewicht stehenden Phasen angegeben.

Bei Stoffen, die vom üblichen Verhalten abweichen, werden zusätzliche Angaben über Umwandlungspunkte, thermische Zersetzung, Homogenitätsbereich und Zusammensetzung des Dampfes am Kopf der Tabelle notiert. Hierzu werden folgende Abkürzungen verwendet: Allotrope Umwandlungstemperatur TPT, Schmelzpunkt MPT, Siedepunkt BPT, Sublimationspunkt SPT, Zersetzungstemperatur DPT, molare Umwandlungswärme L in $kcal \cdot mol^{-1}$.

Falls keine Funktionswerte für hohe Temperaturen angegeben werden, sind keine Daten hierfür verfügbar oder es existieren keine entsprechenden Phasen, z.B. auf Grund thermischer Zersetzung, was aus den Bemerkungen am Kopf der Tabelle ebenfalls zu ersehen ist, vgl. CaO_2, $CdCO_3$.

Bei manchen Stoffen, z.B. As, liegt die Sublimationstemperatur unter dem Schmelzpunkt. Zur Vermeidung von Unklarheiten werden in solchen Fällen beide Temperaturen vermerkt.

Einige Schmelztemperaturen sind genauer als auf 1 K bekannt. Diese Temperaturen werden am Kopf der Tabelle vermerkt, und die Funktionswerte in der Tabelle werden für die exakte Schmelztemperatur berechnet; in der Temperaturspalte wird die Schmelztemperatur aufgerundet angegeben wie für 298,15 K, vgl. Pb.

8. Herkunft der Daten

In den bekannten kritischen Kompilationen thermochemischer Daten werden üblicherweise die Bildungswärme und die Normaltropie der reinen Stoffe bei 25° zusammengestellt, ferner Molwärmen und Umwandlungsenthalpien, in einigen Monographien auch Dampfdrücke. Die Basisdaten der vorliegenden Tabellen sind aus verschiedenen Monographien und Tabellen, teilweise auch aus der Originalliteratur entnommen, und ihre Herkunft ist jeweils durch Literaturangaben ausgewiesen. Hauptsächlich wurden folgende Quellen benutzt: Die Monographie von Kubaschewski, Evans und Alcock, die JANAF-Tables, die eine vollständige thermodynamische Beschreibung für eine große Anzahl von Substanzen geben, die "Selected Values" des National Bureau of Standards (USA), die eine umfassende Darstellung der Werte bei Raumtemperatur enthalten, das entsprechende Werk in der UdSSR, das außerdem Umwandlungsenthalpien auflistet, einige Berichte des National Physical Laboratory (GB) und die ausführliche Zusammenstellung der thermochemischen Eigenschaften von Sulfiden, Seleniden und Telluriden von Mills. Die hier tabellierten Daten von reinen Metallen stammen größtenteils von der neuesten Ausgabe des Werkes "Selected Values of Thermodynamic Properties of Metals and Alloys" von Hultgren und seinen Kollegen.

Im Rahmen der CODATA sind "Schlüsselwerte" von einigen Substanzen akzeptiert worden, die in viele Berechnungen von Gleichgewichten eingehen, wie CO_2, CO, H_2O, HCl, S_2 usw. Diese Werte sind hier mit wenigen Ausnahmen übernommen worden.

Die Basisdaten aus den verschiedenen Quellen sind miteinander verglichen und im Hinblick auf Konsistenz überprüft worden. Bei einigen Daten sind von den erwähnten Kompilationen abweichende Werte ausgewählt, und viele fehlende Daten sind abgeschätzt worden, was jeweils durch einen Stern * gekennzeichnet wird. Besonders wurden die Schätzmethoden für Standardentropien und Molwärmen und die Konsistenzkriterien angewendet, die in der zitierten Monographie von Kubaschewski et al. beschrieben sind. Aus Platzmangel mußte darauf verzichtet werden, die Auswertung der experimentellen Unterlagen für jeden Stoff ausführlich zu beschreiben, wie dies in den JANAF-Tables geschieht. Die Autoren wollen aber gern auf Anforderung Auskunft geben.

9. Genauigkeit

Die Genauigkeit der Daten ist sehr unterschiedlich. In den Tabellen werden die Basisdaten mit derselben Stellenzahl ausgedruckt, die in der jeweils zitierten Literatur angegeben ist. Bei der praktischen Anwendung wird je nach dem vorliegenden Problem und dem Stand der experimentellen Forschung eine unterschiedliche Genauigkeit verlangt. In einigen Fällen genügt schon eine Genauigkeit der molaren freien Enthalpie von $\pm$ 10 kcal · mol^{-1}, um abzuschätzen, ob ein vorgeschlagener Prozeß thermochemisch diskutabel ist oder nicht. Für die Entwicklung und Durchführung von technischen Prozessen ist eine wesentlich größere Genauigkeit der Daten notwendig, um z.B. die Zusammensetzung des Produkts oder die optimalen Arbeitstemperaturen und -drücke mit hinreichender Genauigkeit zu berechnen.

Einer Anregung von F.D. Richardson folgend werden durch die Buchstaben A, B, C, D im Substanzen-Index auf S. LIII folgende Genauigkeiten angezeigt:

A: "Schlüsselwerte" oder gleich gute Werte.

B: "Gute Werte", ± 0,1 in β_i; es besteht keine Notwendigkeit für neue Untersuchungen.

C: "Mittelgute Werte", ± 0,5 in β_i; erneute Untersuchungen sind wünschenswert.

D: "Besser als nichts".

Der Fehler der β-Funktionen ist unterschiedlich in verschiedenen Temperaturbereichen. Er ist durchweg am kleinsten in dem Bereich, in dem experimentelle Untersuchungen vorliegen, der Stoff also aktuell als Reaktionspartner auftritt. Auf diesen Temperaturbereich bezieht sich die Genauigkeitsangabe.

Das Ziel einer kritischen Auswertung thermodynamischer Daten ist neben einer großen Genauigkeit der einzelnen Daten die Konsistenz, d.h. richtige Verknüpfung der Stoffwerte mit den verschiedenen chemischen Gleichgewichten und den Phasendiagrammen. Konsistente Daten müssen richtige Gleichgewichtskonstanten, Umwandlungstemperaturen, Tripelpunkte usw. liefern, die meist genauer gemessen werden können, als man sie, von kalorimetrischen Messungen usw. ausgehend, rechnerisch erhält. Deshalb ist es zulässig, Daten aus kalorimetrischen Messungen, aus EMK-Messungen, Dampfdruckmessungen usw. innerhalb ihrer Meßgenauigkeit den Umwandlungstemperaturen, Tripelpunkten usw. anzupassen, woraus sich eine größere Stellenzahl der Daten ergibt.

Das Differenzensystem der Bildungsenthalpien impliziert, daß z.B. bei Doppeloxiden große Absolutwerte der Bildungsenthalpien auftreten, wenn man sich auf die Elemente bezieht, während die Bildungsenthalpien, bezogen auf die einfachen Oxide, ziemlich klein und genau bekannt sind. Ein Fehler in der freien Standard-Reaktionsenthalpie $d\Delta G_r^o$ ist mit dem relativen Fehler der Gleichgewichtskonstanten dK/K gemäß

$$\frac{dK}{K} = -\frac{d\Delta G_r^o}{RT} \tag{30}$$

verknüpft.

Vollständige Konsistenz aller Daten ist derzeit nicht erreichbar, bleibt aber eine Forderung für die künftige thermochemische Grundlagenforschung und kritische Kompilation der Daten.

10. Anordnung der Stoffe

Die Anordnung der Elemente und Verbindungen in den Tabellen richtet sich nach dem Alphabet und nach dem periodischen System. Die Elemente sind in der alphabetischen Ordnung ihrer chemischen Symbole aufgeführt. Dem Element folgen seine Verbindungen entsprechend dem periodischen System, zunächst die Hydride und Halogenide (Gruppe VIIa), dann die Chalcogenide (Gruppe VIa) usw. über die b-Gruppen zur Gruppe Ia. Der Benutzer der Tabellen wird sich leicht an diese Anordnung gewöhnen. Der Substanzen-Index auf S. LIII erlaubt eine schnelle Orientierung.

Der Substanzen-Index umfaßt die Stoffe des ersten und des vorliegenden Ergänzungsbandes. Es wird für jeden Stoff der Band und die Seitenzahl angegeben. Für Stoffe aus dem ersten Band, deren Daten korrigiert worden sind, werden beide Stellen angegeben. Die Seitennummern für einige Stoffe sind in Klammern gesetzt. Es handelt sich hierbei meist um Stoffe, die zwar präparativ unter besonderen Bedingungen hergestellt werden können, die aber bei höheren Temperaturen instabil sind und deren thermochemische Daten sehr unsicher sind.

Im Substanzen-Index sind schließlich die Fehlergrenzen nach der in Abschnitt 9 erläuterten Bezeichnungsweise angegeben.

11. Danksagung

Die Verfasser danken Herrn Dr. H. Ünal für seine unermüdlich Hilfe bei der Überprüfung der Konsistenz der Daten und Berechnung der Tabellen.

Ferner ist denen zu danken, die sich zum ersten Band kritisch geäußert und Fehler aufgedeckt und uns Informationen für den zweiten Band gegeben haben.

Der Deutschen Forschungsgemeinschaft ist für finanzielle Unterstützung und dem Rechenzentrum der Technischen Hochschule Aachen für bereitwillige Zusammenarbeit zu danken.

12. In den Tabellen benutzte Abkürzungen und Symbole

A, B, C, D:	Temperaturkoeffizienten der Gleichungen für die Molwärme CP
CP:	Molwärme in $cal \cdot mol^{-1}K^{-1}$ $CP = A + B\,10^{-3}T + C\,10^{5}T^{-2} + D\,10^{-6}T^{2}$
BPT:	Siedepunkt in K
BT:	β-Funktion, $\beta(T) = -10^{3}\,G(T)/4{,}575\,T$
DEC.:	Zersetzung
DPT:	Temperatur der thermischen Zersetzung in K
G:	Freie Enthalpie in $kcal \cdot mol^{-1}$
GAS:	"ideales Gas"
H:	Enthalpie in $kcal \cdot mol^{-1}$
L:	Molare Umwandlungswärme in $kcal \cdot mol^{-1}$
LIQ:	flüssige Phase
MPT:	Schmelzpunkt in K
P:	Dampfdruck in Torr
RANGE:	Gültigkeitsbereich der Interpolationsformeln für CP in K
REF.:	Literaturstelle
S:	Entropie in $cal \cdot mol^{-1}K^{-1}$
SOL:	feste Phase

SOL-A:	α-Phase (A : α, B : β, C : γ, D : δ, E : ε, A2 : α_2)	T:	Temperatur in K
SPT:	Sublimationspunkt in K	TPT:	Allotrope Umwandlungstemperatur in K

Im Substanzen-Index bezeichnen die Buchstaben A, B, C, D Fehlergrenzen wie in Abschnitt 9 beschrieben.

Index of substances

In the third column, the number 1 indicates the edition of 1973, the number 2 the present supplement. With regard to "Accuracy" (column 2) see section 9 of the introduction. Comp. = Compound; Acc. = Accuracy; V. = Volume; P. = Page

Selected bibliography

Reference numbers between 1 and 56 omitted in this list are cited in the edition of 1973 but have not been used in this volume.

An asterisk (*) in the reference column of the tables means that the values given have been estimated by the present authors.

1 Kubaschewski, O.; Evans, E.Ll.; Alcock, C.B.: Metallurgical Thermochemistry, London, Pergamon Press 1967. Japanese ed. 1968

2 Landolt-Börnstein: Zahlenwerte und Funktionen, II. Band, Eigenschaften der Materie in ihren Aggregatzuständen, 4. Teil, Kalorische Zustandsgrößen, Berlin, Göttingen, Heidelberg: Springer 1961

3 JANAF Thermochemical Tables, US Department of Commerce/National Bureau of Standards/Institute for Applied Technology, Juni 1971, Supplements 1974, 1975

4 Mčedlov-Petrosjan, O.P.; Babuškin, V.I.; Matveev, G.M.: Thermodynamik der Silikate, Berlin: VEB-Verlag für Bauweisen 1966

5 Kubaschewski, O.: High Temp. - High Pressure 4 (1972) 1

Spencer, P.J.: The Thermodynamic Porperties of Silicates and Alkali Metal Alloys, Teddington, National Physical Laboratory, DCS Reports 21 and 10 (1973, 1971)

Abbasov, A.S.: Evaluation of the Thermodynamik Properties of some III-V Compounds, Teddington, National Physical Laboratory, DCS Report 8 (1970)

9 Holley, C.E.; Storms, E.K.: In Thermodynamics of Nuclear Materials, 1967, pp. 387-426, Intern. Atomic Energy Agency, Vienna 1968

10 Schick, H.L.: Thermodynamics of Certain Refractory Compounds, New York, Academic Press 1966

13 Krasnov, K.S.; Karoseva, N.V.: Opt. i Spektroskopiya, 19 (1965) 30; Opt. Spectry. 19 (1965) 14

16 Brewer, L.; Somayojulu, G.R.; Brackett, E.: Chem. Rev. 63 (1963) 111

19 Wagman, D.D., et al.: Selected Values of Chemical Thermodynamic Properties, National Bureau of Standards, Technical Note 270-3/7, Washington 1968/73

19a Rossini, F.D., et al.: Selected Values of Chemical Thermodynamic Properties, National Bureau of Standards, Washington, 1952, Circular 500

21 Hultgren, R.; Orr, R.L.; Anderson, D.; Kelley, K.K.: Selected Values of Thermodynamic Properties of Metals and Alloys, New York; John Wiley & Sons, 1974

22 Quill, L.L. (Editor): The Chemistry and Metallurgy of Miscellaneous Materials: Thermodynamics, Paper 6, The Thermodynamic Properties of the Halides, by Brewer, L., Bromley, L.A., Gilles, P.W., and Lofgren, N.L., Paper 7, The Fusion and Vaporization Data of the Halides by Brewer, L., New York: McGraw Hill, 1950

27 Rand, M.H.; Kubaschewski, O.: The Thermochemical Properties of Uranium Compounds, Edinburgh: Oliver and Boyd 1963

32 Amosov, V.M.: Izw. VUZ, Tsvetnaya Metallurgiya 6 (1963) 103

33 Schäfer, H.; Kahlenberg, F.: Z. anorg. allg. Chem. 294 (1958) 242, and 305 (1960) 178, 291, 327. Schäfer, H.; Sibbing, E.: Z. anorg. allg. Chem. 305 (1960) 341

40 National Bureau of Standards, Technical News Bulletin, Jan 1971 CODATA Bulletin 10, December 1973, CODATA Special Rep. 3, Sept. 1975

46 Handbook of Chemistry and Physics: 49th Edition, The Chemical Rubber Co. 1968-1969

47 D'Ans Lax: Taschenbuch für Chemiker und Physiker, Berlin, Heidelberg, New-York: Springer 1967

48 Kirschning, H.J.; Plieth, K.; Stranski, I.N.: Z. Kristallographie 106 (1955) 172, Beckers, K.A.; Plieth, K.; Stranski, I.N.: Progress in Inorganic Chemistry 4 (1962) 1, Interscience, New York and London

57 Karapyet'yants, M.Kh.; Karapyet'yants, M.L.: Osnovye termodynamicheskie konstanty, Neorganischeskikh i organicheskikh veshchestv, 1968, Khimiya, Moskau

58 Kelley, K.K.: U.S. Bur. Mines, Bull. No. 584, 1960, Bull. No. 592, 1961

59 King, E.G.; Mah, A.D.; Pankratz, L.B.: INCRA-Series on Metallurgy of Copper, Vol. 2 (1973)

60 Kubaschewski, O.; Catterall, J.A.: Thermochemical Data of Alloys, London: Pergamon 1956

61 Mah, A.D.: U.S. Bur. Mines, Rep. Invest. No. 6727, 1966

62 Glushko, V.P.; Medvedev, V.A. et al. (Editors): "Termicheskie konstanty veshchestv", Part I, 1965, (O, H, halogens, inert gases); Part 2, 1966 (S, Se, Te, Po); Part 3, 1968 (N, P, As, SB, Bi); Part 4, Section 1, 1970, Section 2, 1971, (C, Si, Ge, Sn, Pb); Part 5, 1971, (B, Al, Ga, In, Ti); Part 6, Section 1, 1972, Section 2, 1973, (Groups VIII b, I b, II b), Moscow, Akademiya Nauk SSSR

63 Mills, K.C.: Thermodynamic Data for Inorganic Sulphides Selenides and Tellurides, London: Butterworths 1974, and private communications 1975

64 Weibke, F.; Kubaschewski, O.: Thermochemie der Legierungen. Berlin: Springer 1943

Kubaschewski, O.; Heymer, G.: Trans. Farad. Soc. 56 (1960) 473

65 Atomic Energy Reviews, International Atomic Energy Agency, Vienna, No. 1 Plutonium (M.H. Rand) 1966, No. 2 Niobium (V.I. Lavrentev and Ya.I. Gerassimov) 1968, No. 3 Tantalum (Ya.I. Gerassimov u. V.I. Larentev) 1972, No. 4 Beryllium (P.J. Spencer) 1973, No. 5 Thorium (M.H. Rand) 1974, Techn. Rep. No. 14 (1963)

66 Nat. Bureau of Standards: Sulfates 1966, Carbonates 1969. High Temperature Properties and Decomposition of Inorganic Salts. Stern, K.H., Weise, E.L., Sulfates (1966), Carbonates (1969)

67 Abbasov, A.S., et al.: Dokl. Ak. Nauk. SSSR 156 (1964) 596

68 Adams, G.P.; Margrave, J.L.; Wilson, P.W.: J. Chem. Thermodyn. 2 (1970) 741

69 Akhachinski, V.V.; et al.: Thermodynamics I.A.E.A., Vienna (1962) 309

70 Andon, R.J.L.; Martin, J.F.; Mills, K.C.: J. Chem. Soc. 1971 (A) 1788

71 Andon, R.J.L.; Mills, K.C.: J. Chem. Thermodyn. 3 (1971) 583

72 Baker, F.B., et al.: J. Chem. Thermodyn. 3 (1971) 77

73 Bell, W.E.; Garrison, M.C.; Merten, U.: J. Phys. Chem. 65 (1961) 517

74 Bell, W.E.; Merten, U.; Tagami, M.: J. Phys. Chem. 65 (1961) 510, 517; 66 (1962) 490

75 Bell, W.E.; Tagami, M.: J. Phys. Chem. 1966 70 37 36

76 Bieganski, Z.: Ber. Bunsenges. Phys. Chem. 76 (1972) 1183

77 Blachnik, R.: Ph. D. Thesis, TH Clausthal-Zellerfeld, 1967

78 Blachnik, R.; Igel, R.; Wallbrecht, P.: Z. Naturforsch. 29 (1974) 1198

79 Blachnik, R.; Schneider, A.: J. Chem. Thermodyn. 3 (1971) 227

80 Bokeriya, B.N.; Tsagareishvili, D.Sh.; Gvelesiani, G.G.: Acad. Sci. Georgian SSR, Communications 58, (1970) 369; 58, (1970) 601

81 Chadwick, I.; Irving, R.J.; Sousa-Alonso, A.: J. Chem. Thermodyn. 2 (1970) 611

82 Chatterji, D.; Vest, R.W.: J. Amer. Ceram. Soc. 54 (1971) 73

83 Cubicciotti, D.; Eding, H.: J. Phys. Chem. 69 (1965) 3621

84 Dannöhl, H.D.: Ph.D. Thesis, Stuttgart 1971

85 Deenadas, C.; et al.: J. Phys. Chem. Solids 32 (1971) 1853

86 Denielou, L.; et al.: Compt. rend. 272 (1971) 1855

87 Domalski, E.S.; Armstrong, G.T.: J. Res. Nat. Bur. Stand 71,A (1967) 307

88 Dudchik, G.P.; Polyachenok, O.G.; Novikov, G.I.: Russ. J. Phys. Chem. 45 (1971) 409

89 Dworkin, A.S.; Bredig, M.A.: High Temp. Sci. 3, 1 (1971) 81; J. Phys. Chem. 72 (1968) 1277; J. Phys. Chem. 67 (1963) 697, 2499

90 Eldridge, J.M.; Miller, E.; Komarek, K.L.: Trans. Met. Soc. 233 (1965) 1303; 239 (1967) 570

91 Eremenko, V.N.; et. al.: Russ. J. Phys. Chem. 45 (1971) 1129

92 not used

93 Finogenov, A.D.: Russ. J. Phys. Chem. 45 (1971) 900

94 Fitzgibbon, G.C.; Holley, C.E.: J. Chem. Eng. Data 13 (1968) 63

95 Flowers, R.H.; Rauh, E.G.: J. Inorg. Nucl. Chem. 28 (1966) 1355

96 Franco, J.I.; Kleykamp, H.J.: Ber. Bunsenges. Phys. Chem. 75 (1971) 934, 76 (1972) 691

97 Frederickson, D.R.; Chasanov, M.C.: J. Chem. Thermodyn. 3 (1971) 693

98 Frederickson, D.R.; et al.: High Temp. Sci. 3, 3 (1970) 299

99 Fuger, J.; Brown, D.; Easy, J.F.: J. Chem. Soc. (A) 1969 Inorg. Phys. Theor. p. 2995

100 Glybin, V.P.; Dobrotin, R.B.: Russ. J. Phys. Chem. 45 (1971) 898

101 Gokcen, N.A.; Fujishiro, S.: Trans. AIME 227 (1963) 542

102 Gorelkin, O.S.; et al.: Russ. J. Phys. Chem. 46 (1972) 431

103 not used

104 Gross, P.; Hayman, C.: Fulmer Res. Inst. Private Communication

105 Gross, P.; Hayman, C.; Stuart, M.C.: Trans. Farad. Soc. 65 (1969) 2628/2856

106 Grönvold, F.; Westrum, E.F.: Inorg. Chem. 1 (1962) 36

107 Gupta, S.K.: J. Phys. Chem. 73 (1969) 4086

108 Haraldsen, H.: Z. anorg. allg. Chem. 240 (1939) 337

109 O'Hare, P.A.G.; Wahl, A.C.: J. Chem. Phys. 56 (1972) 4513, 4516

110 Hariharan, A.V.; Eick, H.A.: High Temp. Sci. 3 (1971) 123

111 Hepler, L.G.: J. Am. Chem. Soc. 80 (1958) 6181

112 not used

113 Holleck, H.; Kleykamp, H.: J. Nucl. Mat. 1970 35 158

114 not used

115 Howlett, B.W.; Misra, S.; Bever, M.B.: Trans. Met. Soc. 230 (1964) 1367

116 Huber, E.J.; et al.: J. Phys. Chem. 67 (1963) 1731

117 Huber, E.J.; Fitzgibbon, G.C.; Holley, C.E.: J. Chem. Thermodyn. 3 (1971) 643

118 Huber, E.J.; Holley, C.E.; Whiteman, W.G.: J. Chem. Thermodyn. 1 (1969) 579

119 Johnson, I.; Feder, H.M.: Trans. Met. Soc. AIME 224 (1962) 468

120 Kashkovli, I.Y.; Munier, Z.A.: High Temp. Sci. 4 (1972) 82

121 Kaylor, C.E.; et al.: J. Phys. Chem. 64 (1960) 276; J. Am. Chem. Soc. 81 (1959) 4172

122 Kedrovskii, O.V.; et al.: Russ. J. Phys. Chem. 41 (1967) 205

123 Kerimov, I.G.; et al.: Russ. J. Phys. Chem. 45 (1971) 1118

124 Kirpichev, E.P.; et al.: Russ. J. Phys. Chem. 45 (1971) 860

125 Kleykamp, H.: Z. phys. Chem. Ffm. 67 (1969) 277

126 Kleppa, O.J.; Navrotsky, A.: J. Inorg. Nucl. Chem. 30 (1968) 479

127 Kogan, B.S.: Russ. J. Phys. Chem. 45 (1971) 405, 891

128 Kolbin, N.I.; Samoilov, V.M.: Russ. J. Inorg. Chem. 14 (1969) 330

129 Kornyeyev, V.P.; Gluskova, V.B.; Kelev, E.K.: Izv. Akad. Nauk. Neorg. Mat. 7 (1971) 886

130 Kostryukov, V.N.; Kalininkina, I.N.: Russ. J. Phys. Chem. 38 (1964) 422

131 Kuan, Pan: J. Chinese Soc. Taiwan Ser. II 1 (1954) 1

132 Kubaschewski, P.; Nölting, H.: Ber. Bunsenges. Phys. Chem. 77 (1973) 70

133 Kuznetsov, F.A.; Rezukhina, T.N.: Russ. J. Phys. Chem. 34 (1960) 1146; 35 (1961) 470

134 Krüger, O.L.; Sewage, H.: J. Chem. Phys. 49 (1968) 4540

135 MacLeod, A.C.: J. Inorg. Nucl. Chem. 33 (1971) 2419

136 Leonidov, V.Ya.; Rezukhina, T.N.; Bereznikova, I.A.: Russ. J. Phys. Chem. 34 (1960) 885

137 L'vova, A.S.; Feodos'ev, N.N.: Russ. J. Phys. Chem. 39 (1965) 1091

138 Lim, M.; Searcy, A.W.: J. Phys. Chem. 70 (1966) 1762

139 Lukashenko, G.M.; Polotshaya, R.I.: Russ. J. Phys. Chem. 45 (1971) 1358

140 Lyon, W.G.; Westrum, E.F.; Chavret, M.: J. Chem. Thermodyn. 3 (1971) 571

141 Martin, J.F.; Müller, F.; Kubaschewski, O.: Trans. Farad. Soc. 66 (1970) 1065

142 Mehrotra, G.M.; et al.: Scripta Metall. 7 (1973) 1047

143 Melekh, B.T.; et al.: Russ. J. Phys. Chem. 45 (1971) 1144

144 Mills, K.C.: J. Chem. Thermodyn. 4 (1972) 903

145 Mills, K.C.: High Temp. - High Press. 4 (1972) 371

146 Morozowa, M.P.; Stolyarova, T.A.: Vestn. Leningr. Univ. 19 (16) Ser. Fiz. Khim. (1964) 3-150

147 Muradov, V.G.; Gel'd, P.V.; Kocherov, P.V.: Russ. J. Phys. Chem. 41 (1967) 290

148 Neil, J.M.; Navrotsky, A.; Kleppa, O.J.: Inorg. Chem. 10 (1971) 2076

149 Newland, B.G.; Shelton, R.A.: J. Less Comm. Met. 20 (1970) 245

150 not used

151 Novotny, D.B.; Smith, J.F.: J. Inst. Met. 92 (1963) 122

152 Oetting, F.L.: Chem. Rev. 67 (1967) 261

153 Ogren, I.R.; Magnani, N.J.; Smith, J.F.: Trans. Met. Soc. AIME (1976) 766

154 Opalowski, A.A.; et al.: Russ. J. Phys. Chem. 45 (1971) 1197

155 not used

156 Paukov, I.E.; Khriplovich, L.M.; Popov, A.B.: Russ. J. Phys. Chem. 44 (1970) 309

157 Pfeiffer, M.; Spangenberg, H.J.: Z. phys. Chem. (Leipzig) 232 (1966) 47

158 Predel, B.; Rothacker, D.: Thermochimica Acta 1 (1970) 477

159 Predel, B.; Ruge, H.: Thermochimica Acta 3 (1972) 411

160 Rasch, R.: Sprechsaal 103 (1970) 117

161 Rao, A.V.R.; Tare, V.B.: Scripta Met. 5 (1971) 807

162 Rao, B.S.; Dadape, V.V.: High Temp. Sci. 3 (1971) 1

163 Rezukhina, T.N.: Thesis Moscow 1968

164 Rezukhina, T.N.; Pokarev, B.S.: J. Chem. Thermodyn. 3 (1971) 369

Kuznetsov, F.A.; Rezukhina, A.N.; Golushenko, A.N.: Russ. J. Phys. Chem. 34 (1960) 1010

165 Richelmi, R.; Laffitte, M.: Compt. rend. 265 (1967) 541

166 Roth, W.A.; Berendt, H.; Wirths, G.: Z. Elektrochem. 47 (1941) 185

167 Sandakova, M.I.; et al.: Russ. J. Phys. Chem. 45 (1971) 901

168 Schäfer, H.; Bruderreck, H.; Morcher, B.: Z. anorg. allg. Chem. 352 (1967) 122

169 Schwitzgebel, K.; et al.: J. Chem. Eng. Data 16 (1971) 418

170 Sevastyanova, T.N.; Karpenko, N.V.: Russ. J. Phys. Chem. 45 (1971) 1044

171 Shchukarev, A.A.; Balakireva, I.N.; Voskresenski, A.A.: Zh. Obzhchei. Khim 31 (1961) 1773

172 Shidkovski, A.A.; Balakireva, I.N.; Voskresenski, A.A.: Russ. J. Phys. Chem. 45 (1971) 1060

173 Sirota, N.N.; Smolyarenko, E.M.: Russ. Metall. (1968) 6-159

174 Spedding, F.H.; Henderson, D.C.: J. Chem. Phys. 54 (1971) 1276, 2476

175 Srivastava, R.D.; Farber, M.: J. Phys. Chem. 75 (1971) 1760

176 Stull, D.R.; et al.: J. Chem. Eng. Data 15 (1970) 52

177 Tananaev, I.V.; et al.: Izv. Akad. Nauk. Neorg. Mat. 4 (1968) 2124

178 not used

179 Thomson, W.T.; Flengas, S.N.: Can. J. Chem. 49 (1971) 1550

180 Topol, L.E.; Ransom, L.D.: J. Phys. Chem. 64 (1960) 1339

181 Tsagareishvili, D.Sh.; Gvelesiani, G.G.: Izv. Akad. Nauk., Neorg. Mat. 7 (1971) 1878

182 Tveekrem, J.O.; Chandrasekharaiah, M.S.: J. Electrochem. Soc. 115 (1968) 1021

183 Vasantasree, Miss: Thesis, Imp. Coll. London 1971

184 Vasileff, W.: Grayson-Smith, H.: Canad. J. Res. Chem. 1950 2 BA 367

185 Vasilev, V.P.; et al.: Russ. J. Phys. Chem. 45 (1971) 1169

186 Virmani, Y.; Skinner, H.A.: Manchester Univ., Private Communication 1973

187 Vorob'ev, A.M.; Evseeva, G.V.; Zenkevich, L.V.: Russ. J. Phys. Chem. 45 (1971) 1501

188 Vorob'ev, A.M.; Broier, A.F.: Russ. J. Phys. Chem. 45 (1971) 1358

189 Voronin, G.F.: Russ. J. Phys. Chem. 45 (1971) 1191

190 Westrum, E.F.; et al.: J. Chem. Phys. 35 (1961) 1670; J. Phys. Chem. 67 (1963) 2385

191 Westrum, E.F.: "Lanthanide/Actinide Chemistry" 1967, Progress in the Science and Technology of Rare Earths, Vol.2, Oxford: Pergamon 1966

192 Westrum, E.F.: "Thermodynamics" I.A.E.A. 2 (1966) 497

193 Zhdanov, V.M.: Russ. J. Phys. Chem. 45 (1971) 1356

194 Zumbusch, M.; Heimbrecht, M.; Biltz, W.: Z. anorg. allg. Chem. 242 (1939) 237

195 Mezaki, R.; et al.: "Thermodynamics" I.A.E.A. wien (1962) 775

196 Flotow, A.E.; et al.: J. Chem. Phys. 51 (1969) 583

197 Spencer, P.J.: Private Communication, (1976)

198 Cubicciotti, D.: Inorg. Chem. 7 (1968) 208

199 Gedansky, L.M.; Betrand, G.L.; Hepler, L.G.: J. Chem. Eng. Data 12 (1967) 135

200 Wilcox, D.E.: Thesis, quoted by Sladek, K.J.: J. Chem. Eng. Data 16 (1971)

201 Zmbov, K.F.; Margrave, J.L.: J. Chem. Phys. 45 (1966) 3167

202 Johnson, G.K.; et al.: J. Chem. Phys. 1 (1969) 89

203 Porte, H.A.; Greenberg, E.; Hubbard, W.N.: J. Phys. Chem. 69 (1965) 2308

204 Reznitokii, L.A.; et al.: Russ. J. Phys. Chem. 45 (1971) 529

205 Cubicciotti, D.: J. Electrochem. Soc. 115 (1968) 1138

206 Smith, W.H.: General Electric Res. Lab., Rep. No. 75-RL-1816, Okt. 1957

Seyboldt, A.U.; Oriani, R.A.: J. Metals 8 (1956) 556

Caplan, D.; Fraser, M.J.; Burr, A.A.: Ductile Cr Alloys, Nat. Met. Congr. Philadelphia, Okt. 1955

207 Gattow, G.; Krebs, B.: Z. anorg. allg. Chem. 322 (1963) 113

208 Jordan, A.S.: J. Electrochem. Soc. 118 (1971) 1362

209 Pankratz, L.B.; Weller, W.W.: U.S. Bur. Mines, Rep. Invest. No. 6687 (1965); 7280 (1960)

210 Work, D.E.; von Eick, H.A.: J. Phys. Chem. 74 (1970) 3130

211 Schäfer, H.; Heine, H.: Z. anorg. allg. Chem. 332 (1964) 25

212 Nagamori, M.; Habashi, F.: Metall. Trans. 4 (1974) 523

213 Bernard, O.; Ansara, I.: Rapport LTCM-1974-TM-O2, ENSEEG - B.P.44-38401 St.-Martin d'Hères, France

214 Campbell, G.M.; et al.: Pu 1970

215 Ondracek, G.; Wedemeyer, H.: Sprechsaal 108 (1975) 55

216 Shunk, F.A.: Constitution of Binary Alloys, Second Supplement, New York: McGraw-Hill 1969

217 Schäfer, H.; Tebben, A.; Gerhard, W.: Z. anorg. allg. Chem. 32 (1963) 41

218 Alcock, C.B.; Hooper, G.W.: Proc. Roy. Soc.(A), 254 (1960) 551

219 Münstermann, E.: Diss. Rhein-West, Techn. Hochsch. Aachen 1976

220 Samakhvol, V.V.; et al.: Russ. J. Phys. Chem. 45 (1971) 1171

221 Chart, T.G.: High Temp. - High Pressure 5 (1973) 241; Metal Sci. 8 (1974) 344, 9 (1975) 504

222 Leonov, A.I.; et al.: Russ. J. Inorg. Chem. 14 (1969) 334

223 Stretz, L.A.; Batista, R.G.: Metall. Trans. 5 (1974) 921

224 Rau, H.: J. Chem. Thermodyn. 7 (1975) 27

225 Rau, H.: J. less-Common Metals 24 (1971) 291

226 Martosudirdjo, S.U.; Pratt, J.N.: Thermochim. Acta 10 (1974) 23

227 Sibbing, E.; Schäfer, H.: Z. anorg. allg. Chem. 410 (1974) 67

228 Biltz, W.; Heimbrecht, M.: Z. anorg. allg. Chem. 237 (1938) 132

229 Spear, K.E.; Schäfer, H.; Gilles, P.W.: 3rd Intern. Symposium on High Temperature Technology, (Asilamov in Pacific Grove), California, USA, Butterworths, London, 1967

230 Said, H.; Castanet, R.; Kehiaian, H.V.: J. Less Comm. Metals, in Press

231 Tovopov, N.A.; et al.: Handbook of Phase Diagrams of Silicate Systems Israel Program for Scientific Translations, Jerusalem 1972

232 Jacob, H.T.; Alcock, C.B.: High Temp. High Press. 7 (1975) 43

233 Müller, F.; Kleppa, O.J.: J. inorg. nucl. Chem. 35 (1973) 2673

234 Tretjakov, I.D.; Schmalzried, N.: Ber. Bunsenges. Phys. Chem. 69 (1965) 396

Tables

SILVER

		A	B	C	D	RANGE	REF.
SOL	CP	5.693	1.223			298- 600	21
	CP	4.716	2.294	1.274		600-1234	21
	** MPT= 1233.95 (REF. POINT)						21
LIQ	CP	8.00				1234-2436	21
GAS	CP	4.968				2436-2800	21

PHASE	T	CP	H	S	G	BT	P	
SOL	298	6.058	0	10.20	-3.041	2.230		21
	300	6.060	.011	10.237	-3.060	2.230		
	400	6.182	.623	11.998	-4.176	2.282		
	500	6.304	1.248	13.390	-5.447	2.381		
	600	6.427	1.884	14.550	-6.846	2.494		
	600	6.446	1.884	14.550	-6.846	2.494		
	700	6.582	2.535	15.554	-8.352	2.608		
	800	6.750	3.202	16.443	-9.953	2.719		
	900	6.938	3.886	17.249	-11.638	2.827		
	1000	7.137	4.590	17.990	-13.401	2.929		
	1100	7.345	5.314	18.680	-15.235	3.027		
	1200	7.557	6.059	19.328	-17.135	3.121	.001	
	1234	7.630	6.317	19.540	-17.795	3.152	.003	
			2.7	2.188				21
LIQ	1234	8.000	9.017	21.728	-17.795	3.152	.003	
	1300	8.000	9.545	22.146	-19.244	3.236	.011	
	1400	8.000	10.345	22.738	-21.489	3.355	.060	
	1500	8.000	11.145	23.290	-23.791	3.467	.272	
	1600	8.000	11.945	23.807	-26.146	3.572	1.011	
	1700	8.000	12.745	24.292	-28.551	3.671	3.202	
	1800	8.000	13.545	24.749	-31.003	3.765	8.875	
	1900	8.000	14.345	25.181	-33.500	3.854	21.998	
	2000	8.000	15.145	25.592	-36.039	3.939	49.594	
	2100	8.000	15.945	25.982	-38.618	4.020	103.103	
	2200	8.000	16.745	26.354	-41.235	4.097	199.885	
	2300	8.000	17.545	26.710	-43.888	4.171	364.737	
	2400	8.000	18.345	27.050	-46.576	4.242	631.263	
	2436	8.000	18.633	27.169	-47.552	4.267	760.	
			59.90	24.589				21
GAS	2436	4.968	78.533	51.759	-47.552	4.267		
	2500	4.968	78.851	51.888	-50.869	4.448		
	2600	4.968	79.348	52.083	-56.067	4.714		
	2700	4.968	79.845	52.270	-61.285	4.961		
	2800	4.968	80.341	52.451	-66.521	5.193		

$AgBrO_3$ SILVER BROMATE

		A	B	C	D	RANGE	REF.
SOL	CP	16.0	30.0			298- 400	*
	** DEC.						

PHASE	T	CP	H	S	G	BT	
SOL	298	24.944	-6.3	36.8	-17.272	12.662	62
	300	25.000	-6.254	36.954	-17.340	12.634	
	400	28.000	-3.604	44.557	-21.427	11.709	

SILVER SULFIDE

Ag_2S

		A	B	C	D	RANGE	REF.
SOL-A	CP	15.44	9.55			298- 449	179
SOL-B	CP	19.44	0.7			449- 859	179
SOL-C	CP	19.77				859-1103	179
LIQ	CP	22.25				1103-1300	179
	** DEC.						1

PHASE	T	CP	H	S	G	BT	
SOL-A	298	18.287	-7.6	34.3	-17.827	13.069	1,63
	300	18.305	-7.566	34.413	-17.890	13.035	
	400	19.260	-5.688	39.810	-21.612	11.810	
	449	19.728	-4.733	42.062	-23.619	11.498	
			0.94	2.094			179
SOL-B	449	19.754	-3.793	44.156	-23.619	11.498	
	500	19.790	-2.784	46.283	-25.926	11.334	
	600	19.860	-.802	49.897	-30.740	11.199	
	700	19.930	1.188	52.964	-35.887	11.206	
	800	20.000	3.184	55.630	-41.320	11.290	
	859	20.041	4.365	57.054	-44.644	11.360	
			0.12	.140			179
SOL-C	859	19.770	4.485	57.194	-44.644	11.360	
	900	19.770	5.296	58.116	-47.008	11.417	
	1000	19.770	7.273	60.199	-52.926	11.568	
	1100	19.770	9.250	62.083	-59.041	11.732	
	1103	19.770	9.309	62.137	-59.228	11.737	
			1.88	1.704			179
LIQ	1103	22.250	11.189	63.841	-59.228	11.737	
	1200	22.250	13.348	65.717	-65.513	11.933	
	1300	22.250	15.573	67.498	-72.174	12.135	

Ag_2Te SILVER TELLURIDE

		A	B	C	D	RANGE	REF.
SOL-A	CP	11.758	26.2	0.66		298- 421	63
SOL-B	CP	21.278	-1.720			421- 550	63
	CP	20.08				550- 800	63
	** MPT=	1232					63

PHASE	T	CP	H	S	G	BT	
SOL-A	298	20.312	-8.6	36.7	-19.542	14.327	63
	300	20.351	-8.562	36.826	-19.610	14.288	
	400	22.650	-6.415	42.989	-23.610	12.902	
	421	23.161	-5.934	44.161	-24.525	12.733	
			1.577	3.746			63
SOL-B	421	20.554	-4.357	47.906	-24.525	12.733	
	500	20.418	-2.738	51.430	-28.453	12.439	
	550	20.332	-1.719	53.372	-31.074	12.349	
	550	20.080	-1.719	53.372	-31.074	12.349	
	600	20.080	-.715	55.119	-33.787	12.309	
	700	20.080	1.293	58.214	-39.458	12.321	
	800	20.080	3.301	60.896	-45.416	12.409	

SILVER 1.64-TELLURICE $Ag_{1.64}Te$

		A	B	C	D	RANGE	REF.
SOL	CP	9.563	29.590			298- 418	63
	CP	24.694	-6.494			418- 569	63
	CP	18.88				569- 625	63
	CP	18.52				625- 697	63
	CP	18.76				697- 737	63
	CP	17.45				737- 800	63

PHASE	T	CP	H	S	G	BT	
SOL	298	18.385	-7.5	32.	-17.041	12.493	63
	300	18.440	-7.466	32.114	-17.100	12.459	
	400	21.399	-5.474	37.824	-20.604	11.259	
	418	21.932	-5.084	38.778	-21.293	11.134	
			0.31	.742			63
	418	21.980	-4.774	39.519	-21.293	11.134	
	500	21.447	-2.994	43.410	-24.699	10.797	
	569	20.999	-1.529	46.154	-27.791	10.676	
			0.049	.086			63
	569	18.880	-1.480	46.240	-27.791	10.676	
	600	18.880	-.895	47.242	-29.240	10.652	
	625	18.880	-.423	48.013	-30.431	10.642	
			0.059	.094			63
	625	18.520	-.364	48.107	-30.431	10.642	
	697	18.520	.970	50.126	-33.968	10.653	
			0.526	.755			63
	697	18.760	1.496	50.881	-33.968	10.653	
	700	18.760	1.552	50.962	-34.121	10.655	
	737	18.760	2.246	51.928	-36.025	10.684	
			0.287	.389			63
	737	17.450	2.533	52.317	-36.025	10.684	
	800	17.450	3.632	53.749	-39.367	10.756	

AgP_2 SILVER DIPHOSPHIDE

		A	B	C	D	RANGE	REF.
SOL	CP	15.7	5.5			298- 900	*

PHASE	T	CP	H	S	G	BT	
SOL	298	17.340	-10.7	21.	-16.961	12.435	1
	300	17.350	-10.668	21.107	-17.000	12.386	
	400	17.900	-8.905	26.174	-19.375	10.587	
	500	18.450	-7.088	30.227	-22.202	9.706	
	600	19.000	-5.215	33.640	-25.399	9.253	
	700	19.550	-3.288	36.610	-28.915	9.029	
	800	20.100	-1.305	39.256	-32.710	8.937	
	900	20.650	.732	41.656	-36.758	8.927	

* * * * * * * * * *

AgP_3 SILVER TRIPHOSPHIDE

		A	B	C	D	RANGE	REF.
SOL	CP	22.6	7.5			298- 900	*

PHASE	T	CP	H	S	G	BT	
SOL	298	24.836	-16.5	25.2	-24.013	17.605	1
	300	24.850	-16.454	25.354	-24.060	17.530	
	400	25.600	-13.932	32.605	-26.974	14.740	
	500	26.350	-11.334	38.398	-30.533	13.348	
	600	27.100	-8.662	43.269	-34.623	12.613	
	700	27.850	-5.914	47.503	-39.166	12.230	
	800	28.600	-3.092	51.270	-44.108	12.051	
	900	29.350	-.194	54.682	-49.408	12.000	

SILVER CARBONATE Ag_2CO_3

		A	B	C	D	RANGE	REF.
SOL	CP	18.97	25.85			298- 493	1
	** DPT= 493						

PHASE	T	CP	H	S	G	BT	
SOL	298	26.677	-120.8	40.	-132.726	97.304	1*
	300	26.725	-120.751	40.165	-132.800	96.758	
	400	29.310	-117.949	48.207	-137.232	74.990	
	493	31.714	-115.111	54.577	-142.018	62.966	

* * * * * * * * * *

SILVER CHROMATE Ag_2CrO_4

		A	B	C	D	RANGE	REF.
SOL	CP	31.6	16.0	-2.13		298- 500	19,*

PHASE	T	CP	H	S	G	BT	
SOL	298	33.974	-172.4	52.	-187.904	137.756	131/19
	300	34.033	-172.337	52.210	-188.000	136.976	
	400	36.669	-168.795	62.383	-193.748	105.873	
	500	38.748	-165.021	70.795	-200.419	87.615	

Al

ALUMINIUM

		A	B	C	D	RANGE	REF.
SOL	CP	7.499	-3.918	-0.862	4.960	298- 933	3
LIQ	CP	7.588				933-2767	3
	** BPT=	2766.8					3
GAS	CP	4.971				2767-3200	3

PHASE	T	CP	H	S	G	BT	P	
SOL	298	5.802	0	6.769	-2.018	1.480		3
	300	5.812	.011	6.805	-2.031	1.480		
	400	6.187	.613	8.535	-2.801	1.531		
	500	6.435	1.244	9.942	-3.727	1.629		
	600	6.694	1.900	11.138	-4.782	1.742		
	700	7.011	2.585	12.193	-5.950	1.858		
	800	7.404	3.305	13.154	-7.218	1.972		
	900	7.884	4.069	14.053	-8.579	2.083		
	933	8.063	4.334	14.342	-9.051	2.120		
			2.56	2.743				3
LIQ	933	7.588	6.894	17.085	-9.051	2.120		
	1000	7.588	7.400	17.609	-10.209	2.231		
	1100	7.588	8.159	18.332	-12.006	2.386		
	1200	7.588	8.918	18.993	-13.873	2.527		
	1300	7.588	9.677	19.600	-15.803	2.657		
	1400	7.588	10.436	20.162	-17.792	2.778	.002	
	1500	7.588	11.194	20.686	-19.834	2.890	.013	
	1600	7.588	11.953	21.175	-21.928	2.996	.059	
	1700	7.588	12.712	21.635	-24.068	3.095	.226	
	1800	7.588	13.471	22.069	-26.254	3.188	.742	
	1900	7.588	14.230	22.479	-28.481	3.277	2.138	
	2000	7.588	14.988	22.869	-30.749	3.361	5.521	
	2100	7.588	15.747	23.239	-33.054	3.440	12.986	
	2200	7.588	16.506	23.592	-35.396	3.517	28.178	
	2300	7.588	17.265	23.929	-37.772	3.590	57.013	
	2400	7.588	18.024	24.252	-40.182	3.660	108.517	
	2500	7.588	18.782	24.562	-42.622	3.727	195.750	
	2600	7.588	19.541	24.859	-45.093	3.791	336.758	
	2700	7.588	20.300	25.146	-47.594	3.853	555.477	
	2767	7.588	20.807	25.331	-49.280	3.893	760.	
			69.497	25.118				3
GAS	2767	4.971	90.304	50.449	-49.280	3.893		
	2800	4.971	90.469	50.509	-50.956	3.978		
	2900	4.971	90.966	50.683	-56.015	4.222		
	3000	4.971	91.463	50.852	-61.092	4.451		
	3100	4.971	91.960	51.015	-66.186	4.667		
	3200	4.971	92.457	51.173	-71.295	4.870		

ALUMINIUM HYDRIDE (HEXAGONAL)

		A	B	C	D	RANGE	REF.
SOL	CP	10.8				298- 400	124

PHASE	T	CP	H	S	G	BT	
SOL	298	10.800	-2.75	7.	-4.837	3.546	124/*
	300	10.800	-2.730	7.067	-4.850	3.534	
	400	10.800	-1.650	10.174	-5.720	3.125	

* * * * * * * * * *

ALUMINIUM MONOFLUORIDE (GAS)

		A	B	C	D	RANGE	REF.
GAS	CP	8.019	0.975	-0.604		298- 800	3
	CP	8.907	0.104	-1.831		800-2000	3

PHASE	T	CP	H	S	G	BT	
GAS	298	7.630	-63.4	51.40	-78.725	57.715	3
	300	7.640	-63.386	51.447	-78.820	57.428	
	400	8.031	-62.600	53.705	-84.082	45.947	
	500	8.265	-61.785	55.524	-89.547	39.146	
	600	8.436	-60.949	57.046	-95.177	34.673	
	700	8.578	-60.098	58.358	-100.949	31.522	
	800	8.705	-59.234	59.512	-106.843	29.192	
	800	8.704	-59.234	59.512	-106.843	29.192	
	900	8.775	-58.360	60.541	-112.847	27.407	
	1000	8.828	-57.480	61.468	-118.948	26.000	
	1100	8.870	-56.595	62.312	-125.138	24.866	
	1200	8.905	-55.706	63.085	-131.408	23.936	
	1300	8.934	-54.814	63.799	-137.753	23.161	
	1400	8.959	-53.919	64.462	-144.166	22.508	
	1500	8.982	-53.022	65.081	-150.644	21.952	
	1600	9.002	-52.123	65.661	-157.181	21.473	
	1700	9.020	-51.222	66.208	-163.775	21.058	
	1800	9.038	-50.319	66.724	-170.422	20.695	
	1900	9.054	-49.414	67.213	-177.119	20.376	
	2000	9.069	-48.508	67.678	-183.864	20.094	

$AlF_2(g)$ ALUMINIUM DIFLUORIDE (GAS)

		A	B	C	D	RANGE	REF.
GAS	CP	12.371	1.464	-1.655		298- 800	3A
	CP	13.846	0.025	-3.863		800-2000	3A

PHASE	T	CP	H	S	G	BT	
GAS	298	10.946	-175.	62.9	-193.754	142.044	3A
	300	10.971	-174.980	62.968	-193.870	141.253	
	400	11.922	-173.829	66.271	-200.338	109.474	
	500	12.441	-172.609	68.992	-207.105	90.538	
	600	12.790	-171.347	71.292	-214.122	78.004	
	700	13.058	-170.054	73.285	-221.353	69.119	
	800	13.284	-168.736	75.043	-228.771	62.506	
	800	13.262	-168.736	75.043	-228.771	62.506	
	900	13.392	-167.403	76.613	-236.355	57.403	
	1000	13.485	-166.059	78.029	-244.089	53.353	
	1100	13.554	-164.707	79.318	-251.957	50.066	
	1200	13.608	-163.349	80.500	-259.949	47.350	
	1300	13.650	-161.986	81.591	-268.054	45.070	
	1400	13.684	-160.619	82.604	-276.264	43.133	
	1500	13.712	-159.249	83.549	-284.573	41.468	
	1600	13.735	-157.877	84.434	-292.972	40.024	
	1700	13.755	-156.503	85.268	-301.458	38.760	
	1800	13.772	-155.126	86.054	-310.024	37.647	
	1900	13.786	-153.748	86.799	-318.667	36.660	
	2000	13.799	-152.369	87.507	-327.383	35.780	

ALUMINIUM FLUORIDE

AlF_3

		A	B	C	D	RANGE	REF.
SOL-A	CP	21.740	4.220	-4.487		298- 500	3
	CP	0.907	30.310	14.979		500- 728	3
SOL-B	CP	22.147	2.166	-2.137		728-1549	3
	** BPT= 1549 (89 ALF3 + 11 AL2F6)						3

PHASE	T	CP	H	S	G	ET	P	
SOL-A	298	17.951	-361.0	15.89	-365.738	268.129		3,40
	300	18.020	-360.967	16.001	-365.767	266.497		
	400	20.624	-359.019	21.587	-367.654	200.904		
	500	22.055	-356.879	26.355	-370.057	161.774		
	500	22.054	-356.879	26.355	-370.057	161.774		
	600	23.254	-354.622	30.467	-372.903	135.848		
	700	25.181	-352.205	34.190	-376.138	117.451		
	728	25.799	-351.491	35.189	-377.109	113.226		
			0.135	.185				3
SOL-B	728	23.321	-351.356	35.375	-377.109	113.226		
	800	23.546	-349.669	37.585	-379.737	103.753		
	900	23.833	-347.300	40.375	-383.637	93.172		
	1000	24.099	-344.903	42.900	-387.803	84.766	.004	
	1100	24.353	-342.480	45.209	-392.210	77.935	.093	
	1200	24.598	-340.033	47.338	-396.839	72.284	1.249	
	1300	24.836	-337.561	49.317	-401.672	67.536	11.191	
	1400	25.070	-335.066	51.166	-406.698	63.497	72.970	
	1500	25.301	-332.547	52.903	-411.902	60.022	369.989	
	1549	25.413	-331.305	53.718	-414.514	58.492	760.	
			65.392	42.216				3,*
GAS	1549	19.535	-265.913	95.934	-414.514	58.492		

AlF_3 (g)

ALUMINIUM FLUORIDE (GAS)

		A	B	C	D	RANGE	REF.
GAS	CP	16.587	3.118	-2.291		298- 800	3
	CP	19.751	0.044	-6.817		800-1549	3
	CP	19.751	0.044	-6.817		1549-2000	3

PHASE	T	CP	H	S	G	BT	
GAS	298	14.939	-288.672	66.414	-308.473	226.147	3,*
	300	14.977	-288.644	66.507	-308.596	224.842	
	400	16.402	-287.067	71.033	-315.481	172.394	
	500	17.230	-285.383	74.789	-322.777	141.105	
	600	17.821	-283.629	77.985	-330.420	120.372	
	700	18.302	-281.822	80.769	-338.361	105.655	
	800	18.723	-279.971	83.241	-346.563	94.689	
	800	18.721	-279.971	83.241	-346.563	94.689	
	900	18.949	-278.087	85.460	-355.000	86.217	
	1000	19.113	-276.183	87.465	-363.648	79.486	
	1100	19.236	-274.265	89.293	-372.487	74.016	
	1200	19.330	-272.337	90.971	-381.502	69.490	
	1300	19.405	-270.400	92.521	-390.677	65.688	
	1400	19.465	-268.456	93.961	-400.002	62.452	
	1500	19.514	-266.507	95.306	-409.466	59.667	
	1549	19.535	-265.550	95.934	-414.152	58.441	
	1549	19.535	-265.550	95.934	-414.152	58.441	
	1600	19.555	-264.554	96.567	-419.061	57.249	
	1700	19.590	-262.596	97.753	-428.777	55.130	
	1800	19.620	-260.636	98.874	-438.609	53.262	
	1900	19.646	-258.673	99.936	-448.550	51.602	
	2000	19.669	-256.707	100.944	-458.595	50.120	

ALUMINUM TRIFLUORIDE (DIMERIC GAS) $Al_2F_6(g)$

		A	B	C	D	RANGE	REF.
AS	CP	43.723	0.011	-20.034	2.857	298-1549	3
	CP	43.723	0.011	-20.034	2.857	1549-2000	3

PHASE	T	CP	H	S	G	BT	
AS	298	31.969	-629.45	92.5	-657.029	481.680	3
	300	32.048	-629.391	92.698	-657.200	478.834	
	400	35.670	-625.993	102.447	-666.972	364.466	
	500	38.000	-622.301	110.677	-677.639	296.236	
	600	39.487	-618.421	117.747	-689.069	251.027	
	700	40.475	-614.420	123.913	-701.159	218.941	
	800	41.159	-610.336	129.365	-713.828	195.035	
	900	41.651	-606.194	134.242	-727.012	176.566	
	1000	42.016	-602.010	138.651	-740.661	161.893	
	1100	42.294	-597.794	142.669	-754.729	149.971	
	1200	42.510	-593.553	146.359	-769.183	140.106	
	1300	42.682	-589.293	149.768	-783.992	131.819	
	1400	42.820	-585.018	152.937	-799.129	124.766	
	1500	42.934	-580.730	155.895	-814.572	118.699	
	1549	42.982	-578.625	157.276	-822.245	116.027	
	1549	42.982	-578.625	157.276	-822.245	116.027	
	1600	43.028	-576.432	158.669	-830.302	113.429	
	1700	43.107	-572.125	161.280	-846.300	108.814	
	1800	43.173	-567.811	163.746	-862.553	104.742	
	1900	43.231	-563.491	166.081	-879.045	101.127	
	2000	43.280	-559.165	168.300	-895.765	97.898	

AS CP(T) = A + B*T + C/(T**2) + E/(T**3)

CP(T) = A + B*T + C/(T**2) + E/(T**3)

AlCl (g)

ALUMINIUM MONOCHLORIDE (GAS)

		A	B	C	D	RANGE	REF.
GAS	CP	8.746	0.3	-0.492		298- 800	3
	CP	8.935	0.111	-0.731		800-2000	3

PHASE	T	CP	H	S	G	BT	
GAS	298	8.282	-12.3	54.455	-28.536	20.920	3
	300	8.289	-12.285	54.506	-28.637	20.865	
	400	8.558	-11.441	56.933	-34.214	18.696	
	500	8.699	-10.577	58.859	-40.007	17.489	
	600	8.789	-9.702	60.454	-45.974	16.748	
	700	8.856	-8.820	61.814	-52.089	16.265	
	800	8.909	-7.932	63.000	-58.331	15.938	
	800	8.910	-7.932	63.000	-58.331	15.938	
	900	8.945	-7.039	64.051	-64.685	15.710	
	1000	8.973	-6.143	64.995	-71.138	15.549	
	1100	8.997	-5.244	65.851	-77.681	15.436	
	1200	9.017	-4.344	66.635	-84.306	15.356	
	1300	9.036	-3.441	67.358	-91.006	15.302	
	1400	9.053	-2.537	68.028	-97.776	15.266	
	1500	9.069	-1.630	68.653	-104.610	15.244	
	1600	9.084	-.723	69.239	-111.505	15.233	
	1700	9.098	.186	69.790	-118.457	15.231	
	1800	9.112	1.097	70.311	-125.462	15.235	
	1900	9.126	2.009	70.804	-132.518	15.245	
	2000	9.139	2.922	71.272	-139.622	15.259	

ALUMINIUM CHLORIDE

$AlCl_3$

		A	B	C	D	RANGE	REF.
SOL	CP	15.52	21.0			298- 454	3
	** MPT= 465.7, SPT= 454.3 GAS(AL2CL6)						3

PHASE	T	CP	H	S	G	BT	P	
SOL	298	21.781	-168.65	26.12	-176.438	129.350		3
	300	21.820	-168.610	26.255	-176.486	128.587		
	400	23.920	-166.323	32.820	-179.451	98.060	13.438	
	454	25.060	-164.993	35.936	-181.318	87.238	760.	

$AlCl_3$(g) ALUMINIUM CHLORIDE (GAS)

	A	B	C	D	RANGE	REF.
CP	19.587	0.149	-2.368		298-2000	3

PHASE	T	CP	H	S	G	BT	
GAS	298	16.968	-139.70	75.118	-162.096	118.836	3
	300	17.001	-139.669	75.223	-162.235	118.204	
	400	18.167	-137.902	80.297	-170.021	92.908	
	454	18.507	-136.906	82.632	-174.446	83.932	
	454	18.507	-136.906	82.632	-174.446	83.932	
	500	18.714	-136.055	84.416	-178.263	77.929	
	600	19.019	-134.167	87.858	-186.882	68.081	
	700	19.208	-132.255	90.805	-195.818	61.145	
	800	19.336	-130.327	93.379	-205.030	56.019	
	900	19.429	-128.389	95.662	-214.484	52.091	
	1000	19.499	-126.442	97.712	-224.155	48.996	
	1100	19.555	-124.490	99.574	-234.021	46.502	
	1200	19.601	-122.532	101.277	-244.064	44.456	
	1300	19.641	-120.570	102.848	-254.272	42.753	
	1400	19.675	-118.604	104.305	-264.630	41.316	
	1500	19.705	-116.635	105.663	-275.129	40.092	
	1600	19.733	-114.663	106.936	-285.760	39.038	
	1700	19.758	-112.688	108.133	-296.514	38.125	
	1800	19.782	-110.711	109.263	-307.384	37.327	
	1900	19.805	-108.732	110.333	-318.364	36.625	
	2000	19.826	-106.750	111.349	-329.449	36.005	

ALUMINIUM CHLORIDE (DIMERIC GAS) Al_2Cl_6 (g)

		A	B	C	D	RANGE	REF.
GAS	CP	43.24	0.25	-4.877		298-2000	3

PHASE	T	CP	H	S	G	BT	
GAS	298	37.828	-309.626	113.647	-343.510	251.833	3*
	300	37.896	-309.556	113.881	-343.720	250.434	
	400	40.292	-305.630	125.160	-355.694	194.368	
	454	40.991	-303.422	130.335	-362.633	174.475	
	454	40.991	-303.422	130.335	-362.633	174.475	
	500	41.414	-301.538	134.285	-368.681	161.172	
	600	42.035	-297.363	141.896	-382.501	139.344	
	700	42.420	-293.139	148.407	-397.024	123.973	
	800	42.678	-288.883	154.089	-412.154	112.610	
	900	42.863	-284.606	159.127	-427.820	103.903	
	1000	43.002	-280.312	163.650	-443.963	97.041	
	1100	43.112	-276.006	167.754	-460.536	91.512	
	1200	43.201	-271.690	171.509	-477.502	86.977	
	1300	43.276	-267.366	174.970	-494.828	83.199	
	1400	43.341	-263.036	178.180	-512.488	80.014	
	1500	43.398	-258.698	181.172	-530.457	77.298	
	1600	43.449	-254.356	183.975	-548.716	74.961	
	1700	43.496	-250.009	186.610	-567.246	72.934	
	1800	43.539	-245.657	189.098	-586.033	71.164	
	1900	43.580	-241.301	191.453	-605.061	69.607	
	2000	43.618	-236.941	193.689	-624.319	68.232	

AlBr (g)

ALUMINIUM MONOBROMIDE (GAS)

		A	B	C	D	RANGE	REF.
GAS	CP	8.905	0.14	-0.39		298-3000	3

PHASE	T	CP	H	S	G	BT	
GAS	298	8.508	3.8	57.24	-13.266	9.726	3A
	300	8.514	3.816	57.293	-13.372	9.743	
	400	8.717	4.679	59.774	-19.231	10.509	
	500	8.819	5.556	61.731	-25.309	11.064	
	600	8.881	6.441	63.345	-31.566	11.499	
	700	8.923	7.331	64.717	-37.970	11.856	
	800	8.956	8.225	65.911	-44.503	12.159	
	900	8.983	9.122	66.967	-51.148	12.422	
	1000	9.006	10.022	67.915	-57.893	12.654	
	1100	9.027	10.924	68.774	-64.728	12.862	
	1200	9.046	11.827	69.560	-71.645	13.050	
	1300	9.064	12.733	70.285	-78.638	13.222	
	1400	9.081	13.640	70.958	-85.701	13.380	
	1500	9.098	14.549	71.585	-92.828	13.527	
	1600	9.114	15.460	72.172	-100.016	13.663	
	1700	9.130	16.372	72.725	-107.261	13.791	
	1800	9.145	17.285	73.248	-114.560	13.911	
	1900	9.160	18.201	73.742	-121.910	14.025	
	2000	9.175	19.117	74.213	-129.308	14.132	
	2100	9.190	20.036	74.661	-136.752	14.234	
	2200	9.205	20.955	75.089	-144.239	14.331	
	2300	9.220	21.877	75.498	-151.769	14.423	
	2400	9.234	22.799	75.891	-159.339	14.512	
	2500	9.249	23.724	76.268	-166.947	14.596	
	2600	9.263	24.649	76.631	-174.592	14.678	
	2700	9.278	25.576	76.981	-182.272	14.756	
	2800	9.292	26.505	77.319	-189.987	14.831	
	2900	9.306	27.435	77.645	-197.736	14.904	
	3000	9.321	28.366	77.961	-205.516	14.974	

ALUMINIUM BROMIDE

$AlBr_3$

		A	B	C	D	RANGE	REF.
SOL	CP	11.938	40.53			298- 371	3A
	** MPT=	370.6					3A
LIQ	CP	29.869				371- 527	3A
	** BPT=	527 GAS(AL2BR6 + ALBR3)/527.6 GAS(AL2BR6)					3A

PHASE	T	CP	H	S	G	BT	P	
SOL	298	24.022	-122.16	43.08	-135.004	98.974	.006	3A
	300	24.097	-122.115	43.229	-135.084	98.422	.007	
	371	26.958	-120.313	48.613	-138.329	81.586	4.096	
			2.69	7.258				3A
LIQ	371	29.869	-117.623	55.872	-138.329	81.586	4.096	
	400	29.869	-116.745	58.152	-140.006	76.506	16.437	
	500	29.869	-113.758	64.817	-146.167	63.898	415.681	
	527	29.869	-112.952	66.388	-147.938	61.359	760.	

$AlBr_3(g)$ ALUMINIUM BROMIDE (GAS)

		A	B	C	D	RANGE	REF.
GAS	CP	19.294	0.672	-1.306		298- 527	3A
	CP	19.834	0.017	-1.830		527-2000	3A

PHASE	T	CP	H	S	G	BT	
GAS	298	18.025	-98.2	83.43	-123.075	90.228	3A
	300	18.044	-98.167	83.542	-123.229	89.784	
	400	18.747	-96.323	88.842	-131.859	72.054	
	500	19.108	-94.428	93.067	-140.962	61.623	
	527	19.178	-93.911	94.074	-143.488	59.513	
	527	19.184	-93.911	94.074	-143.488	59.513	
	600	19.336	-92.505	96.573	-150.449	54.808	
	700	19.472	-90.564	99.565	-160.260	50.042	
	800	19.562	-88.612	102.171	-170.349	46.543	
	900	19.623	-86.653	104.479	-180.684	43.882	
	1000	19.668	-84.688	106.549	-191.237	41.800	
	1100	19.701	-82.719	108.425	-201.987	40.137	
	1200	19.727	-80.748	110.141	-212.917	38.783	
	1300	19.748	-78.774	111.721	-224.011	37.665	
	1400	19.764	-76.798	113.185	-235.257	36.730	
	1500	19.778	-74.821	114.549	-246.644	35.941	
	1600	19.790	-72.843	115.826	-258.164	35.268	
	1700	19.800	-70.863	117.026	-269.807	34.691	
	1800	19.808	-68.883	118.158	-281.567	34.191	
	1900	19.816	-66.902	119.229	-293.437	33.757	
	2000	19.822	-64.920	120.245	-305.411	33.378	

ALUMINIUM BROMIDE (DIMERIC GAS) Al_2Br_6 (g)

		A	B	C	D	RANGE	REF.
GAS	CP	43.504	0.112	-3.289		298-2000	3A

PHASE	T	CP	H	S	G	BT	
GAS	298	39.837	-224.008	130.8	-263.006	192.815	3A,*
	300	39.883	-223.934	131.047	-263.248	191.802	
	400	41.493	-219.854	142.774	-276.963	151.346	
	500	42.244	-215.663	152.123	-291.724	127.530	
	527	42.379	-214.521	154.348	-295.862	122.712	
	527	42.379	-214.521	154.348	-295.862	122.712	
	600	42.658	-211.416	159.864	-307.335	111.962	
	700	42.911	-207.137	166.461	-323.659	101.065	
	800	43.080	-202.837	172.202	-340.599	93.060	
	900	43.199	-198.522	177.284	-358.078	86.965	
	1000	43.287	-194.198	181.840	-376.038	82.194	
	1100	43.355	-189.866	185.969	-394.431	78.377	
	1200	43.410	-185.527	189.744	-413.220	75.268	
	1300	43.455	-181.184	193.220	-432.370	72.698	
	1400	43.493	-176.837	196.442	-451.855	70.547	
	1500	43.526	-172.486	199.444	-471.651	68.729	
	1600	43.555	-168.132	202.254	-491.738	67.177	
	1700	43.581	-163.775	204.895	-512.096	65.843	
	1800	43.604	-159.415	207.387	-532.712	64.689	
	1900	43.626	-155.054	209.745	-553.569	63.684	
	2000	43.646	-150.690	211.983	-574.657	62.804	

AlOF(g) ALUMINIUM OXYFLUORIDE (GAS)

		A	B	C	D	RANGE	REF.
GAS	CP	14.26	0.30	-3.27		298-2500	3

PHASE	T	CP	H	S	G	BT	
GAS	298	10.671	-145.	55.97	-161.687	118.536	175*/3
	300	10.717	-144.980	56.036	-161.791	117.881	
	400	12.336	-143.816	59.374	-167.566	91.566	
	500	13.102	-142.540	62.218	-173.649	75.912	
	600	13.532	-141.207	64.648	-179.995	65.572	
	700	13.803	-139.839	66.756	-186.568	58.257	
	800	13.989	-138.449	68.612	-193.338	52.825	
	900	14.126	-137.043	70.268	-200.284	48.642	
	1000	14.233	-135.625	71.762	-207.386	45.330	
	1100	14.320	-134.197	73.122	-214.632	42.649	
	1200	14.393	-132.761	74.372	-222.007	40.438	
	1300	14.457	-131.319	75.526	-229.503	38.588	
	1400	14.513	-129.870	76.600	-237.110	37.019	
	1500	14.565	-128.416	77.603	-244.820	35.675	
	1600	14.612	-126.957	78.544	-252.628	34.512	
	1700	14.657	-125.494	79.431	-260.527	33.498	
	1800	14.699	-124.026	80.270	-268.513	32.606	
	1900	14.739	-122.554	81.066	-276.580	31.818	
	2000	14.778	-121.078	81.823	-284.725	31.117	
	2100	14.816	-119.599	82.545	-292.943	30.491	
	2200	14.852	-118.115	83.235	-301.233	29.929	
	2300	14.888	-116.628	83.896	-309.590	29.422	
	2400	14.923	-115.137	84.531	-318.011	28.963	
	2500	14.958	-113.643	85.141	-326.495	28.546	

ALUMINIUM OXYCHLCRIDE — AlOCl

		A	B	C	D	RANGE	REF.
SOL	CP	13.23	8.21	-1.86		298- 700	3,*
	** DEC.						

PHASE	T	CP	H	S	G	BT	P	
SOL	298	13.585	-189.6	13.	-193.476	141.841		1,3
	300	13.626	-189.575	13.084	-193.500	140.984		
	400	15.351	-188.119	17.259	-195.023	106.570		
	500	16.591	-186.520	20.823	-196.932	86.090		
	600	17.639	-184.807	23.942	-199.173	72.558		
	700	18.597	-182.995	26.734	-201.709	62.985		

* * * * * * * * * *

ALUMINIUM OXYCHLORIDE (GAS) — AlOCl (g)

		A	B	C	D	RANGE	REF.
GAS	CP	14.04	0.83	-2.08	-0.20	298-2000	3

PHASE	T	CP	H	S	G	BT	
GAS	298	11.930	-84.4	59.471	-102.131	74.874	175*/3
	300	11.960	-84.378	59.545	-102.241	74.493	
	400	13.040	-83.121	63.154	-108.382	59.225	
	500	13.573	-81.787	66.127	-114.851	50.208	
	600	13.888	-80.413	68.632	-121.592	44.296	
	700	14.099	-79.013	70.790	-128.566	40.145	
	800	14.251	-77.595	72.683	-135.741	37.088	
	900	14.368	-76.164	74.368	-143.096	34.753	
	1000	14.462	-74.722	75.887	-150.610	32.920	
	1100	14.539	-73.272	77.269	-158.268	31.449	
	1200	14.604	-71.815	78.537	-166.060	30.248	
	1300	14.658	-70.352	79.708	-173.973	29.251	
	1400	14.704	-68.884	80.796	-181.998	28.415	
	1500	14.743	-67.411	81.812	-190.129	27.706	
	1600	14.775	-65.935	82.765	-198.359	27.098	
	1700	14.801	-64.457	83.661	-206.681	26.574	
	1800	14.822	-62.975	84.508	-215.089	26.119	
	1900	14.837	-61.492	85.310	-223.581	25.721	
	2000	14.848	-60.008	86.071	-232.150	25.372	

$AlCl_3 \cdot 6H_2O$ ALUMINIUM CHLORIDE HEXAHYDRATE

		A	B	C	D	RANGE	REF.
SOL	CP	71.0				298- 400	176

PHASE	T	CP	H	S	G	BT	
SOL	298	71.000	-643.7	76.03	-666.368	488.527	19*/176
	300	71.000	-643.569	76.469	-666.509	485.617	
	400	71.000	-636.469	96.895	-675.226	368.976	

* * * * * * * * * *

Al_2O (g) ALUMINIUM SUBOXIDE (GAS)

		A	B	C	D	RANGE	REF.
GAS	CP	13.891	0.013	-6.577	1.030	298-2000	3A

PHASE	T	CP	H	S	G	BT	
GAS	298	10.382	-33.	60.8	-51.128	37.483	3A
	300	10.402	-32.981	60.864	-51.240	37.333	
	400	11.395	-31.889	63.998	-57.488	31.414	
	500	12.091	-30.712	66.621	-64.023	27.988	
	600	12.549	-29.479	68.869	-70.800	25.792	
	700	12.858	-28.207	70.828	-77.787	24.289	
	800	13.075	-26.910	72.560	-84.958	23.213	
	900	13.232	-25.594	74.110	-92.293	22.415	
	1000	13.349	-24.265	75.510	-99.775	21.809	
	1100	13.439	-22.925	76.787	-107.391	21.339	
	1200	13.509	-21.578	77.959	-115.129	20.971	
	1300	13.566	-20.224	79.043	-122.980	20.678	
	1400	13.611	-18.865	80.050	-130.935	20.443	
	1500	13.649	-17.502	80.990	-138.987	20.253	
	1600	13.680	-16.136	81.872	-147.131	20.100	
	1700	13.706	-14.766	82.702	-155.360	19.976	
	1800	13.729	-13.394	83.486	-163.670	19.875	
	1900	13.749	-12.021	84.229	-172.056	19.794	
	2000	13.765	-10.645	84.935	-180.515	19.728	

GAS CP(T) = A + B*T + C/(T**2) + E/(T**3)

ALPHA-ALUMINIUM OXICE Al_2O_3

		A	B	C	D	RANGE	REF.
SOL-A	CP	24.821	6.278	-6.953		298- 800	3
	CP	28.804	2.197	-11.560		800-2327	3
LIQ	CP	34.623				2327-3500	3

PHASE	T	CP	H	S	G	BT	
SOL-A	298	18.871	-400.4	12.174	-404.030	296.202	3
	300	18.979	-400.365	12.291	-404.052	294.391	
	400	22.987	-398.243	18.369	-405.590	221.634	
	500	25.179	-395.826	23.754	-407.702	178.231	
	600	26.656	-393.230	28.482	-410.319	149.479	
	700	27.797	-390.505	32.680	-413.381	129.081	
	800	28.757	-387.677	36.456	-416.841	113.891	
	800	28.755	-387.677	36.456	-416.841	113.891	
	900	29.354	-384.770	39.878	-420.661	102.164	
	1000	29.845	-381.809	42.997	-424.807	92.854	
	1100	30.265	-378.803	45.862	-429.252	85.296	
	1200	30.638	-375.758	48.512	-433.972	79.048	
	1300	30.976	-372.677	50.978	-438.948	73.804	
	1400	31.290	-369.563	53.285	-444.162	69.346	
	1500	31.586	-366.420	55.454	-449.600	65.516	
	1600	31.868	-363.247	57.501	-455.249	62.192	
	1700	32.139	-360.046	59.441	-461.097	59.286	
	1800	32.402	-356.819	61.286	-467.134	56.725	
	1900	32.658	-353.566	63.045	-473.351	54.455	
	2000	32.909	-350.288	64.726	-479.740	52.431	
	2100	33.156	-346.985	66.338	-486.294	50.616	
	2200	33.399	-343.657	67.886	-493.006	48.982	
	2300	33.639	-340.305	69.376	-499.869	47.505	
	2327	33.703	-339.396	69.769	-501.748	47.130	
			28.3	12.162			221
LIQ	2327	34.623	-311.096	81.930	-501.748	47.130	
	2400	34.623	-308.568	83.000	-507.768	46.245	
	2500	34.623	-305.106	84.413	-516.139	45.127	
	2600	34.623	-301.644	85.771	-524.649	44.107	
	2700	34.623	-298.181	87.078	-533.292	43.173	
	2800	34.623	-294.719	88.337	-542.063	42.316	
	2900	34.623	-291.257	89.552	-550.958	41.527	
	3000	34.623	-287.795	90.726	-559.972	40.799	
	3100	34.623	-284.332	91.861	-569.101	40.127	
	3200	34.623	-280.870	92.960	-578.343	39.504	
	3300	34.623	-277.408	94.026	-587.692	38.926	
	3400	34.623	-273.945	95.059	-597.147	38.389	
	3500	34.623	-270.483	96.063	-606.703	37.889	

$AlO_2(g)$ ALUMINIUM DIOXIDE (GAS)

		A	B	C	D	RANGE	REF.
GAS	CP	14.173	0.373	-2.517		298-2000	3

PHASE	T	CP	H	S	G	BT	
GAS	298	11.453	-44.	58.7	-61.501	45.088	3
	300	11.488	-43.979	58.771	-61.610	44.889	
	400	12.749	-42.758	62.274	-67.668	36.977	
	500	13.353	-41.450	65.191	-74.045	32.369	
	600	13.698	-40.096	67.658	-80.691	29.396	
	700	13.920	-38.714	69.787	-87.566	27.343	
	800	14.078	-37.314	71.657	-94.640	25.858	
	900	14.198	-35.900	73.322	-101.890	24.746	
	1000	14.294	-34.475	74.823	-109.299	23.890	
	1100	14.375	-33.042	76.190	-116.850	23.219	
	1200	14.446	-31.601	77.444	-124.533	22.684	
	1300	14.509	-30.153	78.602	-132.336	22.251	
	1400	14.567	-28.699	79.680	-140.251	21.897	
	1500	14.621	-27.240	80.687	-148.270	21.606	
	1600	14.671	-25.775	81.632	-156.386	21.364	
	1700	14.720	-24.305	82.523	-164.594	21.163	
	1800	14.767	-22.831	83.366	-172.889	20.994	
	1900	14.812	-21.352	84.165	-181.266	20.853	
	2000	14.856	-19.869	84.926	-189.721	20.735	

ALUMINIUM OXIDE MONOHYCRATE $Al_2O_3 \cdot H_2O$

		A	B	C	D	RANGE	REF.
SOL-A	CP	28.87	8.4			298- 500	58

PHASE	T	CP	H	S	G	BT	
SOL-A	298	31.374	-479.	16.83	-484.018	354.842	62
	300	31.390	-478.942	17.024	-484.049	352.677	
	400	32.230	-475.761	26.170	-486.229	265.699	
	500	33.070	-472.496	33.452	-489.222	213.867	

* * * * * * * * * *

ALUMINIUM OXIDE TRIHYCRATE $Al_2O_3 \cdot 3H_2O$

		A	B	C	D	RANGE	REF.
SOL	CP	17.30	91.2			298- 500	58

PHASE	T	CP	H	S	G	BT	
SOL	298	44.491	-618.2	33.5	-628.188	460.536	62
	300	44.660	-618.118	33.776	-628.250	457.742	
	400	53.780	-613.196	47.873	-632.345	345.543	
	500	62.900	-607.362	60.853	-637.788	278.814	

AlS(g) ALUMINIUM MONOSULFIDE (GAS)

		A	B	C	D	RANGE	REF.
GAS	CP	8.806	0.166	-0.78		298-2000	63

PHASE	T	CP	H	S	G	BT	
GAS	298	7.978	57.1	55.1	40.672	-29.817	63
	300	7.989	57.115	55.149	40.570	-29.559	
	400	8.385	57.936	57.510	34.932	-19.089	
	500	8.577	58.785	59.404	29.083	-12.714	
	600	8.689	59.649	60.978	23.062	-8.402	
	700	8.763	60.522	62.323	16.895	-5.276	
	800	8.817	61.401	63.497	10.603	-2.897	
	900	8.859	62.285	64.538	4.200	-1.020	
	1000	8.894	63.172	65.473	-2.301	.503	
	1100	8.924	64.063	66.323	-8.891	1.767	
	1200	8.951	64.957	67.100	-15.563	2.835	
	1300	8.976	65.854	67.818	-22.309	3.751	
	1400	8.999	66.752	68.484	-29.125	4.547	
	1500	9.020	67.653	69.105	-36.005	5.247	
	1600	9.041	68.556	69.688	-42.945	5.867	
	1700	9.061	69.461	70.237	-49.941	6.421	
	1800	9.081	70.369	70.755	-56.991	6.921	
	1900	9.100	71.278	71.247	-64.091	7.373	
	2000	9.118	72.188	71.714	-71.240	7.786	

ALUMINIUM SULFIDE Al_2S_3

		A	B	C	D	RANGE	REF.
SOL	CP	24.42	8.62			298-1370	*
LIQ	CP	37.5				1370-1800	*

PHASE	T	CP	H	S	G	BT	
SOL	298	26.990	-172.9	29.5	-181.695	133.204	1/*
	300	27.006	-172.850	29.667	-181.750	132.423	
	400	27.868	-170.106	37.554	-185.128	101.163	
	500	28.730	-167.276	43.865	-189.209	82.714	
	600	29.592	-164.360	49.180	-193.868	70.626	
	700	30.454	-161.358	53.806	-199.022	62.146	
	800	31.316	-158.270	57.929	-204.613	55.905	
	900	32.178	-155.095	61.667	-210.595	51.146	
	1000	33.040	-151.834	65.102	-216.936	47.418	
	1100	33.902	-148.487	68.291	-223.607	44.433	
	1200	34.764	-145.054	71.278	-230.588	42.001	
	1300	35.626	-141.534	74.095	-237.857	39.993	
	1370	36.229	-139.019	75.979	-243.110	38.788	
			13.5	9.854			*
LIQ	1370	37.500	-125.519	85.833	-243.110	38.788	
	1400	37.500	-124.394	86.645	-245.698	38.360	
	1500	37.500	-120.644	89.233	-254.493	37.085	
	1600	37.500	-116.894	91.653	-263.539	36.003	
	1700	37.500	-113.144	93.926	-272.819	35.078	
	1800	37.500	-109.394	96.070	-282.320	34.283	

AlSe (g) ALUMINIUM MONOSELENIDE (GAS)

		A	B	C	D	RANGE	REF.
GAS	CP	8.904	0.02	-0.497		298-1800	63

PHASE	T	CP	H	S	G	BT	
GAS	298	8.351	52.9	58.10	35.577	-26.083	63
	300	8.358	52.915	58.152	35.470	-25.843	
	400	8.601	53.765	60.594	29.527	-16.135	
	500	8.715	54.632	62.527	23.368	-10.215	
	600	8.778	55.507	64.122	17.033	-6.205	
	700	8.817	56.386	65.479	10.551	-3.295	
	800	8.842	57.269	66.658	3.943	-1.077	
	900	8.861	58.155	67.700	-2.776	.674	
	1000	8.874	59.041	68.635	-9.593	2.097	
	1100	8.885	59.929	69.481	-16.500	3.279	
	1200	8.893	60.818	70.254	-23.487	4.278	
	1300	8.901	61.708	70.967	-30.548	5.136	
	1400	8.907	62.598	71.626	-37.679	5.883	
	1500	8.912	63.489	72.241	-44.872	6.539	
	1600	8.917	64.381	72.816	-52.125	7.121	
	1700	8.921	65.273	73.357	-59.434	7.642	
	1800	8.925	66.165	73.867	-66.796	8.111	

* * * * * * * * * *

Al_2Se_3 ALUMINIUM SELENIDE

		A	B	C	D	RANGE	REF.
SOL	CP	25.75	8.2			298-1220	63
	** MPT= 1220						63

PHASE	T	CP	H	S	G	BT	
SOL	298	28.195	-135.5	37.0	-146.532	107.425	63
	300	28.210	-135.448	37.174	-146.600	106.813	
	400	29.030	-132.586	45.402	-150.747	82.375	
	500	29.850	-129.642	51.968	-155.626	68.033	
	600	30.670	-126.616	57.483	-161.106	58.691	
	700	31.490	-123.508	62.272	-167.098	52.178	
	800	32.310	-120.318	66.531	-173.542	47.416	
	900	33.130	-117.046	70.384	-180.391	43.811	
	1000	33.950	-113.692	73.917	-187.609	41.007	
	1100	34.770	-110.256	77.191	-195.166	38.781	
	1200	35.590	-106.738	80.252	-203.040	36.984	
	1220	35.754	-106.024	80.841	-204.651	36.666	

ALUMINIUM MONOTELLURIDE (GAS) AlTe (g)

		A	B	C	D	RANGE	REF.
GAS	CP	8.921	0.013	-0.377		298-2000	63

PHASE	T	CP	H	S	G	BT	
GAS	298	8.501	63.9	60.07	45.990	-33.716	63
	300	8.506	63.916	60.123	45.879	-33.427	
	400	8.691	64.777	62.599	39.737	-21.714	
	500	8.777	65.651	64.548	33.377	-14.591	
	600	8.824	66.531	66.153	26.839	-9.777	
	700	8.853	67.415	67.516	20.154	-6.293	
	800	8.872	68.301	68.699	13.342	-3.645	
	900	8.886	69.189	69.745	6.419	-1.559	
	1000	8.896	70.078	70.682	-.603	.132	
	1100	8.904	70.968	71.530	-7.715	1.533	
	1200	8.910	71.859	72.305	-14.907	2.715	
	1300	8.916	72.750	73.019	-22.174	3.728	
	1400	8.920	73.642	73.679	-29.509	4.607	
	1500	8.924	74.534	74.295	-36.908	5.378	
	1600	8.927	75.427	74.871	-44.367	6.061	
	1700	8.930	76.320	75.412	-51.881	6.671	
	1800	8.933	77.213	75.923	-59.448	7.219	
	1900	8.935	78.106	76.406	-67.065	7.715	
	2000	8.938	79.000	76.864	-74.728	8.167	

Al_2Te_3 ALUMINIUM TELLURIDE

		A	B	C	D	RANGE	REF.
SOL	CP	26.5	8.3			298-1163	*
LIQ	CP	42.2				1163-1200	230*

PHASE	T	CP	H	S	G	BT	
SOL	298	28.975	-76.2	46.8	-90.153	66.093	62/*
	300	28.990	-76.146	46.979	-90.240	65.749	
	400	29.820	-73.206	55.433	-95.379	52.120	
	500	30.650	-70.182	62.176	-101.270	44.271	
	600	31.480	-67.076	67.838	-107.778	39.264	
	700	32.310	-63.886	72.753	-114.813	35.851	
	800	33.140	-60.614	77.121	-122.311	33.418	
	900	33.970	-57.258	81.073	-130.224	31.627	
	1000	34.800	-53.820	84.695	-138.514	30.276	
	1100	35.630	-50.298	88.050	-147.154	29.241	
	1163	36.153	-48.037	90.049	-152.764	28.711	
			11.95	10.275			230
LIQ	1163	42.200	-36.087	100.324	-152.764	28.711	
	1200	42.200	-34.526	101.646	-156.501	28.507	

ALUMINIUM MONONITRIDE

AlN

		A	B	C	D	RANGE	REF.
SOL	CP	7.712	5.422	-1.889		298- 600	3
	CP	12.002	0.280	-6.227		600-1000	3
	CP	11.984	0.093	-4.158		1000-2000	3
	** DPT= 2790						3

PHASE	T	CP	H	S	G	BT	
SOL	298	7.204	-76.0	4.816	-77.436	56.770	3
	300	7.240	-75.987	4.861	-77.445	56.426	
	400	8.700	-75.183	7.162	-78.048	42.649	
	500	9.667	-74.262	9.213	-78.869	34.478	
	600	10.440	-73.256	11.046	-79.883	29.101	
	600	10.440	-73.256	11.046	-79.883	29.101	
	700	10.927	-72.186	12.694	-81.072	25.315	
	800	11.253	-71.076	14.176	-82.417	22.518	
	900	11.485	-69.938	15.516	-83.902	20.377	
	1000	11.659	-68.781	16.735	-85.516	18.692	
	1000	11.661	-68.781	16.735	-85.516	18.692	
	1100	11.743	-67.610	17.851	-87.246	17.336	
	1200	11.807	-66.433	18.875	-89.083	16.226	
	1300	11.859	-65.249	19.822	-91.018	15.304	
	1400	11.902	-64.061	20.703	-93.045	14.527	
	1500	11.939	-62.869	21.525	-95.157	13.866	
	1600	11.970	-61.674	22.297	-97.348	13.299	
	1700	11.998	-60.475	23.023	-99.615	12.808	
	1800	12.023	-59.274	23.710	-101.952	12.380	
	1900	12.046	-58.071	24.361	-104.356	12.005	
	2000	12.066	-56.865	24.979	-106.823	11.675	

AlP

ALUMINIUM MONOPHOSPHIDE

		A	B	C	D	RANGE	REF.
SOL	CP	9.6	1.5			298-2000	*

PHASE	T	CP	H	S	G	BT	
SOL	298	10.047	-39.3	11.3	-42.669	31.281	226
	300	10.050	-39.281	11.362	-42.690	31.104	
	400	10.200	-38.269	14.274	-43.978	24.032	
	500	10.350	-37.241	16.566	-45.524	19.901	
	600	10.500	-36.199	18.466	-47.279	17.224	
	700	10.650	-35.141	20.096	-49.209	15.366	
	800	10.800	-34.069	21.528	-51.291	14.014	
	900	10.950	-32.981	22.809	-53.509	12.996	
	1000	11.100	-31.879	23.970	-55.849	12.207	
	1100	11.250	-30.761	25.035	-58.300	11.585	
	1200	11.400	-29.629	26.021	-60.854	11.084	
	1300	11.550	-28.481	26.939	-63.502	10.677	
	1400	11.700	-27.319	27.800	-66.240	10.342	
	1500	11.850	-26.141	28.613	-69.061	10.063	
	1600	12.000	-24.949	29.382	-71.961	9.831	
	1700	12.150	-23.741	30.114	-74.936	9.635	
	1800	12.300	-22.519	30.813	-77.982	9.470	
	1900	12.450	-21.281	31.482	-81.097	9.330	
	2000	12.600	-20.029	32.125	-84.278	9.211	

ALUMINIUM PHOSPHATE

		A	B	C	D	RANGE	REF.
SOL-A	CP	12.30	33.3			298- 853	62,*
SOL-B	CP	40.0				853- 978	*
SOL-C	CP	39.0				978-1200	*
	** MPT=	2273					62

PHASE	T	CP	H	S	G	BT	
SOL-A	298	22.228	-414.2	21.7	-420.670	308.401	62
	300	22.290	-414.159	21.838	-420.710	306.528	
	400	25.620	-411.763	28.706	-423.246	231.282	
	500	28.950	-409.035	34.781	-426.425	186.415	
	600	32.280	-405.973	40.353	-430.185	156.716	
	700	35.610	-402.579	45.579	-434.484	135.670	
	800	38.940	-398.851	50.552	-439.293	120.025	
	853	40.705	-396.741	53.106	-442.040	113.272	
			0.31	.363			62
SOL-B	853	40.000	-396.431	53.469	-442.040	113.272	
	900	40.000	-394.551	55.615	-444.604	107.979	
	978	40.000	-391.431	58.939	-449.073	100.366	
			0.26	.266			62
SOL-C	978	39.000	-391.171	59.205	-449.073	100.366	
	1000	39.000	-390.313	60.073	-450.385	98.445	
	1100	39.000	-386.413	63.790	-456.581	90.727	
	1200	39.000	-382.513	67.183	-463.133	84.359	

AlAs

ALUMINIUM ARSENIDE

		A	B	C	D	RANGE	REF.
SOL	CP	10.5	1.5			298-2013	*
	** MPT=	2013					1

PHASE	T	CP	H	S	G	BT	
SOL	298	10.947	-29.30	14.4	-33.593	24.628	226
	300	10.950	-29.280	14.468	-33.620	24.495	
	400	11.100	-28.177	17.638	-35.233	19.253	
	500	11.250	-27.060	20.131	-37.125	16.230	
	600	11.400	-25.927	22.196	-39.245	14.297	
	700	11.550	-24.780	23.964	-41.555	12.976	
	800	11.700	-23.617	25.516	-44.030	12.030	
	900	11.850	-22.440	26.903	-46.653	11.330	
	1000	12.000	-21.247	28.159	-49.407	10.799	
	1100	12.150	-20.040	29.310	-52.281	10.389	
	1200	12.300	-18.817	30.374	-55.266	10.067	
	1300	12.450	-17.580	31.364	-58.353	9.811	
	1400	12.600	-16.327	32.292	-61.537	9.608	
	1500	12.750	-15.060	33.167	-64.810	9.444	
	1600	12.900	-13.777	33.994	-68.168	9.313	
	1700	13.050	-12.480	34.781	-71.608	9.207	
	1800	13.200	-11.167	35.531	-75.123	9.122	
	1900	13.350	-9.840	36.249	-78.713	9.055	
	2000	13.500	-8.497	36.937	-82.372	9.002	
	2013	13.519	-8.322	37.025	-82.853	8.996	

ALUMINIUM ANTIMONICE

AlSb

		A	B	C	D	RANGE	REF.
SOL	CP	10.4	2.3			298-1333	*
LIQ	CP	14.1				1333-1700	*

PHASE	T	CP	H	S	G	BT	
SOL	298	11.086	-12.034	15.53	-16.664	12.217	21
	300	11.090	-12.013	15.599	-16.693	12.163	
	400	11.320	-10.893	18.820	-18.421	10.066	
	500	11.550	-9.749	21.371	-20.435	8.933	
	600	11.780	-8.583	23.497	-22.681	8.263	
	700	12.010	-7.393	25.330	-25.125	7.845	
	800	12.240	-6.181	26.949	-27.740	7.579	
	900	12.470	-4.945	28.404	-30.509	7.410	
	1000	12.700	-3.687	29.730	-33.417	7.304	
	1100	12.930	-2.405	30.951	-36.452	7.243	
	1200	13.160	-1.101	32.086	-39.604	7.214	
	1300	13.390	.227	33.148	-42.867	7.207	
	1333	13.466	.670	33.485	-43.966	7.209	
			19.6	14.704			62
LIQ	1333	14.100	20.270	48.189	-43.966	7.209	
	1400	14.100	21.214	48.880	-47.218	7.372	
	1500	14.100	22.624	49.853	-52.155	7.600	
	1600	14.100	24.034	50.763	-57.187	7.812	
	1700	14.100	25.444	51.618	-62.306	8.011	

$3Al_2O_3 \cdot 2SiO_2$ MULLITE

		A	B	C	D	RANGE	REF.
SOL	CP	55.83	151.5	-13.35	-92.25	298- 600	3,5
	CP	120.33	8.39	-55.0	-0.60	600-2023	3,5
	** MPT= 2023						3

PHASE	T	CP	H	S	G	BT	
SOL	298	77.781	-1629.83	65.7	-1649.418	1209.219	3
	300	78.144	-1629.686	66.182	-1649.540	1201.851	
	400	93.326	-1621.051	90.920	-1657.419	905.693	
	500	103.177	-1611.193	112.875	-1667.631	729.019	
	600	109.812	-1600.521	132.314	-1679.910	611.989	
	600	109.870	-1600.521	132.314	-1679.910	611.989	
	700	114.685	-1589.278	149.637	-1694.023	528.969	
	800	118.064	-1577.631	165.183	-1709.778	467.152	
	900	120.605	-1565.692	179.242	-1727.010	419.432	
	1000	122.620	-1553.528	192.057	-1745.585	381.549	
	1100	124.288	-1541.180	203.825	-1765.387	350.797	
	1200	125.715	-1528.678	214.702	-1786.320	325.377	
	1300	126.969	-1516.043	224.815	-1808.302	304.044	
	1400	128.094	-1503.289	234.266	-1831.261	285.911	
	1500	129.121	-1490.427	243.139	-1855.136	270.329	
	1600	130.070	-1477.467	251.503	-1879.872	256.813	
	1700	130.956	-1464.415	259.415	-1905.421	244.991	
	1800	131.790	-1451.278	266.924	-1931.741	234.577	
	1900	132.581	-1438.059	274.071	-1958.794	225.343	
	2000	133.335	-1424.763	280.891	-1986.545	217.109	
	2023	133.504	-1421.694	282.417	-1993.023	215.340	

KAOLINITE $Al_2O_3 \cdot 2SiO_2 \cdot 2H_2O$

		A	B	C	D	RANGE	REF.
SOL	CP	65.49	33.17	-14.90		298- 700	5

PHASE	T	CP	H	S	G	BT	
SOL	298	58.618	-979.47	48.5	-993.930	728.669	19
	300	58.885	-979.361	48.863	-994.020	724.241	
	400	69.445	-972.893	67.399	-999.853	546.368	
	500	76.115	-965.596	83.654	-1007.423	440.404	
	600	81.253	-957.720	98.000	-1016.520	370.317	
	700	85.668	-949.369	110.864	-1026.974	320.679	

* * * * * * * * * *

ALUMINIUM DIBORIDE AlB_2

		A	B	C	D	RANGE	REF.
SOL	CP	12.18	6.85	-3.37		298-1300	*
	** DPT= 1673						62

PHASE	T	CP	H	S	G	BT	
SOL	298	10.431	-16.	8.3	-18.475	13.544	87/*
	300	10.491	-15.981	8.365	-18.490	13.472	
	400	12.814	-14.804	11.735	-19.498	10.654	
	500	14.257	-13.446	14.758	-20.825	9.104	
	600	15.354	-11.964	17.458	-22.438	8.174	
	700	16.287	-10.381	19.896	-24.308	7.590	
	800	17.133	-8.709	22.127	-26.411	7.216	
	900	17.929	-6.956	24.192	-28.728	6.977	
	1000	18.693	-5.124	26.120	-31.245	6.829	
	1100	19.436	-3.218	27.937	-33.948	6.746	
	1200	20.166	-1.237	29.660	-36.829	6.708	
	1300	20.886	.815	31.302	-39.878	6.705	

AlB_{12} ALUMINIUM 12-BORIDE

		A	B	C	D	RANGE	REF.
SOL	CP	50.5	27.5	-20.4		298-2473	*
	** MPT=	2473					62

PHASE	T	CP	H	S	G	BT	
SOL	298	35.750	-48.	28.4	-56.467	41.397	87/*
	300	36.083	-47.934	28.622	-56.520	41.180	
	400	48.750	-43.621	40.942	-59.998	32.786	
	500	56.090	-38.354	52.666	-64.686	28.278	
	600	61.333	-32.471	63.376	-70.497	25.682	
	700	65.587	-26.119	73.159	-77.331	24.147	
	800	69.312	-19.371	82.164	-85.103	23.252	
	900	72.731	-12.267	90.528	-93.742	22.767	
	1000	75.960	-4.831	98.359	-103.191	22.555	
	1100	79.064	2.921	105.746	-113.399	22.533	
	1200	82.083	10.979	112.755	-124.327	22.646	
	1300	85.043	19.336	119.442	-135.939	22.857	
	1400	87.959	27.986	125.852	-148.206	23.139	
	1500	90.843	36.926	132.019	-161.102	23.476	
	1600	93.703	46.154	137.973	-174.603	23.853	
	1700	96.544	55.666	143.739	-188.690	24.261	
	1800	99.370	65.462	149.338	-203.345	24.693	
	1900	102.185	75.540	154.786	-218.553	25.143	
	2000	104.990	85.899	160.098	-234.298	25.606	
	2100	107.787	96.538	165.289	-250.568	26.080	
	2200	110.579	107.456	170.367	-267.352	26.563	
	2300	113.364	118.653	175.344	-284.638	27.050	
	2400	116.146	130.129	180.228	-302.418	27.543	
	2473	118.174	138.682	183.738	-315.703	27.904	

18-ALUMINIUM 4-BORON 33-OXIDE $9\,Al_2O_3 \cdot 2\,B_2O_3$

		A	B	C	D	RANGE	REF.
SOL	CP	234.297	146.606	-68.905	-42.208	298-1600	5

PHASE	T	CP	H	S	G	BT	
SOL	298	196.741	-4236.5	156.3	-4283.101	3140.020	5
	300	197.919	-4236.135	157.521	-4283.391	3120.868	
	400	243.120	-4213.837	221.359	-4302.380	2351.028	
	500	269.486	-4188.113	278.651	-4327.438	1891.776	
	600	287.925	-4160.197	329.496	-4357.895	1587.576	
	700	302.177	-4130.666	374.991	-4393.160	1371.791	
	800	313.802	-4099.849	416.124	-4432.748	1211.133	
	900	323.547	-4067.967	453.664	-4476.265	1087.132	
	1000	331.804	-4035.189	488.192	-4523.381	988.717	
	1100	338.797	-4001.649	520.154	-4573.818	908.856	
	1200	344.660	-3967.467	549.892	-4627.337	842.867	
	1300	349.476	-3932.752	577.677	-4683.731	787.513	
	1400	353.302	-3897.605	603.722	-4742.815	740.486	
	1500	356.176	-3862.123	628.200	-4804.423	700.098	
	1600	358.123	-3826.400	651.254	-4868.407	665.083	

* * * * * * * * * *

4-ALUMINIUM 2-BORON 9-OXIDE $2Al_2O_3 \cdot B_2O_3$

		A	B	C	D	RANGE	REF.
SOL	CP	64.6	25.95	-17.0		298-1308	*
	** MPT= 1308 (PERITEC.)						231

PHASE	T	CP	H	S	G	BT	
SOL	298	53.213	-1121.1	37.3	-1132.221	830.052	5/*
	300	53.496	-1121.001	37.630	-1132.290	824.984	
	400	64.355	-1115.050	54.677	-1136.921	621.268	
	500	70.775	-1108.272	69.775	-1143.159	499.742	
	600	75.448	-1100.951	83.109	-1150.817	419.241	
	700	79.296	-1093.209	95.036	-1159.734	362.134	
	800	82.704	-1085.107	105.850	-1169.787	319.614	
	900	85.856	-1076.677	115.775	-1180.875	286.794	
	1000	88.850	-1067.941	124.977	-1192.918	260.747	
	1100	91.740	-1058.911	133.582	-1205.850	239.613	
	1200	94.559	-1049.595	141.685	-1219.618	222.153	
	1300	97.329	-1040.000	149.364	-1234.173	207.511	
	1308	97.549	-1039.221	149.962	-1235.371	206.442	

As ARSENIC

		A	B	C	D	RANGE	REF.
SOL-A	CP	5.504	1.373			298- 800	21
	** SUBL. TO (AS4 + AS2), MPT= 1090						224

PHASE	T	CP	H	S	G	BT	P	
SOL-A	298	5.913	0	8.534	-2.544	1.865		21
	300	5.916	.011	8.571	-2.560	1.865		
	400	6.053	.609	10.291	-3.507	1.916		
	500	6.190	1.222	11.657	-4.607	2.014		
	600	6.328	1.847	12.798	-5.831	2.124	.056	
	700	6.465	2.487	13.783	-7.161	2.236	3.935	
	800	6.602	3.141	14.656	-8.584	2.345	90.005	

* * * * * * * * * *

As (g) ARSENIC (MONATOMIC GAS)

		A	B	C	D	RANGE	REF.
GAS	CP	4.968				298-1200	21

PHASE	T	CP	H	S	G	BT	
GAS	298	4.968	72.120	41.611	59.714	-43.777	21
	300	4.968	72.129	41.642	59.637	-43.451	
	400	4.968	72.626	43.071	55.398	-30.272	
	500	4.968	73.123	44.180	51.033	-22.310	
	600	4.968	73.620	45.085	46.568	-16.965	
	700	4.968	74.116	45.851	42.021	-13.121	
	800	4.968	74.613	46.514	37.402	-10.219	
	900	4.968	75.110	47.100	32.720	-7.947	
	1000	4.968	75.607	47.623	27.984	-6.117	
	1100	4.968	76.104	48.097	23.197	-4.610	
	1200	4.968	76.600	48.529	18.366	-3.345	

ARSENIC (DIATOMIC GAS) As_2 (g)

		A	B	C	D	RANGE	REF.
GAS	CP	8.891	0.036	-0.483		298-1200	224

PHASE	T	CP	H	S	G	BT	
GAS	298	8.358	45.54	57.19	28.489	-20.886	224
	300	8.365	45.555	57.242	28.383	-20.680	
	400	8.604	46.406	59.686	22.531	-12.312	
	500	8.716	47.272	61.619	16.463	-7.197	
	600	8.778	48.147	63.214	10.219	-3.723	
	700	8.818	49.027	64.570	3.828	-1.195	
	800	8.844	49.910	65.750	-2.689	.735	
	900	8.864	50.796	66.793	-9.318	2.263	
	1000	8.879	51.683	67.727	-16.044	3.507	
	1100	8.891	52.571	68.574	-22.860	4.543	
	1200	8.901	53.461	69.348	-29.757	5.420	

* * * * * * * * * *

ARSENIC (TRIATOMIC GAS) As_3 (g)

		A	B	C	D	RANGE	REF.
GAS	CP	14.841	0.048	-0.662		298-1200	21

PHASE	T	CP	H	S	G	BT	
GAS	298	14.111	62.481	74.121	40.382	-29.605	21
	300	14.120	62.507	74.208	40.245	-29.322	
	400	14.446	63.938	78.322	32.609	-17.819	
	500	14.600	65.391	81.564	24.609	-10.758	
	600	14.686	66.856	84.234	16.315	-5.944	
	700	14.739	68.327	86.502	7.776	-2.428	
	800	14.776	69.803	88.473	-.975	.266	
	900	14.802	71.282	90.215	-9.911	2.407	
	1000	14.823	72.763	91.775	-19.012	4.156	
	1100	14.839	74.246	93.189	-28.262	5.616	
	1200	14.853	75.731	94.481	-37.646	6.857	

$As_4(g)$ ARSENIC (TETRATOMIC GAS)

		A	B	C	D	RANGE	REF.
GAS	CP	19.823	0.031	-1.226		298-1200	224

PHASE	T	CP	H	S	G	BT	
GAS	298	18.453	37.34	78.49	13.938	-10.218	224
	300	18.470	37.374	78.604	13.793	-10.049	
	400	19.069	39.255	84.012	5.651	-3.088	
	500	19.348	41.178	88.301	-2.973	1.299	
	600	19.501	43.121	91.843	-11.985	4.366	
	700	19.594	45.076	94.857	-21.324	6.658	
	800	19.656	47.039	97.477	-30.943	8.454	
	900	19.700	49.007	99.795	-40.809	9.911	
	1000	19.731	50.978	101.872	-50.894	11.124	
	1100	19.756	52.953	103.754	-61.177	12.156	
	1200	19.775	54.929	105.474	-71.640	13.049	

ARSENIC FLUORIDE

		A	B	C	D	RANGE	REF.
LIQ	CP	30.3				298- 331	62
	** BPT=	330.1					
GAS	CP	18.18	1.65	-2.66		330-1000	58

PHASE	T	CP	H	S	G	BT	P	
LIQ	298	30.300	-228.7	43.3	-241.610	177.129	197.048	62
	300	30.300	-228.644	43.487	-241.690	176.095	215.203	
	330	30.300	-227.732	46.385	-243.043	160.934	760.	
			8.023	24.305				62,*
GAS	330	16.284	-219.709	70.689	-243.043	160.934		
	400	17.177	-218.537	73.907	-248.100	135.574		
	500	17.941	-216.778	77.830	-255.692	111.778		
	600	18.431	-214.958	81.147	-263.646	96.046		
	700	18.792	-213.096	84.016	-271.907	84.905		
	800	19.084	-211.201	86.545	-280.437	76.622		
	900	19.337	-209.280	88.808	-289.207	70.238		
	1000	19.564	-207.335	90.857	-298.192	65.179		

AsF_5 (g)

ARSENIC PENTAFLUORIDE (GAS)

		A	B	C	D	RANGE	REF.
GAS	CP	31.30		-7.1		298-2000	*

PHASE	T	CP	H	S	G	BT	
GAS	298	23.313	-295.6	75.8	-318.200	233.278	62/*
	300	23.411	-295.557	75.945	-318.340	231.942	
	400	26.862	-293.018	83.223	-326.308	178.310	
	500	28.460	-290.243	89.409	-334.948	146.425	
	600	29.328	-287.350	94.682	-344.159	125.377	
	700	29.851	-284.389	99.245	-353.861	110.495	
	800	30.191	-281.386	103.255	-363.990	99.451	
	900	30.423	-278.355	106.825	-374.497	90.953	
	1000	30.590	-275.303	110.039	-385.343	84.228	
	1100	30.713	-272.238	112.961	-396.495	78.787	
	1200	30.807	-269.162	115.638	-407.927	74.304	
	1300	30.880	-266.077	118.106	-419.616	70.553	
	1400	30.938	-262.986	120.397	-431.542	67.376	
	1500	30.984	-259.890	122.533	-443.690	64.654	
	1600	31.023	-256.790	124.534	-456.044	62.301	
	1700	31.054	-253.686	126.416	-468.593	60.250	
	1800	31.081	-250.579	128.192	-481.324	58.449	
	1900	31.103	-247.470	129.873	-494.228	56.857	
	2000	31.122	-244.358	131.469	-507.296	55.442	

ARSENIC CHLORIDE $AsCl_3$

		A	B	C	D	RANGE	REF.
LIQ	CP	31.9				298- 309	1
	** MPT= 257, L= 2.4 / BPT= 308.43						1
GAS	CP	19.62	0.24	-1.42		308-1000	58

PHASE	T	CP	H	S	G	BT	P	
LIQ	298	31.900	-72.9	51.7	-88.314	64.745	483.762	62/19
	300	31.900	-72.841	51.897	-88.410	64.415	526.251	
	308	31.900	-72.572	52.781	-88.851	62.968	760.	
			7.959	25.805				1,*
GAS	308	18.201	-64.613	78.586	-88.851	62.968		
	400	18.828	-62.914	83.406	-96.277	52.610		
	500	19.172	-61.012	87.649	-104.837	45.830		
	600	19.370	-59.084	91.163	-113.782	41.451		
	700	19.498	-57.141	94.159	-123.052	38.424		
	800	19.590	-55.186	96.769	-132.601	36.230		
	900	19.661	-53.223	99.081	-142.396	34.583		
	1000	19.718	-51.254	101.155	-152.409	33.314		

$AsBr_3$(g)

ARSENIC BROMIDE (GAS)

		A	B	C	D	RANGE	REF.
GAS	CP	19.90		-1.01		298-1100	62

PHASE	T	CP	H	S	G	BT	
GAS	298	18.764	-31.57	86.94	-57.491	42.148	62
	300	18.778	-31.535	87.056	-57.652	42.005	
	400	19.269	-29.629	92.535	-66.644	36.417	
	500	19.496	-27.690	96.862	-76.121	33.277	
	600	19.619	-25.734	100.429	-85.991	31.326	
	700	19.694	-23.768	103.459	-96.189	30.036	
	800	19.742	-21.796	106.092	-106.670	29.145	
	900	19.775	-19.820	108.420	-117.397	28.512	
	1000	19.799	-17.841	110.505	-128.346	28.054	
	1100	19.817	-15.860	112.392	-139.492	27.718	

* * * * * * * * * *

AsJ_3

ARSENIC IODIDE

		A	B	C	D	RANGE	REF.
SOL	CP	17.02	27.8			298- 414	*
LIQ	CP	30.0				414- 644	*
	** BPT= 644, L= 13.5						62

PHASE	T	CP	H	S	G	BT	
SOL	298	25.309	-15.5	50.92	-30.682	22.493	62
	300	25.360	-15.453	51.077	-30.776	22.423	
	400	28.140	-12.778	58.753	-36.279	19.825	
	414	28.529	-12.381	59.728	-37.109	19.592	
			5.21	12.585			62
LIQ	414	30.000	-7.171	72.312	-37.109	19.592	
	500	30.000	-4.591	77.975	-43.579	19.051	
	600	30.000	-1.591	83.444	-51.658	18.819	
	644	30.000	-.271	85.567	-55.377	18.795	

ARSENIC OXIDE (CLAUDETITE) As_2O_3

		A	B	C	D	RANGE	REF.
SOL	CP	14.3	42.0			298- 582	48
	** TPT= 240 (AR.+CL.)						48
LIQ	CP	36.5				582- 734	*
	** BPT= 734 GAS(AS4O6)						62

PHASE	T	CP	H	S	G	BT	
SOL	298	26.822	-156.16	29.33	-164.905	120.895	48
	300	26.900	-156.110	29.496	-164.959	120.189	
	400	31.100	-153.210	37.810	-168.334	91.986	
	500	35.300	-149.890	45.201	-172.491	75.406	
	582	38.744	-146.855	50.817	-176.430	66.261	
			4.4	7.560			48
LIQ	582	36.500	-142.455	58.377	-176.430	66.261	
	600	36.500	-141.798	59.488	-177.491	64.660	
	700	36.500	-138.148	65.115	-183.728	57.370	
	734	36.500	-136.907	66.846	-185.972	55.381	

* * * * * * * * * *

ARSENIC MONOSULFIDE (GAS) AsS (g)

		A	B	C	D	RANGE	REF.
GAS	CP	8.353	0.391			298-1800	63

PHASE	T	CP	H	S	G	BT	
GAS	298	8.470	48.5	55.5	31.953	-23.425	63
	300	8.470	48.516	55.552	31.850	-23.206	
	400	8.509	49.365	57.995	26.167	-14.299	
	500	8.548	50.218	59.898	20.269	-8.861	
	600	8.588	51.074	61.460	14.199	-5.173	
	700	8.627	51.935	62.786	7.985	-2.493	
	800	8.666	52.800	63.941	1.647	-.450	
	900	8.705	53.668	64.964	-4.799	1.166	
	1000	8.744	54.541	65.883	-11.342	2.479	
	1100	8.783	55.417	66.718	-17.973	3.571	
	1200	8.822	56.297	67.484	-24.684	4.496	
	1300	8.861	57.181	68.192	-31.468	5.291	
	1400	8.900	58.070	68.850	-38.320	5.983	
	1500	8.939	58.962	69.465	-45.236	6.592	
	1600	8.979	59.857	70.043	-52.212	7.133	
	1700	9.018	60.757	70.589	-59.244	7.617	
	1800	9.057	61.661	71.105	-66.329	8.055	

As_2S_2 ARSENIC MONOSULFIDE

		A	B	C	D	RANGE	REF.
SOL	CP	19.82	8.93			298- 580	63
LIQ	CP	35.0				580- 700	63
	** DEC. GAS(AS4S4 + AS(X) + S(Y))						63

PHASE	T	CP	H	S	G	BT	
SOL	298	22.482	-34.0	30.3	-43.034	31.549	63
	300	22.499	-33.958	30.439	-43.090	31.395	
	400	23.392	-31.664	37.034	-46.477	25.398	
	500	24.285	-29.280	42.350	-50.455	22.057	
	580	24.999	-27.309	46.006	-53.992	20.347	
			1.45	2.500			63
LIQ	580	35.000	-25.859	48.506	-53.992	20.347	
	600	35.000	-25.159	49.692	-54.974	20.027	
	700	35.000	-21.659	55.088	-60.220	18.804	

* * * * * * * * * *

As_2S_3 ARSENIC SULFIDE

		A	B	C	D	RANGE	REF.
SOL	CP	25.25	8.71			298- 585	63
LIQ	CP	42.51	4.03			585- 996	63
	** DEC. GAS(AS2S2 + S2 + AS4) / BFT= 996, L= 20.6						63/62

PHASE	T	CP	H	S	G	BT	
SOL	298	27.847	-40.	39.1	-51.658	37.871	63
	300	27.863	-39.948	39.272	-51.730	37.690	
	400	28.734	-37.119	47.407	-56.082	30.646	
	500	29.605	-34.202	53.913	-61.158	26.736	
	585	30.345	-31.654	58.617	-65.945	24.640	
			6.85	11.709			63
LIQ	585	44.868	-24.804	70.327	-65.945	24.640	
	600	44.928	-24.130	71.463	-67.008	24.411	
	700	45.331	-19.617	78.419	-74.511	23.266	
	800	45.734	-15.064	84.499	-82.663	22.586	
	900	46.137	-10.471	89.909	-91.388	22.195	
	996	46.524	-6.023	94.604	-100.249	22.000	

ARSENIC MONOSELENIDE (GAS) — AsSe (g)

		A	B	C	D	RANGE	REF.
GAS	CP	8.43	0.372			298-1800	63

PHASE	T	CP	H	S	G	BT	
GAS	298	8.541	49.5	59.19	31.853	-23.352	63
	300	8.542	49.516	59.243	31.743	-23.128	
	400	8.579	50.372	61.705	25.690	-14.038	
	500	8.616	51.232	63.623	19.420	-8.490	
	600	8.653	52.095	65.198	12.976	-4.727	
	700	8.690	52.962	66.534	6.388	-1.995	
	800	8.728	53.833	67.697	-.325	.089	
	900	8.765	54.708	68.727	-7.147	1.736	
	1000	8.802	55.586	69.653	-14.067	3.075	
	1100	8.839	56.468	70.493	-21.075	4.188	
	1200	8.876	57.354	71.264	-28.163	5.130	
	1300	8.914	58.243	71.976	-35.325	5.940	
	1400	8.951	59.137	72.638	-42.557	6.644	
	1500	8.988	60.034	73.257	-49.852	7.264	
	1600	9.025	60.934	73.838	-57.207	7.815	
	1700	9.062	61.839	74.386	-64.618	8.308	
	1800	9.100	62.747	74.905	-72.083	8.753	

* * * * * * * * * *

ARSENIC SELENIDE — As_2Se_3

		A	B	C	D	RANGE	REF.
SOL	CP	22.9	20.5			298- 650	63
LIQ	CP	46.7				650- 800	63

PHASE	T	CP	H	S	G	BT	
SOL	298	29.012	-24.5	46.5	-38.364	28.125	63
	300	29.050	-24.446	46.680	-38.450	28.015	
	400	31.100	-21.439	55.317	-43.566	23.806	
	500	33.150	-18.226	62.477	-49.465	21.624	
	600	35.200	-14.809	68.703	-56.030	20.412	
	650	36.225	-13.023	71.561	-59.538	20.021	
			9.75	15.000			63
LIQ	650	46.700	-3.273	86.561	-59.538	20.021	
	700	46.700	-.938	90.021	-63.953	19.970	
	800	46.700	3.732	96.257	-73.274	20.020	

AsTe (g)

ARSENIC MONOTELLURIDE (GAS)

		A	B	C	D	RANGE	REF.
GAS	CP	8.698	0.17			298-1800	63

PHASE	T	CP	H	S	G	BT	
GAS	298	8.749	54.7	61.09	36.486	-26.749	63
	300	8.749	54.716	61.144	36.373	-26.501	
	400	8.766	55.592	63.663	30.127	-16.463	
	500	8.783	56.469	65.621	23.659	-10.343	
	600	8.800	57.349	67.224	17.014	-6.198	
	700	8.817	58.229	68.582	10.222	-3.192	
	800	8.834	59.112	69.760	3.304	-.903	
	900	8.851	59.996	70.802	-3.725	.905	
	1000	8.868	60.882	71.735	-10.853	2.372	
	1100	8.885	61.770	72.581	-18.070	3.591	
	1200	8.902	62.659	73.355	-25.367	4.621	
	1300	8.919	63.550	74.068	-32.739	5.505	
	1400	8.936	64.443	74.730	-40.179	6.273	
	1500	8.953	65.337	75.347	-47.683	6.948	
	1600	8.970	66.234	75.925	-55.247	7.547	
	1700	8.987	67.131	76.470	-62.867	8.083	
	1800	9.004	68.031	76.984	-70.540	8.566	

* * * * * * * * * *

As_2Te_3

ARSENIC TELLURIDE

		A	B	C	D	RANGE	REF.
SOL	CP	32.31	10.6	-4.444		298- 648	63
LIQ	CP	40.0				648- 800	79

PHASE	T	CP	H	S	G	BT	
SOL	298	30.471	-9.	54.1	-25.130	18.423	77/62
	300	30.552	-8.944	54.289	-25.230	18.383	
	400	33.772	-5.712	63.564	-31.137	17.015	
	500	35.832	-2.226	71.333	-37.893	16.565	
	600	37.436	1.440	78.013	-45.368	16.527	
	648	38.120	3.253	80.920	-49.183	16.590	
			11.	16.975			79
LIQ	648	40.000	14.253	97.895	-49.183	16.590	
	700	40.000	16.333	100.983	-54.355	16.973	
	800	40.000	20.333	106.324	-64.726	17.685	

GOLD

		A	B	C	D	RANGE	REF.
SOL	CP	5.738	1.046			298- 900	21
	CP	1.092	4.419	13.091		900-1200	21
	CP	-50.581	32.215	276.898		1200-1336	21
	** MPT= 1336.15 (REF. POINT)						
LIQ	CP	7.40				1336-3130	21
GAS	CP	2.635	1.168	14.398		3130-4000	21

PHASE	T	CP	H	S	G	BT	P	
SOL	298	6.050	0	11.352	-3.385	2.481		21
	300	6.052	.011	11.389	-3.406	2.481		
	400	6.156	.622	13.145	-4.636	2.533		
	500	6.261	1.242	14.530	-6.022	2.633		
	600	6.366	1.874	15.681	-7.535	2.745		
	700	6.470	2.516	16.670	-9.153	2.858		
	800	6.575	3.168	17.540	-10.864	2.968		
	900	6.679	3.831	18.321	-12.658	3.074		
	900	6.685	3.831	18.321	-12.658	3.074		
	1000	6.820	4.505	19.031	-14.526	3.175		
	1100	7.035	5.197	19.691	-16.463	3.271		
	1200	7.304	5.914	20.314	-18.463	3.363		
	1200	7.306	5.914	20.314	-18.463	3.363		
	1300	7.683	6.658	20.909	-20.525	3.451		
	1336	7.973	6.940	21.124	-21.284	3.482		
			3.0	2.245				21
LIQ	1336	7.400	9.940	23.369	-21.284	3.482		
	1400	7.400	10.413	23.715	-22.788	3.558		
	1500	7.400	11.153	24.225	-25.185	3.670		
	1600	7.400	11.893	24.703	-27.632	3.775	.003	
	1700	7.400	12.633	25.151	-30.124	3.873	.013	
	1800	7.400	13.373	25.574	-32.661	3.966	.050	
	1900	7.400	14.113	25.974	-35.239	4.054	.167	
	2000	7.400	14.853	26.354	-37.855	4.137	.492	
	2100	7.400	15.593	26.715	-40.509	4.216	1.308	
	2200	7.400	16.333	27.059	-43.198	4.292	3.173	
	2300	7.400	17.073	27.388	-45.920	4.364	7.113	
	2400	7.400	17.813	27.703	-48.675	4.433	14.886	
	2500	7.400	18.553	28.005	-51.460	4.499	29.321	
	2600	7.400	19.293	28.295	-54.275	4.563	54.752	
	2700	7.400	20.033	28.575	-57.119	4.624	97.512	
	2800	7.400	20.773	28.844	-59.990	4.683	166.495	
	2900	7.400	21.513	29.104	-62.888	4.740	273.760	

Au

GOLD (CONTINUED)

PHASE	T	CP	H	S	G	BT	P	
LIQ	3000	7.400	22.253	29.354	-65.810	4.795	435.148	
	3100	7.400	22.993	29.597	-68.758	4.848	670.904	
	3130	7.400	23.215	29.668	-69.647	4.864	760.	
			79.918	25.533				21
GAS	3130	6.438	103.133	55.201	-69.647	4.864		
	3200	6.513	103.586	55.344	-73.516	5.022		
	3300	6.622	104.243	55.547	-79.061	5.237		
	3400	6.731	104.910	55.746	-84.625	5.440		
	3500	6.841	105.589	55.943	-90.210	5.634		
	3600	6.951	106.279	56.137	-95.814	5.817		
	3700	7.062	106.979	56.329	-101.437	5.992		
	3800	7.173	107.691	56.519	-107.080	6.159		
	3900	7.285	108.414	56.706	-112.741	6.319		
	4000	7.397	109.148	56.892	-118.421	6.471		

GOLD OXIDE Au_2O_3

		A	B	C	D	RANGE	REF.
SOL	CP	25.7	5.2			298- 500	*

** METASTABLE

PHASE	T	CP	H	S	G	BT	
SOL	298	27.250	-.82	31.15	-10.107	7.410	215
	300	27.260	-.770	31.319	-10.165	7.406	
	400	27.780	1.982	39.232	-13.710	7.492	
	500	28.300	4.786	45.487	-17.957	7.850	

* * * * * * * * * *

GOLD MONOSULFIDE (GAS) AuS (g)

		A	B	C	D	RANGE	REF.
GAS	CP	8.92	0.012	-0.387		298-2000	63

PHASE	T	CP	H	S	G	BT	
GAS	298	8.488	55.1	63.95	36.033	-26.417	63
	300	8.494	55.116	64.003	35.915	-26.168	
	400	8.683	55.976	66.476	29.386	-16.058	
	500	8.771	56.849	68.424	22.637	-9.896	
	600	8.820	57.729	70.028	15.712	-5.724	
	700	8.849	58.612	71.390	8.640	-2.698	
	800	8.869	59.498	72.573	1.440	-.393	
	900	8.883	60.386	73.618	-5.870	1.426	
	1000	8.893	61.275	74.555	-13.280	2.903	
	1100	8.901	62.165	75.403	-20.778	4.129	
	1200	8.908	63.055	76.178	-28.358	5.165	
	1300	8.913	63.946	76.891	-36.012	6.055	
	1400	8.917	64.838	77.551	-43.734	6.828	
	1500	8.921	65.729	78.167	-51.521	7.508	
	1600	8.924	66.622	78.743	-59.366	8.110	
	1700	8.927	67.514	79.284	-67.268	8.649	
	1800	8.930	68.407	79.794	-75.222	9.134	
	1900	8.932	69.300	80.277	-83.226	9.574	
	2000	8.934	70.194	80.735	-91.277	9.976	

AuSe

GOLD MONOSELENIDE (ALPHA)

		A	B	C	D	RANGE	REF.
SOL-A	CP	10.0	6.67			298- 698	63
	** DPT= 698 (PERITEC.)						225

PHASE	T	CP	H	S	G	BT	
SOL-A	298	11.989	-1.9	19.3	-7.654	5.612	225
	300	12.001	-1.878	19.374	-7.690	5.603	
	400	12.668	-.644	22.918	-9.812	5.362	
	500	13.335	.656	25.816	-12.252	5.356	
	600	14.002	2.023	28.307	-14.961	5.450	
	698	14.656	3.427	30.473	-17.843	5.588	

* * * * * * * * * *

AuSe (ß)

GOLD MONOSELENIDE (BETA)

		A	B	C	D	RANGE	REF.
SOL-B	CP	10.0	6.67			298- 649	63
	** DPT= 649 (PERITEC.)						225

PHASE	T	CP	H	S	G	BT	
SOL-B	298	11.989	-0.9	20.7	-7.072	5.184	225
	300	12.001	-.878	20.774	-7.110	5.180	
	400	12.668	.356	24.318	-9.372	5.121	
	500	13.335	1.656	27.216	-11.952	5.225	
	600	14.002	3.023	29.707	-14.801	5.392	
	649	14.329	3.717	30.819	-16.284	5.485	

GOLD DITELLURIDE $AuTe_2$

		A	B	C	D	RANGE	REF.
SOL	CP	15.20	8.94	0.41		298- 737	70
	** MPT= 737, L= 9.7						63

PHASE	T	CP	H	S	G	BT	
SOL	298	18.327	-4.45	33.85	-14.542	10.661	70
	300	18.338	-4.416	33.963	-14.605	10.641	
	400	19.032	-2.549	39.330	-18.281	9.990	
	500	19.834	-.606	43.662	-22.437	9.809	
	600	20.678	1.419	47.352	-26.992	9.833	
	700	21.542	3.530	50.604	-31.893	9.959	
	737	21.864	4.333	51.722	-33.786	10.020	

* * * * * * * * * *

DIGOLD TRIPHOSPHIDE Au_2P_3

		A	B	C	D	RANGE	REF.
SOL	CP	25.9	9.0			298-1000	*
	** DEC.						1

PHASE	T	CP	H	S	G	BT	
SOL	298	28.583	-23.3	36.	-34.033	24.951	1
	300	28.600	-23.247	36.177	-34.100	24.845	
	400	29.500	-20.342	44.528	-38.153	20.849	
	500	30.400	-17.347	51.207	-42.951	18.776	
	600	31.300	-14.262	56.829	-48.360	17.617	
	700	32.200	-11.087	61.722	-54.292	16.953	
	800	33.100	-7.822	66.080	-60.686	16.581	
	900	34.000	-4.467	70.031	-67.495	16.392	
	1000	34.900	-1.022	73.660	-74.682	16.324	

$AuSn$

GOLD TIN

		A	B	C	D	RANGE	REF.
SOL	CP	11.13	3.8			298- 691	21
LIQ	CP	17.4				691- 900	21

PHASE	T	CP	H	S	G	BT	
SOL	298	12.263	-7.28	23.45	-14.272	10.463	21
	300	12.270	-7.257	23.526	-14.315	10.430	
	400	12.650	-6.011	27.108	-16.854	9.210	
	500	13.030	-4.727	29.971	-19.713	8.618	
	600	13.410	-3.405	32.381	-22.834	8.318	
	691	13.756	-2.169	34.298	-25.869	8.183	
			5.86	8.480			21
LIQ	691	17.400	3.691	42.779	-25.869	8.183	
	700	17.400	3.847	43.004	-26.255	8.198	
	800	17.400	5.587	45.327	-30.674	8.381	
	900	17.400	7.327	47.377	-35.312	8.576	

* * * * * * * * * *

$AuPb_2$

GOLD 2-LEAD

		A	B	C	D	RANGE	REF.
SOL	CP	14.80	17.8			298- 527	58
LIQ	CP	23.10				527- 800	58

PHASE	T	CP	H	S	G	BT	
SOL	298	20.107	-1.5	41.9	-13.992	10.258	1
	300	20.140	-1.463	42.024	-14.070	10.251	
	400	21.920	.640	48.062	-18.585	10.156	
	500	23.700	2.921	53.145	-23.651	10.339	
	527	24.181	3.568	54.404	-25.103	10.412	
			5.73	10.873			58
LIQ	527	23.100	9.298	65.277	-25.103	10.412	
	600	23.100	10.984	68.273	-29.980	10.922	
	700	23.100	13.294	71.834	-36.990	11.550	
	800	23.100	15.604	74.919	-44.331	12.112	

BORON (BETA)

		A	B	C	D	RANGE	REF.
SOL-B	CP	6.648	-0.167	-7.689	1.245	298- 800	3
	CP	5.108	1.128	-2.894		800-1500	3
	CP	7.671	0.017	-23.124		1500-2450	3
LIQ	CP	7.30				2450-3200	3
	** BPT= 3931						3

PHASE	T	CP	H	S	G	BT	P	
SOL-B	298	2.646	0	1.403	-.418	.307		3
	300	2.666	.005	1.419	-.421	.307		
	400	3.721	.326	2.335	-.608	.332		
	500	4.485	.739	3.253	-.888	.388		
	600	4.988	1.214	4.119	-1.257	.458		
	700	5.325	1.731	4.914	-1.709	.534		
	800	5.556	2.276	5.641	-2.238	.611		
	800	5.558	2.276	5.641	-2.238	.611		
	900	5.766	2.842	6.308	-2.836	.689		
	1000	5.947	3.428	6.926	-3.498	.765		
	1100	6.110	4.031	7.500	-4.219	.838		
	1200	6.261	4.649	8.038	-4.997	.910		
	1300	6.403	5.283	8.545	-5.826	.980		
	1400	6.540	5.930	9.025	-6.705	1.047		
	1500	6.671	6.590	9.480	-7.630	1.112		
	1500	6.669	6.590	9.480	-7.630	1.112		
	1600	6.795	7.264	9.915	-8.600	1.175		
	1700	6.900	7.949	10.330	-9.612	1.236		
	1800	6.988	8.643	10.727	-10.665	1.295		
	1900	7.063	9.346	11.107	-11.757	1.353		
	2000	7.127	10.055	11.471	-12.886	1.408		
	2100	7.182	10.771	11.820	-14.051	1.462		
	2200	7.231	11.492	12.155	-15.250	1.515	.003	
	2300	7.273	12.217	12.477	-16.481	1.566	.010	
	2400	7.310	12.946	12.788	-17.745	1.616	.031	
	2450	7.327	13.312	12.939	-18.388	1.640	.055	
			5.39	2.200				
LIQ	2450	7.300	18.702	15.139	-18.388	1.640	.055	3
	2500	7.300	19.067	15.286	-19.149	1.674	.091	
	2600	7.300	19.797	15.573	-20.692	1.740	.239	
	2700	7.300	20.527	15.848	-22.263	1.802	.582	
	2800	7.300	21.257	16.114	-23.861	1.863	1.329	
	2900	7.300	21.987	16.370	-25.485	1.921	2.863	
	3000	7.300	22.717	16.617	-27.135	1.977	5.852	
	3100	7.300	23.447	16.857	-28.808	2.031	11.408	
	3200	7.300	24.177	17.088	-30.506	2.084	21.305	

SOL-B CP(T) = A + B*T + C/(T**2) + E/(T**3)

B_2O_3 BORON OXIDE

		A	B	C	D	RANGE	REF.
SOL	CP	24.564	-20.292	-5.826	34.735	298- 723	3
LIQ	CP	58.751	-34.778	-40.910	11.512	723-1400	3
	CP	30.54				1400-2316	3
	** BPT= 2316, L= 87.55						3

PHASE	T	CP	H	S	G	BT	
SOL	298	15.048	-303.64	12.87	-307.477	225.417	3
	300	15.129	-303.612	12.963	-307.501	224.045	
	400	18.364	-301.923	17.800	-309.043	168.876	
	500	20.771	-299.965	22.160	-311.045	135.976	
	600	23.275	-297.765	26.164	-313.463	114.194	
	700	26.191	-295.296	29.964	-316.271	98.758	
	723	26.935	-294.685	30.823	-316.970	95.827	
			5.26	7.275			3
LIQ	723	31.798	-289.425	38.098	-316.970	95.827	
	800	31.904	-286.971	41.324	-320.030	87.440	
	900	31.725	-283.787	45.074	-324.354	78.774	
	1000	31.394	-280.630	48.400	-329.030	71.919	
	1100	31.044	-277.509	51.375	-334.022	66.373	
	1200	30.754	-274.420	54.064	-339.296	61.803	
	1300	30.574	-271.354	56.517	-344.827	57.978	
	1400	30.538	-268.300	58.781	-350.593	54.737	
	1400	30.540	-268.300	58.781	-350.593	54.737	
	1500	30.540	-265.246	60.888	-356.578	51.960	
	1600	30.540	-262.192	62.859	-362.766	49.558	
	1700	30.540	-259.138	64.710	-369.145	47.463	
	1800	30.540	-256.084	66.456	-375.705	45.623	
	1900	30.540	-253.030	68.107	-382.433	43.996	
	2000	30.540	-249.976	69.674	-389.323	42.549	
	2100	30.540	-246.922	71.164	-396.366	41.256	
	2200	30.540	-243.868	72.584	-403.554	40.095	
	2300	30.540	-240.814	73.942	-410.880	39.048	
	2316	30.540	-240.325	74.154	-412.065	38.890	

BORON OXIDE (AMORPH.) B_2O_3

		A	B	C	D	RANGE	REF.
LIQ	CP	13.082	17.814	-2.972		298- 500	3
	CP	79.731	-40.379	-96.855		500- 723	3
	CP	58.751	-34.778	-40.910	11.512	723-1400	3
	CP	30.54				1400-2316	3

PHASE	T	CP	H	S	G	BT	
LIQ	298	15.050	-299.28	18.739	-304.867	223.504	3
	300	15.124	-299.252	18.832	-304.902	222.151	
	400	18.350	-297.568	23.655	-307.030	167.776	
	500	20.800	-295.607	28.021	-309.617	135.352	
	500	20.800	-295.607	28.021	-309.617	135.352	
	600	28.599	-293.083	32.601	-312.644	113.896	
	700	31.699	-290.041	37.285	-316.140	98.717	
	723	32.008	-289.308	38.315	-317.009	95.839	
	723	31.798	-289.308	38.315	-317.009	95.839	
	800	31.904	-286.853	41.541	-320.086	87.455	
	900	31.725	-283.670	45.290	-324.431	78.793	
	1000	31.394	-280.513	48.616	-329.130	71.941	
	1100	31.044	-277.392	51.592	-334.143	66.397	
	1200	30.754	-274.303	54.280	-339.439	61.829	
	1300	30.574	-271.237	56.734	-344.991	58.006	
	1400	30.538	-268.183	58.997	-350.779	54.766	
	1400	30.540	-268.183	58.997	-350.779	54.766	
	1500	30.540	-265.129	61.104	-356.785	51.991	
	1600	30.540	-262.075	63.075	-362.995	49.590	
	1700	30.540	-259.021	64.927	-369.396	47.496	
	1800	30.540	-255.967	66.672	-375.977	45.656	
	1900	30.540	-252.913	68.324	-382.728	44.030	
	2000	30.540	-249.859	69.890	-389.639	42.584	
	2100	30.540	-246.805	71.380	-396.703	41.291	
	2200	30.540	-243.751	72.801	-403.913	40.130	
	2300	30.540	-240.697	74.158	-411.261	39.084	
	2316	30.540	-240.208	74.370	-412.450	38.926	

BS (g)

BORON MONOSULFIDE (GAS)

		A	B	C	D	RANGE	REF.
GAS	CP	8.783	0.134	-4.264	0.839	298-2000	3A

PHASE	T	CP	H	S	G	BT	
GAS	298	7.192	58.1	51.64	42.704	-31.307	3A
	300	7.193	58.113	51.684	42.608	-31.044	
	400	7.483	58.845	53.787	37.330	-20.399	
	500	7.816	59.610	55.494	31.863	-13.929	
	600	8.067	60.405	56.942	26.240	-9.559	
	700	8.251	61.222	58.200	20.481	-6.395	
	800	8.388	62.054	59.312	14.605	-3.990	
	900	8.492	62.898	60.306	8.623	-2.094	
	1000	8.574	63.752	61.205	2.547	-.557	
	1100	8.641	64.612	62.025	-3.615	.718	
	1200	8.696	65.479	62.780	-9.856	1.795	
	1300	8.743	66.351	63.478	-16.170	2.719	
	1400	8.784	67.228	64.127	-22.550	3.521	
	1500	8.819	68.108	64.734	-28.994	4.225	
	1600	8.851	68.992	65.305	-35.496	4.849	
	1700	8.880	69.878	65.842	-42.053	5.407	
	1800	8.907	70.768	66.350	-48.663	5.909	
	1900	8.932	71.660	66.833	-55.323	6.364	
	2000	8.955	72.554	67.291	-62.029	6.779	

GAS CP(T) = A + B*T + C/(T**2) + E/(T**3)

BORON SULFIDE B_2S_3

		A	B	C	D	RANGE	REF.
SOL	CP	23.43	15.3			298- 836	63
LIQ	CP	36.25				836-1100	63

PHASE	T	CP	H	S	G	BT	
SOL	298	27.992	-60.3	22.	-66.859	49.016	63
	300	28.020	-60.248	22.173	-66.900	48.743	
	400	29.550	-57.370	30.444	-69.547	38.004	
	500	31.080	-54.338	37.202	-72.939	31.886	
	600	32.610	-51.154	43.004	-76.956	28.035	
	700	34.140	-47.816	48.145	-81.518	25.454	
	800	35.670	-44.326	52.804	-86.569	23.653	
	836	36.221	-43.032	54.386	-88.499	23.139	
			11.5	13.756			63
LIQ	836	36.250	-31.532	68.142	-88.499	23.139	
	900	36.250	-29.212	70.816	-92.946	22.573	
	1000	36.250	-25.587	74.636	-100.222	21.906	
	1100	36.250	-21.962	78.090	-107.861	21.433	

* * * * * * * * * *

BORON MONOPHOSPHIDE (CUBIC) BP

		A	B	C	D	RANGE	REF.
SOL	CP	5.25	6.7			298-1100	105,*

PHASE	T	CP	H	S	G	BT	
SOL	298	7.248	-27.6	6.4	-29.508	21.633	105
	300	7.260	-27.587	6.445	-29.520	21.508	
	400	7.930	-26.827	8.625	-30.277	16.545	
	500	8.600	-26.001	10.467	-31.234	13.654	
	600	9.270	-25.107	12.094	-32.363	11.790	
	700	9.940	-24.147	13.573	-33.648	10.507	
	800	10.610	-23.119	14.944	-35.074	9.583	
	900	11.280	-22.025	16.233	-36.634	8.897	
	1000	11.950	-20.863	17.456	-38.319	8.376	
	1100	12.620	-19.635	18.626	-40.123	7.973	

Ba

BARIUM

		A	B	C	D	RANGE	REF.
SOL	CP	-10.618	37.855	5.376		298- 582	21
	** TPT= 582 (LAMDA TRANS.)						
	CP	156.085	-245.833			582- 600	21
	CP	-17.397	31.411	25.687		600- 700	21
	CP	45.890	-31.053	-70.168		700- 800	21
	CP	-12.719	16.201	62.939		800-1002	21
LIQ	CP	5.178	2.026	31.648		1002-1300	21
	CP	9.70				1300-2171	21
GAS	CP	-9.396	6.892	98.475		2171-3000	21

PHASE	T	CP	H	S	G	BT	P	
SOL	298	6.716	0	14.918	-4.448	3.261		21
	300	6.712	.012	14.960	-4.475	3.261		
	400	7.884	.724	16.997	-6.075	3.320		
	500	10.460	1.634	19.018	-7.875	3.443		
	582	13.001	2.594	20.791	-9.506	3.570		
	582	13.010	2.594	20.791	-9.506	3.570		
	600	8.585	2.789	21.121	-9.884	3.601		
	600	8.585	2.789	21.121	-9.884	3.601		
	700	9.833	3.702	22.526	-12.066	3.768		
	700	9.833	3.702	22.526	-12.066	3.768		
	800	10.084	4.709	23.871	-14.387	3.931		
	800	10.076	4.709	23.871	-14.387	3.931		
	900	9.632	5.688	25.025	-16.834	4.088	.004	
	1000	9.776	6.655	26.043	-19.388	4.238	.038	
	1002	9.783	6.675	26.062	-19.440	4.241	.039	
			1.852	1.848				21
LIQ	1002	10.360	8.527	27.911	-19.440	4.241	.039	
	1100	10.022	9.524	28.861	-22.223	4.416	.218	
	1200	9.807	10.515	29.723	-25.153	4.582	.918	
	1300	9.684	11.489	30.502	-28.165	4.736	3.056	
	1300	9.700	11.489	30.502	-28.165	4.736	3.056	

BARIUM (CONTINUED)

PHASE	T	CP	H	S	G	BT	P	
LIQ	1400	9.700	12.459	31.221	-31.251	4.879	8.453	
	1500	9.700	13.429	31.891	-34.407	5.014	20.199	
	1600	9.700	14.399	32.517	-37.628	5.140	42.889	
	1700	9.700	15.369	33.105	-40.909	5.260	82.698	
	1800	9.700	16.339	33.659	-44.248	5.373	147.272	
	1900	9.700	17.309	34.184	-47.640	5.481	245.490	
	2000	9.700	18.279	34.681	-51.084	5.583	387.143	
	2100	9.700	19.249	35.154	-54.576	5.681	582.617	
	2171	9.700	19.937	35.477	-57.083	5.747	760.	
			33.822	15.579				21
GAS	2171	7.656	53.759	51.056	-57.083	5.747		
	2200	7.801	53.983	51.158	-58.565	5.819		
	2300	8.317	54.789	51.516	-63.699	6.054		
	2400	8.854	55.647	51.882	-68.869	6.272		
	2500	9.410	56.561	52.254	-74.075	6.477		
	2600	9.980	57.530	52.634	-79.320	6.668		
	2700	10.563	58.557	53.022	-84.603	6.849		
	2800	11.158	59.643	53.417	-89.924	7.020		
	2900	11.762	60.789	53.819	-95.286	7.182		
	3000	12.374	61.996	54.228	-100.688	7.336		

BaH_2 BARIUM HYDRIDE

		A	B	C	D	RANGE	REF.
SOL	CP	8.9	4.1			298-1100	*

PHASE	T	CP	H	S	G	BT	
SOL	298	10.122	-42.7	13.5	-46.725	34.255	19/*
	300	10.130	-42.681	13.563	-46.750	34.062	
	400	10.540	-41.648	16.533	-48.261	26.372	
	500	10.950	-40.573	18.929	-50.038	21.874	
	600	11.360	-39.458	20.962	-52.035	18.956	
	700	11.770	-38.301	22.744	-54.222	16.931	
	800	12.180	-37.104	24.342	-56.577	15.458	
	900	12.590	-35.865	25.800	-59.086	14.350	
	1000	13.000	-34.586	27.148	-61.734	13.494	
	1100	13.410	-33.265	28.406	-64.512	12.819	

BARIUM FLUORIDE

BaF_2

		A	B	C	D	RANGE	REF.
SOL	CP	20.75	0.9	-3.6		298-1563	19,*
LIQ	CP	22.5				1563-1900	*

PHASE	T	CP	H	S	G	BT	
SOL	298	16.969	-288.5	23.03	-295.366	216.539	19
	300	17.020	-288.469	23.135	-295.409	215.234	
	400	18.860	-286.662	28.320	-297.990	162.836	
	500	19.760	-284.727	32.635	-301.044	131.604	
	600	20.290	-282.722	36.288	-304.495	110.927	
	700	20.645	-280.674	39.444	-308.285	96.264	
	800	20.907	-278.596	42.219	-312.371	85.347	
	900	21.116	-276.495	44.694	-316.719	76.920	
	1000	21.290	-274.374	46.928	-321.302	70.230	
	1100	21.442	-272.237	48.964	-326.098	64.798	
	1200	21.580	-270.086	50.836	-331.089	60.308	
	1300	21.707	-267.922	52.568	-336.260	56.538	
	1400	21.826	-265.745	54.181	-341.599	53.333	
	1500	21.940	-263.557	55.691	-347.093	50.578	
	1563	22.009	-262.172	56.595	-350.630	49.034	
			6.8	4.351			1
LIQ	1563	22.500	-255.372	60.946	-350.630	49.034	
	1600	22.500	-254.540	61.472	-352.895	48.210	
	1700	22.500	-252.290	62.836	-359.111	46.173	
	1800	22.500	-250.040	64.122	-365.460	44.379	
	1900	22.500	-247.790	65.339	-371.933	42.788	

BaO_2 BARIUM PEROXIDE

		A	B	C	D	RANGE	REF.
SOL	CP	14.9	6.7			298-1075	*
	** DPT= 1075 (BAO2 = BAO + 1/2O2)						

PHASE	T	CP	H	S	G	BT	
SOL	298	16.898	-151.6	22.25	-158.234	116.004	122*
	300	16.910	-151.569	22.355	-158.275	115.319	
	400	17.580	-149.844	27.311	-160.769	87.852	
	500	18.250	-148.053	31.306	-163.706	71.565	
	600	18.920	-146.194	34.692	-167.010	60.841	
	700	19.590	-144.269	37.659	-170.630	53.280	
	800	20.260	-142.276	40.319	-174.531	47.686	
	900	20.930	-140.217	42.744	-178.686	43.397	
	1000	21.600	-138.090	44.984	-183.074	40.016	
	1075	22.102	-136.451	46.564	-186.508	37.922	

* * * * * * * * * *

$Ba(OH)_2$ BARIUM HYDROXIDE

		A	B	C	D	RANGE	REF.
SOL	CP	17.5	13.5			298- 681	*
	** MPT= 681						47

PHASE	T	CP	H	S	G	BT	
SOL	298	21.525	-225.5	24.1	-232.685	170.586	19/*
	300	21.550	-225.460	24.233	-232.730	169.567	
	400	22.900	-223.238	30.618	-235.485	128.680	
	500	24.250	-220.880	35.873	-238.816	104.401	
	600	25.600	-218.388	40.413	-242.636	88.392	
	681	26.693	-216.270	43.723	-246.045	78.973	

BARIUM SULFIDE

BaS

		A	B	C	D	RANGE	REF.
SOL	CP	11.4	1.4			298-2000	19,*
	** MPT=	2500					63

PHASE	T	CP	H	S	G	BT	
SOL	298	11.817	-110.	18.7	-115.575	84.730	19
	300	11.820	-109.978	18.773	-115.610	84.233	
	400	11.960	-108.789	22.193	-117.666	64.298	
	500	12.100	-107.586	24.877	-120.024	52.470	
	600	12.240	-106.369	27.095	-122.626	44.673	
	700	12.380	-105.138	28.992	-125.433	39.167	
	800	12.520	-103.893	30.655	-128.417	35.087	
	900	12.660	-102.634	32.137	-131.558	31.951	
	1000	12.800	-101.361	33.478	-134.840	29.473	
	1100	12.940	-100.074	34.705	-138.250	27.471	
	1200	13.080	-98.773	35.837	-141.777	25.825	
	1300	13.220	-97.458	36.889	-145.414	24.450	
	1400	13.360	-96.129	37.874	-149.153	23.287	
	1500	13.500	-94.786	38.801	-152.987	22.293	
	1600	13.640	-93.429	39.676	-156.911	21.436	
	1700	13.780	-92.058	40.508	-160.921	20.691	
	1800	13.920	-90.673	41.299	-165.012	20.038	
	1900	14.060	-89.274	42.056	-169.180	19.463	
	2000	14.200	-87.861	42.780	-173.422	18.953	

BaTe

BARIUM TELLURIDE

		A	B	C	D	RANGE	REF.
SOL	CP	10.2	2.0			298-1500	*

PHASE	T	CP	H	S	G	BT	
SOL	298	10.796	-64.4	23.8	-71.496	52.415	*
	300	10.800	-64.380	23.867	-71.540	52.124	
	400	11.000	-63.290	27.001	-74.090	40.487	
	500	11.200	-62.180	29.477	-76.919	33.626	
	600	11.400	-61.050	31.537	-79.972	29.134	
	700	11.600	-59.900	33.309	-83.216	25.985	
	800	11.800	-58.730	34.871	-86.627	23.669	
	900	12.000	-57.540	36.273	-90.185	21.903	
	1000	12.200	-56.330	37.547	-93.877	20.520	
	1100	12.400	-55.100	38.719	-97.691	19.412	
	1200	12.600	-53.850	39.807	-101.618	18.510	
	1300	12.800	-52.580	40.823	-105.650	17.764	
	1400	13.000	-51.290	41.779	-109.781	17.140	
	1500	13.200	-49.980	42.683	-114.005	16.613	

BARIUM NITRIDE Ba_3N_2

		A	B	C	D	RANGE	REF.
SOL	CP	21.0	23.5			298- 950	*
	CP	43.5				950-1300	*

PHASE	T	CP	H	S	G	BT	
SOL	298	28.007	-81.5	36.4	-92.353	67.705	11/58
	300	28.050	-81.448	36.573	-92.420	67.337	
	400	30.400	-78.526	44.965	-96.512	52.739	
	500	32.750	-75.368	52.001	-101.369	44.314	
	600	35.100	-71.976	58.179	-106.883	38.937	
	700	37.450	-68.348	63.767	-112.985	35.280	
	800	39.800	-64.486	68.921	-119.622	32.684	
	900	42.150	-60.388	73.744	-126.758	30.785	
	950	43.325	-58.251	76.055	-130.503	30.027	
	950	43.500	-58.251	76.055	-130.503	30.027	
	1000	43.500	-56.076	78.286	-134.362	29.369	
	1100	43.500	-51.726	82.432	-142.401	28.296	
	1200	43.500	-47.376	86.217	-150.837	27.475	
	1300	43.500	-43.026	89.699	-159.635	26.841	

* * * * * * * * * *

BARIUM NITRATE $Ba(NO_3)_2$

		A	B	C	D	RANGE	REF.
SOL	CP	30.05	35.7	-4.01		298- 868	58
	** MPT= 868						47

PHASE	T	CP	H	S	G	BT	
SOL	298	36.183	-237.11	51.1	-252.345	184.999	19
	300	36.304	-237.043	51.324	-252.440	183.927	
	400	41.824	-233.123	62.564	-258.148	141.065	
	500	46.296	-228.712	72.389	-264.906	115.806	
	600	50.356	-223.877	81.192	-272.592	99.305	
	700	54.222	-218.647	89.247	-281.120	87.781	
	800	57.983	-213.036	96.734	-290.423	79.350	
	868	60.505	-209.007	101.566	-297.166	74.832	

BaC_2 BARIUM CARBIDE

		A	B	C	D	RANGE	REF.
SOL	CP	17.6	0.9	-2.31		298-1500	*

PHASE	T	CP	H	S	G	BT	
SOL	298	15.270	-17.7	21.	-23.961	17.566	95
	300	15.303	-17.672	21.095	-24.000	17.486	
	400	16.516	-16.073	25.686	-26.347	14.397	
	500	17.126	-14.388	29.444	-29.110	12.726	
	600	17.498	-12.655	32.601	-32.216	11.736	
	700	17.759	-10.892	35.319	-35.615	11.121	
	800	17.959	-9.105	37.704	-39.269	10.729	
	900	18.125	-7.301	39.829	-43.148	10.479	
	1000	18.269	-5.481	41.747	-47.228	10.323	
	1100	18.399	-3.648	43.494	-51.491	10.232	
	1200	18.520	-1.802	45.100	-55.922	10.186	
	1300	18.633	.056	46.587	-60.507	10.174	
	1400	18.742	1.925	47.972	-65.236	10.185	
	1500	18.847	3.804	49.269	-70.099	10.215	

2-BARIUM TIN Ba_2Sn

		A	B	C	D	RANGE	REF.
SOL	CP	14.5	10.0			298-1200	*

PHASE	T	CP	H	S	G	BT	
SOL	298	17.481	-90.	30.3	-99.034	72.604	1/*
	300	17.500	-89.968	30.408	-99.090	72.197	
	400	18.500	-88.168	35.580	-102.399	55.956	
	500	19.500	-86.268	39.815	-106.175	46.415	
	600	20.500	-84.268	43.459	-110.343	40.198	
	700	21.500	-82.168	46.694	-114.853	35.864	
	800	22.500	-79.968	49.630	-119.672	32.697	
	900	23.500	-77.668	52.338	-124.772	30.303	
	1000	24.500	-75.268	54.866	-130.133	28.444	
	1100	25.500	-72.768	57.248	-135.740	26.973	
	1200	26.500	-70.168	59.509	-141.579	25.789	

* * * * * * * * * *

BARIUM 3-TIN $BaSn_3$

		A	B	C	D	RANGE	REF.
SOL	CP	20.25	12.5			298- 973	*
	** MPT=	973					1

PHASE	T	CP	H	S	G	BT	
SOL	298	23.977	-46.5	45.	-59.917	43.926	1*
	300	24.000	-46.456	45.148	-60.000	43.716	
	400	25.250	-43.993	52.224	-64.883	35.455	
	500	26.500	-41.406	57.993	-70.402	30.777	
	600	27.750	-38.693	62.935	-76.454	27.852	
	700	29.000	-35.856	67.306	-82.970	25.908	
	800	30.250	-32.893	71.260	-89.901	24.563	
	900	31.500	-29.806	74.895	-97.211	23.609	
	973	32.412	-27.473	77.387	-102.770	23.087	

Ba_2Pb 2-BARIUM LEAD

		A	B	C	D	RANGE	REF.
SOL	CP	15.0	10.0			298-1201	*
	** MPT= 1201						1

PHASE	T	CP	H	S	G	BT	
SOL	298	17.981	-70.	34.1	-80.167	58.772	1/*
	300	18.000	-69.967	34.211	-80.230	58.455	
	400	19.000	-68.117	39.527	-83.927	45.862	
	500	20.000	-66.167	43.874	-88.104	38.515	
	600	21.000	-64.117	47.608	-92.682	33.764	
	700	22.000	-61.967	50.921	-97.611	30.480	
	800	23.000	-59.717	53.924	-102.856	28.103	
	900	24.000	-57.367	56.690	-108.388	26.324	
	1000	25.000	-54.917	59.271	-114.188	24.959	
	1100	26.000	-52.367	61.701	-120.237	23.892	
	1200	27.000	-49.717	64.006	-126.524	23.046	
	1201	27.010	-49.690	64.028	-126.588	23.039	

* * * * * * * * * *

$BaPb_3$ BARIUM 3-LEAD

		A	B	C	D	RANGE	REF.
SOL	CP	20.7	12.7			298- 890	*
	** MPT= 890						1

PHASE	T	CP	H	S	G	BT	
SOL	298	24.487	-42.	54.	-58.100	42.594	1/*
	300	24.510	-41.955	54.152	-58.200	42.404	
	400	25.780	-39.440	61.377	-63.991	34.968	
	500	27.050	-36.799	67.266	-70.431	30.790	
	600	28.320	-34.030	72.310	-77.416	28.203	
	700	29.590	-31.135	76.771	-84.874	26.502	
	800	30.860	-28.112	80.805	-92.756	25.343	
	890	32.003	-25.283	84.155	-100.181	24.604	

BARIUM TUNGSTATE $BaO \cdot WO_3$

		A	B	C	D	RANGE	REF.
SOL	CP	29.51	8.3			298-1300	*

PHASE	T	CP	H	S	G	BT	
SOL	298	31.985	-389.76	36.	-400.493	293.609	*
	300	32.000	-389.701	36.198	-400.560	291.847	
	400	32.830	-386.459	45.517	-404.666	221.129	
	500	33.660	-383.135	52.932	-409.601	179.061	
	600	34.490	-379.727	59.143	-415.213	151.262	
	700	35.320	-376.237	64.522	-421.402	131.585	
	800	36.150	-372.663	69.292	-428.097	116.966	
	900	36.980	-369.007	73.598	-435.245	105.706	
	1000	37.810	-365.267	77.537	-442.804	96.788	
	1100	38.640	-361.445	81.180	-450.743	89.566	
	1200	39.470	-357.539	84.577	-459.032	83.612	
	1300	40.300	-353.551	87.770	-467.651	78.630	

* * * * * * * * * *

BARIUM URANATE $BaO \cdot UO_3$

		A	B	C	D	RANGE	REF.
SOL	CP	29.04	9.62			298-1100	136

PHASE	T	CP	H	S	G	BT	
SOL	298	31.908	-475.3	40.3	-487.315	357.260	*
	300	31.926	-475.241	40.497	-487.390	355.111	
	400	32.888	-472.000	49.814	-491.926	268.812	
	500	33.850	-468.663	57.256	-497.291	217.395	
	600	34.812	-465.230	63.512	-503.338	183.365	
	700	35.774	-461.701	68.951	-509.967	159.240	
	800	36.736	-458.075	73.791	-517.108	141.286	
	900	37.698	-454.354	78.173	-524.710	127.434	
	1000	38.660	-450.536	82.195	-532.731	116.444	
	1100	39.622	-446.622	85.925	-541.139	107.529	

$BaO \cdot ZrO_2$ BARIUM METAZIRCONATE

		A	B	C	D	RANGE	REF.
SOL	CP	29.35	2.10	-5.25		298-2000	*

PHASE	T	CP	H	S	G	BT	
SOL	298	24.070	-423.2	26.5	-431.101	316.048	5/*
	300	24.147	-423.155	26.649	-431.150	314.135	
	400	26.909	-420.584	34.027	-434.195	237.265	
	500	28.300	-417.817	40.195	-437.915	191.438	
	600	29.152	-414.942	45.435	-442.203	161.094	
	700	29.749	-411.995	49.976	-446.979	139.572	
	800	30.210	-408.997	53.980	-452.181	123.547	
	900	30.592	-405.956	57.561	-457.761	111.174	
	1000	30.925	-402.880	60.802	-463.681	101.351	
	1100	31.226	-399.772	63.763	-469.912	93.375	
	1200	31.505	-396.635	66.492	-476.426	86.781	
	1300	31.769	-393.472	69.025	-483.204	81.245	
	1400	32.022	-390.282	71.388	-490.226	76.538	
	1500	32.267	-387.067	73.606	-497.477	72.492	
	1600	32.505	-383.829	75.696	-504.943	68.981	
	1700	32.738	-380.567	77.674	-512.612	65.910	
	1800	32.968	-377.281	79.552	-520.474	63.203	
	1900	33.195	-373.973	81.340	-528.519	60.802	
	2000	33.419	-370.642	83.049	-536.740	58.660	

BARIUM HAFNATE $BaO \cdot HfO_2$

		A	B	C	D	RANGE	REF.
SOL	CP	29.6	3.2	-4.15		298-1500	*

PHASE	T	CP	H	S	G	GT	
SOL	298	25.886	-429.06	29.2	-437.766	320.934	137/*
	300	25.949	-429.012	29.360	-437.820	318.995	
	400	28.286	-426.286	37.187	-441.161	241.071	
	500	29.540	-423.389	43.645	-445.212	194.628	
	600	30.367	-420.392	49.108	-449.857	163.882	
	700	30.993	-417.323	53.838	-455.009	142.079	
	800	31.512	-414.197	58.012	-460.606	125.849	
	900	31.968	-411.022	61.750	-466.597	113.320	
	1000	32.385	-407.804	65.140	-472.944	103.376	
	1100	32.777	-404.546	68.245	-479.616	95.304	
	1200	33.152	-401.250	71.113	-486.585	88.631	
	1300	33.514	-397.916	73.781	-493.832	83.032	
	1400	33.868	-394.547	76.278	-501.336	78.273	
	1500	34.216	-391.143	78.626	-509.082	74.183	

Be

BERYLLIUM

		A	B	C	D	RANGE	REF.
SOL	CP	4.947	1.660	-1.346		298- 800	3
	CP	4.175	2.363			800-1556	3
LIQ	CP	6.095	0.504			1556-2757	3
GAS	CP	4.99				2757-3000	3

PHASE	T	CP	H	S	G	BT	P	
SOL	298	3.928	0	2.28	-.680	.498		3
	300	3.949	.007	2.304	-.684	.498		
	400	4.770	.448	3.566	-.979	.535		
	500	5.239	.950	4.685	-1.392	.609		
	600	5.569	1.491	5.671	-1.911	.696		
	700	5.834	2.062	6.550	-2.523	.788		
	800	6.065	2.657	7.344	-3.218	.879		
	800	6.065	2.657	7.344	-3.218	.879		
	900	6.302	3.275	8.072	-3.990	.969		
	1000	6.538	3.917	8.748	-4.831	1.056		
	1100	6.774	4.583	9.382	-5.738	1.140		
	1200	7.011	5.272	9.982	-6.706	1.222		
	1300	7.247	5.985	10.552	-7.733	1.300		
	1400	7.483	6.721	11.098	-8.816	1.376	.002	
	1500	7.719	7.482	11.622	-9.952	1.450	.011	
	1556	7.852	7.918	11.908	-10.611	1.491	.027	
			2.8	1.799				3
LIQ	1556	6.879	10.718	13.707	-10.611	1.491	.027	
	1600	6.901	11.021	13.899	-11.218	1.533	.052	
	1700	6.952	11.713	14.319	-12.629	1.624	.201	
	1800	7.002	12.411	14.718	-14.081	1.710	.674	
	1900	7.053	13.114	15.098	-15.572	1.791	1.978	
	2000	7.103	13.822	15.461	-17.101	1.869	5.203	
	2100	7.153	14.534	15.809	-18.664	1.943	12.446	
	2200	7.204	15.252	16.143	-20.262	2.013	27.441	
	2300	7.254	15.975	16.464	-21.892	2.081	56.354	
	2400	7.305	16.703	16.774	-23.554	2.145	108.769	
	2500	7.355	17.436	17.073	-25.247	2.207	198.786	
	2600	7.405	18.174	17.363	-26.969	2.267	346.195	
	2700	7.456	18.917	17.643	-28.719	2.325	577.631	
	2757	7.485	19.343	17.799	-29.729	2.357	760.	
			71.137	25.802				3
GAS	2757	4.990	90.480	43.601	-29.729	2.357		
	2800	4.990	90.695	43.679	-31.606	2.467		
	2900	4.990	91.194	43.854	-35.982	2.712		
	3000	4.990	91.693	44.023	-40.376	2.942		

BERYLLIUM MONOFLUORIDE (GAS)

BeF (g)

		A	B	C	D	RANGE	REF.
GAS	CP	8.732	0.177	-4.514	0.913	298-2000	3A

PHASE	T	CP	H	S	G	BT	
GAS	298	7.152	-40.6	49.149	-55.254	40.508	3A
	300	7.151	-40.587	49.193	-55.345	40.324	
	400	7.408	-39.862	51.277	-60.373	32.991	
	500	7.745	-39.103	52.968	-65.587	28.672	
	600	8.007	-38.315	54.404	-70.958	25.850	
	700	8.201	-37.504	55.654	-76.462	23.876	
	800	8.347	-36.677	56.759	-82.084	22.427	
	900	8.459	-35.836	57.749	-87.810	21.326	
	1000	8.549	-34.986	58.645	-93.630	20.466	
	1100	8.622	-34.127	59.463	-99.536	19.779	
	1200	8.684	-33.261	60.216	-105.521	19.221	
	1300	8.737	-32.390	60.913	-111.578	18.760	
	1400	8.783	-31.514	61.563	-117.702	18.377	
	1500	8.824	-30.634	62.170	-123.889	18.053	
	1600	8.861	-29.750	62.741	-130.135	17.778	
	1700	8.895	-28.862	63.279	-136.436	17.542	
	1800	8.927	-27.971	63.788	-142.790	17.339	
	1900	8.957	-27.077	64.272	-149.193	17.163	
	2000	8.985	-26.179	64.732	-155.643	17.010	

GAS CP(T) = A + B*T + C/(T**2) + E/(T**3)

BeF_2 BERYLLIUM FLUORIDE

		A	B	C	D	RANGE	REF.
SOL-A	CP	4.931	25.030	-0.007		298- 438	3
	CP	4.931	25.030	-0.007		438- 500	3
SOL-B	CP	11.320	8.0			500- 825	3
LIQ	CP	12.558	7.866			825-1447	3
	** BPT=	1447.354					
GAS	CP	14.545	0.135	-6.346		1447-2000	3

PHASE	T	CP	H	S	G	BT	P	
SOL-A	298	12.386	-245.4	12.752	-249.202	182.695		3,65
	300	12.432	-245.377	12.829	-249.226	181.585		
	400	14.939	-244.008	16.749	-250.708	136.999		
	438	15.890	-243.423	18.147	-251.371	125.444		
	438	15.890	-243.423	18.147	-251.371	125.444		
	500	17.443	-242.389	20.351	-252.565	110.411		
			0.053	.106				3
SOL-B	500	15.320	-242.336	20.457	-252.565	110.411		
	600	16.120	-240.764	23.321	-254.757	92.808		
	700	16.920	-239.112	25.866	-257.219	80.318		
	800	17.720	-237.380	28.178	-259.922	71.017		
	825	17.920	-236.935	28.726	-260.634	69.053	.001	
			1.137	1.378				3
LIQ	825	19.047	-235.798	30.104	-260.634	69.053	.001	
	900	19.637	-234.347	31.787	-262.955	63.863	.019	
	1000	20.424	-232.344	33.896	-266.241	58.195	.331	
	1100	21.211	-230.262	35.880	-269.730	53.598	3.394	
	1200	21.997	-228.102	37.759	-273.413	49.802	22.978	
	1300	22.784	-225.863	39.551	-277.279	46.621	113.065	
	1400	23.570	-223.545	41.268	-281.321	43.922	432.853	
	1447	23.943	-222.420	42.059	-283.294	42.783	760.	
			47.646	32.919				3
GAS	1447	14.437	-174.774	74.978	-283.294	42.783		
	1500	14.465	-174.013	75.494	-287.255	41.859		
	1600	14.513	-172.564	76.429	-294.851	40.280		
	1700	14.555	-171.111	77.310	-302.539	38.899		
	1800	14.592	-169.654	78.143	-310.312	37.682		
	1900	14.626	-168.193	78.933	-318.166	36.602		
	2000	14.656	-166.729	79.684	-326.097	35.639		

BERYLLIUM IODIDE (GAS)

$BeJ_2 (g)$

		A	B	C	D	RANGE	REF.
GAS	CP	14.754	0.073	-1.304		298-2000	3

PHASE	T	CP	H	S	G	BT	
GAS	298	13.309	-15.7	70.4	-36.690	26.898	65/3
	300	13.327	-15.675	70.482	-36.820	26.827	
	400	13.968	-14.306	74.417	-44.073	24.084	
	500	14.269	-12.893	77.570	-51.678	22.591	
	600	14.436	-11.457	80.188	-59.569	21.701	
	700	14.539	-10.008	82.421	-67.702	21.140	
	800	14.609	-8.550	84.367	-76.044	20.777	
	900	14.659	-7.086	86.091	-84.569	20.539	
	1000	14.697	-5.619	87.638	-93.256	20.384	
	1100	14.727	-4.147	89.040	-102.091	20.286	
	1200	14.751	-2.673	90.322	-111.060	20.230	
	1300	14.772	-1.197	91.504	-120.152	20.202	
	1400	14.790	.281	92.599	-129.358	20.196	
	1500	14.806	1.761	93.620	-138.670	20.207	
	1600	14.820	3.242	94.576	-148.080	20.230	
	1700	14.833	4.724	95.475	-157.583	20.261	
	1800	14.845	6.208	96.323	-167.173	20.300	
	1900	14.857	7.693	97.126	-176.846	20.345	
	2000	14.867	9.180	97.888	-186.597	20.393	

BeOH(g)

BERYLLIUM HYDROXIDE (GAS)

		A	B	C	D	RANGE	REF.
GAS	CP	8.13	3.73	-0.98	-0.68	298-3000	65

PHASE	T	CP	H	S	G	BT	
GAS	298	8.079	-47.5	53.28	-63.385	46.469	65
	300	8.099	-47.485	53.330	-63.484	46.254	
	400	8.901	-46.632	55.780	-68.943	37.674	
	500	9.433	-45.714	57.826	-74.627	32.624	
	600	9.851	-44.749	59.584	-80.499	29.326	
	700	10.208	-43.745	61.130	-86.536	27.022	
	800	10.526	-42.708	62.514	-92.720	25.333	
	900	10.815	-41.641	63.771	-99.035	24.052	
	1000	11.082	-40.546	64.924	-105.471	23.054	
	1100	11.329	-39.425	65.992	-112.017	22.259	
	1200	11.559	-38.281	66.988	-118.667	21.615	
	1300	11.772	-37.114	67.922	-125.413	21.087	
	1400	11.969	-35.927	68.802	-132.249	20.648	
	1500	12.151	-34.721	69.634	-139.171	20.280	
	1600	12.319	-33.497	70.423	-146.175	19.969	
	1700	12.472	-32.258	71.175	-153.255	19.705	
	1800	12.611	-31.003	71.892	-160.408	19.479	
	1900	12.735	-29.736	72.577	-167.632	19.285	
	2000	12.845	-28.457	73.233	-174.923	19.117	
	2100	12.942	-27.167	73.862	-182.278	18.972	
	2200	13.025	-25.869	74.466	-189.694	18.847	
	2300	13.093	-24.563	75.047	-197.170	18.738	
	2400	13.148	-23.251	75.605	-204.703	18.643	
	2500	13.189	-21.934	76.143	-212.291	18.561	
	2600	13.217	-20.613	76.661	-219.931	18.489	
	2700	13.230	-19.291	77.160	-227.622	18.427	
	2800	13.230	-17.968	77.641	-235.362	18.373	
	2900	13.217	-16.645	78.105	-243.150	18.327	
	3000	13.189	-15.325	78.553	-250.983	18.287	

BERYLLIUM MONOSULFIDE

BeS

		A	B	C	D	RANGE	REF.
SOL	CP	9.97	1.94	-2.30		298-1800	65

PHASE	T	CP	H	S	G	BT	
SOL	298	7.961	-55.8	8.0	-58.185	42.657	65
	300	7.996	-55.785	8.049	-58.200	42.404	
	400	9.308	-54.912	10.553	-59.133	32.313	
	500	10.020	-53.943	12.713	-60.299	26.360	
	600	10.495	-52.916	14.584	-61.666	22.465	
	700	10.859	-51.847	16.230	-63.208	19.737	
	800	11.163	-50.746	17.700	-64.906	17.734	
	900	11.432	-49.616	19.031	-66.744	16.210	
	1000	11.680	-48.460	20.248	-68.708	15.018	
	1100	11.914	-47.280	21.372	-70.790	14.067	
	1200	12.138	-46.078	22.419	-72.980	13.293	
	1300	12.356	-44.853	23.399	-75.272	12.656	
	1400	12.569	-43.607	24.322	-77.658	12.125	
	1500	12.778	-42.339	25.197	-80.135	11.677	
	1600	12.984	-41.051	26.028	-82.696	11.297	
	1700	13.188	-39.743	26.821	-85.339	10.973	
	1800	13.391	-38.414	27.581	-88.059	10.693	

* * * * * * * * * *

BERYLLIUM SULFATE DIHYCRATE

$BeSO_4 \cdot 2H_2O$

		A	B	C	D	RANGE	REF.
SOL	CP	17.3	64.5			298- 500	*

PHASE	T	CP	H	S	G	BT	
SOL	298	36.531	-435.7	39.01	-447.331	327.946	19
	300	36.650	-435.632	39.236	-447.403	325.977	
	400	43.100	-431.645	50.663	-451.910	246.945	
	500	49.550	-427.012	60.974	-457.499	200.000	

$2BeO \cdot SiO_2$ BERYLLIUM ORTHOSILICATE (PHENACITE)

		A	B	C	D	RANGE	REF.
SOL	CP	29.65	18.77	-10.84	-8.0	298-1833	5
	** MPT=	1833					3

PHASE	T	CP	H	S	G	BT	
SOL	298	22.341	-512.5	15.4	-517.092	379.089	5
	300	22.517	-512.459	15.539	-517.120	376.772	
	400	29.103	-509.839	23.031	-519.051	283.634	
	500	32.699	-506.734	29.944	-521.706	228.068	
	600	35.021	-503.340	36.125	-525.015	191.262	
	700	36.657	-499.752	41.653	-528.909	165.155	
	800	37.852	-496.023	46.630	-533.327	145.718	
	900	38.725	-492.192	51.142	-538.220	130.715	
	1000	39.336	-488.287	55.255	-543.542	118.807	
	1100	39.721	-484.333	59.024	-549.259	109.142	
	1200	39.901	-480.350	62.490	-555.337	101.154	
	1300	39.890	-476.359	65.684	-561.748	94.451	
	1400	39.695	-472.378	68.634	-568.466	88.753	
	1500	39.323	-468.426	71.361	-575.467	83.857	
	1600	38.779	-464.519	73.883	-582.731	79.608	
	1700	38.064	-460.676	76.213	-590.238	75.890	
	1800	37.181	-456.912	78.364	-597.968	72.613	
	1833	36.854	-455.690	79.037	-600.565	71.615	

BISMUTH

Bi

		A	B	C	D	RANGE	REF.
SOL	CP	2.832	7.282	0.981		298- 544	21
LIQ	CP	4.545	2.479	4.957	-0.951	544-1200	21
	CP	6.5				1200-1837	21
	** BPT= 1837/GAS (BI + BI2)						21

PHASE	T	CP	H	S	G	BT	P	
SOL	298	6.107	0	13.56	-4.043	2.964		21
	300	6.107	.011	13.598	-4.068	2.964		
	400	6.358	.631	15.379	-5.521	3.017		
	500	6.865	1.291	16.850	-7.134	3.119		
	545	7.128	1.602	17.446	-7.897	3.170		
			2.7	4.958				21
LIQ	545	7.285	4.302	22.405	-7.897	3.170		
	600	7.067	4.700	23.100	-9.160	3.337		
	700	6.826	5.394	24.170	-11.525	3.599		
	800	6.694	6.069	25.072	-13.988	3.822		
	900	6.618	6.734	25.855	-16.536	4.016	.003	
	1000	6.569	7.393	26.550	-19.157	4.187	.031	
	1100	6.531	8.048	27.174	-21.843	4.340	.242	
	1200	6.495	8.700	27.741	-24.590	4.479	1.320	
	1200	6.500	8.700	27.741	-24.590	4.479	1.320	
	1300	6.500	9.350	28.261	-27.390	4.605	5.498	
	1400	6.500	10.000	28.743	-30.240	4.721	18.551	
	1500	6.500	10.650	29.191	-33.137	4.829	52.934	
	1600	6.500	11.300	29.611	-36.078	4.929	131.879	
	1700	6.500	11.950	30.005	-39.059	5.022	293.930	
	1800	6.500	12.600	30.377	-42.078	5.110	597.228	
	1837	6.500	12.840	30.509	-43.204	5.141	760.	

Bi (g)

BISMUTH (MONATOMIC GAS)

		A	B	C	D	RANGE	REF.
GAS	CP	4.968				298-1200	21
	CP	4.952	0.013			1200-1500	21
	CP	4.886	0.057			1500-1837	21
	CP	4.776	0.117			1837-2000	21

PHASE	T	CP	H	S	G	BT	
GAS	298	4.968	50.04	44.67	36.722	-26.921	21*
	300	4.968	50.049	44.701	36.639	-26.695	
	400	4.968	50.546	46.130	32.094	-17.538	
	500	4.968	51.043	47.239	27.424	-11.988	
	600	4.968	51.540	48.144	22.653	-8.252	
	700	4.968	52.036	48.910	17.799	-5.558	
	800	4.968	52.533	49.573	12.874	-3.518	
	900	4.968	53.030	50.159	7.887	-1.916	
	1000	4.968	53.527	50.682	2.845	-.622	
	1100	4.968	54.024	51.156	-2.248	.447	
	1200	4.968	54.520	51.588	-7.385	1.345	
	1200	4.968	54.520	51.588	-7.385	1.345	
	1300	4.969	55.017	51.986	-12.564	2.112	
	1400	4.970	55.514	52.354	-17.781	2.776	
	1500	4.971	56.011	52.697	-23.034	3.356	
	1500	4.971	56.011	52.697	-23.034	3.356	
	1600	4.977	56.509	53.018	-28.320	3.869	
	1700	4.983	57.007	53.320	-33.637	4.325	
	1800	4.989	57.505	53.605	-38.983	4.734	
	1837	4.991	57.690	53.706	-40.968	4.875	
	1837	4.991	57.690	53.706	-40.968	4.875	
	1900	4.998	58.005	53.875	-44.357	5.103	
	2000	5.010	58.505	54.131	-49.758	5.438	

BISMUTH (DIATOMIC GAS) Bi_2 (g)

		A	B	C	D	RANGE	REF.
GAS	CP	8.794	0.182			298- 800	21
	CP	8.93				800-1200	21
	CP	8.94				1200-1837	21
	CP	8.94				1837-2000	21

PHASE	T	CP	H	S	G	BT	
GAS	298	8.848	52.6	65.40	33.101	-24.267	21
	300	8.849	52.616	65.455	32.980	-24.029	
	400	8.867	53.502	68.003	26.301	-14.372	
	500	8.885	54.390	69.983	19.398	-8.480	
	600	8.903	55.279	71.605	12.316	-4.487	
	700	8.921	56.170	72.979	5.085	-1.588	
	800	8.940	57.063	74.171	-2.273	.621	
	800	8.930	57.063	74.171	-2.273	.621	
	900	8.930	57.956	75.223	-9.744	2.367	
	1000	8.930	58.849	76.164	-17.314	3.785	
	1100	8.930	59.742	77.015	-24.974	4.963	
	1200	8.930	60.635	77.792	-32.715	5.959	
	1200	8.940	60.635	77.792	-32.715	5.959	
	1300	8.940	61.529	78.508	-40.530	6.815	
	1400	8.940	62.423	79.170	-48.415	7.559	
	1500	8.940	63.317	79.787	-56.363	8.213	
	1600	8.940	64.211	80.364	-64.371	8.794	
	1700	8.940	65.105	80.906	-72.434	9.313	
	1800	8.940	65.999	81.417	-80.551	9.782	
	1837	8.940	66.330	81.599	-83.567	9.943	
	1837	8.940	66.330	81.599	-83.567	9.943	
	1900	8.940	66.893	81.900	-88.717	10.206	
	2000	8.940	67.787	82.359	-96.930	10.593	

BiF(g)

BISMUTH MONOFLUORIDE (GAS)

		A	B	C	D	RANGE	REF.
GAS	CP	8.85	0.20	-0.61		298-2000	*

PHASE	T	CP	H	S	G	BT	
GAS	298	8.223	-7.023	58.32	-24.411	17.896	62
	300	8.232	-7.008	58.371	-24.519	17.865	
	400	8.549	-6.167	60.789	-30.482	16.657	
	500	8.706	-5.303	62.715	-36.661	16.026	
	600	8.801	-4.427	64.311	-43.014	15.670	
	700	8.866	-3.544	65.673	-49.515	15.461	
	800	8.915	-2.655	66.860	-56.143	15.340	
	900	8.955	-1.761	67.912	-62.882	15.272	
	1000	8.989	-.864	68.858	-69.722	15.240	
	1100	9.020	.036	69.716	-76.651	15.231	
	1200	9.048	.940	70.502	-83.663	15.239	
	1300	9.074	1.846	71.227	-90.749	15.258	
	1400	9.099	2.754	71.901	-97.906	15.286	
	1500	9.123	3.666	72.529	-105.128	15.319	
	1600	9.146	4.579	73.119	-112.411	15.357	
	1700	9.169	5.495	73.674	-119.751	15.397	
	1800	9.191	6.413	74.198	-127.145	15.440	
	1900	9.213	7.333	74.696	-134.589	15.483	
	2000	9.235	8.255	75.169	-142.083	15.528	

BISMUTH FLUORIDE

BiF_3

		A	B	C	D	RANGE	REF.
SOL	CP	24.99	-23.84		43.14	298- 922	205
LIQ	CP	44.12				922-1200	205
GAS	CP	19.715	0.116	-2.119		1200-1800	205

PHASE	T	CP	H	S	G	BT	P	
SOL	298	21.717	-216.	29.3	-224.736	164.758		62/205
	300	21.721	-215.960	29.434	-224.790	163.782		
	400	22.356	-213.763	35.749	-228.063	124.625		
	500	23.855	-211.460	40.883	-231.901	101.378		
	600	26.216	-208.963	45.428	-236.220	86.055		
	700	29.441	-206.188	49.700	-240.978	75.247	.005	
	800	33.528	-203.047	53.889	-246.158	67.256	.296	
	900	38.477	-199.453	58.115	-251.757	61.143	6.102	
	922	39.682	-198.594	59.059	-253.046	59.990	10.714	
			5.160	5.597				205
LIQ	922	44.120	-193.434	64.655	-253.046	59.990	10.714	
	1000	44.120	-189.992	68.238	-258.231	56.444	49.351	
	1100	44.120	-185.580	72.443	-265.268	52.711	230.709	
	1200	44.120	-181.168	76.282	-272.707	49.673	760.	
			30.	25.000				*
GAS	1200	19.707	-151.168	101.282	-272.707	49.673		
	1300	19.740	-149.196	102.861	-282.915	47.569		
	1400	19.769	-147.220	104.325	-293.276	45.789		
	1500	19.795	-145.242	105.690	-303.777	44.266		
	1600	19.818	-143.262	106.968	-314.411	42.952		
	1700	19.839	-141.279	108.170	-325.168	41.809		
	1800	19.858	-139.294	109.305	-336.042	40.807		

BiCl (g) BISMUTH MONOCHLORIDE (GAS)

		A	B	C	D	RANGE	REF.
GAS	CP	8.92	0.20	-0.28		298-1800	*

PHASE	T	CP	H	S	G	BT	
GAS	298	8.665	6.	60.94	-12.169	8.922	62
	300	8.669	6.016	60.994	-12.282	8.949	
	400	8.825	6.892	63.512	-18.513	10.116	
	500	8.908	7.779	65.491	-24.967	10.914	
	600	8.962	8.672	67.120	-31.600	11.512	
	700	9.003	9.571	68.505	-38.382	11.985	
	800	9.036	10.473	69.709	-45.294	12.376	
	900	9.065	11.378	70.775	-52.320	12.707	
	1000	9.092	12.286	71.731	-59.446	12.994	
	1100	9.117	13.196	72.599	-66.663	13.246	
	1200	9.141	14.109	73.394	-73.963	13.472	
	1300	9.163	15.024	74.126	-81.340	13.676	
	1400	9.186	15.942	74.806	-88.787	13.862	
	1500	9.208	16.861	75.440	-96.299	14.033	
	1600	9.229	17.783	76.035	-103.873	14.190	
	1700	9.250	18.707	76.596	-111.505	14.337	
	1800	9.271	19.633	77.125	-119.192	14.474	

* * * * * * * * * *

$BiCl_3$ BISMUTH CHLORIDE

		A	B	C	D	RANGE	REF.
SOL	CP	16.45	32.0			298- 507	*
	** MPT= 506.65						62
LIQ	CP	30.5				507- 712	*
	** BPT= 712.4, L= 17.6						62

PHASE	T	CP	H	S	G	BT	
SOL	298	25.991	-90.5	41.	-102.724	75.309	62
	300	26.050	-90.452	41.161	-102.800	74.900	
	400	29.250	-87.687	49.093	-107.324	58.647	
	500	32.450	-84.602	55.964	-112.584	49.217	
	507	32.663	-84.385	56.394	-112.957	48.732	
			5.65	11.152			62
LIQ	507	30.500	-78.735	67.546	-112.957	48.732	
	600	30.500	-75.888	72.704	-119.510	43.537	
	700	30.500	-72.838	77.405	-127.022	39.663	
	712	30.500	-72.460	77.941	-127.985	39.268	

BISMUTH BROMIDE

$BiBr_3$

		A	B	C	D	RANGE	REF.
SOL	CP	23.3	6.0			298- 493	1,*
	** MPT= 492.2						62
LIQ	CP	37.7				492- 734	1,*
	** BPT= 734, L= 18.02						62

PHASE	T	CP	H	S	G	BT	
SOL	298	25.089	-66.	43.4	-78.940	57.872	198
	300	25.100	-65.954	43.555	-79.020	57.574	
	400	25.700	-63.414	50.858	-83.757	45.769	
	492	26.253	-61.019	56.244	-88.702	39.391	
			5.19	10.544			1
LIQ	492	37.700	-55.829	66.789	-88.702	39.391	
	500	37.700	-55.534	67.382	-89.225	39.006	
	600	37.700	-51.764	74.255	-96.318	35.088	
	700	37.700	-47.994	80.067	-104.041	32.487	
	734	37.700	-46.713	81.855	-106.794	31.802	

* * * * * * * * * *

BISMUTH MONOIODIDE

BiJ

		A	B	C	D	RANGE	REF.
SOL-A	CP	6.33	12.4			298- 564	*
SOL-B	CP	13.1				564- 600	*

PHASE	T	CP	H	S	G	BT	
SOL-A	298	10.027	-13.33	29.5	-22.125	16.221	158*
	300	10.050	-13.311	29.562	-22.180	16.160	
	400	11.290	-12.244	32.623	-25.294	13.822	
	500	12.530	-11.053	35.276	-28.691	12.543	
	564	13.324	-10.226	36.832	-30.999	12.014	
			0.31	.550			158*
SOL-B	564	13.100	-9.916	37.381	-30.999	12.014	
	600	13.100	-9.445	38.192	-32.360	11.789	

BiJ_3 BISMUTH IODIDE

		A	B	C	D	RANGE	REF.
SOL	CP	9.55	26.3	0.71		298- 682	83
LIQ	CP	37.7				682- 815	83
	** DEC. / BPT= 815, L= 18.57						1/62

PHASE	T	CP	H	S	G	BT	
SOL	298	18.190	-36.	53.7	-52.011	38.130	83
	300	18.229	-35.966	53.813	-52.110	37.967	
	400	20.514	-34.032	59.363	-57.777	31.572	
	500	22.984	-31.858	64.203	-63.959	27.960	
	600	25.527	-29.432	68.618	-70.603	25.721	
	682	27.639	-27.253	72.020	-76.371	24.477	
			9.35	13.710			83
LIQ	682	37.700	-17.903	85.730	-76.371	24.477	
	700	37.700	-17.224	86.712	-77.923	24.332	
	800	37.700	-13.454	91.746	-86.851	23.730	
	815	37.700	-12.889	92.447	-88.233	23.664	

* * * * * * * * * *

BiOCl BISMUTH OXYFLUORIDE

		A	B	C	D	RANGE	REF.
SOL	CP	13.73	13.3			298- 700	*

PHASE	T	CP	H	S	G	BT	
SOL	298	17.695	-88.7	24.5	-96.005	70.383	62
	300	17.720	-88.667	24.610	-96.050	69.982	
	400	19.050	-86.829	29.889	-98.785	53.981	
	500	20.380	-84.857	34.283	-101.999	44.590	
	600	21.710	-82.753	38.116	-105.623	38.478	
	700	23.040	-80.515	41.563	-109.609	34.226	

BISMUTH OXIDE

Bi_2O_3

		A	B	C	D	RANGE	REF.
SOL-A	CP	24.74	8.0			298- 978	1
SOL-B	CP	35.0				978-1097	*
LIQ	CP	36.5				1097-1800	*
	** BPT=	2163					47

PHASE	T	CP	H	S	G	BT	
SOL-A	298	27.125	-136.4	36.2	-147.193	107.910	161*/1
	300	27.140	-136.350	36.368	-147.260	107.293	
	400	27.940	-133.596	44.285	-151.310	82.683	
	500	28.740	-130.762	50.606	-156.065	68.225	
	600	29.540	-127.848	55.916	-161.398	58.797	
	700	30.340	-124.854	60.530	-167.225	52.217	
	800	31.140	-121.780	64.634	-173.487	47.401	
	900	31.940	-118.626	68.348	-180.139	43.749	
	978	32.564	-116.110	71.028	-185.575	41.475	
			13.6	13.906			161
SOL-B	978	35.000	-102.510	84.934	-185.575	41.475	
	1000	35.000	-101.740	85.712	-187.452	40.973	
	1097	35.000	-98.345	88.953	-195.926	39.039	
			14.3	13.036			*
LIQ	1097	36.500	-84.045	101.988	-195.926	39.039	
	1100	36.500	-83.936	102.088	-196.232	38.993	
	1200	36.500	-80.286	105.264	-206.602	37.632	
	1300	36.500	-76.636	108.185	-217.277	36.532	
	1400	36.500	-72.986	110.890	-228.232	35.633	
	1500	36.500	-69.336	113.408	-239.448	34.892	
	1600	36.500	-65.686	115.764	-250.908	34.277	
	1700	36.500	-62.036	117.977	-262.596	33.764	
	1800	36.500	-58.386	120.063	-274.499	33.333	

Bi_2S_3 BISMUTH SULFIDE

		A	B	C	D	RANGE	REF.
SOL	CP	26.25	9.8			298-1050	62,*
LIQ	CP	45.				1050-1100	63
	** DEC.						63

PHASE	T	CP	H	S	G	BT	
SOL	298	29.172	-37.2	47.9	-51.481	37.742	62
	300	29.190	-37.146	48.081	-51.570	37.574	
	400	30.170	-34.178	56.612	-56.823	31.051	
	500	31.150	-31.112	63.450	-62.837	27.470	
	600	32.130	-27.948	69.216	-69.477	25.311	
	700	33.110	-24.686	74.242	-76.655	23.936	
	800	34.090	-21.326	78.727	-84.308	23.035	
	900	35.070	-17.868	82.799	-92.387	22.438	
	1000	36.050	-14.312	86.545	-100.857	22.045	
	1050	36.540	-12.497	88.316	-105.229	21.906	
			18.7	17.810			62,*
LIQ	1050	45.000	6.203	106.125	-105.229	21.906	
	1100	45.000	8.453	108.218	-110.588	21.975	

* * * * * * * * * *

$Bi_2(SO_4)_3$ BISMUTH SULFATE

		A	B	C	D	RANGE	REF.
SOL	CP	54.6	40.4			298-1000	*

PHASE	T	CP	H	S	G	BT	
SOL	298	66.645	-608.	74.7	-630.272	462.064	19/*
	300	66.720	-607.877	75.112	-630.410	459.315	
	400	70.760	-601.003	94.860	-638.947	349.151	
	500	74.800	-593.725	111.084	-649.266	283.832	
	600	78.840	-586.043	125.078	-661.090	240.834	
	700	82.880	-577.957	137.535	-674.231	210.533	
	800	86.920	-569.467	148.866	-688.559	188.131	
	900	90.960	-560.573	159.337	-703.976	170.972	
	1000	95.000	-551.275	169.129	-720.404	157.465	

BISMUTH SELENIDE Bi_2Se_3

		A	B	C	D	RANGE	REF.
SOL	CP	20.75	11.7			298- 995	79
LIQ	CP	45.				995-1100	63

PHASE	T	CP	H	S	G	BT	
SOL	298	24.238	-33.45	57.3	-50.534	37.047	115/63
	300	24.260	-33.405	57.450	-50.640	36.896	
	400	25.430	-30.921	64.589	-56.756	31.014	
	500	26.600	-28.319	70.390	-63.514	27.766	
	600	27.770	-25.601	75.343	-70.806	25.795	
	700	28.940	-22.765	79.711	-78.563	24.532	
	800	30.110	-19.813	83.652	-86.734	23.698	
	900	31.280	-16.743	87.266	-95.283	23.141	
	995	32.391	-13.719	90.460	-103.726	22.786	
			20.5	20.603			79
LIQ	995	45.000	6.781	111.063	-103.726	22.786	
	1000	45.000	7.006	111.289	-104.282	22.794	
	1100	45.000	11.506	115.577	-115.629	22.976	

Bi_2Te_3 BISMUTH TELLURIDE

		A	B	C	D	RANGE	REF.
SOL	CP	25.81	13.2			298- 870	63
LIQ	CP	40.				870-1000	63
	** DEC., BI2TE3(S) = 2BITE(G) + 0.5TE2(G)						63

PHASE	T	CP	H	S	G	BT	
SOL	298	29.746	-18.8	62.1	-37.315	27.356	115/63*
	300	29.770	-18.745	62.284	-37.430	27.272	
	400	31.090	-15.702	71.029	-44.114	24.106	
	500	32.410	-12.527	78.108	-51.581	22.549	
	600	33.730	-9.220	84.134	-59.700	21.749	
	700	35.050	-5.781	89.433	-68.384	21.353	
	800	36.370	-2.210	94.199	-77.569	21.194	
	850	37.030	-.375	96.424	-82.335	21.173	
			28.4	32.644			115
LIQ	870	40.000	28.025	129.068	-84.264	21.170	
	900	40.000	29.225	130.424	-88.156	21.410	
	1000	40.000	33.225	134.638	-101.413	22.167	

CARBON TETRAFLUORIDE (GAS) CF_4 (g)

		A	B	C	D	RANGE	REF.
GAS	CP	25.179	0.310	-23.017	4.037	298-2000	3

PHASE	T	CP	H	S	G	BT	
GAS	298	14.611	-223.04	62.454	-241.661	177.166	3
	300	14.649	-223.013	62.544	-241.776	176.158	
	400	17.225	-221.421	67.106	-248.263	135.663	
	500	19.357	-219.586	71.192	-255.182	111.555	
	600	20.840	-217.571	74.861	-262.488	95.624	
	700	21.876	-215.433	78.156	-270.142	84.353	
	800	22.619	-213.206	81.128	-278.108	75.986	
	900	23.170	-210.915	83.825	-286.358	69.547	
	1000	23.591	-208.576	86.289	-294.866	64.452	
	1100	23.921	-206.200	88.554	-303.609	60.330	
	1200	24.186	-203.794	90.647	-312.571	56.935	
	1300	24.404	-201.364	92.592	-321.734	54.096	
	1400	24.586	-198.915	94.407	-331.085	51.692	
	1500	24.741	-196.448	96.109	-340.611	49.634	
	1600	24.874	-193.967	97.710	-350.303	47.856	
	1700	24.992	-191.474	99.222	-360.150	46.307	
	1800	25.096	-188.969	100.653	-370.145	44.948	
	1900	25.189	-186.455	102.012	-380.279	43.748	
	2000	25.274	-183.932	103.307	-390.545	42.683	

GAS CP(T) = A + B*T + C/(T**2) + E/(T**3)

CCl_4 (g) CARBON TETRACHLORIDE (GAS)

		A	B	C	D	RANGE	REF.
GAS	CP	24.90	0.48	-4.74		298-2000	3

PHASE	T	CP	H	S	G	BT	
GAS	298	19.711	-22.94	74.02	-45.009	32.997	3
	300	19.777	-22.903	74.142	-45.146	32.893	
	400	22.129	-20.792	80.201	-52.872	28.892	
	500	23.244	-18.517	85.272	-61.153	26.734	
	600	23.871	-16.159	89.570	-69.901	25.465	
	700	24.269	-13.750	93.282	-79.048	24.683	
	800	24.543	-11.309	96.542	-88.542	24.192	
	900	24.747	-8.844	99.445	-98.344	23.884	
	1000	24.906	-6.361	102.061	-108.422	23.699	
	1100	25.036	-3.864	104.441	-118.749	23.596	
	1200	25.147	-1.354	106.624	-129.303	23.553	
	1300	25.244	1.165	108.641	-140.068	23.551	
	1400	25.330	3.694	110.515	-151.027	23.580	
	1500	25.409	6.231	112.265	-162.167	23.631	
	1600	25.483	8.776	113.907	-173.476	23.699	
	1700	25.552	11.327	115.454	-184.945	23.780	
	1800	25.618	13.886	116.917	-196.564	23.869	
	1900	25.681	16.451	118.304	-208.326	23.966	
	2000	25.741	19.022	119.622	-220.223	24.068	

CALCIUM

		A	B	C	D	RANGE	REF.
SOL-A	CP	5.766	-0.802	0.079	4.879	298- 720	21
SOL-B	CP	-0.090	9.866			720-1112	21
LIQ	CP	7.0				1112-1757	21
GAS	CP	4.979				1757-2800	21

PHASE	T	CP	H	S	G	BT	P	
SOL-A	298	6.049	0	9.90	-2.952	2.164		21
	300	6.052	.011	9.937	-2.970	2.164		
	400	6.275	.626	11.706	-4.056	2.216		
	500	6.616	1.270	13.141	-5.300	2.317		
	600	7.063	1.953	14.385	-6.678	2.433		
	700	7.611	2.686	15.514	-8.173	2.552		
	720	7.733	2.840	15.730	-8.486	2.576		
			0.22	.306				21
SOL-B	720	7.014	3.060	16.035	-8.486	2.576		
	800	7.803	3.652	16.815	-9.800	2.678	.001	
	900	8.789	4.482	17.791	-11.530	2.800	.020	
	1000	9.776	5.410	18.768	-13.358	2.920	.194	
	1100	10.763	6.437	19.746	-15.284	3.037	1.233	
	1112	10.881	6.567	19.864	-15.522	3.051	1.502	
			2.04	1.835				21
LIQ	1112	7.000	8.607	21.698	-15.522	3.051	1.502	
	1200	7.000	9.223	22.231	-17.455	3.179	5.293	
	1300	7.000	9.923	22.792	-19.706	3.313	17.893	
	1400	7.000	10.623	23.311	-22.012	3.437	50.539	
	1500	7.000	11.323	23.793	-24.367	3.551	123.690	
	1600	7.000	12.023	24.245	-26.769	3.657	269.525	
	1700	7.000	12.723	24.670	-29.215	3.756	533.875	
	1757	7.000	13.122	24.900	-30.628	3.810	760.	
			36.72	20.899				21
GAS	1757	4.979	49.842	45.800	-30.628	3.810		
	1800	4.979	50.056	45.920	-32.600	3.959		
	1900	4.979	50.554	46.189	-37.206	4.280		
	2000	4.979	51.052	46.445	-41.838	4.572		
	2100	4.979	51.550	46.688	-46.494	4.839		
	2200	4.979	52.048	46.919	-51.175	5.084		
	2300	4.979	52.546	47.141	-55.878	5.310		
	2400	4.979	53.043	47.352	-60.603	5.519		
	2500	4.979	53.541	47.556	-65.348	5.713		
	2600	4.979	54.039	47.751	-70.113	5.894		
	2700	4.979	54.537	47.939	-74.898	6.063		
	2800	4.979	55.035	48.120	-79.701	6.222		

CaH (g)   CALCIUM MONOHYDRIDE (GAS)

		A	B	C	D	RANGE	REF.
GAS	CP	7.68	0.74	-0.70		298-2000	58

PHASE	T	CP	H	S	G	BT	
GAS	298	7.113	54.7	48.19	40.332	-29.568	19
	300	7.124	54.713	48.234	40.243	-29.321	
	400	7.538	55.449	50.347	35.310	-19.295	
	500	7.770	56.215	52.056	30.187	-13.196	
	600	7.930	57.000	53.488	24.908	-9.074	
	700	8.055	57.800	54.720	19.496	-6.088	
	800	8.163	58.611	55.803	13.969	-3.817	
	900	8.260	59.432	56.770	8.339	-2.025	
	1000	8.350	60.263	57.645	2.618	-.572	
	1100	8.436	61.102	58.445	-3.187	.633	
	1200	8.519	61.950	59.182	-9.069	1.652	
	1300	8.601	62.806	59.867	-15.022	2.526	
	1400	8.680	63.670	60.508	-21.041	3.285	
	1500	8.759	64.542	61.109	-27.122	3.952	
	1600	8.837	65.421	61.677	-33.262	4.544	
	1700	8.914	66.309	62.215	-39.456	5.073	
	1800	8.990	67.204	62.727	-45.704	5.550	
	1900	9.067	68.107	63.215	-52.001	5.982	
	2000	9.142	69.018	63.682	-58.346	6.377	

* * * * * * * * * *

CaH_2 CALCIUM HYDRIDE

		A	B	C	D	RANGE	REF.
SOL	CP	7.1	5.85			298-1053	*
	** TPT= 1053, MPT= 1089						1

PHASE	T	CP	H	S	G	BT	
SOL	298	8.844	-44.5	10.	-47.481	34.810	1/19
	300	8.855	-44.484	10.055	-47.500	34.608	
	400	9.440	-43.569	12.682	-48.642	26.580	
	500	10.025	-42.596	14.852	-50.021	21.867	
	600	10.610	-41.564	16.731	-51.603	18.799	
	700	11.195	-40.474	18.411	-53.361	16.662	
	800	11.780	-39.325	19.944	-55.280	15.104	
	900	12.365	-38.118	21.365	-57.346	13.927	
	1000	12.950	-36.852	22.698	-59.550	13.016	
	1053	13.260	-36.157	23.375	-60.771	12.615	

CALCIUM CHLORIDE

		A	B	C	D	RANGE	REF.
SOL	CP	17.18	3.04	-0.65		298-1045	58*
LIQ	CP	24.70				1045-2273	58
	** BPT≤ 2273						1

PHASE	T	CP	H	S	G	BT	
SOL	298	17.355	-190.2	25.	-197.654	144.904	19
	300	17.370	-190.168	25.107	-197.700	144.044	
	400	17.990	-188.398	30.196	-200.476	109.550	
	500	18.440	-186.575	34.260	-203.705	89.052	
	600	18.823	-184.712	37.657	-207.306	75.521	
	700	19.175	-182.812	40.585	-211.221	65.955	
	800	19.510	-180.877	43.168	-215.411	58.856	
	900	19.836	-178.910	45.485	-219.846	53.393	
	1000	20.155	-176.910	47.591	-224.501	49.071	
	1045	20.297	-176.000	48.481	-226.663	47.410	
			6.8	6.507			58
LIQ	1045	24.700	-169.200	54.988	-226.663	47.410	
	1100	24.700	-167.842	56.255	-229.723	45.648	
	1200	24.700	-165.372	58.405	-235.457	42.888	
	1300	24.700	-162.902	60.382	-241.398	40.588	
	1400	24.700	-160.432	62.212	-247.529	38.646	
	1500	24.700	-157.962	63.916	-253.836	36.989	
	1600	24.700	-155.492	65.510	-260.308	35.561	
	1700	24.700	-153.022	67.008	-266.935	34.321	
	1800	24.700	-150.552	68.420	-273.707	33.237	
	1900	24.700	-148.082	69.755	-280.616	32.283	
	2000	24.700	-145.612	71.022	-287.656	31.438	
	2100	24.700	-143.142	72.227	-294.819	30.686	
	2200	24.700	-140.672	73.376	-302.099	30.015	
	2273	24.700	-138.869	74.182	-307.485	29.569	

CaJ_2 CALCIUM IODIDE

		A	B	C	D	RANGE	REF.
SOL	CP	17.034	4.720			298-1052	3A
LIQ	CP	24.7				1052-2028	3A
GAS	CP	14.892	0.006	-0.359		2028-2500	3A

PHASE	T	CP	H	S	G	BT	P	
SOL	298	18.441	-128.3	34.72	-138.652	101.648		3A
	300	18.450	-128.266	34.834	-138.716	101.068		
	400	18.922	-126.397	40.206	-142.480	77.858		
	500	19.394	-124.481	44.480	-146.721	64.140		
	600	19.866	-122.518	48.057	-151.353	55.138		
	700	20.338	-120.508	51.155	-156.317	48.811		
	800	20.810	-118.451	53.902	-161.572	44.145		
	900	21.282	-116.346	56.380	-167.088	40.580		
	1000	21.754	-114.194	58.647	-172.841	37.779	.002	
	1052	21.999	-113.057	59.756	-175.920	36.552	.010	
			10.	9.506				3A
LIQ	1052	24.700	-103.057	69.261	-175.920	36.552	.010	
	1100	24.700	-101.871	70.363	-179.271	35.623	.031	
	1200	24.700	-99.401	72.512	-186.416	33.956	.217	
	1300	24.700	-96.931	74.490	-193.768	32.580	1.110	
	1400	24.700	-94.461	76.320	-201.309	31.430	4.369	
	1500	24.700	-91.991	78.024	-209.027	30.459	13.991	
	1600	24.700	-89.521	79.618	-216.910	29.633	37.947	
	1700	24.700	-87.051	81.116	-224.948	28.923	89.873	
	1800	24.700	-84.581	82.527	-233.131	28.310	190.313	
	1900	24.700	-82.111	83.863	-241.451	27.777	367.058	
	2000	24.700	-79.641	85.130	-249.901	27.312	654.382	
	2028	24.700	-78.950	85.473	-252.289	27.192	760.	
			42.88	21.144				3A
GAS	2028	14.895	-36.070	106.617	-252.289	27.192		
	2100	14.896	-34.997	107.137	-259.985	27.061		
	2200	14.898	-33.507	107.830	-270.733	26.898		
	2300	14.899	-32.018	108.492	-281.550	26.757		
	2400	14.900	-30.528	109.126	-292.431	26.633		
	2500	14.901	-29.038	109.735	-303.374	26.525		

CALCIUM OXYDICHLORIDE

$CaOCl_2$

		A	B	C	D	RANGE	REF.
SOL	CP	20.0	13.3			298- 500	*

PHASE	T	CP	H	S	G	BT	
SOL	298	23.965	-178.4	27.	-186.450	136.690	19/*
	300	23.990	-178.356	27.148	-186.500	135.884	
	400	25.320	-175.890	34.232	-189.583	103.597	
	500	26.650	-173.292	40.025	-193.304	84.505	

* * * * * * * * * *

CALCIUM PEROXIDE

CaO_2

		A	B	C	D	RANGE	REF.
SOL	CP	19.8				298- 427	*
	** DPT= 426.7						

PHASE	T	CP	H	S	G	BT	
SOL	298	19.800	-157.5	20.	-163.463	119.838	19*
	300	19.800	-157.463	20.122	-163.500	119.126	
	400	19.800	-155.483	25.819	-165.811	90.607	
	427	19.800	-154.955	27.098	-166.517	85.299	

CaS

CALCIUM SULFIDE

		A	B	C	D	RANGE	REF.
SOL	CP	10.8	1.85			298-2000	*

PHASE	T	CP	H	S	G	BT	
SOL	298	11.352	-113.8	13.5	-117.825	86.380	63
	300	11.355	-113.779	13.570	-117.850	85.865	
	400	11.540	-112.634	16.862	-119.379	65.234	
	500	11.725	-111.471	19.457	-121.200	52.983	
	600	11.910	-110.289	21.611	-123.256	44.902	
	700	12.095	-109.089	23.461	-125.512	39.192	
	800	12.280	-107.870	25.088	-127.941	34.957	
	900	12.465	-106.633	26.545	-130.524	31.700	
	1000	12.650	-105.377	27.868	-133.245	29.125	
	1100	12.835	-104.103	29.082	-136.094	27.043	
	1200	13.020	-102.810	30.207	-139.059	25.329	
	1300	13.205	-101.499	31.257	-142.133	23.898	
	1400	13.390	-100.169	32.242	-145.308	22.687	
	1500	13.575	-98.821	33.172	-148.579	21.651	
	1600	13.760	-97.454	34.054	-151.941	20.757	
	1700	13.945	-96.069	34.894	-155.389	19.979	
	1800	14.130	-94.665	35.696	-158.918	19.298	
	1900	14.315	-93.243	36.465	-162.527	18.697	
	2000	14.500	-91.802	37.204	-166.210	18.165	

CALCIUM SULFITE

$CaSO_3$

		A	B	C	D	RANGE	REF.
SOL	CP	18.4	11.6			298-1000	*

PHASE	T	CP	H	S	G	ΦT	
SOL	298	21.859	-280.	24.2	-287.215	210.563	*
	300	21.880	-279.960	24.335	-287.260	209.297	
	400	23.040	-277.714	30.789	-290.029	158.486	
	500	24.200	-275.352	36.054	-293.379	128.253	
	600	25.360	-272.874	40.569	-297.215	108.275	
	700	26.520	-270.280	44.566	-301.475	94.138	
	800	27.680	-267.570	48.183	-306.116	83.638	
	900	28.840	-264.744	51.510	-311.102	75.556	
	1000	30.000	-261.802	54.608	-316.410	69.161	

* * * * * * * * * *

CALCIUM SULFATE

$CaSO_4$

		A	B	C	D	RANGE	REF.
SOL	CP	16.78	23.60			298-1200	1
	** TPT= 1466, MPT= 1673, L= 6.7						1

PHASE	T	CP	H	S	G	ΦT	
SOL	298	23.816	-342.76	25.15	-350.258	256.781	19/*
	300	23.860	-342.716	25.297	-350.305	255.231	
	400	26.220	-340.212	32.485	-353.206	193.009	
	500	28.580	-337.472	38.589	-356.766	155.963	
	600	30.940	-334.496	44.008	-360.901	131.476	
	700	33.300	-331.284	48.955	-365.552	114.146	
	800	35.660	-327.836	53.556	-370.681	101.279	
	900	38.020	-324.152	57.892	-376.255	91.379	
	1000	40.380	-320.232	62.020	-382.252	83.552	
	1100	42.740	-316.076	65.979	-388.653	77.229	
	1200	45.100	-311.684	69.799	-395.443	72.030	

$CaSO_4 \cdot 1/2H_2O$ CALCIUM SULFATE 1/2-HYDRATE (ALPHA)

		A	B	C	D	RANGE	REF.
SOL-A	CP	16.57	39.0			298- 800	58,*

PHASE	T	CP	H	S	G	BT	
SOL-A	298	28.198	-376.85	31.2	-386.152	283.095	19
	300	28.270	-376.798	31.375	-386.210	281.392	
	400	32.170	-373.776	40.042	-389.792	213.001	
	500	36.070	-370.364	47.639	-394.183	172.321	
	600	39.970	-366.562	54.560	-399.298	145.464	
	700	43.870	-362.370	61.014	-405.080	126.489	
	800	47.770	-357.788	67.127	-411.489	112.429	

* * * * * * * * * *

$CaSo_4 \cdot 2H_2O$ CALCIUM SULFATE 2-HYDRATE (GYPSUM)

		A	B	C	D	RANGE	REF.
SOL	CP	21.84	76.0			298- 800	58,*

PHASE	T	CP	H	S	G	BT	
SOL	298	44.499	-483.42	46.4	-497.254	364.546	19/58
	300	44.640	-483.338	46.676	-497.340	362.361	
	400	52.240	-478.494	60.559	-502.717	274.709	
	500	59.840	-472.890	73.032	-509.406	222.691	
	600	67.440	-466.526	84.614	-517.294	188.450	
	700	75.040	-459.402	95.581	-526.308	164.343	
	800	82.640	-451.518	106.097	-536.395	146.556	

CALCIUM SELENIDE

		A	B	C	D	RANGE	REF.
SOL	CP	10.9	2.0			298-2000	*

PHASE	T	CP	H	S	G	BT	
SOL	298	11.496	-88.	16.5	-92.919	68.121	63
	300	11.500	-87.979	16.571	-92.950	67.723	
	400	11.700	-86.819	19.907	-94.781	51.793	
	500	11.900	-85.639	22.539	-96.908	42.364	
	600	12.100	-84.439	24.726	-99.275	36.166	
	700	12.300	-83.219	26.607	-101.843	31.801	
	800	12.500	-81.979	28.262	-104.588	28.576	
	900	12.700	-80.719	29.746	-107.490	26.106	
	1000	12.900	-79.439	31.094	-110.533	24.160	
	1100	13.100	-78.139	32.333	-113.705	22.594	
	1200	13.300	-76.819	33.482	-116.997	21.311	
	1300	13.500	-75.479	34.554	-120.399	20.244	
	1400	13.700	-74.119	35.562	-123.905	19.345	
	1500	13.900	-72.739	36.514	-127.510	18.581	
	1600	14.100	-71.339	37.417	-131.207	17.924	
	1700	14.300	-69.919	38.278	-134.992	17.357	
	1800	14.500	-68.479	39.101	-138.861	16.862	
	1900	14.700	-67.019	39.891	-142.811	16.429	
	2000	14.900	-65.539	40.650	-146.838	16.048	

CaTe

CALCIUM TELLURIDE

		A	B	C	D	RANGE	REF.
SOL	CP	9.20	2.5			298-1500	*

PHASE	T	CP	H	S	G	BT	
SOL	298	9.945	-65.	19.3	-70.754	51.871	*
	300	9.950	-64.982	19.362	-70.790	51.577	
	400	10.200	-63.974	22.258	-72.877	39.824	
	500	10.450	-62.942	24.561	-75.222	32.884	
	600	10.700	-61.884	26.488	-77.777	28.334	
	700	10.950	-60.802	28.157	-80.511	25.140	
	800	11.200	-59.694	29.635	-83.402	22.787	
	900	11.450	-58.562	30.969	-86.433	20.992	
	1000	11.700	-57.404	32.188	-89.592	19.583	
	1100	11.950	-56.222	33.315	-92.868	18.454	
	1200	12.200	-55.014	34.365	-96.253	17.532	
	1300	12.450	-53.782	35.352	-99.739	16.770	
	1400	12.700	-52.524	36.284	-103.321	16.131	
	1500	12.950	-51.242	37.168	-106.994	15.591	

TRICALCIUM DINITRIDE Ca_3N_2

		A	B	C	D	RANGE	REF.
SOL	CP	33.17	3.70	-6.29		298-1468	*
	** MPT= 1468						1

PHASE	T	CP	H	S	G	BT	
SOL	298	27.197	-105.	25.8	-112.692	82.617	1/*
	300	27.291	-104.950	25.969	-112.740	82.142	
	400	30.719	-102.027	34.352	-115.768	63.261	
	500	32.504	-98.858	41.416	-119.566	52.269	
	600	33.643	-95.547	47.449	-124.017	45.179	
	700	34.476	-92.140	52.701	-129.030	40.290	
	800	35.147	-88.658	57.350	-134.537	36.759	
	900	35.723	-85.113	61.523	-140.484	34.119	
	1000	36.241	-81.515	65.314	-146.829	32.094	
	1100	36.720	-77.866	68.791	-153.537	30.509	
	1200	37.173	-74.172	72.006	-160.579	29.249	
	1300	37.608	-70.432	74.999	-167.931	28.235	
	1400	38.029	-66.650	77.801	-175.572	27.412	
	1468	38.310	-64.055	79.611	-180.924	26.939	

* * * * * * * * * *

TRICALCIUM DIPHOSPHIDE Ca_3P_2

		A	B	C	D	RANGE	REF.
SOL	CP	25.8	6.7			298-1100	*

PHASE	T	CP	H	S	G	BT	
SOL	298	27.798	-121.	29.6	-129.825	95.177	19/*
	300	27.810	-120.949	29.772	-129.880	94.630	
	400	28.480	-118.134	37.864	-133.280	72.830	
	500	29.150	-115.253	44.291	-137.398	60.065	
	600	29.820	-112.304	49.665	-142.103	51.768	
	700	30.490	-109.289	54.312	-147.307	45.998	
	800	31.160	-106.206	58.427	-152.948	41.789	
	900	31.830	-103.057	62.136	-158.979	38.611	
	1000	32.500	-99.840	65.524	-165.365	36.145	
	1100	33.170	-96.557	68.653	-172.075	34.193	

$Ca_3(PO_4)_2$ CALCIUM ORTHOPHOSPHATE

		A	B	C	D	RANGE	REF.
SOL-A	CP	48.24	39.68	-5.00		298-1373	58
SOL-B	CP	79.00				1373-2003	58
	** MPT=	2003					19

PHASE	T	CP	H	S	G	BT	
SOL-A	298	54.446	-984.	56.4	-1000.816	733.716	19
	300	54.588	-983.899	56.737	-1000.920	729.268	
	400	60.987	-978.103	73.368	-1007.450	550.519	
	500	66.080	-971.743	87.538	-1015.512	443.940	
	600	70.659	-964.904	99.995	-1024.901	373.370	
	700	74.996	-957.620	111.215	-1035.470	323.332	
	800	79.203	-949.909	121.505	-1047.113	286.096	
	900	83.335	-941.781	131.073	-1059.747	257.376	
	1000	87.420	-933.243	140.065	-1073.309	234.603	
	1100	91.475	-924.298	148.587	-1087.745	216.144	
	1200	95.509	-914.949	156.720	-1103.013	200.913	
	1300	99.528	-905.197	164.523	-1119.078	188.159	
	1373	102.455	-897.825	170.040	-1131.290	180.099	
			3.7	2.695			58
SOL-B	1373	79.000	-894.125	172.735	-1131.290	180.099	
	1400	79.000	-891.992	174.274	-1135.975	177.358	
	1500	79.000	-884.092	179.724	-1153.678	168.113	
	1600	79.000	-876.192	184.823	-1171.908	160.097	
	1700	79.000	-868.292	189.612	-1190.632	153.087	
	1800	79.000	-860.392	194.127	-1209.821	146.912	
	1900	79.000	-852.492	198.399	-1229.449	141.438	
	2000	79.000	-844.592	202.451	-1249.494	136.557	
	2003	79.000	-844.355	202.569	-1250.101	136.418	

CALCIUM PYROPHOSPHATE

$Ca_2P_2O_7$

		A	B	C	D	RANGE	REF.
SOL-A	CP	53.03	14.76	-11.16		298-1413	58
SOL-B	CP	76.15				1413-1626	58
LIQ	CP	96.80				1626-1700	58

PHASE	T	CP	H	S	G	BT	
SOL-A	298	44.876	-797.5	45.23	-810.985	594.548	19,*
	300	45.058	-797.417	45.508	-811.069	590.943	
	400	51.959	-792.527	59.527	-816.338	446.086	
	500	55.946	-787.118	71.581	-822.909	359.741	
	600	58.786	-781.375	82.044	-830.601	302.587	
	700	61.084	-775.379	91.283	-839.277	262.069	
	800	63.094	-769.168	99.573	-848.827	231.920	
	900	64.936	-762.765	107.112	-859.166	208.662	
	1000	66.674	-756.184	114.045	-870.229	190.214	
	1100	68.344	-749.433	120.478	-881.959	175.253	
	1200	69.967	-742.517	126.495	-894.311	162.896	
	1300	71.558	-735.440	132.158	-907.246	152.542	
	1400	73.125	-728.206	137.519	-920.732	143.752	
	1413	73.327	-727.254	138.195	-922.524	142.707	
			1.6	1.132			58
SOL-B	1413	76.150	-725.654	139.328	-922.524	142.707	
	1500	76.150	-719.029	143.878	-934.846	136.225	
	1600	76.150	-711.414	148.792	-949.482	129.711	
	1626	76.150	-709.434	150.020	-953.366	128.159	
			24.1	14.822			58
LIQ	1626	96.800	-685.334	164.842	-953.366	128.159	
	1700	96.800	-678.171	169.150	-965.725	124.169	

$CaHPO_4$ CALCIUM MONOHYDROGEN PHCSPHATE

		A	B	C	D	RANGE	REF.
SOL	CP	33.08	13.17	-9.65		298-1000	*

PHASE	T	CP	H	S	G	BT	
SOL	298	26.151	-433.65	26.62	-441.587	323.735	19
	300	26.309	-433.601	26.782	-441.636	321.775	
	400	32.317	-430.637	35.270	-444.745	243.030	
	500	35.805	-427.219	42.883	-448.660	196.136	
	600	38.301	-423.508	49.642	-453.293	165.134	
	700	40.330	-419.574	55.702	-458.565	143.190	
	800	42.108	-415.450	61.206	-464.415	126.889	
	900	43.742	-411.157	66.261	-470.792	114.339	
	1000	45.285	-406.705	70.950	-477.655	104.405	

CALCIUM ANTIMONIDE Ca_3Sb_2

		A	B	C	D	RANGE	REF.
SOL	CP	25.1	7.05			298-1620	*
	** MPT= 1620						1

PHASE	T	CP	H	S	G	BT	
SOL	298	27.202	-174.	37.6	-185.210	135.781	1/*
	300	27.215	-173.950	37.768	-185.280	134.995	
	400	27.920	-171.193	45.694	-189.471	103.536	
	500	28.625	-168.366	52.000	-194.366	84.969	
	600	29.330	-165.468	57.281	-199.837	72.800	
	700	30.035	-162.500	61.855	-205.799	64.262	
	800	30.740	-159.461	65.912	-212.191	57.976	
	900	31.445	-156.352	69.573	-218.968	53.180	
	1000	32.150	-153.172	72.923	-226.095	49.420	
	1100	32.855	-149.922	76.020	-233.544	46.407	
	1200	33.560	-146.601	78.909	-241.292	43.951	
	1300	34.265	-143.210	81.623	-249.320	41.920	
	1400	34.970	-139.748	84.188	-257.612	40.220	
	1500	35.675	-136.216	86.625	-266.153	38.784	
	1600	36.380	-132.613	88.950	-274.933	37.559	
	1620	36.521	-131.884	89.403	-276.717	37.336	

* * * * * * * * * *

CALCIUM BISMUTHIDE Ca_3Bi_2

		A	B	C	D	RANGE	REF.
SOL	CP	29.7	6.0			298-1203	*
	** MPT= 1203						1

PHASE	T	CP	H	S	G	BT	
SOL	298	31.489	-126.	42.5	-138.671	101.663	1/*
	300	31.500	-125.942	42.695	-138.750	101.093	
	400	32.100	-122.762	51.839	-143.497	78.414	
	500	32.700	-119.522	59.066	-149.055	65.161	
	600	33.300	-116.222	65.081	-155.271	56.565	
	700	33.900	-112.862	70.260	-162.043	50.599	
	800	34.500	-109.442	74.825	-169.302	46.257	
	900	35.100	-105.962	78.924	-176.993	42.986	
	1000	35.700	-102.422	82.653	-185.075	40.453	
	1100	36.300	-98.822	86.084	-193.514	38.453	
	1200	36.900	-95.162	89.268	-202.283	36.846	
	1203	36.918	-95.051	89.360	-202.551	36.803	

Ca_2Si 2-CALCIUM SILICON

		A	B	C	D	RANGE	REF.
SOL	CP	16.1	4.3			298-1200	*

PHASE	T	CP	H	S	G	ΘT	
SOL	298	17.382	-50.	19.4	-55.784	40.896	1/*
	300	17.390	-49.968	19.508	-55.820	40.670	
	400	17.820	-48.207	24.569	-58.035	31.713	
	500	18.250	-46.404	28.592	-60.700	26.535	
	600	18.680	-44.557	31.957	-63.732	23.217	
	700	19.110	-42.668	34.869	-67.076	20.945	
	800	19.540	-40.735	37.449	-70.694	19.315	
	900	19.970	-38.760	39.775	-74.558	18.107	
	1000	20.400	-36.741	41.902	-78.643	17.190	
	1100	20.830	-34.680	43.866	-82.932	16.479	
	1200	21.260	-32.575	45.697	-87.412	15.922	

* * * * * * * * * *

CaSi CALCIUM SILICON

		A	B	C	D	RANGE	REF.
SOL	CP	10.2	3.6			298-1513	*
	** MPT=	1513					1

PHASE	T	CP	H	S	G	ΘT	
SOL	298	11.273	-36.	10.8	-39.220	28.753	1/*
	300	11.280	-35.979	10.870	-39.240	28.590	
	400	11.640	-34.833	14.164	-40.499	22.130	
	500	12.000	-33.651	16.800	-42.051	18.383	
	600	12.360	-32.433	19.020	-43.845	15.973	
	700	12.720	-31.179	20.952	-45.846	14.316	
	800	13.080	-29.889	22.674	-48.029	13.123	
	900	13.440	-28.563	24.236	-50.375	12.234	
	1000	13.800	-27.201	25.670	-52.871	11.557	
	1100	14.160	-25.803	27.002	-55.506	11.029	
	1200	14.520	-24.369	28.250	-58.269	10.614	
	1300	14.880	-22.899	29.426	-61.153	10.282	
	1400	15.240	-21.393	30.542	-64.152	10.016	
	1500	15.600	-19.851	31.606	-67.260	9.801	
	1513	15.647	-19.648	31.741	-67.672	9.776	

CALCIUM 2-SILICON

$CaSi_2$

		A	B	C	D	RANGE	REF.
SOL	CP	14.5	6.5			298-1300	*

PHASE	T	CP	H	S	G	BT	
SOL	298	16.438	-36.	12.1	-39.608	29.037	1/*
	300	16.450	-35.970	12.202	-39.630	28.874	
	400	17.100	-34.292	17.023	-41.101	22.460	
	500	17.750	-32.550	20.909	-43.004	18.800	
	600	18.400	-30.742	24.202	-45.263	16.489	
	700	19.050	-28.870	27.088	-47.831	14.935	
	800	19.700	-26.932	29.674	-50.671	13.845	
	900	20.350	-24.930	32.032	-53.758	13.056	
	1000	21.000	-22.862	34.209	-57.071	12.475	
	1100	21.650	-20.730	36.241	-60.595	12.041	
	1200	22.300	-18.532	38.153	-64.316	11.715	
	1300	22.950	-16.270	39.964	-68.222	11.471	

$CaO{\cdot}SiO_2$ CALCIUM METASILICATE (WOLLASTONITE)

		A	B	C	D	RANGE	REF.
SOL-A	CP	26.64	3.60	-6.52		298-1463	5
SOL-B	CP	32.0				1463-1817	5
LIQ	CP	36.0				1817-2200	*

PHASE	T	CP	H	S	G	ΦT	
SOL-A	298	20.379	-390.6	19.6	-396.444	290.640	5
	300	20.476	-390.562	19.726	-396.480	288.874	
	400	24.005	-388.316	26.165	-398.782	217.914	
	500	25.832	-385.816	31.737	-401.684	175.599	
	600	26.989	-383.171	36.555	-405.104	147.579	
	700	27.829	-380.428	40.781	-408.975	127.705	
	800	28.501	-377.611	44.543	-413.245	112.908	
	900	29.075	-374.731	47.934	-417.871	101.487	
	1000	29.588	-371.798	51.024	-422.822	92.420	
	1100	30.061	-368.815	53.866	-428.068	85.061	
	1200	30.507	-365.786	56.501	-433.588	78.978	
	1300	30.934	-362.714	58.960	-439.362	73.873	
	1400	31.347	-359.600	61.268	-445.375	69.536	
	1463	31.602	-357.617	62.653	-449.279	67.124	
			1.3	.889			5
SOL-B	1463	32.000	-356.317	63.542	-449.279	67.124	
	1500	32.000	-355.133	64.341	-451.645	65.813	
	1600	32.000	-351.933	66.406	-458.183	62.593	
	1700	32.000	-348.733	68.346	-464.922	59.778	
	1800	32.000	-345.533	70.175	-471.849	57.298	
	1817	32.000	-344.989	70.476	-473.044	56.906	
			13.4	7.375			1
LIQ	1817	36.000	-331.589	77.851	-473.044	56.906	
	1900	36.000	-328.601	79.459	-479.573	55.171	
	2000	36.000	-325.001	81.306	-487.312	53.291	
	2100	36.000	-321.401	83.062	-495.831	51.609	
	2200	36.000	-317.801	84.737	-504.222	50.097	

CALCIUM METASILICATE (PSEDCWOLLAST.) $CaO \cdot SiO_2$

		A	B	C	D	RANGE	REF.
SOL	CP	25.85	3.94	-5.65		298-1817	5
LIQ	CP	35.0				1817-2200	5

PHASE	T	CP	H	S	G	BT	
SOL	298	20.669	-388.9	20.9	-395.131	289.678	5
	300	20.754	-388.862	21.028	-395.170	287.920	
	400	23.895	-386.610	27.485	-397.604	217.270	
	500	25.560	-384.130	33.012	-400.636	175.141	
	600	26.645	-381.516	37.774	-404.181	147.243	
	700	27.455	-378.810	41.944	-408.171	127.454	
	800	28.119	-376.030	45.655	-412.554	112.720	
	900	28.698	-373.189	49.001	-417.290	101.345	
	1000	29.225	-370.292	52.052	-422.345	92.316	
	1100	29.717	-367.345	54.861	-427.692	84.986	
	1200	30.186	-364.350	57.467	-433.310	78.927	
	1300	30.638	-361.308	59.901	-439.180	73.843	
	1400	31.078	-358.223	62.188	-445.286	69.522	
	1500	31.509	-355.093	64.347	-451.613	65.809	
	1600	31.933	-351.921	66.394	-458.151	62.589	
	1700	32.352	-348.707	68.342	-464.889	59.774	
	1800	32.768	-345.451	70.203	-471.817	57.294	
	1817	32.838	-344.893	70.512	-473.013	56.902	
			19.8	10.897			5
LIQ	1817	35.000	-325.093	81.409	-473.013	56.902	
	1900	35.000	-322.188	82.972	-479.835	55.201	
	2000	35.000	-318.688	84.767	-488.223	53.358	
	2100	35.000	-315.188	86.475	-496.786	51.708	
	2200	35.000	-311.688	88.103	-505.515	50.225	

$3CaO \cdot 2SiO_2$ TRICALCIUM DISILICATE (RANKINITE)

		A	B	C	D	RANGE	REF.
SOL	CP	61.51	13.34	-12.82		298-1700	5

PHASE	T	CP	H	S	G	BT	
SOL	298	51.066	-941.7	50.4	-956.727	701.394	5
	300	51.268	-941.605	50.717	-956.820	697.137	
	400	58.833	-936.056	66.630	-962.708	526.070	
	500	63.052	-929.945	80.247	-970.069	424.074	
	600	65.953	-923.488	92.012	-978.696	356.538	
	700	68.232	-916.775	102.356	-988.424	308.641	
	800	70.179	-909.853	111.597	-999.130	272.986	
	900	71.933	-902.746	119.965	-1010.715	245.468	
	1000	73.568	-895.470	127.630	-1023.100	223.628	
	1100	75.124	-888.035	134.715	-1036.221	205.906	
	1200	76.628	-880.447	141.316	-1050.027	191.262	
	1300	78.093	-872.711	147.508	-1064.471	178.978	
	1400	79.532	-864.829	153.348	-1079.516	168.543	
	1500	80.950	-856.805	158.884	-1095.130	159.582	
	1600	82.353	-848.640	164.153	-1111.284	151.815	
	1700	83.744	-840.335	169.187	-1127.953	145.028	

* * * * * * * * * *

$2CaO \cdot SiO_2$ (γ) CALCIUM ORTHOSILICATE (GAMMA)

		A	B	C	D	RANGE	REF.
SOL-C	CP	31.86	12.32	-4.64		298-1100	5

PHASE	T	CP	H	S	G	BT	
SOL-C	298	30.313	-553.6	28.8	-562.187	412.149	5
	300	30.400	-553.544	28.988	-562.240	409.647	
	400	33.888	-550.313	38.258	-565.616	309.080	
	500	36.164	-546.805	46.077	-569.843	249.112	
	600	37.963	-543.096	52.834	-574.796	209.398	
	700	39.537	-539.220	58.806	-580.384	181.228	
	800	40.991	-535.193	64.182	-586.538	160.256	
	900	42.375	-531.024	69.090	-593.205	144.069	
	1000	43.716	-526.719	73.625	-600.343	131.223	
	1100	45.029	-522.281	77.853	-607.920	120.799	

CALCIUM ORTHOSILICATE (BETA) $2CaO \cdot SiO_2$ (β)

		A	B	C	D	RANGE	REF.
SOL-B	CP	34.87	9.74	-6.26		298- 970	5
SOL-C	CP	32.16	11.02			970-1710	5
SOL-D	CP	49.0				1710-2403	5
	** MPT=	2403					5

PHASE	T	CP	H	S	G	BT	
SOL-B	298	30.732	-551.1	30.5	-560.194	410.688	5
	300	30.836	-551.043	30.690	-560.250	408.197	
	400	34.853	-547.737	40.174	-563.807	308.091	
	500	37.236	-544.125	48.225	-568.237	248.410	
	600	38.975	-540.310	55.174	-573.415	208.894	
	700	40.410	-536.339	61.293	-579.244	180.873	
	800	41.684	-532.234	66.773	-585.652	160.014	
	900	42.863	-528.006	71.752	-592.582	143.918	
	970	43.652	-524.978	74.992	-597.719	134.690	
			0.44	.454			5
SOL-C	970	42.849	-524.538	75.445	-597.719	134.690	
	1000	43.180	-523.247	76.755	-600.002	131.148	
	1100	44.282	-518.874	80.922	-607.889	120.793	
	1200	45.384	-514.391	84.823	-616.178	112.236	
	1300	46.486	-509.797	88.499	-624.846	105.060	
	1400	47.588	-505.094	91.984	-633.871	98.965	
	1500	48.690	-500.280	95.305	-643.237	93.732	
	1600	49.792	-495.356	98.483	-652.928	89.198	
	1700	50.894	-490.321	101.534	-662.930	85.237	
	1710	51.004	-489.812	101.833	-663.946	84.868	
			3.39	1.982			5
SOL-D	1710	49.000	-486.422	103.816	-663.946	84.868	
	1800	49.000	-482.012	106.329	-673.404	81.773	
	1900	49.000	-477.112	108.978	-684.170	78.708	
	2000	49.000	-472.212	111.492	-695.195	75.978	
	2100	49.000	-467.312	113.882	-706.465	73.533	
	2200	49.000	-462.412	116.162	-717.968	71.333	
	2300	49.000	-457.512	118.340	-729.694	69.346	
	2400	49.000	-452.612	120.425	-741.633	67.544	
	2403	49.000	-452.465	120.487	-741.994	67.493	

$3CaO \cdot SiO_2$ TRICALCIUM SILICATE

		A	B	C	D	RANGE	REF.
SOL	CP	49.85	8.62	-10.15		298-1800	5
	** MPT=	2343					5

PHASE	T	CP	H	S	G	BT	
SOL	298	41.002	-699.5	40.3	-711.515	521.625	5
	300	41.158	-699.424	40.554	-711.590	518.463	
	400	46.954	-694.983	53.290	-716.299	391.420	
	500	50.100	-690.118	64.134	-722.185	315.709	
	600	52.203	-684.997	73.464	-729.076	265.601	
	700	53.813	-679.693	81.637	-736.839	230.082	
	800	55.160	-674.243	88.913	-745.373	203.654	
	900	56.355	-668.666	95.480	-754.598	183.266	
	1000	57.455	-662.975	101.475	-764.450	167.093	
	1100	58.493	-657.177	107.000	-774.877	153.975	
	1200	59.489	-651.278	112.132	-785.837	143.140	
	1300	60.455	-645.281	116.932	-797.293	134.055	
	1400	61.400	-639.188	121.447	-809.214	126.341	
	1500	62.329	-633.001	125.715	-821.574	119.719	
	1600	63.246	-626.722	129.767	-834.350	113.982	
	1700	64.153	-620.352	133.629	-847.521	108.971	
	1800	65.053	-613.892	137.321	-861.070	104.562	

CALCIUM ORTHOSILICATE 7/6-HYDRATE $2\,CaO \cdot SiO_2 \cdot \frac{7}{6}\,H_2O$

		A	B	C	D	RANGE	REF.
SOL	CP	41.04	22.4	-7.4		298- 900	4

PHASE	T	CP	H	S	G	BT	
SOL	298	39.394	-637.1	38.4	-648.549	475.463	4,*
	300	39.538	-637.027	38.644	-648.620	472.583	
	400	45.375	-632.756	50.892	-653.112	356.892	
	500	49.280	-628.014	61.457	-658.742	287.975	
	600	52.424	-622.924	70.728	-665.361	242.390	
	700	55.210	-617.541	79.021	-672.855	210.103	
	800	57.804	-611.889	86.564	-681.140	186.104	
	900	60.286	-605.983	93.517	-690.149	167.614	

* * * * * * * * * *

TRICALCIUM DISILICATE TRIHYDRATE $3\,CaO \cdot 2\,SiO_2 \cdot 3\,H_2O$

		A	B	C	D	RANGE	REF.
SOL	CP	81.54	45.1	-14.67		298- 900	4

PHASE	T	CP	H	S	G	BT	
SOL	298	78.484	-1143.1	74.6	-1165.342	854.334	4,*
	300	78.770	-1142.955	75.086	-1165.480	849.166	
	400	90.411	-1134.445	99.488	-1174.240	641.661	
	500	98.222	-1124.995	120.543	-1185.266	518.149	
	600	104.525	-1114.849	139.023	-1198.263	436.526	
	700	110.116	-1104.113	155.562	-1213.006	378.769	
	800	115.328	-1092.838	170.609	-1229.326	335.881	
	900	120.319	-1081.055	184.483	-1247.089	302.875	

4 CaO·3 SiO_2·3/2 H_2O

4-CALCIUM 3-SILICATE 3/2-HYDRATE

		A	B	C	D	RANGE	REF.
SOL	CP	87.95	39.5	-13.48		298-1000	4

PHASE	T	CP	H	S	G	BT	
SOL	298	84.563	-1439.8	78.95	-1463.339	1072.801	4,*
	300	84.822	-1439.643	79.474	-1463.485	1066.292	
	400	95.325	-1430.589	105.449	-1472.769	804.792	
	500	102.308	-1420.691	127.508	-1484.445	648.938	
	600	107.906	-1410.172	146.670	-1498.174	545.783	
	700	112.849	-1399.131	163.680	-1513.707	472.664	
	800	117.444	-1387.614	179.052	-1530.856	418.267	
	900	121.836	-1375.649	193.140	-1549.475	376.314	
	1000	126.102	-1363.251	206.198	-1569.450	343.049	

* * * * * * * * * *

6 CaO·6 SiO_2· H_2O

6-CALCIUM 6-SILICATE HYDRATE

		A	B	C	D	RANGE	REF.
SOL	CP	132.25	65.2	-18.35		298- 900	4

PHASE	T	CP	H	S	G	BT	
SOL	298	131.047	-2396.4	121.3	-2432.566	1783.358	4,*
	300	131.421	-2396.157	122.112	-2432.791	1772.525	
	400	146.861	-2382.179	162.218	-2447.066	1337.195	
	500	157.510	-2366.938	196.184	-2465.030	1077.609	
	600	166.273	-2350.739	225.695	-2486.155	905.703	
	700	174.145	-2333.712	251.925	-2510.060	783.781	
	800	181.543	-2315.925	275.666	-2536.458	693.021	
	900	188.665	-2297.413	297.462	-2565.128	622.982	

5-CALCIUM 6-SILICATE 3-HYDRATE $5\ CaO \cdot 6\ SiO_2 \cdot 3\ H_2O$

		A	B	C	D	RANGE	REF.
SOL	CP	143.55	74.7	-20.82		298- 900	4

PHASE	T	CP	H	S	G	BT	
SOL	298	142.400	-2374.7	122.65	-2411.268	1767.745	4,*
	300	142.827	-2374.436	123.532	-2411.496	1757.010	
	400	160.417	-2359.202	167.239	-2426.097	1325.736	
	500	172.572	-2342.526	204.399	-2444.725	1068.732	
	600	182.587	-2324.757	236.768	-2466.818	898.659	
	700	191.591	-2306.042	265.600	-2491.962	778.130	
	800	200.057	-2286.456	291.740	-2519.848	688.483	
	900	208.210	-2266.041	315.776	-2550.240	619.366	

* * * * * * * * * *

$5\ CaO \cdot 6\ SiO_2 \cdot 1\frac{1}{2}\ H_2O$

5-CALCIUM 6-SILICATE 11/2-HYDRATE

		A	B	C	D	RANGE	REF.
SOL	CP	110.6	189.1			298- 700	4

PHASE	T	CP	H	S	G	BT	
SOL	298	166.980	-2556.0	146.15	-2599.575	1905.796	4,*
	300	167.330	-2555.691	147.184	-2599.846	1894.241	
	400	186.240	-2538.012	197.912	-2617.177	1430.151	
	500	205.150	-2518.443	241.501	-2639.193	1153.746	
	600	224.060	-2496.982	280.576	-2665.328	970.976	
	700	242.970	-2473.631	316.535	-2695.205	841.594	

$5\,CaO \cdot 6\,SiO_2 \cdot 2\frac{1}{2}\,H_2O$

5-CALCIUM 6-SLICATE 21/2-HYDRATE

		A	B	C	D	RANGE	REF.
SOL	CP	132.2	270.			298- 700	4

PHASE	T	CP	H	S	G	BT	
SOL	298	212.700	-2911.0	193.15	-2968.588	2176.326	4,*
	300	213.200	-2910.606	194.467	-2968.946	2163.167	
	400	240.200	-2887.936	259.499	-2991.736	1634.828	
	500	267.200	-2862.566	315.998	-3020.565	1320.466	
	600	294.200	-2834.496	367.101	-3054.757	1112.844	
	700	321.200	-2803.726	414.480	-3093.862	966.077	

* * * * * * * * * *

$CaO \cdot 2\,SiO_2 \cdot 2\,H_2O$

CALCIUM 2-SILICATE 2-HYDRATE

		A	B	C	D	RANGE	REF.
SOL	CP	44.81	18.7	-10.35		298-1000	4

PHASE	T	CP	H	S	G	BT	
SOL	298	38.742	-750.2	40.9	-762.394	558.925	4,*
	300	38.920	-750.128	41.140	-762.470	555.534	
	400	45.821	-745.855	53.386	-767.209	419.240	
	500	50.020	-741.050	64.090	-773.095	337.965	
	600	53.155	-735.886	73.498	-779.984	284.147	
	700	55.788	-730.436	81.894	-787.761	245.983	
	800	58.153	-724.737	89.500	-796.337	217.578	
	900	60.362	-718.810	96.478	-805.640	195.662	
	1000	62.475	-712.668	102.948	-815.615	178.277	

2-CALCIUM 3-SLICATE 5/2-HYDRATE $2\,CaO \cdot 3\,SiO_2 \cdot 5/2\,H_2O$

		A	B	C	D	RANGE	REF.
SOL	CP	79.47	36.3	-17.55		298-1000	4

PHASE	T	CP	H	S	G	BT	
SOL	298	70.550	-1175.7	64.9	-1195.050	876.113	4,*
	300	70.860	-1175.569	65.337	-1195.170	870.798	
	400	83.021	-1167.814	87.564	-1202.840	657.289	
	500	90.600	-1159.111	106.953	-1212.588	530.093	
	600	96.375	-1149.753	123.999	-1224.152	445.957	
	700	101.298	-1139.864	139.233	-1237.327	386.363	
	800	105.768	-1129.508	153.055	-1251.952	342.063	
	900	109.973	-1118.719	165.757	-1267.901	307.930	
	1000	114.015	-1107.519	177.555	-1285.073	280.890	

* * * * * * * * * *

CALCIUM MAGNESIUM DIOXIDE $CaO \cdot MgO$

		A	B	C	D	RANGE	REF.
SOL	CP	23.38	1.83	-4.36		298-1800	*

PHASE	T	CP	H	S	G	BT	
SOL	298	19.021	-297.1	15.85	-301.826	221.274	5
	300	19.085	-297.065	15.968	-301.855	219.931	
	400	21.387	-295.026	21.817	-303.753	165.985	
	500	22.551	-292.824	26.727	-306.187	133.852	
	600	23.267	-290.530	30.906	-309.074	112.595	
	700	23.771	-288.177	34.532	-312.350	97.533	
	800	24.163	-285.780	37.733	-315.966	86.330	
	900	24.489	-283.347	40.598	-319.885	77.689	
	1000	24.774	-280.883	43.194	-324.077	70.836	
	1100	25.033	-278.393	45.567	-328.517	65.279	
	1200	25.273	-275.878	47.756	-333.184	60.689	
	1300	25.501	-273.339	49.788	-338.063	56.841	
	1400	25.720	-270.778	51.685	-343.137	53.573	
	1500	25.931	-268.195	53.467	-348.396	50.768	
	1600	26.138	-265.592	55.147	-353.827	48.337	
	1700	26.340	-262.968	56.738	-359.422	46.213	
	1800	26.539	-260.324	58.249	-365.172	44.344	

$CaO \cdot MgO \cdot 2SiO_2$ DIOPSIDE

		A	B	C	D	RANGE	REF.
SOL	CP	44.46	29.58	-13.36	-10.50	298-1665	5
LIQ	CP	84.98				1665-1900	5

PHASE	T	CP	H	S	G	BT	
SOL	298	37.317	-765.4	34.2	-775.597	568.604	5
	300	37.545	-765.331	34.432	-775.660	565.144	
	400	46.262	-761.092	46.565	-779.718	426.076	
	500	51.281	-756.197	57.469	-784.931	343.139	
	600	54.717	-750.888	67.139	-791.171	288.223	
	700	57.294	-745.282	75.775	-798.324	249.282	
	800	59.316	-739.447	83.563	-806.298	220.300	
	900	60.928	-733.432	90.646	-815.014	197.939	
	1000	62.204	-727.273	97.134	-824.407	180.198	
	1100	63.189	-721.001	103.112	-834.423	165.807	
	1200	63.908	-714.644	108.642	-845.015	153.919	
	1300	64.378	-708.227	113.778	-856.139	143.949	
	1400	64.610	-701.776	118.559	-867.758	135.481	
	1500	64.611	-695.313	123.018	-879.840	128.210	
	1600	64.386	-688.861	127.182	-892.352	121.906	
	1665	64.120	-684.684	129.741	-900.703	118.243	
			30.7	18.438			5
LIQ	1665	84.980	-653.984	148.179	-900.703	118.243	
	1700	84.980	-651.010	149.947	-905.920	116.480	
	1800	84.980	-642.512	154.804	-921.160	111.859	
	1900	84.980	-634.014	159.399	-936.872	107.779	

MONTICELLITE $CaO \cdot MgO \cdot SiO_2$

		A	B	C	D	RANGE	REF.
SOL	CP	36.0	7.65	-7.85		298-1400	5

PHASE	T	CP	H	S	G	BT	
SOL	298	29.450	-540.6	26.4	-548.471	402.094	5
	300	29.573	-540.545	26.583	-548.520	399.650	
	400	34.154	-537.332	35.796	-551.650	301.448	
	500	36.685	-533.780	43.711	-555.636	242.901	
	600	38.409	-530.021	50.560	-560.357	204.137	
	700	39.753	-526.111	56.585	-565.720	176.650	
	800	40.893	-522.077	61.970	-571.653	156.189	
	900	41.916	-517.936	66.846	-578.097	140.400	
	1000	42.865	-513.696	71.312	-585.008	127.871	
	1100	43.766	-509.364	75.440	-592.348	117.705	
	1200	44.635	-504.944	79.286	-600.087	109.305	
	1300	45.481	-500.438	82.892	-608.198	102.261	
	1400	46.309	-495.849	86.293	-616.658	96.278	

* * * * * * * * * *

AKERMANITE $2CaO \cdot MgO \cdot 2SiO_2$

		A	B	C	D	RANGE	REF.
SOL	CP	60.22	11.29	-11.48		298-1727	5
	** MPT=	1727					5

PHASE	T	CP	H	S	G	BT	
SOL	298	50.672	-926.2	50.	-941.107	689.943	5
	300	50.851	-926.106	50.314	-941.200	685.756	
	400	57.561	-920.646	65.977	-947.036	517.506	
	500	61.273	-914.690	79.252	-954.316	417.187	
	600	63.805	-908.429	90.659	-962.825	350.756	
	700	65.780	-901.947	100.648	-972.400	303.638	
	800	67.458	-895.283	109.544	-982.918	268.557	
	900	68.964	-888.461	117.577	-994.280	241.477	
	1000	70.362	-881.494	124.916	-1006.410	219.980	
	1100	71.690	-874.391	131.685	-1019.245	202.532	
	1200	72.971	-867.157	137.978	-1032.732	188.111	
	1300	74.218	-859.798	143.869	-1046.827	176.011	
	1400	75.440	-852.315	149.414	-1061.494	165.729	
	1500	76.645	-844.710	154.660	-1076.700	156.896	
	1600	77.836	-836.986	159.644	-1092.417	149.237	
	1700	79.016	-829.143	164.399	-1108.621	142.542	
	1727	79.333	-827.006	165.646	-1113.077	140.877	

$2CaO{\cdot}5MgO{\cdot}8SiO_2{\cdot}H_2O$ TREMOLITE

		A	B	C	D	RANGE	REF.
SOL	CP	182.6	64.8	-40.2		298-1100	5

PHASE	T	CP	H	S	G	BT	
SOL	298	156.697	-2953.12	131.2	-2992.237	2193.664	5
	300	157.373	-2952.829	132.171	-2992.481	2180.314	
	400	183.395	-2935.651	181.411	-3008.216	1643.834	
	500	198.920	-2916.485	224.115	-3028.543	1323.953	
	600	210.313	-2896.001	261.430	-3052.860	1112.153	
	700	219.756	-2874.487	294.577	-3080.690	961.964	
	800	228.159	-2852.084	324.478	-3111.667	850.182	
	900	235.957	-2828.875	351.806	-3145.500	763.935	
	1000	243.380	-2804.905	377.053	-3181.959	695.510	
	1100	250.558	-2780.207	400.588	-3220.854	640.011	

* * * * * * * * * *

$3CaO{\cdot}MgO{\cdot}2SiO_2$ MERWINITE

		A	B	C	D	RANGE	REF.
SOL	CP	72.90	12.04	-14.40		298-1700	5

PHASE	T	CP	H	S	G	BT	
SOL	298	60.291	-1090.9	60.5	-1108.938	812.983	5
	300	60.512	-1090.788	60.874	-1109.050	808.051	
	400	68.716	-1084.277	79.550	-1116.097	609.889	
	500	73.160	-1077.165	95.401	-1124.865	491.744	
	600	76.124	-1069.693	109.016	-1135.102	413.516	
	700	78.389	-1061.963	120.927	-1146.612	358.037	
	800	80.282	-1054.027	131.521	-1159.244	316.733	
	900	81.958	-1045.914	141.075	-1172.882	284.853	
	1000	83.500	-1037.640	149.791	-1187.431	259.548	
	1100	84.954	-1029.217	157.818	-1202.817	239.010	
	1200	86.348	-1020.651	165.270	-1218.976	222.036	
	1300	87.700	-1011.949	172.236	-1235.855	207.794	
	1400	89.021	-1003.112	178.783	-1253.409	195.692	
	1500	90.320	-994.145	184.970	-1271.600	185.297	
	1600	91.601	-985.049	190.840	-1290.392	176.283	
	1700	92.870	-975.825	196.431	-1309.758	168.403	

ANORTHITE $CaO \cdot Al_2O_3 \cdot 2SiO_2$

		A	B	C	D	RANGE	REF.
SOL	CP	70.99	10.37	-32.35		298-1826	5
LIQ	CP	91.0				1826-2100	*

PHASE	T	CP	H	S	G	BT	
SOL	298	37.690	-1009.2	48.4	-1023.630	750.442	5
	300	38.157	-1009.130	48.635	-1023.720	745.880	
	400	54.919	-1004.364	62.231	-1029.256	562.435	
	500	63.235	-998.416	75.470	-1036.151	452.962	
	600	68.226	-991.825	87.473	-1044.308	380.440	
	700	71.647	-984.822	98.261	-1053.604	328.994	
	800	74.231	-977.523	108.004	-1063.926	290.690	
	900	76.329	-969.992	116.872	-1075.176	261.124	
	1000	78.125	-962.267	125.009	-1087.276	237.656	
	1100	79.723	-954.373	132.531	-1100.157	218.611	
	1200	81.187	-946.327	139.532	-1113.765	202.872	
	1300	82.557	-938.139	146.085	-1128.049	189.668	
	1400	83.857	-929.818	152.251	-1142.969	178.449	
	1500	85.107	-921.369	158.079	-1158.488	168.814	
	1600	86.318	-912.797	163.611	-1174.575	160.461	
	1700	87.500	-904.106	168.879	-1191.201	153.160	
	1800	88.658	-895.298	173.914	-1208.343	146.733	
	1826	88.955	-892.989	175.187	-1212.881	145.187	
			39.9	21.851			5
LIQ	1826	91.000	-853.089	197.038	-1212.881	145.187	
	1900	91.000	-846.355	200.653	-1227.596	141.225	
	2000	91.000	-837.255	205.321	-1247.897	136.382	
	2100	91.000	-828.155	209.761	-1268.653	132.048	

$CaO \cdot Al_2O_3 \cdot SiO_2$ CALCIUM-ALUMINIUM PYROXENE

		A	B	C	D	RANGE	REF.
SOL	CP	55.74	5.05	-17.62		298-1700	5

PHASE	T	CP	H	S	G	ET	
SOL	298	37.424	-786.8	34.6	-797.116	584.380	5
	300	37.677	-786.731	34.832	-797.180	580.823	
	400	46.747	-782.448	47.090	-801.284	437.860	
	500	51.217	-777.528	58.051	-806.553	352.592	
	600	53.876	-772.263	67.642	-812.848	296.120	
	700	55.679	-766.781	76.090	-820.044	256.064	
	800	57.027	-761.143	83.616	-828.036	226.239	
	900	58.110	-755.384	90.398	-836.742	203.216	
	1000	59.028	-749.526	96.569	-846.095	184.939	
	1100	59.839	-743.582	102.234	-856.039	170.102	
	1200	60.576	-737.561	107.472	-866.527	157.837	
	1300	61.262	-731.468	112.348	-877.521	147.545	
	1400	61.911	-725.310	116.912	-888.987	138.796	
	1500	62.532	-719.087	121.205	-900.895	131.278	
	1600	63.132	-712.804	125.260	-913.220	124.757	
	1700	63.715	-706.461	129.105	-925.940	119.054	

GEHLENITE $2CaO \cdot Al_2O_3 \cdot SiO_2$

		A	B	C	D	RANGE	REF.
SOL	CP	65.84	6.67	-18.69		298-1863	5
	** MPT= 1863						5

PHASE	T	CP	H	S	G	BT	
SOL	298	46.803	-953.	47.4	-967.132	709.022	5
	300	47.074	-952.913	47.690	-967.220	704.714	
	400	56.827	-947.653	62.756	-972.755	531.560	
	500	61.699	-941.704	76.012	-979.709	428.288	
	600	64.650	-935.376	87.541	-987.900	359.891	
	700	66.695	-928.803	97.668	-997.171	311.373	
	800	68.256	-922.053	106.680	-1007.397	275.245	
	900	69.536	-915.161	114.795	-1018.477	247.353	
	1000	70.641	-908.151	122.180	-1030.331	225.209	
	1100	71.632	-901.037	128.960	-1042.893	207.232	
	1200	72.546	-893.827	135.233	-1056.107	192.369	
	1300	73.405	-886.529	141.074	-1069.925	179.895	
	1400	74.224	-879.148	146.544	-1084.309	169.291	
	1500	75.014	-871.686	151.692	-1099.223	160.178	
	1600	75.782	-864.146	156.558	-1114.638	152.273	
	1700	76.532	-856.530	161.175	-1130.526	145.359	
	1800	77.269	-848.840	165.570	-1146.865	139.267	
	1863	77.728	-843.957	168.236	-1157.381	135.791	

$3CaO \cdot Al_2O_3 \cdot 3SiO_2$ GROSSULAR

		A	B	C	D	RANGE	REF.
SOL	CP	109.06	11.76	-31.41		298-1700	5

PHASE	T	CP	H	S	G	BT	
SOL	298	77.232	-1587.8	57.7	-1605.003	1176.657	5
	300	77.688	-1587.657	58.179	-1605.110	1169.479	
	400	94.133	-1578.957	83.095	-1612.195	880.981	
	500	102.376	-1569.092	105.074	-1621.629	708.909	
	600	107.391	-1558.586	124.214	-1633.115	594.942	
	700	110.882	-1547.664	141.045	-1646.395	514.097	
	800	113.560	-1536.436	156.032	-1661.262	453.897	
	900	115.766	-1524.967	169.539	-1677.552	407.420	
	1000	117.679	-1513.293	181.837	-1695.130	370.520	
	1100	119.400	-1501.438	193.135	-1713.886	340.564	
	1200	120.991	-1489.417	203.593	-1733.729	315.798	
	1300	122.489	-1477.243	213.337	-1754.581	295.011	
	1400	123.921	-1464.922	222.467	-1776.376	277.342	
	1500	125.304	-1452.460	231.064	-1799.057	262.158	
	1600	126.649	-1439.862	239.194	-1822.573	248.985	
	1700	127.965	-1427.131	246.912	-1846.882	237.465	

SPHENE $CaO \cdot TiO_2SiO_2$

		A	B	C	D	RANGE	REF.
SOL	CP	42.39	5.54	-9.63		298-1673	5
LIQ	CP	66.8				1673-2000	5

PHASE	T	CP	H	S	G	BT	
SOL	298	33.209	-621.95	30.9	-631.163	462.717	5
	300	33.352	-621.888	31.106	-631.220	459.905	
	400	38.587	-618.258	41.514	-634.864	346.920	
	500	41.308	-614.251	50.444	-639.473	279.551	
	600	43.039	-610.029	58.138	-644.911	234.940	
	700	44.303	-605.659	64.871	-651.069	203.300	
	800	45.317	-601.176	70.856	-657.861	179.743	
	900	46.187	-596.600	76.244	-665.220	161.559	
	1000	46.967	-591.942	81.152	-673.094	147.124	
	1100	47.688	-587.209	85.662	-681.437	135.407	
	1200	48.369	-582.405	89.841	-690.215	125.722	
	1300	49.022	-577.536	93.739	-699.396	117.595	
	1400	49.655	-572.602	97.395	-708.955	110.688	
	1500	50.272	-567.605	100.842	-718.868	104.753	
	1600	50.878	-562.548	104.106	-729.117	99.606	
	1673	51.314	-558.818	106.385	-736.800	96.264	
			29.6	17.693			5
LIQ	1673	66.800	-529.218	124.078	-736.800	96.264	
	1700	66.800	-527.414	125.148	-740.165	95.167	
	1800	66.800	-520.734	128.966	-752.872	91.423	
	1900	66.800	-514.054	132.577	-765.951	88.116	
	2000	66.800	-507.374	136.004	-779.382	85.178	

Ca_2Sn 2-CALCIUM TIN

		A	B	C	D	RANGE	REF.
SOL	CP	15.8	4.4			298-1395	*
LIQ	CP	22.9				1395-1700	*

PHASE	T	CP	H	S	G	ΦT	
SOL	298	17.112	-75.	24.	-82.156	60.230	1/*
	300	17.120	-74.968	24.106	-82.200	59.891	
	400	17.560	-73.234	29.091	-84.871	46.378	
	500	18.000	-71.456	33.057	-87.985	38.463	
	600	18.440	-69.634	36.378	-91.461	33.319	
	700	18.880	-67.768	39.253	-95.246	29.741	
	800	19.320	-65.858	41.803	-99.301	27.131	
	900	19.760	-63.904	44.104	-103.598	25.160	
	1000	20.200	-61.906	46.209	-108.115	23.632	
	1100	20.640	-59.864	48.155	-112.834	22.421	
	1200	21.080	-57.778	49.969	-117.742	21.447	
	1300	21.520	-55.648	51.674	-122.825	20.651	
	1395	21.938	-53.584	53.206	-127.807	20.026	
			15.	10.753			*
LIQ	1395	22.900	-38.584	63.959	-127.807	20.026	
	1400	22.900	-38.470	64.041	-128.127	20.004	
	1500	22.900	-36.180	65.621	-134.611	19.615	
	1600	22.900	-33.890	67.099	-141.248	19.296	
	1700	22.900	-31.600	68.487	-148.028	19.033	

CALCIUM TIN

CaSn

		A	B	C	D	RANGE	REF.
SOL	CP	10.4	3.7			298-1260	*
LIQ	CP	15.3				1260-1700	*

PHASE	T	CP	H	S	G	BT	
SOL	298	11.503	-38.	16.9	-43.039	31.552	1/*
	300	11.510	-37.979	16.971	-43.070	31.381	
	400	11.880	-36.809	20.333	-44.942	24.559	
	500	12.250	-35.603	23.024	-47.115	20.597	
	600	12.620	-34.359	25.290	-49.533	18.045	
	700	12.990	-33.079	27.263	-52.163	16.288	
	800	13.360	-31.761	29.022	-54.979	15.021	
	900	13.730	-30.407	30.617	-57.962	14.077	
	1000	14.100	-29.015	32.082	-61.098	13.355	
	1100	14.470	-27.587	33.444	-64.375	12.792	
	1200	14.840	-26.121	34.719	-67.784	12.347	
	1260	15.062	-25.224	35.448	-69.889	12.124	
			10.	7.937			*
LIQ	1260	15.300	-15.224	43.385	-69.889	12.124	
	1300	15.300	-14.612	43.863	-71.634	12.044	
	1400	15.300	-13.082	44.997	-76.077	11.878	
	1500	15.300	-11.552	46.052	-80.630	11.749	
	1600	15.300	-10.022	47.040	-85.286	11.651	
	1700	15.300	-8.492	47.967	-90.036	11.577	

Ca_2Pb 2-CALCIUM LEAD

		A	B	C	D	RANGE	REF.
SOL	CP	14.92	5.26			298-1383	*
	** MPT= 1383						1

PHASE	T	CP	H	S	G	BT	
SOL	298	16.488	-51.5	25.2	-59.013	43.264	1/147*
	300	16.498	-51.469	25.302	-59.060	43.031	
	400	17.024	-49.793	30.120	-61.841	33.793	
	500	17.550	-48.065	33.976	-65.052	28.438	
	600	18.076	-46.283	37.222	-68.616	24.997	
	700	18.602	-44.449	40.048	-72.483	22.633	
	800	19.128	-42.563	42.566	-76.616	20.933	
	900	19.654	-40.624	44.849	-80.988	19.669	
	1000	20.180	-38.632	46.947	-85.579	18.706	
	1100	20.706	-36.588	48.895	-90.373	17.958	
	1200	21.232	-34.491	50.720	-95.354	17.369	
	1300	21.758	-32.341	52.440	-100.513	16.900	
	1383	22.195	-30.517	53.800	-104.923	16.583	

* * * * * * * * * *

CaPb CALCIUM LEAD

		A	B	C	D	RANGE	REF.
SOL	CP	10.4	5.4			298-1223	*
	** MPT= 1223						1

PHASE	T	CP	H	S	G	BT	
SOL	298	12.010	-28.6	19.3	-34.354	25.186	1/*
	300	12.020	-28.578	19.374	-34.390	25.057	
	400	12.560	-27.349	22.906	-36.511	19.952	
	500	13.100	-26.066	25.767	-38.949	17.027	
	600	13.640	-24.729	28.203	-41.651	15.173	
	700	14.180	-23.338	30.346	-44.580	13.920	
	800	14.720	-21.893	32.275	-47.713	13.036	
	900	15.260	-20.394	34.040	-51.030	12.393	
	1000	15.800	-18.841	35.676	-54.516	11.916	
	1100	16.340	-17.234	37.207	-58.161	11.557	
	1200	16.880	-15.573	38.652	-61.955	11.285	
	1223	17.004	-15.183	38.973	-62.848	11.232	

CALCIUM 2-ALUMINIUM $CaAl_2$

		A	B	C	D	RANGE	REF.
SOL	CP	19.8	1.9	-2.68		298-1353	*
LIQ	CP	22.5				1353-1800	*

PHASE	T	CP	H	S	G	BT	
SOL	298	17.352	-51.8	20.4	-57.882	42.435	21,*
	300	17.392	-51.768	20.507	-57.920	42.200	
	400	18.885	-49.945	25.742	-60.242	32.919	
	500	19.678	-48.013	30.049	-63.038	27.557	
	600	20.196	-46.018	33.685	-66.229	24.127	
	700	20.583	-43.978	36.829	-69.758	21.782	
	800	20.901	-41.904	39.598	-73.582	20.104	
	900	21.179	-39.799	42.077	-77.668	18.863	
	1000	21.432	-37.669	44.321	-81.990	17.921	
	1100	21.669	-35.514	46.375	-86.526	17.193	
	1200	21.894	-33.335	48.270	-91.260	16.623	
	1300	22.111	-31.135	50.031	-96.176	16.171	
	1353	22.224	-29.960	50.917	-98.851	15.970	
			15.1	11.160			*
LIQ	1353	22.500	-14.860	62.078	-98.851	15.970	
	1400	22.500	-13.803	62.846	-101.787	15.892	
	1500	22.500	-11.553	64.398	-108.150	15.760	
	1600	22.500	-9.303	65.850	-114.663	15.664	
	1700	22.500	-7.053	67.214	-121.317	15.598	
	1800	22.500	-4.803	68.500	-128.103	15.556	

$CaAl_4$ CALCIUM 4-ALUMINIUM

		A	B	C	D	RANGE	REF.
SOL	CP	29.7	7.8	-2.7		298-1050	*
LIQ	CP	36.5				1050-1700	*

PHASE	T	CP	H	S	G	BT	
SOL	298	28.988	-52.2	33.	-62.039	45.482	21,*
	300	29.040	-52.146	33.179	-62.100	45.246	
	400	31.132	-49.128	41.847	-65.867	35.993	
	500	32.520	-45.942	48.951	-70.418	30.784	
	600	33.630	-42.633	54.981	-75.622	27.549	
	700	34.609	-39.221	60.240	-81.388	25.414	
	800	35.518	-35.714	64.921	-87.651	23.948	
	900	36.387	-32.118	69.155	-94.358	22.916	
	1000	37.230	-28.437	73.032	-101.470	22.179	
	1050	37.645	-26.565	74.859	-105.167	21.893	
			18.8	17.905			*
LIQ	1050	36.500	-7.765	92.764	-105.167	21.893	
	1100	36.500	-5.940	94.462	-109.848	21.828	
	1200	36.500	-2.290	97.638	-119.456	21.759	
	1300	36.500	1.360	100.559	-129.367	21.752	
	1400	36.500	5.010	103.264	-139.560	21.789	
	1500	36.500	8.660	105.782	-150.014	21.860	
	1600	36.500	12.310	108.138	-160.711	21.955	
	1700	36.500	15.960	110.351	-171.637	22.068	

CALCIUM ZINC

$CaZn$

		A	B	C	D	RANGE	REF.
SOL	CP	9.75	6.8			298- 750	*
LIQ	CP	15.2				750-1000	*

PHASE	T	CP	H	S	G	BT	
SOL	298	11.777	-17.5	15.9	-22.241	16.305	1/*
	300	11.790	-17.478	15.973	-22.270	16.226	
	400	12.470	-16.265	19.458	-24.048	13.141	
	500	13.150	-14.984	22.313	-26.141	11.428	
	600	13.830	-13.635	24.771	-28.498	10.382	
	700	14.510	-12.218	26.954	-31.086	9.707	
	750	14.850	-11.484	27.967	-32.459	9.460	
			5.3	7.067			*
LIQ	750	15.200	-6.184	35.033	-32.459	9.460	
	800	15.200	-5.424	36.014	-34.236	9.354	
	900	15.200	-3.904	37.805	-37.928	9.212	
	1000	15.200	-2.384	39.406	-41.790	9.135	

* * * * * * * * * *

CALCIUM 2-ZINC

$CaZn_2$

		A	B	C	D	RANGE	REF.
SOL	CP	15.9	6.65			298- 963	*
LIQ	CP	23.1				963-1000	*

PHASE	T	CP	H	S	G	BT	
SOL	298	17.883	-22.5	24.3	-29.745	21.807	1/*
	300	17.895	-22.467	24.411	-29.790	21.705	
	400	18.560	-20.644	29.650	-32.504	17.762	
	500	19.225	-18.755	33.863	-35.686	15.601	
	600	19.890	-16.799	37.427	-39.255	14.301	
	700	20.555	-14.777	40.543	-43.157	13.476	
	800	21.220	-12.688	43.331	-47.353	12.938	
	900	21.885	-10.533	45.869	-51.815	12.584	
	963	22.304	-9.141	47.363	-54.752	12.427	
			10.3	10.696			*
LIQ	963	23.100	1.159	58.059	-54.752	12.427	
	1000	23.100	2.014	58.930	-56.916	12.441	

$CaO \cdot UO_3$ CALCIUM URANATE

		A	B	C	D	RANGE	REF.
SOL-A	CP	27.63	11.9			298-1025	136
SOL-B	CP	27.01	12.58			1025-1200	136

PHASE	T	CP	H	S	G	BT	
SOL-A	298	31.178	-477.6	34.2	-487.797	357.613	*
	300	31.200	-477.542	34.393	-487.860	355.454	
	400	32.390	-474.363	43.532	-491.775	268.730	
	500	33.580	-471.064	50.887	-496.508	217.053	
	600	34.770	-467.647	57.115	-501.916	182.847	
	700	35.960	-464.110	62.564	-507.905	158.596	
	800	37.150	-460.455	67.443	-514.409	140.549	
	900	38.340	-456.680	71.888	-521.379	126.625	
	1000	39.530	-452.787	75.989	-528.775	115.579	
	1025	39.827	-451.795	76.968	-530.687	113.168	
			0.22	.215			136
SOL-B	1025	39.904	-451.575	77.183	-530.687	113.168	
	1100	40.848	-448.547	80.034	-536.584	106.624	
	1200	42.106	-444.399	83.642	-544.769	99.229	

CALCIUM HAFNATE $CaO \cdot HfO_2$

		A	B	C	D	RANGE	REF.
SOL	CP	28.90	3.24	4.65		298-2000	*

PHASE	T	CP	H	S	G	BT	
SOL	298	35.097	-425.1	24.	-432.256	316.895	137/*
	300	35.039	-425.035	24.217	-432.300	314.973	
	400	33.102	-421.644	33.985	-435.238	237.835	
	500	32.380	-418.376	41.281	-439.016	191.920	
	600	32.136	-415.153	47.158	-443.448	161.547	
	700	32.117	-411.941	52.109	-448.417	140.021	
	800	32.219	-408.725	56.403	-453.848	124.002	
	900	32.390	-405.495	60.207	-459.682	111.641	
	1000	32.605	-402.246	63.631	-465.876	101.831	
	1100	32.848	-398.973	66.749	-472.398	93.869	
	1200	33.111	-395.676	69.619	-479.218	87.289	
	1300	33.387	-392.351	72.280	-486.315	81.768	
	1400	33.673	-388.998	74.764	-493.668	77.075	
	1500	33.967	-385.616	77.098	-501.262	73.044	
	1600	34.266	-382.204	79.299	-509.083	69.547	
	1700	34.569	-378.763	81.386	-517.118	66.489	
	1800	34.876	-375.290	83.370	.357	63.796	
	1900	35.185	-371.787	85.264	-533.790	61.408	
	2000	35.496	-368.253	87.077	-542.407	59.279	

Cd

CADMIUM

		A	B	C	D	RANGE	REF.
SOL	CP	5.331	2.906			298- 594	21
LIQ	CP	7.10				594-1040	21
GAS	CP	4.968				1040-1200	21

PHASE	T	CP	H	S	G	BT	P	
SOL	298	6.197	0	12.38	-3.691	2.706		21
	300	6.203	.011	12.418	-3.714	2.706		
	400	6.493	.646	14.243	-5.051	2.760		
	500	6.784	1.310	15.723	-6.551	2.864	.002	
	594	7.057	1.961	16.914	-8.086	2.976	.131	
			1.48	2.492				21
LIQ	594	7.100	3.441	19.406	-8.086	2.976	.131	
	600	7.100	3.483	19.477	-8.203	2.988	.161	
	700	7.100	4.193	20.572	-10.207	3.187	2.902	
	800	7.100	4.903	21.520	-12.313	3.364	24.865	
	900	7.100	5.613	22.356	-14.507	3.523	130.234	
	1000	7.100	6.323	23.104	-16.781	3.668	484.002	
	1040	7.100	6.607	23.383	-17.711	3.722	760.	
			23.28	22.385				21
GAS	1040	4.968	29.887	45.767	-17.711	3.722		
	1100	4.968	30.185	46.046	-20.465	4.067		
	1200	4.968	30.682	46.478	-25.092	4.570		

CADMIUM FLUORIDE CdF_2

		A	B	C	D	RANGE	REF.
SOL	CP	14.35	5.5			298-1345	*
	** MPT= 1345, L= 5.4						62

PHASE	T	CP	H	S	G	BT	
SOL	298	15.990	-167.4	20.	-173.363	127.096	62
	300	16.000	-167.370	20.099	-173.400	126.339	
	400	16.550	-165.743	24.777	-175.654	95.986	
	500	17.100	-164.060	28.529	-178.325	77.956	
	600	17.650	-162.323	31.696	-181.340	66.062	
	700	18.200	-160.530	34.458	-184.651	57.658	
	800	18.750	-158.683	36.924	-188.222	51.427	
	900	19.300	-156.780	39.164	-192.028	46.637	
	1000	19.850	-154.823	41.226	-196.049	42.852	
	1100	20.400	-152.810	43.144	-200.268	39.795	
	1200	20.950	-150.743	44.942	-204.674	37.281	
	1300	21.500	-148.620	46.641	-209.254	35.183	
	1345	21.747	-147.647	47.377	-211.369	34.350	

* * * * * * * * * *

CADMIUM BROMIDE $CdBr_2$

		A	B	C	D	RANGE	REF.
SOL	CP	19.11	5.04	-2.03		298- 842	180*
	** MPT= 841.15						62
LIQ	CP	24.3				841-1138	180
	** BPT= 1138, L= 24.5						62

PHASE	T	CP	H	S	G	BT	
SOL	298	18.329	-75.35	33.18	-85.243	62.493	62
	300	18.366	-75.316	33.293	-85.304	62.152	
	400	19.857	-73.398	38.802	-88.919	48.589	
	500	20.818	-71.362	43.342	-93.032	40.670	
	600	21.570	-69.241	47.206	-97.564	35.543	
	700	22.224	-67.051	50.581	-102.457	31.993	
	800	22.825	-64.798	53.588	-107.668	29.418	
	841	23.062	-63.854	54.739	-109.897	28.558	
			7.97	9.475			180
LIQ	841	24.300	-55.884	64.214	-109.897	28.558	
	900	24.300	-54.454	65.857	-113.725	27.620	
	1000	24.300	-52.024	68.417	-120.441	26.326	
	1100	24.300	-49.594	70.733	-127.401	25.316	
	1138	24.300	-48.670	71.559	-130.104	24.990	

CdJ_2

CADMIUM IODIDE

		A	B	C	D	RANGE	REF.
SOL-A	CP	16.15	9.0			298- 595	62,*
SOL-B	CP	21.5				595- 661	180
	** MPT=	661.15					62
LIQ	CP	24.4				661-1017	180
	** BPT=	1017					62

PHASE	T	CP	H	S	G	BT	
SOL-A	298	18.833	-48.8	37.84	-60.082	44.047	62
	300	18.850	-48.765	37.957	-60.152	43.827	
	400	19.750	-46.835	43.503	-64.236	35.102	
	500	20.650	-44.815	48.006	-68.818	30.085	
	595	21.505	-42.813	51.671	-73.557	27.022	
			0.6	1.008			*
SOL-B	595	21.500	-42.213	52.679	-73.557	27.022	
	600	21.500	-42.105	52.859	-73.821	26.893	
	661	21.500	-40.791	54.946	-77.118	25.496	
			4.95	7.487			180
LIQ	661	24.400	-35.841	62.433	-77.118	25.496	
	700	24.400	-34.893	63.826	-79.571	24.846	
	800	24.400	-32.453	67.084	-86.120	23.530	
	900	24.400	-30.013	69.958	-92.975	22.580	
	1000	24.400	-27.573	72.529	-100.101	21.880	
	1017	24.400	-27.158	72.940	-101.338	21.780	

CADMIUM OXIDE

CdO

		A	B	C	D	RANGE	REF.
SOL	CP	11.53	1.525	-1.17		298-1500	145

PHASE	T	CP	H	S	G	BT	
SOL	298	10.668	-62.	13.1	-65.906	48.317	188/1
	300	10.687	-61.980	13.166	-65.930	48.036	
	400	11.409	-60.871	16.351	-67.412	36.837	
	500	11.824	-59.708	18.945	-69.181	30.243	
	600	12.120	-58.510	21.128	-71.187	25.933	
	700	12.359	-57.286	23.015	-73.396	22.918	
	800	12.567	-56.040	24.679	-75.783	20.706	
	900	12.758	-54.773	26.170	-78.326	19.023	
	1000	12.938	-53.488	27.524	-81.012	17.708	
	1100	13.111	-52.186	28.765	-83.828	16.657	
	1200	13.279	-50.866	29.913	-86.762	15.804	
	1300	13.443	-49.530	30.983	-89.808	15.100	
	1400	13.605	-48.178	31.985	-92.956	14.513	
	1500	13.765	-46.809	32.929	-96.203	14.019	

CdSe

CADMIUM SELENIDE

		A	B	C	D	RANGE	REF.
SOL	CP	11.19	2.23			298-1537	63
	** MPT= 1537, L= 10.5						63/62

PHASE	T	CP	H	S	G	BT	
SOL	298	11.855	-34.6	19.9	-40.533	29.716	63
	300	11.859	-34.578	19.973	-40.570	29.559	
	400	12.082	-33.381	23.416	-42.747	23.359	
	500	12.305	-32.162	26.135	-45.229	19.772	
	600	12.528	-30.920	28.399	-47.959	17.471	
	700	12.751	-29.656	30.347	-50.899	15.893	
	800	12.974	-28.370	32.064	-54.021	14.760	
	900	13.197	-27.061	33.605	-57.306	13.918	
	1000	13.420	-25.730	35.007	-60.737	13.276	
	1100	13.643	-24.377	36.296	-64.303	12.778	
	1200	13.866	-23.002	37.493	-67.993	12.385	
	1300	14.089	-21.604	38.612	-71.799	12.072	
	1400	14.312	-20.184	39.664	-75.714	11.821	
	1500	14.535	-18.742	40.659	-79.730	11.618	
	1537	14.618	-18.202	41.014	-81.241	11.553	

* * * * * * * * * *

$CdSeO_3$

CADMIUM SELENITE

		A	B	C	D	RANGE	REF.
SOL	CP	18.75	13.0			298- 953	*
LIQ	CP	34.1				953-1000	*

PHASE	T	CP	H	S	G	BT	
SOL	298	22.626	-139.1	28.7	-147.657	108.250	62
	300	22.650	-139.058	28.840	-147.710	107.621	
	400	23.950	-136.728	35.534	-150.942	82.482	
	500	25.250	-134.268	41.018	-154.777	67.662	
	600	26.550	-131.678	45.737	-159.120	57.967	
	700	27.850	-128.958	49.927	-163.907	51.181	
	800	29.150	-126.108	53.731	-169.093	46.200	
	900	30.450	-123.128	57.239	-174.643	42.415	
	953	31.139	-121.496	59.001	-177.724	40.763	
			14.0	14.690			62
LIQ	953	34.100	-107.496	73.691	-177.724	40.763	
	1000	34.100	-105.893	75.333	-181.226	39.612	

CADMIUM TELLURIDE

CdTe

		A	B	C	D	RANGE	REF.
SOL-B	CP	12.55	4.54	-1.76		298-1364	21
	** MPT= 1364, L= 10.6						21/62

PHASE	T	CP	H	S	G	BT	
SOL-B	298	11.924	-24.33	22.2	-30.949	22.689	21
	300	11.956	-24.308	22.274	-30.990	22.579	
	400	13.266	-23.041	25.910	-33.405	18.254	
	500	14.116	-21.669	28.967	-36.153	15.805	
	600	14.785	-20.223	31.602	-39.184	14.275	
	700	15.369	-18.715	33.925	-42.463	13.259	
	800	15.907	-17.151	36.013	-45.961	12.558	
	900	16.419	-15.535	37.916	-49.659	12.061	
	1000	16.914	-13.868	39.672	-53.540	11.703	
	1100	17.399	-12.152	41.307	-57.590	11.444	
	1200	17.876	-10.388	42.841	-61.798	11.256	
	1300	18.348	-8.577	44.291	-66.155	11.123	
	1364	18.648	-7.393	45.180	-69.018	11.060	

Cd_3As_2 CADMIUM ARSENIDE

		A	B	C	D	RANGE	REF.
SOL	CP	32.55	2.85	-3.07		298- 994	*
	** MPT= 994, L= 29.3						62

PHASE	T	CP	H	S	G	BT	
SOL	298	29.946	-9.1	49.5	-23.858	17.491	173
	300	29.994	-9.045	49.685	-23.950	17.450	
	400	31.771	-5.946	58.588	-29.381	16.055	
	500	32.747	-2.716	65.791	-35.611	15.568	
	600	33.407	.594	71.823	-42.500	15.483	
	700	33.918	3.961	77.013	-49.948	15.597	
	800	34.350	7.375	81.571	-57.882	15.815	
	900	34.736	10.829	85.639	-66.246	16.089	
	994	35.072	14.110	89.107	-74.462	16.374	

* * * * * * * * * *

CdSb CADMIUM MONOANTIMONIDE

		A	B	C	D	RANGE	REF.
SOL	CP	10.64	4.64			298- 729	213
LIQ	CP	34.212	-18.027			729- 900	213

PHASE	T	CP	H	S	G	BT	
SOL	298	12.023	-3.19	22.85	-10.003	7.333	213
	300	12.032	-3.168	22.924	-10.045	7.319	
	400	12.496	-1.941	26.449	-12.521	6.842	
	500	12.960	-.669	29.288	-15.312	6.694	
	600	13.424	.651	31.691	-18.364	6.690	
	700	13.888	2.016	33.796	-21.641	6.757	
	729	14.023	2.421	34.362	-22.629	6.785	
			8.5	11.660			213
LIQ	729	21.070	10.921	46.022	-22.629	6.785	
	800	19.790	12.372	47.922	-25.966	7.094	
	900	17.988	14.260	50.149	-30.873	7.498	

CADMIUM CARBONATE

$CdCO_3$

		A	B	C	D	RANGE	REF.
SOL	CP	10.3	31.5			298- 560	*
	** DPT= 560, CDCO3 = CDO + CO2						

PHASE	T	CP	H	S	G	BT	
SOL	298	19.692	-179.7	22.1	-186.289	136.572	66
	300	19.750	-179.664	22.222	-186.330	135.760	
	400	22.900	-177.531	28.335	-188.865	103.205	
	500	26.050	-175.084	33.783	-191.975	83.924	
	560	27.940	-173.464	36.841	-194.095	75.759	

* * * * * * * * * *

CADMIUM METASILICATE

$CdO \cdot SiO_2$

		A	B	C	D	RANGE	REF.
SOL	CP	21.0	10.2	-2.5		298-1100	5
	** MPT= 1514						62

PHASE	T	CP	H	S	G	BT	
SOL	298	21.229	-284.5	23.3	-291.447	213.665	5
	300	21.282	-284.461	23.431	-291.490	212.379	
	400	23.517	-282.212	29.885	-294.166	160.746	
	500	25.100	-279.778	35.310	-297.433	130.025	
	600	26.426	-277.200	40.006	-301.204	109.728	
	700	27.630	-274.497	44.171	-305.417	95.368	
	800	28.769	-271.677	47.935	-310.025	84.706	
	900	29.871	-268.744	51.388	-314.993	76.501	
	1000	30.950	-265.703	54.591	-320.294	70.010	
	1100	32.013	-262.555	57.591	-325.905	64.760	

$CdO \cdot Al_2O_3$ CADMIUM ALUMINATE

		A	B	C	D	RANGE	REF.
SOL	CP	38.25	5.70	-7.9		298-1200	*

PHASE	T	CP	H	S	G	BT	
SOL	298	31.062	-458.8	29.9	-467.715	342.890	126/*
	300	31.182	-458.742	30.093	-467.770	340.816	
	400	35.592	-455.376	39.746	-471.275	257.527	
	500	37.940	-451.690	47.963	-475.671	207.944	
	600	39.476	-447.815	55.024	-480.829	175.165	
	700	40.628	-443.807	61.199	-486.646	151.958	
	800	41.576	-439.696	66.688	-493.046	134.712	
	900	42.405	-435.496	71.633	-499.966	121.425	
	1000	43.160	-431.217	76.141	-507.358	110.898	
	1100	43.867	-426.866	80.288	-515.182	102.371	
	1200	44.541	-422.445	84.134	-523.405	95.338	

* * * * * * * * * *

$CdO \cdot Ga_2O_3$ CADMIUM GALLIUM TETRACXIDE

		A	B	C	D	RANGE	REF.
SOL	CP	38.51	5.22	-6.19		298-1200	*

PHASE	T	CP	H	S	G	BT	
SOL	298	33.103	-324.1	33.4	-334.058	244.904	5/*
	300	33.198	-324.039	33.605	-334.120	243.439	
	400	36.729	-320.521	43.701	-338.001	184.700	
	500	38.644	-316.744	52.120	-342.804	149.860	
	600	39.923	-312.813	59.285	-348.384	126.916	
	700	40.901	-308.770	65.515	-354.630	110.735	
	800	41.719	-304.638	71.032	-361.463	98.760	
	900	42.444	-300.429	75.988	-368.818	89.573	
	1000	43.111	-296.151	80.495	-376.646	82.327	
	1100	43.740	-291.808	84.633	-384.905	76.484	
	1200	44.344	-287.404	88.465	-393.562	71.687	

CADMIUM TUNGSTATE $CdO \cdot WO_3$

		A	B	C	D	RANGE	REF.
SOL	CP	27.4	7.9			298-1300	*

PHASE	T	CP	H	S	G	BT	
SOL	298	29.755	-282.36	37.	-293.392	215.091	5/*
	300	29.770	-282.305	37.184	-293.460	213.814	
	400	30.560	-279.288	45.857	-297.631	162.640	
	500	31.350	-276.193	52.761	-302.573	132.272	
	600	32.140	-273.018	58.546	-308.146	112.257	
	700	32.930	-269.765	63.560	-314.257	98.129	
	800	33.720	-266.432	68.009	-320.840	87.661	
	900	34.510	-263.021	72.026	-327.844	79.622	
	1000	35.300	-259.530	75.703	-335.233	73.275	
	1100	36.090	-255.961	79.104	-342.976	68.152	
	1200	36.880	-252.312	82.279	-351.047	63.943	
	1300	37.670	-248.585	85.262	-359.425	60.433	

$CdO \cdot TiO_2$ CADMIUM TITANATE

		A	B	C	D	RANGE	REF.
SOL-A	CP	27.75	2.3	-4.35		298-1100	*
SOL-B	CP	27.75	2.3	-4.35		1100-1600	*

PHASE	T	CP	H	S	G	BT	
SOL-A	298	23.542	-294.42	25.1	-301.904	221.331	148/*
	300	23.607	-294.376	25.246	-301.950	220.000	
	400	25.951	-291.883	32.402	-304.844	166.581	
	500	27.160	-289.222	38.335	-308.390	134.815	
	600	27.922	-286.466	43.358	-312.481	113.836	
	700	28.472	-283.645	47.706	-317.039	98.997	
	800	28.910	-280.775	51.537	-322.005	87.979	
	900	29.283	-277.865	54.964	-327.333	79.498	
	1000	29.615	-274.920	58.067	-332.987	72.784	
	1100	29.920	-271.943	60.904	-338.937	67.350	
			3.58	3.255			148
SOL-B	1100	29.920	-268.363	64.159	-338.937	67.350	
	1200	30.208	-265.356	66.774	-345.486	62.930	
	1300	30.483	-262.322	69.203	-352.286	59.233	
	1400	30.748	-259.260	71.472	-359.321	56.100	
	1500	31.007	-256.172	73.602	-366.576	53.417	
	1600	31.260	-253.059	75.612	-374.037	51.098	

CERIUM

Ce

		A	B	C	D	RANGE	REF.
SOL-C	CP	5.420	3.381			298- 600	21
	CP	5.099	3.918			600- 999	21
SOL-D	CP	8.99				999-1071	21
LIQ	CP	9.01				1071-3000	21
	** BPT= 3699 / L= 98.99						21

PHASE	T	CP	H	S	G	BT	P	
SOL-C	298	6.428	0	16.6	-4.949	3.628		21
	300	6.434	.012	16.640	-4.980	3.628		
	400	6.772	.672	18.537	-6.743	3.684		
	500	7.110	1.366	20.085	-8.676	3.793		
	600	7.449	2.094	21.411	-10.752	3.917		
	600	7.450	2.094	21.411	-10.752	3.917		
	700	7.842	2.859	22.589	-12.953	4.045		
	800	8.233	3.663	23.661	-15.266	4.171		
	900	8.625	4.506	24.654	-17.683	4.295		
	999	9.013	5.379	25.574	-20.170	4.413		
			0.715	.716				21
SOL-D	999	8.990	6.094	26.290	-20.170	4.413		
	1000	8.990	6.103	26.299	-20.196	4.414		
	1071	8.990	6.741	26.915	-22.085	4.507		
			1.305	1.218				21
LIQ	1071	9.010	8.046	28.134	-22.085	4.507		
	1100	9.010	8.307	28.374	-22.905	4.551		
	1200	9.010	9.208	29.158	-25.782	4.696		
	1300	9.010	10.109	29.880	-28.734	4.831		
	1400	9.010	11.010	30.547	-31.756	4.958		
	1500	9.010	11.911	31.169	-34.842	5.077		
	1600	9.010	12.812	31.750	-37.988	5.190		
	1700	9.010	13.713	32.297	-41.191	5.296		
	1800	9.010	14.614	32.812	-44.447	5.397		
	1900	9.010	15.515	33.299	-47.752	5.494	.002	
	2000	9.010	16.416	33.761	-51.106	5.585	.008	
	2100	9.010	17.317	34.201	-54.504	5.673	.027	
	2200	9.010	18.218	34.620	-57.945	5.757	.079	
	2300	9.010	19.119	35.020	-61.427	5.838	.211	
	2400	9.010	20.020	35.404	-64.948	5.915	.519	
	2500	9.010	20.921	35.771	-68.507	5.990	1.190	
	2600	9.010	21.822	36.125	-72.102	6.062	2.560	
	2700	9.010	22.723	36.465	-75.732	6.131	5.205	
	2800	9.010	23.624	36.793	-79.395	6.198	10.061	
	2900	9.010	24.525	37.109	-83.090	6.263	18.582	
	3000	9.010	25.426	37.414	-86.816	6.325	32.948	

CeH_2

CERIUM DIHYDRIDE

		A	B	C	D	RANGE	REF.
SOL	CP	8.40	4.6			298-1200	76,*

PHASE	T	CP	H	S	G	BT	
SOL	298	9.771	-49.	13.33	-52.974	38.836	19/76
	300	9.780	-48.982	13.390	-52.999	38.615	
	400	10.240	-47.981	16.267	-54.488	29.775	
	500	10.700	-46.934	18.601	-56.235	24.583	
	600	11.160	-45.841	20.593	-58.197	21.201	
	700	11.620	-44.702	22.348	-60.345	18.843	
	800	12.080	-43.517	23.929	-62.660	17.120	
	900	12.540	-42.286	25.379	-65.127	15.817	
	1000	13.000	-41.009	26.724	-67.733	14.805	
	1100	13.460	-39.686	27.984	-70.469	14.003	
	1200	13.920	-38.317	29.175	-73.327	13.357	

CERIUM FLUORIDE

CeF_3

		A	B	C	D	RANGE	REF.
SOL	CP	19.2	4.0			298-1710	*
LIQ	CP	32.0				1710-2400	138

PHASE	T	CP	H	S	G	BT	
SOL	298	20.393	-425.	27.55	-433.214	317.597	*/1
	300	20.400	-424.962	27.676	-433.265	315.676	
	400	20.800	-422.902	33.600	-436.342	238.438	
	500	21.200	-420.802	38.284	-439.944	192.325	
	600	21.600	-418.662	42.185	-443.973	161.739	
	700	22.000	-416.482	45.544	-448.363	140.004	
	800	22.400	-414.262	48.508	-453.069	123.789	
	900	22.800	-412.002	51.170	-458.055	111.246	
	1000	23.200	-409.702	53.592	-463.295	101.267	
	1100	23.600	-407.362	55.822	-468.767	93.148	
	1200	24.000	-404.982	57.893	-474.454	86.421	
	1300	24.400	-402.562	59.830	-480.341	80.764	
	1400	24.800	-400.102	61.653	-486.416	75.943	
	1500	25.200	-397.602	63.377	-492.668	71.791	
	1600	25.600	-395.062	65.017	-499.089	68.182	
	1700	26.000	-392.482	66.581	-505.669	65.017	
	1710	26.040	-392.222	66.733	-506.336	64.722	
			13.5	7.895			138
LIQ	1710	32.000	-378.722	74.628	-506.336	64.722	
	1800	32.000	-375.842	76.269	-513.127	62.310	
	1900	32.000	-372.642	77.999	-520.841	59.918	
	2000	32.000	-369.442	79.641	-528.724	57.784	
	2100	32.000	-366.242	81.202	-536.766	55.870	
	2200	32.000	-363.042	82.691	-544.962	54.144	
	2300	32.000	-359.842	84.113	-553.302	52.583	
	2400	32.000	-356.642	85.475	-561.782	51.164	

$CeCl_3$ CERIUM CHLORIDE

		A	B	C	D	RANGE	REF.
SOL	CP	23.3	3.25	-1.2		298-1090	*
LIQ	CP	34.7				1090-1300	*

PHASE	T	CP	H	S	G	BT	
SOL	298	22.919	-252.8	35.8	-263.474	193.157	1
	300	22.942	-252.758	35.942	-263.540	192.015	
	400	23.850	-250.414	42.678	-267.485	146.167	
	500	24.445	-247.998	48.067	-272.031	118.921	
	600	24.917	-245.529	52.567	-277.069	100.936	
	700	25.330	-243.016	56.440	-282.524	88.220	
	800	25.712	-240.464	59.847	-288.342	78.782	
	900	26.077	-237.874	62.897	-294.481	71.519	
	1000	26.430	-235.249	65.663	-300.912	65.773	
	1090	26.741	-232.856	67.954	-306.926	61.548	
			12.8	11.743			89
LIQ	1090	34.700	-220.056	79.697	-306.926	61.548	
	1100	34.700	-219.709	80.014	-307.724	61.147	
	1200	34.700	-216.239	83.033	-315.879	57.537	
	1300	34.700	-212.769	85.810	-324.323	54.531	

CERIUM BROMIDE $CeBr_3$

		A	B	C	D	RANGE	REF.
SOL	CP	22.64	5.86	-0.07		298-1005	89
LIQ	CP	36.49				1005-1200	89

PHASE	T	CP	H	S	G	BT	
SOL	298	24.308	-212.	43.5	-224.970	164.929	*
	300	24.320	-211.955	43.650	-225.050	163.971	
	400	24.940	-209.492	50.733	-229.785	125.565	
	500	25.542	-206.968	56.363	-235.149	102.797	
	600	26.137	-204.384	61.072	-241.027	87.806	
	700	26.728	-201.740	65.145	-247.342	77.234	
	800	27.317	-199.038	68.753	-254.040	69.410	
	900	27.905	-196.277	72.004	-261.081	63.408	
	1000	28.493	-193.457	74.975	-268.432	58.674	
	1005	28.522	-193.315	75.117	-268.807	58.463	
			12.4	12.338			89
LIQ	1005	36.490	-180.915	87.455	-268.807	58.463	
	1100	36.490	-177.448	90.751	-277.274	55.097	
	1200	36.490	-173.799	93.926	-286.511	52.188	

CeJ_3 CERIUM IODIDE

		A	B	C	D	RANGE	REF.
SOL	CP	19.15	14.52	0.9		298-1033	19
LIQ	CP	36.5				1033-1300	89*

PHASE	T	CP	H	S	G	BT	
SOL	298	24.492	-155.3	51.3	-170.595	125.066	19/*
	300	24.506	-155.255	51.452	-170.690	124.364	
	400	25.520	-152.756	58.631	-176.209	96.289	
	500	26.770	-150.143	64.458	-182.372	79.725	
	600	28.112	-147.399	69.456	-189.073	68.879	
	700	29.498	-144.519	73.893	-196.245	61.279	
	800	30.907	-141.499	77.924	-203.838	55.694	
	900	32.329	-138.337	81.646	-211.819	51.444	
	1000	33.760	-135.033	85.127	-220.160	48.122	
	1033	34.234	-133.911	86.230	-222.987	47.183	
			12.2	11.810			89*
LIQ	1033	36.500	-121.711	98.041	-222.987	47.183	
	1100	36.500	-119.266	100.334	-229.634	45.630	
	1200	36.500	-115.616	103.510	-239.828	43.685	
	1300	36.500	-111.966	106.432	-250.327	42.089	

CERIUM OXIDE Ce_2O_3

		A	B	C	D	RANGE	REF.
SOL	CP	25.77	9.9	-2.2		298-1200	163*
	** MPT= 1960						47

PHASE	T	CP	H	S	G	BT	
SOL	298	26.247	-435.4	36.	-446.133	327.069	164/1
	300	26.296	-435.351	36.163	-446.200	325.100	
	400	28.355	-432.611	44.031	-450.224	246.024	
	500	29.840	-429.699	50.524	-454.961	198.890	
	600	31.099	-426.651	56.078	-460.298	167.686	
	700	32.251	-423.482	60.960	-466.154	145.559	
	800	33.346	-420.202	65.338	-472.473	129.091	
	900	34.408	-416.814	69.327	-479.209	116.383	
	1000	35.450	-413.321	73.007	-486.328	106.301	
	1100	36.478	-409.725	76.434	-493.802	98.123	
	1200	37.497	-406.026	79.651	-501.608	91.368	

* * * * * * * * * *

CERIUM 1.72-OXIDE $CeO_{1.72}$

		A	B	C	D	RANGE	REF.
SOL	CP	13.95	4.75	-1.8		298-1200	*

PHASE	T	CP	H	S	G	BT	
SOL	298	13.341	-238.	16.5	-242.919	178.089	*/191
	300	13.375	-237.975	16.583	-242.950	177.013	
	400	14.725	-236.564	20.633	-244.817	133.780	
	500	15.605	-235.045	24.019	-247.055	108.002	
	600	16.300	-233.449	26.927	-249.605	90.931	
	700	16.908	-231.788	29.486	-252.428	78.822	
	800	17.469	-230.069	31.781	-255.494	69.807	
	900	18.003	-228.295	33.869	-258.778	62.848	
	1000	18.520	-226.469	35.793	-262.262	57.325	
	1100	19.026	-224.592	37.582	-265.932	52.843	
	1200	19.525	-222.664	39.259	-269.775	49.139	

$CeO_{1.83}$ CERIUM 1.83-OXIDE

		A	B	C	D	RANGE	REF.
SOL	CP	14.52	4.52	-1.52		298-1200	*

PHASE	T	CP	H	S	G	BT	
SOL	298	14.158	-247.	16.2	-251.830	184.621	*/191
	300	14.187	-246.974	16.288	-251.860	183.505	
	400	15.378	-245.490	20.547	-253.709	138.639	
	500	16.172	-243.911	24.068	-255.945	111.889	
	600	16.810	-242.261	27.075	-258.506	94.173	
	700	17.374	-240.551	29.709	-261.348	81.607	
	800	17.898	-238.787	32.064	-264.438	72.251	
	900	18.400	-236.972	34.201	-267.753	65.028	
	1000	18.888	-235.108	36.165	-271.273	59.295	
	1100	19.366	-233.195	37.988	-274.981	54.641	
	1200	19.838	-231.235	39.693	-278.866	50.795	

* * * * * * * * * *

CeO_2 CERIUM DIOXIDE

		A	B	C	D	RANGE	REF.
SOL	CP	15.49	4.23	-1.815		298-1300	191,133
	** MPT > 2873						1

PHASE	T	CP	H	S	G	BT	
SOL	298	14.709	-260.5	14.9	-264.942	194.234	1,*
	300	14.742	-260.473	14.991	-264.970	193.057	
	400	16.048	-258.927	19.429	-266.699	145.737	
	500	16.879	-257.278	23.104	-268.831	117.522	
	600	17.524	-255.557	26.241	-271.302	98.835	
	700	18.081	-253.776	28.985	-274.066	85.579	
	800	18.590	-251.943	31.433	-277.089	75.707	
	900	19.073	-250.059	33.650	-280.345	68.086	
	1000	19.538	-248.129	35.684	-283.813	62.036	
	1100	19.993	-246.152	37.568	-287.476	57.124	
	1200	20.440	-244.130	39.326	-291.322	53.064	
	1300	20.882	-242.064	40.980	-295.338	49.658	

CERIUM MONOSULFIDE

CeS

		A	B	C	D	RANGE	REF.
SOL	CP	12.558	3.238	-1.408		298-2000	63,*
	** MPT= 2723						63

PHASE	T	CP	H	S	G	BT	
SOL	298	11.939	-109.1	18.7	-114.675	84.071	63
	300	11.965	-109.078	18.774	-114.710	83.577	
	400	12.973	-107.826	22.368	-116.773	63.811	
	500	13.614	-106.495	25.336	-119.163	52.093	
	600	14.110	-105.108	27.863	-121.826	44.381	
	700	14.537	-103.675	30.071	-124.725	38.946	
	800	14.928	-102.202	32.038	-127.832	34.927	
	900	15.298	-100.690	33.818	-131.126	31.846	
	1000	15.655	-99.143	35.448	-134.591	29.419	
	1100	16.003	-97.560	36.957	-138.212	27.464	
	1200	16.346	-95.942	38.364	-141.979	25.861	
	1300	16.684	-94.291	39.686	-145.882	24.528	
	1400	17.019	-92.605	40.934	-149.913	23.406	
	1500	17.352	-90.887	42.120	-154.067	22.451	
	1600	17.684	-89.135	43.250	-158.336	21.631	
	1700	18.014	-87.350	44.332	-162.715	20.921	
	1800	18.343	-85.532	45.371	-167.201	20.304	
	1900	18.671	-83.681	46.372	-171.788	19.763	
	2000	18.999	-81.798	47.338	-176.474	19.287	

* * * * * * * * * *

TRICERIUM TETRASULFIDE

Ce_3S_4

		A	B	C	D	RANGE	REF.
SOL	CP	40.11	9.48			298-1200	*
	** MPT= 2323						63

PHASE	T	CP	H	S	G	BT	
SOL	298	42.936	-395.	61.	-413.187	302.915	63
	300	42.954	-394.921	61.266	-413.300	301.130	
	400	43.902	-390.578	73.753	-420.079	229.551	
	500	44.850	-386.140	83.651	-427.966	187.089	
	600	45.798	-381.608	91.912	-436.755	159.109	
	700	46.746	-376.981	99.043	-446.310	139.363	
	800	47.694	-372.259	105.347	-456.536	124.737	
	900	48.642	-367.442	111.019	-467.359	113.505	
	1000	49.590	-362.530	116.193	-478.723	104.639	
	1100	50.538	-357.524	120.964	-490.584	97.483	
	1200	51.486	-352.423	125.402	-502.905	91.604	

Ce_2S_3 CERIUM SULFIDE

		A	B	C	D	RANGE	REF.
SOL	CP	29.86	3.04			298-2160	63
	** MPT= 2160						63

PHASE	T	CP	H	S	G	BT	
SOL	298	30.766	-284.	43.1	-296.850	217.626	63
	300	30.772	-283.943	43.290	-296.930	216.343	
	400	31.076	-280.851	52.185	-301.724	164.877	
	500	31.380	-277.728	59.152	-307.304	134.340	
	600	31.684	-274.575	64.900	-313.515	114.213	
	700	31.988	-271.391	69.807	-320.256	100.002	
	800	32.292	-268.177	74.098	-327.455	89.469	
	900	32.596	-264.933	77.919	-335.060	81.375	
	1000	32.900	-261.658	81.369	-343.027	74.979	
	1100	33.204	-258.353	84.519	-351.323	69.811	
	1200	33.508	-255.017	87.421	-359.922	65.560	
	1300	33.812	-251.651	90.115	-368.801	62.009	
	1400	34.116	-248.255	92.632	-377.940	59.007	
	1500	34.420	-244.828	94.996	-387.322	56.440	
	1600	34.724	-241.371	97.227	-396.934	54.226	
	1700	35.028	-237.883	99.342	-406.764	52.300	
	1800	35.332	-234.365	101.352	-416.799	50.613	
	1900	35.636	-230.817	103.271	-427.031	49.126	
	2000	35.940	-227.238	105.106	-437.451	47.809	
	2100	36.244	-223.629	106.867	-448.050	46.635	
	2160	36.426	-221.449	107.891	-454.493	45.992	

* * * * * * * * * *

$Ce_2(SO_4)_3$ CERIUM SULFATE

		A	B	C	D	RANGE	REF.
SOL	CP	53.	47.5			298-1000	*
	** MPT= 1673-1773						47

PHASE	T	CP	H	S	G	BT	
SOL	298	67.162	-944.5	68.7	-964.983	707.447	81/*
	300	67.250	-944.376	69.116	-965.110	703.177	
	400	72.000	-937.413	89.113	-973.058	531.726	
	500	76.750	-929.976	105.689	-982.820	429.648	
	600	81.500	-922.063	120.103	-994.125	362.158	
	700	86.250	-913.676	133.023	-1006.791	314.377	
	800	91.000	-904.813	144.850	-1020.693	278.878	
	900	95.750	-895.476	155.842	-1035.734	251.544	
	1000	100.500	-885.663	166.176	-1051.839	229.910	

CERIUM MONOTELLURICE

		A	B	C	D	RANGE	REF.
SOL	CP	8.932	0.006	-0.271		298-2090	63
	** MPT= 2090						63

PHASE	T	CP	H	S	G	BT	
SOL	298	8.629	-72.	23.4	-78.977	57.899	63
	300	8.633	-71.984	23.453	-79.020	57.574	
	400	8.765	-71.113	25.958	-81.496	44.533	
	500	8.827	-70.233	27.921	-84.194	36.806	
	600	8.860	-69.349	29.533	-87.069	31.719	
	700	8.881	-68.462	30.901	-90.092	28.132	
	800	8.894	-67.573	32.088	-93.243	25.476	
	900	8.904	-66.683	33.136	-96.505	23.438	
	1000	8.911	-65.792	34.074	-99.867	21.829	
	1100	8.916	-64.901	34.924	-103.317	20.530	
	1200	8.920	-64.009	35.700	-106.849	19.462	
	1300	8.924	-63.117	36.414	-110.455	18.572	
	1400	8.927	-62.224	37.076	-114.130	17.819	
	1500	8.929	-61.331	37.692	-117.869	17.176	
	1600	8.931	-60.438	38.268	-121.667	16.621	
	1700	8.933	-59.545	38.809	-125.521	16.139	
	1800	8.934	-58.652	39.320	-129.428	15.717	
	1900	8.936	-57.758	39.803	-133.384	15.345	
	2000	8.937	-56.865	40.261	-137.388	15.015	
	2090	8.938	-56.060	40.655	-141.029	14.749	

CeN

CERIUM MONONITRIDE

		A	B	C	D	RANGE	REF.
SOL	CP	11.1	1.65	-1.73		298-2000	*

PHASE	T	CP	H	S	G	BT	
SOL	298	9.646	-78.	10.6	-81.160	59.500	1/*
	300	9.673	-77.982	10.660	-81.180	59.148	
	400	10.679	-76.959	13.598	-82.398	45.026	
	500	11.233	-75.861	16.045	-83.883	36.670	
	600	11.609	-74.718	18.128	-85.594	31.182	
	700	11.902	-73.542	19.940	-87.500	27.322	
	800	12.150	-72.339	21.546	-89.576	24.474	
	900	12.371	-71.113	22.990	-91.804	22.296	
	1000	12.577	-69.865	24.304	-94.169	20.583	
	1100	12.772	-68.598	25.512	-96.661	19.207	
	1200	12.960	-67.311	26.632	-99.269	18.082	
	1300	13.143	-66.006	27.676	-101.985	17.147	
	1400	13.322	-64.682	28.657	-104.802	16.363	
	1500	13.498	-63.341	29.582	-107.714	15.696	
	1600	13.672	-61.983	30.459	-110.717	15.125	
	1700	13.845	-60.607	31.293	-113.805	14.633	
	1800	14.017	-59.214	32.089	-116.974	14.204	
	1900	14.187	-57.804	32.851	-120.221	13.830	
	2000	14.357	-56.377	33.583	-123.543	13.502	

DICERIUM TRICARBIDE Ce_2C_3

		A	B	C	D	RANGE	REF.
SOL	CP	29.25	2.85	-4.8		298-1500	*

PHASE	T	CP	H	S	G	BT	
SOL	298	24.700	-42.2	41.5	-54.573	40.009	72/*
	300	24.772	-42.154	41.653	-54.650	39.818	
	400	27.390	-39.529	49.186	-59.204	32.352	
	500	28.755	-36.716	55.458	-64.445	28.173	
	600	29.627	-33.794	60.783	-70.264	25.597	
	700	30.265	-30.799	65.400	-76.578	23.912	
	800	30.780	-27.745	69.476	-83.326	22.767	
	900	31.222	-24.645	73.127	-90.459	21.969	
	1000	31.620	-21.502	76.438	-97.940	21.408	
	1100	31.988	-18.322	79.469	-105.737	21.011	
	1200	32.337	-15.105	82.267	-113.826	20.733	
	1300	32.671	-11.855	84.869	-122.184	20.544	
	1400	32.995	-8.572	87.302	-130.794	20.421	
	1500	33.312	-5.256	89.589	-139.640	20.348	

* * * * * * * * * *

CERIUM DICARBIDE CeC_2

		A	B	C	D	RANGE	REF.
SOL	CP	16.4	2.2	-2.2		298-1500	*
	** MPT > 2573						47

PHASE	T	CP	H	S	G	BT	
SOL	298	14.581	-23.2	21.5	-29.610	21.708	72/*
	300	14.616	-23.173	21.590	-29.650	21.603	
	400	15.905	-21.639	25.994	-32.037	17.506	
	500	16.620	-20.010	29.626	-34.823	15.223	
	600	17.109	-18.323	32.701	-37.943	13.823	
	700	17.491	-16.592	35.368	-41.350	12.912	
	800	17.816	-14.826	37.726	-45.007	12.297	
	900	18.108	-13.030	39.841	-48.887	11.873	
	1000	18.380	-11.205	41.763	-52.969	11.578	
	1100	18.638	-9.354	43.527	-57.234	11.373	
	1200	18.887	-7.478	45.160	-61.670	11.233	
	1300	19.130	-5.577	46.681	-66.263	11.141	
	1400	19.368	-3.652	48.108	-71.003	11.086	
	1500	19.602	-1.704	49.452	-75.881	11.057	

$CeSi_2$ CERIUM 2-SILICON

		A	B	C	D	RANGE	REF.
SOL	CP	17.236	2.692	-1.254	0.548	298-1500	*
	** MPT=	1700					1

PHASE	T	CP	H	S	G	BT	
SOL	298	16.677	-45.	24.1	-52.185	38.258	*
	300	16.700	-44.969	24.203	-52.230	38.055	
	400	17.617	-43.249	29.145	-54.907	30.004	
	500	18.217	-41.456	33.144	-58.028	25.367	
	600	18.700	-39.609	36.509	-61.515	22.410	
	700	19.133	-37.717	39.425	-65.315	20.395	
	800	19.544	-35.783	42.007	-69.389	18.959	
	900	19.948	-33.809	44.332	-73.708	17.901	
	1000	20.351	-31.794	46.455	-78.249	17.104	
	1100	20.757	-29.739	48.413	-82.993	16.491	
	1200	21.168	-27.642	50.237	-87.927	16.016	
	1300	21.588	-25.505	51.948	-93.037	15.643	
	1400	22.015	-23.325	53.563	-98.313	15.349	
	1500	22.451	-21.101	55.097	-103.747	15.118	

CERIUM HEXABORIDE

CeB_6

		A	B	C	D	RANGE	REF.
SOL	CP	33.48	7.12	-9.70		298-2463	*
	** MPT= 2463-2563						216

PHASE	T	CP	H	S	G	BT	
SOL	298	24.691	-84.	17.7	-89.277	65.451	*
	300	24.838	-83.954	17.853	-89.310	65.071	
	400	30.265	-81.165	25.839	-91.501	50.001	
	500	33.160	-77.982	32.931	-94.447	41.288	
	600	35.058	-74.566	39.154	-98.058	35.722	
	700	36.484	-70.986	44.670	-102.255	31.930	
	800	37.660	-67.277	49.620	-106.973	29.228	
	900	38.690	-63.459	54.117	-112.163	27.241	
	1000	39.630	-59.542	58.242	-117.784	25.745	
	1100	40.510	-55.535	62.061	-123.802	24.600	
	1200	41.350	-51.441	65.622	-130.188	23.714	
	1300	42.162	-47.265	68.964	-136.919	23.021	
	1400	42.953	-43.009	72.118	-143.974	22.478	
	1500	43.729	-38.675	75.108	-151.337	22.053	
	1600	44.493	-34.264	77.954	-158.991	21.720	
	1700	45.248	-29.777	80.675	-166.924	21.462	
	1800	45.997	-25.215	83.282	-175.122	21.266	
	1900	46.739	-20.578	85.789	-183.577	21.119	
	2000	47.477	-15.867	88.205	-192.277	21.014	
	2100	48.212	-11.082	90.539	-201.215	20.944	
	2200	48.944	-6.225	92.799	-210.383	20.902	
	2300	49.673	-1.294	94.991	-219.773	20.886	
	2400	50.400	3.710	97.120	-229.379	20.891	
	2463	50.857	6.899	98.432	-235.539	20.903	

$CeAl_2$

CERIUM 2-ALUMINIUM

		A	B	C	D	RANGE	REF.
SOL	CP	20.44	1.37	-2.75		298-1738	*
LIQ	CP	23.2				1738-2000	*

PHASE	T	CP	H	S	G	BT	
SOL	298	17.755	-42.	27.1	-50.080	36.714	1*
	300	17.795	-41.967	27.210	-50.130	36.525	
	400	19.269	-40.104	32.559	-53.128	29.032	
	500	20.025	-38.136	36.947	-56.610	24.748	
	600	20.498	-36.108	40.643	-60.494	22.038	
	700	20.838	-34.041	43.830	-64.722	20.210	
	800	21.106	-31.943	46.630	-69.247	18.920	
	900	21.333	-29.821	49.130	-74.038	17.981	
	1000	21.535	-27.677	51.388	-79.065	17.282	
	1100	21.720	-25.515	53.449	-84.309	16.753	
	1200	21.893	-23.334	55.347	-89.750	16.348	
	1300	22.058	-21.136	57.105	-95.373	16.036	
	1400	22.218	-18.922	58.746	-101.167	15.795	
	1500	22.373	-16.693	60.284	-107.119	15.609	
	1600	22.525	-14.448	61.733	-113.221	15.467	
	1700	22.674	-12.188	63.103	-119.463	15.360	
	1738	22.730	-11.325	63.605	-121.871	15.327	
			19.5	11.220			*
LIQ	1738	23.200	8.175	74.825	-121.871	15.327	
	1800	23.200	9.613	75.638	-126.535	15.366	
	1900	23.200	11.933	76.892	-134.162	15.434	
	2000	23.200	14.253	78.082	-141.911	15.509	

CERIUM 4-ALUMINIUM $CeAl_4$

		A	B	C	D	RANGE	REF.
SOL	CP	30.35	7.3	-2.75		298-1278	*
	** TPT= 1278 / MPT= 1518						1

PHASE	T	CP	H	S	G	BT	
SOL	298	29.433	-42.	38.7	-53.538	39.250	1,*
	300	29.484	-41.946	38.882	-53.610	39.060	
	400	31.551	-38.884	47.675	-57.954	31.669	
	500	32.900	-35.658	54.868	-63.092	27.581	
	600	33.966	-32.313	60.963	-68.891	25.097	
	700	34.899	-28.869	66.271	-75.259	23.500	
	800	35.760	-25.336	70.987	-82.126	22.439	
	900	36.580	-21.719	75.247	-89.441	21.722	
	1000	37.375	-18.021	79.143	-97.163	21.238	
	1100	38.153	-14.244	82.741	-105.260	20.916	
	1200	38.919	-10.391	86.094	-113.703	20.711	
	1278	39.511	-7.332	88.563	-120.516	20.612	

* * * * * * * * * *

MONOCERIUM ALUMINATE $CeAlO_3$

		A	B	C	D	RANGE	REF.
SOL	CP	25.63	6.8	-4.5		298-1800	*

PHASE	T	CP	H	S	G	BT	
SOL	298	22.595	-422.2	26.2	-430.012	315.249	222,*
	300	22.670	-422.158	26.340	-430.060	313.341	
	400	25.537	-419.732	33.300	-433.052	236.640	
	500	27.230	-417.088	39.192	-436.684	190.900	
	600	28.460	-414.301	44.270	-440.863	160.606	
	700	29.472	-411.403	48.735	-445.518	139.116	
	800	30.367	-408.411	52.730	-450.595	123.113	
	900	31.194	-405.332	56.355	-456.052	110.759	
	1000	31.980	-402.173	59.683	-461.856	100.952	
	1100	32.738	-398.937	62.767	-467.980	92.992	
	1200	33.477	-395.626	65.647	-474.402	86.412	
	1300	34.204	-392.242	68.355	-481.104	80.892	
	1400	34.920	-388.786	70.916	-488.069	76.201	
	1500	35.630	-385.258	73.350	-495.283	72.172	
	1600	36.334	-381.660	75.672	-502.735	68.680	
	1700	37.034	-377.991	77.896	-510.414	65.627	
	1800	37.731	-374.253	80.032	-518.311	62.940	

Co

COBALT

		A	B	C	D	RANGE	REF.
SOL-A	CP	5.146	3.314	-0.185		298- 700	21
SOL-B	CP	1.062	7.171	6.030		700-1000	21
	CP	66.757	-80.333	-103.640	32.785	1000-1394	21
	** CURIE-PT= 1394						21
	CP	-196.485	91.344	1551.779		1394-1500	21
	CP	-13.422	8.642	224.048		1500-1768	21
LIQ	CP	9.68				1768-3201	21
GAS	CP	6.27				3201-3600	21

PHASE	T	CP	H	S	G	BT	P	
SOL-A	298	5.926	0	7.18	-2.141	1.569		21
	300	5.935	.011	7.217	-2.154	1.569		
	400	6.356	.626	8.984	-2.967	1.621		
	500	6.729	1.281	10.442	-3.941	1.723		
	600	7.083	1.971	11.701	-5.049	1.839		
	700	7.428	2.697	12.819	-6.276	1.960		
			0.108	.154				21
SOL-B	700	7.312	2.805	12.973	-6.276	1.960		
	800	7.741	3.557	13.976	-7.624	2.083		
	900	8.260	4.356	14.917	-9.069	2.203		
	1000	8.836	5.211	15.817	-10.606	2.318		
	1000	8.845	5.211	15.817	-10.606	2.318		
	1100	9.495	6.126	16.689	-12.232	2.431		
	1200	10.371	7.117	17.551	-13.944	2.540		
	1300	11.598	8.212	18.427	-15.742	2.647		
	1394	13.148	9.372	19.287	-17.514	2.746		
	1394	10.704	9.372	19.287	-17.514	2.746		
	1400	10.569	9.436	19.333	-17.630	2.753		
	1500	9.499	10.422	20.014	-19.599	2.856		
	1500	9.499	10.422	20.014	-19.599	2.856		
	1600	9.157	11.353	20.615	-21.631	2.955		
	1700	9.022	12.260	21.165	-23.720	3.050	.002	
	1768	9.025	12.873	21.518	-25.171	3.112	.005	
			3.87	2.189				21
LIQ	1768	9.680	16.743	23.707	-25.171	3.112	.005	
	1800	9.680	17.053	23.881	-25.933	3.149	.009	

COBALT (CONTINUED)

PHASE	T	CP	H	S	G	BT	P	
LIQ	1900	9.680	18.021	24.404	-28.347	3.261	.036	
	2000	9.680	18.989	24.901	-30.813	3.367	.124	
	2100	9.680	19.957	25.373	-33.326	3.469	.382	
	2200	9.680	20.925	25.823	-35.886	3.565	1.059	
	2300	9.680	21.893	26.254	-38.490	3.658	2.678	
	2400	9.680	22.861	26.666	-41.137	3.747	6.246	
	2500	9.680	23.829	27.061	-43.823	3.832	13.575	
	2600	9.680	24.797	27.440	-46.548	3.913	27.725	
	2700	9.680	25.765	27.806	-49.311	3.992	53.575	
	2800	9.680	26.733	28.158	-52.109	4.068	98.546	
	2900	9.680	27.701	28.498	-54.942	4.141	173.442	
	3000	9.680	28.669	28.826	-57.808	4.212	293.392	
	3100	9.680	29.637	29.143	-60.707	4.280	478.867	
	3200	9.680	30.605	29.450	-63.636	4.347	760.	
	3201	9.680	30.615	29.453	-63.666	4.347	760.	
			90.01	28.119				21
GAS	3201	6.270	120.625	57.573	-63.666	4.347		
	3300	6.270	121.245	57.764	-69.375	4.595		
	3400	6.270	121.872	57.951	-75.161	4.832		
	3500	6.270	122.499	58.133	-80.965	5.056		
	3600	6.270	123.126	58.309	-86.787	5.269		

CoF_2 COBALT DIFLUORIDE

		A	B	C	D	RANGE	REF.
SOL	CP	15.3	3.74			298-1523	19,*
	** MPT= 1523						1

PHASE	T	CP	H	S	G	BT	
SOL	298	16.415	-165.4	19.59	-171.241	125.540	19
	300	16.422	-165.370	19.692	-171.277	124.792	
	400	16.796	-163.709	24.467	-173.496	94.806	
	500	17.170	-162.010	28.255	-176.138	77.000	
	600	17.544	-160.275	31.419	-179.126	65.255	
	700	17.918	-158.502	34.151	-182.407	56.958	
	800	18.292	-156.691	36.568	-185.946	50.805	
	900	18.666	-154.843	38.744	-189.713	46.075	
	1000	19.040	-152.958	40.730	-193.688	42.336	
	1100	19.414	-151.035	42.563	-197.854	39.315	
	1200	19.788	-149.075	44.268	-202.197	36.830	
	1300	20.162	-147.078	45.867	-206.704	34.755	
	1400	20.536	-145.043	47.374	-211.367	33.000	
	1500	20.910	-142.970	48.804	-216.176	31.501	
	1523	20.996	-142.489	49.123	-217.303	31.187	

* * * * * * * * * *

CoF_3 COBALT TRIFLUORIDE

		A	B	C	D	RANGE	REF.
SOL	CP	20.55	5.83	-0.61		298-1200	*

PHASE	T	CP	H	S	G	BT	
SOL	298	21.602	-193.8	22.8	-200.598	147.062	19/*
	300	21.621	-193.760	22.934	-200.640	146.186	
	400	22.501	-191.552	29.280	-203.264	111.073	
	500	23.221	-189.265	34.380	-206.455	90.254	
	600	23.879	-186.910	38.673	-210.113	76.544	
	700	24.507	-184.490	42.401	-214.171	66.876	
	800	25.119	-182.009	45.713	-218.580	59.721	
	900	25.722	-179.467	48.707	-223.303	54.233	
	1000	26.319	-176.865	51.448	-228.313	49.904	
	1100	26.913	-174.203	53.984	-233.586	46.415	
	1200	27.504	-171.482	56.351	-239.104	43.553	

COBALT MONOCHLORIDE (GAS)

CoCl (g)

		A	B	C	D	RANGE	REF.
GAS	CP	8.873	0.113	-0.455		298-2000	3A

PHASE	T	CP	H	S	G	BT	
GAS	298	8.395	46.1	58.7	28.599	-20.966	3A
	300	8.401	46.116	58.752	28.490	-20.758	
	400	8.634	46.969	61.205	22.487	-12.288	
	500	8.747	47.839	63.145	16.266	-7.111	
	600	8.814	48.717	64.747	9.869	-3.595	
	700	8.859	49.601	66.109	3.324	-1.038	
	800	8.892	50.488	67.294	-3.347	.914	
	900	8.919	51.379	68.343	-10.130	2.460	
	1000	8.940	52.272	69.284	-17.012	3.718	
	1100	8.960	53.167	70.137	-23.984	4.766	
	1200	8.977	54.064	70.917	-31.037	5.653	
	1300	8.993	54.962	71.636	-38.165	6.417	
	1400	9.008	55.862	72.303	-45.363	7.082	
	1500	9.022	56.764	72.925	-52.624	7.668	
	1600	9.036	57.667	73.508	-59.946	8.189	
	1700	9.049	58.571	74.056	-67.325	8.656	
	1800	9.062	59.477	74.574	-74.757	9.078	
	1900	9.075	60.383	75.064	-82.239	9.461	
	2000	9.088	61.292	75.530	-89.769	9.811	

$CoCl_2$

COBALT DICHLORIDE

		A	B	C	D	RANGE	REF.
SOL	CP	19.615	1.607	-1.184		298-1013	3A
LIQ	CP	23.7				1013-1342	3A
	** BPT= 1342 GAS(COCL2 + CO2CL4)						3A

PHASE	T	CP	H	S	G	BT	P	
SOL	298	18.762	-74.7	26.12	-82.488	60.473		3A
	300	18.782	-74.665	26.236	-82.536	60.136		
	400	19.518	-72.746	31.752	-85.447	46.692		
	500	19.945	-70.772	36.156	-88.850	38.841		
	600	20.250	-68.761	39.821	-92.654	33.754		
	700	20.498	-66.723	42.962	-96.797	30.225		
	800	20.716	-64.663	45.713	-101.233	27.659	.013	
	900	20.915	-62.581	48.165	-105.929	25.727	.426	
	1000	21.104	-60.480	50.378	-110.858	24.231	6.708	
	1013	21.128	-60.205	50.651	-111.515	24.062	9.216	
			10.7	10.563				3A
LIQ	1013	23.700	-49.505	61.214	-111.515	24.062	9.216	
	1100	23.700	-47.443	63.167	-116.927	23.234	39.713	
	1200	23.700	-45.073	65.229	-123.348	22.468	158.721	
	1300	23.700	-42.703	67.126	-129.967	21.852	498.089	
	1342	23.700	-41.708	67.879	-132.802	21.630	760.	

COBALT DICHLORIDE (GAS) $CoCl_2(g)$

		A	B	C	D	RANGE	REF.
GAS	CP	13.961	1.789	-0.174	-0.511	298-2000	3A

PHASE	T	CP	H	S	G	ΘT	
GAS	298	14.253	-22.575	71.3	-43.833	32.135	3A,*
	300	14.258	-22.549	71.388	-43.965	32.033	
	400	14.486	-21.111	75.523	-51.320	28.044	
	500	14.658	-19.653	78.775	-59.041	25.810	
	600	14.802	-18.180	81.460	-67.056	24.429	
	700	14.927	-16.693	83.752	-75.320	23.519	
	800	15.038	-15.195	85.752	-83.797	22.895	
	900	15.136	-13.686	87.529	-92.463	22.456	
	1000	15.222	-12.168	89.129	-101.297	22.141	
	1100	15.296	-10.642	90.583	-110.284	21.914	
	1200	15.360	-9.109	91.917	-119.410	21.750	
	1300	15.413	-7.571	93.148	-128.664	21.633	
	1342	15.432	-6.923	93.639	-132.586	21.595	
	1342	15.432	-6.923	93.639	-132.586	21.595	
	1400	15.455	-6.027	94.292	-138.036	21.551	
	1500	15.487	-4.480	95.360	-147.519	21.496	
	1600	15.508	-2.930	96.360	-157.106	21.463	
	1700	15.519	-1.379	97.300	-166.789	21.445	
	1800	15.520	.173	98.188	-176.564	21.441	
	1900	15.511	1.725	99.027	-186.425	21.447	
	2000	15.491	3.275	99.822	-196.368	21.461	

$Co_2Cl_4(g)$ COBALT DICHLORIDE (DIMERIC GAS)

		A	B	C	D	RANGE	REF.
GAS	CP	31.754	0.384	-1.308		298-2000	3A

PHASE	T	CP	H	S	G	BT	
GAS	298	30.397	-83.8	107.6	-115.881	84.954	3A
	300	30.416	-83.744	107.788	-116.080	84.576	
	400	31.090	-80.664	116.644	-127.321	69.575	
	500	31.423	-77.537	123.621	-139.347	60.917	
	600	31.621	-74.384	129.368	-152.005	55.375	
	700	31.756	-71.214	134.254	-165.192	51.582	
	800	31.857	-68.034	138.501	-178.834	48.862	
	900	31.938	-64.844	142.258	-192.876	46.843	
	1000	32.007	-61.646	145.627	-207.273	45.306	
	1100	32.068	-58.443	148.680	-221.991	44.111	
	1200	32.124	-55.233	151.473	-237.000	43.169	
	1300	32.176	-52.018	154.046	-252.278	42.417	
	1342	32.197	-50.666	155.070	-258.770	42.147	
	1342	32.197	-50.666	155.070	-258.770	42.147	
	1400	32.225	-48.798	156.432	-267.803	41.812	
	1500	32.272	-45.573	158.657	-283.559	41.320	
	1600	32.317	-42.344	160.742	-299.530	40.919	
	1700	32.362	-39.110	162.702	-315.703	40.592	
	1800	32.405	-35.871	164.553	-332.067	40.324	
	1900	32.447	-32.629	166.306	-348.611	40.105	
	2000	32.489	-29.382	167.972	-365.325	39.926	

COBALT TRICHLORIDE (GAS) $CoCl_3$ (g)

		A	B	C	D	RANGE	REF.
GAS	CP	21.458	-1.539	-2.499	0.490	298-2000	3A

PHASE	T	CP	H	S	G	BT	
GAS	298	18.231	-39.1	79.8	-62.892	46.108	3A
	300	18.264	-39.066	79.913	-63.040	45.931	
	400	19.359	-37.177	85.342	-71.313	38.969	
	500	19.811	-35.215	89.717	-80.073	35.005	
	600	20.017	-33.222	93.350	-89.232	32.507	
	700	20.111	-31.215	96.443	-98.726	30.828	
	800	20.150	-29.202	99.132	-108.507	29.647	
	900	20.161	-27.186	101.506	-118.541	28.790	
	1000	20.159	-25.170	103.630	-128.800	28.153	
	1100	20.151	-23.155	105.551	-139.261	27.672	
	1200	20.143	-21.140	107.304	-149.905	27.305	
	1300	20.138	-19.126	108.916	-160.717	27.023	
	1342	20.136	-18.280	109.556	-165.305	26.924	
	1342	20.136	-18.280	109.556	-165.305	26.924	
	1400	20.136	-17.112	110.408	-171.684	26.805	
	1500	20.141	-15.098	111.798	-182.795	26.637	
	1600	20.152	-13.084	113.098	-194.040	26.508	
	1700	20.171	-11.068	114.320	-205.412	26.411	
	1800	20.198	-9.049	115.474	-216.902	26.339	
	1900	20.234	-7.028	116.567	-228.505	26.288	
	2000	20.278	-5.002	117.606	-240.214	26.253	

$CoBr_2$ COBALT DIBROMIDE

		A	B	C	D	RANGE	REF.
SOL	CP	16.3	4.9			298- 951	*
	** MPT= 951						1

PHASE	T	CP	H	S	G	BT	
SOL	298	17.761	-52.8	32.	-62.341	45.703	19/*
	300	17.770	-52.767	32.110	-62.400	45.465	
	400	18.260	-50.966	37.289	-65.881	36.001	
	500	18.750	-49.115	41.416	-69.823	30.524	
	600	19.240	-47.216	44.878	-74.143	27.010	
	700	19.730	-45.267	47.881	-78.784	24.601	
	800	20.220	-43.270	50.547	-83.708	22.871	
	900	20.710	-41.223	52.957	-88.885	21.587	
	951	20.960	-40.161	54.106	-91.615	21.057	

* * * * * * * * * *

CoJ_2 COBALT DIIODIDE

		A	B	C	D	RANGE	REF.
SOL	CP	15.8	7.7			298- 793	*
	** MPT= 793						1

PHASE	T	CP	H	S	G	BT	
SOL	298	18.096	-21.2	36.6	-32.112	23.542	19/*
	300	18.110	-21.167	36.712	-32.180	23.446	
	400	18.880	-19.317	42.027	-36.128	19.742	
	500	19.650	-17.391	46.323	-40.552	17.728	
	600	20.420	-15.387	49.974	-45.371	16.529	
	700	21.190	-13.307	53.179	-50.532	15.779	
	793	21.906	-11.303	55.866	-55.605	15.327	

COBALT HYDROXIDE $Co(OH)_2$

		A	B	C	D	RANGE	REF.
SOL	CP	19.8	11.4			298- 500	*

PHASE	T	CP	H	S	G	BT	
SOL	298	23.199	-129.4	22.3	-136.049	99.740	199
	300	23.220	-129.357	22.444	-136.090	99.155	
	400	24.360	-126.978	29.280	-138.690	75.787	
	500	25.500	-124.485	34.838	-141.904	62.035	

* * * * * * * * * *

COBALT 0.89-SULFIDE $CoS_{0.89}$

		A	B	C	D	RANGE	REF.
SOL	CP	9.62	3.71			298-1000	63

PHASE	T	CP	H	S	G	BT	
SOL	298	10.726	-22.6	12.5	-26.327	19.301	63
	300	10.733	-22.580	12.566	-26.350	19.199	
	400	11.104	-21.488	15.705	-27.770	15.175	
	500	11.475	-20.359	18.223	-29.471	12.883	
	600	11.846	-19.193	20.347	-31.402	11.440	
	700	12.217	-17.990	22.201	-33.531	10.470	
	800	12.588	-16.750	23.857	-35.835	9.791	
	900	12.959	-15.473	25.361	-38.297	9.301	
	1000	13.330	-14.158	26.746	-40.904	8.941	

$CoS_{1.333}$ COBALT 1.333-SULFIDE

		A	B	C	D	RANGE	REF.
SOL	CP	11.42	6.10			298- 900	63
	** DPT= 900						63

PHASE	T	CP	H	S	G	BT	
SOL	298	13.239	-28.6	14.7	-32.983	24.180	63
	300	13.250	-28.575	14.782	-33.010	24.051	
	400	13.860	-27.220	18.677	-34.691	18.957	
	500	14.470	-25.803	21.836	-36.721	16.053	
	600	15.080	-24.326	24.528	-39.043	14.223	
	700	15.690	-22.787	26.898	-41.616	12.995	
	800	16.300	-21.188	29.033	-44.414	12.135	
	900	16.910	-19.527	30.988	-47.417	11.516	

* * * * * * * * * *

CoS_2 COBALT DISULFIDE

		A	B	C	D	RANGE	REF.
SOL	CP	14.50	6.05			298-1000	63

PHASE	T	CP	H	S	G	BT	
SOL	298	16.304	-36.6	16.5	-41.519	30.439	63
	300	16.315	-36.570	16.601	-41.550	30.273	
	400	16.920	-34.908	21.377	-43.459	23.748	
	500	17.525	-33.186	25.218	-45.795	20.020	
	600	18.130	-31.403	28.467	-48.483	17.662	
	700	18.735	-29.560	31.307	-51.475	16.073	
	800	19.340	-27.656	33.848	-54.734	14.955	
	900	19.945	-25.692	36.161	-58.237	14.144	
	1000	20.550	-23.667	38.293	-61.961	13.543	

COBALT(2) SULFATE

$CoSO_4$

		A	B	C	D	RANGE	REF.
SOL	CP	30.09	9.91			298-1000	2
	** MPT= 1262						47

PHASE	T	CP	H	S	G	BT	
SOL	298	33.045	-212.	27.1	-220.080	161.345	66
	300	33.063	-211.939	27.304	-220.130	160.386	
	400	34.054	-208.583	36.952	-223.364	122.057	
	500	35.045	-205.128	44.657	-227.457	99.435	
	600	36.036	-201.574	51.134	-232.255	84.610	
	700	37.027	-197.921	56.764	-237.655	74.209	
	800	38.018	-194.169	61.773	-243.587	66.554	
	900	39.009	-190.317	66.308	-249.994	60.715	
	1000	40.000	-186.367	70.469	-256.836	56.139	

* * * * * * * * * *

COBALT SELENITE

$CoSeO_3$

		A	B	C	D	RANGE	REF.
SOL	CP	19.1	14.3			298- 932	*
LIQ	CP	34.5				932-1100	*

PHASE	T	CP	H	S	G	BT	
SOL	298	23.364	-139.1	27.2	-147.210	107.922	62
	300	23.390	-139.057	27.345	-147.260	107.293	
	400	24.820	-136.646	34.269	-150.354	82.161	
	500	26.250	-134.093	39.961	-154.073	67.355	
	600	27.680	-131.396	44.874	-158.320	57.676	
	700	29.110	-128.557	49.248	-163.030	50.907	
	800	30.540	-125.574	53.228	-168.157	45.945	
	900	31.970	-122.449	56.908	-173.666	42.178	
	932	32.428	-121.418	58.033	-175.505	41.161	
			3.9	4.185			62
LIQ	932	34.500	-117.518	62.218	-175.505	41.161	
	1000	34.500	-115.172	64.647	-179.820	39.305	
	1100	34.500	-111.722	67.935	-186.451	37.049	

Co_3N TRICOBALT NITRIDE

		A	B	C	D	RANGE	REF.
SOL	CP	17.5	15.0			298- 600	*

** METASTABLE

PHASE	T	CP	H	S	G	BT	
SOL	298	21.972	2.	23.6	-5.036	3.692	*
	300	22.000	2.041	23.736	-5.080	3.701	
	400	23.500	4.316	30.270	-7.793	4.258	
	500	25.000	6.741	35.675	-11.097	4.851	
	600	26.500	9.316	40.366	-14.904	5.429	

* * * * * * * * * *

Co_2P DICOBALT PHOSPHIDE

		A	B	C	D	RANGE	REF.
SOL	CP	13.85	5.5			298-1600	*

PHASE	T	CP	H	S	G	BT	
SOL	298	15.490	-44.8	18.5	-50.316	36.887	64,*
	300	15.500	-44.771	18.596	-50.350	36.685	
	400	16.050	-43.194	23.130	-52.446	28.659	
	500	16.600	-41.561	26.771	-54.947	24.020	
	600	17.150	-39.874	29.846	-57.781	21.050	
	700	17.700	-38.131	32.531	-60.903	19.017	
	800	18.250	-36.334	34.930	-64.278	17.562	
	900	18.800	-34.481	37.112	-67.882	16.486	
	1000	19.350	-32.574	39.121	-71.695	15.671	
	1100	19.900	-30.611	40.991	-75.701	15.042	
	1200	20.450	-28.594	42.746	-79.889	14.552	
	1300	21.000	-26.521	44.405	-84.247	14.165	
	1400	21.550	-24.394	45.981	-88.767	13.859	
	1500	22.100	-22.211	47.487	-93.441	13.616	
	1600	22.650	-19.974	48.930	-98.263	13.424	

COBALT MONOPHOSPHICE

CoP

		A	B	C	D	RANGE	REF.
SOL	CP	9.78	3.5			298-1500	*

PHASE	T	CP	H	S	G	BT	
SOL	298	10.824	-30.	12.	-33.578	24.617	64,*
	300	10.830	-29.980	12.067	-33.600	24.481	
	400	11.180	-28.879	15.231	-34.972	19.110	
	500	11.530	-27.744	17.763	-36.625	16.011	
	600	11.880	-26.573	19.896	-38.511	14.030	
	700	12.230	-25.368	21.754	-40.595	12.676	
	800	12.580	-24.127	23.409	-42.855	11.709	
	900	12.930	-22.852	24.911	-45.272	10.995	
	1000	13.280	-21.541	26.292	-47.833	10.455	
	1100	13.630	-20.196	27.574	-50.527	10.040	
	1200	13.980	-18.815	28.775	-53.345	9.717	
	1300	14.330	-17.400	29.908	-56.280	9.463	
	1400	14.680	-15.949	30.983	-59.325	9.262	
	1500	15.030	-14.464	32.007	-62.475	9.104	

* * * * * * * * * *

COBALT TRIPHOSPHIDE

CoP_3

		A	B	C	D	RANGE	REF.
SOL	CP	22.4	6.0			298-1400	*

PHASE	T	CP	H	S	G	BT	
SOL	298	24.189	-48.9	23.5	-55.907	40.986	64,*
	300	24.200	-48.855	23.650	-55.950	40.765	
	400	24.800	-46.405	30.694	-58.683	32.067	
	500	25.400	-43.895	36.292	-62.041	27.122	
	600	26.000	-41.325	40.976	-65.911	24.011	
	700	26.600	-38.695	45.029	-70.216	21.925	
	800	27.200	-36.005	48.620	-74.901	20.465	
	900	27.800	-33.255	51.859	-79.928	19.412	
	1000	28.400	-30.445	54.819	-85.264	18.637	
	1100	29.000	-27.575	57.554	-90.884	18.059	
	1200	29.600	-24.645	60.103	-96.768	17.626	
	1300	30.200	-21.655	62.496	-102.900	17.301	
	1400	30.800	-18.605	64.756	-109.263	17.059	

$CoSb_{(0.98)}$ COBALT 0.98-ANTIMONIDE

		A	B	C	D	RANGE	REF.
SOL	CP	10.10	6.2			298-1475	*
	** COSB(0.84-0.98), MPT= 1475						21

PHASE	T	CP	H	S	G	BT	
SOL	298	11.949	-10.	16.9	-15.039	11.025	21
	300	11.960	-9.978	16.974	-15.070	10.980	
	400	12.580	-8.751	20.500	-16.951	9.263	
	500	13.200	-7.462	23.373	-19.149	8.371	
	600	13.820	-6.111	25.835	-21.612	7.873	
	700	14.440	-4.698	28.012	-24.306	7.590	
	800	15.060	-3.223	29.980	-27.207	7.434	
	900	15.680	-1.686	31.790	-30.297	7.358	
	1000	16.300	-.087	33.474	-33.561	7.336	
	1100	16.920	1.574	35.057	-36.988	7.350	
	1200	17.540	3.297	36.556	-40.570	7.390	
	1300	18.160	5.082	37.984	-44.297	7.448	
	1400	18.780	6.929	39.352	-48.164	7.520	
	1475	19.245	8.355	40.345	-51.153	7.580	

* * * * * * * * * *

$CoSb_2$ COBALT DIANTIMONIDE

		A	B	C	D	RANGE	REF.
SOL	CP	15.50	7.93			298-1200	*
	** DPT= 1192 (PERITEC.)						21

PHASE	T	CP	H	S	G	BT	
SOL	298	17.864	-12.	28.75	-20.572	15.082	21
	300	17.879	-11.967	28.861	-20.625	15.027	
	400	18.672	-10.139	34.113	-23.784	12.997	
	500	19.465	-8.233	38.364	-27.415	11.985	
	600	20.258	-6.246	41.983	-31.436	11.452	
	700	21.051	-4.181	45.166	-35.797	11.178	
	800	21.844	-2.036	48.028	-40.459	11.054	
	900	22.637	.188	50.647	-45.394	11.025	
	1000	23.430	2.491	53.073	-50.582	11.056	
	1100	24.223	4.874	55.343	-56.004	11.128	
	1200	25.016	7.336	57.485	-61.646	11.229	

COBALT TRIANTIMONICE $CoSb_3$

		A	B	C	D	RANGE	REF.
SOL	CP	21.0	9.65			298-1132	*
	** DPT= 1132 (PERITEC.)						21

PHASE	T	CP	H	S	G	BT	
SOL	298	23.877	-15.	38.7	-26.538	19.456	21
	300	23.895	-14.956	38.848	-26.610	19.388	
	400	24.860	-12.518	45.854	-30.860	16.863	
	500	25.825	-9.984	51.505	-35.736	15.622	
	600	26.790	-7.353	56.299	-41.132	14.984	
	700	27.755	-4.626	60.501	-46.977	14.669	
	800	28.720	-1.802	64.270	-53.218	14.540	
	900	29.685	1.118	67.709	-59.820	14.528	
	1000	30.650	4.135	70.886	-66.751	14.590	
	1100	31.615	7.248	73.853	-73.990	14.702	
	1132	31.924	8.265	74.764	-76.368	14.746	

* * * * * * * * * *

COBALT CARBONATE $CoCO_3$

		A	B	C	D	RANGE	REF.
SOL	CP	22.0	9.3	-4.3		298- 800	*
	** DPT= 443, COCO3 = COO + CO2						

PHASE	T	CP	H	S	G	BT	
SOL	298	19.936	-170.	21.2	-176.321	129.264	166/130
	300	20.012	-169.963	21.324	-176.360	128.496	
	400	23.032	-167.796	27.537	-178.811	97.711	
	500	24.930	-165.392	32.893	-181.839	79.492	
	600	26.386	-162.824	37.571	-185.367	67.529	
	700	27.632	-160.122	41.734	-189.336	59.121	
	800	28.768	-157.301	45.499	-193.700	52.924	

$2CoO \cdot SiO_2$ COBALT ORTHOSILICATE

		A	B	C	D	RANGE	REF.
SOL	CP	37.62	5.27	-6.38		298-1690	5
LIQ	CP	58.0				1690-2000	5

PHASE	T	CP	H	S	G	BT	
SOL	298	32.014	-336.2	37.9	-347.500	254.759	5
	300	32.112	-336.141	38.098	-347.570	253.239	
	400	35.740	-332.726	47.897	-351.885	192.287	
	500	37.703	-329.046	56.101	-357.096	156.108	
	600	39.010	-325.207	63.097	-363.065	132.264	
	700	40.007	-321.254	69.188	-369.686	115.437	
	800	40.839	-317.211	74.586	-376.879	102.973	
	900	41.575	-313.089	79.440	-384.585	93.403	
	1000	42.252	-308.897	83.855	-392.753	85.848	
	1100	42.890	-304.640	87.913	-401.344	79.750	
	1200	43.501	-300.320	91.671	-410.325	74.741	
	1300	44.093	-295.941	95.176	-419.670	70.562	
	1400	44.672	-291.502	98.465	-429.353	67.034	
	1500	45.241	-287.006	101.567	-439.357	64.023	
	1600	45.803	-282.454	104.504	-449.661	61.429	
	1690	46.303	-278.309	107.025	-459.181	59.389	
			24.	14.201			5
LIQ	1690	58.000	-254.309	121.226	-459.181	59.389	
	1700	58.000	-253.729	121.568	-460.395	59.196	
	1800	58.000	-247.929	124.883	-472.719	57.404	
	1900	58.000	-242.129	128.019	-485.366	55.837	
	2000	58.000	-236.329	130.994	-498.318	54.461	

COBALT TITANATE

$CoO \cdot TiO_2$

		A	B	C	D	RANGE	REF.
SOL	CP	29.51	2.32	-3.95		298-1700	*

PHASE	T	CP	H	S	G	BT	
SOL	298	25.758	-291.5	23.15	-298.402	218.764	5
	300	25.817	-291.452	23.310	-298.445	217.446	
	400	27.969	-288.749	31.071	-301.178	164.578	
	500	29.090	-285.891	37.444	-304.613	133.164	
	600	29.805	-282.944	42.814	-308.633	112.435	
	700	30.328	-279.937	47.450	-313.152	97.783	
	800	30.749	-276.882	51.528	-318.105	86.914	
	900	31.110	-273.789	55.171	-323.443	78.553	
	1000	31.435	-270.661	58.466	-329.127	71.940	
	1100	31.736	-267.503	61.476	-335.126	66.592	
	1200	32.020	-264.315	64.250	-341.415	62.188	
	1300	32.292	-261.099	66.824	-347.970	58.507	
	1400	32.556	-257.857	69.226	-354.774	55.390	
	1500	32.814	-254.588	71.481	-361.810	52.723	
	1600	33.068	-251.294	73.607	-369.066	50.419	
	1700	33.317	-247.975	75.619	-376.528	48.412	

Co_2B DICOBALT BORIDE

		A	B	C	D	RANGE	REF.
SOL	CP	16.34	5.34	-3.41		298-1500	*

PHASE	T	CP	H	S	G	BT	
SOL	298	14.096	-30.	14.3	-34.264	25.119	*
	300	14.153	-29.974	14.387	-34.290	24.984	
	400	16.345	-28.437	18.793	-35.954	19.647	
	500	17.646	-26.733	22.590	-38.028	16.624	
	600	18.597	-24.919	25.895	-40.456	14.738	
	700	19.382	-23.019	28.822	-43.195	13.488	
	800	20.079	-21.046	31.456	-46.211	12.626	
	900	20.725	-19.005	33.859	-49.478	12.017	
	1000	21.339	-16.902	36.074	-52.976	11.579	
	1100	21.932	-14.738	38.136	-56.688	11.264	
	1200	22.511	-12.516	40.069	-60.599	11.038	
	1300	23.080	-10.236	41.894	-64.698	10.878	
	1400	23.642	-7.900	43.625	-68.975	10.769	
	1500	24.198	-5.508	45.275	-73.420	10.699	

* * * * * * * * * *

CoB COBALT MONOBORIDE

		A	B	C	D	RANGE	REF.
SOL	CP	10.26	3.5	-2.69		298-1300	*
	** MPT=	1733					62

PHASE	T	CP	H	S	G	BT	
SOL	298	8.277	-22.5	7.3	-24.676	18.091	*
	300	8.321	-22.485	7.351	-24.690	17.989	
	400	9.979	-21.560	9.999	-25.560	13.967	
	500	10.934	-20.511	12.336	-26.679	11.663	
	600	11.613	-19.382	14.392	-28.018	10.207	
	700	12.161	-18.193	16.225	-29.550	9.227	
	800	12.640	-16.953	17.880	-31.257	8.540	
	900	13.078	-15.666	19.395	-33.122	8.044	
	1000	13.491	-14.338	20.794	-35.132	7.679	
	1100	13.888	-12.969	22.099	-37.277	7.407	
	1200	14.273	-11.561	23.324	-39.549	7.204	
	1300	14.651	-10.114	24.481	-41.940	7.052	

COBALT TIN

CoSn

		A	B	C	D	RANGE	REF.
SOL-D	CP	10.9	4.5			298-1209	*
	** DPT= 1209 (PERITEC.)						21

PHASE	T	CP	H	S	G	BT	
SOL-D	298	12.242	-7.	17.1	-12.098	8.870	21/*
	300	12.250	-6.977	17.176	-12.130	8.838	
	400	12.700	-5.730	20.761	-14.034	7.669	
	500	13.150	-4.437	23.644	-16.259	7.108	
	600	13.600	-3.100	26.081	-18.748	6.830	
	700	14.050	-1.717	28.211	-21.465	6.703	
	800	14.500	-.290	30.117	-24.383	6.662	
	900	14.950	1.183	31.851	-27.483	6.675	
	1000	15.400	2.700	33.449	-30.749	6.721	
	1100	15.850	4.263	34.938	-34.169	6.790	
	1200	16.300	5.870	36.336	-37.733	6.873	
	1209	16.340	6.017	36.458	-38.061	6.881	

CoAl

COBALT ALUMINIUM

		A	B	C	D	RANGE	REF.
SOL	CP	10.2	3.0			298-1918	*
LIQ	CP	17.0				1918-2100	*

PHASE	T	CP	H	S	G	BT	
SOL	298	11.094	-26.4	13.	-30.276	22.196	64/*
	300	11.100	-26.379	13.069	-30.300	22.077	
	400	11.400	-25.254	16.303	-31.776	17.364	
	500	11.700	-24.099	18.879	-33.539	14.662	
	600	12.000	-22.914	21.039	-35.538	12.946	
	700	12.300	-21.699	22.911	-37.737	11.784	
	800	12.600	-20.454	24.573	-40.113	10.960	
	900	12.900	-19.179	26.074	-42.647	10.357	
	1000	13.200	-17.874	27.449	-45.324	9.907	
	1100	13.500	-16.539	28.721	-48.133	9.564	
	1200	13.800	-15.174	29.909	-51.065	9.301	
	1300	14.100	-13.779	31.025	-54.112	9.098	
	1400	14.400	-12.354	32.081	-57.268	8.941	
	1500	14.700	-10.899	33.085	-60.527	8.820	
	1600	15.000	-9.414	34.043	-63.884	8.727	
	1700	15.300	-7.899	34.962	-67.334	8.658	
	1800	15.600	-6.354	35.845	-70.875	8.607	
	1900	15.900	-4.779	36.696	-74.502	8.571	
	1918	15.954	-4.493	36.846	-75.164	8.566	
			15.	7.821			*
LIQ	1918	17.000	10.507	44.667	-75.164	8.566	
	2000	17.000	11.901	45.379	-78.856	8.618	
	2100	17.000	13.601	46.208	-83.436	8.684	

2-COBALT 5-ALUMINIUM — Co_2Al_5

		A	B	C	D	RANGE	REF.
SOL	CP	35.2	14.0			298-1443	*
	** DPT= 1443 (PERITEC.)						21

PHASE	T	CP	H	S	G	BT	
SOL	298	39.374	-70.	46.3	-83.804	61.439	64/*
	300	39.400	-69.927	46.544	-83.890	61.122	
	400	40.800	-65.917	58.070	-89.145	48.713	
	500	42.200	-61.767	67.325	-95.429	41.718	
	600	43.600	-57.477	75.142	-102.563	37.363	
	700	45.000	-53.047	81.969	-110.425	34.481	
	800	46.400	-48.477	88.069	-118.932	32.495	
	900	47.800	-43.767	93.615	-128.020	31.092	
	1000	49.200	-38.917	98.723	-137.641	30.085	
	1100	50.600	-33.927	103.478	-147.753	29.360	
	1200	52.000	-28.797	107.941	-158.327	28.839	
	1300	53.400	-23.527	112.159	-169.333	28.471	
	1400	54.800	-18.117	116.167	-180.751	28.220	
	1443	55.402	-15.748	117.834	-185.783	28.142	

* * * * * * * * * *

COBALT 3-ALUMINIUM — $CoAl_3$

		A	B	C	D	RANGE	REF.
SOL	CP	20.05	8.5			298-1200	*

PHASE	T	CP	H	S	G	BT	
SOL	298	22.584	-36.4	26.5	-44.301	32.478	1/*
	300	22.600	-36.358	26.640	-44.350	32.313	
	400	23.450	-34.056	33.258	-47.359	25.879	
	500	24.300	-31.668	38.582	-50.959	22.277	
	600	25.150	-29.196	43.087	-55.048	20.054	
	700	26.000	-26.638	47.028	-59.558	18.597	
	800	26.850	-23.996	50.555	-64.440	17.607	
	900	27.700	-21.268	53.767	-69.658	16.918	
	1000	28.550	-18.456	56.729	-75.185	16.434	
	1100	29.400	-15.558	59.490	-80.998	16.095	
	1200	30.250	-12.576	62.085	-87.078	15.861	

Co_2Al_9 2-COBALT 9-ALUMINIUM

		A	B	C	D	RANGE	REF.
SOL	CP	54.9	25.8			298-1216	*
	** DPT= 1216 (PERITEC.)						21

PHASE	T	CP	H	S	G	BT	
SOL	298	62.592	-79.2	73.1	-100.995	74.041	1/*
	300	62.640	-79.084	73.487	-101.130	73.683	
	400	65.220	-72.691	91.861	-109.436	59.801	
	500	67.800	-66.040	106.692	-119.386	52.191	
	600	70.380	-59.131	119.281	-130.700	47.614	
	700	72.960	-51.964	130.324	-143.191	44.712	
	800	75.540	-44.539	140.235	-156.727	42.822	
	900	78.120	-36.856	149.281	-171.209	41.581	
	1000	80.700	-28.915	157.645	-186.561	40.778	
	1100	83.280	-20.716	165.458	-202.720	40.282	
	1200	85.860	-12.259	172.815	-219.637	40.007	
	1216	86.273	-10.882	173.955	-222.411	39.979	

COBALT TUNGSTATE

$CoO \cdot WO_3$

		A	B	C	D	RANGE	REF.
SOL-A	CP	27.6	11.59			298- 986	163
SOL-B	CP	29.25	10.02			986-1500	163

PHASE	T	CP	H	S	G	BT	
SOL-A	298	31.056	-273.36	30.2	-282.364	207.006	5/*
	300	31.077	-273.303	30.392	-282.420	205.771	
	400	32.236	-270.137	39.491	-285.933	156.248	
	500	33.395	-266.855	46.809	-290.260	126.890	
	600	34.554	-263.458	53.000	-295.258	107.562	
	700	35.713	-259.945	58.414	-300.834	93.937	
	800	36.872	-256.315	63.258	-306.922	83.858	
	900	38.031	-252.570	67.668	-313.471	76.131	
	986	39.028	-249.257	71.183	-319.443	70.815	
			0.445	.451			163
SOL-B	986	39.130	-248.812	71.635	-319.443	70.815	
	1000	39.270	-248.263	72.187	-320.450	70.044	
	1100	40.272	-244.286	75.977	-327.861	65.149	
	1200	41.274	-240.208	79.524	-335.638	61.136	
	1300	42.276	-236.031	82.868	-343.759	57.799	
	1400	43.278	-231.753	86.037	-352.205	54.989	
	1500	44.280	-227.375	89.057	-360.961	52.599	

$CoO \cdot Fe_2O_3$ COBALT 2-IRON 4-OXICE

		A	B	C	D	RANGE	REF.
SOL	CP	39.709	14.961	-5.97		298- 740	204
	CP	6248.59	5556.8	1971.43		740- 784	204
	CP	157.189	-129.577			784- 855	204
	CP	46.4				855-1000	204*

PHASE	T	CP	H	S	G	BT	
SOL	298	37.454	-261.2	34.1	-271.367	198.944	5*
	300	37.564	-261.131	34.332	-271.430	197.763	
	400	41.962	-257.134	45.801	-275.454	150.521	
	500	44.801	-252.788	55.486	-280.531	122.636	
	600	47.027	-248.193	63.857	-286.507	104.374	
	700	48.963	-243.392	71.254	-293.270	91.575	
	740	49.690	-241.419	73.995	-296.175	87.483	
	740	49.605	-241.419	73.995	-296.175	87.483	
	784	55.605	-239.256	76.833	-299.493	83.499	
	784	55.601	-239.256	76.833	-299.493	83.499	
	800	53.527	-238.383	77.936	-300.731	82.167	
	855	46.401	-235.635	81.260	-305.112	78.001	
	855	46.400	-235.635	81.260	-305.112	78.001	
	900	46.400	-233.547	83.640	-308.823	75.003	
	1000	46.400	-228.907	88.529	-317.436	69.385	

CHROMIUM

		A	B	C	D	RANGE	REF.
SOL	CP	4.234	5.489	-0.090	-2.159	298-1000	21
	CP	4.318	3.712	-3.991		1000-2130	21
LIQ	CP	9.4				2130-2945	21
GAS	CP	7.358				2945-3100	21

PHASE	T	CP	H	S	G	BT	P	
SOL	298	5.577	0	5.65	-1.685	1.235		21
	300	5.586	.010	5.685	-1.695	1.235		
	400	6.028	.592	7.354	-2.350	1.284		
	500	6.403	1.214	8.740	-3.157	1.380		
	600	6.725	1.871	9.937	-4.092	1.491		
	700	7.000	2.557	10.995	-5.139	1.605		
	800	7.229	3.269	11.945	-6.287	1.718		
	900	7.414	4.002	12.808	-7.525	1.828		
	1000	7.555	4.750	13.597	-8.846	1.934		
	1000	7.631	4.750	13.597	-8.846	1.934		
	1100	8.071	5.536	14.345	-10.244	2.035		
	1200	8.495	6.364	15.065	-11.714	2.134		
	1300	8.907	7.234	15.762	-13.256	2.229		
	1400	9.311	8.145	16.437	-14.866	2.321		
	1500	9.709	9.096	17.093	-16.542	2.411		
	1600	10.101	10.087	17.732	-18.284	2.498	.003	
	1700	10.490	11.116	18.356	-20.088	2.583	.015	
	1800	10.876	12.185	18.966	-21.955	2.666	.067	
	1900	11.260	13.292	19.565	-23.881	2.747	.253	
	2000	11.642	14.437	20.152	-25.867	2.827	.829	
	2100	12.023	15.620	20.729	-27.911	2.905	2.409	
	2130	12.137	15.982	20.901	-28.536	2.928	3.249	
			4.047	1.900				21
LIQ	2130	9.400	20.029	22.801	-28.536	2.928	3.249	
	2200	9.400	20.687	23.104	-30.142	2.995	6.127	
	2300	9.400	21.627	23.522	-32.474	3.086	14.144	
	2400	9.400	22.567	23.922	-34.846	3.174	30.365	
	2500	9.400	23.507	24.306	-37.258	3.258	61.176	
	2600	9.400	24.447	24.675	-39.707	3.338	116.532	
	2700	9.400	25.387	25.030	-42.192	3.416	211.236	
	2800	9.400	26.327	25.371	-44.712	3.490	366.374	
	2900	9.400	27.267	25.701	-47.266	3.563	610.895	
	2945	9.400	27.690	25.846	-48.426	3.594	760.	
			82.28	27.939				21
GAS	2945	7.358	109.970	53.785	-48.426	3.594		
	3000	7.358	110.375	53.921	-51.388	3.744		
	3100	7.358	111.111	54.162	-56.792	4.004		

CrF_2 CHROMIUM DIFLUORIDE

		A	B	C	D	RANGE	REF.
SOL	CP	14.1	5.0			298-1373	*
	** MPT=	1373					1

PHASE	T	CP	H	S	G	BT	
SOL	298	15.591	-186.3	20.	-192.263	140.952	19/*
	300	15.600	-186.271	20.096	-192.300	140.109	
	400	16.100	-184.686	24.653	-194.547	106.310	
	500	16.600	-183.051	28.299	-197.201	86.208	
	600	17.100	-181.366	31.370	-200.188	72.928	
	700	17.600	-179.631	34.043	-203.462	63.532	
	800	18.100	-177.846	36.426	-206.987	56.554	
	900	18.600	-176.011	38.587	-210.739	51.181	
	1000	19.100	-174.126	40.572	-214.699	46.929	
	1100	19.600	-172.191	42.416	-218.849	43.487	
	1200	20.100	-170.206	44.143	-223.178	40.652	
	1300	20.600	-168.171	45.772	-227.675	38.281	
	1373	20.965	-166.654	46.907	-231.058	36.784	

CHROMIUM TRIFLUORIDE CrF_3

		A	B	C	D	RANGE	REF.
SOL	CP	16.4	8.2			298-1373	*
	** MPT=	1373					1

PHASE	T	CP	H	S	G	BT	
SOL	298	18.845	-266.	22.45	-272.693	199.917	1
	300	18.860	-265.965	22.567	-272.735	198.714	
	400	19.680	-264.038	28.105	-275.280	150.426	
	500	20.500	-262.029	32.584	-278.321	121.670	
	600	21.320	-259.938	36.394	-281.775	102.650	
	700	22.140	-257.765	39.742	-285.585	89.176	
	800	22.960	-255.510	42.752	-289.712	79.156	
	900	23.780	-253.173	45.504	-294.127	71.433	
	1000	24.600	-250.754	48.052	-298.806	65.313	
	1100	25.420	-248.253	50.435	-303.731	60.354	
	1200	26.240	-245.670	52.682	-308.888	56.264	
	1300	27.060	-243.005	54.815	-314.264	52.840	
	1373	27.659	-241.008	56.309	-318.320	50.676	

* * * * * * * * * *

CHROMIUM TETRAFLUORIDE CrF_4

		A	B	C	D	RANGE	REF.
SOL	CP	19.2	16.0			298- 600	*

PHASE	T	CP	H	S	G	BT	
SOL	298	23.970	-286.5	30.8	-295.683	216.771	1/*
	300	24.000	-286.456	30.948	-295.740	215.476	
	400	25.600	-283.976	38.072	-299.204	163.500	
	500	27.200	-281.336	43.956	-303.314	132.596	
	600	28.800	-278.536	49.057	-307.970	112.193	

$CrCl_2$ CHROMIUM DICHLORIDE

		A	B	C	D	RANGE	REF.
SOL	CP	15.25	5.96			298-1088	19*
LIQ	CP	24.0				1088-1573	*
	** BPT=	1573					1

PHASE	T	CP	H	S	G	BT	
SOL	298	17.027	-94.5	27.56	-102.717	75.304	19
	300	17.038	-94.468	27.665	-102.768	74.877	
	400	17.634	-92.735	32.649	-105.794	57.811	
	500	18.230	-90.942	36.647	-109.265	47.766	
	600	18.826	-89.089	40.024	-113.103	41.203	
	700	19.422	-87.176	42.971	-117.256	36.614	
	800	20.018	-85.204	45.603	-121.687	33.248	
	900	20.614	-83.173	47.995	-126.369	30.691	
	1000	21.210	-81.082	50.198	-131.280	28.695	
	1088	21.734	-79.192	52.009	-135.778	27.278	
			7.7	7.077			1
LIQ	1088	24.000	-71.492	59.086	-135.778	27.278	
	1100	24.000	-71.204	59.349	-136.488	27.121	
	1200	24.000	-68.804	61.437	-142.529	25.962	
	1300	24.000	-66.404	63.358	-148.770	25.014	
	1400	24.000	-64.004	65.137	-155.196	24.230	
	1500	24.000	-61.604	66.793	-161.793	23.576	
	1573	24.000	-59.852	67.933	-166.711	23.166	

CHROMIUM TRICHLCRICE — $CrCl_3$

		A	B	C	D	RANGE	REF.
SOL	CP	19.0	9.85			298-1218	19,*
	** SPT= 1218, L= 56.8, MPT= 1150						1

PHASE	T	CP	H	S	G	BT	
SOL	298	21.937	-133.	29.4	-141.766	103.931	19/1
	300	21.955	-132.959	29.536	-141.820	103.330	
	400	22.940	-130.715	35.987	-145.109	79.295	
	500	23.925	-128.371	41.211	-148.977	65.127	
	600	24.910	-125.930	45.661	-153.326	55.856	
	700	25.895	-123.389	49.574	-158.091	49.365	
	800	26.880	-120.751	53.097	-163.228	44.598	
	900	27.865	-118.013	56.319	-168.701	40.972	
	1000	28.850	-115.178	59.306	-174.484	38.139	
	1100	29.835	-112.243	62.102	-180.556	35.878	
	1200	30.820	-109.211	64.740	-186.899	34.044	
	1218	30.997	-108.654	65.201	-188.069	33.750	

* * * * * * * * * *

CHROMIUM DIBROMIDE — $CrBr_2$

		A	B	C	D	RANGE	REF.
SOL	CP	15.75	5.3			298-1115	*
	** MPT= 1115						1

PHASE	T	CP	H	S	G	BT	
SOL	298	17.330	-72.2	32.2	-81.800	59.969	19/*
	300	17.340	-72.168	32.307	-81.860	59.643	
	400	17.870	-70.407	37.368	-85.355	46.642	
	500	18.400	-68.594	41.413	-89.300	39.038	
	600	18.930	-66.727	44.814	-93.616	34.104	
	700	19.460	-64.808	47.772	-98.248	30.679	
	800	19.990	-62.835	50.405	-103.160	28.186	
	900	20.520	-60.810	52.790	-108.321	26.308	
	1000	21.050	-58.731	54.980	-113.711	24.855	
	1100	21.580	-56.600	57.011	-119.312	23.708	
	1115	21.659	-56.276	57.304	-120.169	23.557	

CrJ_2 CHROMIUM DIIODIDE

		A	B	C	D	RANGE	REF.
SOL	CP	16.0	5.4			298-1066	*
	** MPT=	1066					1

PHASE	T	CP	H	S	G	BT	
SOL	298	17.610	-37.8	40.4	-49.845	36.542	1/*
	300	17.620	-37.767	40.509	-49.920	36.372	
	400	18.160	-35.978	45.652	-54.239	29.639	
	500	18.700	-34.135	49.762	-59.016	25.800	
	600	19.240	-32.238	53.219	-64.170	23.377	
	700	19.780	-30.287	56.226	-69.645	21.747	
	800	20.320	-28.282	58.902	-75.404	20.602	
	900	20.860	-26.223	61.327	-81.417	19.774	
	1000	21.400	-24.110	63.553	-87.663	19.161	
	1066	21.756	-22.686	64.932	-91.903	18.844	

* * * * * * * * * *

CrJ_3 CHROMIUM TRIIODIDE

		A	B	C	D	RANGE	REF.
SOL	CP	25.2	5.0			298- 900	*

PHASE	T	CP	H	S	G	BT	
SOL	298	26.691	-49.	47.7	-63.222	46.349	19/*
	300	26.700	-48.951	47.865	-63.310	46.128	
	400	27.200	-46.256	55.615	-68.502	37.433	
	500	27.700	-43.511	61.738	-74.380	32.516	
	600	28.200	-40.716	66.832	-80.815	29.441	
	700	28.700	-37.871	71.217	-87.723	27.392	
	800	29.200	-34.976	75.082	-95.041	25.968	
	900	29.700	-32.031	78.550	-102.726	24.949	

CHROMIUM DIOXIDE DICHLORIDE (GAS) CrO_2Cl_2 (g)

		A	B	C	D	RANGE	REF.
GAS	CP	25.48		-4.75		298-1500	19
	** MPT= 178, BPT= 390						1

PHASE	T	CP	H	S	G	BT	
GAS	298	20.137	-128.6	78.8	-152.094	111.503	19
	300	20.202	-128.563	78.925	-152.240	110.922	
	400	22.511	-126.411	85.100	-160.451	87.678	
	500	23.580	-124.100	90.252	-169.226	73.979	
	600	24.161	-121.710	94.607	-178.475	65.018	
	700	24.511	-119.275	98.360	-188.127	58.744	
	800	24.738	-116.812	101.648	-198.131	54.134	
	900	24.894	-114.330	104.572	-208.445	50.624	
	1000	25.005	-111.835	107.201	-219.036	47.877	
	1100	25.087	-109.330	109.588	-229.877	45.678	
	1200	25.150	-106.818	111.774	-240.946	43.888	
	1300	25.199	-104.301	113.789	-252.226	42.409	
	1400	25.238	-101.779	115.658	-263.699	41.171	
	1500	25.269	-99.253	117.400	-275.353	40.124	

CrO_2 CHROMIUM DIOXIDE

		A	B	C	D	RANGE	REF.
SOL	CP	22.6	4.10			298- 600	*

PHASE	T	CP	H	S	G	BT	
SOL	298	23.822	-139.0	12.2	-142.637	104.570	165/*
	300	23.830	-138.956	12.347	-142.660	103.942	
	400	24.240	-136.552	19.259	-144.256	78.828	
	500	24.650	-134.108	24.712	-146.464	64.028	
	600	25.060	-131.622	29.243	-149.168	54.342	

* * * * * * * * * *

Cr_5O_{12} 5-CHROMIUM 12-OXIDE

		A	B	C	D	RANGE	REF.
SOL	CP	87.7	17.75	-16.28		298- 700	*
	** DPT= 643-705						165

PHASE	T	CP	H	S	G	BT	
SOL	298	74.678	-701.5	71.	-722.669	529.802	165/*
	300	74.936	-701.362	71.463	-722.800	526.631	
	400	84.625	-693.327	94.511	-731.131	399.525	
	500	90.063	-684.572	114.024	-741.584	324.190	
	600	93.828	-675.369	130.793	-753.845	274.625	
	700	96.803	-665.633	145.488	-767.674	239.711	

8-CHROMIUM 21-OXIDE

Cr_8O_{21}

		A	B	C	D	RANGE	REF.
SOL	CP	146.9	33.3	-28.8		298- 700	*
	** DPT= 600-640						165

PHASE	T	CP	H	S	G	BT	
SOL	298	124.430	-1118.	122.6	-1154.553	846.424	165,*
	300	124.890	-1117.769	123.371	-1154.781	841.370	
	400	142.220	-1104.314	161.962	-1169.099	638.852	
	500	152.030	-1089.565	194.831	-1186.981	518.899	
	600	158.880	-1074.004	223.184	-1207.915	440.042	
	700	164.332	-1057.835	248.098	-1231.504	384.544	

* * * * * * * * * *

CHROMIUM TRIOXIDE

CrO_3

		A	B	C	D	RANGE	REF.
SOL	CP	19.73	5.18	-4.18		298- 470	*
LIQ	CP	30.0				470- 600	*
	** DPT= 490-580						165

PHASE	T	CP	H	S	G	BT	
SOL	298	16.572	-138.2	17.2	-143.328	105.077	165*/1
	300	16.640	-138.169	17.303	-143.360	104.452	
	400	19.189	-136.363	22.481	-145.356	79.429	
	470	20.272	-134.980	25.665	-147.043	68.384	
			3.4	7.234			165
LIQ	470	30.000	-131.580	32.899	-147.043	68.384	
	500	30.000	-130.680	34.755	-148.058	64.725	
	600	30.000	-127.680	40.225	-151.815	55.306	

CrS

CHROMIUM MONOSULFIDE

		A	B	C	D	RANGE	REF.
SOL-A	CP	7.85	11.167			298- 450	63,*
	** TPT= 450 (LAMDA-TR.), L= 0.057						63
SOL-B	CP	12.343	1.172			450-1840	63
	** MPT= 1840						63

PHASE	T	CP	H	S	G	BT	
SOL-A	298	11.179	-37.2	15.3	-41.762	30.616	63
	300	11.200	-37.179	15.369	-41.790	30.448	
	400	12.317	-36.003	18.744	-43.501	23.771	
	450	12.875	-35.374	20.227	-44.476	21.603	
			0.057	.127			63*
SOL-B	450	12.870	-35.317	20.354	-44.476	21.603	
	500	12.929	-34.672	21.713	-45.528	19.903	
	600	13.046	-33.373	24.080	-47.821	17.421	
	700	13.163	-32.062	26.100	-50.333	15.717	
	800	13.281	-30.740	27.866	-53.033	14.490	
	900	13.398	-29.406	29.437	-55.899	13.576	
	1000	13.515	-28.061	30.854	-58.915	12.878	
	1100	13.632	-26.703	32.148	-62.066	12.333	
	1200	13.749	-25.334	33.339	-65.341	11.902	
	1300	13.867	-23.953	34.444	-68.731	11.556	
	1400	13.984	-22.561	35.476	-72.228	11.277	
	1500	14.101	-21.157	36.445	-75.824	11.049	
	1600	14.218	-19.741	37.359	-79.515	10.863	
	1700	14.335	-18.313	38.224	-83.294	10.710	
	1800	14.453	-16.874	39.047	-87.158	10.584	
	1840	14.499	-16.295	39.365	-88.727	10.540	

CHROMIUM 1.17-SULFIDE

$CrS_{1.17}$

		A	B	C	D	RANGE	REF.
SOL-A	CP	14.46				298- 301	63
SOL-B	CP	12.48	1.8			301- 590	63
SOL-C	CP	12.48	1.8			590-1500	63

PHASE	T	CP	H	S	G	BT	
SOL-A	298	14.460	-39.5	16.7	-44.479	32.608	63
	300	14.460	-39.473	16.789	-44.510	32.430	
	301	14.460	-39.459	16.838	-44.527	32.334	
			0.018	.060			63
SOL-B	301	13.022	-39.441	16.897	-44.527	32.334	
	400	13.200	-38.143	20.624	-46.393	25.351	
	500	13.380	-36.814	23.589	-48.608	21.250	
	590	13.542	-35.602	25.817	-50.834	18.833	
			0.065	.110			63
SOL-C	590	13.542	-35.537	25.927	-50.834	18.833	
	600	13.560	-35.402	26.155	-51.095	18.614	
	700	13.740	-34.037	28.258	-53.818	16.805	
	800	13.920	-32.654	30.105	-56.738	15.502	
	900	14.100	-31.253	31.755	-59.832	14.531	
	1000	14.280	-29.834	33.250	-63.084	13.789	
	1100	14.460	-28.397	34.619	-66.478	13.210	
	1200	14.640	-26.942	35.885	-70.004	12.751	
	1300	14.820	-25.469	37.064	-73.652	12.384	
	1400	15.000	-23.978	38.169	-77.414	12.087	
	1500	15.180	-22.469	39.210	-81.284	11.845	

$Cr_2(SO_4)_3$ CHROMIUM SULFATE

		A	B	C	D	RANGE	REF.
SOL	CP	85.58	19.0	-21.45		298-1200	184,*

PHASE	T	CP	H	S	G	BT	
SOL	298	67.115	-695.7	61.85	-714.141	523.550	183*/184*
	300	67.447	-695.576	62.266	-714.255	520.405	
	400	79.774	-688.140	83.572	-721.569	394.300	
	500	86.500	-679.800	102.156	-730.878	319.509	
	600	91.022	-670.912	118.348	-741.920	270.281	
	700	94.502	-661.629	132.650	-754.484	235.592	
	800	97.428	-652.029	145.465	-768.401	209.946	
	900	100.032	-642.154	157.093	-783.538	190.295	
	1000	102.435	-632.030	167.758	-799.788	174.817	
	1100	104.707	-621.672	177.629	-817.063	162.357	
	1200	106.890	-611.091	186.833	-835.291	152.148	

DICHROMIUM NITRIDE — Cr_2N

		A	B	C	D	RANGE	REF.
SOL	CP	15.24	6.8			298-1000	206,*
	** CRN(0.38-0.5)						

PHASE	T	CP	H	S	G	BT	
SOL	298	17.267	-27.3	17.65	-32.562	23.872	206,*
	300	17.280	-27.268	17.757	-32.595	23.749	
	400	17.960	-25.506	22.821	-34.634	18.926	
	500	18.640	-23.676	26.902	-37.127	16.230	
	600	19.320	-21.778	30.360	-39.994	14.570	
	700	20.000	-19.812	33.390	-43.185	13.485	
	800	20.680	-17.778	36.105	-46.662	12.749	
	900	21.360	-15.676	38.580	-50.398	12.240	
	1000	22.040	-13.506	40.865	-54.371	11.884	

* * * * * * * * * *

CHROMIUM HEXACARBONYL — $Cr(CO)_6$

		A	B	C	D	RANGE	REF.
SOL	CP	48.7	18.0			298- 424	1
	** DPT= 424						1

PHASE	T	CP	H	S	G	BT	
SOL	298	54.067	-222.	76.5	-244.808	179.474	186/*
	300	54.100	-221.900	76.835	-244.950	178.470	
	400	55.900	-216.400	92.645	-253.458	138.502	
	424	56.332	-215.053	95.914	-255.721	131.828	

CrB

CHROMIUM MONOBORIDE

		A	B	C	D	RANGE	REF.
SOL	CP	10.15	3.80	-2.42		298-1200	195
	** MPT= 1823						47

PHASE	T	CP	H	S	G	BT	
SOL	298	8.561	-18.	5.75	-19.714	14.453	*
	300	8.601	-17.984	5.803	-19.725	14.372	
	400	10.157	-17.038	8.515	-20.444	11.171	
	500	11.082	-15.973	10.888	-21.417	9.362	
	600	11.758	-14.829	12.970	-22.612	8.237	
	700	12.316	-13.625	14.826	-24.003	7.495	
	800	12.812	-12.368	16.503	-25.571	6.987	
	900	13.271	-11.064	18.039	-27.299	6.630	
	1000	13.708	-9.715	19.460	-29.175	6.377	
	1100	14.130	-8.323	20.786	-31.188	6.197	
	1200	14.542	-6.889	22.034	-33.329	6.071	

* * * * * * * * * *

CrB_2

CHROMIUM DIBORIDE

		A	B	C	D	RANGE	REF.
SOL	CP	9.613	10.72			298-1200	195
	** MPT= 2123-2173						47

PHASE	T	CP	H	S	G	BT	
SOL	298	12.809	-22.5	6.25	-24.363	17.861	*
	300	12.829	-22.476	6.329	-24.375	17.760	
	400	13.901	-21.140	10.167	-25.206	13.774	
	500	14.973	-19.696	13.384	-26.388	11.536	
	600	16.045	-18.145	16.209	-27.870	10.153	
	700	17.117	-16.487	18.762	-29.621	9.249	
	800	18.189	-14.722	21.118	-31.616	8.638	
	900	19.261	-12.849	23.322	-33.839	8.218	
	1000	20.333	-10.870	25.407	-36.277	7.929	
	1100	21.405	-8.783	27.395	-38.918	7.733	
	1200	22.477	-6.589	29.304	-41.753	7.605	

3-CHROMIUM 2-SILICCN Cr_3Si

		A	B	C	D	RANGE	REF.
SOL	CP	19.65	10.132	-1.032		298-2043	221*
	** MPT=	2043					221

PHASE	T	CP	H	S	G	GT	
SOL	298	21.510	-25.2	20.96	-31.449	23.056	221
	300	21.543	-25.160	21.093	-31.488	22.942	
	400	23.058	-22.927	27.508	-33.930	18.541	
	500	24.303	-20.557	32.790	-36.952	16.154	
	600	25.443	-18.069	37.323	-40.463	14.741	
	700	26.532	-15.470	41.327	-44.399	13.864	
	800	27.594	-12.764	44.940	-48.716	13.310	
	900	28.641	-9.952	48.250	-53.377	12.964	
	1000	29.679	-7.036	51.322	-58.358	12.756	
	1100	30.710	-4.016	54.199	-63.635	12.645	
	1200	31.737	-.894	56.915	-69.192	12.603	
	1300	32.761	2.331	59.496	-75.014	12.613	
	1400	33.782	5.658	61.961	-81.088	12.660	
	1500	34.802	9.087	64.327	-87.403	12.736	
	1600	35.821	12.618	66.605	-93.950	12.835	
	1700	36.839	16.251	68.807	-100.721	12.950	
	1800	37.856	19.986	70.942	-107.709	13.079	
	1900	38.872	23.822	73.016	-114.908	13.219	
	2000	39.888	27.760	75.036	-122.311	13.367	
	2043	40.325	29.485	75.889	-125.555	13.433	

Cr_5Si_3 5-CHROMIUM 3-SILICCN

		A	B	C	D	RANGE	REF.
SOL	CP	47.48	11.78	-6.12		298-1300	221
	CP	140.29	-129.8		53.9	1300-1920	221
	** MPT= 1920						221

PHASE	T	CP	H	S	G	BT	
SOL	298	44.108	-53.3	43.42	-66.246	48.566	221
	300	44.214	-53.218	43.693	-66.326	48.325	
	400	48.367	-48.568	57.043	-71.385	39.008	
	500	50.922	-43.596	68.127	-77.659	33.950	
	600	52.848	-38.404	77.588	-84.957	30.950	
	700	54.477	-33.036	85.859	-93.138	29.083	
	800	55.948	-27.514	93.231	-102.099	27.896	
	900	57.326	-21.850	99.901	-111.760	27.143	
	1000	58.648	-16.050	106.010	-122.060	26.680	
	1100	59.932	-10.121	111.660	-132.947	26.418	
	1200	61.191	-4.065	116.929	-144.380	26.299	
	1300	62.432	2.116	121.876	-156.322	26.284	
	1300	62.641	2.116	121.876	-156.322	26.284	
	1400	64.214	8.450	126.569	-168.746	26.346	
	1500	66.865	14.995	131.084	-181.630	26.467	
	1600	70.594	21.859	135.512	-194.960	26.634	
	1700	75.401	29.150	139.931	-208.732	26.838	
	1800	81.286	36.975	144.402	-222.948	27.073	
	1900	88.249	45.443	148.979	-237.616	27.336	
	1920	89.771	47.223	149.911	-240.605	27.391	

CHROMIUM SILICON

		A	B	C	D	RANGE	REF.
SOL	CP	12.43	2.09	-2.01		298-1700	221
	** MPT= 1730 (PERITEC.)						221

PHASE	T	CP	H	S	G	BT	
SOL	298	10.792	-13.1	10.46	-16.219	11.890	221
	300	10.824	-13.080	10.527	-16.238	11.831	
	400	12.010	-11.931	13.823	-17.461	9.541	
	500	12.671	-10.695	16.580	-18.985	8.299	
	600	13.126	-9.404	18.932	-20.763	7.564	
	700	13.483	-8.073	20.983	-22.761	7.107	
	800	13.788	-6.709	22.804	-24.952	6.818	
	900	14.063	-5.316	24.444	-27.316	6.634	
	1000	14.319	-3.897	25.939	-29.836	6.522	
	1100	14.563	-2.453	27.315	-32.500	6.458	
	1200	14.798	-.985	28.593	-35.296	6.429	
	1300	15.028	.507	29.786	-38.215	6.425	
	1400	15.253	2.021	30.908	-41.251	6.440	
	1500	15.476	3.557	31.968	-44.395	6.469	
	1600	15.695	5.116	32.974	-47.643	6.509	
	1700	15.913	6.696	33.932	-50.988	6.556	

$CrSi_2$ CHROMIUM 2-SILICON

		A	B	C	D	RANGE	REF.
SOL	CP	15.68	5.38	-1.86		298-1730	221
	** MPT=	1730					221

PHASE	T	CP	H	S	G	BT	
SOL	298	15.192	-19.14	13.96	-23.302	17.083	221
	300	15.227	-19.112	14.054	-23.328	16.997	
	400	16.669	-17.511	18.651	-24.971	13.645	
	500	17.626	-15.793	22.478	-27.033	11.818	
	600	18.391	-13.992	25.762	-29.449	10.728	
	700	19.066	-12.118	28.648	-32.172	10.046	
	800	19.693	-10.180	31.235	-35.168	9.609	
	900	20.292	-8.180	33.590	-38.411	9.329	
	1000	20.874	-6.122	35.758	-41.880	9.154	
	1100	21.444	-4.006	37.774	-45.558	9.053	
	1200	22.007	-1.833	39.664	-49.431	9.004	
	1300	22.564	.395	41.448	-53.487	8.993	
	1400	23.117	2.679	43.140	-57.717	9.011	
	1500	23.667	5.019	44.754	-62.113	9.051	
	1600	24.215	7.413	46.299	-66.666	9.107	
	1700	24.762	9.862	47.783	-71.370	9.177	
	1730	24.925	10.607	48.218	-72.810	9.199	

CESIUM

		A	B	C	D	RANGE	REF.
SOL	CP	10.625	-1.735	-2.136		298- 301	3
	** MPT= 301.55						3
LIQ	CP	7.144	0.216	0.484		301- 700	3
	CP	7.396				700- 952	3
	** BPT= 951.6 GAS(CS + CS2)						3*

PHASE	T	CP	H	S	G	BT	P	
SOL	298	7.705	0	20.351	-6.068	4.448		3
	300	7.731	.014	20.399	-6.105	4.448		
	302	7.753	.026	20.439	-6.137	4.448		
			0.499	1.655				3
LIQ	302	7.741	.525	22.093	-6.137	4.448		
	400	7.533	1.276	24.248	-8.424	4.603	.002	
	500	7.446	2.024	25.918	-10.935	4.780	.200	
	600	7.408	2.766	27.272	-13.597	4.953	3.747	
	700	7.394	3.506	28.413	-16.382	5.116	29.803	
	700	7.396	3.506	28.413	-16.382	5.116	29.803	
	800	7.396	4.246	29.400	-19.274	5.266	139.292	
	900	7.396	4.985	30.271	-22.259	5.406	457.278	
	952	7.396	5.367	30.684	-23.831	5.474	760.	

Cs (g)

CESIUM (GAS)

		A	B	C	D	RANGE	REF.
GAS	CP	4.968				298- 952	3
	CP	4.968				952-1100	3
	CP	4.256	0.372	3.735		1100-2000	3

PHASE	T	CP	H	S	G	BT	
GAS	298	4.968	18.515	41.942	6.010	-4.406	3*
	300	4.968	18.524	41.973	5.932	-4.322	
	400	4.968	19.021	43.402	1.660	-.907	
	500	4.968	19.518	44.511	-2.737	1.197	
	600	4.968	20.015	45.416	-7.235	2.636	
	700	4.968	20.511	46.182	-11.816	3.690	
	800	4.968	21.008	46.845	-16.468	4.500	
	900	4.968	21.505	47.431	-21.183	5.145	
	952	4.968	21.761	47.708	-23.637	5.429	
	952	4.968	21.761	47.708	-23.637	5.429	
	1000	4.968	22.002	47.954	-25.952	5.673	
	1100	4.968	22.499	48.428	-30.772	6.115	
	1100	4.974	22.499	48.428	-30.772	6.115	
	1200	4.962	22.995	48.860	-35.636	6.491	
	1300	4.961	23.491	49.257	-40.543	6.817	
	1400	4.967	23.988	49.625	-45.487	7.102	
	1500	4.980	24.485	49.968	-50.467	7.354	
	1600	4.997	24.984	50.290	-55.480	7.579	
	1700	5.018	25.485	50.593	-60.524	7.782	
	1800	5.041	25.987	50.881	-65.598	7.966	
	1900	5.066	26.493	51.154	-70.700	8.133	
	2000	5.093	27.001	51.414	-75.828	8.287	

CESIUM (DIATOMIC GAS)

		A	B	C	D	RANGE	REF.
GAS	CP	8.886	0.673			298-2000	3

PHASE	T	CP	H	S	G	BT	
GAS	298	9.087	25.40	67.87	5.165	-3.786	3
	300	9.088	25.417	67.926	5.039	-3.671	
	400	9.155	26.329	70.550	-1.891	1.033	
	500	9.222	27.248	72.600	-9.052	3.957	
	600	9.290	28.173	74.287	-16.399	5.974	
	700	9.357	29.106	75.725	-23.901	7.463	
	800	9.424	30.045	76.978	-31.538	8.617	
	900	9.492	30.991	78.092	-39.292	9.543	
	952	9.526	31.481	78.622	-43.336	9.954	
	952	9.526	31.481	78.622	-43.336	9.954	
	1000	9.559	31.943	79.096	-47.153	10.307	
	1100	9.626	32.902	80.010	-55.109	10.951	
	1200	9.694	33.868	80.851	-63.152	11.503	
	1300	9.761	34.841	81.629	-71.277	11.984	
	1400	9.828	35.821	82.355	-79.476	12.408	
	1500	9.895	36.807	83.035	-87.746	12.786	
	1600	9.963	37.800	83.676	-96.082	13.126	
	1700	10.030	38.799	84.282	-104.480	13.434	
	1800	10.097	39.806	84.857	-112.937	13.714	
	1900	10.165	40.819	85.405	-121.451	13.972	
	2000	10.232	41.839	85.928	-130.018	14.210	

Cs_2O CESIUM OXIDE

		A	B	C	D	RANGE	REF.
SOL	CP	16.1	8.0			298- 763	*
	** MPT= 763						1

PHASE	T	CP	H	S	G	BT	
SOL	298	18.485	-75.9	30.5	-84.994	62.310	1
	300	18.500	-75.866	30.614	-85.050	61.967	
	400	19.300	-73.976	36.046	-88.394	48.303	
	500	20.100	-72.006	40.439	-92.225	40.317	
	600	20.900	-69.956	44.174	-96.460	35.140	
	700	21.700	-67.826	47.456	-101.045	31.552	
	763	22.204	-66.443	49.347	-104.095	29.820	

* * * * * * * * * *

Cs_2O_3 CESIUM TRIOXIDE

		A	B	C	D	RANGE	REF.
SOL	CP	28.7	11.5			298- 775	*
	** MPT= 775						1

PHASE	T	CP	H	S	G	BT	
SOL	298	32.129	-135.	56.	-151.696	111.211	16
	300	32.150	-134.941	56.199	-151.800	110.601	
	400	33.300	-131.668	65.605	-157.910	86.290	
	500	34.450	-128.281	73.160	-164.860	72.070	
	600	35.600	-124.778	79.542	-172.503	62.843	
	700	36.750	-121.161	85.116	-180.742	56.438	
	775	37.612	-118.372	88.900	-187.269	52.817	

CESIUM DIOXIDE CsO_2

		A	B	C	D	RANGE	REF.
SOL	CP	17.3	7.4			298- 705	*
	** MPT= 705						1

PHASE	T	CP	H	S	G	BT	
SOL	298	19.506	-76.	35.	-86.435	63.367	16
	300	19.520	-75.964	35.121	-86.500	63.024	
	400	20.260	-73.975	40.838	-90.310	49.350	
	500	21.000	-71.912	45.438	-94.631	41.369	
	600	21.740	-69.775	49.332	-99.374	36.202	
	700	22.480	-67.564	52.739	-104.481	32.625	
	705	22.517	-67.451	52.899	-104.745	32.475	

Cs_2SO_4 CESIUM SULFATE

		A	B	C	D	RANGE	REF.
SOL-A	CP	28.75	21.0	-3.4		298- 933	*
SOL-B	CP	32.9	13.4			933-1392	*
LIQ	CP	47.8				1392-1800	*

PHASE	T	CP	H	S	G	BT	
SOL-A	298	31.186	-339.4	49.6	-354.188	259.662	19/*
	300	31.272	-339.342	49.793	-354.280	258.128	
	400	35.025	-336.016	59.338	-359.751	196.585	
	500	37.890	-332.366	67.471	-366.101	160.044	
	600	40.406	-328.449	74.604	-373.212	135.961	
	700	42.756	-324.290	81.011	-380.998	118.969	
	800	45.019	-319.901	86.869	-389.396	106.392	
	900	47.230	-315.288	92.299	-398.357	96.747	
	933	47.952	-313.717	94.013	-401.431	94.046	
			2.34	2.508			*
SOL-B	933	45.402	-311.377	96.521	-401.431	94.046	
	1000	46.300	-308.305	99.700	-408.006	89.182	
	1100	47.640	-303.608	104.176	-418.202	83.100	
	1200	48.980	-298.777	108.379	-428.832	78.111	
	1300	50.320	-293.812	112.352	-439.870	73.959	
	1392	51.553	-289.126	115.835	-450.368	70.719	
			9.1	6.537			*
LIQ	1392	47.800	-280.026	122.372	-450.368	70.719	
	1400	47.800	-279.644	122.646	-451.348	70.468	
	1500	47.800	-274.864	125.944	-463.779	67.582	
	1600	47.800	-270.084	129.029	-476.530	65.100	
	1700	47.800	-265.304	131.927	-489.579	62.948	
	1800	47.800	-260.524	134.659	-502.909	61.070	

CESIUM CARBONATE Cs_2CO_3

		A	B	C	D	RANGE	REF.
SOL	CP	27.59	16.57	-2.62		298-1065	156,*
	** MPT= 1065						66

PHASE	T	CP	H	S	G	BT	
SOL	298	29.583	-274.2	48.87	-288.771	211.703	200/156
	300	29.650	-274.145	49.053	-288.861	210.464	
	400	32.580	-271.025	58.011	-294.229	160.781	
	500	34.827	-267.651	65.529	-300.416	131.329	
	600	36.804	-264.068	72.056	-307.302	111.950	
	700	38.654	-260.294	77.870	-314.803	98.299	
	800	40.437	-256.339	83.148	-322.858	88.213	
	900	42.180	-252.208	88.012	-331.419	80.490	
	1000	43.898	-247.904	92.545	-340.449	74.415	
	1065	45.006	-245.015	95.344	-346.556	71.127	

Cu

COPPER

		A	B	C	D	RANGE	REF.
SOL	CP	5.940	0.905	-0.332		298-1357	3
	** MPT=	1356.6					3
LIQ	CP	7.5				1357-2848	3
GAS	CP	5.84				2848-3400	3

PHASE	T	CP	H	S	G	BT	P	
SOL	298	5.836	0	7.913	-2.359	1.730		3
	300	5.843	.011	7.949	-2.374	1.730		
	400	6.094	.609	9.668	-3.258	1.780		
	500	6.260	1.227	11.046	-4.296	1.878		
	600	6.391	1.860	12.200	-5.460	1.989		
	700	6.506	2.505	13.194	-6.731	2.102		
	800	6.612	3.161	14.069	-8.095	2.212		
	900	6.714	3.827	14.854	-9.542	2.317		
	1000	6.812	4.503	15.566	-11.063	2.418		
	1100	6.908	5.189	16.220	-12.653	2.514		
	1200	7.003	5.885	16.825	-14.306	2.606		
	1300	7.097	6.590	17.390	-16.017	2.693		
	1357	7.150	6.993	17.693	-17.010	2.741		
			3.17	2.337				3
LIQ	1357	7.500	10.163	20.030	-17.010	2.741		
	1400	7.500	10.488	20.266	-17.884	2.792		
	1500	7.500	11.238	20.783	-19.937	2.905	.006	
	1600	7.500	11.988	21.267	-22.040	3.011	.027	
	1700	7.500	12.738	21.722	-24.189	3.110	.109	
	1800	7.500	13.488	22.151	-26.383	3.204	.376	
	1900	7.500	14.238	22.556	-28.619	3.292	1.132	
	2000	7.500	14.988	22.941	-30.894	3.376	3.042	
	2100	7.500	15.738	23.307	-33.206	3.456	7.418	
	2200	7.500	16.488	23.656	-35.555	3.533	16.640	
	2300	7.500	17.238	23.989	-37.937	3.605	34.714	
	2400	7.500	17.988	24.308	-40.352	3.675	67.978	
	2500	7.500	18.738	24.615	-42.798	3.742	125.927	
	2600	7.500	19.488	24.909	-45.275	3.806	222.118	
	2700	7.500	20.238	25.192	-47.780	3.868	375.136	
	2800	7.500	20.988	25.465	-50.313	3.928	609.541	
	2848	7.500	21.348	25.592	-51.538	3.955	760.	
			72.743	25.542				3
GAS	2848	5.840	94.091	51.134	-51.538	3.955		
	2900	5.840	94.395	51.240	-54.200	4.085		
	3000	5.840	94.979	51.438	-59.334	4.323		
	3100	5.840	95.563	51.629	-64.487	4.547		
	3200	5.840	96.147	51.814	-69.659	4.758		
	3300	5.840	96.731	51.994	-74.850	4.958		
	3400	5.840	97.315	52.168	-80.058	5.147		

COPPER(1) BROMIDE

		A	B	C	D	RANGE	REF.
SOL-A	CP	13.56	2.24	-1.0		298- 665	59
SOL-B	CP	14.0				665- 745	59
SOL-C	CP	14.10				743- 761	59
LIQ	CP	15.50				761-2000	59

PHASE	T	CP	H	S	G	GT	P	
SOL-A	298	13.103	-25.0	22.96	-31.846	23.347		59
	300	13.121	-24.976	23.041	-31.888	23.234		
	400	13.831	-23.625	26.923	-34.394	18.794		
	500	14.280	-22.218	30.060	-37.248	16.283		
	600	14.626	-20.772	32.696	-40.389	14.714		
	665	14.823	-19.815	34.210	-42.564	13.991		
			1.8	2.707				59
SOL-B	665	14.000	-18.015	36.917	-42.564	13.991		
	700	14.000	-17.525	37.635	-43.869	13.698		
	745	14.000	-16.895	38.507	-45.583	13.374		
			0.37	.498				59
SOL-C	743	14.100	-16.525	39.005	-45.506	13.387		
	761	14.100	-16.271	39.343	-46.211	13.273		
			1.72	2.260				59
LIQ	761	15.500	-14.551	41.603	-46.211	13.273		
	800	15.500	-13.947	42.378	-47.849	13.073		
	900	15.500	-12.397	44.203	-52.179	12.673		
	1000	15.500	-10.847	45.836	-56.683	12.390	.011	
	1100	15.500	-9.297	47.314	-61.341	12.189	.089	
	1200	15.500	-7.747	48.662	-66.141	12.048	.505	
	1300	15.500	-6.197	49.903	-71.070	11.950	2.141	
	1400	15.500	-4.647	51.052	-76.119	11.884	7.261	
	1500	15.500	-3.097	52.121	-81.278	11.844	20.603	
	1600	15.500	-1.547	53.121	-86.541	11.822	50.633	
	1700	15.500	.003	54.061	-91.900	11.816	110.628	
	1800	15.500	1.553	54.947	-97.351	11.822	219.297	
	1900	15.500	3.103	55.785	-102.888	11.836	400.735	
	2000	15.500	4.653	56.580	-108.507	11.859	683.676	

$CuBr_2$ COPPER BROMIDE

		A	B	C	D	RANGE	REF.
SOL	CP	18.25	1.24	-0.46		298- 600	59

PHASE	T	CP	H	S	G	BT	
SOL	298	18.102	-33.4	30.8	-42.583	31.218	59
	300	18.111	-33.367	30.912	-42.640	31.067	
	400	18.458	-31.536	36.174	-46.006	25.140	
	500	18.686	-29.679	40.319	-49.838	21.787	
	600	18.866	-27.801	43.742	-54.046	19.689	

* * * * * * * * * *

$CuBr_3$ (g) COPPER TRIBROMIDE (GAS)

		A	B	C	D	RANGE	REF.
GAS	CP	31.76	0.08	-1.09		298-1300	59

PHASE	T	CP	H	S	G	BT	
GAS	298	30.558	-36.2	108.	-68.400	50.145	59
	300	30.573	-36.143	108.189	-68.600	49.982	
	400	31.111	-33.055	117.069	-79.883	43.652	
	500	31.364	-29.930	124.041	-91.951	40.197	
	600	31.505	-26.786	129.773	-104.650	38.124	
	700	31.594	-23.631	134.637	-117.877	36.808	
	800	31.654	-20.469	138.860	-131.556	35.944	
	900	31.697	-17.301	142.591	-145.633	35.369	
	1000	31.731	-14.129	145.932	-160.062	34.986	
	1100	31.758	-10.955	148.958	-174.808	34.736	
	1200	31.780	-7.778	151.722	-189.844	34.580	
	1300	31.800	-4.599	154.267	-205.146	34.493	

COPPER HYDROXIDE

$Cu(OH)_2$

		A	B	C	D	RANGE	REF.
SOL	CP	20.79	5.56	-1.29		298- 700	59

PHASE	T	CP	H	S	G	BT	
SOL	298	20.997	-105.9	20.8	-112.102	82.184	59
	300	21.025	-105.861	20.930	-112.140	81.705	
	400	22.208	-103.695	27.153	-114.556	62.599	
	500	23.054	-101.430	32.203	-117.532	51.380	
	600	23.768	-99.089	36.471	-120.971	44.070	
	700	24.419	-96.679	40.184	-124.808	38.972	

Cu_2SO_4 COPPER(1) SULFATE

		A	B	C	D	RANGE	REF.
SOL	CP	23.8	16.6			298-1000	*

PHASE	T	CP	H	S	G	BT	
SOL	298	28.749	-179.2	43.6	-192.199	140.905	212/*
	300	28.780	-179.147	43.778	-192.280	140.095	
	400	30.440	-176.186	52.285	-197.100	107.705	
	500	32.100	-173.059	59.256	-202.687	88.606	
	600	33.760	-169.766	65.255	-208.919	76.109	
	700	35.420	-166.307	70.584	-215.715	67.358	
	800	37.080	-162.682	75.422	-223.019	60.934	
	900	38.740	-158.891	79.885	-230.787	56.050	
	1000	40.400	-154.934	84.052	-238.986	52.237	

* * * * * * * * * *

$CuSO_4 \cdot H_2O$ COPPER SULFATE 1-HYDRATE

		A	B	C	D	RANGE	REF.
SOL-A	CP	31.26	16.94	-4.45		298- 600	59

PHASE	T	CP	H	S	G	BT	
SOL-A	298	31.305	-258.8	34.68	-269.140	197.311	59
	300	31.398	-258.742	34.874	-269.204	196.141	
	400	35.255	-255.394	44.479	-273.186	149.282	
	500	37.950	-251.728	52.648	-278.052	121.553	
	600	40.188	-247.819	59.770	-283.680	103.344	

COPPER SULFATE 3-HYDRATE $CuSO_4 \cdot 3\,H_2O$

		A	B	C	D	RANGE	REF.
SOL-A	CP	48.83	17.16	-4.4		298- 600	59

PHASE	T	CP	H	S	G	BT	
SOL-A	298	48.997	-401.8	53.1	-417.632	306.174	59
	300	49.089	-401.709	53.403	-417.730	304.357	
	400	52.944	-396.592	68.097	-423.831	231.602	
	500	55.650	-391.157	80.215	-431.264	188.531	
	600	57.904	-385.477	90.564	-439.816	160.224	

* * * * * * * * * *

COPPER SULFATE 5-HYDRATE $CuSO_4 \cdot 5\,H_2O$

		A	B	C	D	RANGE	REF.
SOL	CP	67.15	16.94	-4.44		298- 600	59

PHASE	T	CP	H	S	G	BT	
SOL	298	67.206	-544.1	72.0	-565.567	414.627	59
	300	67.299	-543.976	72.416	-565.700	412.168	
	400	71.151	-537.038	92.349	-573.977	313.649	
	500	73.844	-529.782	108.527	-584.046	255.321	
	600	76.081	-522.284	122.193	-595.599	216.976	

Cu_2Se (β) DICOPPER(1) SELENIDE

		A	B	C	D	RANGE	REF.
SOL-B	CP	14.0	18.5			298- 395	132
	** CUSE(0.50-0.54)						1
SOL-A	CP	20.1				395- 800	132
	** MPT= 1386						19

PHASE	T	CP	H	S	G	BT	
SOL-B	298	19.516	-15.8	31.7	-25.251	18.512	63*
	300	19.550	-15.764	31.821	-25.310	18.441	
	395	21.307	-13.823	37.430	-28.608	15.831	
			1.63	4.127			132
SOL-A	395	20.100	-12.193	41.556	-28.608	15.831	
	400	20.100	-12.093	41.809	-28.816	15.747	
	500	20.100	-10.083	46.294	-33.230	14.527	
	600	20.100	-8.073	49.959	-38.048	13.861	
	700	20.100	-6.063	53.057	-43.203	13.490	
	800	20.100	-4.053	55.741	-48.646	13.291	

* * * * * * * * * *

CuSe COPPER SELENIDE

		A	B	C	D	RANGE	REF.
SOL-A	CP	13.1				298- 326	63
SOL-B	CP	15.0				326- 655	*
	** DPT= 655						

PHASE	T	CP	H	S	G	BT	
SOL-A	298	13.100	-10.0	18.7	-15.575	11.419	63
	300	13.100	-9.976	18.781	-15.610	11.373	
	326	13.100	-9.635	19.870	-16.113	10.803	
			0.330	1.012			63
SOL-B	326	15.000	-9.305	20.882	-16.113	10.803	
	400	15.000	-8.195	23.951	-17.775	9.713	
	500	15.000	-6.695	27.298	-20.344	8.894	
	600	15.000	-5.195	30.033	-23.215	8.457	
	655	15.000	-4.370	31.348	-24.903	8.310	

COPPER SELENITE $CuO \cdot SeO_2$

		A	B	C	D	RANGE	REF.
SOL	CP	19.0	14.7			298- 800	*

PHASE	T	CP	H	S	G	BT	
SOL	298	23.383	-102.6	24.7	-109.964	80.617	*/62
	300	23.410	-102.557	24.845	-110.010	80.153	
	400	24.880	-100.142	31.781	-112.854	61.669	
	500	26.350	-97.581	37.490	-116.326	50.853	
	600	27.820	-94.872	42.425	-120.327	43.835	
	700	29.290	-92.017	46.823	-124.793	38.967	
	800	30.760	-89.014	50.830	-129.679	35.431	

Cu_2Te COPPER(1) TELLURIDE

		A	B	C	D	RANGE	REF.
SOL-Z	CP	14.3	12.8			298- 433	132
SOL-E	CP	14.45	12.8			433- 531	132
SOL-D	CP	27.0				531- 590	132
SOL-C	CP	32.0				590- 633	132
SOL-B	CP	26.12	-4.1			633- 841	132
SOL-A	CP	21.				841-1128	132,*
	** MPT=	1128					62

PHASE	T	CP	H	S	G	BT	
SOL-Z	298	18.116	-10.	32.2	-19.600	14.369	144
	300	18.140	-9.966	32.312	-19.660	14.324	
	400	19.420	-8.088	37.706	-23.171	12.662	
	433	19.842	-7.441	39.262	-24.441	12.338	
			0.053	.122			132
SOL-E	433	19.992	-7.388	39.384	-24.441	12.338	
	500	20.850	-6.019	42.321	-27.180	11.882	
	531	21.247	-5.367	43.587	-28.512	11.736	
			0.455	.857			132
SOL-D	531	27.000	-4.912	44.444	-28.512	11.736	
	590	27.000	-3.319	47.289	-31.219	11.566	
			0.23	.390			132
SOL-C	590	32.000	-3.089	47.678	-31.219	11.566	
	600	32.000	-2.769	48.216	-31.699	11.548	
	633	32.000	-1.713	49.930	-33.318	11.505	
			0.585	.924			132
SOL-B	633	23.525	-1.128	50.854	-33.318	11.505	
	700	23.250	.439	53.207	-36.806	11.493	
	800	22.840	2.744	56.285	-42.284	11.553	
	841	22.672	3.677	57.422	-44.615	11.596	
			0.477	.567			132
SOL-A	841	21.000	4.154	57.989	-44.615	11.596	
	900	21.000	5.393	59.413	-48.079	11.677	
	1000	21.000	7.493	61.626	-54.133	11.832	
	1100	21.000	9.593	63.627	-60.397	12.001	
	1128	21.000	10.181	64.155	-62.186	12.050	

COPPER TELLURIDE

CuTe

		A	B	C	D	RANGE	REF.
SOL	CP	9.88	9.65	0.227		298- 613	63
	** DPT= 613						63

PHASE	T	CP	H	S	G	BT	
SOL	298	13.013	-6.0	20.7	-12.172	8.923	63
	300	13.027	-5.976	20.781	-12.210	8.896	
	400	13.882	-4.631	24.643	-14.488	7.917	
	500	14.796	-3.198	27.838	-17.117	7.483	
	600	15.733	-1.671	30.618	-20.042	7.301	
	613	15.856	-1.466	30.957	-20.443	7.289	

Cu_3P TRICOPPER PHOSPHIDE

		A	B	C	D	RANGE	REF.
SOL	CP	18.6	8.0			298-1303	*
	** MPT=	1303					1

PHASE	T	CP	H	S	G	BT	
SOL	298	20.985	-29.223	28.5	-37.720	27.653	64*/*
	300	21.000	-29.184	28.630	-37.773	27.521	
	400	21.800	-27.044	34.781	-40.956	22.381	
	500	22.600	-24.824	39.731	-44.690	19.537	
	600	23.400	-22.524	43.922	-48.878	17.806	
	700	24.200	-20.144	47.590	-53.457	16.692	
	800	25.000	-17.684	50.873	-58.383	15.952	
	900	25.800	-15.144	53.864	-63.622	15.452	
	1000	26.600	-12.524	56.624	-69.148	15.114	
	1100	27.400	-9.824	59.197	-74.940	14.891	
	1200	28.200	-7.044	61.615	-80.982	14.751	
	1300	29.000	-4.184	63.904	-87.259	14.672	
	1303	29.024	-4.097	63.971	-87.451	14.670	

* * * * * * * * * *

CuP_2 COPPER DIPHOSPHIDE

		A	B	C	D	RANGE	REF.
SOL	CP	15.4	5.33			298-1100	*

PHASE	T	CP	H	S	G	BT	
SOL	298	16.989	-24.323	19.5	-30.137	22.094	108*/*
	300	16.999	-24.292	19.605	-30.173	21.984	
	400	17.532	-22.565	24.568	-32.392	17.701	
	500	18.065	-20.785	28.538	-35.054	15.324	
	600	18.598	-18.952	31.879	-38.079	13.872	
	700	19.131	-17.066	34.786	-41.415	12.932	
	800	19.664	-15.126	37.375	-45.026	12.302	
	900	20.197	-13.133	39.722	-48.882	11.872	
	1000	20.730	-11.086	41.877	-52.964	11.577	
	1100	21.263	-8.987	43.878	-57.253	11.377	

TRICOPPER ARSENIDE Cu_3As

		A	B	C	D	RANGE	REF.
SOL	CP	20.1	7.2			298-1098	*
	** MPT=	1098					1

PHASE	T	CP	H	S	G	BT	
SOL	298	22.247	-25.6	32.8	-35.379	25.937	64/*
	300	22.260	-25.559	32.938	-35.440	25.822	
	400	22.980	-23.297	39.440	-39.073	21.351	
	500	23.700	-20.963	44.645	-43.285	18.923	
	600	24.420	-18.557	49.030	-47.975	17.477	
	700	25.140	-16.079	52.848	-53.073	16.572	
	800	25.860	-13.529	56.252	-58.531	15.992	
	900	26.580	-10.907	59.340	-64.313	15.619	
	1000	27.300	-8.213	62.178	-70.390	15.386	
	1098	28.006	-5.503	64.762	-76.612	15.251	

* * * * * * * * * *

COPPER CARBONATE $CuCO_3$

		A	B	C	D	RANGE	REF.
SOL	CP	22.0	9.3	-4.3		298- 800	*

PHASE	T	CP	H	S	G	BT	
SOL	298	19.936	-142.5	21.	-148.761	109.060	166/19
	300	20.012	-142.463	21.124	-148.800	108.415	
	400	23.032	-140.296	27.337	-151.231	82.640	
	500	24.930	-137.892	32.693	-154.239	67.427	
	600	26.386	-135.324	37.371	-157.747	57.467	
	700	27.632	-132.622	41.534	-161.696	50.491	
	800	28.768	-129.801	45.299	-166.040	45.366	

$Cu_2O \cdot Fe_2O_3$ 2-COPPER 2-IRON 4-OXIDE

		A	B	C	D	RANGE	REF.
SOL-A	CP	46.84	3.6	-8.6		298-1091	59
SOL-B	CP	43.74	7.2			1091-1470	59
LIQ	CP	60.568				1470-1600	59

PHASE	T	CP	H	S	G	BT	
SOL-A	298	38.239	-245.2	42.48	-257.865	189.046	59
	300	38.364	-245.129	42.717	-257.944	187.938	
	400	42.905	-241.036	54.462	-262.820	143.618	
	500	45.200	-236.620	64.306	-268.773	117.496	
	600	46.611	-232.024	72.681	-275.633	100.413	
	700	47.605	-227.311	79.944	-283.272	88.453	
	800	48.376	-222.511	86.353	-291.593	79.670	
	900	49.018	-217.640	92.089	-300.520	72.986	
	1000	49.580	-212.710	97.283	-309.993	67.758	
	1091	50.045	-208.177	101.622	-319.046	63.920	
			0.18	.165			59
SOL-B	1091	51.595	-207.997	101.787	-319.046	63.920	
	1100	51.660	-207.532	102.211	-319.964	63.579	
	1200	52.380	-202.330	106.737	-330.414	60.185	
	1300	53.100	-197.056	110.958	-341.301	57.386	
	1400	53.820	-191.710	114.919	-352.597	55.050	
	1470	54.324	-187.925	117.557	-360.734	53.639	
			30.76	20.925			59
LIQ	1470	60.568	-157.165	138.482	-360.734	53.639	
	1500	60.568	-155.348	139.706	-364.907	53.174	
	1600	60.568	-149.291	143.615	-379.075	51.786	

COPPER 2-IRON 4-OXICE

$CuO \cdot Fe_2O_3$

		A	B	C	D	RANGE	REF.
SOL-A	CP	33.37	28.15	-5.6		298- 675	59
SOL-B	CP	54.3				675- 795	59
SOL-C	CP	39.68	9.8			795-1338	59
LIQ	CP	54.0				1358-1500	59

PHASE	T	CP	H	S	G	BT	
SOL-A	298	35.463	-231.35	35.076	-241.808	177.274	59
	300	35.593	-231.284	35.296	-241.873	176.228	
	400	41.130	-227.429	46.350	-245.969	134.409	
	500	45.205	-223.105	55.981	-251.095	109.768	
	600	48.704	-218.406	64.538	-257.129	93.672	
	675	51.142	-214.661	70.416	-262.192	84.903	
			0.18	.267			59
SOL-B	675	54.300	-214.481	70.683	-262.192	84.903	
	700	54.300	-213.124	72.658	-263.984	82.431	
	795	54.300	-207.965	79.568	-271.222	74.570	
							59
SOL-C	795	47.471	-207.965	79.568	-271.222	74.570	
	800	47.520	-207.728	79.866	-271.620	74.213	
	900	48.500	-202.927	85.519	-279.894	67.977	
	1000	49.480	-198.028	90.680	-288.708	63.106	
	1100	50.460	-193.031	95.442	-298.017	59.218	
	1200	51.440	-187.936	99.875	-307.785	56.063	
	1300	52.420	-182.743	104.031	-317.983	53.465	
	1338	52.792	-180.744	105.546	-321.965	52.597	
			3.12	2.297			59
LIQ	1358	54.000	-177.624	107.844	-324.076	52.162	
	1400	54.000	-175.356	109.489	-328.640	51.310	
	1500	54.000	-169.956	113.214	-339.777	49.512	

$CuFeS_2$ COPPER IRON DISULFICE

		A	B	C	D	RANGE	REF.
SOL-A	CP	20.79	12.8	-1.34		298- 830	59
SOL-B	CP	-344.64	440.96			830- 930	59
SOL-C	CP	41.22				930-1200	59

PHASE	T	CP	H	S	G	BT	
SOL-A	298	23.099	-45.5	29.87	-54.406	39.886	59
	300	23.141	-45.457	30.013	-54.461	39.680	
	400	25.072	-43.042	36.948	-57.821	31.596	
	500	26.654	-40.454	42.717	-61.812	27.022	
	600	28.098	-37.716	47.705	-66.339	24.167	
	700	29.477	-34.836	52.141	-71.335	22.275	
	800	30.821	-31.821	56.165	-76.753	20.971	
	830	31.219	-30.891	57.307	-78.455	20.661	
			2.41	2.904			59
SOL-B	830	21.357	-28.481	60.210	-78.455	20.661	
	900	52.224	-25.905	63.172	-82.761	20.100	
	930	65.453	-24.140	65.100	-84.684	19.903	
							59
SOL-C	930	41.220	-24.140	65.100	-84.684	19.903	
	1000	41.220	-21.255	68.092	-89.347	19.529	
	1100	41.220	-17.133	72.020	-96.355	19.147	
	1200	41.220	-13.011	75.607	-103.739	18.896	

5-COPPER IRON 4-SULFIDE

Cu_5FeS_4

		A	B	C	D	RANGE	REF.
SOL-A	CP	49.76	35.080	-1.35		298- 485	59
SOL-B	CP	-34.31	247.			485- 540	59
SOL-C	CP	45.18	28.08	62.36		540-1200	59

PHASE	T	CP	H	S	G	BT	
SOL-A	298	58.700	-90.9	86.6	-116.720	85.569	59
	300	58.784	-90.791	86.963	-116.880	85.159	
	400	62.948	-84.700	104.458	-126.483	69.117	
	485	66.200	-79.210	116.893	-135.903	61.249	
			1.43	2.948			59
SOL-B	485	85.485	-77.780	119.842	-135.903	61.249	
	500	89.190	-76.470	122.502	-137.721	60.206	
	540	99.070	-72.705	129.741	-142.765	57.788	
							59
SOL-C	540	81.729	-72.705	129.741	-142.765	57.788	
	600	79.350	-67.879	138.218	-150.809	54.940	
	700	77.563	-60.051	150.288	-165.253	51.601	
	800	77.388	-52.313	160.620	-180.810	49.402	
	900	78.151	-44.542	169.772	-197.338	47.927	
	1000	79.496	-36.664	178.072	-214.736	46.937	
	1100	81.222	-28.631	185.727	-232.930	46.285	
	1200	83.207	-20.411	192.878	-251.864	45.877	

Dy

DYSPROSIUM

		A	B	C	D	RANGE	REF.
SOL-A	CP	6.749	-0.098			298- 400	21
	CP	5.483	1.811	0.813		400- 800	21
	CP	4.520	-0.888	8.028	3.156	800-1657	21
SOL-B	CP	6.7				1657-1682	21
LIQ	CP	11.93				1682-2500	21
	** BPT= 2835, L= 55.						21

PHASE	T	CP	H	S	G	BT	P	
SOL-A	298	6.720	0	17.90	-5.337	3.913		21
	300	6.720	.012	17.942	-5.370	3.913		
	400	6.710	.684	19.873	-7.265	3.970		
	400	6.716	.684	19.873	-7.265	3.970		
	500	6.714	1.354	21.369	-9.330	4.079		
	600	6.795	2.029	22.600	-11.531	4.201		
	700	6.917	2.715	23.656	-13.845	4.323		
	800	7.059	3.413	24.589	-16.258	4.442		
	800	7.084	3.413	24.589	-16.258	4.442		
	900	7.268	4.130	25.432	-18.759	4.556		
	1000	7.591	4.872	26.214	-21.342	4.665		
	1100	8.025	5.652	26.957	-24.001	4.769		
	1200	8.557	6.480	27.677	-26.733	4.869		
	1300	9.174	7.366	28.386	-29.536	4.966	.002	
	1400	9.872	8.317	29.091	-32.410	5.060	.011	
	1500	10.646	9.343	29.798	-35.354	5.152	.054	
	1600	11.492	10.449	30.512	-38.370	5.242	.217	
	1657	12.006	11.119	30.923	-40.120	5.292	.440	
			0.995	.600				21
SOL-B	1657	6.700	12.114	31.523	-40.120	5.292	.440	
	1682	6.700	12.281	31.624	-40.910	5.316	.589	
			2.643	1.571				21
LIQ	1682	11.930	14.924	33.195	-40.910	5.316	.589	
	1700	11.930	15.139	33.322	-41.508	5.337	.715	
	1800	11.930	16.332	34.004	-44.875	5.449	1.960	
	1900	11.930	17.525	34.649	-48.308	5.557	4.785	
	2000	11.930	18.718	35.261	-51.804	5.662	10.599	
	2100	11.930	19.911	35.843	-55.359	5.762	21.611	
	2200	11.930	21.104	36.398	-58.971	5.859	41.039	
	2300	11.930	22.297	36.928	-62.638	5.953	73.283	
	2400	11.930	23.490	37.436	-66.356	6.043	124.048	
	2500	11.930	24.683	37.923	-70.124	6.131	200.382	

DYSPROSIUM CHLORIDE

$DyCl_3$

		A	B	C	D	RANGE	REF.
SOL	CP	22.60	4.30	-0.34		298- 924	89
LIQ	CP	34.60				924-1000	89

PHASE	T	CP	H	S	G	BT	
SOL	298	23.500	-239.	35.1	-249.465	182.887	19/*
	300	23.512	-238.957	35.245	-249.530	181.807	
	400	24.107	-236.574	42.094	-253.412	138.477	
	500	24.614	-234.138	47.529	-257.902	112.744	
	600	25.086	-231.653	52.059	-262.888	95.770	
	700	25.541	-229.121	55.960	-268.293	83.776	
	800	25.987	-226.545	59.400	-274.065	74.881	
	900	26.428	-223.924	62.486	-280.162	68.042	
	924	26.533	-223.289	63.183	-281.670	66.631	
			6.1	6.602			89
LIQ	924	34.600	-217.189	69.785	-281.670	66.631	
	1000	34.600	-214.559	72.520	-287.079	62.749	

Dy_2O_3 DYSPROSIUM OXIDE

		A	B	C	D	RANGE	REF.
SOL	CP	30.7	4.6	-4.0		298-2000	*

PHASE	T	CP	H	S	G	BT	
SOL	298	27.572	-445.3	35.8	-455.974	334.283	117/1
	300	27.636	-445.249	35.971	-456.040	332.270	
	400	30.040	-442.351	44.290	-460.067	251.403	
	500	31.400	-439.274	51.151	-464.850	203.213	
	600	32.349	-436.085	56.964	-470.263	171.316	
	700	33.104	-432.811	62.009	-476.217	148.702	
	800	33.755	-429.467	66.472	-482.645	131.870	
	900	34.346	-426.062	70.483	-489.496	118.882	
	1000	34.900	-422.599	74.130	-496.730	108.575	
	1100	35.429	-419.083	77.482	-504.313	100.211	
	1200	35.942	-415.514	80.587	-512.218	93.300	
	1300	36.443	-411.895	83.483	-520.423	87.503	
	1400	36.936	-408.226	86.202	-528.909	82.577	
	1500	37.422	-404.508	88.767	-537.658	78.347	
	1600	37.904	-400.741	91.198	-546.658	74.680	
	1700	38.382	-396.927	93.510	-555.894	71.475	
	1800	38.857	-393.065	95.717	-565.356	68.653	
	1900	39.329	-389.156	97.831	-575.034	66.153	
	2000	39.800	-385.199	99.860	-584.919	63.926	

ERBIUM

		A	B	C	D	RANGE	REF.
SOL	CP	6.784	-0.478	-0.045	1.368	298-1795	21
LIQ	CP	9.25				1795-2500	21
	** BPT= 3136, L= 62.469						21

PHASE	T	CP	H	S	G	BT	P	
SOL	298	6.712	0	17.49	-5.215	3.823		21
	300	6.714	.012	17.532	-5.247	3.823		
	400	6.784	.687	19.472	-7.102	3.881		
	500	6.869	1.370	20.995	-9.128	3.990		
	600	6.977	2.062	22.256	-11.292	4.114		
	700	7.111	2.766	23.342	-13.573	4.238		
	800	7.270	3.485	24.301	-15.956	4.360		
	900	7.456	4.221	25.168	-18.430	4.476		
	1000	7.669	4.977	25.964	-20.987	4.587		
	1100	7.910	5.756	26.706	-23.621	4.694		
	1200	8.177	6.560	27.406	-26.327	4.796		
	1300	8.472	7.392	28.072	-29.101	4.893		
	1400	8.794	8.255	28.711	-31.941	4.987	.001	
	1500	9.143	9.152	29.330	-34.843	5.077	.006	
	1600	9.520	10.085	29.932	-37.806	5.165	.027	
	1700	9.923	11.056	30.521	-40.829	5.250	.104	
	1795	10.332	12.018	31.071	-43.755	5.328	.319	
			4.757	2.650				21
LIQ	1795	9.250	16.775	33.721	-43.755	5.328	.319	
	1800	9.250	16.822	33.747	-43.923	5.334	.336	
	1900	9.250	17.747	34.247	-47.323	5.444	.893	
	2000	9.250	18.672	34.722	-50.772	5.549	2.146	
	2100	9.250	19.597	35.173	-54.267	5.648	4.720	
	2200	9.250	20.522	35.603	-57.806	5.743	9.630	
	2300	9.250	21.447	36.015	-61.387	5.834	18.406	
	2400	9.250	22.372	36.408	-65.008	5.921	33.234	
	2500	9.250	23.297	36.786	-68.668	6.004	57.093	

$ErCl_3$ ERBIUM CHLORIDE

		A	B	C	D	RANGE	REF.
SOL	CP	22.84	4.2	-0.25		298-1049	89
LIQ	CP	35.5				1049-1200	*

PHASE	T	CP	H	S	G	BT	
SOL	298	23.811	-229.1	35.1	-239.565	175.630	1
	300	23.822	-229.056	35.247	-239.630	174.594	
	400	24.364	-226.646	42.177	-243.517	133.069	
	500	24.840	-224.185	47.666	-248.018	108.423	
	600	25.291	-221.679	52.235	-253.019	92.175	
	700	25.729	-219.128	56.166	-258.444	80.701	
	800	26.161	-216.533	59.630	-264.237	72.196	
	900	26.589	-213.895	62.736	-270.358	65.661	
	1000	27.015	-211.215	65.560	-276.775	60.497	
	1049	27.223	-209.886	66.857	-280.019	58.347	
			7.8	7.436			89
LIQ	1049	35.500	-202.086	74.293	-280.019	58.347	
	1100	35.500	-200.276	75.978	-283.852	56.404	
	1200	35.500	-196.726	79.067	-291.606	53.116	

ERBIUM OXIDE (CUBIC) Er_2O_3

		A	B	C	D	RANGE	REF.
SOL	CP	27.05	7.0	-3.5		298-2000	*
	** TPT= 2543, MPT= 2673						1

PHASE	T	CP	H	S	G	BT	
SOL	298	25.200	-453.6	36.6	-464.512	340.542	1
	300	25.261	-453.553	36.756	-464.580	338.492	
	400	27.662	-450.895	44.387	-468.650	256.093	
	500	29.150	-448.050	50.729	-473.415	206.957	
	600	30.278	-445.077	56.147	-478.765	174.414	
	700	31.236	-442.000	60.888	-484.622	151.326	
	800	32.103	-438.832	65.116	-490.926	134.133	
	900	32.918	-435.581	68.945	-497.632	120.858	
	1000	33.700	-432.250	72.454	-504.704	110.318	
	1100	34.461	-428.842	75.702	-512.114	101.761	
	1200	35.207	-425.358	78.732	-519.837	94.688	
	1300	35.943	-421.801	81.580	-527.854	88.752	
	1400	36.671	-418.170	84.270	-536.148	83.708	
	1500	37.394	-414.467	86.825	-544.704	79.374	
	1600	38.113	-410.691	89.261	-553.509	75.616	
	1700	38.829	-406.844	91.593	-562.552	72.331	
	1800	39.542	-402.926	93.833	-571.824	69.438	
	1900	40.253	-398.936	95.990	-581.316	66.876	
	2000	40.962	-394.875	98.072	-591.020	64.592	

$EuCl_3$ EUROPIUM CHLORIDE

		A	B	C	D	RANGE	REF.
SOL	CP	21.63	6.25			298- 896	*
LIQ	CP	34.0				896-1200	*

PHASE	T	CP	H	S	G	BT	
SOL	298	23.493	-220.4	30.	-229.344	168.137	88
	300	23.505	-220.357	30.145	-229.400	167.140	
	400	24.130	-217.975	36.993	-232.772	127.198	
	500	24.755	-215.531	42.445	-236.753	103.498	
	600	25.380	-213.024	47.013	-241.232	87.880	
	700	26.005	-210.455	50.972	-246.135	76.857	
	800	26.630	-207.823	54.486	-251.411	68.692	
	896	27.230	-205.237	57.537	-256.791	62.644	
			7.9	8.817			*
LIQ	896	34.000	-197.337	66.354	-256.791	62.644	
	900	34.000	-197.201	66.505	-257.056	62.430	
	1000	34.000	-193.801	70.088	-263.889	57.681	
	1100	34.000	-190.401	73.328	-271.063	53.862	
	1200	34.000	-187.001	76.287	-278.545	50.737	

EUROPIUM OXIDE (MONOCL.) Eu_2O_3

		A	B	C	D	RANGE	REF.
SOL-A	CP	29.60	6.48	-2.08		298- 895	1
SOL-B	CP	31.06	4.16			895-2000	1

** TPT= 895 (MONOCL.-CUB), 2513 / MPT= 2593 1

PHASE	T	CP	H	S	G	BT	
SOL-A	298	29.192	-412.4	35.	-422.835	309.988	1/191
	300	29.233	-412.346	35.181	-422.900	308.124	
	400	30.892	-409.332	43.839	-426.868	233.261	
	500	32.008	-406.185	50.858	-431.614	188.684	
	600	32.910	-402.938	56.775	-437.003	159.200	
	700	33.712	-399.606	61.909	-442.943	138.312	
	800	34.459	-396.197	66.460	-449.365	122.777	
	895	35.140	-392.891	70.365	-455.867	111.333	
			0.13	.145			1
SOL-B	895	34.783	-392.761	70.510	-455.867	111.333	
	900	34.804	-392.587	70.704	-456.221	110.800	
	1000	35.220	-389.086	74.392	-463.478	101.307	
	1100	35.636	-385.543	77.769	-471.089	93.609	
	1200	36.052	-381.959	80.887	-479.023	87.254	
	1300	36.468	-378.333	83.789	-487.259	81.927	
	1400	36.884	-374.665	86.507	-495.775	77.404	
	1500	37.300	-370.956	89.066	-504.555	73.523	
	1600	37.716	-367.205	91.486	-513.584	70.162	
	1700	38.132	-363.413	93.785	-522.848	67.226	
	1800	38.548	-359.579	95.977	-532.337	64.643	
	1900	38.964	-355.703	98.072	-542.040	62.357	
	2000	39.380	-351.786	100.081	-551.949	60.322	

EuS

EUROPIUM MONOSULFIDE

		A	B	C	D	RANGE	REF.
SOL	CP	11.65	1.15			298-2400	110

PHASE	T	CP	H	S	G	BT	
SOL	298	11.993	-106.7	23.6	-113.736	83.382	110
	300	11.995	-106.678	23.674	-113.780	82.900	
	400	12.110	-105.473	27.141	-116.329	63.568	
	500	12.225	-104.256	29.855	-119.183	52.102	
	600	12.340	-103.028	32.094	-122.284	44.548	
	700	12.455	-101.788	34.005	-125.591	39.217	
	800	12.570	-100.537	35.676	-129.077	35.267	
	900	12.685	-99.274	37.163	-132.721	32.233	
	1000	12.800	-98.000	38.505	-136.505	29.837	
	1100	12.915	-96.714	39.731	-140.418	27.902	
	1200	13.030	-95.417	40.860	-144.448	26.311	
	1300	13.145	-94.108	41.907	-148.587	24.983	
	1400	13.260	-92.788	42.885	-152.827	23.861	
	1500	13.375	-91.456	43.804	-157.162	22.902	
	1600	13.490	-90.113	44.671	-161.586	22.075	
	1700	13.605	-88.758	45.492	-166.095	21.356	
	1800	13.720	-87.392	46.273	-170.683	20.727	
	1900	13.835	-86.014	47.018	-175.348	20.172	
	2000	13.950	-84.625	47.731	-180.086	19.682	
	2100	14.065	-83.224	48.414	-184.893	19.245	
	2200	14.180	-81.812	49.071	-189.768	18.854	
	2300	14.295	-80.388	49.704	-194.707	18.504	
	2400	14.410	-78.953	50.315	-199.708	18.188	

IRON

		A	B	C	D	RANGE	REF.
SOL-A	CP	6.734	-1.749	-0.692	5.985	298- 800	21
	CP	-62.967	61.140	148.0		800-1000	21
	CP	-153.419	166.429			1000-1042	21
	** CURIE-PT= 1042						
	CP	465.166	-427.222			1042-1060	21
	CP	-134.305	79.862	696.012		1060-1184	21
SOL-C	CP	5.734	1.998			1184-1665	21
SOL-D	CP	5.888	2.367			1665-1809	21
LIQ	CP	11.				1809-3135	21
GAS	CP	6.468				3135-3600	21

PHASE	T	CP	H	S	G	BT	P	
SOL-A	298	5.966	0	6.52	-1.944	1.425		21
	300	5.979	.011	6.557	-1.956	1.425		
	400	6.559	.639	8.361	-2.705	1.478		
	500	7.079	1.321	9.880	-3.619	1.582		
	600	7.647	2.057	11.220	-4.675	1.703		
	700	8.301	2.853	12.446	-5.859	1.829		
	800	9.057	3.720	13.603	-7.162	1.957		
	800	9.070	3.720	13.603	-7.162	1.957		
	900	10.331	4.676	14.727	-8.578	2.083		
	1000	12.973	5.832	15.943	-10.110	2.210		
	1000	13.010	5.832	15.943	-10.110	2.210		
	1042	20.000	6.525	16.621	-10.793	2.264		
	1042	20.001	6.525	16.621	-10.793	2.264		
	1060	12.311	6.816	16.898	-11.095	2.288		
	1060	12.294	6.816	16.898	-11.095	2.288		
	1100	11.065	7.282	17.329	-11.780	2.341		
	1184	9.901	8.150	18.090	-13.268	2.450		
			0.215	.182				21
SOL-C	1184	8.100	8.365	18.272	-13.268	2.450		

Fe

IRON (CONTINUED)

PHASE	T	CP	H	S	G	BT	P	
SOL-C	1200	8.132	8.495	18.381	-13.562	2.470		
	1300	8.331	9.318	19.039	-15.433	2.595		
	1400	8.531	10.161	19.664	-17.368	2.712		
	1500	8.731	11.024	20.260	-19.365	2.822		
	1600	8.931	11.908	20.829	-21.420	2.926		
	1665	9.061	12.492	21.188	-22.785	2.991	.003	
			0.200	.120				21
SOL-D	1665	9.829	12.692	21.308	-22.785	2.991	.003	
	1700	9.912	13.038	21.513	-23.535	3.026	.005	
	1800	10.149	14.041	22.086	-25.715	3.123	.023	
	1809	10.170	14.132	22.137	-25.914	3.131	.026	
			3.300	1.824				21
LIQ	1809	11.000	17.432	23.961	-25.914	3.131	.026	
	1900	11.000	18.433	24.501	-28.119	3.235	.086	
	2000	11.000	19.533	25.065	-30.597	3.344	.281	
	2100	11.000	20.633	25.602	-33.131	3.448	.816	
	2200	11.000	21.733	26.114	-35.717	3.549	2.138	
	2300	11.000	22.833	26.603	-38.353	3.645	5.127	
	2400	11.000	23.933	27.071	-41.037	3.737	11.375	
	2500	11.000	25.033	27.520	-43.767	3.827	23.575	
	2600	11.000	26.133	27.951	-46.540	3.913	46.015	
	2700	11.000	27.233	28.366	-49.356	3.996	85.165	
	2800	11.000	28.333	28.767	-52.213	4.076	150.348	
	2900	11.000	29.433	29.153	-55.109	4.154	254.448	
	3000	11.000	30.533	29.525	-58.043	4.229	414.638	
	3100	11.000	31.633	29.886	-61.014	4.302	653.068	
	3135	11.000	32.018	30.010	-62.062	4.327	760.	
			83.55	26.651				21
GAS	3135	6.468	115.568	56.660	-62.062	4.327		
	3200	6.468	115.989	56.793	-65.749	4.491		
	3300	6.468	116.635	56.992	-71.439	4.732		
	3400	6.468	117.282	57.185	-77.147	4.960		
	3500	6.468	117.929	57.373	-82.875	5.176		
	3600	6.468	118.576	57.555	-88.622	5.381		

IRON TRIBROMIDE

$FeBr_3$

		A	B	C	D	RANGE	REF.
SOL	CP	17.8	18.0			298- 600	*

PHASE	T	CP	H	S	G	BT	
SOL	298	23.167	-63.9	41.5	-76.273	55.917	1,*
	300	23.200	-63.857	41.643	-76.350	55.629	
	400	25.000	-61.447	48.564	-80.873	44.193	
	500	26.800	-58.857	54.336	-86.025	37.607	
	600	28.600	-56.087	59.381	-91.716	33.412	

$Fe_{1-y}O$ WUESTITE

		A	B	C	D	RANGE	REF.
SOL	$\overline{CP}$	9.474	2.376	-0.639		298- 800	*
	$\overline{CP}$	9.282	2.513			800-1645	*
LIQ	$\overline{CP}$	16.3				1650-3000	*
	** DPT=	3687					3

PHASE	T	$\overline{CP}$	$\overline{H}$	$\overline{S}$	$\overline{G}$	BT	
SOL	298	9.464	-62.382	16.083	-67.177	49.249	*
	300	9.477	-62.364	16.142	-67.207	48.967	
	400	10.025	-61.387	18.949	-68.967	37.687	
	500	10.406	-60.365	21.229	-70.979	31.029	
	600	10.722	-59.308	23.155	-73.201	26.667	
	700	11.007	-58.221	24.829	-75.602	23.607	
	800	11.275	-57.107	26.317	-78.161	21.355	
	800	11.292	-57.107	26.317	-78.161	21.355	
	900	11.544	-55.965	27.661	-80.861	19.638	
	1000	11.795	-54.798	28.891	-83.689	18.293	
	1100	12.046	-53.606	30.027	-86.636	17.215	
	1200	12.298	-52.389	31.086	-89.692	16.337	
	1300	12.549	-51.147	32.080	-92.851	15.612	
	1400	12.800	-49.879	33.019	-96.106	15.005	
	1500	13.051	-48.587	33.911	-99.453	14.492	
	1600	13.303	-47.269	34.761	-102.887	14.056	
	1645	13.416	-46.668	35.132	-104.459	13.880	
			7.558	4.581			*
LIQ	1650	16.300	-39.110	39.712	-104.635	13.861	
	1700	16.300	-38.295	40.199	-106.633	13.710	
	1800	16.300	-36.665	41.130	-110.700	13.443	
	1900	16.300	-35.035	42.012	-114.857	13.213	
	2000	16.300	-33.405	42.848	-119.101	13.016	
	2100	16.300	-31.775	43.643	-123.426	12.847	
	2200	16.300	-30.145	44.401	-127.828	12.700	
	2300	16.300	-28.515	45.126	-132.305	12.574	
	2400	16.300	-26.885	45.820	-136.852	12.464	
	2500	16.300	-25.255	46.485	-141.468	12.369	
	2600	16.300	-23.625	47.124	-146.148	12.287	
	2700	16.300	-21.995	47.740	-150.892	12.215	
	2800	16.300	-20.365	48.332	-155.696	12.154	
	2900	16.300	-18.735	48.904	-160.558	12.102	
	3000	16.300	-17.105	49.457	-165.476	12.057	

On page 248 are tabulated the partial heat, $\partial H/\partial n_{FeO} = \bar{H}$, the partial entropy, $\partial S/\partial n_{FeO} = \bar{S}$, and the chemical potential, $\partial G/\partial n_{FeO} = \bar{G}$, of the stoichiometric combination, FeO, in the reference state of wuestite of composition $Fe_{0.95}O$, that is $y = 0.05$. Wuestite of this composition has no congruent melting point but a melting range between the solidus and liquidus temperatures of 1645 and 1650 K, respectively.

The integral enthalpy of wuestite, which is needed for heat balance calculations, is given with sufficient accuracy by

$$H_{Fe_{1-y}O} = \bar{H}_{FeO} - y \cdot H_{Fe}$$

where $\bar{H}_{FeO}$ is the partial enthalpy of FeO, tabulated on p. 248, and H_{Fe} the enthalpy of Fe.

In the table below, the compositions of the phase boundaries Fe/wuestite (y_1) and wuestite/magnetite (y_2) are listed. Values of a_{FeO} at the latter boundary are also given and are defined by

$$RT \ln a_{FeO} = \bar{G}(y) - \bar{G}(y_o) \; ; \; y_o = 0.05$$

Since activity values, a_{FeO}, along the phase boundary Fe/wuestite are very close to unity, i.e. within 1 - 2 %, they have not been tabulated.

T(K)	y_1	y_2	a_{FeO}
900	0.0514	0.0721	0.95
1000	0.0480	0.0909	0.87
1100	0.0465	0.1043	0.86
1200	0.0461	0.1153	0.82
1300	0.0461	0.1252	0.74
1400	0.0461	0.1343	0.70
1500	0.0469	0.1440	0.67
1600	0.0484	0.1553	0.67

$Fe_2O_3 \cdot H_2O$ IRON TRIOXIDE HYDRATE

		A	B	C	D	RANGE	REF.
SOL	CP	42.				298- 400	*

PHASE	T	CP	H	S	G	BT	
SOL	298	42.000	-267.2	28.4	-275.667	202.097	19/*
	300	42.000	-267.122	28.660	-275.720	200.889	
	400	42.000	-262.922	40.742	-279.219	152.579	

IRON MONOSULFIDE

FeS

		A	B	C	D	RANGE	REF.
SOL-A	CP	0.12	40.			298- 411	63
	** HOMOGENEITY RANGE...						
SOL-B	CP	17.40				411- 598	1
SOL-C	CP	12.20	2.38			598-1468	1
LIQ	CP	17.0				1468-2000	1
	** DEC.						63

PHASE	T	CP	H	S	G	BT	
SOL-A	298	12.046	-24.	14.41	-28.296	20.745	62,63
	300	12.120	-23.978	14.485	-28.323	20.636	
	400	16.120	-22.566	18.519	-29.973	16.379	
	411	16.560	-22.386	18.963	-30.180	16.050	
			0.57	1.387			1
SOL-B	411	17.400	-21.816	20.349	-30.180	16.050	
	500	17.400	-20.267	23.760	-32.147	14.053	
	598	17.400	-18.562	26.874	-34.633	12.659	
			0.12	.201			1
SOL-C	598	13.623	-18.442	27.075	-34.633	12.659	
	600	13.628	-18.415	27.121	-34.687	12.636	
	700	13.866	-17.040	29.239	-37.508	11.712	
	800	14.104	-15.642	31.106	-40.527	11.073	
	900	14.342	-14.219	32.781	-43.722	10.619	
	1000	14.580	-12.773	34.305	-47.078	10.290	
	1100	14.818	-11.303	35.705	-50.579	10.051	
	1200	15.056	-9.810	37.005	-54.216	9.875	
	1300	15.294	-8.292	38.219	-57.977	9.748	
	1400	15.532	-6.751	39.362	-61.857	9.658	
	1468	15.694	-5.689	40.102	-64.559	9.613	
			7.73	5.266			1
LIQ	1468	17.000	2.041	45.368	-64.559	9.613	
	1500	17.000	2.585	45.734	-66.017	9.620	
	1600	17.000	4.285	46.831	-70.645	9.651	
	1700	17.000	5.985	47.862	-75.381	9.692	
	1800	17.000	7.685	48.834	-80.216	9.741	
	1900	17.000	9.385	49.753	-85.146	9.795	
	2000	17.000	11.085	50.625	-90.165	9.854	

FeS_2 IRON DISULFIDE

		A	B	C	D	RANGE	REF.
SOL	CP	17.869	1.333	-3.045		298-1016	63

PHASE	T	CP	H	S	G	BT	
SOL	298	14.841	-41.	12.65	-44.772	32.823	63
	300	14.886	-40.973	12.742	-44.795	32.638	
	400	16.499	-39.393	17.276	-46.303	25.302	
	500	17.317	-37.698	21.054	-48.225	21.082	
	600	17.823	-35.939	24.259	-50.495	18.395	
	700	18.181	-34.138	27.035	-53.062	16.569	
	800	18.460	-32.306	29.481	-55.891	15.271	
	900	18.693	-30.448	31.669	-58.950	14.317	
	1000	18.897	-28.568	33.649	-62.218	13.599	
	1016	18.928	-28.266	33.950	-62.758	13.502	

* * * * * * * * * *

$FeSe_{0.96}$ IRON 0.96-SELENIDE

		A	B	C	D	RANGE	REF.
SOL-A	CP	12.98	5.05	-1.242		298- 731	63
	** TPT=	730.8					63
SOL-B	CP	16.219	-2.885			731- 800	63
	CP	39.47	-65.95		42.5	800-1000	63*

PHASE	T	CP	H	S	G	BT	
SOL-A	298	13.088	-16.	16.54	-20.931	15.345	63
	300	13.115	-15.976	16.621	-20.962	15.273	
	400	14.224	-14.605	20.558	-22.828	12.474	
	500	15.008	-13.141	23.820	-25.051	10.951	
	600	15.665	-11.607	26.616	-27.576	10.046	
	700	16.262	-10.010	29.076	-30.363	9.481	
	731	16.438	-9.507	29.780	-31.270	9.353	
			2.29	3.134			63
SOL-B	731	14.111	-7.217	32.913	-31.270	9.353	
	800	13.911	-6.247	34.181	-33.592	9.178	
	800	13.910	-6.247	34.181	-33.592	9.178	
	900	14.540	-4.832	35.847	-37.094	9.009	
	1000	16.020	-3.311	37.448	-40.759	8.909	

IRON 0.9-TELLURIDE

$FeTe_{0.9}$

		A	B	C	D	RANGE	REF.
SOL-B	CP	12.633	1.011	-0.946		298-1200	63
	** FETE(0.92-0.98), MPT= 1200						63

PHASE	T	CP	H	S	G	BT	
SOL-B	298	11.870	-5.55	19.14	-11.257	8.252	63
	300	11.885	-5.528	19.213	-11.292	8.227	
	400	12.446	-4.308	22.719	-13.396	7.320	
	500	12.760	-3.047	25.533	-15.813	6.913	
	600	12.977	-1.759	27.879	-18.487	6.735	
	700	13.148	-.453	29.893	-21.378	6.675	
	800	13.294	.869	31.658	-24.457	6.682	
	900	13.426	2.206	33.232	-27.703	6.728	
	1000	13.549	3.554	34.653	-31.098	6.797	
	1100	13.667	4.915	35.950	-34.629	6.881	
	1200	13.781	6.288	37.144	-38.285	6.974	

* * * * * * * * * *

IRON DITELLURIDE

$FeTe_2$

		A	B	C	D	RANGE	REF.
SOL-E	CP	17.284	4.912	-1.002		298- 933	63
	** MPT= 933						63

PHASE	T	CP	H	S	G	BT	
SOL-E	298	17.621	-17.3	23.94	-24.438	17.916	63
	300	17.644	-17.267	24.049	-24.482	17.838	
	400	18.623	-15.451	29.269	-27.158	14.841	
	500	19.339	-13.551	33.504	-30.303	13.247	
	600	19.953	-11.586	37.086	-33.837	12.327	
	700	20.518	-9.562	40.204	-37.705	11.774	
	800	21.057	-7.483	42.979	-41.867	11.439	
	900	21.581	-5.351	45.490	-46.292	11.243	
	933	21.752	-4.636	46.270	-47.806	11.200	

Fe_4N TETRAIRON NITRIDE

		A	B	C	D	RANGE	REF.
SOL-A	CP	26.84	8.16			298- 753	58
	** METASTABLE						
SOL-B	CP	26.84	8.16			753- 900	58

PHASE	T	CP	H	S	G	BT	
SOL-A	298	29.273	-2.65	37.2	-13.741	10.074	62/1
	300	29.288	-2.596	37.381	-13.810	10.062	
	400	30.104	.374	45.919	-17.994	9.833	
	500	30.920	3.425	52.724	-22.937	10.027	
	600	31.736	6.558	58.433	-28.502	10.363	
	700	32.552	9.772	63.387	-34.598	10.804	
	753	32.984	11.509	65.778	-38.022	11.037	
			2.	2.656			62
SOL-B	753	32.984	13.509	68.434	-38.022	11.037	
	800	33.368	15.068	70.443	-41.286	11.280	
	900	34.184	18.446	74.420	-48.532	11.787	

IRON PENTACARBONYL $Fe(CO)_5$

		A	B	C	D	RANGE	REF.
LIQ	CP	57.5				298- 380	62
	** BPT=	379.842					
GAS	CP	47.26	9.4	-8.9		380-1000	62,2

PHASE	T	CP	H	S	G	BT	P	
LIQ	298	57.500	-182.6	80.7	-206.661	151.507	29.150	62
	300	57.500	-182.494	81.056	-206.810	150.681	32.199	
	380	57.500	-177.903	94.624	-213.845	123.057	760.	
			8.355	21.996				62*
GAS	380	44.662	-169.548	116.620	-213.845	123.057		
	400	45.457	-168.639	118.950	-216.219	118.153		
	500	48.400	-163.935	129.435	-228.653	99.957		
	600	50.428	-158.989	138.447	-242.057	88.181		
	700	52.024	-153.864	146.345	-256.305	80.033		
	800	53.389	-148.592	153.383	-271.298	74.125		
	900	54.621	-143.190	159.743	-286.959	69.693		
	1000	55.770	-137.670	165.558	-303.228	66.279		

$2FeO \cdot SiO_2$ IRON ORTHOSILICATE (FAYALITE)

		A	B	C	D	RANGE	REF.
SOL	CP	36.51	9.36	-6.70		298-1493	5
LIQ	CP	57.5				1493-1700	5
	** DEC.						1

PHASE	T	CP	H	S	G	BT	
SOL	298	31.764	-350.464	33.137	-360.344	264.175	5*
	300	31.874	-350.405	33.334	-360.405	262.590	
	400	36.066	-346.985	43.145	-364.243	199.040	
	500	38.510	-343.248	51.474	-368.985	161.305	
	600	40.265	-339.305	58.657	-374.499	136.430	
	700	41.695	-335.205	64.974	-380.687	118.872	
	800	42.951	-330.972	70.625	-387.472	105.867	
	900	44.107	-326.618	75.752	-394.795	95.882	
	1000	45.200	-322.153	80.456	-402.608	88.002	
	1100	46.252	-317.580	84.813	-410.874	81.644	
	1200	47.277	-312.903	88.882	-419.561	76.423	
	1300	48.282	-308.125	92.706	-428.643	72.071	
	1400	49.272	-303.247	96.320	-438.096	68.399	
	1493	50.184	-298.623	99.518	-447.203	65.472	
			22.	14.735			5
LIQ	1493	57.500	-276.623	114.254	-447.203	65.472	
	1500	57.500	-276.220	114.523	-448.004	65.283	
	1600	57.500	-270.470	118.234	-459.644	62.793	
	1700	57.500	-264.720	121.719	-471.643	60.642	

2-IRON TITANIUM Fe_2Ti

		A	B	C	D	RANGE	REF.
SOL	CP	19.35	3.30	-3.15		298-1803	*
	** MPT= 1803						1

PHASE	T	CP	H	S	G	BT	
SOL	298	16.790	-20.9	17.8	-26.207	19.213	*
	300	16.840	-20.869	17.904	-26.240	19.118	
	400	18.701	-19.081	23.035	-28.295	15.462	
	500	19.740	-17.155	27.328	-30.819	13.473	
	600	20.455	-15.143	30.994	-33.740	12.291	
	700	21.017	-13.069	34.191	-37.002	11.554	
	800	21.498	-10.943	37.029	-40.566	11.084	
	900	21.931	-8.771	39.587	-44.399	10.783	
	1000	22.335	-6.557	41.918	-48.476	10.596	
	1100	22.720	-4.305	44.065	-52.776	10.487	
	1200	23.091	-2.014	46.058	-57.284	10.434	
	1300	23.454	.313	47.921	-61.984	10.422	
	1400	23.809	2.677	49.672	-66.864	10.439	
	1500	24.160	5.075	51.327	-71.915	10.479	
	1600	24.507	7.508	52.897	-77.127	10.536	
	1700	24.851	9.976	54.393	-82.492	10.606	
	1800	25.193	12.479	55.823	-88.003	10.686	
	1803	25.203	12.554	55.865	-88.171	10.689	

FeTi

IRON TITANIUM

		A	B	C	D	RANGE	REF.
SOL	CP	12.67	2.3	-1.94		298-1590	*
	** MPT= 1590 (PERIT.)						1

PHASE	T	CP	H	S	G	BT	
SOL	298	11.173	-9.7	12.6	-13.457	9.865	1/*
	300	11.204	-9.679	12.669	-13.480	9.822	
	400	12.377	-8.493	16.073	-14.923	8.154	
	500	13.044	-7.220	18.912	-16.676	7.290	
	600	13.511	-5.891	21.333	-18.691	6.809	
	700	13.884	-4.521	23.445	-20.932	6.536	
	800	14.207	-3.116	25.320	-23.372	6.386	
	900	14.500	-1.680	27.011	-25.990	6.312	
	1000	14.776	-.216	28.553	-28.769	6.288	
	1100	15.040	1.274	29.974	-31.696	6.298	
	1200	15.295	2.791	31.293	-34.761	6.332	
	1300	15.545	4.333	32.527	-37.952	6.381	
	1400	15.791	5.900	33.688	-41.264	6.442	
	1500	16.034	7.491	34.786	-44.688	6.512	
	1590	16.250	8.944	35.727	-47.861	6.580	

IRON TITANIUM TRIOXIDE (ILMENITE) $FeO \cdot TiO_2$

		A	B	C	D	RANGE	REF.
SOL	CP	27.87	4.36	-4.79		298-1743	1
LIQ	CP	47.60				1743-2000	1
	** DEC.						1

PHASE	T	CP	H	S	G	BT	
SOL	298	23.781	-295.282	25.3	-302.825	222.007	5,*
	300	23.856	-295.238	25.447	-302.872	220.672	
	400	26.620	-292.698	32.737	-305.792	167.100	
	500	28.134	-289.954	38.853	-309.380	135.248	
	600	29.155	-287.087	44.078	-313.533	114.220	
	700	29.944	-284.130	48.633	-318.174	99.352	
	800	30.610	-281.102	52.676	-323.243	88.318	
	900	31.203	-278.011	56.316	-328.695	79.829	
	1000	31.751	-274.863	59.632	-334.495	73.114	
	1100	32.270	-271.662	62.683	-340.613	67.683	
	1200	32.769	-268.409	65.513	-347.025	63.210	
	1300	33.255	-265.108	68.155	-353.709	59.472	
	1400	33.730	-261.759	70.637	-360.650	56.308	
	1500	34.197	-258.362	72.980	-367.832	53.600	
	1600	34.659	-254.920	75.202	-375.242	51.263	
	1700	35.116	-251.431	77.316	-382.869	49.228	
	1743	35.312	-249.917	78.196	-386.212	48.433	
			21.7	12.450			1
LIQ	1743	47.600	-228.217	90.646	-386.212	48.433	
	1800	47.600	-225.503	92.178	-391.423	47.532	
	1900	47.600	-220.743	94.751	-400.771	46.105	
	2000	47.600	-215.983	97.193	-410.369	44.849	

Fe_2B DIIRON BORIDE

		A	B	C	D	RANGE	REF.
SOL	CP	17.02	3.3			298-1662	*
	** MPT= 1662						62

PHASE	T	CP	H	S	G	BT	
SOL	298	18.004	-17.	13.54	-21.037	15.423	102/*
	300	18.010	-16.967	13.651	-21.062	15.346	
	400	18.340	-15.149	18.878	-22.700	12.405	
	500	18.670	-13.299	23.006	-24.802	10.842	
	600	19.000	-11.415	26.439	-27.278	9.938	
	700	19.330	-9.499	29.392	-30.073	9.391	
	800	19.660	-7.549	31.995	-33.145	9.056	
	900	19.990	-5.567	34.330	-36.463	8.856	
	1000	20.320	-3.551	36.453	-40.004	8.744	
	1100	20.650	-1.503	38.405	-43.748	8.693	
	1200	20.980	.579	40.216	-47.681	8.685	
	1300	21.310	2.693	41.908	-51.788	8.707	
	1400	21.640	4.841	43.500	-56.059	8.752	
	1500	21.970	7.021	45.004	-60.485	8.814	
	1600	22.300	9.235	46.432	-65.057	8.888	
	1662	22.505	10.624	47.284	-67.962	8.938	

IRON MONOBORIDE

		A	B	C	D	RANGE	REF.
SOL	CP	11.541	1.539			298-1923	*
	** MPT=	1923					62

PHASE	T	CP	H	S	G	BT	
SOL	298	12.000	-17.	6.62	-18.974	13.910	102/*
	300	12.003	-16.978	6.694	-18.986	13.833	
	400	12.157	-15.770	10.168	-19.837	10.840	
	500	12.310	-14.546	12.897	-20.995	9.178	
	600	12.464	-13.308	15.156	-22.401	8.161	
	700	12.618	-12.054	17.089	-24.016	7.499	
	800	12.772	-10.784	18.783	-25.811	7.052	
	900	12.926	-9.499	20.297	-27.766	6.743	
	1000	13.080	-8.199	21.667	-29.865	6.528	
	1100	13.234	-6.883	22.920	-32.096	6.378	
	1200	13.388	-5.552	24.079	-34.446	6.274	
	1300	13.542	-4.206	25.156	-36.909	6.206	
	1400	13.696	-2.844	26.165	-39.475	6.163	
	1500	13.849	-1.466	27.116	-42.140	6.141	
	1600	14.003	-.074	28.014	-44.897	6.133	
	1700	14.157	1.334	28.868	-47.741	6.138	
	1800	14.311	2.758	29.681	-50.669	6.153	
	1900	14.465	4.196	30.459	-53.676	6.175	
	1923	14.500	4.530	30.634	-54.379	6.181	

$FeO \cdot Al_2O_3$ IRON DIALUMINIUM TETROXIDE

		A	B	C	D	RANGE	REF.
SOL	CP	37.14	6.25	-7.49		298-2053	*
	** MPT= 2053						5

PHASE	T	CP	H	S	G	BT	
SOL	298	30.578	-472.182	25.4	-479.755	351.717	5*
	300	30.693	-472.125	25.590	-479.802	349.583	
	400	34.959	-468.817	35.079	-482.848	263.851	
	500	37.269	-465.196	43.148	-486.770	212.796	
	600	38.809	-461.388	50.087	-491.440	179.031	
	700	39.986	-457.446	56.161	-496.759	155.116	
	800	40.970	-453.397	61.567	-502.650	137.336	
	900	41.840	-449.256	66.443	-509.055	123.632	
	1000	42.641	-445.031	70.893	-515.925	112.770	
	1100	43.396	-440.729	74.993	-523.222	103.969	
	1200	44.120	-436.353	78.800	-530.914	96.706	
	1300	44.822	-431.906	82.360	-538.974	90.622	
	1400	45.508	-427.389	85.707	-547.378	85.461	
	1500	46.182	-422.805	88.869	-556.109	81.036	
	1600	46.847	-418.153	91.871	-565.147	77.206	
	1700	47.506	-413.435	94.731	-574.478	73.864	
	1800	48.159	-408.652	97.465	-584.089	70.928	
	1900	48.808	-403.804	100.086	-593.967	68.331	
	2000	49.453	-398.891	102.606	-604.103	66.022	
	2053	49.794	-396.261	103.904	-609.575	64.900	

3-IRON 2-MOLYBDENUM — Fe_3Mo_2

		A	B	C	D	RANGE	REF.
SOL	CP	24.3	11.81	1.45		298-1600	*

PHASE	T	CP	H	S	G	BT	
SOL	298	29.452	-1.0	35.	-11.435	8.383	*
	300	29.454	-.946	35.182	-11.500	8.379	
	400	29.930	2.019	43.706	-15.464	8.450	
	500	30.785	5.053	50.473	-20.184	8.824	
	600	31.789	8.181	56.173	-25.523	9.298	
	700	32.863	11.413	61.153	-31.395	9.803	
	800	33.975	14.754	65.614	-37.737	10.311	
	900	35.108	18.208	69.681	-44.504	10.809	
	1000	36.255	21.776	73.439	-51.662	11.292	
	1100	37.411	25.460	76.948	-59.184	11.760	
	1200	38.573	29.259	80.253	-67.045	12.212	
	1300	39.739	33.174	83.387	-75.229	12.649	
	1400	40.908	37.207	86.375	-83.718	13.071	
	1500	42.079	41.356	89.237	-92.499	13.479	
	1600	43.253	45.623	91.990	-101.562	13.875	

* * * * * * * * * *

3-IRON 2-TUNGSTEN — Fe_3W_2

		A	B	C	D	RANGE	REF.
SOL	CP	28.29	9.90	-1.60		298-1300	21

PHASE	T	CP	H	S	G	BT	
SOL	298	29.442	-7.5	35.	-17.935	13.149	*
	300	29.482	-7.445	35.182	-18.000	13.115	
	400	31.250	-4.403	43.922	-21.972	12.007	
	500	32.600	-1.209	51.045	-26.731	11.686	
	600	33.786	2.111	57.095	-32.145	11.711	
	700	34.893	5.546	62.387	-38.125	11.905	
	800	35.960	9.089	67.116	-44.604	12.187	
	900	37.002	12.737	71.412	-51.534	12.516	
	1000	38.030	16.489	75.364	-58.875	12.869	
	1100	39.048	20.343	79.036	-66.597	13.233	
	1200	40.059	24.298	82.477	-74.675	13.602	
	1300	41.065	28.354	85.723	-83.086	13.970	

$FeO \cdot Cr_2O_3$ IRON DICHROMIUM TETROXIDE

		A	B	C	D	RANGE	REF.
SOL	CP	38.96	5.34	-7.62		298-2000	*
	** MPT= 2423						5

PHASE	T	CP	H	S	G	BT	
SOL	298	31.980	-338.088	34.9	-348.493	255.487	5*
	300	32.095	-338.029	35.098	-348.558	253.959	
	400	36.333	-334.581	44.988	-352.576	192.665	
	500	38.582	-330.826	53.359	-357.505	156.286	
	600	40.047	-326.890	60.530	-363.208	132.316	
	700	41.143	-322.828	66.789	-369.581	115.404	
	800	42.041	-318.668	72.343	-376.542	102.880	
	900	42.825	-314.424	77.341	-384.031	93.268	
	1000	43.538	-310.105	81.891	-391.996	85.682	
	1100	44.204	-305.718	86.072	-400.397	79.562	
	1200	44.839	-301.265	89.945	-409.200	74.535	
	1300	45.451	-296.751	93.559	-418.377	70.345	
	1400	46.047	-292.176	96.949	-427.904	66.808	
	1500	46.631	-287.542	100.146	-437.760	63.790	
	1600	47.206	-282.850	103.174	-447.928	61.192	
	1700	47.774	-278.100	106.053	-458.390	58.938	
	1800	48.337	-273.295	108.799	-469.134	56.968	
	1900	48.895	-268.433	111.428	-480.146	55.237	
	2000	49.449	-263.516	113.950	-491.416	53.707	

GALLIUM

		A	B	C	D	RANGE	REF.
SOL	CP	-12.344	62.366			298- 303	21
	** MPT=	302.8					21
LIQ	CP	5.829	0.547	0.742		303- 700	21
	CP	6.35				700-2478	21
GAS	CP	5.081				2478-2800	21

PHASE	T	CP	H	S	G	BT	P	
SOL	298	6.250	0	9.758	-2.909	2.133		21
	300	6.366	.012	9.797	-2.927	2.133		
	303	6.540	.030	9.857	-2.955	2.133		
			1.336	4.412				21
LIQ	303	6.804	1.366	14.269	-2.955	2.133		
	400	6.512	2.011	16.118	-4.437	2.424		
	500	6.399	2.655	17.557	-6.123	2.677		
	600	6.363	3.293	18.719	-7.939	2.892		
	700	6.363	3.929	19.700	-9.861	3.079		
	700	6.350	3.929	19.700	-9.861	3.079		
	800	6.350	4.564	20.548	-11.874	3.244		
	900	6.350	5.199	21.296	-13.967	3.392		
	1000	6.350	5.834	21.965	-16.131	3.526		
	1100	6.350	6.469	22.570	-18.358	3.648		
	1200	6.350	7.104	23.123	-20.643	3.760		
	1300	6.350	7.739	23.631	-22.981	3.864	.007	
	1400	6.350	8.374	24.102	-25.368	3.961	.042	
	1500	6.350	9.009	24.540	-27.800	4.051	.192	
	1600	6.350	9.644	24.949	-30.275	4.136	.721	
	1700	6.350	10.279	25.334	-32.790	4.216	2.306	
	1800	6.350	10.914	25.697	-35.341	4.292	6.470	
	1900	6.350	11.549	26.041	-37.928	4.363	16.258	
	2000	6.350	12.184	26.366	-40.549	4.432	37.199	
	2100	6.350	12.819	26.676	-43.201	4.497	78.545	
	2200	6.350	13.454	26.972	-45.884	4.559	154.744	
	2300	6.350	14.089	27.254	-48.595	4.618	287.053	
	2400	6.350	14.724	27.524	-51.334	4.675	505.189	
	2478	6.350	15.219	27.727	-53.489	4.718	760.	
			61.836	24.954				21
GAS	2478	5.081	77.055	52.681	-53.489	4.718		
	2500	5.081	77.167	52.726	-54.648	4.778		
	2600	5.081	77.675	52.925	-59.931	5.038		
	2700	5.081	78.183	53.117	-65.233	5.281		
	2800	5.081	78.691	53.302	-70.554	5.508		

GaCl (g) GALLIUM MONOCHLORIDE (GAS)

		A	B	C	D	RANGE	REF.
GAS	CP	9.082		-0.481		298-2000	*

PHASE	T	CP	H	S	G	BT	
GAS	298	8.541	-19.54	57.36	-36.642	26.863	62
	300	8.548	-19.524	57.413	-36.748	26.775	
	400	8.781	-18.656	59.909	-42.620	23.289	
	500	8.890	-17.772	61.881	-48.713	21.295	
	600	8.948	-16.880	63.508	-54.984	20.031	
	700	8.984	-15.983	64.890	-61.406	19.174	
	800	9.007	-15.083	66.091	-67.956	18.567	
	900	9.023	-14.182	67.153	-74.620	18.123	
	1000	9.034	-13.279	68.104	-81.383	17.789	
	1100	9.042	-12.375	68.966	-88.237	17.534	
	1200	9.049	-11.471	69.753	-95.174	17.336	
	1300	9.054	-10.566	70.477	-102.186	17.181	
	1400	9.057	-9.660	71.148	-109.267	17.060	
	1500	9.061	-8.754	71.773	-116.414	16.964	
	1600	9.063	-7.848	72.358	-123.621	16.888	
	1700	9.065	-6.941	72.908	-130.884	16.829	
	1800	9.067	-6.035	73.426	-138.201	16.782	
	1900	9.069	-5.128	73.916	-145.569	16.746	
	2000	9.070	-4.221	74.381	-152.984	16.720	

* * * * * * * * * *

$GaCl_3$ GALLIUM CHLORIDE

		A	B	C	D	RANGE	REF.
SOL	CP	28.3				298- 351	*
LIQ	CP	30.6				351- 575	*
	** BPT= 575, L= 15.0						1

PHASE	T	CP	H	S	G	BT	
SOL	298	28.300	-125.4	32.3	-135.030	98.993	62/*
	300	28.300	-125.348	32.475	-135.090	98.426	
	351	28.300	-123.904	36.918	-136.863	85.229	
			2.75	7.835			62
LIQ	351	30.600	-121.154	44.753	-136.863	85.229	
	400	30.600	-119.655	48.752	-139.156	76.041	
	500	30.600	-116.595	55.580	-144.385	63.119	
	575	30.600	-114.300	59.857	-148.718	56.533	

GALLIUM BROMIDE $GaBr_3$

		A	B	C	D	RANGE	REF.
SOL	CP	18.78	18.5			298- 396	*
LIQ	CP	30.0				396- 587	*
	** BPT= 587, L= 14.0						1

PHASE	T	CP	H	S	G	BT	
SOL	298	24.296	-92.4	43.	-105.220	77.139	1/*
	300	24.330	-92.355	43.150	-105.300	76.721	
	396	26.106	-89.934	50.140	-109.790	60.600	
			2.8	7.071			*
LIQ	396	30.000	-87.134	57.211	-109.790	60.600	
	400	30.000	-87.014	57.513	-110.019	60.120	
	500	30.000	-84.014	64.207	-116.118	50.762	
	587	30.000	-81.404	69.019	-121.918	45.398	

* * * * * * * * * *

GALLIUM IODIDE GaJ_3

		A	B	C	D	RANGE	REF.
SOL	CP	28.0				298- 486	*
LIQ	CP	30.7				486- 618	*
	** BPT=618.2						62

PHASE	T	CP	H	S	G	BT	
SOL	298	28.000	-57.2	48.7	-71.720	52.579	62/*
	300	28.000	-57.148	48.873	-71.810	52.321	
	400	28.000	-54.348	56.928	-77.120	42.142	
	486	28.000	-51.940	62.381	-82.257	36.995	
			5.3	10.905			62
LIQ	486	30.700	-46.640	73.286	-82.257	36.995	
	500	30.700	-46.210	74.158	-83.290	36.411	
	600	30.700	-43.140	79.756	-90.994	33.149	
	618	30.700	-42.582	80.673	-92.454	32.689	

Ga_2O_3 GALLIUM OXIDE

		A	B	C	D	RANGE	REF.
SOL	CP	26.98	3.69	-5.02		298-2068	145
	** MPT= 2068						1

PHASE	T	CP	H	S	G	BT	
SOL	298	22.433	-260.3	20.31	-266.355	195.270	62
	300	22.509	-260.258	20.449	-266.393	194.093	
	400	25.318	-257.850	27.360	-268.793	146.882	
	500	26.817	-255.237	33.184	-271.829	118.832	
	600	27.800	-252.503	38.165	-275.402	100.329	
	700	28.539	-249.685	42.508	-279.441	87.257	
	800	29.148	-246.800	46.360	-283.888	77.565	
	900	29.681	-243.858	49.825	-288.700	70.115	
	1000	30.168	-240.865	52.977	-293.842	64.228	
	1100	30.624	-237.825	55.874	-299.287	59.471	
	1200	31.059	-234.741	58.558	-305.010	55.557	
	1300	31.480	-231.614	61.060	-310.992	52.290	
	1400	31.890	-228.445	63.408	-317.217	49.526	
	1500	32.292	-225.236	65.622	-323.669	47.165	
	1600	32.688	-221.987	67.719	-330.337	45.128	
	1700	33.079	-218.698	69.712	-337.210	43.357	
	1800	33.467	-215.371	71.614	-344.277	41.807	
	1900	33.852	-212.005	73.434	-351.530	40.441	
	2000	34.234	-208.601	7[illegible].180	-358.961	39.231	
	2068	34.494	-206.264	76.329	-364.113	38.485	

DIGALLIUM SULFIDE (GAS) $Ga_2S(g)$

		A	B	C	D	RANGE	REF.
GAS	CP	13.385	0.275	-2.21		298-2000	63

PHASE	T	CP	H	S	G	BT	
GAS	298	10.981	5.	69.3	-15.662	11.482	63
	300	11.012	5.020	69.368	-15.790	11.505	
	400	12.114	6.184	72.709	-22.899	12.513	
	500	12.638	7.425	75.475	-30.313	13.251	
	600	12.936	8.705	77.807	-37.980	13.836	
	700	13.126	10.008	79.817	-45.863	14.321	
	800	13.260	11.328	81.579	-53.935	14.736	
	900	13.360	12.659	83.147	-62.173	15.100	
	1000	13.439	13.999	84.558	-70.559	15.423	
	1100	13.505	15.347	85.842	-79.080	15.714	
	1200	13.562	16.700	87.020	-87.724	15.979	
	1300	13.612	18.059	88.108	-96.481	16.222	
	1400	13.657	19.422	89.118	-105.343	16.447	
	1500	13.699	20.790	90.062	-114.303	16.656	
	1600	13.739	22.162	90.947	-123.353	16.852	
	1700	13.776	23.538	91.781	-132.490	17.035	
	1800	13.812	24.917	92.570	-141.708	17.208	
	1900	13.846	26.300	93.317	-151.003	17.372	
	2000	13.880	27.686	94.028	-160.370	17.527	

GaS — GALLIUM MONOSULFIDE

		A	B	C	D	RANGE	REF.
SOL	CP	9.88	3.75			298-1233	63
	** MPT= 1233						63

PHASE	T	CP	H	S	G	BT	
SOL	298	10.998	-50.	13.8	-54.114	39.672	63
	300	11.005	-49.980	13.868	-54.140	39.446	
	400	11.380	-48.860	17.085	-55.695	30.434	
	500	11.755	-47.704	19.665	-57.536	25.152	
	600	12.130	-46.509	21.841	-59.614	21.717	
	700	12.505	-45.278	23.739	-61.895	19.327	
	800	12.880	-44.008	25.434	-64.355	17.583	
	900	13.255	-42.702	26.972	-66.977	16.266	
	1000	13.630	-41.357	28.388	-69.746	15.245	
	1100	14.005	-39.976	29.705	-72.651	14.436	
	1200	14.380	-38.556	30.940	-75.684	13.786	
	1233	14.504	-38.080	31.331	-76.711	13.599	

* * * * * * * * * *

Ga_2S_3 — DIGALLIUM SULFIDE

		A	B	C	D	RANGE	REF.
SOL	CP	21.63	11.3			298-1363	63
	** MPT= 1363						63

PHASE	T	CP	H	S	G	BT	
SOL	298	24.999	-122.8	33.4	-132.758	97.327	63,120
	300	25.020	-122.754	33.555	-132.820	96.772	
	400	26.150	-120.195	40.907	-136.558	74.622	
	500	27.280	-117.524	46.864	-140.956	61.620	
	600	28.410	-114.739	51.937	-145.902	53.152	
	700	29.540	-111.842	56.402	-151.323	47.252	
	800	30.670	-108.831	60.420	-157.167	42.942	
	900	31.800	-105.708	64.098	-163.396	39.683	
	1000	32.930	-102.471	67.507	-169.978	37.154	
	1100	34.060	-99.122	70.698	-176.890	35.149	
	1200	35.190	-95.659	73.710	-184.112	33.536	
	1300	36.320	-92.084	76.572	-191.627	32.220	
	1363	37.032	-89.773	78.307	-196.506	31.513	

GALLIUM MONOSELENIDE

GaSe

		A	B	C	D	RANGE	REF.
SOL	CP	10.67	3.1			298-1233	63
	** MPT= 1233						63

PHASE	T	CP	H	S	G	BT	
SOL	298	11.594	-38.0	16.8	-43.009	31.531	63
	300	11.600	-37.979	16.872	-43.040	31.359	
	400	11.910	-36.803	20.251	-44.904	24.537	
	500	12.220	-35.597	22.942	-47.068	20.576	
	600	12.530	-34.359	25.198	-49.478	18.025	
	700	12.840	-33.091	27.152	-52.097	16.268	
	800	13.150	-31.791	28.887	-54.901	15.000	
	900	13.460	-30.461	30.454	-57.869	14.054	
	1000	13.770	-29.099	31.888	-60.987	13.331	
	1100	14.080	-27.707	33.215	-64.243	12.766	
	1200	14.390	-26.283	34.453	-67.627	12.318	
	1233	14.492	-25.806	34.845	-68.771	12.191	

* * * * * * * * * *

DIGALLIUM SELENIDE

Ga_2Se_3

		A	B	C	D	RANGE	REF.
SOL	CP	25.27	8.44			298-1293	63
	** MPT= 1293						62

PHASE	T	CP	H	S	G	BT	
SOL	298	27.786	-97.	46.4	-110.834	81.255	63*
	300	27.802	-96.949	46.572	-110.920	80.816	
	400	28.646	-94.126	54.686	-116.000	63.388	
	500	29.490	-91.219	61.168	-121.804	53.247	
	600	30.334	-88.228	66.620	-128.200	46.703	
	700	31.178	-85.153	71.359	-135.104	42.187	
	800	32.022	-81.993	75.577	-142.455	38.922	
	900	32.866	-78.748	79.398	-150.206	36.480	
	1000	33.710	-75.419	82.904	-158.324	34.606	
	1100	34.554	-72.006	86.157	-166.779	33.140	
	1200	35.398	-68.509	89.200	-175.548	31.976	
	1293	36.183	-65.180	91.871	-183.969	31.100	

$Ga_2(SeO_4)_3$ GALLIUM SELENATE

		A	B	C	D	RANGE	REF.
SOL	CP	43.8	75.0			298- 800	*

PHASE	T	CP	H	S	G	BT	
SOL	298	66.161	-472.8	72.5	-494.416	362.465	177/*
	300	66.300	-472.677	72.910	-494.550	360.328	
	400	73.800	-465.672	93.010	-502.877	274.796	
	500	81.300	-457.917	110.284	-513.059	224.288	
	600	88.800	-449.412	125.770	-524.874	191.211	
	700	96.300	-440.157	140.021	-538.172	168.048	
	800	103.800	-430.152	153.370	-552.848	151.051	

GALLIUM MONOTELLURICE — GaTe

		A	B	C	D	RANGE	REF.
SOL	CP	10.82	3.34			298-1108	63
	** MPT= 1108						62,63

PHASE	T	CP	H	S	G	BT	
SOL	298	11.816	-28.6	20.4	-34.682	25.426	1/123
	300	11.822	-28.578	20.473	-34.720	25.297	
	400	12.156	-27.379	23.920	-36.947	20.190	
	500	12.490	-26.147	26.668	-39.481	17.259	
	600	12.824	-24.881	28.975	-42.266	15.398	
	700	13.158	-23.582	30.977	-45.266	14.135	
	800	13.492	-22.250	32.756	-48.454	13.239	
	900	13.826	-20.884	34.364	-51.811	12.583	
	1000	14.160	-19.484	35.838	-55.323	12.092	
	1100	14.494	-18.052	37.203	-58.975	11.719	
	1108	14.521	-17.936	37.308	-59.273	11.693	

* * * * * * * * * *

DIGALLIUM TELLURIDE — Ga_2Te_3

		A	B	C	D	RANGE	REF.
SOL	CP	26.35	15.5			298- 943	*
	** TPT= 943, MPT= 1065						62

PHASE	T	CP	H	S	G	BT	
SOL	298	30.971	-65.	53.2	-80.862	59.281	1/67*
	300	31.000	-64.943	53.392	-80.960	58.987	
	400	32.550	-61.765	62.522	-86.774	47.417	
	500	34.100	-58.433	69.952	-93.409	40.834	
	600	35.650	-54.945	76.306	-100.729	36.695	
	700	37.200	-51.303	81.918	-108.645	33.925	
	800	38.750	-47.505	86.987	-117.094	31.993	
	900	40.300	-43.553	91.640	-126.029	30.608	
	943	40.966	-41.805	93.536	-130.010	30.135	

GaN

GALLIUM NITRIDE

		A	B	C	D	RANGE	REF.
SOL	CP	9.1	2.15			298-1773	*
	** MPT= 1773						62

PHASE	T	CP	H	S	G	BT	
SOL	298	9.741	-26.2	7.1	-28.317	20.760	62/*
	300	9.745	-26.182	7.160	-28.330	20.641	
	400	9.960	-25.197	9.993	-29.194	15.953	
	500	10.175	-24.190	12.239	-30.309	13.250	
	600	10.390	-23.162	14.113	-31.629	11.523	
	700	10.605	-22.112	15.731	-33.123	10.343	
	800	10.820	-21.041	17.161	-34.769	9.500	
	900	11.035	-19.948	18.448	-36.551	8.877	
	1000	11.250	-18.834	19.621	-38.455	8.405	
	1100	11.465	-17.698	20.704	-40.472	8.042	
	1200	11.680	-16.541	21.711	-42.593	7.758	
	1300	11.895	-15.362	22.654	-44.812	7.535	
	1400	12.110	-14.162	23.543	-47.122	7.357	
	1500	12.325	-12.940	24.386	-49.519	7.216	
	1600	12.540	-11.697	25.188	-51.998	7.104	
	1700	12.755	-10.432	25.955	-54.556	7.015	
	1773	12.912	-9.495	26.495	-56.470	6.962	

GALLIUM PHOSPHIDE

GaP

		A	B	C	D	RANGE	REF.
SOL	CP	10.0	1.63			298-1790	62*
	** MPT=	1790					62

PHASE	T	CP	H	S	G	BT	
SOL	298	10.486	-29.2	12.5	-32.927	24.139	226
	300	10.489	-29.181	12.565	-32.950	24.007	
	400	10.652	-28.124	15.605	-34.365	18.779	
	500	10.815	-27.050	17.999	-36.050	15.759	
	600	10.978	-25.961	19.985	-37.952	13.826	
	700	11.141	-24.855	21.690	-40.037	12.502	
	800	11.304	-23.732	23.188	-42.283	11.553	
	900	11.467	-22.594	24.529	-44.670	10.849	
	1000	11.630	-21.439	25.746	-47.185	10.314	
	1100	11.793	-20.268	26.862	-49.816	9.899	
	1200	11.956	-19.080	27.895	-52.554	9.573	
	1300	12.119	-17.877	28.858	-55.392	9.314	
	1400	12.282	-16.657	29.762	-58.324	9.106	
	1500	12.445	-15.420	30.615	-61.343	8.939	
	1600	12.608	-14.168	31.424	-64.445	8.804	
	1700	12.771	-12.899	32.193	-67.627	8.695	
	1790	12.918	-11.743	32.855	-70.554	8.615	

GaAs

GALLIUM ARSENIDE

		A	B	C	D	RANGE	REF.
SOL	CP	10.80	1.45			298-1238	5*
LIQ	CP	14.1				1238-1400	*

PHASE	T	CP	H	S	G	BT	
SOL	298	11.232	-19.5	15.35	-24.077	17.651	5*
	300	11.235	-19.479	15.419	-24.105	17.563	
	400	11.380	-18.348	18.671	-25.817	14.108	
	500	11.525	-17.203	21.226	-27.816	12.160	
	600	11.670	-16.043	23.340	-30.048	10.946	
	700	11.815	-14.869	25.150	-32.474	10.140	
	800	11.960	-13.680	26.737	-35.070	9.582	
	900	12.105	-12.477	28.155	-37.816	9.184	
	1000	12.250	-11.259	29.437	-40.697	8.895	
	1100	12.395	-10.027	30.612	-43.700	8.684	
	1200	12.540	-8.780	31.696	-46.816	8.528	
	1238	12.595	-8.303	32.088	-48.028	8.480	
			21.	16.963			5*
LIQ	1238	14.100	12.697	49.051	-48.028	8.480	
	1300	14.100	13.571	49.740	-51.091	8.590	
	1400	14.100	14.981	50.785	-56.118	8.762	

* * * * * * * * * *

GaSb

GALLIUM ANTIMONIDE

		A	B	C	D	RANGE	REF.
SOL	CP	10.9	3.0			298- 985	5*
	** TPT= 985 / MPT= 1216						62/1

PHASE	T	CP	H	S	G	BT	
SOL	298	11.794	-10.	18.48	-15.510	11.371	5
	300	11.800	-9.978	18.553	-15.544	11.325	
	400	12.100	-8.783	21.989	-17.579	9.606	
	500	12.400	-7.558	24.721	-19.919	8.708	
	600	12.700	-6.303	27.008	-22.508	8.200	
	700	13.000	-5.018	28.989	-25.310	7.903	
	800	13.300	-3.703	30.744	-28.298	7.732	
	900	13.600	-2.358	32.328	-31.453	7.639	
	985	13.855	-1.191	33.567	-34.254	7.601	

GADOLINIUM

		A	B	C	D	RANGE	REF.
SOL-A	CP	-5.461	18.068	7.941		298- 500	21
	CP	5.721	2.060			500-1100	21
	CP	4.763	2.915			1100-1533	21
SOL-B	CP	6.76				1533-1585	21
LIQ	CP	8.88				1585-2400	21
	** BPT= 3539, L= 85.89						21

PHASE	T	CP	H	S	G	BT	P	
SOL-A	298	8.859	0	16.24	-4.842	3.550		21
	300	8.783	.016	16.295	-4.872	3.550		
	400	6.729	.764	18.460	-6.620	3.617		
	500	6.749	1.428	19.942	-8.543	3.734		
	500	6.751	1.428	19.942	-8.543	3.734		
	600	6.957	2.114	21.191	-10.601	3.862		
	700	7.163	2.820	22.279	-12.776	3.989		
	800	7.369	3.546	23.249	-15.053	4.113		
	900	7.575	4.294	24.129	-17.422	4.231		
	1000	7.781	5.061	24.938	-19.876	4.345		
	1100	7.987	5.850	25.689	-22.408	4.453		
	1100	7.969	5.850	25.689	-22.408	4.453		
	1200	8.261	6.661	26.395	-25.012	4.556		
	1300	8.552	7.502	27.067	-27.686	4.655		
	1400	8.844	8.372	27.712	-30.425	4.750		
	1500	9.135	9.271	28.332	-33.227	4.842		
	1533	9.232	9.574	28.532	-34.166	4.871		
			0.935	.610				21
SOL-B	1533	6.760	10.509	29.142	-34.166	4.871		
	1585	6.760	10.860	29.367	-35.687	4.921		
			2.403	1.516				21
LIQ	1585	8.880	13.263	30.883	-35.687	4.921		
	1600	8.880	13.397	30.967	-36.151	4.939		
	1700	8.880	14.285	31.505	-39.275	5.050	.001	
	1800	8.880	15.173	32.013	-42.451	5.155	.005	
	1900	8.880	16.061	32.493	-45.676	5.255	.017	
	2000	8.880	16.949	32.949	-48.949	5.350	.054	
	2100	8.880	17.837	33.382	-52.265	5.440	.157	
	2200	8.880	18.725	33.795	-55.624	5.527	.409	
	2300	8.880	19.613	34.190	-59.024	5.609	.978	
	2400	8.880	20.501	34.568	-62.462	5.689	2.174	

$GdCl_3$ GADOLINIUM CHLORIDE

		A	B	C	D	RANGE	REF.
SOL	CP	20.67	8.20	0.34		298- 875	89
LIQ	CP	33.345				875-1000	89

PHASE	T	CP	H	S	G	BT	
SOL	298	23.497	-240.1	34.9	-250.505	183.650	1
	300	23.508	-240.057	35.045	-250.570	182.565	
	400	24.162	-237.674	41.894	-254.432	139.034	
	500	24.906	-235.221	47.365	-258.904	113.182	
	600	25.684	-232.692	51.974	-263.876	96.130	
	700	26.479	-230.084	55.993	-269.279	84.084	
	800	27.283	-227.396	59.581	-275.061	75.153	
	875	27.889	-225.327	62.053	-279.623	69.851	
			9.7	11.086			89
LIQ	875	33.345	-215.627	73.139	-279.623	69.851	
	900	33.345	-214.793	74.078	-281.463	68.358	
	1000	33.345	-211.459	77.591	-289.050	63.180	

GADOLINIUM IODIDE

GdJ_3

		A	B	C	D	RANGE	REF.
SOL-A	CP	24.29	1.80	-1.00		298-1013	89
SOL-B	CP	30.64				1013-1203	89
LIQ	CP	37.25				1203-1300	89

PHASE	T	CP	H	S	G	BT	
SOL-A	298	23.702	-142.	51.8	-157.444	115.425	19/*
	300	23.719	-141.956	51.947	-157.540	114.783	
	400	24.385	-139.547	58.871	-163.096	89.124	
	500	24.790	-137.087	64.359	-169.267	73.997	
	600	25.092	-134.593	68.907	-175.937	64.094	
	700	25.346	-132.071	72.794	-183.026	57.151	
	800	25.574	-129.524	76.194	-190.479	52.044	
	900	25.787	-126.956	79.218	-198.253	48.149	
	1000	25.990	-124.367	81.946	-206.313	45.096	
	1013	26.016	-124.029	82.281	-207.381	44.747	
			0.14	.138			89
SOL-B	1013	30.640	-123.889	82.420	-207.381	44.747	
	1100	30.640	-121.224	84.944	-214.662	42.655	
	1200	30.640	-118.160	87.610	-223.292	40.673	
	1203	30.640	-118.068	87.687	-223.555	40.619	
			12.9	10.723			89
LIQ	1203	37.250	-105.168	98.410	-223.555	40.619	
	1300	37.250	-101.555	101.299	-233.243	39.217	

GdOCl

GADOLINIUM OXYCHLORIDE

		A	B	C	D	RANGE	REF.
SOL	CP	16.0	3.92	-0.74		298-1000	*

PHASE	T	CP	H	S	G	BT	
SOL	298	16.336	-235.1	22.8	-241.898	177.340	100/*
	300	16.354	-235.070	22.901	-241.940	176.277	
	400	17.105	-233.394	27.716	-244.481	133.596	
	500	17.664	-231.655	31.595	-247.452	108.176	
	600	18.146	-229.864	34.859	-250.779	91.359	
	700	18.593	-228.027	37.690	-254.410	79.441	
	800	19.020	-226.146	40.201	-258.307	70.576	
	900	19.437	-224.223	42.465	-262.442	63.738	
	1000	19.846	-222.259	44.535	-266.793	58.315	

GERMANIUM

		A	B	C	D	RANGE	REF.
SOL	CP	6.158	0.019	-0.517		298- 600	21
	CP	5.420	0.990			600- 900	21
	CP	4.745	1.758			900-1210	21
	** MPT= 1210.4						21
LIQ	CP	6.6				1210-2600	21
	** BPT= 3107, L= 79.104						21

PHASE	T	CP	H	S	G	BT	P	
SOL	298	5.582	0	7.43	-2.215	1.624		21
	300	5.589	.010	7.465	-2.229	1.624		
	400	5.842	.584	9.112	-3.061	1.673		
	500	5.961	1.175	10.430	-4.041	1.766		
	600	6.026	1.774	11.523	-5.140	1.872		
	600	6.014	1.774	11.523	-5.140	1.872		
	700	6.113	2.380	12.458	-6.340	1.980		
	800	6.212	2.997	13.280	-7.628	2.084		
	900	6.311	3.623	14.018	-8.993	2.184		
	900	6.327	3.623	14.018	-8.993	2.184		
	1000	6.503	4.264	14.694	-10.429	2.280		
	1100	6.679	4.923	15.322	-11.930	2.371		
	1200	6.855	5.600	15.910	-13.492	2.458		
	1210	6.873	5.672	15.970	-13.658	2.466		
			8.83	7.295				21
LIQ	1210	6.600	14.502	23.265	-13.658	2.466		
	1300	6.600	15.093	23.736	-15.764	2.651		
	1400	6.600	15.753	24.225	-18.162	2.836		
	1500	6.600	16.413	24.680	-20.608	3.003		
	1600	6.600	17.073	25.106	-23.097	3.155	.004	
	1700	6.600	17.733	25.507	-25.628	3.295	.016	
	1800	6.600	18.393	25.884	-28.198	3.424	.058	
	1900	6.600	19.053	26.241	-30.804	3.544	.184	
	2000	6.600	19.713	26.579	-33.445	3.655	.521	
	2100	6.600	20.373	26.901	-36.120	3.760	1.327	
	2200	6.600	21.033	27.208	-38.825	3.857	3.089	
	2300	6.600	21.693	27.502	-41.561	3.950	6.645	
	2400	6.600	22.353	27.782	-44.325	4.037	13.337	
	2500	6.600	23.013	28.052	-47.117	4.120	25.187	
	2600	6.600	23.673	28.311	-49.935	4.198	45.065	

GeH_4 (g) GERMANIUM TETRAHYDRIDE (GAS)

		A	B	C	D	RANGE	REF.
GAS	CP	14.95	5.28	-5.17		298-2000	62

PHASE	T	CP	H	S	G	BT	
GAS	298	10.708	21.7	51.9	6.226	-4.564	62
	300	10.790	21.720	51.966	6.130	-4.466	
	400	13.831	22.969	55.539	.753	-.412	
	500	15.522	24.443	58.821	-4.968	2.172	
	600	16.682	26.056	61.759	-10.999	4.007	
	700	17.591	27.771	64.401	-17.310	5.405	
	800	18.366	29.570	66.802	-23.871	6.522	
	900	19.064	31.442	69.006	-30.663	7.447	
	1000	19.713	33.381	71.048	-37.667	8.233	
	1100	20.331	35.383	72.956	-44.868	8.916	
	1200	20.927	37.446	74.751	-52.255	9.518	
	1300	21.508	39.568	76.449	-59.815	10.057	
	1400	22.078	41.748	78.064	-67.542	10.545	
	1500	22.640	43.984	79.606	-75.426	10.991	
	1600	23.196	46.275	81.085	-83.461	11.402	
	1700	23.747	48.623	82.508	-91.641	11.783	
	1800	24.294	51.025	83.881	-99.961	12.139	
	1900	24.839	53.481	85.209	-108.416	12.472	
	2000	25.381	55.992	86.497	-117.001	12.787	

GERMANIUM DIFLUORICE

GeF_2

		A	B	C	D	RANGE	REF.
SOL	CP	20.0				298- 385	*
LIQ	CP	21.8				385- 500	*

PHASE	T	CP	H	S	G	BT	
SOL	298	20.000	-157.35	20.1	-163.343	119.750	66/*
	300	20.000	-157.313	20.224	-163.380	119.038	
	385	20.000	-155.613	25.213	-165.320	93.858	
			2.3	5.974			*
LIQ	385	21.800	-153.313	31.187	-165.320	93.858	
	400	21.800	-152.986	32.020	-165.794	90.598	
	500	21.800	-150.806	36.885	-169.248	73.988	

* * * * * * * * * *

GERMANIUM TETRAFLUORIDE (GAS)

GeF_4 (g)

		A	B	C	D	RANGE	REF.
GAS	CP	23.33	2.0	-3.91		298-1000	*

PHASE	T	CP	H	S	G	BT	
GAS	298	19.528	-284.87	72.35	-306.441	224.658	104/62
	300	19.586	-284.834	72.471	-306.575	223.370	
	400	21.686	-282.757	78.432	-314.130	171.655	
	500	22.766	-280.529	83.398	-322.228	140.865	
	600	23.444	-278.216	87.613	-330.784	120.504	
	700	23.932	-275.847	91.265	-339.732	106.083	
	800	24.319	-273.433	94.487	-349.023	95.361	
	900	24.647	-270.985	97.371	-358.618	87.096	
	1000	24.939	-268.505	99.983	-368.488	80.544	

$GeCl_4$ GERMANIUM TETRACHLORIDE

		A	B	C	D	RANGE	REF.
LIQ	CP	36.25				298- 357	*
	** BPT=	356.27					62
GAS	CP	25.57		-2.28		356-1000	62,*

PHASE	T	CP	H	S	G	BT	P	
LIQ	298	36.250	-129.	59.33	-146.689	107.541	82.841	62/*
	300	36.250	-128.933	59.554	-146.799	106.958	90.400	
	356	36.250	-126.893	65.786	-150.331	92.231	760.	
			7.654	21.484				62,*
GAS	356	23.774	-119.239	87.270	-150.331	92.231		
	400	24.145	-118.191	90.044	-154.209	84.267		
	500	24.658	-115.748	95.494	-163.495	71.473		
	600	24.937	-113.267	100.016	-173.277	63.124		
	700	25.105	-110.764	103.874	-183.476	57.291		
	800	25.214	-108.248	107.234	-194.035	53.015		
	900	25.289	-105.723	110.208	-204.910	49.766		
	1000	25.342	-103.191	112.875	-216.066	47.228		

GERMANIUM TETRABROMIDE (GAS) $GeBr_4$ (g)

		A	B	C	D	RANGE	REF.
GAS	CP	25.66	0.16	-1.22		298-1000	58

PHASE	T	CP	H	S	G	BT	
GAS	298	24.335	-71.4	94.7	-99.635	73.044	62
	300	24.352	-71.355	94.851	-99.810	72.721	
	400	24.961	-68.885	101.952	-109.666	59.927	
	500	25.252	-66.373	107.557	-120.151	52.525	
	600	25.417	-63.839	112.176	-131.145	47.776	
	700	25.523	-61.291	116.103	-142.563	44.516	
	800	25.597	-58.735	119.516	-154.348	42.172	
	900	25.653	-56.172	122.535	-166.454	40.426	
	1000	25.698	-53.605	125.240	-178.845	39.092	

* * * * * * * * * *

GERMANIUM TETRAIODIDE (GAS) GeJ_4 (g)

		A	B	C	D	RANGE	REF.
GAS	CP	26.10	0.24	-1.10		298- 800	62

PHASE	T	CP	H	S	G	BT	
GAS	298	24.934	-9.	102.5	-39.560	29.002	62
	300	24.950	-8.954	102.654	-39.750	28.962	
	400	25.508	-6.427	109.919	-50.395	27.538	
	500	25.780	-3.861	115.644	-61.683	26.965	
	600	25.938	-1.275	120.359	-73.490	26.772	
	700	26.044	1.325	124.366	-85.732	26.770	
	800	26.120	3.933	127.849	-98.346	26.870	

GeO (g)

GERMANIUM MONOXIDE (GAS)

		A	B	C	D	RANGE	REF.
GAS	CP	8.85		-1.35		298-2000	62

PHASE	T	CP	H	S	G	BT	
GAS	298	7.331	-7.33	53.49	-23.278	17.066	62
	300	7.350	-7.316	53.535	-23.377	17.032	
	400	8.006	-6.544	55.753	-28.845	15.762	
	500	8.310	-5.726	57.576	-34.515	15.088	
	600	8.475	-4.886	59.107	-40.351	14.700	
	700	8.574	-4.034	60.422	-46.329	14.466	
	800	8.639	-3.173	61.571	-52.430	14.325	
	900	8.683	-2.306	62.591	-58.639	14.241	
	1000	8.715	-1.436	63.508	-64.944	14.196	
	1100	8.738	-.564	64.340	-71.338	14.175	
	1200	8.756	.311	65.101	-77.810	14.173	
	1300	8.770	1.187	65.802	-84.356	14.183	
	1400	8.781	2.065	66.453	-90.969	14.203	
	1500	8.790	2.944	67.059	-97.645	14.229	
	1600	8.797	3.823	67.626	-104.379	14.259	
	1700	8.803	4.703	68.160	-111.169	14.294	
	1800	8.808	5.584	68.663	-118.010	14.330	
	1900	8.813	6.465	69.140	-124.901	14.369	
	2000	8.816	7.346	69.592	-131.838	14.408	

GERMANIUM DIOXIDE (TETRAGONAL)

		A	B	C	D	RANGE	REF.
SOL-A	CP	15.92	2.77	-4.24		298-1308	71
	**	TPT= 1308 (TO HEXAGONAL)					1
SOL-B	CP	16.47	2.35	-4.23		1308-1389	71
	**	MPT= 1389, L= 10.5					1

PHASE	T	CP	H	S	G	BT	
SOL-A	298	11.976	-138.6	9.5	-141.432	103.687	71
	300	12.040	-138.578	9.574	-141.450	103.060	
	400	14.378	-137.242	13.401	-142.602	77.925	
	500	15.609	-135.738	16.753	-144.114	63.001	
	600	16.404	-134.135	19.674	-145.939	53.165	
	700	16.994	-132.463	22.248	-148.037	46.226	
	800	17.473	-130.739	24.550	-150.379	41.087	
	900	17.890	-128.971	26.632	-152.940	37.144	
	1000	18.266	-127.163	28.537	-155.700	34.033	
	1100	18.617	-125.318	30.295	-158.642	31.524	
	1200	18.950	-123.440	31.929	-161.755	29.463	
	1300	19.270	-121.529	33.458	-165.025	27.747	
	1308	19.295	-121.375	33.577	-165.293	27.622	
			5.369	4.105			*
SOL-B	1308	19.297	-116.006	37.681	-165.293	27.622	
	1389	19.515	-114.434	38.847	-168.393	26.499	

GeO_2 GERMANIUM DIOXIDE (HEX.)

		A	B	C	D	RANGE	REF.
SOL-B	CP	16.47	2.35	-4.23		298-1389	71
	** MPT= 1389, L= 10.5						1

PHASE	T	CP	H	S	G	BT	
SOL-B	298	12.412	-133.449	13.21	-137.388	100.721	*/58
	300	12.475	-133.426	13.287	-137.412	100.118	
	400	14.766	-132.049	17.232	-138.942	75.925	
	500	15.953	-130.508	20.666	-140.841	61.570	
	600	16.705	-128.873	23.646	-143.060	52.117	
	700	17.252	-127.174	26.264	-145.558	45.451	
	800	17.689	-125.426	28.597	-148.303	40.520	
	900	18.063	-123.638	30.702	-151.270	36.738	
	1000	18.397	-121.815	32.623	-154.438	33.757	
	1100	18.705	-119.959	34.391	-157.789	31.354	
	1200	18.996	-118.074	36.031	-161.312	29.383	
	1300	19.275	-116.161	37.563	-164.992	27.741	
	1308	19.297	-116.006	37.681	-165.293	27.622	
	1308	19.297	-116.006	37.681	-165.293	27.622	
	1389	19.515	-114.434	38.847	-168.393	26.499	

GERMANIUM MONOSULFICE

		A	B	C	D	RANGE	REF.
SOL	CP	9.99	4.81			298- 938	63
LIQ	CP	14.5				938-1111	63
	** BPT=	1111.43					
GAS	CP	8.78	0.1	-0.68		1111-2000	63

PHASE	T	CP	H	S	G	BT	P	
SOL	298	11.424	-18.2	15.77	-22.902	16.790		63
	300	11.433	-18.179	15.841	-22.931	16.708		
	400	11.914	-17.012	19.196	-24.690	13.492		
	500	12.395	-15.796	21.906	-26.749	11.694		
	600	12.876	-14.533	24.208	-29.057	10.586		
	700	13.357	-13.221	26.229	-31.581	9.861	.038	
	800	13.838	-11.861	28.044	-34.296	9.371	1.347	
	900	14.319	-10.453	29.702	-37.185	9.031	20.728	
	938	14.502	-9.906	30.298	-38.325	8.931	49.828	
			5.6	5.970				63
LIQ	938	14.500	-4.306	36.268	-38.325	8.931	49.828	
	1000	14.500	-3.407	37.196	-40.603	8.875	148.545	
	1100	14.500	-1.957	38.578	-44.392	8.821	653.529	
	1111	14.500	-1.791	38.728	-44.834	8.817	760.	
			32.022	28.812				*
GAS	1111	8.836	30.231	67.539	-44.834	8.817		
	1200	8.853	31.014	68.218	-50.847	9.262		
	1300	8.870	31.901	68.927	-57.704	9.702		
	1400	8.885	32.788	69.585	-64.630	10.091		
	1500	8.900	33.678	70.198	-71.620	10.436		
	1600	8.913	34.568	70.773	-78.669	10.747		
	1700	8.926	35.460	71.314	-85.773	11.028		
	1800	8.939	36.354	71.824	-92.930	11.285		
	1900	8.951	37.248	72.308	-100.137	11.520		
	2000	8.963	38.144	72.767	-107.391	11.737		

GeS_2 GERMANIUM DISULFIDE

		A	B	C	D	RANGE	REF.
SOL	CP	13.49	7.42			298-1113	63
	** DEC.						63
LIQ	CP	22.4				1113-1200	*

PHASE	T	CP	H	S	G	BT	
SOL	298	15.702	-37.5	20.9	-43.731	32.060	63
	300	15.716	-37.471	20.997	-43.770	31.891	
	400	16.458	-35.862	25.620	-46.110	25.197	
	500	17.200	-34.179	29.372	-48.865	21.362	
	600	17.942	-32.422	32.574	-51.966	18.931	
	700	18.684	-30.591	35.395	-55.368	17.289	
	800	19.426	-28.685	37.939	-59.036	16.130	
	900	20.168	-26.706	40.269	-62.948	15.288	
	1000	20.910	-24.652	42.433	-67.085	14.663	
	1100	21.652	-22.524	44.461	-71.430	14.194	
	1113	21.748	-22.242	44.715	-72.010	14.142	
			10.	8.985			63
LIQ	1113	22.400	-12.242	53.700	-72.010	14.142	
	1200	22.400	-10.293	55.386	-76.756	13.981	

GERMANIUM MONOSELENIDE

GeSe

		A	B	C	D	RANGE	REF.
SOL	CP	10.78	3.92			298- 948	63
LIQ	CP	15.2				948-1125	63
	** BPT=	1125.39					
GAS	CP	8.84	0.04	-0.42		1125-2000	*

PHASE	T	CP	H	S	G	BT	P	
SOL	298	11.949	-16.5	18.7	-22.075	16.184		63
	300	11.956	-16.478	18.774	-22.110	16.109		
	400	12.348	-15.263	22.267	-24.170	13.207		
	500	12.740	-14.008	25.065	-26.541	11.602		
	600	13.132	-12.715	27.422	-29.168	10.626		
	700	13.524	-11.382	29.476	-32.015	9.997	.030	
	800	13.916	-10.010	31.307	-35.056	9.578	1.079	
	900	14.308	-8.599	32.969	-38.271	9.295	16.860	
	948	14.496	-7.907	33.717	-39.871	9.193	50.737	
			5.9	6.224				63
LIQ	948	15.200	-2.007	39.941	-39.871	9.193	50.737	
	1000	15.200	-1.217	40.753	-41.970	9.174	125.171	
	1100	15.200	.303	42.201	-46.118	9.164	547.040	
	1125	15.200	.689	42.548	-47.194	9.166	760.	
			31.744	28.207				63,*
GAS	1125	8.852	32.433	70.755	-47.194	9.166		
	1200	8.859	33.094	71.324	-52.495	9.562		
	1300	8.867	33.980	72.033	-59.663	10.032		
	1400	8.875	34.867	72.691	-66.900	10.445		
	1500	8.881	35.755	73.303	-74.200	10.812		
	1600	8.888	36.643	73.876	-81.559	11.142		
	1700	8.893	37.532	74.415	-88.974	11.440		
	1800	8.899	38.422	74.924	-96.441	11.711		
	1900	8.904	39.312	75.405	-103.958	11.959		
	2000	8.909	40.203	75.862	-111.521	12.188		

$GeSe_2$ GERMANIUM DISELENICE

		A	B	C	D	RANGE	REF.
SOL	CP	15.03	6.63			298-1013	63
	** MPT=	1013					63

PHASE	T	CP	H	S	G	BT	
SOL	298	17.007	-27.	26.9	-35.020	25.674	63
	300	17.019	-26.969	27.005	-35.070	25.552	
	400	17.682	-25.233	31.992	-38.030	20.782	
	500	18.345	-23.432	36.009	-41.437	18.114	
	600	19.008	-21.564	39.412	-45.212	16.471	
	700	19.671	-19.631	42.392	-49.305	15.396	
	800	20.334	-17.630	45.062	-53.680	14.667	
	900	20.997	-15.564	47.495	-58.310	14.161	
	1000	21.660	-13.431	49.742	-63.173	13.808	
	1013	21.746	-13.149	50.022	-63.821	13.771	

* * * * * * * * * *

GeTe GERMANIUM MONOTELLURIDE

		A	B	C	D	RANGE	REF.
SOL	CP	11.50	3.0			298- 997	193,63
	** MPT=	997					63

PHASE	T	CP	H	S	G	BT	
SOL	298	12.394	-7.8	21.5	-14.210	10.418	62/193
	300	12.400	-7.777	21.577	-14.250	10.383	
	400	12.700	-6.522	25.185	-16.596	9.069	
	500	13.000	-5.237	28.051	-19.263	8.421	
	600	13.300	-3.922	30.448	-22.191	8.084	
	700	13.600	-2.577	32.521	-25.341	7.913	
	800	13.900	-1.202	34.356	-28.687	7.838	
	900	14.200	.203	36.011	-32.207	7.822	
	997	14.491	1.594	37.479	-35.772	7.843	

GERMANIUM PHOSPHIDE

GeP

		A	B	C	D	RANGE	REF.
SOL	CP	10.35	2.7	-1.25		298-1700	*

PHASE	T	CP	H	S	G	BT	
SOL	298	9.749	-6.5	14.6	-10.853	7.957	1
	300	9.771	-6.482	14.660	-10.880	7.927	
	400	10.649	-5.457	17.604	-12.498	6.830	
	500	11.200	-4.363	20.043	-14.384	6.288	
	600	11.623	-3.221	22.124	-16.495	6.009	
	700	11.985	-2.040	23.943	-18.800	5.870	
	800	12.315	-.825	25.565	-21.277	5.813	
	900	12.626	.422	27.034	-23.908	5.806	
	1000	12.925	1.700	28.380	-26.680	5.832	
	1100	13.217	3.007	29.625	-29.581	5.878	
	1200	13.503	4.343	30.787	-32.602	5.938	
	1300	13.786	5.708	31.879	-35.736	6.009	
	1400	14.066	7.100	32.911	-38.976	6.085	
	1500	14.344	8.521	33.891	-42.316	6.166	
	1600	14.621	9.969	34.826	-45.753	6.250	
	1700	14.897	11.445	35.721	-49.280	6.336	

HF(g)

HYDROGEN FLUORIDE (GAS)

		A	B	C	D	RANGE	REF.
GAS	CP	6.43	0.82	0.26		298-2500	1

** MPT= 190,L= 0.94,BPT= 292.65/GAS(HF+H2F2+...) 1

PHASE	T	CP	H	S	G	BT	
GAS	298	6.967	-65.14	41.508	-77.516	56.828	3,19
	300	6.965	-65.127	41.551	-77.592	56.534	
	400	6.920	-64.434	43.546	-81.852	44.728	
	500	6.944	-63.741	45.092	-86.287	37.721	
	600	6.994	-63.044	46.362	-90.861	33.101	
	700	7.057	-62.342	47.445	-95.553	29.837	
	800	7.127	-61.632	48.392	-100.346	27.417	
	900	7.200	-60.916	49.236	-105.228	25.556	
	1000	7.276	-60.192	49.998	-110.190	24.085	
	1100	7.353	-59.461	50.695	-115.226	22.896	
	1200	7.432	-58.722	51.338	-120.328	21.918	
	1300	7.511	-57.974	51.936	-125.492	21.100	
	1400	7.591	-57.219	52.496	-130.714	20.408	
	1500	7.672	-56.456	53.022	-135.990	19.816	
	1600	7.752	-55.685	53.520	-141.317	19.306	
	1700	7.833	-54.906	53.993	-146.693	18.861	
	1800	7.914	-54.118	54.443	-152.115	18.472	
	1900	7.995	-53.323	54.873	-157.581	18.128	
	2000	8.076	-52.519	55.285	-163.089	17.824	
	2100	8.158	-51.708	55.681	-168.637	17.553	
	2200	8.239	-50.888	56.062	-174.225	17.310	
	2300	8.321	-50.060	56.430	-179.849	17.092	
	2400	8.403	-49.224	56.786	-185.510	16.895	
	2500	8.484	-48.379	57.131	-191.206	16.717	

WATER (GAS) H_2O (g)

		A	B	C	D	RANGE	REF.
GAS	CP	7.17	2.56	0.08		298-2500	1

PHASE	T	CP	H	S	G	BT	
GAS	298	8.023	-57.795	45.106	-71.243	52.230	40
	300	8.027	-57.780	45.156	-71.327	51.969	
	400	8.244	-56.967	47.494	-75.964	41.511	
	500	8.482	-56.131	49.359	-80.810	35.327	
	600	8.728	-55.270	50.927	-85.826	31.266	
	700	8.978	-54.385	52.291	-90.989	28.412	
	800	9.230	-53.474	53.506	-96.280	26.306	
	900	9.484	-52.539	54.608	-101.686	24.696	
	1000	9.738	-51.578	55.621	-107.198	23.431	
	1100	9.993	-50.591	56.561	-112.808	22.416	
	1200	10.248	-49.579	57.441	-118.508	21.586	
	1300	10.503	-48.542	58.271	-124.294	20.899	
	1400	10.758	-47.479	59.059	-130.161	20.322	
	1500	11.014	-46.390	59.810	-136.105	19.833	
	1600	11.269	-45.276	60.529	-142.122	19.416	
	1700	11.525	-44.136	61.220	-148.210	19.056	
	1800	11.780	-42.971	61.886	-154.365	18.745	
	1900	12.036	-41.780	62.530	-160.586	18.474	
	2000	12.292	-40.564	63.153	-166.871	18.237	
	2100	12.548	-39.322	63.759	-173.216	18.029	
	2200	12.804	-38.054	64.349	-179.622	17.846	
	2300	13.060	-36.761	64.924	-186.086	17.685	
	2400	13.315	-35.442	65.485	-192.606	17.542	
	2500	13.571	-34.098	66.034	-199.182	17.415	

$H_2S_2(g)$ DIHYDROGEN DISULFIDE (GAS)

		A	B	C	D	RANGE	REF.
GAS	CP	12.28	3.866	-1.005		298-1000	63

PHASE	T	CP	H	S	G	BT	
GAS	298	12.302	3.76	63.66	-15.220	11.158	63
	300	12.323	3.783	63.736	-15.338	11.175	
	400	13.198	5.062	67.411	-21.902	11.968	
	500	13.811	6.414	70.425	-28.798	12.589	
	600	14.320	7.821	72.989	-35.972	13.105	
	700	14.781	9.277	75.232	-43.386	13.547	
	800	15.216	10.777	77.234	-51.011	13.937	
	900	15.635	12.319	79.050	-58.826	14.287	
	1000	16.045	13.903	80.719	-66.816	14.605	

* * * * * * * * * *

H_2CS_3 TRITHIO-CARBONIC ACID

		A	B	C	D	RANGE	REF.
LIQ	CP	35.5				298- 331	207
	** DEC.						

PHASE	T	CP	H	S	G	BT	
LIQ	298	35.500	6.1	53.3	-9.791	7.178	207
	300	35.500	6.166	53.520	-9.890	7.206	
	331	35.500	7.266	57.011	-11.604	7.663	

HYDROGEN SELENIDE (GAS) $H_2Se(g)$

		A	B	C	D	RANGE	REF.
GAS	CP	7.59	3.50	-0.31		298-2000	63

PHASE	T	CP	H	S	G	BT	
GAS	298	8.285	7.0	52.3	-8.593	6.300	63
	300	8.296	7.015	52.351	-8.690	6.332	
	400	8.796	7.871	54.809	-14.053	7.679	
	500	9.216	8.772	56.818	-19.637	8.585	
	600	9.604	9.713	58.533	-25.407	9.256	
	700	9.977	10.692	60.042	-31.337	9.785	
	800	10.342	11.708	61.398	-37.410	10.221	
	900	10.702	12.760	62.637	-43.613	10.592	
	1000	11.059	13.849	63.783	-49.934	10.915	
	1100	11.414	14.972	64.853	-56.367	11.201	
	1200	11.768	16.131	65.862	-62.903	11.458	
	1300	12.122	17.326	66.818	-69.537	11.692	
	1400	12.474	18.556	67.729	-76.265	11.907	
	1500	12.826	19.821	68.602	-83.082	12.107	
	1600	13.178	21.121	69.441	-89.984	12.293	
	1700	13.529	22.456	70.250	-96.969	12.468	
	1800	13.880	23.827	71.033	-104.033	12.633	
	1900	14.231	25.232	71.793	-111.175	12.790	
	2000	14.582	26.673	72.532	-118.391	12.939	

H_2Te (g)

HYDROGEN TELLURIDE (GAS)

		A	B	C	D	RANGE	REF.
GAS	CP	8.48	2.88	-0.74		298-2000	63

PHASE	T	CP	H	S	G	BT	
GAS	298	8.506	23.8	54.7	7.491	-5.492	1
	300	8.522	23.816	54.753	7.390	-5.384	
	400	9.169	24.703	57.300	1.783	-.974	
	500	9.624	25.643	59.397	-4.055	1.773	
	600	10.002	26.625	61.186	-10.087	3.675	
	700	10.345	27.643	62.754	-16.285	5.085	
	800	10.668	28.694	64.157	-22.632	6.184	
	900	10.981	29.776	65.431	-29.112	7.070	
	1000	11.286	30.889	66.604	-35.715	7.807	
	1100	11.587	32.033	67.694	-42.430	8.431	
	1200	11.885	33.207	68.715	-49.251	8.971	
	1300	12.180	34.410	69.678	-56.171	9.445	
	1400	12.474	35.643	70.591	-63.185	9.865	
	1500	12.767	36.905	71.462	-70.288	10.242	
	1600	13.059	38.196	72.295	-77.476	10.584	
	1700	13.350	39.517	73.096	-84.746	10.896	
	1800	13.641	40.866	73.867	-92.095	11.183	
	1900	13.932	42.245	74.612	-99.519	11.449	
	2000	14.221	43.652	75.334	-107.016	11.696	

PHOSPHORIC ACID

H_3PO_4

		A	B	C	D	RANGE	REF.
SOL	CP	25.5				298- 316	3
	** MPT=	315.51					3
LIQ	CP	48.00				316-1200	3

PHASE	T	CP	H	S	G	BT	
SOL	298	25.500	-305.67	26.42	-313.547	229.867	1,3
	300	25.500	-305.623	26.578	-313.596	228.485	
	316	25.500	-305.227	27.863	-314.018	217.546	
			3.1	9.810			3
LIQ	316	48.000	-302.127	37.673	-314.032	217.218	
	400	48.000	-298.095	48.988	-317.690	173.601	
	500	48.000	-293.295	59.699	-323.145	141.265	
	600	48.000	-288.495	68.450	-329.565	120.060	
	700	48.000	-283.695	75.849	-336.790	105.165	
	800	48.000	-278.895	82.259	-344.703	94.181	
	900	48.000	-274.095	87.913	-353.217	85.784	
	1000	48.000	-269.295	92.970	-362.265	79.184	
	1100	48.000	-264.495	97.545	-371.795	73.879	
	1200	48.000	-259.695	101.721	-381.761	69.538	

HCN (g)

HYDROGEN CYANIDE (GAS)

		A	B	C	D	RANGE	REF.
GAS	CP	8.350	3.860	-0.829		298- 800	3
	CP	11.772	1.19	-9.069		800-2000	3

PHASE	T	CP	H	S	G	BT	
GAS	298	8.568	-32.3	48.21	-46.674	34.217	3,62
	300	8.587	-32.284	48.263	-46.763	34.071	
	400	9.376	-31.383	50.850	-51.723	28.264	
	500	9.948	-30.416	53.006	-56.919	24.882	
	600	10.436	-29.396	54.863	-62.314	22.701	
	700	10.883	-28.330	56.506	-67.884	21.197	
	800	11.308	-27.220	57.987	-73.610	20.112	
	800	11.307	-27.220	57.987	-73.610	20.112	
	900	11.723	-26.068	59.344	-79.478	19.302	
	1000	12.055	-24.878	60.597	-85.475	18.683	
	1100	12.331	-23.659	61.759	-91.594	18.200	
	1200	12.570	-22.413	62.843	-97.825	17.819	
	1300	12.782	-21.146	63.857	-104.160	17.513	
	1400	12.975	-19.858	64.812	-110.594	17.267	
	1500	13.154	-18.551	65.713	-117.121	17.067	
	1600	13.322	-17.227	66.568	-123.735	16.904	
	1700	13.481	-15.887	67.380	-130.433	16.771	
	1800	13.634	-14.531	68.155	-137.210	16.662	
	1900	13.782	-13.160	68.896	-144.063	16.573	
	2000	13.925	-11.775	69.607	-150.988	16.501	

METABORIC ACID HBO_2

		A	B	C	D	RANGE	REF.
SOL	CP	12.450	8.240	-1.660		298- 509	3
	** MPT= 509						3

PHASE	T	CP	H	S	G	BT	P	
SOL	298	13.039	-191.87	11.7	-195.358	143.221		3
	300	13.078	-191.846	11.781	-195.380	142.353		
	400	14.708	-190.451	15.783	-196.764	107.521		
	500	15.906	-188.918	19.198	-198.517	86.783		
	509	16.003	-188.774	19.483	-198.691	85.324		

* * * * * * * * * *

METABORIC ACID (GAS) $HBO_2(g)$

		A	B	C	D	RANGE	REF.
GAS	CP	9.607	6.825	-1.382		298- 800	3
	CP	16.202	1.26	-15.123		800-2000	3

PHASE	T	CP	H	S	G	BT	
GAS	298	10.087	-134.0	57.273	-151.076	110.757	3
	300	10.119	-133.981	57.335	-151.182	110.151	
	400	11.473	-132.897	60.446	-157.075	85.833	
	500	12.467	-131.698	63.117	-163.256	71.369	
	600	13.318	-130.408	65.466	-169.688	61.817	
	700	14.102	-129.037	67.579	-176.342	55.064	
	800	14.851	-127.589	69.511	-183.198	50.054	
	800	14.847	-127.589	69.511	-183.198	50.054	
	900	15.469	-126.072	71.297	-190.239	46.203	
	1000	15.950	-124.500	72.953	-197.453	43.159	
	1100	16.338	-122.885	74.492	-204.826	40.701	
	1200	16.664	-121.234	75.928	-212.348	38.679	
	1300	16.945	-119.553	77.273	-220.009	36.992	
	1400	17.194	-117.846	78.538	-227.800	35.566	
	1500	17.420	-116.115	79.732	-235.714	34.348	
	1600	17.627	-114.363	80.863	-243.744	33.298	
	1700	17.821	-112.590	81.938	-251.885	32.386	
	1800	18.003	-110.799	82.962	-260.130	31.588	
	1900	18.177	-108.990	83.940	-268.475	30.886	
	2000	18.344	-107.164	84.876	-276.916	30.264	

Hf

HAFNIUM

		A	B	C	D	RANGE	REF.
SOL-A	CP	5.607	1.822			298-2013	21
SOL-B	CP	8.8				2013-2500	21
LIQ	CP	8.0				2500-3700	21
	** BPT= 4876, L= 137.539						21

PHASE	T	CP	H	S	G	BT	P	
SOL-A	298	6.150	0	10.41	-3.104	2.275		21
	300	6.154	.011	10.448	-3.123	2.275		
	400	6.336	.636	12.243	-4.261	2.329		
	500	6.518	1.279	13.677	-5.560	2.431		
	600	6.700	1.939	14.881	-6.989	2.546		
	700	6.882	2.619	15.928	-8.531	2.664		
	800	7.065	3.316	16.859	-10.171	2.779		
	900	7.247	4.032	17.701	-11.900	2.890		
	1000	7.429	4.765	18.474	-13.709	2.996		
	1100	7.611	5.517	19.191	-15.593	3.098		
	1200	7.793	6.288	19.861	-17.545	3.196		
	1300	7.976	7.076	20.492	-19.563	3.289		
	1400	8.158	7.883	21.090	-21.643	3.379		
	1500	8.340	8.708	21.659	-23.780	3.465		
	1600	8.522	9.551	22.203	-25.974	3.548		
	1700	8.704	10.412	22.725	-28.220	3.628		
	1800	8.887	11.292	23.227	-30.518	3.706		
	1900	9.069	12.189	23.713	-32.865	3.781		
	2000	9.251	13.105	24.183	-35.260	3.854		
	2013	9.275	13.226	24.243	-35.575	3.863		
			1.61	.800				21
SOL-B	2013	8.800	14.836	25.042	-35.575	3.863		
	2100	8.800	15.601	25.415	-37.770	3.931		
	2200	8.800	16.481	25.824	-40.332	4.007		
	2300	8.800	17.361	26.215	-42.934	4.080		
	2400	8.800	18.241	26.590	-45.574	4.151		
	2500	8.800	19.121	26.949	-48.251	4.219	.001	
			5.75	2.300				21
LIQ	2500	8.000	24.871	29.249	-48.251	4.219	.001	
	2600	8.000	25.671	29.563	-51.192	4.304	.003	
	2700	8.000	26.471	29.865	-54.164	4.385	.008	
	2800	8.000	27.271	30.156	-57.165	4.463	.021	
	2900	8.000	28.071	30.436	-60.194	4.537	.049	
	3000	8.000	28.871	30.708	-63.252	4.608	.108	
	3100	8.000	29.671	30.970	-66.336	4.677	.227	
	3200	8.000	30.471	31.224	-69.445	4.744	.456	
	3300	8.000	31.271	31.470	-72.580	4.807	.877	
	3400	8.000	32.071	31.709	-75.739	4.869	1.623	
	3500	8.000	32.871	31.941	-78.922	4.929	2.900	
	3600	8.000	33.671	32.166	-82.127	4.986	5.017	
	3700	8.000	34.471	32.385	-85.355	5.042	8.424	

HAFNIUM TETRAFLUORIDE

HfF_4

		A	B	C	D	RANGE	REF.
SOL	CP	31.9	0.75	-9.0		298-1100	121

PHASE	T	CP	H	S	G	BT	
SOL	298	21.999	-461.4	32.5	-471.090	345.365	1/*
	300	22.125	-461.359	32.636	-471.150	343.279	
	400	26.575	-458.893	39.701	-474.773	259.439	
	500	28.675	-456.119	45.882	-479.060	209.425	
	600	29.850	-453.188	51.223	-483.922	176.292	
	700	30.588	-450.163	55.884	-489.282	152.781	
	800	31.094	-447.078	60.003	-495.080	135.268	
	900	31.464	-443.949	63.688	-501.268	121.741	
	1000	31.750	-440.788	67.018	-507.806	110.996	
	1100	31.981	-437.601	70.055	-514.662	102.268	

HfN

HAFNIUM NITRIDE

		A	B	C	D	RANGE	REF.
SOL	CP	11.10	1.68	-1.72		298-1700	*
	** MPT= 3573						47

PHASE	T	CP	H	S	G	BT	
SOL	298	9.666	-88.2	12.1	-91.808	67.306	1/*
	300	9.693	-88.182	12.160	-91.830	66.907	
	400	10.697	-87.157	15.103	-93.198	50.928	
	500	11.252	-86.057	17.554	-94.834	41.458	
	600	11.630	-84.912	19.641	-96.697	35.226	
	700	11.925	-83.734	21.457	-98.753	30.836	
	800	12.175	-82.528	23.066	-100.981	27.590	
	900	12.400	-81.300	24.513	-103.361	25.103	
	1000	12.608	-80.049	25.830	-105.879	23.143	
	1100	12.806	-78.778	27.041	-108.524	21.565	
	1200	12.997	-77.488	28.164	-111.285	20.270	
	1300	13.182	-76.179	29.212	-114.154	19.194	
	1400	13.364	-74.852	30.195	-117.125	18.286	
	1500	13.544	-73.506	31.123	-120.191	17.514	
	1600	13.721	-72.143	32.003	-123.348	16.851	
	1700	13.896	-70.762	32.840	-126.591	16.277	

* * * * * * * * * *

HfB_2

HAFNIUM DIBORIDE

		A	B	C	D	RANGE	REF.
SOL	CP	15.57	4.4	-4.5		298-1500	19,*
	** MPT= 2420						

PHASE	T	CP	H	S	G	BT	
SOL	298	11.820	-80.3	10.2	-83.341	61.099	19
	300	11.890	-80.278	10.273	-83.360	60.736	
	400	14.517	-78.942	14.099	-84.582	46.219	
	500	15.970	-77.412	17.507	-86.166	37.668	
	600	16.960	-75.763	20.511	-88.069	32.084	
	700	17.732	-74.027	23.185	-90.257	28.183	
	800	18.387	-72.221	25.596	-92.698	25.327	
	900	18.974	-70.352	27.797	-95.369	23.162	
	1000	19.520	-68.427	29.824	-98.251	21.476	
	1100	20.038	-66.449	31.709	-101.329	20.135	
	1200	20.537	-64.420	33.474	-104.589	19.051	
	1300	21.024	-62.342	35.137	-108.020	18.162	
	1400	21.500	-60.216	36.713	-111.614	17.426	
	1500	21.970	-58.042	38.212	-115.360	16.810	

MERCURY SULFIDE (REC.)

HgS

		A	B	C	D	RANGE	REF.
SOL-R	CP	10.46	3.72			298- 618	63
SOL-B	CP	10.52	3.63			618-1098	63
	** DEC., MPT= 1098						63

PHASE	T	CP	H	S	G	BT	P	
SOL-R	298	11.569	-12.75	19.7	-18.624	13.653		63
	300	11.576	-12.729	19.772	-18.660	13.596		
	400	11.948	-11.552	23.153	-20.813	11.373		
	500	12.320	-10.339	25.859	-23.268	10.172		
	600	12.692	-9.088	28.138	-25.971	9.461		
	618	12.759	-8.859	28.514	-26.481	9.366		
			0.95	1.537				63
SOL-B	618	12.763	-7.909	30.051	-26.481	9.366		
	700	13.061	-6.851	31.660	-29.012	9.059	.014	
	800	13.424	-5.526	33.427	-32.268	8.816	.536	
	900	13.787	-4.166	35.029	-35.692	8.668	8.868	
	1000	14.150	-2.769	36.501	-39.270	8.584	81.407	
	1098	14.506	-1.365	37.840	-42.913	8.543	472.039	

Hg_2SO_4 MERCURY(1) SULFATE

		A	B	C	D	RANGE	REF.
SOL	CP	31.8				298- 400	58,*
	** DISP.						66

PHASE	T	CP	H	S	G	BT	
SOL	298	31.800	-177.61	48.	-191.921	140.701	66
	300	31.800	-177.551	48.197	-192.010	139.898	
	400	31.800	-174.371	57.345	-197.309	107.819	

* * * * * * * * * *

$HgSO_4$ MERCURY SULFATE

		A	B	C	D	RANGE	REF.
SOL	CP	14.0	35.			298- 800	*
	** DEC.						66

PHASE	T	CP	H	S	G	BT	
SOL	298	24.435	-168.3	33.5	-178.288	130.706	66/*
	300	24.500	-168.255	33.651	-178.350	129.945	
	400	28.000	-165.630	41.179	-182.101	99.509	
	500	31.500	-162.655	47.803	-186.556	81.555	
	600	35.000	-159.330	53.855	-191.643	69.815	
	700	38.500	-155.655	59.514	-197.314	61.613	
	800	42.000	-151.630	64.883	-203.536	55.611	

MERCURY SELENIDE

HgSe

		A	B	C	D	RANGE	REF.
SOL	CP	11.7	3.7			298-1043	63
	** DEC., MPT= 1043						63

PHASE	T	CP	H	S	G	BT	P	
SOL	298	12.803	-10.4	24.1	-17.585	12.892		63
	300	12.810	-10.376	24.179	-17.630	12.845		
	400	13.180	-9.077	27.915	-20.243	11.062		
	500	13.550	-7.740	30.896	-23.188	10.137		
	600	13.920	-6.367	33.399	-26.406	9.620		
	700	14.290	-4.956	35.573	-29.857	9.323		
	800	14.660	-3.509	37.505	-33.513	9.156	.002	
	900	15.030	-2.024	39.253	-37.352	9.072	.067	
	1000	15.400	-.503	40.856	-41.359	9.040	.936	
	1043	15.559	.163	41.507	-43.129	9.039	2.461	

* * * * * * * * * *

MERCURY SELENITE

$HgSeO_3$

		A	B	C	D	RANGE	REF.
SOL	CP	20.0	14.8			298- 633	*
	** TPT= 633, MPT= 738						62

PHASE	T	CP	H	S	G	BT	
SOL	298	24.413	-87.3	38.8	-98.868	72.482	62,*
	300	24.440	-87.255	38.951	-98.940	72.088	
	400	25.920	-84.737	46.185	-103.211	56.399	
	500	27.400	-82.071	52.128	-108.135	47.272	
	600	28.880	-79.257	57.254	-113.609	41.388	
	633	29.368	-78.296	58.813	-115.525	39.891	

HgTe

MERCURY TELLURIDE

		A	B	C	D	RANGE	REF.
SOL	CP	12.45	2.17			298- 943	63
	** DEC., MPT= 943, L= 8.5						63

PHASE	T	CP	H	S	G	BT	P	
SOL	298	13.097	-7.6	27.0	-15.650	11.473		63
	300	13.101	-7.576	27.081	-15.700	11.439		
	400	13.318	-6.255	30.880	-18.607	10.168		
	500	13.535	-4.912	33.875	-21.850	9.552		
	600	13.752	-3.548	36.362	-25.365	9.240		
	700	13.969	-2.162	38.498	-29.110	9.090		
	800	14.186	-.754	40.377	-33.056	9.032		
	900	14.403	.675	42.061	-37.179	9.030	.020	
	943	14.496	1.297	42.735	-39.002	9.040	.068	

HOLMIUM

Ho

		A	B	C	D	RANGE	REF.
SOL-A	CP	6.853	0.015	-0.327		298- 700	21
	CP	7.186	-3.626	1.043	3.945	700-1701	21
SOL-B	CP	6.7				1701-1743	21
LIQ	CP	10.50				1743-2400	21
	** BPT= 2968 / L= 57.18						21/*

PHASE	T	CP	H	S	G	BT	P	
SOL-A	298	6.490	0	17.93	-5.346	3.919		21
	300	6.494	.012	17.970	-5.379	3.919		
	400	6.655	.671	19.864	-7.275	3.975		
	500	6.730	1.340	21.358	-9.339	4.082		
	600	6.771	2.015	22.589	-11.538	4.203		
	700	6.797	2.694	23.634	-13.850	4.325		
	700	6.794	2.694	23.634	-13.850	4.325		
	800	6.973	3.381	24.552	-16.260	4.443		
	900	7.247	4.092	25.388	-18.758	4.556		
	1000	7.609	4.834	26.170	-21.336	4.664		
	1100	8.057	5.616	26.915	-23.991	4.767		
	1200	8.588	6.448	27.639	-26.719	4.867		
	1300	9.201	7.337	28.350	-29.518	4.963		
	1400	9.895	8.291	29.057	-32.388	5.057	.004	
	1500	10.670	9.318	29.765	-35.329	5.148	.023	
	1600	11.524	10.427	30.481	-38.342	5.238	.095	
	1700	12.459	11.626	31.207	-41.426	5.326	.331	
	1701	12.469	11.638	31.214	-41.457	5.327	.335	
			1.121	.659				
SOL-B	1701	6.700	12.759	31.873	-41.457	5.327	.335	21
	1743	6.700	13.041	32.037	-42.799	5.367	.537	
			3.34	1.916				*
LIQ	1743	10.500	16.381	33.953	-42.799	5.367	.537	
	1800	10.500	16.979	34.291	-44.744	5.433	.954	
	1900	10.500	18.029	34.859	-48.202	5.545	2.386	
	2000	10.500	19.079	35.397	-51.715	5.652	5.413	
	2100	10.500	20.129	35.909	-55.280	5.754	11.289	
	2200	10.500	21.179	36.398	-58.896	5.852	21.908	
	2300	10.500	22.229	36.865	-62.559	5.945	39.946	
	2400	10.500	23.279	37.311	-66.268	6.035	68.982	

HoF_3

HOLMIUM FLUORIDE

		A	B	C	D	RANGE	REF.
SOL-A	CP	25.54	0.115	-5.04		298-1343	174
SOL-B	CP	39.04	-3.93	-51.14		1343-1416	174
LIQ	CP	22.85	0.039	0.85		1416-1800	174

PHASE	T	CP	H	S	G	BT	
SOL-A	298	19.905	-408.	26.5	-415.901	304.905	19/*
	300	19.974	-407.963	26.623	-415.950	303.060	
	400	22.436	-405.825	32.757	-418.928	228.922	
	500	23.581	-403.518	37.901	-422.468	184.686	
	600	24.209	-401.126	42.261	-426.482	155.367	
	700	24.592	-398.684	46.024	-430.901	134.551	
	800	24.844	-396.211	49.325	-435.671	119.036	
	900	25.021	-393.718	52.262	-440.754	107.044	
	1000	25.151	-391.209	54.905	-446.114	97.511	
	1100	25.250	-388.689	57.307	-451.727	89.762	
	1200	25.328	-386.159	59.508	-457.569	83.346	
	1300	25.391	-383.623	61.538	-463.622	77.952	
	1343	25.415	-382.531	62.364	-466.286	75.890	
							174
SOL-B	1343	30.927	-382.531	62.364	-466.286	75.890	
	1400	30.929	-380.768	63.650	-469.878	73.361	
	1416	30.925	-380.273	64.001	-470.899	72.690	
			13.46	9.506			174
LIQ	1416	22.948	-366.813	73.507	-470.899	72.690	
	1500	22.946	-364.886	74.830	-477.130	69.527	
	1600	22.946	-362.591	76.310	-484.688	66.214	
	1700	22.946	-360.297	77.701	-492.389	63.309	
	1800	22.946	-358.002	79.013	-500.225	60.744	

HOLMIUM CHLORIDE $HoCl_3$

		A	B	C	D	RANGE	REF.
SOL	CP	22.84	3.10	-0.23		298- 993	89
LIQ	CP	34.35				993-1100	89

PHASE	T	CP	H	S	G	BT	
SOL	298	23.506	-240.3	35.1	-250.765	183.840	19/*
	300	23.514	-240.257	35.245	-250.830	182.754	
	400	23.936	-237.883	42.070	-254.711	139.186	
	500	24.298	-235.471	47.451	-259.197	113.310	
	600	24.636	-233.024	51.911	-264.171	96.237	
	700	24.963	-230.544	55.733	-269.558	84.171	
	800	25.284	-228.032	59.088	-275.302	75.219	
	900	25.602	-225.488	62.084	-281.363	68.334	
	993	25.895	-223.093	64.616	-287.257	63.231	
			7.3	7.351			89
LIQ	993	34.350	-215.793	71.967	-287.257	63.231	
	1000	34.350	-215.553	72.209	-287.761	62.899	
	1100	34.350	-212.118	75.483	-295.148	58.648	

* * * * * * * * * *

HOLMIUM OXIDE Ho_2O_3

		A	B	C	D	RANGE	REF.
SOL	CP	30.08	1.66	-2.75		298-1500	181
	** TPT= 2543, MPT= 2663						1

PHASE	T	CP	H	S	G	BT	
SOL	298	27.481	-449.6	37.8	-460.870	337.872	1
	300	27.522	-449.549	37.970	-460.940	335.840	
	400	29.025	-446.712	46.121	-465.161	254.186	
	500	29.810	-443.767	52.690	-470.112	205.513	
	600	30.312	-440.759	58.172	-475.663	173.283	
	700	30.681	-437.709	62.874	-481.721	150.420	
	800	30.978	-434.626	66.991	-488.218	133.393	
	900	31.234	-431.515	70.654	-495.104	120.244	
	1000	31.465	-428.379	73.957	-502.337	109.800	
	1100	31.679	-425.222	76.966	-509.885	101.318	
	1200	31.881	-422.044	79.732	-517.722	94.303	
	1300	32.075	-418.846	82.291	-525.825	88.411	
	1400	32.264	-415.629	84.675	-534.174	83.400	
	1500	32.448	-412.394	86.907	-542.755	79.090	

In

INDIUM

		A	B	C	D	RANGE	REF.
SOL	CP	-2.152	18.483	2.693		298- 430	21
	** MPT=	629.76					21
LIQ	CP	7.136	-0.208			430- 900	21
	CP	6.95				900-2346	21
GAS	CP	5.789				2346-2800	21

PHASE	T	CP	H	S	G	BT	P	
SOL	298	6.388	0	13.82	-4.120	3.021		21
	300	6.385	.012	13.860	-4.146	3.021		
	400	6.924	.668	15.743	-5.629	3.076		
	430	7.249	.879	16.251	-6.105	3.105		
			0.78	1.815				21
LIQ	430	7.047	1.659	18.066	-6.105	3.105		
	500	7.032	2.153	19.132	-7.413	3.241		
	600	7.011	2.855	20.412	-9.392	3.422		
	700	6.990	3.555	21.491	-11.489	3.587		
	800	6.970	4.253	22.424	-13.685	3.739		
	900	6.949	4.949	23.243	-15.970	3.878		
	900	6.950	4.949	23.243	-15.970	3.878		
	1000	6.950	5.644	23.976	-18.331	4.007		
	1100	6.950	6.339	24.638	-20.762	4.126		
	1200	6.950	7.034	25.243	-23.257	4.236	.008	
	1300	6.950	7.729	25.799	-25.809	4.340	.049	
	1400	6.950	8.424	26.314	-28.415	4.436	.232	
	1500	6.950	9.119	26.793	-31.071	4.528	.890	
	1600	6.950	9.814	27.242	-33.773	4.614	2.882	
	1700	6.950	10.509	27.663	-36.518	4.695	8.118	
	1800	6.950	11.204	28.061	-39.305	4.773	20.356	
	1900	6.950	11.899	28.436	-42.130	4.847	46.285	
	2000	6.950	12.594	28.793	-44.991	4.917	96.828	
	2100	6.950	13.289	29.132	-47.888	4.984	188.601	
	2200	6.950	13.984	29.455	-50.817	5.049	345.376	
	2300	6.950	14.679	29.764	-53.778	5.111	599.418	
	2346	6.950	14.999	29.902	-55.151	5.138	760.	
			55.319	23.580				21
GAS	2346	5.789	70.318	53.482	-55.151	5.138		
	2400	5.789	70.631	53.614	-58.042	5.286		
	2500	5.789	71.210	53.850	-63.416	5.545		
	2600	5.789	71.788	54.077	-68.812	5.785		
	2700	5.789	72.367	54.296	-74.231	6.009		
	2800	5.789	72.946	54.506	-79.671	6.219		

INDIUM MONOCHLORIDE

InCl

		A	B	C	D	RANGE	REF.
SOL-A	CP	8.4	10.0			298- 393	*
SOL-B	CP	14.0				393- 498	*
LIQ	CP	15.0				498- 881	*
	** BPT= 881, L= 23.1						1/62

PHASE	T	CP	H	S	G	BT	
SOL-A	298	11.381	-44.5	22.7	-51.268	37.586	62
	300	11.400	-44.479	22.770	-51.310	37.384	
	393	12.330	-43.375	25.969	-53.581	29.801	
			1.65	4.198			62
SOL-B	393	14.000	-41.725	30.167	-53.581	29.801	
	400	14.000	-41.627	30.414	-53.793	29.395	
	498	14.000	-40.255	33.482	-56.930	24.987	
			2.2	4.418			62
LIQ	498	15.000	-38.055	37.900	-56.930	24.987	
	500	15.000	-38.025	37.960	-57.005	24.920	
	600	15.000	-36.525	40.695	-60.942	22.201	
	700	15.000	-35.025	43.007	-65.130	20.337	
	800	15.000	-33.525	45.010	-69.534	18.998	
	881	15.000	-32.310	46.457	-73.239	18.171	

$InCl_2$ INDIUM DICHLORIDE

		A	B	C	D	RANGE	REF.
SOL	CP	14.0	12.0			298- 509	*
	** TPT= 463, MPT= 509 / BPT= 796, L= 37.						62

PHASE	T	CP	H	S	G	BT	
SOL	298	17.576	-86.7	29.2	-95.406	69.944	62
	300	17.600	-86.667	29.309	-95.460	69.552	
	400	18.800	-84.847	34.536	-98.662	53.914	
	463	19.556	-83.639	37.340	-100.928	47.647	
	463	19.556	-83.639	37.340	-100.928	47.647	
	500	20.000	-82.907	38.860	-102.338	44.738	
	509	20.108	-82.727	39.218	-102.689	44.098	

* * * * * * * * * *

$InCl_3$ INDIUM TRICHLORIDE

		A	B	C	D	RANGE	REF.
SOL	CP	18.8	13.3			298- 856	*
	** MPT= 856						62

PHASE	T	CP	H	S	G	BT	
SOL	298	22.765	-128.4	33.7	-138.448	101.499	62/*
	300	22.790	-128.358	33.841	-138.510	100.918	
	400	24.120	-126.012	40.579	-142.244	77.729	
	500	25.450	-123.534	46.104	-146.586	64.081	
	600	26.780	-120.922	50.862	-151.440	55.169	
	700	28.110	-118.178	55.090	-156.741	48.943	
	800	29.440	-115.300	58.930	-162.445	44.384	
	856	30.185	-113.631	60.947	-165.802	42.337	

INDIUM MONOBROMIDE — InBr

		A	B	C	D	RANGE	REF.
SOL	CP	10.4	6.0			298- 558	*
	** MPT=	558.35					62
LIQ	CP	14.500				558-1091	*
	** BPT=	1091.58					
GAS	CP	8.98	0.1	-0.175		1091-1300	62,*

PHASE	T	CP	H	S	G	BT	P	
SOL	298	12.189	-41.9	27.3	-50.039	36.685		62/*
	300	12.200	-41.877	27.375	-50.090	36.495		
	400	12.800	-40.627	30.967	-53.014	28.970		
	500	13.400	-39.317	33.888	-56.261	24.595	.003	
	558	13.750	-38.525	35.386	-58.283	22.816	.060	
			5.8	10.394				62
LIQ	558	14.500	-32.725	45.780	-58.271	22.826	.059	
	600	14.500	-32.116	46.833	-60.216	21.937	.243	
	700	14.500	-30.666	49.068	-65.014	20.301	3.442	
	800	14.500	-29.216	51.004	-70.020	19.131	23.931	
	900	14.500	-27.766	52.712	-75.207	18.265	104.058	
	1000	14.500	-26.316	54.240	-80.556	17.608	327.086	
	1092	14.500	-24.988	55.510	-85.582	17.137	760.	
			19.7	18.047				62
GAS	1092	9.074	-5.288	73.557	-85.582	17.137		
	1100	9.076	-5.212	73.627	-86.202	17.129		
	1200	9.088	-4.304	74.417	-93.605	17.050		
	1300	9.100	-3.394	75.145	-101.083	16.996		

* * * * * * * * * *

INDIUM TRIBROMIDE — IrBr$_3$

		A	B	C	D	RANGE	REF.
SOL	CP	19.6	13.0			298- 693	*
	** SPT= 682, L= 29.5 / MPT= 692.85						62

PHASE	T	CP	H	S	G	BT	
SOL	298	23.476	-98.2	42.7	-110.931	81.326	62/*
	300	23.500	-98.157	42.845	-111.010	80.882	
	400	24.800	-95.742	49.784	-115.655	63.200	
	500	26.100	-93.197	55.457	-120.925	52.864	
	600	27.400	-90.522	60.331	-126.720	46.164	
	693	28.607	-87.921	64.358	-132.512	41.805	

InJ

INDIUM MONOIODIDE

		A	B	C	D	RANGE	REF.
SOL	CP	11.5	3.0			298- 638	1
LIQ	CP	14.5				638- 986	*
GAS	CP	9.25		-0.4		986-2000	*

PHASE	T	CP	H	S	G	BT	P	
SOL	298	12.394	-27.8	31.1	-37.072	27.179		1,*
	300	12.400	-27.777	31.177	-37.130	27.053		
	400	12.700	-26.522	34.785	-40.436	22.096		
	500	13.000	-25.237	37.651	-44.063	19.262	.003	
	600	13.300	-23.922	40.048	-47.951	17.468	.388	
	638	13.414	-23.414	40.868	-49.488	16.955	1.601	
			5.36	8.401				62
LIQ	638	14.500	-18.054	49.269	-49.488	16.955	1.601	
	700	14.500	-17.155	50.614	-52.585	16.420	7.857	
	800	14.500	-15.705	52.550	-57.746	15.778	58.515	
	900	14.500	-14.255	54.258	-63.088	15.322	268.743	
	986	14.500	-13.008	55.581	-67.812	15.033	760.	
			21.077	21.376				62,*
GAS	986	9.209	8.069	76.958	-67.812	15.033		
	1000	9.210	8.197	77.088	-68.890	15.058		
	1100	9.217	9.119	77.966	-76.643	15.230		
	1200	9.222	10.041	78.768	-84.481	15.388		
	1300	9.226	10.963	79.506	-92.395	15.535		
	1400	9.230	11.886	80.190	-100.380	15.672		
	1500	9.232	12.809	80.827	-108.431	15.801		
	1600	9.234	13.732	81.423	-116.544	15.921		
	1700	9.236	14.656	81.983	-124.715	16.035		
	1800	9.238	15.580	82.511	-132.940	16.143		
	1900	9.239	16.503	83.010	-141.216	16.246		
	2000	9.240	17.427	83.484	-149.541	16.343		

INDIUM TRIIODIDE

InJ_3

		A	B	C	D	RANGE	REF.
SOL	CP	39.2				298- 480	*
LIQ	CP	32.5				480- 720	*
	** BPT= 720, L= 22.8						62

PHASE	T	CP	H	S	G	BT	
SOL	298	39.200	-56.1	48.6	-70.590	51.751	62,*
	300	39.200	-56.027	48.842	-70.680	51.497	
	400	39.200	-52.107	60.120	-76.155	41.615	
	480	39.200	-48.971	67.267	-81.259	37.003	
			4.8	10.000			62
LIQ	480	32.500	-44.171	77.267	-81.259	37.003	
	500	32.500	-43.521	78.593	-82.818	36.205	
	600	32.500	-40.271	84.519	-90.983	33.145	
	700	32.500	-37.021	89.529	-99.692	31.129	
	720	32.500	-36.371	90.444	-101.491	30.811	

In_2O_3 INDIUM OXIDE

		A	B	C	D	RANGE	REF.
SOL-A	CP	29.6	1.9	-5.51		298-1523	*
SOL-B	CP	29.6	1.9	-5.51		1523-2183	*
	** MPT= 2183						62

PHASE	T	CP	H	S	G	BT	
SOL-A	298	23.968	-221.3	25.8	-228.992	167.878	62
	300	24.048	-221.256	25.949	-229.040	166.878	
	400	26.916	-218.688	33.315	-232.014	126.784	
	500	28.346	-215.918	39.490	-235.663	103.022	
	600	29.209	-213.037	44.740	-239.881	87.388	
	700	29.806	-210.085	49.290	-244.588	76.374	
	800	30.259	-207.081	53.300	-249.721	68.230	
	900	30.630	-204.036	56.886	-255.234	61.988	
	1000	30.949	-200.957	60.130	-261.087	57.068	
	1100	31.235	-197.847	63.094	-267.251	53.105	
	1200	31.497	-194.711	65.823	-273.698	49.854	
	1300	31.744	-191.548	68.354	-280.409	47.147	
	1400	31.979	-188.362	70.715	-287.363	44.865	
	1500	32.205	-185.153	72.929	-294.547	42.921	
	1523	32.256	-184.412	73.420	-296.230	42.515	
SOL-B	1523	32.256	-184.412	73.420	-296.230	42.515	
	1600	32.425	-181.921	75.015	-301.945	41.249	
	1700	32.639	-178.668	76.987	-309.546	39.800	
	1800	32.850	-175.394	78.859	-317.339	38.535	
	1900	33.057	-172.098	80.640	-325.315	37.425	
	2000	33.262	-168.782	82.341	-333.464	36.444	
	2100	33.465	-165.446	83.969	-341.780	35.574	
	2183	33.632	-162.661	85.269	-348.804	34.925	

INDIUM MONOSULFIDE

InS

		A	B	C	D	RANGE	REF.
SOL	CP	10.16	4.5			298- 965	63
LIQ	CP	14.5				965-1800	*

PHASE	T	CP	H	S	G	BT	P	
SOL	298	11.502	-32.	16.5	-36.919	27.066		63
	300	11.510	-31.979	16.571	-36.950	26.922		
	400	11.960	-30.805	19.944	-38.783	21.193		
	500	12.410	-29.587	22.661	-40.917	17.887		
	600	12.860	-28.323	24.964	-43.301	15.775		
	700	13.310	-27.015	26.980	-45.901	14.333		
	800	13.760	-25.661	28.786	-48.690	13.303		
	900	14.210	-24.263	30.433	-51.652	12.545		
	965	14.502	-23.330	31.434	-53.663	12.155		
			8.6	8.912				63
LIQ	965	14.500	-14.730	40.346	-53.663	12.155		
	1000	14.500	-14.222	40.863	-55.085	12.040		
	1100	14.500	-12.772	42.245	-59.241	11.772		
	1200	14.500	-11.322	43.506	-63.530	11.572		
	1300	14.500	-9.872	44.667	-67.939	11.423		
	1400	14.500	-8.422	45.741	-72.460	11.313	.003	
	1500	14.500	-6.972	46.742	-77.085	11.233	.017	
	1600	14.500	-5.522	47.678	-81.806	11.176	.079	
	1700	14.500	-4.072	48.557	-86.618	11.137	.300	
	1800	14.500	-2.622	49.385	-91.516	11.113	.980	

In_5S_6 5-INDIUM 6-SULFIDE

		A	B	C	D	RANGE	REF.
SOL	CP	61.3	14.3	-2.55		298-1043	*
	** DPT= 1043						63

PHASE	T	CP	H	S	G	BT	
SOL	298	62.695	-185.	89.5	-211.684	155.190	63
	300	62.757	-184.884	89.888	-211.850	154.354	
	400	65.426	-178.466	108.333	-221.799	121.202	
	500	67.430	-171.820	123.155	-233.397	102.032	
	600	69.172	-164.988	135.605	-246.352	89.746	
	700	70.790	-157.990	146.391	-260.463	81.331	
	800	72.342	-150.833	155.945	-275.589	75.298	
	900	73.855	-143.523	164.554	-291.621	70.825	
	1000	75.345	-136.062	172.412	-308.475	67.426	
	1043	75.980	-132.809	175.598	-315.957	66.215	

INDIUM SULFIDE In_2S_3

		A	B	C	D	RANGE	REF.
SOL-A	CP	30.82	0.78	-2.54		298- 660	63
SOL-B	CP	23.37	13.24			660-1100	63
SOL-C	CP	38.10				1100-1363	63
	** MPT=	1363					62

PHASE	T	CP	H	S	G	BT	
SOL-A	298	28.195	-85.	39.1	-96.658	70.862	63
	300	28.232	-84.948	39.275	-96.730	70.477	
	400	29.544	-82.050	47.602	-101.091	55.241	
	500	30.194	-79.060	54.271	-106.196	46.424	
	600	30.582	-76.020	59.813	-111.908	40.768	
	660	30.752	-74.180	62.736	-115.585	38.280	
			0.26	.394			63
SOL-B	660	32.108	-73.920	63.130	-115.585	38.280	
	700	32.638	-72.625	65.035	-118.149	36.893	
	800	33.962	-69.295	69.479	-124.878	34.120	
	900	35.286	-65.832	73.556	-132.033	32.066	
	1000	36.610	-62.238	77.342	-139.580	30.509	
	1100	37.934	-58.510	80.894	-147.493	29.308	
			0.96	.873			63
SOL-C	1100	38.100	-57.550	81.766	-147.493	29.308	
	1200	38.100	-53.740	85.081	-155.838	28.386	
	1300	38.100	-49.930	88.131	-164.501	27.659	
	1363	38.100	-47.530	89.934	-170.110	27.280	

$In_2(SO_4)_3$ INDIUM SULFATE

		A	B	C	D	RANGE	REF.
SOL	CP	47.85	60.0			298- 943	209,*
	** TPT= 943						62

PHASE	T	CP	H	S	G	BT	
SOL	298	65.739	-651.4	72.2	-672.926	493.335	62
	300	65.850	-651.278	72.607	-673.060	490.390	
	400	71.850	-644.393	92.373	-681.342	372.318	
	500	77.850	-636.908	109.050	-691.433	302.266	
	600	83.850	-628.823	123.774	-703.088	256.134	
	700	89.850	-620.138	137.150	-716.143	223.620	
	800	95.850	-610.853	149.540	-730.485	199.586	
	900	101.850	-600.968	161.176	-746.026	181.184	
	943	104.430	-596.533	165.989	-753.061	174.553	

INDIUM MONOSELENIDE

InSe

		A	B	C	D	RANGE	REF.
SOL	CP	10.86	3.9			298- 933	63
LIQ	CP	14.7				933-1200	*

PHASE	T	CP	H	S	G	BT	P	
SOL	298	12.023	-28.2	19.5	-34.014	24.936		63
	300	12.030	-28.178	19.574	-34.050	24.809		
	400	12.420	-26.955	23.089	-36.191	19.776		
	500	12.810	-25.694	25.902	-38.645	16.894		
	600	13.200	-24.393	28.272	-41.356	15.066		
	700	13.590	-23.054	30.336	-44.289	13.830		
	800	13.980	-21.675	32.176	-47.416	12.955		
	900	14.370	-20.258	33.845	-50.719	12.318		
	933	14.499	-19.781	34.365	-51.844	12.146		
			8.3	8.896				*
LIQ	933	14.700	-11.481	43.261	-51.844	12.146		
	1000	14.700	-10.497	44.281	-54.777	11.973		
	1100	14.700	-9.027	45.682	-59.276	11.779		
	1200	14.700	-7.557	46.961	-63.909	11.641		

InSe (g)

INDIUM MONOSELENIDE (GAS)

		A	B	C	D	RANGE	REF.
GAS	CP	8.716	0.137			298-2000	63

PHASE	T	CP	H	S	G	BT	
GAS	298	8.757	56.0	63.0	37.217	-27.284	63
	300	8.757	56.016	63.054	37.100	-27.031	
	400	8.771	56.893	65.575	30.662	-16.755	
	500	8.784	57.770	67.534	24.003	-10.493	
	600	8.798	58.649	69.137	17.167	-6.254	
	700	8.812	59.530	70.494	10.184	-3.180	
	800	8.826	60.412	71.672	3.075	-.840	
	900	8.839	61.295	72.712	-4.146	1.007	
	1000	8.853	62.180	73.644	-11.464	2.506	
	1100	8.867	63.066	74.488	-18.871	3.750	
	1200	8.880	63.953	75.260	-26.359	4.801	
	1300	8.894	64.842	75.972	-33.921	5.703	
	1400	8.908	65.732	76.631	-41.552	6.487	
	1500	8.921	66.623	77.246	-49.246	7.176	
	1600	8.935	67.516	77.823	-57.000	7.787	
	1700	8.949	68.410	78.365	-64.810	8.333	
	1800	8.963	69.306	78.877	-72.672	8.825	
	1900	8.976	70.203	79.362	-80.584	9.271	
	2000	8.990	71.101	79.822	-88.543	9.677	

* * * * * * * * * *

In_2Se_3

INDIUM SELENIDE

		A	B	C	D	RANGE	REF.
SOL-A	CP	14.32	64.65			298- 470	63
SOL-B	CP	14.32	64.65			470- 600	63
	** TPT= 920, 1020 / MPT= 1173						63

PHASE	T	CP	H	S	G	BT	
SOL-A	298	33.595	-78.	48.1	-92.341	67.697	63
	300	33.715	-77.938	48.308	-92.430	67.344	
	400	40.180	-74.243	58.893	-97.800	53.443	
	470	44.705	-71.272	65.728	-102.164	47.513	
			0.336	.715			63
SOL-B	470	44.705	-70.936	66.443	-102.164	47.513	
	500	46.645	-69.566	69.268	-104.200	45.552	
	600	53.110	-64.578	78.344	-111.584	40.650	

DIINDIUM TELLURIDE In_2Te

		A	B	C	D	RANGE	REF.
SOL	CP	13.5	8.8			298- 733	*
	** MPT= 733						62

PHASE	T	CP	H	S	G	BT	
SOL	298	16.124	-19.1	37.5	-30.281	22.199	62/*
	300	16.140	-19.070	37.600	-30.350	22.113	
	400	17.020	-17.412	42.363	-34.358	18.775	
	500	17.900	-15.666	46.256	-38.794	16.959	
	600	18.780	-13.832	49.597	-43.591	15.880	
	700	19.660	-11.910	52.558	-48.701	15.207	
	733	19.950	-11.257	53.471	-50.451	15.044	

* * * * * * * * * *

INDIUM MONOTELLURICE $InTe$

		A	B	C	D	RANGE	REF.
SOL	CP	10.0	4.7			298- 968	123,*
LIQ	CP	14.8				968-1200	*

PHASE	T	CP	H	S	G	BT	
SOL	298	11.401	-17.2	25.26	-24.731	18.131	62/123
	300	11.410	-17.179	25.331	-24.778	18.053	
	400	11.880	-16.014	28.677	-27.485	15.019	
	500	12.350	-14.803	31.379	-30.492	13.330	
	600	12.820	-13.544	33.672	-33.748	12.294	
	700	13.290	-12.239	35.684	-37.217	11.621	
	800	13.760	-10.886	37.489	-40.877	11.169	
	900	14.230	-9.487	39.137	-44.710	10.859	
	968	14.550	-8.508	40.185	-47.407	10.705	
			8.6	8.884			62
LIQ	968	14.800	.092	49.069	-47.407	10.705	
	1000	14.800	.565	49.550	-48.985	10.707	
	1100	14.800	2.045	50.961	-54.012	10.733	
	1200	14.800	3.525	52.249	-59.173	10.778	

In_2Te_3 INDIUM TELLURIDE

		A	B	C	D	RANGE	REF.
SOL-A	CP	26.5	10.0			298- 883	*
SOL-B	CP	26.5	10.0			883- 943	*
LIQ	CP	37.0				943-1200	*

PHASE	T	CP	H	S	G	BT	
SOL-A	298	29.481	-45.8	57.	-62.795	46.036	62/*
	300	29.500	-45.745	57.182	-62.900	45.829	
	400	30.500	-42.745	65.806	-69.068	37.742	
	500	31.500	-39.645	72.719	-76.005	33.226	
	600	32.500	-36.445	78.551	-83.576	30.447	
	700	33.500	-33.145	83.636	-91.691	28.631	
	800	34.500	-29.745	88.174	-100.285	27.400	
	883	35.330	-26.847	91.620	-107.748	26.672	
SOL-B	883	35.330	-26.847	91.620	-107.748	26.672	
	900	35.500	-26.245	92.296	-109.312	26.548	
	943	35.930	-24.710	93.962	-113.316	26.266	
			19.5	20.679			62
LIQ	943	37.000	-5.210	114.641	-113.316	26.266	
	1000	37.000	-3.101	116.813	-119.913	26.211	
	1100	37.000	.599	120.339	-131.774	26.185	
	1200	37.000	4.299	123.559	-143.971	26.224	

INDIUM NITRIDE

		A	B	C	D	RANGE	REF.
SOL	CP	9.1	2.9			298-1473	*
	** MPT= 1473						62

PHASE	T	CP	H	S	G	BT	
SOL	298	9.965	-33.	10.4	-36.101	26.466	187/*
	300	9.970	-32.982	10.462	-36.120	26.317	
	400	10.260	-31.970	13.370	-37.318	20.392	
	500	10.550	-30.930	15.690	-38.775	16.951	
	600	10.840	-29.860	17.639	-40.444	14.734	
	700	11.130	-28.762	19.332	-42.294	13.207	
	800	11.420	-27.634	20.837	-44.304	12.105	
	900	11.710	-26.478	22.199	-46.457	11.283	
	1000	12.000	-25.292	23.448	-48.740	10.654	
	1100	12.290	-24.078	24.605	-51.143	10.163	
	1200	12.580	-22.834	25.687	-53.658	9.774	
	1300	12.870	-21.562	26.705	-56.278	9.463	
	1400	13.160	-20.260	27.670	-58.998	9.211	
	1473	13.372	-19.292	28.344	-61.042	9.058	

InP

INDIUM PHOSPHIDE

		A	B	C	D	RANGE	REF.
SOL-A	CP	9.8	3.5			298- 910	62,*
SOL-B	CP	13.200				910-1328	*
LIQ	CP	14.000				1328-1700	*

PHASE	T	CP	H	S	G	BT	
SOL-A	298	10.844	-18.0	14.28	-22.258	16.317	226
	300	10.850	-17.980	14.347	-22.284	16.236	
	400	11.200	-16.877	17.516	-23.884	13.051	
	500	11.550	-15.740	20.053	-25.767	11.264	
	600	11.900	-14.567	22.190	-27.881	10.157	
	700	12.250	-13.360	24.051	-30.195	9.429	
	800	12.600	-12.117	25.709	-32.685	8.930	
	900	12.950	-10.840	27.213	-35.332	8.581	
	910	12.985	-10.710	27.357	-35.605	8.552	
			0.09	.099			62
SOL-B	910	13.200	-10.620	27.456	-35.605	8.552	
	1000	13.200	-9.432	28.701	-38.133	8.335	
	1100	13.200	-8.112	29.959	-41.067	8.160	
	1200	13.200	-6.792	31.107	-44.121	8.037	
	1300	13.200	-5.472	32.164	-47.285	7.950	
	1328	13.200	-5.103	32.445	-48.190	7.932	
			15.	11.295			62
LIQ	1328	14.000	9.897	43.740	-48.190	7.932	
	1400	14.000	10.905	44.479	-51.366	8.020	
	1500	14.000	12.305	45.445	-55.863	8.140	
	1600	14.000	13.705	46.349	-60.453	8.259	
	1700	14.000	15.105	47.198	-65.131	8.374	

INDIUM ARSENIDE

InAs

		A	B	C	D	RANGE	REF.
SOL	CP	10.9	1.8			298-1215	5*
LIQ	CP	14.3				1215-1800	*

PHASE	T	CP	H	S	G	BT	
SOL	298	11.437	-13.8	17.85	-19.122	14.019	5
	300	11.440	-13.779	17.921	-19.155	13.956	
	400	11.620	-12.626	21.236	-21.120	11.541	
	500	11.800	-11.455	23.849	-23.379	10.220	
	600	11.980	-10.266	26.016	-25.875	9.426	
	700	12.160	-9.059	27.876	-28.572	8.922	
	800	12.340	-7.834	29.512	-31.443	8.591	
	900	12.520	-6.591	30.976	-34.469	8.371	
	1000	12.700	-5.330	32.304	-37.634	8.226	
	1100	12.880	-4.051	33.523	-40.926	8.132	
	1200	13.060	-2.754	34.651	-44.335	8.076	
	1215	13.087	-2.558	34.814	-44.856	8.070	
			18.4	15.144			62
LIQ	1215	14.300	15.842	49.958	-44.856	8.070	
	1300	14.300	17.058	50.925	-49.144	8.263	
	1400	14.300	18.488	51.985	-54.291	8.476	
	1500	14.300	19.918	52.971	-59.539	8.676	
	1600	14.300	21.348	53.894	-64.883	8.864	
	1700	14.300	22.778	54.761	-70.316	9.041	
	1800	14.300	24.208	55.578	-75.833	9.209	

InSb

INDIUM ANTIMONIDE

		A	B	C	D	RANGE	REF.
SOL-A	CP	10.6	3.6			298- 799	5
	** MPT=	798.35					62
LIQ	CP	14.8				798-1200	*

PHASE	T	CP	H	S	G	BT	
SOL-A	298	11.673	-7.44	20.95	-13.686	10.034	5
	300	11.680	-7.418	21.022	-13.725	10.000	
	400	12.040	-6.232	24.432	-16.005	8.746	
	500	12.400	-5.010	27.157	-18.589	8.126	
	600	12.760	-3.752	29.450	-21.422	7.804	
	700	13.120	-2.458	31.444	-24.469	7.641	
	798	13.474	-1.151	33.191	-27.649	7.570	
			11.8	14.780			5
LIQ	798	14.800	10.649	47.972	-27.649	7.570	
	800	14.800	10.674	48.002	-27.728	7.576	
	900	14.800	12.154	49.745	-32.617	7.922	
	1000	14.800	13.634	51.305	-37.671	8.234	
	1100	14.800	15.114	52.715	-42.873	8.519	
	1200	14.800	16.594	54.003	-48.210	8.781	

IRIDIUM

		A	B	C	D	RANGE	REF.
SOL	CP	5.584	1.281			298- 600	21
	CP	5.099	1.913	0.397	-0.036	600-2716	21
LIQ	CP	10.				2716-4701	21
GAS	CP	7.553				4701-5000	21

PHASE	T	CP	H	S	G	BT	P	
SOL	298	5.966	0	8.486	-2.530	1.855		21
	300	5.968	.011	8.523	-2.546	1.855		
	400	6.096	.614	10.257	-3.489	1.906		
	500	6.224	1.230	11.632	-4.585	2.005		
	600	6.353	1.859	12.778	-5.807	2.116		
	600	6.344	1.859	12.778	-5.807	2.116		
	700	6.501	2.501	13.767	-7.136	2.228		
	800	6.668	3.160	14.646	-8.557	2.338		
	900	6.841	3.835	15.442	-10.062	2.444		
	1000	7.016	4.528	16.171	-11.643	2.545		
	1100	7.193	5.238	16.848	-13.295	2.642		
	1200	7.370	5.967	17.482	-15.012	2.734		
	1300	7.549	6.712	18.079	-16.790	2.823		
	1400	7.727	7.476	18.645	-18.626	2.908		
	1500	7.905	8.258	19.184	-20.518	2.990		
	1600	8.083	9.057	19.700	-22.462	3.069		
	1700	8.261	9.874	20.195	-24.457	3.145		
	1800	8.438	10.709	20.672	-26.501	3.218		
	1900	8.615	11.562	21.133	-28.591	3.289		
	2000	8.791	12.432	21.580	-30.727	3.358		
	2100	8.967	13.320	22.013	-32.907	3.425		
	2200	9.142	14.226	22.434	-35.129	3.490		
	2300	9.316	15.149	22.844	-37.393	3.554		
	2400	9.490	16.089	23.244	-39.698	3.615		
	2500	9.663	17.046	23.635	-42.042	3.676		
	2600	9.835	18.021	24.018	-44.424	3.735	.002	
	2700	10.007	19.013	24.392	-46.845	3.792	.006	
	2716	10.035	19.174	24.451	-47.236	3.801	.007	
			6.247	2.300				21
LIQ	2716	10.000	25.421	26.751	-47.236	3.801	.007	
	2800	10.000	26.261	27.056	-49.496	3.864	.017	
	2900	10.000	27.261	27.407	-52.219	3.936	.043	
	3000	10.000	28.261	27.746	-54.977	4.006	.102	
	3100	10.000	29.261	28.074	-57.768	4.073	.228	
	3200	10.000	30.261	28.391	-60.591	4.139	.484	
	3300	10.000	31.261	28.699	-63.446	4.202	.981	
	3400	10.000	32.261	28.997	-66.331	4.264	1.904	

IRIDIUM (CONTINUED)

PHASE	T	CP	H	S	G	BT	P	
LIQ	3500	10.000	33.261	29.287	-69.245	4.324	3.553	
	3600	10.000	34.261	29.569	-72.188	4.383	6.400	
	3700	10.000	35.261	29.843	-75.158	4.440	11.154	
	3800	10.000	36.261	30.110	-78.156	4.496	18.863	
	3900	10.000	37.261	30.369	-81.180	4.550	31.025	
	4000	10.000	38.261	30.623	-84.230	4.603	49.733	
	4100	10.000	39.261	30.870	-87.304	4.654	77.848	
	4200	10.000	40.261	31.111	-90.403	4.705	119.196	
	4300	10.000	41.261	31.346	-93.526	4.754	178.801	
	4400	10.000	42.261	31.576	-96.672	4.802	263.138	
	4500	10.000	43.261	31.800	-99.841	4.850	380.421	
	4600	10.000	44.261	32.020	-103.032	4.896	540.911	
	4700	10.000	45.261	32.235	-106.245	4.941	760.	
	4701	10.000	45.271	32.237	-106.277	4.942	760.	
			144.391	30.715				21
GAS	4701	7.553	189.662	62.952	-106.277	4.942		
	4800	7.553	190.410	63.110	-112.518	5.124		
	4900	7.553	191.165	63.266	-118.836	5.301		
	5000	7.553	191.920	63.418	-125.171	5.472		

IRIDIUM TRICHLORIDE $IrCl_3$

		A	B	C	D	RANGE	REF.
SOL	CP	20.3	4.5	-1.0		298-1100	*
	** DEC.						

PHASE	T	CP	H	S	G	BT	
SOL	298	20.517	-60.8	27.45	-68.984	50.574	75
	300	20.539	-60.762	27.577	-69.035	50.299	
	400	21.475	-58.658	33.624	-72.107	39.403	
	500	22.150	-56.475	38.491	-75.721	33.102	
	600	22.722	-54.231	42.581	-79.780	29.064	
	700	23.246	-51.932	46.124	-84.219	26.298	
	800	23.744	-49.583	49.260	-88.991	24.315	
	900	24.227	-47.184	52.085	-94.061	22.844	
	1000	24.700	-44.738	54.662	-99.400	21.727	
	1100	25.167	-42.244	57.038	-104.986	20.862	

* * * * * * * * * *

IRIDIUM TRIBROMIDE $IrBr_3$

		A	B	C	D	RANGE	REF.
SOL	CP	23.72	4.86			298- 800	128
	** DEC.						1

PHASE	T	CP	H	S	G	BT	
SOL	298	25.169	-33.1	42.	-45.622	33.447	128,*
	300	25.178	-33.053	42.156	-45.700	33.297	
	400	25.664	-30.511	49.466	-50.298	27.485	
	500	26.150	-27.921	55.245	-55.543	24.281	
	600	26.636	-25.281	60.055	-61.314	22.337	
	700	27.122	-22.593	64.198	-67.532	21.087	
	800	27.608	-19.857	67.851	-74.138	20.256	

IrJ IRIDIUM MONOIODIDE

		A	B	C	D	RANGE	REF.
SOL	CP	11.6	3.0			298- 800	*
	** DEC.						1

PHASE	T	CP	H	S	G	BT	
SOL	298	12.494	-11.	26.	-18.752	13.747	1/*
	300	12.500	-10.977	26.077	-18.800	13.698	
	400	12.800	-9.712	29.714	-21.598	11.802	
	500	13.100	-8.417	32.603	-24.718	10.806	
	600	13.400	-7.092	35.018	-28.103	10.238	
	700	13.700	-5.737	37.106	-31.711	9.902	
	800	14.000	-4.352	38.955	-35.516	9.704	

* * * * * * * * * *

IrJ_2 IRIDIUM DIIODIDE

		A	B	C	D	RANGE	REF.
SOL	CP	17.8	4.0			298- 800	*
	** DISP.						1

PHASE	T	CP	H	S	G	BT	
SOL	298	18.993	-20.	38.	-31.330	22.968	1/*
	300	19.000	-19.965	38.118	-31.400	22.878	
	400	19.400	-18.045	43.638	-35.500	19.399	
	500	19.800	-16.085	48.010	-40.090	17.526	
	600	20.200	-14.085	51.656	-45.078	16.422	
	700	20.600	-12.045	54.799	-50.404	15.739	
	800	21.000	-9.965	57.576	-56.026	15.308	

IRIDIUM TRIOXIDE (GAS) IrO_3 (g)

		A	B	C	D	RANGE	REF.
GAS	CP	19.64	0.42	-1.32		298-1800	*

PHASE	T	CP	H	S	G	BT	
GAS	298	18.280	3.2	69.4	-17.492	12.823	75
	300	18.299	3.234	69.513	-17.620	12.838	
	400	18.983	5.103	74.884	-24.851	13.580	
	500	19.322	7.019	79.160	-32.561	14.234	
	600	19.525	8.963	82.703	-40.659	14.812	
	700	19.665	10.922	85.723	-49.084	15.327	
	800	19.770	12.894	88.356	-57.791	15.790	
	900	19.855	14.876	90.690	-66.745	16.210	
	1000	19.928	16.865	92.786	-75.921	16.595	
	1100	19.993	18.861	94.688	-85.296	16.949	
	1200	20.052	20.863	96.430	-94.853	17.277	
	1300	20.108	22.871	98.038	-104.578	17.583	
	1400	20.161	24.885	99.530	-114.457	17.870	
	1500	20.211	26.903	100.922	-124.480	18.139	
	1600	20.260	28.927	102.228	-134.639	18.393	
	1700	20.308	30.955	103.458	-144.923	18.634	
	1800	20.355	32.989	104.620	-155.328	18.862	

Ir_2S_3 DIIRIDIUM TRISULFICE

		A	B	C	D	RANGE	REF.
SOL	CP	26.36	7.88	-2.30		298-1400	63

PHASE	T	CP	H	S	G	BT	
SOL	298	26.122	-58.4	23.2	-65.317	47.885	62
	300	26.168	-58.352	23.362	-65.360	47.621	
	400	28.074	-55.631	31.174	-68.101	37.214	
	500	29.380	-52.756	37.585	-71.549	31.278	
	600	30.449	-49.763	43.039	-75.586	27.536	
	700	31.407	-46.670	47.805	-80.134	25.022	
	800	32.305	-43.484	52.058	-85.130	23.260	
	900	33.168	-40.210	55.913	-90.532	21.987	
	1000	34.010	-36.851	59.452	-96.303	21.050	
	1100	34.838	-33.408	62.732	-102.414	20.350	
	1200	35.656	-29.884	65.799	-108.842	19.825	
	1300	36.468	-26.277	68.685	-115.567	19.431	
	1400	37.275	-22.590	71.417	-122.574	19.137	

* * * * * * * * * *

IrS_2 IRIDIUM DISULFIDE

		A	B	C	D	RANGE	REF.
SOL	CP	16.39	3.77	-1.57		298-1400	63

PHASE	T	CP	H	S	G	BT	
SOL	298	15.748	-34.6	14.7	-38.983	28.579	62
	300	15.777	-34.571	14.798	-39.010	28.423	
	400	16.917	-32.931	19.508	-40.734	22.259	
	500	17.647	-31.201	23.366	-42.883	18.747	
	600	18.216	-29.407	26.635	-45.388	16.535	
	700	18.709	-27.560	29.481	-48.196	15.050	
	800	19.161	-25.666	32.009	-51.273	14.009	
	900	19.589	-23.729	34.290	-54.590	13.258	
	1000	20.003	-21.749	36.376	-58.125	12.705	
	1100	20.407	-19.728	38.301	-61.860	12.292	
	1200	20.805	-17.668	40.094	-65.781	11.982	
	1300	21.198	-15.567	41.775	-69.875	11.749	
	1400	21.588	-13.428	43.360	-74.132	11.574	

POTASSIUM CHLORIDE (GAS) $KCl(g)$

		A	B	C	D	RANGE	REF.
GAS	CP	8.878	0.227	-0.201		298-2500	3

PHASE	T	CP	H	S	G	BT	
GAS	298	8.720	-51.31	57.12	-68.340	50.102	3
	300	8.723	-51.294	57.174	-68.446	49.870	
	400	8.843	-50.415	59.702	-74.296	40.599	
	500	8.911	-49.527	61.683	-80.368	35.134	
	600	8.958	-48.633	63.312	-86.621	31.556	
	700	8.996	-47.736	64.696	-93.023	29.047	
	800	9.028	-46.834	65.899	-99.554	27.200	
	900	9.057	-45.930	66.964	-106.198	25.792	
	1000	9.085	-45.023	67.920	-112.943	24.687	
	1100	9.111	-44.113	68.787	-119.779	23.801	
	1200	9.136	-43.201	69.581	-126.698	23.078	
	1300	9.161	-42.286	70.313	-133.693	22.479	
	1400	9.186	-41.368	70.993	-140.759	21.976	
	1500	9.210	-40.449	71.628	-147.890	21.551	
	1600	9.233	-39.527	72.223	-155.083	21.186	
	1700	9.257	-38.602	72.783	-162.334	20.872	
	1800	9.280	-37.675	73.313	-169.639	20.600	
	1900	9.304	-36.746	73.816	-176.995	20.362	
	2000	9.327	-35.814	74.293	-184.401	20.153	
	2100	9.350	-34.881	74.749	-191.853	19.969	
	2200	9.373	-33.944	75.184	-199.350	19.806	
	2300	9.396	-33.006	75.602	-206.890	19.662	
	2400	9.419	-32.065	76.002	-214.470	19.533	
	2500	9.442	-31.122	76.387	-222.090	19.418	

K_2S POTASSIUM SULFIDE

		A	B	C	D	RANGE	REF.
SOL-A	CP	16.78	4.74			298- 820	89
	** TPT=	820-1100					89
SOL	CP	16.78	4.74			820-1100	89
SOL-B	CP	34.02				1100-1221	89
LIQ	CP	24.13				1221-1500	89

PHASE	T	CP	H	S	G	BT	
SOL-A	298	18.193	-99.	27.0	-107.050	78.480	63
	300	18.202	-98.966	27.113	-107.100	78.033	
	400	18.676	-97.122	32.414	-110.088	60.157	
	500	19.150	-95.231	36.632	-113.547	49.638	
	600	19.624	-93.292	40.166	-117.392	42.766	
	700	20.098	-91.306	43.226	-121.565	37.959	
	800	20.572	-89.273	45.941	-126.026	34.433	
	820	20.667	-88.860	46.450	-126.949	33.840	
							89
SOL	820	20.667	-88.860	46.450	-126.949	33.840	
	900	21.046	-87.192	48.391	-130.744	31.753	
	1000	21.520	-85.064	50.633	-135.697	29.661	
	1100	21.994	-82.888	52.707	-140.865	27.991	
							89
SOL-B	1100	34.020	-82.888	52.707	-140.865	27.991	
	1200	34.020	-79.486	55.667	-146.286	26.646	
	1221	34.020	-78.772	56.257	-147.461	26.398	
			3.86	3.161			89
LIQ	1221	24.130	-74.912	59.418	-147.461	26.398	
	1300	24.130	-73.005	60.931	-152.216	25.593	
	1400	24.130	-70.592	62.719	-158.399	24.731	
	1500	24.130	-68.179	64.384	-164.755	24.008	

POTASSIUM SULFITE

K_2SO_3

		A	B	C	D	RANGE	REF.
SOL	CP	20.72	19.8			298- 973	160
LIQ	CP	39.98				973-1200	160

PHASE	T	CP	H	S	G	ET	
SOL	298	26.623	-272.	38.4	-283.449	207.802	160
	300	26.660	-271.951	38.565	-283.520	206.572	
	400	28.640	-269.186	46.506	-287.788	157.261	
	500	30.620	-266.223	53.109	-292.777	127.990	
	600	32.600	-263.062	58.867	-298.382	108.700	
	700	34.580	-259.703	64.041	-304.531	95.092	
	800	36.560	-256.146	68.788	-311.176	85.021	
	900	38.540	-252.391	73.208	-318.278	77.299	
	973	39.985	-249.525	76.269	-323.735	72.725	
			6.	6.166			160
LIQ	973	39.980	-243.525	82.436	-323.735	72.725	
	1000	39.980	-242.445	83.530	-325.975	71.251	
	1100	39.980	-238.447	87.341	-334.522	66.472	
	1200	39.980	-234.449	90.819	-343.432	62.556	

K_2HPO_4 DIPOTASSIUM MONOHYDROGEN PHOSPHATE

		A	B	C	D	RANGE	REF.
SOL	CP	39.33	9.91	-10.09		298- 900	127

PHASE	T	CP	H	S	G	BT	
SOL	298	30.934	-424.4	41.99	-436.919	320.314	127
	300	31.092	-424.343	42.182	-436.997	318.395	
	400	36.988	-420.904	52.035	-441.718	241.376	
	500	40.249	-417.029	60.667	-447.363	195.568	
	600	42.473	-412.887	68.212	-453.815	165.324	
	700	44.208	-408.551	74.894	-460.976	143.943	
	800	45.681	-404.054	80.896	-468.771	128.079	
	900	47.003	-399.419	86.354	-477.137	115.880	

* * * * * * * * * *

KH_2PO_4 POTASSIUM DIHYDROGEN PHOSPHATE

		A	B	C	D	RANGE	REF.
SOL	CP	34.54	9.31	-8.49		298-1100	127

PHASE	T	CP	H	S	G	BT	
SOL	298	27.765	-373.36	32.15	-382.946	280.744	127
	300	27.900	-373.309	32.322	-383.005	279.057	
	400	32.958	-370.236	41.126	-386.687	211.304	
	500	35.799	-366.788	48.809	-391.192	171.013	
	600	37.768	-363.105	55.519	-396.416	144.414	
	700	39.324	-359.248	61.461	-402.271	125.611	
	800	40.661	-355.247	66.802	-408.688	111.663	
	900	41.871	-351.120	71.662	-415.615	100.939	
	1000	43.001	-346.875	76.132	-423.008	92.461	
	1100	44.079	-342.521	80.282	-430.831	85.610	

POTASSIUM BISMUTHIDE K_3Bi

		A	B	C	D	RANGE	REF.
SOL	CP	19.8	16.0			298- 948	*
	** MPT= 944						21

PHASE	T	CP	H	S	G	BT	
SOL	298	24.570	-54.1	47.3	-68.202	50.001	21/*
	300	24.600	-54.055	47.452	-68.290	49.756	
	400	26.200	-51.515	54.748	-73.414	40.117	
	500	27.800	-48.815	60.766	-79.198	34.622	
	600	29.400	-45.955	65.976	-85.540	31.162	
	700	31.000	-42.935	70.629	-92.375	28.845	
	800	32.600	-39.755	74.872	-99.653	27.227	
	900	34.200	-36.415	78.805	-107.339	26.069	
	948	34.968	-34.754	80.601	-111.165	25.631	

$K_2O \cdot SiO_2$ POTASSIUM METASILICATE

		A	B	C	D	RANGE	REF.
SOL	CP	32.42	5.85	-5.15		298-1249	5
LIQ	CP	40.0				1249-1800	5

PHASE	T	CP	H	S	G	GT	
SOL	298	28.371	-370.0	34.9	-380.405	278.882	5
	300	28.453	-369.947	35.076	-380.470	277.210	
	400	31.541	-366.930	43.736	-384.424	210.068	
	500	33.285	-363.682	50.976	-389.170	170.129	
	600	34.499	-360.290	57.157	-394.584	143.746	
	700	35.464	-356.790	62.550	-400.575	125.082	
	800	36.295	-353.202	67.340	-407.074	111.222	
	900	37.049	-349.534	71.660	-414.027	100.553	
	1000	37.755	-345.793	75.600	-421.393	92.108	
	1100	38.429	-341.984	79.230	-429.137	85.273	
	1200	39.082	-338.108	82.602	-437.231	79.641	
	1249	39.397	-336.185	84.173	-441.317	77.232	
			11.5	9.207			5
LIQ	1249	40.000	-324.685	93.380	-441.317	77.232	
	1300	40.000	-322.645	94.981	-446.120	75.010	
	1400	40.000	-318.645	97.945	-455.769	71.158	
	1500	40.000	-314.645	100.705	-465.703	67.862	
	1600	40.000	-310.645	103.286	-475.904	65.014	
	1700	40.000	-306.645	105.711	-486.355	62.534	
	1800	40.000	-302.645	107.998	-497.041	60.357	

POTASSIUM TETRASILICATE $K_2O \cdot 4SiO_2$

		A	B	C	D	RANGE	REF.
SOL-A	CP	60.52	38.09			298- 865	5
SOL-B	CP	93.54	3.87			865-1043	5
LIQ	CP	98.0				1043-1800	5

PHASE	T	CP	H	S	G	BT	
SOL-A	298	71.877	-1031.5	63.5	-1050.433	770.091	5
	300	71.947	-1031.367	63.945	-1050.550	765.428	
	400	75.756	-1023.982	85.164	-1058.048	578.168	
	500	79.565	-1016.216	102.478	-1067.455	466.647	
	600	83.374	-1008.069	117.321	-1078.461	392.882	
	700	87.183	-999.541	130.459	-1090.862	340.628	
	800	90.992	-990.632	142.350	-1104.512	301.779	
	865	93.468	-984.637	149.553	-1114.001	281.500	
			0.77	.890			5
SOL-B	865	96.888	-983.867	150.443	-1114.001	281.500	
	900	97.023	-980.474	154.289	-1119.334	271.848	
	1000	97.410	-970.752	164.531	-1135.284	248.149	
	1043	97.576	-966.560	168.636	-1142.447	239.420	
			11.7	11.218			5
LIQ	1043	98.000	-954.860	179.854	-1142.447	239.420	
	1100	98.000	-949.274	185.068	-1152.849	229.081	
	1200	98.000	-939.474	193.595	-1171.788	213.440	
	1300	98.000	-929.674	201.439	-1191.545	200.344	
	1400	98.000	-919.874	208.702	-1212.057	189.236	
	1500	98.000	-910.074	215.463	-1233.269	179.711	
	1600	98.000	-900.274	221.788	-1255.135	171.467	
	1700	98.000	-890.474	227.729	-1277.614	164.271	
	1800	98.000	-880.674	233.331	-1300.670	157.944	

$K_2O{\cdot}4B_2O_3$ DIPOTASSIUM OCTABORATE

		A	B	C	D	RANGE	REF.
SOL	CP	51.625	87.5	-0.801		298- 700	3
	** REF. STATE B2O3 (LIQ)						3
	CP	119.144	21.678	-106.549		700-1130	3
LIQ	CP	158.1				1130-1300	3

PHASE	T	CP	H	S	G	BT	
SOL	298	76.812	-1420.92	70.2	-1441.850	1057.047	3
	300	76.985	-1420.778	70.676	-1441.980	1050.623	
	400	86.124	-1412.619	94.083	-1450.253	792.488	
	500	95.055	-1403.560	114.262	-1460.691	638.553	
	600	103.902	-1393.611	132.376	-1473.037	536.625	
	700	112.712	-1382.780	149.054	-1487.118	464.362	
	700	112.574	-1382.780	149.054	-1487.118	464.362	
	800	119.838	-1371.143	164.583	-1502.809	410.604	
	900	125.500	-1358.866	179.037	-1519.999	369.156	
	1000	130.167	-1346.076	192.508	-1538.584	336.303	
	1100	134.184	-1332.854	205.107	-1558.472	309.681	
	1130	135.296	-1328.811	208.733	-1564.679	302.661	
			29.9	26.460			3
LIQ	1130	158.100	-1298.911	235.193	-1564.679	302.661	
	1200	158.100	-1287.844	244.695	-1581.479	288.065	
	1300	158.100	-1272.034	257.350	-1606.589	270.129	

KALIOPHILITE $K_2O \cdot Al_2O_3 \cdot 2SiO$

		A	B	C	D	RANGE	REF.
SOL	CP	57.26				298-1400	5

PHASE	T	CP	H	S	G	BT	
SOL	298	57.260	-1007.8	63.6	-1026.762	752.738	5
	300	57.260	-1007.694	63.954	-1026.880	748.182	
	400	57.260	-1001.968	80.427	-1034.139	565.103	
	500	57.260	-996.242	93.204	-1042.844	455.888	
	600	57.260	-990.516	103.644	-1052.702	383.498	
	700	57.260	-984.790	112.470	-1063.519	332.090	
	800	57.260	-979.064	120.116	-1075.157	293.759	
	900	57.260	-973.338	126.861	-1087.513	264.120	
	1000	57.260	-967.612	132.894	-1100.506	240.548	
	1100	57.260	-961.886	138.351	-1114.072	221.376	
	1200	57.260	-956.160	143.333	-1128.160	205.494	
	1300	57.260	-950.434	147.917	-1142.726	192.135	
	1400	57.260	-944.708	152.160	-1157.732	180.754	

* * * * * * * * * *

LEUCITE $K_2O \cdot Al_2O_3 \cdot 4SiO_2$

		A	B	C	D	RANGE	REF.
SOL	CP	78.46				298-1400	5

PHASE	T	CP	H	S	G	BT	
SOL	298	78.460	1450.4	88.	1424.163	-1044.080	5
	300	78.460	1450.545	88.485	1424.000	-1037.522	
	400	78.460	1458.391	111.057	1413.968	-772.660	
	500	78.460	1466.237	128.565	1401.955	-612.876	
	600	78.460	1474.083	142.870	1388.361	-505.778	
	700	78.460	1481.929	154.964	1373.454	-428.869	
	800	78.460	1489.775	165.441	1357.422	-370.880	
	900	78.460	1497.621	174.682	1340.407	-325.539	
	1000	78.460	1505.467	182.949	1322.518	-289.075	
	1100	78.460	1513.313	190.427	1303.843	-259.085	
	1200	78.460	1521.159	197.254	1284.454	-233.963	
	1300	78.460	1529.005	203.534	1264.411	-212.595	
	1400	78.460	1536.851	209.349	1243.763	-194.186	

$K_2O \cdot Al_2O_3 \cdot 6SiO_2$ MICROCLINE

		A	B	C	D	RANGE	REF.
SOL	CP	136.74	18.48	-45.52		298-1400	5

PHASE	T	CP	H	S	G	BT	
SOL	298	91.042	-1891.4	105.	-1922.706	1409.571	5
	300	91.706	-1891.231	105.565	-1922.901	1401.020	
	400	115.682	-1880.703	135.687	-1934.978	1057.365	
	500	127.772	-1868.474	162.927	-1949.937	852.432	
	600	135.184	-1855.301	186.924	-1967.455	716.741	
	700	140.386	-1841.509	208.173	-1987.230	620.525	
	800	144.411	-1827.262	227.191	-2009.015	548.911	
	900	147.752	-1812.650	244.398	-2032.608	493.651	
	1000	150.668	-1797.726	260.120	-2057.845	449.802	
	1100	153.306	-1782.525	274.605	-2084.591	414.226	
	1200	155.755	-1767.071	288.051	-2112.732	384.833	
	1300	158.071	-1751.379	300.610	-2142.172	360.180	
	1400	160.290	-1735.460	312.406	-2172.828	339.239	

* * * * * * * * * *

$K_2O \cdot Al_2O_3 \cdot 6SiO_2$ (α) ANDULARIA

		A	B	C	D	RANGE	REF.
SOL	CP	136.74	18.48	-45.52		298-1400	5

PHASE	T	CP	H	S	G	BT	
SOL	298	91.042	-1890.	112.	-1923.393	1410.075	5
	300	91.706	-1889.831	112.565	-1923.601	1401.530	
	400	115.682	-1879.303	142.687	-1936.378	1058.130	
	500	127.772	-1867.074	169.927	-1952.037	853.350	
	600	135.184	-1853.901	193.924	-1970.255	717.761	
	700	140.386	-1840.109	215.173	-1990.730	621.618	
	800	144.411	-1825.862	234.191	-2013.215	550.059	
	900	147.752	-1811.250	251.398	-2037.508	494.841	
	1000	150.668	-1796.326	267.120	-2063.445	451.026	
	1100	153.306	-1781.125	281.605	-2090.891	415.478	
	1200	155.755	-1765.671	295.051	-2119.732	386.108	
	1300	158.071	-1749.979	307.610	-2149.872	361.475	
	1400	160.290	-1734.060	319.406	-2181.228	340.551	

SANIDINE $K_2O \cdot Al_2O_3 \cdot 6SiO_2$

		A	B	C	D	RANGE	REF.
SOL	CP	136.74	18.48	-45.52		298-1473	5
LIQ	CP	183.0				1473-1900	5

PHASE	T	CP	H	S	G	BT	
SOL	298	91.042	-1888.8	113.8	-1922.729	1409.588	5
	300	91.706	-1888.631	114.365	-1922.941	1401.050	
	400	115.682	-1878.103	144.487	-1935.898	1057.868	
	500	127.772	-1865.874	171.727	-1951.737	853.218	
	600	135.184	-1852.701	195.724	-1970.135	717.718	
	700	140.386	-1838.909	216.973	-1990.790	621.636	
	800	144.411	-1824.662	235.991	-2013.455	550.124	
	900	147.752	-1810.050	253.198	-2037.928	494.943	
	1000	150.668	-1795.126	268.920	-2064.045	451.157	
	1100	153.306	-1779.925	283.405	-2091.671	415.633	
	1200	155.755	-1764.471	296.851	-2120.692	386.283	
	1300	158.071	-1748.779	309.410	-2151.012	361.667	
	1400	160.290	-1732.860	321.206	-2182.548	340.757	
	1473	161.863	-1721.101	329.393	-2206.297	327.394	
			29.4	19.959			5
LIQ	1473	183.000	-1691.701	349.352	-2206.297	327.394	
	1500	183.000	-1686.760	352.676	-2215.775	322.882	
	1600	183.000	-1668.460	364.487	-2251.639	307.601	
	1700	183.000	-1650.160	375.581	-2288.648	294.265	
	1800	183.000	-1631.860	386.041	-2326.735	282.542	
	1900	183.000	-1613.560	395.936	-2365.838	272.170	

K_2CrO_4 POTASSIUM CHROMATE

		A	B	C	D	RANGE	REF.
SOL-A	CP	29.57	17.9			298- 939	58,*
SOL-B	CP	35.5	12.0			939-1244	*
LIQ	CP	50.0				1244-1600	*

PHASE	T	CP	H	S	G	BT	
SOL-A	298	34.907	-331.2	47.8	-345.452	253.257	172,111/58
	300	34.940	-331.135	48.016	-345.540	251.760	
	400	36.730	-327.552	58.313	-350.877	191.736	
	500	38.520	-323.789	66.701	-357.140	156.127	
	600	40.310	-319.848	73.882	-364.177	132.669	
	700	42.100	-315.727	80.231	-371.889	116.125	
	800	43.890	-311.428	85.969	-380.203	103.881	
	900	45.680	-306.949	91.242	-389.067	94.491	
	939	46.378	-305.154	93.194	-392.664	91.404	
			3.28	3.493			*
SOL-B	939	46.768	-301.874	96.688	-392.664	91.404	
	1000	47.500	-298.999	99.654	-398.653	87.137	
	1100	48.700	-294.189	104.237	-408.850	81.242	
	1200	49.900	-289.259	108.526	-419.491	76.410	
	1244	50.428	-287.052	110.333	-424.306	74.553	
			6.8	5.466			*
LIQ	1244	50.000	-280.252	115.799	-424.306	74.553	
	1300	50.000	-277.452	118.001	-430.853	72.443	
	1400	50.000	-272.452	121.706	-442.840	69.140	
	1500	50.000	-267.452	125.156	-455.185	66.329	
	1600	50.000	-262.452	128.383	-467.864	63.916	

LANTHANUM

		A	B	C	D	RANGE	REF.
SOL-A	CP	6.320	0.556			298- 550	21
SOL-B	CP	4.220	3.593	0.933		550-1134	21
SOL-C	CP	9.45				1134-1193	21
LIQ	CP	8.2				1193-3000	21
	** BPT= 3730, L= 98.882						21

PHASE	T	CP	H	S	G	BT	P	
SOL-A	298	6.486	0	13.6	-4.055	2.973		21
	300	6.487	.012	13.640	-4.080	2.973		
	400	6.542	.663	15.514	-5.542	3.028		
	500	6.598	1.320	16.980	-7.169	3.134		
	550	6.626	1.651	17.610	-8.034	3.193		
			0.087	.158				21
SOL-B	550	6.505	1.738	17.768	-8.034	3.193		
	600	6.635	2.067	18.340	-8.937	3.256		
	700	6.926	2.744	19.384	-10.824	3.380		
	800	7.240	3.452	20.329	-12.811	3.500		
	900	7.569	4.193	21.201	-14.888	3.616		
	1000	7.906	4.966	22.015	-17.049	3.727		
	1100	8.249	5.774	22.785	-19.289	3.833		
	1134	8.367	6.057	23.038	-20.068	3.868		
			0.746	.658				21
SOL-C	1134	9.450	6.803	23.696	-20.068	3.868		
	1193	9.450	7.360	24.175	-21.481	3.936		
			1.481	1.241				21
LIQ	1193	8.200	8.841	25.416	-21.481	3.936		
	1200	8.200	8.899	25.464	-21.659	3.945		
	1300	8.200	9.719	26.121	-24.238	4.075		
	1400	8.200	10.539	26.728	-26.881	4.197		
	1500	8.200	11.359	27.294	-29.583	4.311		
	1600	8.200	12.179	27.823	-32.339	4.418		
	1700	8.200	12.999	28.321	-35.146	4.519		
	1800	8.200	13.819	28.789	-38.002	4.615		
	1900	8.200	14.639	29.233	-40.903	4.706	.002	
	2000	8.200	15.459	29.653	-43.848	4.792	.007	
	2100	8.200	16.279	30.053	-46.833	4.875	.023	
	2200	8.200	17.099	30.435	-49.858	4.954	.069	
	2300	8.200	17.919	30.799	-52.920	5.029	.184	
	2400	8.200	18.739	31.148	-56.017	5.102	.457	
	2500	8.200	19.559	31.483	-59.149	5.171	1.052	
	2600	8.200	20.379	31.805	-62.313	5.239	2.272	
	2700	8.200	21.199	32.114	-65.509	5.303	4.630	
	2800	8.200	22.019	32.412	-68.736	5.366	8.966	
	2900	8.200	22.839	32.700	-71.991	5.426	16.583	
	3000	8.200	23.659	32.978	-75.275	5.485	29.432	

LaH_2 LANTHANUM DIHYDRIDE

		A	B	C	D	RANGE	REF.
SOL	CP	9.38	3.62			298-1200	76,*

PHASE	T	CP	H	S	G	BT	
SOL	298	10.459	-45.2	12.35	-48.882	35.836	19/76
	300	10.466	-45.181	12.415	-48.905	35.632	
	400	10.828	-44.116	15.475	-50.306	27.490	
	500	11.190	-43.015	17.930	-51.980	22.724	
	600	11.552	-41.878	20.002	-53.879	19.628	
	700	11.914	-40.705	21.810	-55.972	17.478	
	800	12.276	-39.495	23.425	-58.235	15.911	
	900	12.638	-38.249	24.892	-60.652	14.730	
	1000	13.000	-36.968	26.242	-63.210	13.816	
	1100	13.362	-35.649	27.498	-65.897	13.094	
	1200	13.724	-34.295	28.676	-68.707	12.515	

LANTHANUM FLUORIDE LaF_3

		A	B	C	D	RANGE	REF.
SOL-A	CP	18.578	4.769			298-1766	*
LIQ	CP	36.5				1766-2000	*

PHASE	T	CP	H	S	G	BT	
SOL-A	298	20.000	-426.	27.1	-434.080	318.232	93/*
	300	20.009	-425.963	27.224	-434.130	316.306	
	400	20.486	-423.938	33.045	-437.156	238.883	
	500	20.962	-421.866	37.668	-440.700	192.656	
	600	21.439	-419.746	41.532	-444.665	161.991	
	700	21.916	-417.578	44.872	-448.989	140.199	
	800	22.393	-415.363	47.830	-453.627	123.942	
	900	22.870	-413.099	50.495	-458.545	111.365	
	1000	23.347	-410.788	52.929	-463.718	101.359	
	1100	23.824	-408.430	55.177	-469.125	93.219	
	1200	24.301	-406.024	57.270	-474.748	86.475	
	1300	24.778	-403.570	59.234	-480.574	80.803	
	1400	25.255	-401.068	61.088	-486.591	75.971	
	1500	25.731	-398.519	62.847	-492.789	71.809	
	1600	26.208	-395.922	64.523	-499.158	68.191	
	1700	26.685	-393.277	66.126	-505.691	65.020	
	1766	27.000	-391.506	67.148	-510.089	63.134	
			12.009	6.800			174
LIQ	1766	36.500	-379.497	73.948	-510.089	63.134	
	1800	36.500	-378.256	74.644	-512.615	62.248	
	1900	36.500	-374.606	76.618	-520.179	59.842	
	2000	36.500	-370.956	78.490	-527.935	57.698	

$LaBr_3$ LANTHANUM BROMIDE

		A	B	C	D	RANGE	REF.
SOL	CP	22.64	6.0			298-1061	89
LIQ	CP	34.5				1061-1859	89
	** BPT=	1859					1

PHASE	T	CP	H	S	G	BT	
SOL	298	24.429	-208.	43.5	-220.970	161.997	93/*
	300	24.440	-207.955	43.651	-221.050	161.057	
	400	25.040	-205.481	50.764	-225.787	123.381	
	500	25.640	-202.947	56.416	-231.155	101.051	
	600	26.240	-200.353	61.144	-237.039	86.353	
	700	26.840	-197.699	65.234	-243.363	75.991	
	800	27.440	-194.985	68.857	-250.070	68.325	
	900	28.040	-192.211	72.124	-257.122	62.446	
	1000	28.640	-189.377	75.109	-264.486	57.811	
	1061	29.006	-187.619	76.816	-269.120	55.442	
			12.9	12.158			89
LIQ	1061	34.500	-174.719	88.974	-269.120	55.442	
	1100	34.500	-173.373	90.219	-272.614	54.171	
	1200	34.500	-169.923	93.221	-281.789	51.328	
	1300	34.500	-166.473	95.983	-291.251	48.970	
	1400	34.500	-163.023	98.539	-300.978	46.991	
	1500	34.500	-159.573	100.920	-310.953	45.312	
	1600	34.500	-156.123	103.146	-321.157	43.874	
	1700	34.500	-152.673	105.238	-331.577	42.633	
	1800	34.500	-149.223	107.210	-342.201	41.554	
	1859	34.500	-147.188	108.323	-348.559	40.983	

LANTHANUM IODIDE

LaJ_3

		A	B	C	D	RANGE	REF.
SOL	CP	23.222	4.735	-0.653		298-1051	89
LIQ	CP	36.273				1051-1745	89
	** BPT=	1745					1

PHASE	T	CP	H	S	G	BT	
SOL	298	23.899	-157.	51.3	-172.295	126.313	1/*
	300	23.917	-156.956	51.448	-172.390	125.603	
	400	24.708	-154.522	58.443	-177.900	97.213	
	500	25.328	-152.020	64.025	-184.032	80.451	
	600	25.882	-149.459	68.693	-190.674	69.462	
	700	26.403	-146.844	72.722	-197.750	61.748	
	800	26.908	-144.179	76.280	-205.203	56.066	
	900	27.403	-141.463	79.478	-212.994	51.729	
	1000	27.892	-138.698	82.391	-221.089	48.326	
	1051	28.139	-137.270	83.784	-225.327	46.862	
			13.4	12.750			89
LIQ	1051	36.273	-123.870	96.534	-225.327	46.862	
	1100	36.273	-122.092	98.187	-230.098	45.722	
	1200	36.273	-118.465	101.343	-240.077	43.730	
	1300	36.273	-114.838	104.247	-250.358	42.095	
	1400	36.273	-111.210	106.935	-260.919	40.737	
	1500	36.273	-107.583	109.437	-271.739	39.598	
	1600	36.273	-103.956	111.778	-282.801	38.634	
	1700	36.273	-100.328	113.977	-294.090	37.813	
	1745	36.273	-98.696	114.925	-299.240	37.483	

LaOCl

LANTHANUM OXYCHLORIDE

		A	B	C	D	RANGE	REF.
SOL	CP	16.85	2.93	-1.1		298-1200	*

PHASE	T	CP	H	S	G	BT	
SOL	298	16.486	-243.9	19.8	-249.803	183.135	100/*
	300	16.507	-243.869	19.902	-249.840	182.033	
	400	17.334	-242.174	24.775	-252.084	137.751	
	500	17.875	-240.412	28.704	-254.764	111.372	
	600	18.302	-238.602	32.002	-257.804	93.918	
	700	18.677	-236.753	34.852	-261.150	81.546	
	800	19.022	-234.868	37.369	-264.763	72.340	
	900	19.351	-232.949	39.628	-268.615	65.237	
	1000	19.670	-230.998	41.684	-272.682	59.603	
	1100	19.982	-229.015	43.573	-276.946	55.031	
	1200	20.290	-227.002	45.325	-281.392	51.255	

LANTHANUM MONOSULFICE

		A	B	C	D	RANGE	REF.
SOL	CP	11.11	1.3			298-2600	63
	** MPT=	2600					63

PHASE	T	CP	H	S	G	BT	
SOL	298	11.498	-109.	17.	-114.069	83.626	63
	300	11.500	-108.979	17.071	-114.100	83.133	
	400	11.630	-107.822	20.397	-115.981	63.378	
	500	11.760	-106.653	23.006	-118.156	51.653	
	600	11.890	-105.470	25.162	-120.567	43.923	
	700	12.020	-104.275	27.005	-123.178	38.463	
	800	12.150	-103.066	28.618	-125.961	34.416	
	900	12.280	-101.845	30.057	-128.896	31.304	
	1000	12.410	-100.610	31.357	-131.967	28.845	
	1100	12.540	-99.363	32.546	-135.164	26.858	
	1200	12.670	-98.102	33.643	-138.474	25.223	
	1300	12.800	-96.829	34.662	-141.890	23.857	
	1400	12.930	-95.542	35.615	-145.404	22.702	
	1500	13.060	-94.243	36.512	-149.011	21.714	
	1600	13.190	-92.930	37.359	-152.705	20.861	
	1700	13.320	-91.605	38.163	-156.481	20.120	
	1800	13.450	-90.266	38.928	-160.336	19.470	
	1900	13.580	-88.915	39.658	-164.265	18.897	
	2000	13.710	-87.550	40.358	-168.266	18.390	
	2100	13.840	-86.173	41.030	-172.336	17.938	
	2200	13.970	-84.782	41.677	-176.472	17.533	
	2300	14.100	-83.379	42.301	-180.671	17.170	
	2400	14.230	-81.962	42.904	-184.931	16.843	
	2500	14.360	-80.533	43.487	-189.251	16.547	
	2600	14.490	-79.090	44.053	-193.628	16.278	

La_2S_3 LANTHANUM SULFIDE

		A	B	C	D	RANGE	REF.
SOL	CP	27.85	3.5			298-2400	63
	** MPT=	2400					63

PHASE	T	CP	H	S	G	BT	
SOL	298	28.894	-292.	39.43	-303.756	222.689	63
	300	28.900	-291.947	39.609	-303.829	221.369	
	400	29.250	-289.039	47.971	-308.227	168.430	
	500	29.600	-286.097	54.535	-313.364	136.990	
	600	29.950	-283.119	59.963	-319.097	116.247	
	700	30.300	-280.107	64.606	-325.331	101.586	
	800	30.650	-277.059	68.675	-331.999	90.710	
	900	31.000	-273.977	72.305	-339.051	82.344	
	1000	31.350	-270.859	75.589	-346.448	75.726	
	1100	31.700	-267.707	78.594	-354.160	70.375	
	1200	32.050	-264.519	81.367	-362.159	65.967	
	1300	32.400	-261.297	83.946	-370.427	62.283	
	1400	32.750	-258.039	86.360	-378.943	59.164	
	1500	33.100	-254.747	88.632	-387.694	56.495	
	1600	33.450	-251.419	90.779	-396.665	54.189	
	1700	33.800	-248.057	92.817	-405.846	52.182	
	1800	34.150	-244.659	94.759	-415.226	50.422	
	1900	34.500	-241.227	96.615	-424.795	48.869	
	2000	34.850	-237.759	98.394	-434.546	47.491	
	2100	35.200	-234.257	100.102	-444.471	46.263	
	2200	35.550	-230.719	101.748	-454.564	45.163	
	2300	35.900	-227.147	103.336	-464.819	44.174	
	2400	36.250	-223.539	104.871	-475.230	43.281	

LANTHANUM MONOSELENIDE

LaSe

		A	B	C	D	RANGE	REF.
SOL	CP	11.34	1.40			298-2200	63
	** MPT=	2250					63

PHASE	T	CP	H	S	G	BT	P	
SOL	298	11.757	-86.0	19.4	-91.784	67.289		63
	300	11.760	-85.978	19.473	-91.820	66.900		
	400	11.900	-84.795	22.875	-93.945	51.336		
	500	12.040	-83.598	25.545	-96.371	42.129		
	600	12.180	-82.387	27.753	-99.039	36.080		
	700	12.320	-81.162	29.641	-101.911	31.822		
	800	12.460	-79.923	31.295	-104.960	28.677		
	900	12.600	-78.670	32.771	-108.164	26.269		
	1000	12.740	-77.403	34.106	-111.509	24.374		
	1100	12.880	-76.122	35.327	-114.982	22.848		
	1200	13.020	-74.827	36.453	-118.571	21.598		
	1300	13.160	-73.518	37.501	-122.270	20.558		
	1400	13.300	-72.195	38.481	-126.069	19.683		
	1500	13.440	-70.858	39.404	-129.964	18.938		
	1600	13.580	-69.507	40.276	-133.948	18.299		
	1700	13.720	-68.142	41.103	-138.018	17.746		
	1800	13.860	-66.763	41.891	-142.168	17.264		
	1900	14.000	-65.370	42.644	-146.395	16.841		
	2000	14.140	-63.963	43.366	-150.695	16.469	.002	
	2100	14.280	-62.542	44.059	-155.067	16.140	.008	
	2200	14.420	-61.107	44.727	-159.506	15.848	.032	

LaSe (g)

LANTHANUM MONOSELENIDE (GAS)

		A	B	C	D	RANGE	REF.
GAS	CP	8.925	0.009	-0.343		298-2000	63

PHASE	T	CP	H	S	G	BT	
GAS	298	8.542	44.80	62.85	26.061	-19.106	63
	300	8.547	44.816	62.903	25.945	-18.903	
	400	8.714	45.680	65.388	19.525	-10.669	
	500	8.792	46.556	67.342	12.885	-5.633	
	600	8.835	47.437	68.949	6.068	-2.211	
	700	8.861	48.322	70.313	-.897	.280	
	800	8.879	49.209	71.497	-7.989	2.183	
	900	8.891	50.098	72.544	-15.192	3.690	
	1000	8.900	50.987	73.481	-22.494	4.917	
	1100	8.907	51.878	74.330	-29.885	5.938	
	1200	8.912	52.769	75.105	-37.357	6.805	
	1300	8.916	53.660	75.819	-44.904	7.550	
	1400	8.920	54.552	76.479	-52.519	8.200	
	1500	8.923	55.444	77.095	-60.198	8.772	
	1600	8.926	56.337	77.671	-67.937	9.281	
	1700	8.928	57.229	78.212	-75.731	9.737	
	1800	8.931	58.122	78.723	-83.578	10.149	
	1900	8.933	59.015	79.205	-91.475	10.523	
	2000	8.934	59.909	79.664	-99.419	10.865	

LANTHANUM SELENIDE

La_2Se_3

		A	B	C	D	RANGE	REF.
SOL	CP	28.85	3.9			298-1900	63
	** MPT=	1900					63

PHASE	T	CP	H	S	G	BT	
SOL	298	30.013	-223.	48.33	-237.410	174.049	63
	300	30.020	-222.944	48.516	-237.499	173.041	
	400	30.410	-219.923	57.205	-242.805	132.680	
	500	30.800	-216.862	64.033	-248.879	108.800	
	600	31.190	-213.763	69.683	-255.573	93.105	
	700	31.580	-210.624	74.520	-262.789	82.057	
	800	31.970	-207.447	78.763	-270.457	73.895	
	900	32.360	-204.230	82.551	-278.526	67.644	
	1000	32.750	-200.975	85.980	-286.955	62.722	
	1100	33.140	-197.680	89.120	-295.712	58.761	
	1200	33.530	-194.347	92.020	-304.771	55.514	
	1300	33.920	-190.974	94.719	-314.110	52.814	
	1400	34.310	-187.563	97.248	-323.709	50.540	
	1500	34.700	-184.112	99.628	-333.554	48.605	
	1600	35.090	-180.623	101.880	-343.631	46.944	
	1700	35.480	-177.094	104.019	-353.927	45.506	
	1800	35.870	-173.527	106.058	-364.431	44.254	
	1900	36.260	-169.920	108.008	-375.135	43.156	

La_2Te_3 LANTHANUM TELLURIDE

		A	B	C	D	RANGE	REF.
SOL	CP	30.63	3.2			298-1760	63
	** MPT=	1760					63

PHASE	T	CP	H	S	G	BT	
SOL	298	31.584	-187.5	55.36	-204.006	149.560	63
	300	31.590	-187.442	55.555	-204.108	148.713	
	400	31.910	-184.267	64.687	-210.141	114.831	
	500	32.230	-181.060	71.842	-216.981	94.855	
	600	32.550	-177.821	77.746	-224.468	81.774	
	700	32.870	-174.550	82.788	-232.501	72.600	
	800	33.190	-171.247	87.198	-241.005	65.848	
	900	33.510	-167.912	91.126	-249.925	60.698	
	1000	33.830	-164.545	94.673	-259.218	56.660	
	1100	34.150	-161.146	97.912	-268.849	53.423	
	1200	34.470	-157.715	100.898	-278.792	50.782	
	1300	34.790	-154.252	103.669	-289.022	48.595	
	1400	35.110	-150.757	106.259	-299.519	46.763	
	1500	35.430	-147.230	108.692	-310.268	45.212	
	1600	35.750	-143.671	110.989	-321.253	43.887	
	1700	36.070	-140.080	113.166	-332.462	42.747	
	1760	36.262	-137.910	114.421	-339.290	42.137	

LANTHANUM NITRIDE

		A	B	C	D	RANGE	REF.
SOL	CP	11.1	1.65	-1.73		298-1800	*

PHASE	T	CP	H	S	G	BT	
SOL	298	9.646	-71.5	10.6	-74.660	54.735	1/*
	300	9.673	-71.482	10.660	-74.680	54.412	
	400	10.679	-70.459	13.598	-75.898	41.474	
	500	11.233	-69.361	16.045	-77.383	33.829	
	600	11.609	-68.218	18.128	-79.094	28.814	
	700	11.902	-67.042	19.940	-81.000	25.293	
	800	12.150	-65.839	21.546	-83.076	22.698	
	900	12.371	-64.613	22.990	-85.304	20.717	
	1000	12.577	-63.365	24.304	-87.669	19.163	
	1100	12.772	-62.098	25.512	-90.161	17.916	
	1200	12.960	-60.811	26.632	-92.769	16.898	
	1300	13.143	-59.506	27.676	-95.485	16.055	
	1400	13.322	-58.182	28.657	-98.302	15.348	
	1500	13.498	-56.841	29.582	-101.214	14.749	
	1600	13.672	-55.483	30.459	-104.217	14.237	
	1700	13.845	-54.107	31.293	-107.305	13.797	
	1800	14.017	-52.714	32.089	-110.474	13.415	

$LaAl_2$ LANTHANUM 2-ALUMINIUM

		A	B	C	D	RANGE	REF.
SOL	CP	16.6	3.4			298-1697	*
	** MPT= 1697						1

PHASE	T	CP	H	S	G	BT	
SOL	298	17.614	-36.	23.6	-43.036	31.551	1/85
	300	17.620	-35.967	23.709	-43.080	31.388	
	400	17.960	-34.188	28.824	-45.718	24.983	
	500	18.300	-32.375	32.869	-48.810	21.338	
	600	18.640	-30.528	36.235	-52.270	19.042	
	700	18.980	-28.647	39.134	-56.041	17.499	
	800	19.320	-26.732	41.691	-60.085	16.417	
	900	19.660	-24.783	43.986	-64.371	15.633	
	1000	20.000	-22.800	46.075	-68.875	15.055	
	1100	20.340	-20.783	47.997	-73.580	14.621	
	1200	20.680	-18.732	49.781	-78.470	14.293	
	1300	21.020	-16.647	51.450	-83.533	14.045	
	1400	21.360	-14.528	53.020	-88.757	13.857	
	1500	21.700	-12.375	54.506	-94.134	13.717	
	1600	22.040	-10.188	55.917	-99.656	13.614	
	1697	22.370	-8.035	57.224	-105.143	13.543	

* * * * * * * * * *

LaMg LANTHANUM MAGNESIUM

		A	B	C	D	RANGE	REF.
SOL	CP	12.12	4.76			298-1000	*

PHASE	T	CP	H	S	G	BT	
SOL	298	13.539	-4.3	22.25	-10.934	8.016	153*
	300	13.548	-4.275	22.334	-10.975	7.996	
	400	14.024	-2.896	26.296	-13.415	7.331	
	500	14.500	-1.470	29.477	-16.209	7.086	
	600	14.976	.004	32.163	-19.294	7.029	
	700	15.452	1.525	34.507	-22.630	7.066	
	800	15.928	3.094	36.601	-26.187	7.155	
	900	16.404	4.711	38.505	-29.944	7.272	
	1000	16.880	6.375	40.258	-33.883	7.406	

LiF

LITHIUM FLUORIDE

			A	B	C	D	RANGE	REF.
SOL	CP		10.203	4.163	-1.267		298-1122	3
	**	MPT=	1121.3					3
LIQ	CP		15.34				1121-1990	3
	**	BPT=	1990 GAS(69LIF+27LI2F2+4LI3F3)					3

PHASE	T	CP	H	S	G	BT	P	
SOL	298	10.019	-147.45	8.523	-149.991	109.961		3
	300	10.044	-147.431	8.585	-150.007	109.295		
	400	11.076	-146.371	11.629	-151.022	82.526		
	500	11.778	-145.227	14.179	-152.316	66.586		
	600	12.349	-144.020	16.378	-153.847	56.046		
	700	12.859	-142.759	18.321	-155.583	48.582		
	800	13.335	-141.449	20.069	-157.504	43.034		
	900	13.793	-140.093	21.666	-159.592	38.759		
	1000	14.239	-138.691	23.143	-161.834	35.373	.001	
	1100	14.678	-137.245	24.521	-164.217	32.631	.022	
	1121	14.770	-136.931	24.803	-164.743	32.114	.038	
			6.474	5.774				3
LIQ	1121	15.340	-130.457	30.577	-164.743	32.114	.038	
	1200	15.340	-129.250	31.617	-167.191	30.454	.180	
	1300	15.340	-127.716	32.845	-170.414	28.653	.962	
	1400	15.340	-126.182	33.982	-173.757	27.128	3.964	
	1500	15.340	-124.648	35.040	-177.208	25.823	13.321	
	1600	15.340	-123.114	36.030	-180.762	24.694	37.987	
	1700	15.340	-121.580	36.960	-184.412	23.711	94.724	
	1800	15.340	-120.046	37.837	-188.153	22.848	211.413	
	1900	15.340	-118.512	38.666	-191.978	22.085	430.081	
	1990	15.340	-117.131	39.376	-195.490	21.472	760.	

LiF(g)

LITHIUM FLUORIDE (GAS)

		A	B	C	D	RANGE	REF.
GAS	CP	8.934	0.141	-3.127	0.536	298-1990	3
	CP	8.934	0.141	-3.127	0.536	1990-2500	3

PHASE	T	CP	H	S	G	BT	
GAS	298	7.481	-81.475	47.84	-95.738	70.188	3
	300	7.487	-81.461	47.886	-95.827	69.819	
	400	7.874	-80.693	50.093	-100.730	55.044	
	500	8.182	-79.889	51.885	-105.832	46.265	
	600	8.398	-79.060	53.397	-111.098	40.473	
	700	8.551	-78.212	54.704	-116.505	36.379	
	800	8.663	-77.351	55.853	-122.034	33.343	
	900	8.748	-76.480	56.879	-127.671	31.007	
	1000	8.816	-75.602	57.804	-133.406	29.160	
	1100	8.871	-74.717	58.647	-139.229	27.666	
	1200	8.917	-73.828	59.421	-145.133	26.436	
	1300	8.957	-72.934	60.136	-151.112	25.408	
	1400	8.991	-72.037	60.801	-157.159	24.537	
	1500	9.022	-71.136	61.423	-163.270	23.792	
	1600	9.051	-70.232	62.006	-169.442	23.148	
	1700	9.076	-69.326	62.556	-175.671	22.587	
	1800	9.100	-68.417	63.075	-181.952	22.095	
	1900	9.123	-67.506	63.568	-188.285	21.661	
	1990	9.142	-66.684	63.990	-194.025	21.311	
	1990	9.142	-66.684	63.990	-194.025	21.311	
	2000	9.145	-66.593	64.036	-194.665	21.275	
	2100	9.165	-65.677	64.483	-201.091	20.931	
	2200	9.185	-64.760	64.910	-207.561	20.622	
	2300	9.204	-63.840	65.318	-214.073	20.344	
	2400	9.222	-62.919	65.710	-220.624	20.093	
	2500	9.240	-61.996	66.087	-227.214	19.866	

GAS CP(T) = A + B*T + C/(T**2) + E/(T**3)

CP(T) = A + B*T + C/(T**2) + E/(T**3)

LITHIUM FLUORIDE (DIMERIC GAS) $Li_2F_2(g)$

		A	B	C	D	RANGE	REF.
GAS	CP	19.923	-0.017	-6.899	0.777	298-1990	3
	CP	19.923	-0.017	-6.899	0.777	1990-2500	3

PHASE	T	CP	H	S	G	BT	
GAS	298	15.089	-225.33	61.8	-243.756	178.702	3
	300	15.130	-225.302	61.893	-243.870	177.683	
	400	16.818	-223.696	66.501	-250.297	136.774	
	500	17.776	-221.962	70.366	-257.146	112.413	
	600	18.356	-220.154	73.663	-264.351	96.303	
	700	18.730	-218.298	76.522	-271.864	84.891	
	800	18.983	-216.412	79.041	-279.644	76.406	
	900	19.163	-214.504	81.288	-287.663	69.863	
	1000	19.294	-212.581	83.314	-295.894	64.676	
	1100	19.393	-210.646	85.158	-304.319	60.471	
	1200	19.468	-208.703	86.848	-312.921	56.998	
	1300	19.528	-206.753	88.409	-321.685	54.087	
	1400	19.576	-204.798	89.858	-330.599	51.616	
	1500	19.614	-202.838	91.210	-339.653	49.494	
	1600	19.645	-200.875	92.477	-348.838	47.655	
	1700	19.671	-198.909	93.669	-358.146	46.049	
	1800	19.693	-196.941	94.794	-367.570	44.635	
	1900	19.711	-194.971	95.859	-377.103	43.383	
	1990	19.725	-193.196	96.771	-385.771	42.373	
	1990	19.725	-193.196	96.771	-385.771	42.373	
	2000	19.726	-192.999	96.870	-386.740	42.267	
	2100	19.739	-191.026	97.833	-396.475	41.267	
	2200	19.750	-189.051	98.752	-406.305	40.368	
	2300	19.760	-187.076	99.630	-416.224	39.556	
	2400	19.768	-185.099	100.471	-426.230	38.819	
	2500	19.775	-183.122	101.278	-436.317	38.148	

GAS CP(T) = A + B*T + C/(T**2) + E/(T**3)

CP(T) = A + B*T + C/(T**2) + E/(T**3)

$Li_3F_3(g)$ LITHIUM FLUORIDE (TRIMERIC GAS)

		A	B	C	D	RANGE	REF.
GAS	CP	31.861	-0.022	-10.859	1.288	298-1990	3
	CP	31.861	-0.022	-10.859	1.288	1990-2500	3

PHASE	T	CP	H	S	G	BT	
GAS	298	24.498	-362.62	76.	-385.279	282.455	3
	300	24.559	-362.575	76.152	-385.420	280.816	
	400	27.078	-359.981	83.595	-393.419	214.983	
	500	28.537	-357.194	89.808	-402.098	175.781	
	600	29.428	-354.292	95.096	-411.350	149.854	
	700	30.005	-351.319	99.679	-421.094	131.489	
	800	30.398	-348.298	103.713	-431.268	117.833	
	900	30.677	-345.243	107.310	-441.822	107.304	
	1000	30.882	-342.165	110.553	-452.718	98.955	
	1100	31.036	-339.068	113.504	-463.923	92.185	
	1200	31.155	-335.958	116.210	-475.411	86.596	
	1300	31.248	-332.838	118.708	-487.158	81.910	
	1400	31.323	-329.709	121.026	-499.146	77.931	
	1500	31.384	-326.574	123.190	-511.358	74.515	
	1600	31.433	-323.433	125.217	-523.780	71.555	
	1700	31.474	-320.288	127.124	-536.398	68.968	
	1800	31.508	-317.138	128.924	-549.201	66.691	
	1900	31.537	-313.986	130.628	-562.179	64.674	
	1990	31.559	-311.147	132.088	-574.002	63.048	
	1990	31.559	-311.147	132.088	-574.002	63.048	
	2000	31.562	-310.831	132.246	-575.324	62.877	
	2100	31.582	-307.674	133.787	-588.626	61.267	
	2200	31.600	-304.515	135.256	-602.079	59.819	
	2300	31.616	-301.354	136.661	-615.675	58.510	
	2400	31.629	-298.192	138.007	-629.409	57.323	
	2500	31.640	-295.028	139.299	-643.275	56.243	

GAS CP(T) = A + B*T + C/(T**2) + E/(T**3)

CP(T) = A + B*T + C/(T**2) + E/(T**3)

LITHIUM SULFIDE

Li_2S

		A	B	C	D	RANGE	REF.
SOL	CP	15.85	4.82			298-1223	*
	** MPT=	1223					63

PHASE	T	CP	H	S	G	BT	
SOL	298	17.287	-106.7	14.5	-111.023	81.393	63/*
	300	17.296	-106.668	14.607	-111.050	80.911	
	400	17.778	-104.914	19.649	-112.774	61.625	
	500	18.260	-103.112	23.668	-114.946	50.250	
	600	18.742	-101.262	27.039	-117.486	42.800	
	700	19.224	-99.364	29.965	-120.339	37.577	
	800	19.706	-97.418	32.563	-123.468	33.734	
	900	20.188	-95.423	34.912	-126.844	30.806	
	1000	20.670	-93.380	37.064	-130.444	28.512	
	1100	21.152	-91.289	39.057	-134.251	26.677	
	1200	21.634	-89.150	40.918	-138.251	25.182	
	1223	21.745	-88.651	41.330	-139.197	24.878	

Li_2SO_4 LITHIUM SULFATE

		A	B	C	D	RANGE	REF.
SOL-A	CP	18.9	32.3			298- 859	*
SOL-B	CP	33.5	13.5			859-1132	*
LIQ	CP	46.8				1132-1300	*

PHASE	T	CP	H	S	G	BT	
SOL-A	298	28.530	-342.8	29.	-351.446	257.652	19/1
	300	28.590	-342.747	29.177	-351.500	256.102	
	400	31.820	-339.727	37.844	-354.864	193.915	
	500	35.050	-336.383	45.291	-359.029	156.952	
	600	38.280	-332.717	51.967	-363.897	132.567	
	700	41.510	-328.727	58.111	-369.405	115.349	
	800	44.740	-324.415	63.864	-375.506	102.597	
	859	46.646	-321.719	67.115	-379.370	96.534	
			6.5	7.567			19
SOL-B	859	45.096	-315.219	74.682	-379.370	96.534	
	900	45.650	-313.358	76.797	-382.476	92.890	
	1000	47.000	-308.726	81.677	-390.403	85.334	
	1100	48.350	-303.958	86.220	-398.800	79.245	
	1132	48.782	-302.404	87.612	-401.582	77.542	
			3.3	2.915			1
LIQ	1132	46.800	-299.104	90.528	-401.582	77.542	
	1200	46.800	-295.922	93.258	-407.831	74.286	
	1300	46.800	-291.242	97.004	-417.347	70.172	

LITHIUM SELENIDE Li_2Se

		A	B	C	D	RANGE	REF.
SOL	CP	15.8	4.3			298-1375	63
	** MPT=	1375					63

PHASE	T	CP	H	S	G	BT	
SOL	298	17.082	-96.	17.	-101.069	74.095	63
	300	17.090	-95.968	17.106	-101.100	73.661	
	400	17.520	-94.238	22.081	-103.070	56.323	
	500	17.950	-92.464	26.037	-105.483	46.113	
	600	18.380	-90.648	29.347	-108.256	39.438	
	700	18.810	-88.788	32.213	-111.337	34.766	
	800	19.240	-86.886	34.753	-114.688	31.336	
	900	19.670	-84.940	37.044	-118.280	28.726	
	1000	20.100	-82.952	39.138	-122.090	26.686	
	1100	20.530	-80.920	41.074	-126.102	25.058	
	1200	20.960	-78.846	42.879	-130.301	23.734	
	1300	21.390	-76.728	44.574	-134.674	22.644	
	1375	21.712	-75.112	45.783	-138.063	21.947	

* * * * * * * * * *

LITHIUM TELLURIDE Li_2Te

		A	B	C	D	RANGE	REF.
SOL	CP	16.5	4.0			298-1100	*

PHASE	T	CP	H	S	G	BT	
SOL	298	17.693	-85.	18.5	-90.516	66.359	63
	300	17.700	-84.967	18.609	-90.550	65.975	
	400	18.100	-83.177	23.756	-92.680	50.645	
	500	18.500	-81.347	27.838	-95.266	41.646	
	600	18.900	-79.477	31.246	-98.225	35.783	
	700	19.300	-77.567	34.190	-101.500	31.694	
	800	19.700	-75.617	36.793	-105.052	28.703	
	900	20.100	-73.627	39.137	-108.850	26.436	
	1000	20.500	-71.597	41.275	-112.872	24.672	
	1100	20.900	-69.527	43.248	-117.100	23.269	

$2Li_2O{\cdot}SiO_2$ LITHIUM ORTHOSILICATE

		A	B	C	D	RANGE	REF.
SOL	CP	32.82	16.33	-2.35		298-1528	5
LIQ	CP	68.6				1528-2200	*

PHASE	T	CP	H	S	G	BT	
SOL	298	35.045	-556.9	29.	-565.546	414.612	5
	300	35.108	-556.835	29.217	-565.600	412.095	
	400	37.883	-553.177	39.721	-569.066	310.965	
	500	40.045	-549.278	48.413	-573.484	250.704	
	600	41.965	-545.176	55.886	-578.708	210.823	
	700	43.771	-540.889	62.492	-584.633	182.555	
	800	45.517	-536.424	68.451	-591.185	161.526	
	900	47.227	-531.787	73.911	-598.306	145.308	
	1000	48.915	-526.979	78.974	-605.954	132.449	
	1100	50.589	-522.004	83.715	-614.090	122.025	
	1200	52.253	-516.862	88.188	-622.688	113.422	
	1300	53.910	-511.554	92.436	-631.721	106.216	
	1400	55.562	-506.080	96.492	-641.169	100.104	
	1500	57.211	-500.441	100.381	-651.013	94.865	
	1528	57.672	-498.833	101.444	-653.839	93.531	
			7.44	4.869			5
LIQ	1528	68.600	-491.393	106.313	-653.839	93.531	
	1600	68.600	-486.454	109.471	-661.608	90.384	
	1700	68.600	-479.594	113.630	-672.765	86.501	
	1800	68.600	-472.734	117.551	-684.326	83.100	
	1900	68.600	-465.874	121.260	-696.269	80.100	
	2000	68.600	-459.014	124.779	-708.572	77.440	
	2100	68.600	-452.154	128.126	-721.219	75.068	
	2200	68.600	-445.294	131.317	-734.192	72.945	

EUCRYPTITE $Li_2O \cdot Al_2O_3 \cdot 2SiO_2 (\beta)$

		A	B	C	D	RANGE	REF.
SOL-B	CP	73.74	13.60	-21.0		298-1300	5
SOL-C	CP	62.0	24.0			1300-1600	5

PHASE	T	CP	H	S	G	BT	
SOL-B	298	54.171	-1011.1	49.6	-1025.888	752.097	5
	300	54.487	-1010.999	49.936	-1025.980	747.527	
	400	66.055	-1004.899	67.406	-1031.862	563.859	
	500	72.140	-997.963	82.858	-1039.392	454.379	
	600	76.067	-990.541	96.379	-1048.369	381.919	
	700	78.974	-982.783	108.332	-1058.616	330.559	
	800	81.339	-974.764	119.036	-1069.994	292.348	
	900	83.387	-966.526	128.737	-1082.390	262.875	
	1000	85.240	-958.093	137.620	-1095.714	239.500	
	1100	86.964	-949.482	145.826	-1109.891	220.545	
	1200	88.602	-940.703	153.464	-1124.860	204.893	
	1300	90.177	-931.764	160.618	-1140.568	191.773	
			0.6	.462			5
SOL-C	1300	93.200	-931.164	161.080	-1140.568	191.773	
	1400	95.600	-921.724	168.075	-1157.029	180.645	
	1500	98.000	-912.044	174.752	-1174.172	171.100	
	1600	100.400	-902.124	181.154	-1191.970	162.837	

$Li_2O \cdot Al_2O_3 \cdot 4SiO_2(\alpha)$ ALPHA-SPODUMENE

		A	B	C	D	RANGE	REF.
SOL-A	CP	88.66	32.88	-20.00		298-1200	5

PHASE	T	CP	H	S	G	BT	
SOL-A	298	75.964	-1456.3	61.8	-1474.726	1081.148	5
	300	76.302	-1456.159	62.271	-1474.840	1074.565	
	400	89.312	-1447.809	86.204	-1482.291	809.995	
	500	97.100	-1438.463	107.026	-1491.976	652.230	
	600	102.832	-1428.456	125.256	-1503.609	547.763	
	700	107.594	-1417.929	141.474	-1516.961	473.680	
	800	111.839	-1406.954	156.123	-1531.852	418.539	
	900	115.783	-1395.571	169.525	-1548.144	375.991	
	1000	119.540	-1383.803	181.920	-1565.723	342.235	
	1100	123.175	-1371.667	193.485	-1584.500	314.853	
	1200	126.727	-1359.171	204.355	-1604.397	292.240	

* * * * * * * * * *

$Li_2O \cdot Al_2O_3 \cdot 4SiO_2(\beta)$ BETA-SPODUMENE

		A	B	C	D	RANGE	REF.
SOL	CP	99.04	21.80	-24.64		298-1700	5

PHASE	T	CP	H	S	G	BT	
SOL	298	77.821	-1442.9	73.8	-1464.903	1073.948	5
	300	78.202	-1442.756	74.283	-1465.040	1067.425	
	400	92.360	-1434.142	98.966	-1473.728	805.316	
	500	100.084	-1424.489	120.474	-1484.726	649.061	
	600	105.276	-1414.207	139.205	-1497.730	545.621	
	700	109.271	-1403.473	155.744	-1512.494	472.285	
	800	112.630	-1392.374	170.560	-1528.822	417.711	
	900	115.618	-1380.959	184.001	-1546.560	375.607	
	1000	118.376	-1369.258	196.327	-1565.585	342.204	
	1100	120.984	-1357.289	207.733	-1585.795	315.111	
	1200	123.489	-1345.065	218.368	-1607.106	292.733	
	1300	125.922	-1332.594	228.349	-1629.447	273.972	
	1400	128.303	-1319.882	237.768	-1652.757	258.042	
	1500	130.645	-1306.934	246.700	-1676.984	244.369	
	1600	132.957	-1293.754	255.206	-1702.083	232.525	
	1700	135.247	-1280.344	263.335	-1728.013	222.181	

LITHIUM ALUMINIUM

		A	B	C	D	RANGE	REF.
SOL	CP	10.5	4.0			298- 993	*
	** MPT= 993						1

PHASE	T	CP	H	S	G	BT	
SOL	298	11.693	-11.7	11.2	-15.039	11.026	*
	300	11.700	-11.678	11.272	-15.060	10.973	
	400	12.100	-10.488	14.693	-16.366	8.943	
	500	12.500	-9.258	17.436	-17.976	7.859	
	600	12.900	-7.988	19.750	-19.839	7.227	
	700	13.300	-6.678	21.769	-21.917	6.844	
	800	13.700	-5.328	23.571	-24.185	6.608	
	900	14.100	-3.938	25.208	-26.625	6.466	
	993	14.472	-2.610	26.612	-29.036	6.391	

$LiAlF_4(g)$ LITHIUM TETRAFLUOROALUMINATE (GAS)

		A	B	C	D	RANGE	REF.
GAS	CP	28.706	2.968	-4.911		298- 800	3
	CP	31.689	0.039	-9.424		800-2200	3

PHASE	T	CP	H	S	G	BT	
GAS	298	24.066	-443.	78.46	-466.393	341.921	3
	300	24.140	-442.955	78.609	-466.538	339.918	
	400	26.824	-440.390	85.970	-474.778	259.442	
	500	28.226	-437.632	92.120	-483.692	211.450	
	600	29.123	-434.761	97.351	-493.172	179.662	
	700	29.781	-431.815	101.892	-503.139	157.108	
	800	30.313	-428.809	105.904	-513.533	140.309	
	800	30.248	-428.809	105.904	-513.533	140.309	
	900	30.561	-425.768	109.486	-524.305	127.336	
	1000	30.786	-422.700	112.718	-535.418	117.031	
	1100	30.953	-419.613	115.660	-546.839	108.662	
	1200	31.081	-416.511	118.359	-558.542	101.738	
	1300	31.182	-413.397	120.851	-570.504	95.923	
	1400	31.263	-410.275	123.165	-582.707	90.977	
	1500	31.329	-407.145	125.325	-595.132	86.722	
	1600	31.383	-404.010	127.348	-607.767	83.028	
	1700	31.429	-400.869	129.252	-620.598	79.794	
	1800	31.468	-397.724	131.050	-633.614	76.942	
	1900	31.502	-394.576	132.752	-646.805	74.410	
	2000	31.531	-391.424	134.369	-660.161	72.149	
	2100	31.557	-388.269	135.908	-673.676	70.120	
	2200	31.580	-385.112	137.376	-687.341	68.290	

TRILITHIUM HEXAFLUOROALUMINATE Li_3AlF_6

		A	B	C	D	RANGE	REF.
SOL-B	CP	49.22	26.25	-7.72		298- 748	3
SOL-C	CP	68.0				748- 848	3
SOL-D	CP	70.5				848- 978	3
SOL-E	CP	73.0				978-1058	3
LIQ	CP	86.0				1058-2000	3

PHASE	T	CP	H	S	G	BT	
SOL-B	298	48.362	-808.7	44.907	-822.089	602.689	3
	300	48.517	-808.610	45.207	-822.172	599.033	
	400	54.895	-803.413	60.115	-827.459	452.163	
	500	59.257	-797.696	72.855	-834.123	364.644	
	600	62.826	-791.587	83.982	-841.976	306.731	
	700	66.019	-785.143	93.910	-850.880	265.692	
	748	67.475	-781.939	98.336	-855.494	249.991	
			0.5	.668			3
SOL-C	748	68.000	-781.439	99.004	-855.494	249.991	
	800	68.000	-777.903	103.575	-860.763	235.181	
	848	68.000	-774.639	107.537	-865.830	223.175	
			0.3	.354			3
SOL-D	848	70.500	-774.339	107.891	-865.830	223.175	
	900	70.500	-770.673	112.086	-871.551	211.670	
	978	70.500	-765.174	117.946	-880.525	196.794	
			0.1	.102			3
SOL-E	978	73.000	-765.074	118.048	-880.525	196.794	
	1000	73.000	-763.468	119.672	-883.140	193.036	
	1058	73.000	-759.234	123.788	-890.202	183.913	
			20.6	19.471			3
LIQ	1058	86.000	-738.634	143.259	-890.202	183.913	
	1100	86.000	-735.022	146.607	-896.289	178.100	
	1200	86.000	-726.422	154.090	-911.329	165.998	
	1300	86.000	-717.822	160.973	-927.087	155.878	
	1400	86.000	-709.222	167.347	-943.507	147.308	
	1500	86.000	-700.622	173.280	-960.542	139.970	
	1600	86.000	-692.022	178.830	-978.150	133.627	
	1700	86.000	-683.422	184.044	-996.297	128.100	
	1800	86.000	-674.822	188.960	-1014.949	123.248	
	1900	86.000	-666.222	193.609	-1034.080	118.962	
	2000	86.000	-657.622	198.021	-1053.663	115.154	

Li_2BeF_4 DILITHIUM TETRAFLUOROBERYLLATE

		A	B	C	D	RANGE	REF.
SOL	CP	21.99	35.36	-0.24		298- 745	65
LIQ	CP	55.5				745-2000	65

PHASE	T	CP	H	S	G	BT	
SOL	298	32.263	-544.18	29.8	-553.065	405.462	3,65
	300	32.331	-544.120	30.000	-553.120	403.002	
	400	35.984	-540.704	39.804	-556.625	304.167	
	500	39.574	-536.925	48.220	-561.035	245.261	
	600	43.139	-532.790	55.750	-566.240	206.280	
	700	46.693	-528.298	62.667	-572.165	178.662	
	745	48.290	-526.161	65.625	-575.052	168.717	
			10.606	14.236			3,65
LIQ	745	55.500	-515.555	79.862	-575.052	168.717	
	800	55.500	-512.502	83.815	-579.554	158.348	
	900	55.500	-506.952	90.352	-588.269	142.870	
	1000	55.500	-501.402	96.199	-597.602	130.623	
	1100	55.500	-495.852	101.489	-607.490	120.713	
	1200	55.500	-490.302	106.318	-617.884	112.547	
	1300	55.500	-484.752	110.760	-628.741	105.715	
	1400	55.500	-479.202	114.873	-640.025	99.926	
	1500	55.500	-473.652	118.703	-651.706	94.966	
	1600	55.500	-468.102	122.284	-663.757	90.677	
	1700	55.500	-462.552	125.649	-676.156	86.937	
	1800	55.500	-457.002	128.821	-688.881	83.653	
	1900	55.500	-451.452	131.822	-701.914	80.749	
	2000	55.500	-445.902	134.669	-715.240	78.168	

LITHIUM TRIFLUOROBERYLLATE $LiBeF_3$

		A	B	C	D	RANGE	REF.
SOL	CP	18.43	20.29	-2.09		298- 633	3,65
	** MPT= 633						3

PHASE	T	CP	H	S	G	BT	P	
SOL	298	22.128	-394.5	21.3	-400.851	293.871		3,65
	300	22.195	-394.459	21.437	-400.890	292.088		
	400	25.240	-392.080	28.260	-403.384	220.428		
	500	27.739	-389.428	34.166	-406.512	177.710		
	600	30.023	-386.539	39.428	-410.196	149.434		
	633	30.752	-385.536	41.055	-411.524	142.102		

* * * * * * * * * *

LITHIUM TRIFLUOROBERYLLATE (GAS) $LiBeF_3(g)$

		A	B	C	D	RANGE	REF.
GAS	CP	12.860	21.070	-3.260	-9.790	298-1000	3
	CP	25.650	0.050	-18.800		1000-1800	3

PHASE	T	CP	H	S	G	BT	
GAS	298	14.604	-212.	64.	-231.082	169.410	3
	300	14.678	-211.973	64.091	-231.200	168.452	
	400	17.684	-210.342	68.762	-237.847	129.971	
	500	19.643	-208.470	72.931	-244.936	107.076	
	600	21.072	-206.431	76.645	-252.418	91.955	
	700	22.147	-204.267	79.978	-260.252	81.265	
	800	22.941	-202.011	82.990	-268.403	73.334	
	900	23.491	-199.687	85.726	-276.841	67.235	
	1000	23.814	-197.320	88.220	-285.540	62.413	
	1000	23.820	-197.320	88.220	-285.540	62.413	
	1100	24.151	-194.921	90.507	-294.478	58.515	
	1200	24.404	-192.492	92.619	-303.636	55.307	
	1300	24.603	-190.042	94.581	-312.997	52.627	
	1400	24.761	-187.573	96.410	-322.547	50.359	
	1500	24.889	-185.090	98.123	-332.275	48.419	
	1600	24.996	-182.596	99.733	-342.169	46.744	
	1700	25.084	-180.092	101.251	-352.218	45.287	
	1800	25.160	-177.580	102.687	-362.416	44.009	

Lu LUTETIUM

		A	B	C	D	RANGE	REF.
SOL	CP	6.550	-1.285	0.062	1.979	298-1936	21
LIQ	CP	11.45				1936-3000	21
	** BPT= 3668 / L= 84.704						21/*

PHASE	T	CP	H	S	G	BT	P	
SOL	298	6.413	0	12.18	-3.631	2.662		21
	300	6.411	.012	12.220	-3.654	2.662		
	400	6.391	.651	14.060	-4.972	2.717		
	500	6.427	1.292	15.489	-6.452	2.821		
	600	6.509	1.938	16.667	-8.062	2.937		
	700	6.633	2.595	17.679	-9.780	3.054		
	800	6.798	3.266	18.575	-11.594	3.168		
	900	7.004	3.956	19.388	-13.493	3.277		
	1000	7.250	4.669	20.138	-15.469	3.381		
	1100	7.536	5.407	20.842	-17.519	3.481		
	1200	7.862	6.177	21.512	-19.637	3.577		
	1300	8.228	6.981	22.155	-21.820	3.669		
	1400	8.633	7.824	22.779	-24.067	3.758		
	1500	9.078	8.709	23.390	-26.376	3.843		
	1600	9.563	9.641	23.991	-28.745	3.927		
	1700	10.087	10.623	24.586	-31.174	4.008		
	1800	10.651	11.660	25.179	-33.662	4.088	.002	
	1900	11.254	12.755	25.770	-36.209	4.166	.007	
	1936	11.481	13.164	25.984	-37.141	4.193	.011	
			4.813	2.486				*
LIQ	1936	11.450	17.977	28.470	-37.141	4.193	.011	
	2000	11.450	18.710	28.842	-38.975	4.260	.024	
	2100	11.450	19.855	29.401	-41.887	4.360	.074	
	2200	11.450	21.000	29.934	-44.854	4.456	.204	
	2300	11.450	22.145	30.443	-47.873	4.550	.512	
	2400	11.450	23.290	30.930	-50.942	4.640	1.183	
	2500	11.450	24.435	31.397	-54.059	4.726	2.545	
	2600	11.450	25.580	31.846	-57.221	4.811	5.139	
	2700	11.450	26.725	32.278	-60.427	4.892	9.812	
	2800	11.450	27.870	32.695	-63.676	4.971	17.826	
	2900	11.450	29.015	33.097	-66.966	5.047	30.971	
	3000	11.450	30.160	33.485	-70.295	5.122	51.699	

LUTETIUM OXIDE

Lu_2O_3

		A	B	C	D	RANGE	REF.
SOL	CP	26.915	5.500	-3.748		298-1500	*

PHASE	T	CP	H	S	G	BT	
SOL	298	24.339	-449.8	26.3	-457.641	335.505	1
	300	24.401	-449.755	26.451	-457.690	333.472	
	400	26.772	-447.183	33.833	-460.716	251.758	
	500	28.166	-444.432	39.967	-464.415	203.023	
	600	29.174	-441.563	45.195	-468.680	170.739	
	700	30.000	-438.603	49.756	-473.432	147.832	
	800	30.729	-435.566	53.810	-478.614	130.769	
	900	31.402	-432.459	57.469	-484.181	117.591	
	1000	32.040	-429.286	60.811	-490.097	107.125	
	1100	32.655	-426.052	63.894	-496.334	98.626	
	1200	33.255	-422.756	66.761	-502.869	91.597	
	1300	33.843	-419.401	69.446	-509.681	85.697	
	1400	34.424	-415.988	71.975	-516.753	80.680	
	1500	34.998	-412.516	74.370	-524.071	76.367	

Mg

MAGNESIUM

		A	B	C	D	RANGE	REF.
SOL	CP	5.112	2.815			298- 922	21
LIQ	CP	7.8				922-1363	21
GAS	CP	4.968				1363-2000	21

PHASE	T	CP	H	S	G	BT	P	
SOL	298	5.951	0	7.81	-2.329	1.707		21
	300	5.956	.011	7.847	-2.343	1.707		
	400	6.238	.621	9.599	-3.219	1.759		
	500	6.519	1.259	11.021	-4.252	1.859		
	600	6.801	1.925	12.235	-5.416	1.973		
	700	7.082	2.619	13.304	-6.694	2.090	.008	
	800	7.364	3.341	14.268	-8.074	2.206	.183	
	900	7.645	4.092	15.152	-9.545	2.318	1.971	
	922	7.707	4.260	15.337	-9.881	2.342	3.098	
			2.14	2.321				21
LIQ	922	7.800	6.400	17.658	-9.881	2.342	3.098	
	1000	7.800	7.009	18.292	-11.283	2.466	11.891	
	1100	7.800	7.789	19.035	-13.150	2.613	49.884	
	1200	7.800	8.569	19.714	-15.088	2.748	163.012	
	1300	7.800	9.349	20.338	-17.091	2.874	439.938	
	1363	7.800	9.840	20.707	-18.384	2.948	760.	
			30.449	22.340				21
GAS	1363	4.968	40.289	43.047	-18.384	2.948		
	1400	4.968	40.473	43.180	-19.979	3.119		
	1500	4.968	40.970	43.523	-24.314	3.543		
	1600	4.968	41.467	43.844	-28.683	3.918		
	1700	4.968	41.964	44.145	-33.082	4.254		
	1800	4.968	42.460	44.429	-37.511	4.555		
	1900	4.968	42.957	44.697	-41.968	4.828		
	2000	4.968	43.454	44.952	-46.450	5.077		

MAGNESIUM HYDRIDE

MgH_2

		A	B	C	D	RANGE	REF.
SOL	CP	6.50	11.8	-1.40		298- 560	3
	** DPT= 560						3

PHASE	T	CP	H	S	G	BT	
SOL	298	8.443	-18.2	7.43	-20.415	14.967	3
	300	8.484	-18.184	7.482	-20.429	14.885	
	400	10.345	-17.238	10.192	-21.315	11.647	
	500	11.840	-16.127	12.665	-22.459	9.818	
	560	12.662	-15.392	14.053	-23.261	9.079	

* * * * * * * * * *

MAGNESIUM IODIDE

MgJ_2

		A	B	C	D	RANGE	REF.
SOL	CP	16.7	4.3			298- 923	*
	** MPT= 923						1

PHASE	T	CP	H	S	G	BT	
SOL	298	17.982	-87.5	31.	-96.743	70.924	19/*
	300	17.990	-87.467	31.111	-96.800	70.528	
	400	18.420	-85.646	36.346	-100.184	54.746	
	500	18.850	-83.783	40.502	-104.034	45.479	
	600	19.280	-81.876	43.977	-108.262	39.440	
	700	19.710	-79.927	46.981	-112.814	35.227	
	800	20.140	-77.934	49.641	-117.647	32.144	
	900	20.570	-75.899	52.038	-122.733	29.808	
	923	20.669	-75.424	52.558	-123.936	29.350	

MgS

MAGNESIUM SULFIDE

		A	B	C	D	RANGE	REF.
SOL	CP	10.30	1.97			298-2000	63

PHASE	T	CP	H	S	G	BT	
SOL	298	10.887	-82.7	12.03	-86.287	63.256	19/176
	300	10.891	-82.680	12.097	-86.309	62.885	
	400	11.088	-81.581	15.257	-87.684	47.915	
	500	11.285	-80.462	17.753	-89.339	39.055	
	600	11.482	-79.324	19.828	-91.221	33.232	
	700	11.679	-78.166	21.613	-93.295	29.132	
	800	11.876	-76.988	23.185	-95.536	26.103	
	900	12.073	-75.791	24.595	-97.926	23.783	
	1000	12.270	-74.574	25.877	-100.451	21.956	
	1100	12.467	-73.337	27.056	-103.098	20.486	
	1200	12.664	-72.080	28.149	-105.859	19.282	
	1300	12.861	-70.804	29.171	-108.726	18.281	
	1400	13.058	-69.508	30.131	-111.691	17.438	
	1500	13.255	-68.192	31.039	-114.750	16.721	
	1600	13.452	-66.857	31.900	-117.897	16.106	
	1700	13.649	-65.502	32.722	-121.129	15.574	
	1800	13.846	-64.127	33.507	-124.441	15.111	
	1900	14.043	-62.733	34.261	-127.829	14.706	
	2000	14.240	-61.319	34.987	-131.292	14.349	

MAGNESIUM SELENIDE — MgSe

		A	B	C	D	RANGE	REF.
SOL	CP	10.25	4.08			298-1900	*

PHASE	T	CP	H	S	G	BT	
SOL	298	11.466	-65.2	14.7	-69.583	51.012	1/*
	300	11.474	-65.179	14.771	-69.610	50.718	
	400	11.882	-64.011	18.128	-71.262	38.941	
	500	12.290	-62.802	20.823	-73.214	32.006	
	600	12.698	-61.553	23.100	-75.413	27.473	
	700	13.106	-60.263	25.088	-77.824	24.301	
	800	13.514	-58.932	26.864	-80.423	21.974	
	900	13.922	-57.560	28.480	-83.192	20.204	
	1000	14.330	-56.147	29.968	-86.115	18.823	
	1100	14.738	-54.694	31.353	-89.182	17.721	
	1200	15.146	-53.200	32.652	-92.383	16.827	
	1300	15.554	-51.665	33.881	-95.710	16.092	
	1400	15.962	-50.089	35.049	-99.157	15.481	
	1500	16.370	-48.472	36.164	-102.718	14.968	
	1600	16.778	-46.815	37.233	-106.388	14.534	
	1700	17.186	-45.117	38.263	-110.163	14.164	
	1800	17.594	-43.378	39.256	-114.039	13.848	
	1900	18.002	-41.598	40.219	-118.013	13.576	

* * * * * * * * * *

MAGNESIUM SELENITE — $MgSeO_3$

		A	B	C	D	RANGE	REF.
SOL	CP	18.3	11.3			298- 900	*

PHASE	T	CP	H	S	G	BT	
SOL	298	21.669	-215.15	22.4	-221.829	162.627	19/*
	300	21.690	-215.110	22.534	-221.870	161.654	
	400	22.820	-212.884	28.929	-224.456	122.653	
	500	23.950	-210.546	34.142	-227.617	99.505	
	600	25.080	-208.094	38.609	-231.260	84.248	
	700	26.210	-205.530	42.560	-235.322	73.481	
	800	27.340	-202.852	46.133	-239.759	65.508	
	900	28.470	-200.062	49.419	-244.539	59.390	

MgTe

MAGNESIUM TELLURIDE

		A	B	C	D	RANGE	REF.
SOL	CP	9.0	2.5			298-1500	*

PHASE	T	CP	H	S	G	BT	
SOL	298	9.745	-50.	17.8	-55.307	40.547	1/*
	300	9.750	-49.982	17.860	-55.340	40.321	
	400	10.000	-48.994	20.699	-57.274	31.297	
	500	10.250	-47.982	22.958	-59.461	25.994	
	600	10.500	-46.944	24.849	-61.854	22.533	
	700	10.750	-45.882	26.486	-64.422	20.116	
	800	11.000	-44.794	27.938	-67.145	18.346	
	900	11.250	-43.682	29.248	-70.005	17.002	
	1000	11.500	-42.544	30.446	-72.991	15.954	
	1100	11.750	-41.382	31.554	-76.091	15.120	
	1200	12.000	-40.194	32.587	-79.299	14.444	
	1300	12.250	-38.982	33.557	-82.606	13.889	
	1400	12.500	-37.744	34.474	-86.008	13.428	
	1500	12.750	-36.482	35.345	-89.500	13.042	

MAGNESIUM ANTIMONIDE Mg_3Sb_2

		A	B	C	D	RANGE	REF.
SOL-A	CP	27.0	9.6			298-1198	5
SOL-B	CP	38.4				1198-1518	5
	** MPT= 1518, L= 4.0						5

PHASE	T	CP	H	S	G	BT	
SOL-A	298	29.862	-71.74	31.15	-81.027	59.403	5
	300	29.880	-71.685	31.335	-81.085	59.078	
	400	30.840	-68.649	40.062	-84.674	46.270	
	500	31.800	-65.517	47.047	-89.040	38.925	
	600	32.760	-62.289	52.930	-94.047	34.261	
	700	33.720	-58.965	58.052	-99.601	31.101	
	800	34.680	-55.545	62.617	-105.638	28.863	
	900	35.640	-52.029	66.757	-112.110	27.228	
	1000	36.600	-48.417	70.562	-118.979	26.006	
	1100	37.560	-44.709	74.095	-126.214	25.080	
	1198	38.501	-40.982	77.340	-133.636	24.382	
			17.51	14.616			5
SOL-B	1198	38.400	-23.472	91.957	-133.636	24.382	
	1200	38.400	-23.395	92.021	-133.820	24.375	
	1300	38.400	-19.555	95.094	-143.177	24.074	
	1400	38.400	-15.715	97.940	-152.831	23.861	
	1500	38.400	-11.875	100.589	-162.759	23.717	
	1518	38.400	-11.184	101.047	-164.574	23.697	

Mg_3P_2 TRIMAGNESIUM DIPHOSPHIDE

		A	B	C	D	RANGE	REF.
SOL	CP	19.2	16.6			298-1200	*

PHASE	T	CP	H	S	G	BT	
SOL	298	24.149	-111.	18.5	-116.516	85.420	*
	300	24.180	-110.955	18.649	-116.550	84.918	
	400	25.840	-108.454	25.833	-118.787	64.911	
	500	27.500	-105.787	31.777	-121.676	53.192	
	600	29.160	-102.954	36.938	-125.117	45.580	
	700	30.820	-99.955	41.558	-129.046	40.295	
	800	32.480	-96.790	45.781	-133.415	36.452	
	900	34.140	-93.459	49.703	-138.192	33.562	
	1000	35.800	-89.962	53.386	-143.348	31.333	
	1100	37.460	-86.299	56.876	-148.863	29.580	
	1200	39.120	-82.470	60.206	-154.718	28.182	

MAGNESIUM ORTHOPHOSPHATE $Mg_3(PO_4)_2$

		A	B	C	D	RANGE	REF.
SOL	CP	29.025	80.255	0.26	-26.005	298-1621	3
LIQ	CP	113.445				1621-2000	3

PHASE	T	CP	H	S	G	BT	
SOL	298	50.934	-903.6	45.22	-917.082	672.330	40
	300	51.050	-903.506	45.535	-917.166	668.245	
	400	57.129	-898.093	61.064	-922.519	504.109	
	500	62.755	-892.095	74.425	-929.308	406.255	
	600	67.888	-885.559	86.328	-937.356	341.477	
	700	72.514	-878.534	97.147	-946.537	295.562	
	800	76.626	-871.073	107.104	-956.756	261.409	
	900	80.223	-863.226	116.342	-967.934	235.078	
	1000	83.301	-855.046	124.958	-980.004	214.209	
	1100	85.861	-846.583	133.022	-992.908	197.299	
	1200	87.902	-837.891	140.584	-1006.592	183.350	
	1300	89.423	-829.020	147.684	-1021.009	171.670	
	1400	90.425	-820.024	154.350	-1036.114	161.766	
	1500	90.908	-810.953	160.609	-1051.865	153.277	
	1600	90.870	-801.859	166.477	-1068.223	145.932	
	1621	90.796	-799.952	167.662	-1071.731	144.515	
			29.	17.890			3
LIQ	1621	113.445	-770.952	185.552	-1071.731	144.515	
	1700	113.445	-761.990	190.950	-1086.605	139.711	
	1800	113.445	-750.645	197.435	-1106.027	134.308	
	1900	113.445	-739.301	203.568	-1126.080	129.546	
	2000	113.445	-727.956	209.387	-1146.730	125.326	

$MgCO_3$ MAGNESIUM CARBONATE

		A	3	C	D	RANGE	REF.
SOL	CP	18.62	13.80	-4.16		298- 729	1

** DPT= 729, MGCO3 = MGO + CO2

PHASE	T	CP	H	S	G	BT	
SOL	298	18.055	-265.7	15.74	-270.393	198.230	66
	300	18.138	-265.667	15.852	-270.422	197.029	
	400	21.540	-263.668	21.577	-272.299	148.797	
	500	23.856	-261.393	26.644	-274.715	120.094	
	600	25.744	-258.911	31.165	-277.610	101.133	
	700	27.431	-256.251	35.262	-280.934	87.723	
	729	27.897	-255.449	36.385	-281.973	84.545	

SERPENTINE $3MgO \cdot 2SiO_2 \cdot 2H_2O$

		A	B	C	D	RANGE	REF.
SOL	CP	75.82	31.60	-17.58		298-1000	5

PHASE	T	CP	H	S	G	BT	
SOL	298	65.465	-1043.04	53.1	-1058.872	776.278	5
	300	65.767	-1042.919	53.506	-1058.970	771.563	
	400	77.472	-1035.696	74.205	-1065.378	582.174	
	500	84.588	-1027.571	92.306	-1073.724	469.387	
	600	89.897	-1018.837	108.215	-1083.766	394.814	
	700	94.352	-1009.619	122.415	-1095.310	342.017	
	800	98.353	-999.981	135.279	-1108.204	302.788	
	900	102.090	-989.957	147.081	-1122.330	272.576	
	1000	105.662	-979.569	158.023	-1137.592	248.654	

* * * * * * * * * *

ANTHOPHYLLITE $7MgO \cdot 8SiO_2 \cdot H_2O$

		A	B	C	D	RANGE	REF.
SOL	CP	199.0	34.1	-52.3		298-1100	5

PHASE	T	CP	H	S	G	BT	
SOL	298	150.332	-2888.72	133.6	-2928.553	2146.976	5
	300	151.119	-2888.441	134.532	-2928.801	2133.917	
	400	179.952	-2871.706	182.479	-2944.698	1609.124	
	500	195.130	-2852.886	224.411	-2965.092	1296.215	
	600	204.932	-2832.854	260.907	-2989.399	1089.034	
	700	212.197	-2811.983	293.066	-3017.129	942.117	
	800	218.108	-2790.459	321.798	-3047.898	832.759	
	900	223.233	-2768.387	347.789	-3081.398	748.366	
	1000	227.870	-2745.829	371.552	-3117.381	681.395	
	1100	232.188	-2722.824	393.475	-3155.647	627.054	

$3MgO \cdot 4SiO_2 \cdot H_2O$ TALC

		A	B	C	D	RANGE	REF.
SOL	CP	84.37	41.89	-17.75		298-1100	5

PHASE	T	CP	H	S	G	GT	
SOL	298	76.892	-1413.82	62.3	-1432.395	1050.115	5
	300	77.215	-1413.677	62.777	-1432.510	1043.723	
	400	90.032	-1405.253	86.923	-1440.023	786.898	
	500	98.215	-1395.819	107.942	-1449.790	633.788	
	600	104.573	-1385.670	126.429	-1461.527	532.432	
	700	110.071	-1374.932	142.969	-1475.011	460.581	
	800	115.109	-1363.671	158.000	-1490.070	407.123	
	900	119.880	-1351.919	171.835	-1506.571	365.895	
	1000	124.485	-1339.700	184.705	-1524.405	333.203	
	1100	128.982	-1327.026	196.781	-1543.486	306.704	

2-MAGNESIUM GERMANIUM

Mg_2Ge

		A	B	C	D	RANGE	REF.
SOL	CP	17.55	4.3	-0.76		298-1388	5
LIQ	CP	22.35				1388-1600	*

PHASE	T	CP	H	S	G	BT	
SOL	298	17.977	-27.53	17.43	-32.727	23.993	5
	300	17.996	-27.497	17.541	-32.759	23.868	
	400	18.795	-25.655	22.835	-34.789	19.010	
	500	19.396	-23.744	27.096	-37.292	16.303	
	600	19.919	-21.778	30.679	-40.185	14.640	
	700	20.405	-19.761	33.787	-43.412	13.556	
	800	20.871	-17.698	36.542	-46.931	12.823	
	900	21.326	-15.588	39.027	-50.712	12.316	
	1000	21.774	-13.433	41.297	-54.729	11.963	
	1100	22.217	-11.233	43.393	-58.965	11.717	
	1200	22.657	-8.989	45.345	-63.403	11.549	
	1300	23.095	-6.702	47.176	-68.030	11.438	
	1388	23.479	-4.652	48.701	-72.249	11.378	
			20.4	14.697			*
LIQ	1388	22.350	15.748	63.398	-72.249	11.378	
	1400	22.350	16.016	63.591	-73.011	11.399	
	1500	22.350	18.251	65.133	-79.448	11.577	
	1600	22.350	20.486	66.575	-86.034	11.753	

$2MgO \cdot 2Al_2O_3 \cdot 5SiO_2$ CORDIERITE

		A	B	C	D	RANGE	REF.
SOL	CP	149.7	21.8	-48.0		298-1700	5

PHASE	T	CP	H	S	G	BT	
SOL	298	102.202	-2177.8	97.3	-2206.810	1617.853	5
	300	102.907	-2177.610	97.934	-2206.991	1608.008	
	400	128.420	-2165.877	131.514	-2218.483	1212.286	
	500	141.400	-2152.326	161.698	-2233.175	976.252	
	600	149.447	-2137.757	188.239	-2250.700	819.927	
	700	155.164	-2122.513	211.726	-2270.721	709.047	
	800	159.640	-2106.765	232.748	-2292.964	626.493	
	900	163.394	-2090.609	251.773	-2317.205	562.770	
	1000	166.700	-2074.101	269.162	-2343.264	512.189	
	1100	169.713	-2057.279	285.194	-2370.992	471.136	
	1200	172.527	-2040.165	300.083	-2400.264	437.207	
	1300	175.200	-2022.778	313.998	-2430.976	408.739	
	1400	177.771	-2005.129	327.077	-2463.036	384.549	
	1500	180.267	-1987.226	339.427	-2496.367	363.769	
	1600	182.705	-1969.077	351.139	-2530.900	345.751	
	1700	185.099	-1950.687	362.288	-2566.576	330.000	

2-MAGNESIUM LEAD

Mg_2Pb

		A	B	C	D	RANGE	REF.
SOL-A	CP	15.75	8.25			298- 710	*
SOL-B	CP	22.2				710- 822	*
LIQ	CP	22.7				822-1100	*

PHASE	T	CP	H	S	G	BT	
SOL-A	298	18.210	-11.5	28.5	-19.997	14.660	90*
	300	18.225	-11.466	28.613	-20.050	14.608	
	400	19.050	-9.603	33.969	-23.190	12.672	
	500	19.875	-7.656	38.308	-26.810	11.720	
	600	20.700	-5.628	42.005	-30.830	11.231	
	700	21.525	-3.516	45.258	-35.197	10.990	
	710	21.607	-3.301	45.564	-35.651	10.975	
			0.53	.746			90*
SOL-B	710	22.200	-2.771	46.310	-35.651	10.975	
	800	22.200	-.773	48.960	-39.940	10.913	
	822	22.200	-.284	49.562	-41.024	10.909	
			9.6	11.679			90*
LIQ	822	22.700	9.316	61.241	-41.024	10.909	
	900	22.700	11.086	63.298	-45.882	11.143	
	1000	22.700	13.356	65.690	-52.334	11.439	
	1100	22.700	15.626	67.854	-59.013	11.726	

MgCe

MAGNESIUM CERIUM

		A	B	C	D	RANGE	REF.
SOL	CP	10.65	6.18			298-1013	*
	** MPT=	1013					1

PHASE	T	CP	H	S	G	BT	
SOL	298	12.493	-3.85	25.25	-11.378	8.342	153*
	300	12.504	-3.827	25.327	-11.425	8.324	
	400	13.122	-2.546	29.009	-14.149	7.732	
	500	13.740	-1.202	32.004	-17.204	7.521	
	600	14.358	.202	34.563	-20.536	7.481	
	700	14.976	1.669	36.823	-24.107	7.528	
	800	15.594	3.198	38.863	-27.893	7.621	
	900	16.212	4.788	40.736	-31.874	7.741	
	1000	16.830	6.440	42.476	-36.036	7.877	
	1013	16.910	6.659	42.694	-36.589	7.895	

MANGANESE

		A	B	C	D	RANGE	REF.
SOL-A	CP	4.958	4.476			298- 600	21
	CP	5.738	3.217			600- 980	21
SOL-B	CP	7.991	1.015			980-1360	21
SOL-C	CP	7.580	2.0			1360-1410	21
SOL-D	CP	8.026	1.975			1410-1517	21
LIQ	CP	11.				1517-2335	21
GAS	CP	5.008				2335-2600	21

PHASE	T	CP	H	S	G	BT	P	
SOL-A	298	6.293	0	7.65	-2.281	1.672		21
	300	6.301	.012	7.689	-2.295	1.672		
	400	6.748	.664	9.563	-3.161	1.727		
	500	7.196	1.361	11.117	-4.197	1.835		
	600	7.644	2.103	12.468	-5.378	1.959		
	600	7.668	2.103	12.468	-5.378	1.959		
	700	7.990	2.886	13.675	-6.686	2.088		
	800	8.312	3.701	14.762	-8.109	2.215		
	900	8.633	4.549	15.760	-9.635	2.340		
	980	8.891	5.249	16.506	-10.926	2.437		
			0.532	.543				21
SOL-B	980	8.986	5.781	17.049	-10.926	2.437		
	1000	9.006	5.961	17.231	-11.269	2.463		
	1100	9.107	6.867	18.094	-13.036	2.590		
	1200	9.209	7.783	18.891	-14.886	2.711	.004	
	1300	9.310	8.709	19.632	-16.812	2.827	.032	
	1360	9.371	9.269	20.053	-18.003	2.893	.094	
			0.507	.373				21
SOL-C	1360	10.300	9.776	20.426	-18.003	2.893	.094	
	1400	10.380	10.190	20.726	-18.826	2.939	.183	
	1410	10.400	10.294	20.800	-19.034	2.951	.215	
			0.449	.318				21
SOL-D	1410	10.811	10.743	21.118	-19.034	2.951	.215	
	1500	10.988	11.724	21.792	-20.965	3.055	.814	
	1517	11.022	11.911	21.916	-21.336	3.074	1.027	
			2.882	1.900				21
LIQ	1517	11.000	14.793	23.816	-21.336	3.074	1.027	
	1600	11.000	15.706	24.402	-23.338	3.188	2.820	

Mn

MANGANESE (CONTINUED)

PHASE	T	CP	H	S	G	BT	P	
LIQ	1700	11.000	16.806	25.069	-25.812	3.319	8.273	
	1800	11.000	17.906	25.698	-28.350	3.443	21.322	
	1900	11.000	19.006	26.293	-30.950	3.561	49.298	
	2000	11.000	20.106	26.857	-33.608	3.673	103.985	
	2100	11.000	21.206	27.393	-36.320	3.780	202.817	
	2200	11.000	22.306	27.905	-39.086	3.883	369.843	
	2300	11.000	23.406	28.394	-41.901	3.982	636.275	
	2335	11.000	23.791	28.560	-42.897	4.016	760.	
			54.032	23.140				21
GAS	2335	5.008	77.823	51.700	-42.897	4.016		
	2400	5.008	78.148	51.838	-46.262	4.213		
	2500	5.008	78.649	52.042	-51.456	4.499		
	2600	5.008	79.150	52.239	-56.671	4.764		

MANGANESE DIFLUORIDE

MnF_2

		A	B	C	D	RANGE	REF.
SOL	CP	14.79	5.70	-0.47		298-1203	19,*
LIQ	CP	22.0				1203-2093	*
	** BPT=	2093					1

PHASE	T	CP	H	S	G	BT	
SOL	298	15.961	-192.	22.05	-198.574	145.578	1*/19
	300	15.978	-191.970	22.149	-198.615	144.710	
	400	16.776	-190.331	26.859	-201.075	109.877	
	500	17.452	-188.619	30.677	-203.958	89.162	
	600	18.079	-186.842	33.915	-207.191	75.479	
	700	18.684	-185.004	36.747	-210.727	65.801	
	800	19.277	-183.106	39.281	-214.531	58.615	
	900	19.862	-181.149	41.585	-218.576	53.085	
	1000	20.443	-179.134	43.708	-222.842	48.709	
	1100	21.021	-177.060	45.683	-227.312	45.169	
	1200	21.597	-174.929	47.537	-231.974	42.254	
	1203	21.615	-174.865	47.591	-232.117	42.175	
			5.5	4.572			16
LIQ	1203	22.000	-169.365	52.163	-232.117	42.175	
	1300	22.000	-167.231	53.869	-237.261	39.892	
	1400	22.000	-165.031	55.500	-242.730	37.897	
	1500	22.000	-162.831	57.017	-248.357	36.190	
	1600	22.000	-160.631	58.437	-254.130	34.717	
	1700	22.000	-158.431	59.771	-260.041	33.435	
	1800	22.000	-156.231	61.028	-266.082	32.311	
	1900	22.000	-154.031	62.218	-272.245	31.319	
	2000	22.000	-151.831	63.346	-278.523	30.440	
	2093	22.000	-149.785	64.346	-284.461	29.707	

MnF_3 MANGANESE TRIFLUORIDE

		A	B	C	D	RANGE	REF.
SOL	CP	19.6	7.4			298-1000	*

PHASE	T	CP	H	S	G	BT	
SOL	298	21.806	-240.	25.	-247.454	181.413	1,*
	300	21.820	-239.960	25.135	-247.500	180.328	
	400	22.560	-237.741	31.513	-250.346	136.801	
	500	23.300	-235.448	36.627	-253.761	110.934	
	600	24.040	-233.081	40.941	-257.645	93.860	
	700	24.780	-230.640	44.702	-261.931	81.790	
	800	25.520	-228.125	48.059	-266.572	72.834	
	900	26.260	-225.536	51.108	-271.533	65.946	
	1000	27.000	-222.873	53.913	-276.785	60.500	

MANGANESE MONOSULFIDE (GREEN)

MnS

		A	B	C	D	RANGE	REF.
SOL	CP	11.4	1.80			298-1803	63
LIQ	CP	16.				1803-2200	63

PHASE	T	CP	H	S	G	BT	
SOL	298	11.937	-51.0	19.2	-56.724	41.586	63
	300	11.940	-50.978	19.274	-56.760	41.355	
	400	12.120	-49.775	22.733	-58.868	32.168	
	500	12.300	-48.554	25.457	-61.283	26.790	
	600	12.480	-47.315	27.716	-63.944	23.295	
	700	12.660	-46.058	29.653	-66.815	20.863	
	800	12.840	-44.783	31.355	-69.867	19.089	
	900	13.020	-43.490	32.878	-73.080	17.749	
	1000	13.200	-42.179	34.259	-76.438	16.708	
	1100	13.380	-40.850	35.526	-79.928	15.882	
	1200	13.560	-39.503	36.698	-83.540	15.217	
	1300	13.740	-38.138	37.790	-87.265	14.673	
	1400	13.920	-36.755	38.815	-91.096	14.223	
	1500	14.100	-35.354	39.781	-95.026	13.847	
	1600	14.280	-33.935	40.697	-99.050	13.531	
	1700	14.460	-32.498	41.568	-103.164	13.264	
	1800	14.640	-31.043	42.400	-107.363	13.037	
	1803	14.645	-30.999	42.424	-107.490	13.031	
			6.240	3.461			63
LIQ	1803	16.000	-24.759	45.885	-107.490	13.031	
	1900	16.000	-23.207	46.724	-111.982	12.883	
	2000	16.000	-21.607	47.544	-116.696	12.754	
	2100	16.000	-20.007	48.325	-121.489	12.645	
	2200	16.000	-18.407	49.069	-126.359	12.554	

MnS_2

MANGANESE DISULFIDE

		A	B	C	D	RANGE	REF.
SOL	CP	16.66	4.22	-1.04		298- 700	63

PHASE	T	CP	H	S	G	BT	
SOL	298	16.748	-53.5	23.88	-60.620	44.442	63
	300	16.770	-53.469	23.984	-60.664	44.200	
	400	17.698	-51.742	28.946	-63.320	34.601	
	500	18.354	-49.938	32.968	-66.422	29.037	
	600	18.903	-48.075	36.364	-69.893	25.462	
	700	19.402	-46.159	39.316	-73.680	23.007	

* * * * * * * * * *

MnSe

MANGANESE SELENIDE

		A	B	C	D	RANGE	REF.
SOL	CP	11.65	1.84			298-1500	*

PHASE	T	CP	H	S	G	BT	
SOL	298	12.199	-37.7	21.7	-44.170	32.382	63
	300	12.202	-37.677	21.775	-44.210	32.211	
	400	12.386	-36.448	25.311	-46.572	25.449	
	500	12.570	-35.200	28.095	-49.248	21.529	
	600	12.754	-33.934	30.403	-52.176	19.008	
	700	12.938	-32.649	32.382	-55.317	17.273	
	800	13.122	-31.346	34.122	-58.644	16.023	
	900	13.306	-30.025	35.678	-62.136	15.091	
	1000	13.490	-28.685	37.090	-65.775	14.377	
	1100	13.674	-27.327	38.384	-69.550	13.820	
	1200	13.858	-25.950	39.582	-73.449	13.379	
	1300	14.042	-24.555	40.698	-77.463	13.024	
	1400	14.226	-23.142	41.746	-81.586	12.738	
	1500	14.410	-21.710	42.733	-85.810	12.504	

MANGANESE TELLURIDE — MnTe

		A	B	C	D	RANGE	REF.
SOL-A	CP	-19.5	123.164			298- 307	63,*
	** HOMOGENEITY RANGE....						
SOL-B	CP	13.55	0.66			307-1438	63
	** DEC., MPT= 1438						63

PHASE	T	CP	H	S	G	BT	
SOL-A	298	17.221	-26.602	22.4	-33.281	24.399	146*/63
	300	17.449	-26.570	22.507	-33.322	24.278	
	307	18.311	-26.445	22.920	-33.481	23.838	
SOL-B	307	13.753	-26.445	22.920	-33.481	23.838	
	400	13.814	-25.163	26.567	-35.790	19.557	
	500	13.880	-23.778	29.656	-38.606	16.877	
	600	13.946	-22.387	32.193	-41.702	15.192	
	700	14.012	-20.989	34.347	-45.032	14.062	
	800	14.078	-19.585	36.223	-48.563	13.268	
	900	14.144	-18.173	37.885	-52.270	12.694	
	1000	14.210	-16.756	39.378	-56.134	12.270	
	1100	14.276	-15.331	40.736	-60.141	11.950	
	1200	14.342	-13.901	41.981	-64.277	11.708	
	1300	14.408	-12.463	43.131	-68.534	11.523	
	1400	14.474	-11.019	44.201	-72.901	11.382	
	1438	14.499	-10.468	44.589	-74.588	11.338	

* * * * * * * * * *

MANGANESE DITELLURIDE — $MnTe_2$

		A	B	C	D	RANGE	REF.
SOL	CP	18.32	1.0			298- 700	63

PHASE	T	CP	H	S	G	BT	
SOL	298	18.618	-30.	34.66	-40.334	29.570	63
	300	18.620	-29.966	34.775	-40.398	29.434	
	400	18.720	-28.099	40.146	-44.157	24.129	
	500	18.820	-26.222	44.333	-48.388	21.153	
	600	18.920	-24.335	47.774	-52.999	19.307	
	700	19.020	-22.438	50.698	-57.926	18.088	

Mn_4N TETRAMANGENESE MONONITRIDE

		A	B	C	D	RANGE	REF.
SOL	CP	22.16	30.5			298- 800	58
	** MNN(0.2-0.25)						1

PHASE	T	CP	H	S	G	BT	
SOL	298	31.254	-30.5	35.9	-41.204	30.207	1/*
	300	31.310	-30.442	36.094	-41.270	30.069	
	400	34.360	-27.159	45.519	-45.366	24.790	
	500	37.410	-23.570	53.513	-50.327	22.001	
	600	40.460	-19.677	60.604	-56.039	20.415	
	700	43.510	-15.478	67.070	-62.427	19.493	
	800	46.560	-10.975	73.079	-69.438	18.972	

* * * * * * * * * *

Mn_5N_2 PENTAMANGANESE DINITRIDE

		A	B	C	D	RANGE	REF.
SOL	CP	30.55	38.4			298- 800	58
	** MNN(0.4-0.5)						1

PHASE	T	CP	H	S	G	BT	
SOL	298	41.999	-48.8	44.8	-62.157	45.569	1/60
	300	42.070	-48.722	45.060	-62.240	45.348	
	400	45.910	-44.323	57.689	-67.399	36.830	
	500	49.750	-39.540	68.346	-73.713	32.224	
	600	53.590	-34.373	77.756	-81.027	29.518	
	700	57.430	-28.822	86.305	-89.236	27.864	
	800	61.270	-22.887	94.224	-98.267	26.849	

MANGANESE MONOPHOSPHIDE

MnP

		A	B	C	D	RANGE	REF.
SOL	CP	10.75	2.5			298-1420	*
	** MPT= 1420						1

PHASE	T	CP	H	S	G	BT	
SOL	298	11.495	-23.	12.5	-26.727	19.594	171/*
	300	11.500	-22.979	12.571	-26.750	19.490	
	400	11.750	-21.816	15.914	-28.182	15.400	
	500	12.000	-20.629	18.562	-29.910	13.075	
	600	12.250	-19.416	20.772	-31.880	11.614	
	700	12.500	-18.179	22.680	-34.054	10.634	
	800	12.750	-16.916	24.365	-36.408	9.948	
	900	13.000	-15.629	25.881	-38.922	9.453	
	1000	13.250	-14.316	27.264	-41.580	9.089	
	1100	13.500	-12.979	28.538	-44.371	8.817	
	1200	13.750	-11.616	29.724	-47.285	8.613	
	1300	14.000	-10.229	30.834	-50.313	8.460	
	1400	14.250	-8.816	31.881	-53.449	8.345	
	1420	14.300	-8.531	32.083	-54.089	8.326	

* * * * * * * * * *

MANGANESE TRIPHOSPHIDE

MnP_3

		A	B	C	D	RANGE	REF.
SOL	CP	22.4	7.2			298-1000	*

PHASE	T	CP	H	S	G	BT	
SOL	298	24.547	-41.6	23.1	-48.487	35.547	64,*
	300	24.560	-41.555	23.252	-48.530	35.359	
	400	25.280	-39.063	30.416	-51.229	27.994	
	500	26.000	-36.499	36.134	-54.566	23.854	
	600	26.720	-33.863	40.938	-58.426	21.284	
	700	27.440	-31.155	45.111	-62.733	19.589	
	800	28.160	-28.375	48.822	-67.433	18.424	
	900	28.880	-25.523	52.181	-72.485	17.604	
	1000	29.600	-22.599	55.261	-77.859	17.018	

MnAs

MANGANESE ARSENIDE

		A	B	C	D	RANGE	REF.
SOL	CP	11.65	2.4			298-1208	*
	** TPT= 313 / 400, MPT= 1208						1

PHASE	T	CP	H	S	G	BT	
SOL	298	12.366	-13.7	16.	-18.470	13.541	1/*
	300	12.370	-13.677	16.077	-18.500	13.479	
	400	12.610	-12.428	19.668	-20.295	11.090	
	500	12.850	-11.155	22.508	-22.409	9.796	
	600	13.090	-9.858	24.872	-24.781	9.028	
	700	13.330	-8.537	26.908	-27.372	8.547	
	800	13.570	-7.192	28.703	-30.155	8.239	
	900	13.810	-5.823	30.315	-33.107	8.041	
	1000	14.050	-4.430	31.783	-36.213	7.915	
	1100	14.290	-3.013	33.133	-39.460	7.841	
	1200	14.530	-1.572	34.387	-42.836	7.803	
	1208	14.549	-1.456	34.483	-43.112	7.801	

DIMANGANESE ANTIMONIDE Mn_2Sb

		A	B	C	D	RANGE	REF.
SOL	CP	16.9	7.0			298-1200	*

PHASE	T	CP	H	S	G	BT	
SOL	298	18.987	-7.8	32.7	-17.550	12.866	91
	300	19.000	-7.765	32.817	-17.610	12.831	
	400	19.700	-5.830	38.379	-21.182	11.575	
	500	20.400	-3.825	42.850	-25.250	11.038	
	600	21.100	-1.750	46.632	-29.729	10.830	
	700	21.800	.395	49.937	-34.561	10.792	
	800	22.500	2.610	52.894	-39.705	10.848	
	900	23.200	4.895	55.584	-45.130	10.961	
	1000	23.900	7.250	58.065	-50.814	11.107	
	1100	24.600	9.675	60.375	-56.738	11.274	
	1200	25.300	12.170	62.546	-62.885	11.454	

* * * * * * * * * *

MANGANESE ANTIMONICE $MnSb$

		A	B	C	D	RANGE	REF.
SOL	CP	11.0	4.85			298-1200	*

PHASE	T	CP	H	S	G	BT	
SOL	298	12.446	-6.5	22.1	-13.089	9.596	91
	300	12.455	-6.477	22.177	-13.130	9.567	
	400	12.940	-5.207	25.827	-15.538	8.491	
	500	13.425	-3.889	28.766	-18.272	7.988	
	600	13.910	-2.522	31.257	-21.276	7.751	
	700	14.395	-1.107	33.437	-24.513	7.654	
	800	14.880	.357	35.391	-27.956	7.638	
	900	15.365	1.869	37.172	-31.586	7.671	
	1000	15.850	3.430	38.816	-35.386	7.735	
	1100	16.335	5.039	40.349	-39.345	7.818	
	1200	16.820	6.697	41.791	-43.453	7.915	

MnBi

MANGANESE BISMUTHIDE

		A	B	C	D	RANGE	REF.
SOL	CP	10.30	7.35			298- 718	*
	** MPT= 718						1

PHASE	T	CP	H	S	G	BT	
SOL	298	12.491	-4.7	22.5	-11.408	8.364	1/*
	300	12.505	-4.677	22.577	-11.450	8.342	
	400	13.240	-3.390	26.275	-13.900	7.596	
	500	13.975	-2.029	29.309	-16.683	7.293	
	600	14.710	-.595	31.922	-19.748	7.194	
	700	15.445	.913	34.244	-23.058	7.200	
	718	15.577	1.192	34.638	-23.678	7.208	

* * * * * * * * * *

Mn_7C_3

HEPTAMANGANESE TRICARBIDE

		A	B	C	D	RANGE	REF.
SOL	CP	58.95	13.10	-9.50		298-1500	*

PHASE	T	CP	H	S	G	BT	
SOL	298	52.169	-26.5	57.1	-43.524	31.909	101
	300	52.324	-26.403	57.423	-43.630	31.789	
	400	58.252	-20.842	73.383	-50.195	27.429	
	500	61.700	-14.832	86.779	-58.221	25.452	
	600	64.171	-8.533	98.256	-67.487	24.585	
	700	66.181	-2.013	108.303	-77.825	24.301	
	800	67.946	4.695	117.257	-89.111	24.347	
	900	69.567	11.572	125.355	-101.248	24.590	
	1000	71.100	18.605	132.765	-114.159	24.953	
	1100	72.575	25.790	139.611	-127.782	25.391	
	1200	74.010	33.119	145.987	-142.066	25.877	
	1300	75.418	40.591	151.967	-156.966	26.392	
	1400	76.805	48.202	157.607	-172.448	26.924	
	1500	78.178	55.951	162.953	-188.478	27.465	

3-MANGANESE SILICON

Mn_3Si

		A	B	C	D	RANGE	REF.
SOL	CP	24.111	12.447	-3.518		298-1000	221
	** MPT=	1348					221

PHASE	T	CP	H	S	G	BT	
SOL	298	23.865	-19.02	24.771	-26.405	19.358	221
	300	23.936	-18.976	24.919	-26.451	19.272	
	400	26.891	-16.422	32.245	-29.320	16.022	
	500	28.927	-13.627	38.474	-32.864	14.367	
	600	30.602	-10.648	43.900	-36.988	13.475	
	700	32.106	-7.512	48.731	-41.624	12.997	
	800	33.519	-4.230	53.112	-46.719	12.765	
	900	34.879	-.810	57.138	-52.235	12.686	
	1000	36.206	2.744	60.882	-58.138	12.708	

Mn_5Si_3 5-MANGANESE 3-SILICCN

		A	B	C	D	RANGE	REF.
SOL	CP	48.126	12.945	-4.685		298-1573	221
LIQ	CP	77.591				1573-2000	221

PHASE	T	CP	H	S	G	BT	
SOL	298	46.715	-48.0	57.097	-65.023	47.670	221
	300	46.804	-47.913	57.386	-65.129	47.453	
	400	50.376	-43.038	71.387	-71.593	39.122	
	500	52.724	-37.877	82.893	-79.324	34.677	
	600	54.592	-32.509	92.676	-88.115	32.100	
	700	56.231	-26.966	101.217	-97.818	30.544	
	800	57.750	-21.267	108.825	-108.327	29.598	
	900	59.198	-15.419	115.711	-119.559	29.037	
	1000	60.602	-9.428	122.022	-131.450	28.732	
	1100	61.978	-3.299	127.862	-143.948	28.604	
	1200	63.335	2.967	133.313	-157.010	28.599	
	1300	64.677	9.367	138.436	-170.600	28.684	
	1400	66.010	15.902	143.278	-184.687	28.835	
	1500	67.335	22.569	147.877	-199.247	29.034	
	1573	68.299	27.520	151.100	-210.160	29.203	
			41.3	26.256			221
LIQ	1573	77.591	68.820	177.355	-210.160	29.203	
	1600	77.591	70.915	178.676	-214.967	29.367	
	1700	77.591	78.674	183.380	-233.072	29.967	
	1800	77.591	86.433	187.815	-251.634	30.557	
	1900	77.591	94.192	192.010	-270.627	31.133	
	2000	77.591	101.951	195.990	-290.029	31.697	

MANGANESE SILICON

		A	B	C	D	RANGE	REF.
SOL	CP	11.788	3.054	-1.530		298-1548	221
LIQ	CP	18.815				1548-2000	221

PHASE	T	CP	H	S	G	BT	
SOL	298	10.977	-14.48	11.096	-17.788	13.041	221
	300	11.004	-14.460	11.164	-17.809	12.975	
	400	12.053	-13.301	14.489	-19.097	10.435	
	500	12.703	-12.062	17.252	-20.688	9.044	
	600	13.195	-10.766	19.614	-22.534	8.209	
	700	13.614	-9.425	21.680	-24.601	7.682	
	800	13.992	-8.045	23.523	-26.863	7.340	
	900	14.348	-6.627	25.191	-29.300	7.116	
	1000	14.689	-5.175	26.721	-31.896	6.972	
	1100	15.021	-3.690	28.136	-34.640	6.883	
	1200	15.347	-2.172	29.457	-37.520	6.834	
	1300	15.668	-.621	30.698	-40.529	6.814	
	1400	15.986	.962	31.871	-43.658	6.816	
	1500	16.301	2.576	32.985	-46.901	6.834	
	1548	16.452	3.362	33.501	-48.497	6.848	
			14.197	9.171			221
LIQ	1548	18.815	17.559	42.672	-48.497	6.848	
	1600	18.815	18.538	43.293	-50.732	6.931	
	1700	18.815	20.419	44.434	-55.119	7.087	
	1800	18.815	22.301	45.510	-59.617	7.239	
	1900	18.815	24.182	46.527	-64.219	7.388	
	2000	18.815	26.064	47.492	-68.920	7.532	

$MnSi_{1.7}$ MANGANESE 1.7-SILICCN

		A	B	C	D	RANGE	REF.
SOL	CP	17.191	1.103	-3.123		298-1425	221
	** MPT=	1425					221

PHASE	T	CP	H	S	G	BT	
SOL	298	14.007	-18.069	13.425	-22.072	16.181	221
	300	14.052	-18.043	13.512	-22.097	16.100	
	400	15.680	-16.546	17.809	-23.669	12.934	
	500	16.493	-14.933	21.404	-25.635	11.206	
	600	16.985	-13.257	24.457	-27.932	10.175	
	700	17.326	-11.541	27.103	-30.513	9.528	
	800	17.585	-9.795	29.434	-33.342	9.110	
	900	17.798	-8.025	31.518	-36.391	8.838	
	1000	17.982	-6.236	33.403	-39.639	8.664	
	1100	18.146	-4.430	35.124	-43.066	8.558	
	1200	18.298	-2.607	36.710	-46.659	8.499	
	1300	18.440	-.770	38.180	-50.404	8.475	
	1400	18.576	1.080	39.552	-54.292	8.476	
	1425	18.609	1.545	39.881	-55.285	8.480	

MANGANESE METASILICATE (RHODONITE) $MnO \cdot SiO_2$

		A	B	C	D	RANGE	REF.
SOL	CP	26.42	3.88	-6.16		298-1564	5
LIQ	CP	36.25				1564-3000	*

PHASE	T	CP	H	S	G	BT	
SOL	298	20.647	-315.6	24.5	-322.905	236.727	5
	300	20.740	-315.562	24.628	-322.950	235.301	
	400	24.122	-313.297	31.119	-325.745	178.003	
	500	25.896	-310.789	36.710	-329.144	143.888	
	600	27.037	-308.139	41.538	-333.062	121.334	
	700	27.879	-305.391	45.772	-337.431	105.365	
	800	28.561	-302.568	49.541	-342.200	93.497	
	900	29.152	-299.682	52.939	-347.327	84.354	
	1000	29.684	-296.740	56.039	-352.778	77.110	
	1100	30.179	-293.746	58.891	-358.527	71.242	
	1200	30.648	-290.705	61.538	-364.550	66.403	
	1300	31.100	-287.617	64.009	-370.828	62.350	
	1400	31.538	-284.485	66.329	-377.347	58.914	
	1500	31.966	-281.310	68.520	-384.090	55.969	
	1564	32.236	-279.255	69.861	-388.518	54.298	
			16.	10.230			*
LIQ	1564	36.250	-263.255	80.091	-388.518	54.298	
	1600	36.250	-261.950	80.916	-391.417	53.472	
	1700	36.250	-258.325	83.114	-399.619	51.381	
	1800	36.250	-254.700	85.186	-408.035	49.549	
	1900	36.250	-251.075	87.146	-416.653	47.932	
	2000	36.250	-247.450	89.005	-425.461	46.498	
	2100	36.250	-243.825	90.774	-434.451	45.220	
	2200	36.250	-240.200	92.460	-443.613	44.075	
	2300	36.250	-236.575	94.072	-452.940	43.045	
	2400	36.250	-232.950	95.614	-462.425	42.115	
	2500	36.250	-229.325	97.094	-472.061	41.273	
	2600	36.250	-225.700	98.516	-481.842	40.508	
	2700	36.250	-222.075	99.884	-491.763	39.811	
	2800	36.250	-218.450	101.202	-501.817	39.174	
	2900	36.250	-214.825	102.474	-512.001	38.591	
	3000	36.250	-211.200	103.703	-522.311	38.055	

$2MnO \cdot SiO_2$ MANGANESE ORTHOSILICATE (TEPHROITE)

		A	B	C	D	RANGE	REF.
SOL	CP	38.02	4.66	-7.44		298-1618	5
LIQ	CP	58.1				1618-3000	5

PHASE	T	CP	H	S	G	BT	
SOL	298	31.040	-413.5	34.	-423.637	310.576	5
	300	31.151	-413.442	34.192	-423.700	308.707	
	400	35.234	-410.097	43.788	-427.612	233.668	
	500	37.374	-406.458	51.901	-432.408	189.031	
	600	38.749	-402.647	58.844	-437.954	159.546	
	700	39.764	-398.720	64.896	-444.147	138.686	
	800	40.585	-394.701	70.261	-450.910	123.199	
	900	41.295	-390.606	75.084	-458.181	111.277	
	1000	41.936	-386.444	79.468	-465.912	101.839	
	1100	42.531	-382.221	83.493	-474.063	94.200	
	1200	43.095	-377.939	87.218	-482.601	87.905	
	1300	43.638	-373.602	90.689	-491.498	82.639	
	1400	44.164	-369.212	93.943	-500.731	78.178	
	1500	44.679	-364.770	97.007	-510.280	74.358	
	1600	45.185	-360.276	99.907	-520.127	71.056	
	1618	45.276	-359.462	100.413	-521.930	70.509	
			21.42	13.239			5
LIQ	1618	58.100	-338.042	113.651	-521.930	70.509	
	1700	58.100	-333.278	116.524	-531.368	68.321	
	1800	58.100	-327.468	119.845	-543.188	65.961	
	1900	58.100	-321.658	122.986	-555.331	63.886	
	2000	58.100	-315.848	125.966	-567.780	62.052	
	2100	58.100	-310.038	128.801	-580.520	60.424	
	2200	58.100	-304.228	131.504	-593.536	58.970	
	2300	58.100	-298.418	134.086	-606.817	57.668	
	2400	58.100	-292.608	136.559	-620.350	56.498	
	2500	58.100	-286.798	138.931	-634.125	55.443	
	2600	58.100	-280.988	141.210	-648.133	54.488	
	2700	58.100	-275.178	143.402	-662.364	53.622	
	2800	58.100	-269.368	145.515	-676.811	52.835	
	2900	58.100	-263.558	147.554	-691.465	52.117	
	3000	58.100	-257.748	149.524	-706.319	51.462	

MANGANESE 2-TIN

$MnSn_2$

		A	B	C	D	RANGE	REF.
SOL	CP	22.08	0.3	-2.72		298- 800	*

PHASE	T	CP	H	S	G	BT	
SOL	298	19.110	-6.55	31.28	-15.876	11.639	139
	300	19.148	-6.515	31.398	-15.934	11.610	
	400	20.500	-4.523	37.119	-19.370	10.585	
	500	21.142	-2.437	41.770	-23.322	10.196	
	600	21.504	-.303	45.660	-27.699	10.091	
	700	21.735	1.859	48.993	-32.436	10.128	
	800	21.895	4.041	51.906	-37.484	10.242	

* * * * * * * * * *

MANGANESE MONOBORICE

MnB

		A	B	C	D	RANGE	REF.
SOL	CP	10.15	3.8	-2.42		298-1200	*

PHASE	T	CP	H	S	G	BT	
SOL	298	8.561	-18.	7.75	-20.311	14.890	*
	300	8.601	-17.984	7.803	-20.325	14.809	
	400	10.157	-17.038	10.515	-21.244	11.609	
	500	11.082	-15.973	12.888	-22.417	9.800	
	600	11.758	-14.829	14.970	-23.812	8.675	
	700	12.316	-13.625	16.826	-25.403	7.932	
	800	12.812	-12.368	18.503	-27.171	7.424	
	900	13.271	-11.064	20.039	-29.099	7.067	
	1000	13.708	-9.715	21.460	-31.175	6.814	
	1100	14.130	-8.323	22.786	-33.388	6.634	
	1200	14.542	-6.889	24.034	-35.729	6.508	

MnB_2 MANGANESE DIBORIDE

		A	B	C	D	RANGE	REF.
SOL	CP	9.62	10.72			298-1200	*

PHASE	T	CP	H	S	G	BT	
SOL	298	12.816	-22.5	8.25	-24.960	18.298	*
	300	12.836	-22.476	8.329	-24.975	18.197	
	400	13.908	-21.139	12.169	-26.007	14.211	
	500	14.980	-19.695	15.387	-27.388	11.973	
	600	16.052	-18.143	18.213	-29.071	10.591	
	700	17.124	-16.484	20.768	-31.022	9.687	
	800	18.196	-14.718	23.125	-33.218	9.076	
	900	19.268	-12.845	25.330	-35.642	8.656	
	1000	20.340	-10.865	27.416	-38.280	8.367	
	1100	21.412	-8.777	29.404	-41.122	8.171	
	1200	22.484	-6.582	31.313	-44.158	8.043	

* * * * * * * * * *

$MnO \cdot MoO_3$ MANGANESE MOLYBDATE

		A	B	C	D	RANGE	REF.
SOL	CP	26.0	12.26			298-1000	*

PHASE	T	CP	H	S	G	BT	
SOL	298	29.655	-284.7	32.5	-294.390	215.823	5/*
	300	29.678	-284.645	32.684	-294.450	214.536	
	400	30.904	-281.616	41.389	-298.172	162.935	
	500	32.130	-278.464	48.417	-302.673	132.316	
	600	33.356	-275.190	54.383	-307.820	112.138	
	700	34.582	-271.793	59.617	-313.525	97.900	
	800	35.808	-268.274	64.315	-319.726	87.357	
	900	37.034	-264.632	68.603	-326.375	79.265	
	1000	38.260	-260.867	72.569	-333.436	72.882	

MOLYBDENUM

		A	B	C	D	RANGE	REF.
SOL	CP	6.111	0.680	-0.522		298- 700	3
	CP	8.105	-2.847	-2.200	1.663	700-1500	3
	CP	3.984	2.317			1500-2000	3
	CP	49.318	-30.263	-251.864	6.534	2000-2892	3
LIQ	CP	10.				2892-3800	3
	** BPT= 4919 / L= 140.812						3

PHASE	T	CP	H	S	G	BT	P	
SOL	298	5.727	0	6.837	-2.038	1.494		3
	300	5.735	.011	6.872	-2.051	1.494		
	400	6.057	.602	8.572	-2.827	1.545		
	500	6.242	1.218	9.945	-3.755	1.641		
	600	6.374	1.849	11.095	-4.808	1.752		
	700	6.480	2.492	12.086	-5.968	1.864		
	700	6.478	2.492	12.086	-5.968	1.864		
	800	6.548	3.143	12.955	-7.221	1.973		
	900	6.618	3.801	13.730	-8.556	2.078		
	1000	6.701	4.467	14.432	-9.965	2.178		
	1100	6.804	5.142	15.075	-11.441	2.273		
	1200	6.931	5.829	15.672	-12.978	2.364		
	1300	7.084	6.529	16.233	-14.574	2.450		
	1400	7.266	7.246	16.765	-16.224	2.533		
	1500	7.478	7.983	17.273	-17.926	2.612		
	1500	7.459	7.983	17.273	-17.926	2.612		
	1600	7.691	8.741	17.762	-19.678	2.688		
	1700	7.923	9.522	18.235	-21.478	2.762		
	1800	8.155	10.325	18.694	-23.324	2.832		
	1900	8.386	11.152	19.142	-25.216	2.901		
	2000	8.618	12.003	19.578	-27.152	2.967		
	2000	8.631	12.003	19.578	-27.152	2.967		
	2100	8.869	12.877	20.004	-29.132	3.032		
	2200	9.160	13.778	20.423	-31.153	3.095		
	2300	9.517	14.712	20.838	-33.216	3.157		
	2400	9.950	15.684	21.252	-35.321	3.217		
	2500	10.468	16.704	21.668	-37.467	3.276		
	2600	11.078	17.781	22.091	-39.654	3.334	.002	
	2700	11.786	18.923	22.522	-41.885	3.391	.005	
	2800	12.596	20.142	22.965	-44.159	3.447	.013	
	2892	13.434	21.338	23.385	-46.291	3.499	.031	
			6.652	2.300				3

Mo

MOLYBDENUM (CONTINUED)

PHASE	T	CP	H	S	G	BT	P
LIQ	2892	10.000	27.990	25.685	-46.291	3.499	.031
	2900	10.000	28.070	25.713	-46.497	3.505	.033
	3000	10.000	29.070	26.052	-49.085	3.576	.076
	3100	10.000	30.070	26.380	-51.707	3.646	.164
	3200	10.000	31.070	26.697	-54.361	3.713	.337
	3300	10.000	32.070	27.005	-57.046	3.778	.661
	3400	10.000	33.070	27.303	-59.761	3.842	1.246
	3500	10.000	34.070	27.593	-62.506	3.904	2.262
	3600	10.000	35.070	27.875	-65.280	3.964	3.968
	3700	10.000	36.070	28.149	-68.081	4.022	6.748
	3800	10.000	37.070	28.416	-70.909	4.079	11.152

MOLYBDENUM DIOXIDE DICHLORIDE

MoO_2Cl_2

		A	B	C	D	RANGE	REF.
SOL	CP	18.03	24.07	-0.23		298- 642	*

PHASE	T	CP	H	S	G	BT	
SOL	298	24.948	-171.4	34.	-181.537	133.088	19/*
	300	24.995	-171.354	34.154	-181.600	132.313	
	400	27.514	-168.728	41.692	-185.405	101.314	
	500	29.973	-165.853	48.097	-189.901	83.017	
	600	32.408	-162.734	53.777	-195.000	71.038	
	642	33.427	-161.351	56.004	-197.306	67.176	

Mo_2S_3 MOLYBDENUM SESQUISULFIDE

		A	B	C	D	RANGE	REF.
SOL	CP	26.36	7.88	-2.30		298-1500	63

PHASE	T	CP	H	S	G	BT	
SOL	298	26.122	-98.	27.	-106.050	77.747	63
	300	26.168	-97.952	27.162	-106.100	77.304	
	400	28.074	-95.231	34.974	-109.221	59.684	
	500	29.380	-92.356	41.385	-113.049	49.420	
	600	30.449	-89.363	46.839	-117.466	42.793	
	700	31.407	-86.270	51.605	-122.394	38.218	
	800	32.305	-83.084	55.858	-127.770	34.910	
	900	33.168	-79.810	59.713	-133.552	32.435	
	1000	34.010	-76.451	63.252	-139.703	30.536	
	1100	34.838	-73.008	66.532	-146.194	29.050	
	1200	35.656	-69.484	69.599	-153.002	27.869	
	1300	36.468	-65.877	72.485	-160.107	26.920	
	1400	37.275	-62.190	75.217	-167.494	26.150	
	1500	38.078	-58.423	77.816	-175.146	25.522	

MOLYBDENUM DISULFICE MoS_2

		A	B	C	D	RANGE	REF.
SOL	CP	17.35	1.78	-2.20		298-1458	97
LIQ	CP	23.0				1458-2000	*

PHASE	T	CP	H	S	G	BT	
SOL	298	15.406	-65.8	14.96	-70.260	51.509	63
	300	15.440	-65.771	15.055	-70.288	51.212	
	400	16.687	-64.158	19.690	-72.033	39.363	
	500	17.360	-62.452	23.492	-74.198	32.436	
	600	17.807	-60.693	26.699	-76.712	27.946	
	700	18.147	-58.895	29.470	-79.524	24.832	
	800	18.430	-57.065	31.912	-82.595	22.567	
	900	18.680	-55.210	34.098	-85.898	20.862	
	1000	18.910	-53.330	36.078	-89.408	19.543	
	1100	19.126	-51.428	37.891	-93.108	18.501	
	1200	19.333	-49.505	39.564	-96.981	17.665	
	1300	19.534	-47.562	41.119	-101.017	16.985	
	1400	19.730	-45.598	42.574	-105.202	16.425	
	1458	19.842	-44.451	43.377	-107.695	16.145	
			10.9	7.476			*
LIQ	1458	23.000	-33.551	50.853	-107.695	16.145	
	1500	23.000	-32.585	51.506	-109.844	16.006	
	1600	23.000	-30.285	52.991	-115.070	15.720	
	1700	23.000	-27.985	54.385	-120.439	15.486	
	1800	23.000	-25.685	55.700	-125.944	15.294	
	1900	23.000	-23.385	56.943	-131.577	15.137	
	2000	23.000	-21.085	58.123	-137.331	15.009	

Mo_3Si 3-MOLYBDENUM SILICCN

		A	B	C	D	RANGE	REF.
SOL	CP	20.516	5.421	0.077		298-2298	221
	** MPT= 2298						221

PHASE	T	CP	H	S	G	BT	
SOL	298	22.219	-27.82	25.37	-35.384	25.941	221
	300	22.228	-27.779	25.507	-35.431	25.815	
	400	22.733	-25.531	31.970	-38.319	20.939	
	500	23.257	-23.232	37.099	-41.781	18.265	
	600	23.790	-20.879	41.386	-45.711	16.653	
	700	24.326	-18.474	45.094	-50.039	15.625	
	800	24.865	-16.014	48.377	-54.716	14.950	
	900	25.404	-13.501	51.337	-59.704	14.500	
	1000	25.945	-10.933	54.042	-64.975	14.202	
	1100	26.485	-8.312	56.540	-70.506	14.010	
	1200	27.027	-5.636	58.868	-76.277	13.894	
	1300	27.568	-2.906	61.052	-82.274	13.833	
	1400	28.109	-.122	63.115	-88.484	13.815	
	1500	28.651	2.716	65.073	-94.894	13.828	
	1600	29.193	5.608	66.939	-101.495	13.865	
	1700	29.734	8.554	68.725	-108.279	13.922	
	1800	30.276	11.555	70.440	-115.238	13.994	
	1900	30.818	14.609	72.092	-122.365	14.077	
	2000	31.360	17.718	73.686	-129.654	14.170	
	2100	31.902	20.881	75.229	-137.101	14.270	
	2200	32.444	24.099	76.726	-144.699	14.376	
	2298	32.975	27.304	78.151	-152.288	14.485	

5-MOLYBDENUM 3-SILICON Mo_5Si_3

		A	B	C	D	RANGE	REF.
SOL	CP	43.824	8.375	-2.868		298-2463	221
	** MPT= 2463						221

PHASE	T	CP	H	S	G	BT	
SOL	298	43.095	-74.0	49.556	-88.775	65.083	221
	300	43.150	-73.920	49.823	-88.867	64.748	
	400	45.381	-69.484	62.571	-94.512	51.646	
	500	46.864	-64.868	72.864	-101.300	44.284	
	600	48.052	-60.120	81.517	-109.030	39.720	
	700	49.101	-55.262	89.004	-117.565	36.710	
	800	50.076	-50.303	95.625	-126.802	34.645	
	900	51.007	-45.248	101.577	-136.667	33.192	
	1000	51.912	-40.102	106.998	-147.100	32.153	
	1100	52.799	-34.866	111.988	-158.053	31.406	
	1200	53.675	-29.542	116.619	-169.486	30.872	
	1300	54.542	-24.132	120.950	-181.367	30.495	
	1400	55.403	-18.634	125.024	-193.667	30.237	
	1500	56.259	-13.051	128.875	-206.364	30.071	
	1600	57.112	-7.383	132.533	-219.436	29.978	
	1700	57.962	-1.629	136.021	-232.865	29.941	
	1800	58.810	4.210	139.358	-246.635	29.950	
	1900	59.657	10.133	142.561	-260.732	29.995	
	2000	60.502	16.141	145.642	-275.143	30.070	
	2100	61.346	22.234	148.614	-289.857	30.170	
	2200	62.190	28.410	151.488	-304.863	30.289	
	2300	63.032	34.671	154.271	-320.151	30.425	
	2400	63.874	41.017	156.971	-335.714	30.575	
	2463	64.404	45.058	158.633	-345.656	30.675	

$MoSi_2$ MOLYBDENUM 2-SILICCN

		A	B	C	D	RANGE	REF.
SOL	CP	16.212	2.861	-1.570		298-2303	221
	** MPT= 2303						221

PHASE	T	CP	H	S	G	BT	
SOL	298	15.299	-31.48	15.539	-36.113	26.475	221
	300	15.326	-31.452	15.634	-36.142	26.333	
	400	16.375	-29.861	20.202	-37.942	20.733	
	500	17.014	-28.190	23.929	-40.154	17.554	
	600	17.492	-26.464	27.075	-42.709	15.559	
	700	17.894	-24.694	29.802	-45.555	14.225	
	800	18.255	-22.886	32.216	-48.659	13.295	
	900	18.593	-21.043	34.386	-51.991	12.627	
	1000	18.916	-19.168	36.362	-55.529	12.138	
	1100	19.229	-17.261	38.179	-59.258	11.775	
	1200	19.536	-15.322	39.866	-63.161	11.505	
	1300	19.838	-13.353	41.441	-67.227	11.303	
	1400	20.137	-11.355	42.922	-71.446	11.155	
	1500	20.434	-9.326	44.322	-75.809	11.047	
	1600	20.728	-7.268	45.650	-80.308	10.971	
	1700	21.021	-5.180	46.915	-84.937	10.921	
	1800	21.313	-3.064	48.125	-89.689	10.891	
	1900	21.604	-.918	49.285	-94.560	10.878	
	2000	21.895	1.257	50.401	-99.545	10.879	
	2100	22.184	3.461	51.476	-104.639	10.891	
	2200	22.474	5.694	52.515	-109.839	10.913	
	2300	22.763	7.956	53.520	-115.141	10.942	
	2303	22.771	8.024	53.550	-115.301	10.943	

AMMONIA (GAS) $NH_3(g)$

		A	B	C	D	RANGE	REF.
GAS	CP	6.165	7.558	0.084		298- 800	3
	CP	12.601	2.50	-15.231		800-2000	3

PHASE	T	CP	H	S	G	BT	
GAS	298	8.513	-10.98	46.049	-24.710	18.115	40
	300	8.526	-10.964	46.102	-24.795	18.065	
	400	9.241	-10.076	48.651	-29.537	16.140	
	500	9.978	-9.115	50.792	-34.512	15.087	
	600	10.723	-8.080	52.677	-39.687	14.458	
	700	11.473	-6.971	54.387	-45.041	14.064	
	800	12.225	-5.786	55.968	-50.560	13.814	
	800	12.221	-5.786	55.968	-50.560	13.814	
	900	12.971	-4.525	57.452	-56.232	13.657	
	1000	13.578	-3.196	58.851	-62.047	13.562	
	1100	14.092	-1.812	60.170	-67.999	13.512	
	1200	14.543	-.380	61.416	-74.079	13.493	
	1300	14.950	1.095	62.596	-80.280	13.498	
	1400	15.324	2.609	63.718	-86.596	13.520	
	1500	15.674	4.159	64.787	-93.022	13.555	
	1600	16.006	5.743	65.810	-99.552	13.600	
	1700	16.324	7.360	66.789	-106.182	13.653	
	1800	16.631	9.008	67.731	-112.909	13.711	
	1900	16.929	10.686	68.638	-119.728	13.774	
	2000	17.220	12.393	69.514	-126.635	13.840	

* * * * * * * * * *

AMMONIUM SULFATE $(NH_4)_2SO_4$

		A	B	C	D	RANGE	REF.
SOL	CP	24.75	67.1			298- 600	2
	** DEC.						

PHASE	T	CP	H	S	G	BT	
SOL	298	44.756	-282.1	52.6	-297.783	218.310	62
	300	44.880	-282.017	52.877	-297.880	217.035	
	400	51.590	-277.194	66.707	-303.877	166.053	
	500	58.300	-271.699	78.940	-311.169	136.030	
	600	65.010	-265.534	90.163	-319.631	116.441	

Na_2S SODIUM SULFIDE

		A	B	C	D	RANGE	REF.
SOL	CP	19.81	1.64			298-1251	63
LIQ	CP	22.0				1251-2200	*

PHASE	T	CP	H	S	G	BT	
SOL	298	20.299	-89.5	19.	-95.165	69.767	63
	300	20.302	-89.462	19.126	-95.200	69.363	
	400	20.466	-87.424	24.989	-97.419	53.235	
	500	20.630	-85.369	29.573	-100.156	43.784	
	600	20.794	-83.298	33.349	-103.307	37.635	
	700	20.958	-81.210	36.567	-106.807	33.351	
	800	21.122	-79.106	39.376	-110.607	30.221	
	900	21.286	-76.986	41.873	-114.672	27.850	
	1000	21.450	-74.849	44.124	-118.974	26.005	
	1100	21.614	-72.696	46.176	-123.490	24.539	
	1200	21.778	-70.526	48.064	-128.203	23.352	
	1251	21.862	-69.414	48.972	-130.678	22.833	
			6.3	5.036			*
LIQ	1251	22.000	-63.114	54.008	-130.678	22.833	
	1300	22.000	-62.036	54.853	-133.345	22.420	
	1400	22.000	-59.836	56.484	-138.913	21.688	
	1500	22.000	-57.636	58.002	-144.638	21.077	
	1600	22.000	-55.436	59.422	-150.510	20.561	
	1700	22.000	-53.236	60.755	-156.520	20.125	
	1800	22.000	-51.036	62.013	-162.659	19.752	
	1900	22.000	-48.836	63.202	-168.920	19.433	
	2000	22.000	-46.636	64.331	-175.297	19.158	
	2100	22.000	-44.436	65.404	-181.784	18.921	
	2200	22.000	-42.236	66.428	-188.376	18.716	

SODIUM MONOSULFIDE

NaS

		A	B	C	D	RANGE	REF.
SOL	CP	9.8	6.66			298- 753	*
LIQ	CP	14.9				753-1500	*

PHASE	T	CP	H	S	G	ET	
SOL	298	11.786	-48.1	10.7	-51.290	37.602	63
	300	11.798	-48.078	10.773	-51.310	37.384	
	400	12.464	-46.865	14.258	-52.568	28.726	
	500	13.130	-45.585	17.111	-54.141	23.668	
	600	13.796	-44.239	19.564	-55.977	20.392	
	700	14.462	-42.826	21.740	-58.045	18.125	
	753	14.815	-42.050	22.809	-59.225	17.192	
			2.26	3.001			63
LIQ	753	14.900	-39.790	25.810	-59.225	17.192	
	800	14.900	-39.090	26.712	-60.460	16.519	
	900	14.900	-37.600	28.467	-63.220	15.354	
	1000	14.900	-36.110	30.037	-66.147	14.458	
	1100	14.900	-34.620	31.457	-69.223	13.755	
	1200	14.900	-33.130	32.754	-72.434	13.194	
	1300	14.900	-31.640	33.946	-75.770	12.740	
	1400	14.900	-30.150	35.050	-79.221	12.369	
	1500	14.900	-28.660	36.078	-82.778	12.062	

Na_2S_3 DISODIUM TRISULFIDE

		A	B	C	D	RANGE	REF.
SOL	CP	26.05	13.1			298- 626	*
LIQ	CP	37.2				626-1000	*

PHASE	T	CP	H	S	G	BT	
SOL	298	29.956	-103.4	24.3	-110.645	81.116	63
	300	29.980	-103.345	24.485	-110.690	80.649	
	400	31.290	-100.281	33.289	-113.597	62.075	
	500	32.600	-97.087	40.412	-117.293	51.276	
	600	33.910	-93.761	46.472	-121.644	44.315	
	626	34.251	-92.875	47.918	-122.871	42.903	
			4.7	7.508			63
LIQ	626	37.200	-88.175	55.426	-122.871	42.903	
	700	37.200	-85.422	59.582	-127.129	39.697	
	800	37.200	-81.702	64.549	-133.342	36.432	
	900	37.200	-77.982	68.931	-140.020	34.006	
	1000	37.200	-74.262	72.850	-147.112	32.156	

* * * * * * * * * *

NaS_2 SODIUM DISULFIDE

		A	B	C	D	RANGE	REF.
SOL	CP	16.26	6.45			298- 458	*
LIQ	CP	22.4				458-1000	*

PHASE	T	CP	H	S	G	BT	
SOL	298	18.183	-49.2	20.	-55.163	40.441	63
	300	18.195	-49.166	20.113	-55.200	40.219	
	400	18.840	-47.315	25.435	-57.489	31.415	
	458	19.214	-46.211	28.011	-59.040	28.177	
			2.06	4.498			63
LIQ	458	22.400	-44.151	32.509	-59.040	28.177	
	500	22.400	-43.210	34.474	-60.447	26.425	
	600	22.400	-40.970	38.558	-64.105	23.353	
	700	22.400	-38.730	42.011	-68.138	21.277	
	800	22.400	-36.490	45.002	-72.492	19.807	
	900	22.400	-34.250	47.641	-77.127	18.731	
	1000	22.400	-32.010	50.001	-82.011	17.926	

DISODIUM SULFITE

Na_2SO_3

		A	B	C	D	RANGE	REF.
SOL	CP	25.6	10.4			298-1000	*
LIQ	CP	43.5				1000-2000	*

PHASE	T	CP	H	S	G	BT	
SOL	298	28.701	-260.38	34.9	-270.785	198.518	19A/*
	300	28.720	-260.327	35.078	-270.850	197.341	
	400	29.760	-257.403	43.482	-274.796	150.162	
	500	30.800	-254.375	50.235	-279.492	122.182	
	600	31.840	-251.243	55.942	-284.808	103.755	
	700	32.880	-248.007	60.928	-290.657	90.759	
	800	33.920	-244.667	65.387	-296.976	81.141	
	900	34.960	-241.223	69.442	-303.721	73.763	
	1000	36.000	-237.675	73.179	-310.854	67.946	
			6.0	6.000			*
LIQ	1000	43.500	-231.675	79.179	-310.854	67.946	
	1100	43.500	-227.325	83.325	-318.983	63.385	
	1200	43.500	-222.975	87.110	-327.507	59.655	
	1300	43.500	-218.625	90.592	-336.395	56.561	
	1400	43.500	-214.275	93.816	-345.617	53.961	
	1500	43.500	-209.925	96.817	-355.150	51.752	
	1600	43.500	-205.575	99.624	-364.974	49.860	
	1700	43.500	-201.225	102.262	-375.070	48.225	
	1800	43.500	-196.875	104.748	-385.421	46.803	
	1900	43.500	-192.525	107.100	-396.015	45.558	
	2000	43.500	-188.175	109.331	-406.837	44.463	

Na_2Te SODIUM TELLURIDE

		A	B	C	D	RANGE	REF.
SOL	CP	17.5	3.3			298-1226	*
LIQ	CP	23.0				1226-1500	*

PHASE	T	CP	H	S	G	BT	
SOL	298	18.484	-83.	22.6	-89.738	65.789	189*
	300	18.490	-82.966	22.714	-89.780	65.414	
	400	18.820	-81.100	28.079	-92.332	50.455	
	500	19.150	-79.202	32.314	-95.359	41.687	
	600	19.480	-77.270	35.834	-98.771	35.982	
	700	19.810	-75.306	38.862	-102.509	32.009	
	800	20.140	-73.308	41.529	-106.531	29.107	
	900	20.470	-71.278	43.920	-110.806	26.911	
	1000	20.800	-69.214	46.094	-115.308	25.204	
	1100	21.130	-67.118	48.092	-120.019	23.849	
	1200	21.460	-64.988	49.945	-124.922	22.754	
	1226	21.546	-64.429	50.405	-126.226	22.504	
			3.3	2.692			189*
LIQ	1226	23.000	-61.129	53.097	-126.226	22.504	
	1300	23.000	-59.427	54.445	-130.206	21.893	
	1400	23.000	-57.127	56.150	-135.737	21.192	
	1500	23.000	-54.827	57.736	-141.432	20.609	

SODIUM MONOTELLURIDE

NaTe

		A	B	C	D	RANGE	REF.
SOL	CP	11.2	4.3			298- 650	*
LIQ	CP	16.0				650-1500	*

PHASE	T	CP	H	S	G	BT	
SOL	298	12.482	-41.4	20.	-47.363	34.723	189*
	300	12.490	-41.377	20.077	-47.400	34.536	
	400	12.920	-40.106	23.729	-49.598	27.103	
	500	13.350	-38.793	26.658	-52.122	22.786	
	600	13.780	-37.436	29.130	-54.915	20.005	
	650	13.995	-36.742	30.242	-56.399	18.966	
			0.4	.615			189*
LIQ	650	16.000	-36.342	30.857	-56.399	18.966	
	700	16.000	-35.542	32.043	-57.972	18.102	
	800	16.000	-33.942	34.180	-61.286	16.745	
	900	16.000	-32.342	36.064	-64.800	15.738	
	1000	16.000	-30.742	37.750	-68.492	14.971	
	1100	16.000	-29.142	39.275	-72.344	14.375	
	1200	16.000	-27.542	40.667	-76.342	13.906	
	1300	16.000	-25.942	41.948	-80.474	13.531	
	1400	16.000	-24.342	43.133	-84.729	13.229	
	1500	16.000	-22.742	44.237	-89.098	12.983	

$NaTe_3$ SODIUM TRITELLURIDE

		A	B	C	D	RANGE	REF.
SOL	CP	20.0	13.33			298- 728	*
LIQ	CP	34.0				728-1500	*

PHASE	T	CP	H	S	G	BT	
SOL	298	23.974	-50.3	34.7	-60.646	44.461	189*
	300	23.999	-50.256	34.848	-60.710	44.233	
	400	25.332	-47.789	41.935	-64.563	35.280	
	500	26.665	-45.189	47.731	-69.055	30.188	
	600	27.998	-42.456	52.710	-74.082	26.988	
	700	29.331	-39.590	57.126	-79.578	24.849	
	728	29.704	-38.763	58.284	-81.194	24.378	
			8.8	12.088			189*
LIQ	728	34.000	-29.963	70.372	-81.194	24.378	
	800	34.000	-27.515	73.578	-86.378	23.601	
	900	34.000	-24.115	77.583	-93.940	22.815	
	1000	34.000	-20.715	81.165	-101.880	22.269	
	1100	34.000	-17.315	84.406	-110.162	21.890	
	1200	34.000	-13.915	87.364	-118.752	21.631	
	1300	34.000	-10.515	90.086	-127.627	21.459	
	1400	34.000	-7.115	92.605	-136.763	21.352	
	1500	34.000	-3.715	94.951	-146.142	21.296	

* * * * * * * * * *

Na_3As SODIUM ARSENIDE

		A	B	C	D	RANGE	REF.
SOL	CP	19.7	12.3			298-1000	*

PHASE	T	CP	H	S	G	BT	
SOL	298	23.367	-52.	33.	-61.839	45.335	*
	300	23.390	-51.957	33.145	-61.900	45.100	
	400	24.620	-49.556	40.042	-65.573	35.832	
	500	25.850	-47.033	45.668	-69.867	30.543	
	600	27.080	-44.386	50.490	-74.680	27.206	
	700	28.310	-41.617	54.756	-79.946	24.964	
	800	29.540	-38.724	58.617	-85.618	23.393	
	900	30.770	-35.709	62.167	-91.659	22.261	
	1000	32.000	-32.570	65.473	-98.043	21.430	

SODIUM ORTHOSILICATE $2Na_2O \cdot SiO_2$

		A	B	C	D	RANGE	REF.
SOL	CP	38.86	17.74			298-1393	5
LIQ	CP	62.0				1393-2000	5

PHASE	T	CP	H	S	G	BT	
SOL	298	44.149	-503.5	46.8	-517.453	379.355	5
	300	44.182	-503.418	47.073	-517.540	377.079	
	400	45.956	-498.911	60.027	-522.922	265.750	
	500	47.730	-494.227	70.472	-529.463	231.459	
	600	49.504	-489.365	79.331	-536.964	195.615	
	700	51.278	-484.326	87.095	-545.293	170.271	
	800	53.052	-479.110	94.058	-554.356	151.463	
	900	54.826	-473.716	100.409	-564.084	136.997	
	1000	56.600	-468.145	106.278	-574.422	125.557	
	1100	58.374	-462.396	111.755	-585.327	116.309	
	1200	60.148	-456.470	116.911	-596.763	108.700	
	1300	61.922	-450.366	121.795	-608.700	102.346	
	1393	63.572	-444.531	126.130	-620.230	97.322	
			13.8	9.907			
LIQ	1393	62.000	-430.731	136.037	-620.230	97.322	5
	1400	62.000	-430.297	136.347	-621.183	96.984	
	1500	62.000	-424.097	140.625	-635.034	92.537	
	1600	62.000	-417.897	144.626	-649.299	88.702	
	1700	62.000	-411.697	148.385	-663.951	85.368	
	1800	62.000	-405.497	151.929	-678.969	82.449	
	1900	62.000	-399.297	155.281	-694.331	79.877	
	2000	62.000	-393.097	158.461	-710.019	77.598	

$Na_2O \cdot 2SiO_2$ SODIUM DISILICATE

		A	B	C	D	RANGE	REF.
SOL-A	CP	44.38	16.86	-10.67		298- 951	3,5
SOL-B	CP	70.0				951- 980	3,5
SOL-C	CP	70.0				980-1147	3,5
LIQ	CP	62.43				1147-2000	3,5

PHASE	T	CP	H	S	G	BT	
SOL-A	298	37.404	-590.36	39.21	-602.050	441.374	3,5
	300	37.582	-590.291	39.442	-602.123	438.705	
	400	44.455	-586.152	51.302	-606.672	331.515	
	500	48.542	-581.489	61.691	-612.334	267.687	
	600	51.532	-576.479	70.816	-618.968	225.489	
	700	54.004	-571.199	78.950	-626.464	195.617	
	800	56.201	-565.687	86.307	-634.733	173.424	
	900	58.237	-559.964	93.045	-643.705	156.334	
	951	59.234	-556.969	96.283	-648.533	149.060	
			0.1	.105			3,5
SOL-B	951	70.000	-556.869	96.388	-648.533	149.060	
	980	70.000	-554.839	98.490	-651.359	145.279	
			0.15	.153			3,5
SOL-C	980	70.000	-554.689	98.643	-651.359	145.279	
	1000	70.000	-553.289	100.058	-653.346	142.808	
	1100	70.000	-546.289	106.729	-663.691	131.881	
	1147	70.000	-542.999	109.658	-668.776	127.446	
			8.5	7.411			3,5
LIQ	1147	62.430	-534.499	117.069	-668.776	127.446	
	1200	62.430	-531.190	119.889	-675.056	122.961	
	1300	62.430	-524.947	124.886	-687.298	115.561	
	1400	62.430	-518.704	129.512	-700.021	109.293	
	1500	62.430	-512.461	133.820	-713.190	103.926	
	1600	62.430	-506.218	137.849	-726.776	99.286	
	1700	62.430	-499.975	141.634	-740.752	95.243	
	1800	62.430	-493.732	145.202	-755.095	91.693	
	1900	62.430	-487.489	148.577	-769.786	88.557	
	2000	62.430	-481.246	151.780	-784.805	85.771	

SODIUM MONOTITANATE

$Na_2O \cdot TiO_2$

		A	B	C	D	RANGE	REF.
SOL-A	CP	25.18	20.729			298- 560	1
SOL-B	CP	25.95	17.00			560-1303	1
LIQ	CP	46.9				1303-2000	1

PHASE	T	CP	H	S	G	BT	
SOL-A	298	31.360	-376.7	29.1	-385.376	282.526	*/5
	300	31.399	-376.642	29.294	-385.430	280.823	
	400	33.472	-373.398	38.611	-388.843	212.482	
	500	35.544	-369.948	46.302	-393.099	171.847	
	560	36.788	-367.778	50.400	-396.002	154.567	
			0.4	.714			1
SOL-B	560	35.470	-367.378	51.114	-396.002	154.567	
	600	36.150	-365.945	53.585	-398.096	145.026	
	700	37.850	-362.245	59.285	-403.745	126.072	
	800	39.550	-358.375	64.450	-409.935	112.004	
	900	41.250	-354.335	69.206	-416.621	101.183	
	1000	42.950	-350.125	73.640	-423.766	92.626	
	1100	44.650	-345.745	77.814	-431.340	85.711	
	1200	46.350	-341.195	81.772	-439.321	80.022	
	1300	48.050	-336.475	85.549	-447.689	75.273	
	1303	48.101	-336.331	85.660	-447.945	75.143	
			16.8	12.893			1
LIQ	1303	46.900	-319.531	98.553	-447.945	75.143	
	1400	46.900	-314.982	101.920	-457.670	71.455	
	1500	46.900	-310.292	105.156	-468.026	68.201	
	1600	46.900	-305.602	108.183	-478.695	65.395	
	1700	46.900	-300.912	111.026	-489.657	62.958	
	1800	46.900	-296.222	113.707	-500.895	60.825	
	1900	46.900	-291.532	116.243	-512.393	58.947	
	2000	46.900	-286.842	118.649	-524.139	57.283	

$Na_2O \cdot 2TiO_2$ SODIUM DITITANATE

		A	B	C	D	RANGE	REF.
SOL	CP	49.32	7.06	-4.6		298-1258	1
LIQ	CP	68.5				1258-2000	1,*

PHASE	T	CP	H	S	G	BT	
SOL	298	46.250	-607.	41.5	-619.373	454.074	*/5
	300	46.327	-606.914	41.786	-619.450	451.330	
	400	49.269	-602.119	55.563	-624.344	341.171	
	500	51.010	-597.099	66.757	-630.477	275.618	
	600	52.278	-591.932	76.174	-637.636	232.290	
	700	53.323	-586.651	84.313	-645.670	201.614	
	800	54.249	-581.271	91.495	-654.467	178.816	
	900	55.106	-575.803	97.934	-663.944	161.249	
	1000	55.920	-570.251	103.783	-674.034	147.330	
	1100	56.706	-564.620	109.149	-684.684	136.053	
	1200	57.473	-558.911	114.117	-695.851	126.749	
	1258	57.911	-555.565	116.840	-702.549	122.069	
			26.2	20.827			1
LIQ	1258	68.500	-529.365	137.666	-702.549	122.069	
	1300	68.500	-526.488	139.916	-708.378	119.105	
	1400	68.500	-519.638	144.992	-722.627	112.822	
	1500	68.500	-512.788	149.718	-737.365	107.448	
	1600	68.500	-505.938	154.139	-752.560	102.809	
	1700	68.500	-499.088	158.292	-768.184	98.770	
	1800	68.500	-492.238	162.207	-784.211	95.229	
	1900	68.500	-485.388	165.911	-800.619	92.105	
	2000	68.500	-478.538	169.425	-817.387	89.332	

SODIUM TRITITANATE $Na_2O{\cdot}3TiO_2$

		A	B	C	D	RANGE	REF.
SOL	CP	63.46	10.64	-5.64		298-1401	58
LIQ	CP	94.15				1401-2000	58,*

PHASE	T	CP	H	S	G	BT	
SOL	298	60.288	-834.	55.9	-850.667	623.639	*/5
	300	60.385	-833.888	56.273	-850.770	619.869	
	400	64.191	-827.640	74.223	-857.329	468.486	
	500	66.524	-821.097	88.813	-865.504	378.362	
	600	68.277	-814.354	101.102	-875.015	318.767	
	700	69.757	-807.451	111.741	-885.669	276.556	
	800	71.091	-800.407	121.144	-897.323	245.170	
	900	72.340	-793.235	129.590	-909.866	220.975	
	1000	73.536	-785.941	137.274	-923.215	201.796	
	1100	74.698	-778.529	144.337	-937.300	186.249	
	1200	75.836	-771.002	150.886	-952.066	173.418	
	1300	76.958	-763.363	157.001	-967.463	162.667	
	1400	78.068	-755.611	162.744	-983.453	153.545	
	1401	78.079	-755.533	162.800	-983.616	153.460	
			37.1	26.481			58,*
LIQ	1401	94.150	-718.433	189.281	-983.616	153.460	
	1500	94.150	-709.112	195.710	-1002.677	146.110	
	1600	94.150	-699.697	201.786	-1022.555	139.693	
	1700	94.150	-690.282	207.494	-1043.022	134.108	
	1800	94.150	-680.867	212.875	-1064.043	129.210	
	1900	94.150	-671.452	217.966	-1085.587	124.888	
	2000	94.150	-662.037	222.795	-1107.627	121.052	

Na_3AlF_6 CRYOLITE

		A	B	C	D	RANGE	REF.
SOL-A	CP	41.174	37.871			298- 838	3
SOL-B	CP	67.4				838-1153	3
SOL-C	CP	85.				1153-1285	3
LIQ	CP	94.70				1285-2500	3

PHASE	T	CP	H	S	G	BT	
SOL-A	298	52.465	-791.	56.99	-807.992	592.353	3
	300	52.535	-790.903	57.315	-808.097	588.778	
	400	56.322	-785.460	72.947	-814.639	445.158	
	500	60.109	-779.638	85.922	-822.599	359.606	
	600	63.897	-773.438	97.216	-831.768	303.012	
	700	67.684	-766.859	107.350	-842.004	262.921	
	800	71.471	-759.901	116.635	-853.209	233.117	
	838	72.910	-757.158	119.985	-857.705	223.719	
			1.97	2.351			3
SOL-B	838	67.400	-755.188	122.336	-857.705	223.719	
	900	67.400	-751.009	127.146	-865.441	210.186	
	1000	67.400	-744.269	134.248	-878.517	192.026	
	1100	67.400	-737.529	140.672	-892.268	177.301	
	1153	67.400	-733.957	143.843	-899.808	170.581	
			0.09	.078			3
SOL-C	1153	85.000	-733.867	143.921	-899.808	170.581	
	1200	85.000	-729.872	147.317	-906.653	165.146	
	1285	85.000	-722.647	153.135	-919.425	156.395	
			25.64	19.953			3
LIQ	1285	94.700	-697.007	173.088	-919.425	156.395	
	1300	94.700	-695.587	174.187	-922.030	155.028	
	1400	94.700	-686.117	181.205	-939.804	146.730	
	1500	94.700	-676.647	187.739	-958.254	139.636	
	1600	94.700	-667.177	193.850	-977.337	133.516	
	1700	94.700	-657.707	199.592	-997.012	128.192	
	1800	94.700	-648.237	205.004	-1017.245	123.527	
	1900	94.700	-638.767	210.125	-1038.003	119.414	
	2000	94.700	-629.297	214.982	-1059.261	115.766	
	2100	94.700	-619.827	219.602	-1080.992	112.515	
	2200	94.700	-610.357	224.008	-1103.174	109.605	
	2300	94.700	-600.887	228.218	-1125.787	106.989	
	2400	94.700	-591.417	232.248	-1148.812	104.628	
	2500	94.700	-581.947	236.114	-1172.231	102.490	

JADEITE $Na_2O \cdot Al_2O_3 \cdot 4SiO_2$

		A	B	C	D	RANGE	REF.
SOL	CP	96.32	22.84	-23.74		298-1200	5

PHASE	T	CP	H	S	G	BT	
SOL	298	76.424	-1444.1	63.8	-1463.122	1072.642	5
	300	76.794	-1443.958	64.274	-1463.240	1066.113	
	400	90.619	-1435.505	88.497	-1470.904	803.773	
	500	98.244	-1426.032	109.604	-1480.834	647.359	
	600	103.430	-1415.936	127.998	-1492.734	543.801	
	700	107.463	-1405.384	144.255	-1506.363	470.371	
	800	110.883	-1394.463	158.833	-1521.530	415.718	
	900	113.945	-1383.219	172.073	-1538.085	373.548	
	1000	116.786	-1371.681	184.227	-1555.908	340.089	
	1100	119.482	-1359.867	195.485	-1574.900	312.946	
	1200	122.079	-1347.788	205.993	-1594.980	290.525	

* * * * * * * * * *

LOW ALBITE $Na_2O \cdot Al_2O_3 \cdot 6SiO_2$

		A	B	C	D	RANGE	REF.
SOL	CP	123.40	27.80	-30.02		298-1391	5
	** MPT= 1391						5

PHASE	T	CP	H	S	G	BT	
SOL	298	97.918	-1874.7	100.4	-1904.634	1396.322	5
	300	98.384	-1874.518	101.007	-1904.821	1387.847	
	400	115.757	-1863.707	131.991	-1916.503	1047.270	
	500	125.292	-1851.617	158.929	-1931.082	844.189	
	600	131.741	-1838.749	182.373	-1948.173	709.717	
	700	136.733	-1825.317	203.069	-1967.465	614.353	
	800	140.949	-1811.428	221.609	-1988.715	543.365	
	900	144.714	-1797.142	238.431	-2011.730	488.580	
	1000	148.198	-1782.494	253.861	-2036.355	445.105	
	1100	151.499	-1767.508	268.141	-2062.464	409.829	
	1200	154.675	-1752.198	281.460	-2089.951	380.683	
	1300	157.764	-1736.576	293.964	-2118.728	356.238	
	1391	160.518	-1722.094	304.730	-2145.973	337.214	

$Na_2O \cdot Al_2O_3 \cdot 6SiO_2$ HIGH ALBITE

		A	B	C	D	RANGE	REF.
SOL-A	CP	135.18	19.52	-41.06		298- 973	5
SOL-B	CP	135.18	19.52	-41.06		973-1391	5

PHASE	T	CP	H	S	G	BT	
SOL-A	298	94.610	-1869.5	109.4	-1902.118	1394.477	5
	300	95.414	-1869.324	109.988	-1902.321	1386.026	
	400	117.325	-1858.545	140.849	-1914.884	1046.385	
	500	128.516	-1846.201	168.347	-1930.374	843.880	
	600	135.486	-1832.978	192.436	-1948.440	709.814	
	700	140.464	-1819.169	213.713	-1968.768	614.760	
	800	144.380	-1804.920	232.734	-1991.107	544.018	
	900	147.679	-1790.313	249.934	-2015.254	489.436	
	973	149.836	-1779.453	261.536	-2033.927	456.911	
			3.4	3.494			5
SOL-B	973	149.836	-1776.053	265.030	-2033.927	456.911	
	1000	150.594	-1771.997	269.142	-2041.139	446.151	
	1100	153.259	-1756.803	283.621	-2068.786	411.085	
	1200	155.753	-1741.351	297.065	-2097.828	382.118	
	1300	158.126	-1725.656	309.626	-2128.170	357.826	
	1391	160.210	-1711.171	320.394	-2156.840	338.922	

SODIUM 2-CADMIUM $NaCd_2$

		A	B	C	D	RANGE	REF.
SOL	CP	17.1	8.1			298- 655	*
LIQ	CP	21.3				655- 900	*

PHASE	T	CP	H	S	G	BT	
SOL	298	19.515	-7.	32.8	-16.779	12.301	5,*
	300	19.530	-6.964	32.921	-16.840	12.270	
	400	20.340	-4.970	38.650	-20.430	11.164	
	500	21.150	-2.896	43.276	-24.534	10.725	
	600	21.960	-.740	47.204	-29.063	10.587	
	655	22.405	.480	49.149	-31.713	10.583	
			7.4	11.298			5*
LIQ	655	21.300	7.880	60.447	-31.713	10.583	
	700	21.300	8.838	61.862	-34.465	10.762	
	800	21.300	10.968	64.706	-40.797	11.147	
	900	21.300	13.098	67.215	-47.395	11.511	

* * * * * * * * * *

SODIUM CHROMITE $Na_2O \cdot Cr_2O_3$

		A	B	C	D	RANGE	REF.
SOL	CP	45.2	7.2	-4.10		298-1100	*

PHASE	T	CP	H	S	G	BT	
SOL	298	42.734	-418.4	38.8	-429.968	315.218	5/*
	300	42.804	-418.321	39.065	-430.040	313.326	
	400	45.517	-413.891	51.791	-434.607	237.490	
	500	47.160	-409.252	62.136	-440.320	192.489	
	600	48.381	-404.472	70.846	-446.980	162.834	
	700	49.403	-399.582	78.383	-454.450	141.905	
	800	50.319	-394.595	85.041	-462.628	126.401	
	900	51.174	-389.520	91.017	-471.435	114.496	
	1000	51.990	-384.362	96.451	-480.813	105.096	
	1100	52.781	-379.123	101.444	-490.711	97.508	

Na_2CrO_4 SODIUM CHROMATE

		A	B	C	D	RANGE	REF.
SOL-A	CP	24.15	33.46			298- 694	86
SOL-B	CP	35.84	12.33			694-1070	86
LIQ	CP	48.9				1070-1600	86

PHASE	T	CP	H	S	G	BT	
SOL-A	298	34.126	-317.6	44.4	-330.838	242.543	172/*
	300	34.188	-317.537	44.611	-330.920	241.108	
	400	37.534	-313.951	54.905	-335.913	183.559	
	500	40.880	-310.030	63.640	-341.850	149.443	
	600	44.226	-305.775	71.389	-348.608	126.997	
	694	47.371	-301.470	78.049	-355.636	112.009	
			2.29	3.300			86
SOL-B	694	44.397	-299.180	81.349	-355.636	112.009	
	700	44.471	-298.913	81.731	-356.125	111.202	
	800	45.704	-294.404	87.750	-364.604	99.619	
	900	46.937	-289.772	93.204	-373.656	90.748	
	1000	48.170	-285.017	98.213	-383.230	83.766	
	1070	49.033	-281.615	101.501	-390.221	79.714	
			5.79	5.411			86
LIQ	1070	48.900	-275.825	106.913	-390.221	79.714	
	1100	48.900	-274.358	108.265	-393.449	78.182	
	1200	48.900	-269.468	112.520	-404.491	73.678	
	1300	48.900	-264.578	116.434	-415.941	69.936	
	1400	48.900	-259.688	120.058	-427.768	66.787	
	1500	48.900	-254.798	123.431	-439.945	64.109	
	1600	48.900	-249.908	126.587	-452.447	61.810	

SODIUM MOLYBDATE $Na_2O{\cdot}MoO_3$

		A	B	C	D	RANGE	REF.
SOL-A	CP	29.958	18.78			298- 718	86
SOL-B	CP	-51.493	121.04			718- 866	86
SOL-C	CP	-140.97	213.06			866- 915	86
SOL-D	CP	264.2	-228.			915- 962	86
LIQ	CP	50.9				962-1600	86,*

PHASE	T	CP	H	S	G	BT	
SOL-A	298	35.557	-350.4	38.1	-361.760	265.213	*/5
	300	35.592	-350.334	38.320	-361.830	263.629	
	400	37.470	-346.681	48.816	-366.208	200.113	
	500	39.348	-342.840	57.379	-371.530	162.417	
	600	41.226	-338.811	64.719	-377.643	137.575	
	700	43.104	-334.595	71.215	-384.446	120.046	
	718	43.442	-333.816	72.314	-385.738	117.429	
			5.2	7.242			86
SOL-B	718	35.414	-328.616	79.556	-385.738	117.429	
	800	45.339	-325.305	83.913	-392.436	107.223	
	866	53.328	-322.049	87.820	-398.101	100.481	
			0.5	.577			86
SOL-C	866	43.540	-321.549	88.397	-398.101	100.481	
	900	50.784	-319.946	90.212	-401.137	97.422	
	915	53.980	-319.160	91.078	-402.497	96.150	
			1.98	2.164			86
SOL-D	915	55.580	-317.180	93.242	-402.497	96.150	
	962	44.864	-314.820	95.760	-406.941	92.462	
			5.12	5.322			86
LIQ	962	50.900	-309.700	101.082	-406.941	92.462	
	1000	50.900	-307.765	103.054	-410.819	89.797	
	1100	50.900	-302.675	107.905	-421.371	83.730	
	1200	50.900	-297.585	112.334	-432.387	78.759	
	1300	50.900	-292.495	116.408	-443.826	74.624	
	1400	50.900	-287.405	120.181	-455.658	71.141	
	1500	50.900	-282.315	123.692	-467.854	68.175	
	1600	50.900	-277.225	126.977	-480.389	65.627	

NbF_5 NIOBIUM PENTAFLUORICE

		A	B	C	D	RANGE	REF.
SOL	CP	38.0				298- 352	*
LIQ	CP	42.5				352- 506	*
GAS	CP	31.23	0.26	-5.9		506-1500	*

PHASE	T	CP	H	S	G	BT	P	
SOL	298	38.000	-433.5	38.3	-444.919	326.178	.019	65
	300	38.000	-433.430	38.535	-444.990	324.219	.023	
	352	38.000	-431.454	44.609	-447.156	277.668	1.944	
			2.92	8.295				65
LIQ	352	42.500	-428.534	52.905	-447.156	277.668	1.944	
	400	42.500	-426.494	58.338	-449.829	245.808	22.886	
	500	42.500	-422.244	67.821	-456.154	199.412	654.737	
	506	42.500	-421.989	68.328	-456.563	197.224	760.	
			12.5	24.704				1
GAS	506	29.057	-409.489	93.032	-456.563	197.224		
	600	29.747	-406.722	98.045	-465.549	169.599		
	700	30.208	-403.723	102.668	-475.590	148.506		
	800	30.516	-400.686	106.723	-486.064	132.804		
	900	30.736	-397.623	110.330	-496.920	120.685		
	1000	30.900	-394.540	113.578	-508.118	111.064		
	1100	31.028	-391.444	116.529	-519.625	103.254		
	1200	31.132	-388.336	119.233	-531.415	96.797		
	1300	31.219	-385.218	121.729	-543.465	91.377		
	1400	31.293	-382.092	124.045	-555.755	86.769		
	1500	31.358	-378.960	126.206	-568.269	82.808		

NIOBIUM 2.33-CHLORICE $NbCl_{2.33}$

		A	B	C	D	RANGE	REF.
SOL	CP	19.5	3.6	-1.4		298- 800	*

PHASE	T	CP	H	S	G	BT	
SOL	298	18.998	-113.5	31.2	-122.802	90.029	65/*
	300	19.024	-113.465	31.318	-122.860	89.516	
	400	20.065	-111.505	36.947	-126.284	69.008	
	500	20.740	-109.463	41.501	-130.214	56.924	
	600	21.271	-107.362	45.331	-134.561	49.020	
	700	21.734	-105.211	48.645	-139.263	43.486	
	800	22.161	-103.016	51.575	-144.277	39.420	

* * * * * * * * * *

NIOBIUM PENTABROMICE $NbBr_5$

		A	B	C	D	RANGE	REF.
SOL	CP	27.91	31.2			298- 500	32
LIQ	CP	44.				500- 545	*
	** BPT= 545, L= 12.						1

PHASE	T	CP	H	S	G	BT	
SOL	298	37.212	-131.53	73.	-153.295	112.383	32
	300	37.270	-131.461	73.230	-153.430	111.789	
	400	40.390	-127.578	84.380	-161.330	88.158	
	500	43.510	-123.383	93.728	-170.247	74.425	
			8.5	17.000			*
LIQ	500	44.000	-114.883	110.728	-170.247	74.425	
	545	44.000	-112.903	114.519	-175.316	70.313	

NbJ_5 NIOBIUM PENTAIODIDE

		A	B	C	D	RANGE	REF.
SOL	CP	30.98	20.87			298- 600	32
LIQ	CP	44.				600- 620	32
	** BPT= 620, L= 14						32

PHASE	T	CP	H	S	G	BT	
SOL	298	37.202	-64.7	82.	-89.148	65.356	32
	300	37.241	-64.631	82.230	-89.300	65.064	
	400	39.328	-60.803	93.230	-98.095	53.604	
	500	41.415	-56.766	102.230	-107.880	47.161	
	600	43.502	-52.520	109.965	-118.499	43.169	
			9.	15.000			32
LIQ	600	44.000	-43.520	124.965	-118.499	43.169	
	620	44.000	-42.640	126.408	-121.012	42.663	

* * * * * * * * * *

Nb_5Si_3 5-NIOBIUM 3-SILICON

		A	B	C	D	RANGE	REF.
SOL	CP	45.21	7.357	-3.604		298-2000	5

PHASE	T	CP	H	S	G	BT	
SOL	298	43.349	-108.	60.	-125.889	92.292	5/*
	300	43.413	-107.920	60.268	-126.000	91.803	
	400	45.900	-103.442	73.134	-132.695	72.511	
	500	47.447	-98.770	83.553	-140.546	61.441	
	600	48.623	-93.964	92.311	-149.351	54.408	
	700	49.624	-89.051	99.883	-158.969	49.639	
	800	50.532	-84.042	106.569	-169.298	46.256	
	900	51.386	-78.946	112.571	-180.260	43.779	
	1000	52.207	-73.766	118.028	-191.794	41.922	
	1100	53.005	-68.506	123.041	-203.851	40.507	
	1200	53.788	-63.166	127.687	-216.390	39.415	
	1300	54.561	-57.748	132.023	-229.378	38.567	
	1400	55.326	-52.254	136.094	-242.786	37.906	
	1500	56.085	-46.683	139.937	-256.589	37.390	
	1600	56.840	-41.037	143.581	-270.767	36.990	
	1700	57.592	-35.315	147.050	-285.300	36.683	
	1800	58.341	-29.519	150.363	-300.171	36.451	
	1900	59.088	-23.647	153.537	-315.367	36.280	
	2000	59.834	-17.701	156.587	-330.875	36.161	

NIOBIUM 2-SILICON

$NbSi_2$

		A	B	C	D	RANGE	REF.
SOL	CP	15.10	3.67	-0.668		298-2000	5

PHASE	T	CP	H	S	G	BT	
SOL	298	15.443	-33.	16.7	-37.979	27.843	5/*
	300	15.459	-32.971	16.796	-38.010	27.694	
	400	16.150	-31.389	21.344	-39.926	21.818	
	500	16.668	-29.747	25.006	-42.250	18.470	
	600	17.116	-28.057	28.085	-44.908	16.360	
	700	17.533	-26.325	30.755	-47.853	14.942	
	800	17.932	-24.551	33.122	-51.049	13.948	
	900	18.321	-22.739	35.257	-54.470	13.229	
	1000	18.703	-20.887	37.207	-58.094	12.698	
	1100	19.082	-18.998	39.007	-61.906	12.301	
	1200	19.458	-17.071	40.684	-65.892	12.002	
	1300	19.831	-15.107	42.256	-70.039	11.776	
	1400	20.204	-13.105	43.739	-74.340	11.607	
	1500	20.575	-11.066	45.146	-78.785	11.480	
	1600	20.946	-8.990	46.486	-83.367	11.389	
	1700	21.316	-6.877	47.766	-88.080	11.325	
	1800	21.685	-4.727	48.995	-92.918	11.283	
	1900	22.054	-2.540	50.178	-97.877	11.260	
	2000	22.423	-.316	51.318	-102.952	11.252	

NbB_2 NIOBIUM DIBORIDE

		A	B	C	D	RANGE	REF.
SOL	CP	11.23	9.21	-2.25		298-1400	195

PHASE	T	CP	H	S	G	BT	
SOL	298	11.445	-60.	9.	-62.683	45.954	190
	300	11.493	-59.979	9.071	-62.700	45.683	
	400	13.508	-58.721	12.676	-63.791	34.859	
	500	14.935	-57.296	15.850	-65.221	28.512	
	600	16.131	-55.741	18.680	-66.950	24.390	
	700	17.218	-54.073	21.250	-68.948	21.529	
	800	18.246	-52.300	23.616	-71.193	19.452	
	900	19.241	-50.425	25.823	-73.666	17.891	
	1000	20.215	-48.452	27.901	-76.353	16.689	
	1100	21.175	-46.383	29.873	-79.243	15.746	
	1200	22.126	-44.218	31.756	-82.325	14.995	
	1300	23.070	-41.958	33.564	-85.592	14.391	
	1400	24.009	-39.604	35.309	-89.036	13.901	

NIOBIUM 2-CHROMIUM

$NbCr_2$

		A	B	C	D	RANGE	REF.
SOL	CP	17.75	5.68	-1.76		298-1500	141

PHASE	T	CP	H	S	G	BT	
SOL	298	17.464	-5.	20.	-10.963	8.037	141
	300	17.498	-4.968	20.108	-11.000	8.015	
	400	18.922	-3.141	25.355	-13.282	7.258	
	500	19.886	-1.198	29.686	-16.041	7.012	
	600	20.669	.831	33.382	-19.198	6.994	
	700	21.367	2.933	36.621	-22.702	7.089	
	800	22.019	5.103	39.518	-26.511	7.244	
	900	22.645	7.336	42.147	-30.597	7.431	
	1000	23.254	9.631	44.565	-34.934	7.636	
	1100	23.853	11.986	46.809	-39.504	7.850	
	1200	24.444	14.401	48.910	-44.291	8.068	
	1300	25.030	16.875	50.890	-49.282	8.286	
	1400	25.612	19.407	52.766	-54.465	8.504	
	1500	26.192	21.997	54.553	-59.832	8.719	

* * * * * * * * * *

NIOBIUM 2-IRON

$NbFe_2$

		A	B	C	D	RANGE	REF.
SOL	CP	16.0	6.3			298-1300	*
	** NBFE(1.7-2.3), MPT(NBFE2)= 1928						21

PHASE	T	CP	H	S	G	BT	
SOL	298	17.878	-14.7	18.	-20.067	14.711	65
	300	17.890	-14.667	18.111	-20.100	14.645	
	400	18.520	-12.846	23.344	-22.184	12.122	
	500	19.150	-10.963	27.544	-24.735	10.813	
	600	19.780	-9.016	31.091	-27.671	10.081	
	700	20.410	-7.007	34.187	-30.938	9.661	
	800	21.040	-4.934	36.954	-34.498	9.426	
	900	21.670	-2.799	39.468	-38.320	9.307	
	1000	22.300	-.600	41.784	-42.385	9.264	
	1100	22.930	1.661	43.939	-46.672	9.274	
	1200	23.560	3.986	45.961	-51.168	9.320	
	1300	24.190	6.373	47.872	-55.861	9.392	

Nd

NEODYMIUM

		A	B	C	D	RANGE	REF.
SOL-A	CP	6.171	0.523	-0.063	3.333	298-1128	21
SOL-B	CP	10.65				1128-1289	21
LIQ	CP	11.66				1289-2800	21
	** BPT= 3341, L= 65.249						21

PHASE	T	CP	H	S	G	BT	P	
SOL-A	298	6.552	0	16.99	-5.066	3.714		21
	300	6.558	.012	17.031	-5.097	3.714		
	400	6.874	.683	18.959	-6.900	3.771		
	500	7.241	1.389	20.532	-8.877	3.881		
	600	7.667	2.134	21.889	-11.000	4.007		
	700	8.157	2.924	23.106	-13.250	4.137		
	800	8.713	3.767	24.231	-15.618	4.267		
	900	9.334	4.669	25.293	-18.094	4.395		
	1000	10.021	5.636	26.311	-20.675	4.519		
	1100	10.774	6.675	27.301	-23.356	4.641		
	1128	10.997	6.980	27.575	-24.124	4.675		
			0.724	.642				21
SOL-B	1128	10.650	7.704	28.216	-24.124	4.675		
	1200	10.650	8.471	28.875	-26.180	4.769		
	1289	10.650	9.419	29.637	-28.784	4.881		
			1.707	1.324				21
LIQ	1289	11.660	11.126	30.962	-28.784	4.881		
	1300	11.660	11.254	31.061	-29.125	4.897		
	1400	11.660	12.420	31.925	-32.275	5.039		
	1500	11.660	13.586	32.729	-35.508	5.174	.002	
	1600	11.660	14.752	33.482	-38.819	5.303	.009	
	1700	11.660	15.918	34.189	-42.203	5.426	.032	
	1800	11.660	17.084	34.855	-45.655	5.544	.105	
	1900	11.660	18.250	35.485	-49.172	5.657	.300	
	2000	11.660	19.416	36.084	-52.751	5.765	.766	
	2100	11.660	20.582	36.652	-56.388	5.869	1.778	
	2200	11.660	21.748	37.195	-60.081	5.969	3.806	
	2300	11.660	22.914	37.713	-63.826	6.066	7.596	
	2400	11.660	24.080	38.209	-67.623	6.159	14.259	
	2500	11.660	25.246	38.685	-71.468	6.249	25.364	
	2600	11.660	26.412	39.143	-75.359	6.335	43.029	
	2700	11.660	27.578	39.583	-79.296	6.419	69.997	
	2800	11.660	28.744	40.007	-83.275	6.501	109.691	

NEODYMIUM DIHYDRIDE

NdH_2

		A	B	C	D	RANGE	REF.
SOL	CP	9.14	3.85			298-1200	*

PHASE	T	CP	H	S	G	BT	
SOL	298	10.288	-46.	14.08	-50.198	36.801	19/76
	300	10.295	-45.981	14.144	-50.224	36.593	
	400	10.680	-44.932	17.158	-51.795	28.304	
	500	11.065	-43.845	19.583	-53.636	23.448	
	600	11.450	-42.719	21.634	-55.700	20.291	
	700	11.835	-41.555	23.428	-57.955	18.097	
	800	12.220	-40.352	25.033	-60.379	16.497	
	900	12.605	-39.111	26.495	-62.956	15.290	
	1000	12.990	-37.831	27.843	-65.674	14.355	
	1100	13.375	-36.513	29.099	-68.522	13.616	
	1200	13.760	-35.156	30.279	-71.491	13.022	

NdF_3 NEODYMIUM FLUORIDE

		A	B	C	D	RANGE	REF.
SOL	CP	17.92	8.75	4.85		298-1650	174
LIQ	CP	44.09	-1.06	-28.5		1650-1900	174,*

PHASE	T	CP	H	S	G	BT	
SOL	298	25.985	-399.	27.6	-407.229	298.547	201/*
	300	25.934	-398.952	27.761	-407.280	296.743	
	400	24.451	-396.450	34.970	-410.437	224.283	
	500	24.235	-394.021	40.389	-414.216	181.078	
	600	24.517	-391.586	44.828	-418.483	152.453	
	700	25.035	-389.110	48.644	-423.161	132.135	
	800	25.678	-386.575	52.028	-428.197	116.994	
	900	26.394	-383.972	55.093	-433.556	105.296	
	1000	27.155	-381.295	57.913	-439.208	96.002	
	1100	27.946	-378.540	60.538	-445.132	88.451	
	1200	28.757	-375.705	63.004	-451.310	82.206	
	1300	29.582	-372.788	65.338	-457.728	76.961	
	1400	30.417	-369.788	67.561	-464.374	72.502	
	1500	31.261	-366.705	69.688	-471.237	68.668	
	1600	32.109	-363.536	71.733	-478.309	65.343	
	1650	32.536	-361.920	72.728	-481.920	63.841	
			13.075	7.924			174
LIQ	1650	41.294	-348.845	80.652	-481.920	63.841	
	1700	41.302	-346.780	81.885	-485.984	62.486	
	1800	41.302	-342.650	84.246	-494.292	60.023	
	1900	41.287	-338.520	86.478	-502.829	57.846	

NEODYMIUM CHLORIDE

$NdCl_3$

		A	B	C	D	RANGE	REF.
SOL	CP	18.65	14.6			298-1032	89,*
LIQ	CP	35.0				1032-1947	89,*
	** BPT=	1947					1

PHASE	T	CP	H	S	G	BT	
SOL	298	23.003	-248.8	34.4	-259.056	189.919	19/*
	300	23.030	-248.757	34.542	-259.120	188.794	
	400	24.490	-246.381	41.368	-262.928	143.677	
	500	25.950	-243.859	46.989	-267.354	116.876	
	600	27.410	-241.191	51.850	-272.301	99.199	
	700	28.870	-238.377	56.184	-277.707	86.716	
	800	30.330	-235.417	60.135	-283.525	77.466	
	900	31.790	-232.311	63.791	-289.724	70.364	
	1000	33.250	-229.059	67.216	-296.276	64.760	
	1032	33.717	-227.988	68.271	-298.444	63.211	
			11.6	11.240			89
LIQ	1032	35.000	-216.388	79.511	-298.444	63.211	
	1100	35.000	-214.008	81.745	-303.927	60.393	
	1200	35.000	-210.508	84.790	-312.256	56.877	
	1300	35.000	-207.008	87.592	-320.877	53.952	
	1400	35.000	-203.508	90.186	-329.768	51.486	
	1500	35.000	-200.008	92.600	-338.908	49.386	
	1600	35.000	-196.508	94.859	-348.283	47.580	
	1700	35.000	-193.008	96.981	-357.876	46.014	
	1800	35.000	-189.508	98.982	-367.675	44.648	
	1900	35.000	-186.008	100.874	-377.668	43.448	
	1947	35.000	-184.363	101.729	-382.430	42.933	

NdJ_3 NEODYMIUM IODIDE

		A	B	C	D	RANGE	REF.
SOL-A	CP	21.715	8.56	-0.148		298- 847	89
SOL-B	CP	28.057				847-1060	89
LIQ	CP	37.224				1060-1710	89
	** BPT=	1710					1

PHASE	T	CP	H	S	G	BT	
SOL-A	298	24.101	-150.2	51.4	-165.525	121.349	1/*
	300	24.119	-150.155	51.549	-165.620	120.670	
	400	25.046	-147.697	58.616	-171.143	93.521	
	500	25.936	-145.147	64.301	-177.298	77.507	
	600	26.810	-142.510	69.107	-183.974	67.022	
	700	27.677	-139.786	73.305	-191.099	59.672	
	800	28.540	-136.975	77.057	-198.620	54.268	
	847	28.945	-135.624	78.698	-202.281	52.201	
			3.3	3.896			89
SOL-B	847	28.057	-132.324	82.594	-202.281	52.201	
	900	28.057	-130.837	84.297	-206.704	50.201	
	1000	28.057	-128.031	87.253	-215.284	47.057	
	1060	28.057	-126.348	88.888	-220.569	45.483	
			9.9	9.340			89
LIQ	1060	37.224	-116.448	98.228	-220.569	45.483	
	1100	37.224	-114.959	99.606	-224.526	44.615	
	1200	37.224	-111.236	102.845	-234.651	42.741	
	1300	37.224	-107.514	105.825	-245.086	41.208	
	1400	37.224	-103.792	108.583	-255.808	39.939	
	1500	37.224	-100.069	111.152	-266.796	38.877	
	1600	37.224	-96.347	113.554	-278.033	37.983	
	1700	37.224	-92.624	115.811	-289.502	37.223	
	1710	37.224	-92.252	116.029	-290.662	37.154	

NEODYMIUM OXYCHLORIDE — NdOCl

		A	B	C	D	RANGE	REF.
SOL	CP	16.4	4.57	-0.95		298-1100	*

PHASE	T	CP	H	S	G	BT	
SOL	298	16.694	-241.7	22.6	-248.438	182.135	100/*
	300	16.715	-241.669	22.703	-248.480	181.042	
	400	17.634	-239.948	27.647	-251.007	137.162	
	500	18.305	-238.150	31.657	-253.979	111.029	
	600	18.878	-236.290	35.046	-257.318	93.741	
	700	19.405	-234.376	37.996	-260.973	81.491	
	800	19.908	-232.410	40.620	-264.907	72.379	
	900	20.396	-230.395	42.993	-269.089	65.353	
	1000	20.875	-228.331	45.167	-273.499	59.781	
	1100	21.348	-226.220	47.179	-278.117	55.264	

* * * * * * * * * *

NEODYMIUM MONOSULFIDE — NdS

		A	B	C	D	RANGE	REF.
SOL	CP	11.04	2.0			298-2000	63
	** MPT=	2400					63

PHASE	T	CP	H	S	G	BT	
SOL	298	11.636	-108.	18.6	-113.546	83.242	63
	300	11.640	-107.978	18.672	-113.580	82.754	
	400	11.840	-106.804	22.048	-115.624	63.182	
	500	12.040	-105.610	24.712	-117.966	51.570	
	600	12.240	-104.396	26.924	-120.551	43.917	
	700	12.440	-103.162	28.826	-123.341	38.514	
	800	12.640	-101.908	30.500	-126.309	34.511	
	900	12.840	-100.634	32.001	-129.435	31.435	
	1000	13.040	-99.340	33.364	-132.704	29.006	
	1100	13.240	-98.026	34.616	-136.104	27.045	
	1200	13.440	-96.692	35.777	-139.624	25.433	
	1300	13.640	-95.338	36.860	-143.257	24.087	
	1400	13.840	-93.964	37.879	-146.994	22.950	
	1500	14.040	-92.570	38.840	-150.831	21.979	
	1600	14.240	-91.156	39.753	-154.761	21.142	
	1700	14.440	-89.722	40.622	-158.780	20.415	
	1800	14.640	-88.268	41.453	-162.884	19.779	
	1900	14.840	-86.794	42.250	-167.069	19.220	
	2000	15.040	-85.300	43.016	-171.333	18.725	

Nd_2S_3 NEODYMIUM SULFIDE

		A	B	C	D	RANGE	REF.
SOL	CP	28.33	3.19			298-2000	63
	** MPT= 2480						63

PHASE	T	CP	H	S	G	BT	
SOL	298	29.281	-276.	44.28	-289.202	212.019	63
	300	29.287	-275.946	44.461	-289.284	210.772	
	400	29.606	-273.001	52.930	-294.173	160.750	
	500	29.925	-270.025	59.571	-299.810	131.064	
	600	30.244	-267.016	65.055	-306.049	111.493	
	700	30.563	-263.976	69.741	-312.795	97.672	
	800	30.882	-260.904	73.843	-319.978	87.426	
	900	31.201	-257.799	77.499	-327.548	79.550	
	1000	31.520	-254.663	80.803	-335.466	73.326	
	1100	31.839	-251.495	83.822	-343.699	68.296	
	1200	32.158	-248.296	86.606	-352.223	64.157	
	1300	32.477	-245.064	89.192	-361.014	60.700	
	1400	32.796	-241.800	91.611	-370.056	57.776	
	1500	33.115	-238.505	93.885	-379.331	55.276	
	1600	33.434	-235.177	96.032	-388.828	53.119	
	1700	33.753	-231.818	98.068	-398.534	51.242	
	1800	34.072	-228.427	100.007	-408.439	49.598	
	1900	34.391	-225.003	101.857	-418.533	48.149	
	2000	34.710	-221.548	103.630	-428.807	46.864	

* * * * * * * * * *

$Nd_2(SO_4)_3$ NEODYMIUM SULFATE

		A	B	C	D	RANGE	REF.
SOL	CP	51.0	47.5			298-1100	*

PHASE	T	CP	H	S	G	BT	
SOL	298	65.162	-932.	68.9	-952.543	698.326	1/*
	300	65.250	-931.879	69.303	-952.670	694.113	
	400	70.000	-925.117	88.725	-960.607	524.922	
	500	74.750	-917.879	104.855	-970.307	424.178	
	600	79.500	-910.167	118.904	-981.509	357.563	
	700	84.250	-901.979	131.516	-994.040	310.395	
	800	89.000	-893.317	143.076	-1007.777	275.349	
	900	93.750	-884.179	153.833	-1022.629	248.362	
	1000	98.500	-874.567	163.956	-1038.523	227.000	
	1100	103.250	-864.479	173.567	-1055.403	209.717	

NEODYMIUM MONOSELENIDE

NdSe

		A	B	C	D	RANGE	REF.
SOL	CP	11.35	1.50			298-2000	63
	** MPT=	2100					63

PHASE	T	CP	H	S	G	BT	
SOL	298	11.797	-86.	21.6	-92.440	67.769	63
	300	11.800	-85.978	21.673	-92.480	67.381	
	400	11.950	-84.791	25.088	-94.826	51.817	
	500	12.100	-83.588	27.771	-97.474	42.611	
	600	12.250	-82.371	29.990	-100.365	36.563	
	700	12.400	-81.138	31.890	-103.461	32.306	
	800	12.550	-79.891	33.555	-106.735	29.163	
	900	12.700	-78.628	35.042	-110.166	26.756	
	1000	12.850	-77.351	36.388	-113.739	24.861	
	1100	13.000	-76.058	37.620	-117.440	23.336	
	1200	13.150	-74.751	38.757	-121.260	22.087	
	1300	13.300	-73.428	39.816	-125.189	21.049	
	1400	13.450	-72.091	40.807	-129.221	20.175	
	1500	13.600	-70.738	41.740	-133.348	19.431	
	1600	13.750	-69.371	42.623	-137.567	18.793	
	1700	13.900	-67.988	43.461	-141.871	18.241	
	1800	14.050	-66.591	44.259	-146.258	17.760	
	1900	14.200	-65.178	45.023	-150.722	17.339	
	2000	14.350	-63.751	45.755	-155.261	16.968	

Nd_2Se_3 NEODYMIUM SELENIDE

		A	B	C	D	RANGE	REF.
SOL	CP	30.11	3.35			298-1830	63
	** MPT=	1830					63

PHASE	T	CP	H	S	G	BT	
SOL	298	31.109	-225.	44.28	-238.202	174.630	63
	300	31.115	-224.942	44.472	-238.284	173.613	
	400	31.450	-221.814	53.470	-243.202	132.897	
	500	31.785	-218.652	60.523	-248.914	108.815	
	600	32.120	-215.457	66.348	-255.266	92.993	
	700	32.455	-212.228	71.325	-262.156	81.860	
	800	32.790	-208.966	75.680	-269.510	73.637	
	900	33.125	-205.670	79.562	-277.276	67.341	
	1000	33.460	-202.341	83.069	-285.410	62.385	
	1100	33.795	-198.978	86.274	-293.880	58.396	
	1200	34.130	-195.582	89.229	-302.657	55.129	
	1300	34.465	-192.152	91.974	-311.718	52.412	
	1400	34.800	-188.689	94.540	-321.046	50.124	
	1500	35.135	-185.192	96.953	-330.621	48.178	
	1600	35.470	-181.662	99.231	-340.432	46.507	
	1700	35.805	-178.098	101.391	-350.464	45.061	
	1800	36.140	-174.501	103.447	-360.706	43.802	
	1830	36.240	-173.415	104.046	-363.819	43.455	

NEODYMIUM MONOTELLURIDE

NdTe

		A	B	C	D	RANGE	REF.
SOL	CP	11.56	1.46			298-2028	63
	** MPT= 2028						63

PHASE	T	CP	H	S	G	BT	
SOL	298	11.995	-72.	23.3	-78.947	57.877	63
	300	11.998	-71.978	23.374	-78.990	57.552	
	400	12.144	-70.771	26.846	-81.509	44.540	
	500	12.290	-69.549	29.571	-84.335	36.868	
	600	12.436	-68.313	31.825	-87.408	31.843	
	700	12.582	-67.062	33.753	-90.689	28.318	
	800	12.728	-65.796	35.443	-94.150	25.724	
	900	12.874	-64.516	36.950	-97.771	23.745	
	1000	13.020	-63.222	38.314	-101.536	22.194	
	1100	13.166	-61.912	39.562	-105.430	20.950	
	1200	13.312	-60.588	40.714	-109.445	19.935	
	1300	13.458	-59.250	41.785	-113.570	19.095	
	1400	13.604	-57.897	42.788	-117.800	18.392	
	1500	13.750	-56.529	43.731	-122.126	17.796	
	1600	13.896	-55.147	44.623	-126.544	17.287	
	1700	14.042	-53.750	45.470	-131.049	16.850	
	1800	14.188	-52.338	46.277	-135.637	16.471	
	1900	14.334	-50.912	47.048	-140.303	16.141	
	2000	14.480	-49.472	47.787	-145.045	15.852	
	2028	14.521	-49.065	47.989	-146.386	15.778	

Nd_2Te_3 NEODYMIUM TELLURIDE

		A	B	C	D	RANGE	REF.
SOL	CP	30.73	3.34			298-1650	63
	** MPT= 1650						63

PHASE	T	CP	H	S	G	BT	
SOL	298	31.726	-190.	60.56	-208.056	152.530	63
	300	31.732	-189.941	60.756	-208.168	151.671	
	400	32.066	-186.751	69.931	-214.724	117.335	
	500	32.400	-183.528	77.122	-222.089	97.088	
	600	32.734	-180.271	83.059	-230.107	83.828	
	700	33.068	-176.981	88.130	-238.672	74.527	
	800	33.402	-173.658	92.567	-247.712	67.681	
	900	33.736	-170.301	96.521	-257.169	62.458	
	1000	34.070	-166.911	100.092	-267.003	58.361	
	1100	34.404	-163.487	103.355	-277.178	55.078	
	1200	34.738	-160.030	106.363	-287.666	52.398	
	1300	35.072	-156.539	109.157	-298.443	50.180	
	1400	35.406	-153.015	111.768	-309.491	48.320	
	1500	35.740	-149.458	114.222	-320.792	46.746	
	1600	36.074	-145.867	116.540	-332.331	45.400	
	1650	36.241	-144.060	117.652	-338.186	44.800	

NEODYMIUM DIZIRCONATE $Nd_2O_3 \cdot 2ZrO_2$

		A	B	C	D	RANGE	REF.
SOL	CP	61.0	10.7	-9.6		298-1400	*

PHASE	T	CP	H	S	G	BT	
SOL	298	53.391	-967.24	62.	-985.725	722.653	129/*
	300	53.543	-967.141	62.331	-985.840	718.281	
	400	59.280	-961.467	78.616	-992.913	542.575	
	500	62.510	-955.365	92.218	-1001.474	437.803	
	600	64.753	-948.997	103.823	-1011.290	368.412	
	700	66.531	-942.430	113.942	-1022.189	319.185	
	800	68.060	-935.699	122.928	-1034.041	282.525	
	900	69.445	-928.822	131.025	-1046.745	254.219	
	1000	70.740	-921.813	138.410	-1060.222	231.743	
	1100	71.977	-914.676	145.210	-1074.408	213.494	
	1200	73.173	-907.419	151.525	-1089.248	198.406	
	1300	74.342	-900.043	157.428	-1104.699	185.742	
	1400	75.490	-892.551	162.979	-1120.722	174.976	

Ni

NICKEL

		A	B	C	D	RANGE	REF.
SOL	CP	4.561	5.616			298- 500	21
	CP	-60.030	85.191	62.011		500- 631	21
	** CURIE-PT= 631						
	CP	111.662	-162.222			631- 640	21
	CP	-92.184	96.612	156.437		640- 700	21
	CP	-2.599	13.066	13.498	-3.941	700-1400	21
	CP	8.65				1400-1726	21
LIQ	CP	10.30				1726-3187	21
GAS	CP	5.352				3187-3600	21

PHASE	T	CP	H	S	G	BT	P	
SOL	298	6.235	0	7.14	-2.129	1.561		21
	300	6.246	.012	7.179	-2.142	1.561		
	400	6.807	.664	9.052	-2.957	1.616		
	500	7.369	1.373	10.632	-3.943	1.724		
	500	7.370	1.373	10.632	-3.943	1.724		
	600	8.310	2.123	11.996	-5.075	1.849		
	631	9.300	2.395	12.438	-5.453	1.889		
	631	9.300	2.395	12.438	-5.453	1.889		
	640	7.840	2.472	12.559	-5.566	1.901		
	640	7.840	2.472	12.559	-5.566	1.901		
	700	7.370	2.920	13.229	-6.340	1.980		
	700	7.371	2.920	13.229	-6.340	1.980		
	800	7.441	3.659	14.215	-7.713	2.107		
	900	7.635	4.412	15.102	-9.180	2.229		
	1000	7.876	5.187	15.919	-10.731	2.346		
	1100	8.121	5.987	16.681	-12.362	2.456		
	1200	8.343	6.811	17.397	-14.066	2.562		
	1300	8.525	7.655	18.073	-15.840	2.663		
	1400	8.658	8.514	18.710	-17.679	2.760		
	1400	8.650	8.514	18.710	-17.679	2.760		
	1500	8.650	9.379	19.306	-19.580	2.853		

NICKEL (CONTINUED)

PHASE	T	CP	H	S	G	BT	P	
SOL	1600	8.650	10.244	19.865	-21.539	2.943		
	1700	8.650	11.109	20.389	-23.552	3.028	.002	
	1726	8.650	11.334	20.520	-24.084	3.050	.003	
			4.176	2.419				21
LIQ	1726	10.300	15.510	22.940	-24.084	3.050	.003	
	1800	10.300	16.272	23.372	-25.798	3.133	.010	
	1900	10.300	17.302	23.929	-28.163	3.240	.041	
	2000	10.300	18.332	24.457	-30.583	3.342	.142	
	2100	10.300	19.362	24.960	-33.054	3.440	.437	
	2200	10.300	20.392	25.439	-35.574	3.534	1.207	
	2300	10.300	21.422	25.897	-38.141	3.625	3.038	
	2400	10.300	22.452	26.335	-40.753	3.712	7.052	
	2500	10.300	23.482	26.756	-43.407	3.795	15.241	
	2600	10.300	24.512	27.160	-46.103	3.876	30.927	
	2700	10.300	25.542	27.549	-48.839	3.954	59.343	
	2800	10.300	26.572	27.923	-51.612	4.029	108.335	
	2900	10.300	27.602	28.285	-54.423	4.102	189.154	
	3000	10.300	28.632	28.634	-57.269	4.173	317.315	
	3100	10.300	29.662	28.971	-60.149	4.241	513.465	
	3187	10.300	30.558	29.257	-62.682	4.299	760.	
			88.253	27.692				21
GAS	3187	5.352	118.811	56.948	-62.682	4.299		
	3200	5.352	118.881	56.970	-63.423	4.332		
	3300	5.352	119.416	57.135	-69.128	4.579		
	3400	5.352	119.951	57.294	-74.850	4.812		
	3500	5.352	120.487	57.449	-80.587	5.033		
	3600	5.352	121.022	57.600	-86.339	5.242		

NiF_2 NICKEL FLUORIDE

		A	B	C	D	RANGE	REF.
SOL	CP	15.0	4.3			298-1747	*
	** MPT= 1747, L= 68. / BPT= 2013						62/1

PHASE	T	CP	H	S	G	BT	
SOL	298	16.282	-157.2	17.6	-162.447	119.093	62/1
	300	16.290	-157.170	17.701	-162.480	118.383	
	400	16.720	-155.519	22.446	-164.498	89.889	
	500	17.150	-153.826	26.223	-166.937	72.978	
	600	17.580	-152.089	29.388	-169.722	61.830	
	700	18.010	-150.310	32.130	-172.801	53.958	
	800	18.440	-148.487	34.563	-176.138	48.125	
	900	18.870	-146.622	36.760	-179.706	43.644	
	1000	19.300	-144.713	38.770	-183.484	40.106	
	1100	19.730	-142.762	40.630	-187.455	37.249	
	1200	20.160	-140.767	42.365	-191.606	34.901	
	1300	20.590	-138.730	43.996	-195.924	32.942	
	1400	21.020	-136.649	45.537	-200.402	31.288	
	1500	21.450	-134.526	47.002	-205.029	29.877	
	1600	21.880	-132.359	48.400	-209.800	28.661	
	1700	22.310	-130.150	49.740	-214.707	27.606	
	1747	22.512	-129.097	50.351	-217.060	27.158	

NICKEL MONOCHLORIDE (GAS) NiCl (g)

		A	B	C	D	RANGE	REF.
GAS	CP	8.91	0.13	-0.47		298-1800	*

PHASE	T	CP	H	S	G	BT	
GAS	298	8.420	38.7	68.	18.426	-13.508	162,*
	300	8.427	38.716	68.052	18.300	-13.333	
	400	8.668	39.572	70.514	11.366	-6.211	
	500	8.787	40.445	72.462	4.214	-1.842	
	600	8.857	41.328	74.071	-3.115	1.135	
	700	8.905	42.216	75.440	-10.592	3.307	
	800	8.941	43.108	76.632	-18.197	4.972	
	900	8.969	44.004	77.687	-25.914	6.294	
	1000	8.993	44.902	78.633	-33.731	7.373	
	1100	9.014	45.802	79.491	-41.638	8.274	
	1200	9.033	46.705	80.276	-49.627	9.039	
	1300	9.051	47.609	81.000	-57.691	9.700	
	1400	9.068	48.515	81.671	-65.825	10.277	
	1500	9.084	49.423	82.298	-74.024	10.787	
	1600	9.100	50.332	82.884	-82.283	11.241	
	1700	9.115	51.243	83.436	-90.599	11.649	
	1800	9.129	52.155	83.958	-98.969	12.018	

$NiCl_2$ NICKEL CHLORIDE

		A	B	C	D	RANGE	REF.
SOL	CP	17.50	3.16	-1.19		298-1260	1
	** MPT= 1303, L= 18.5						1
GAS	CP	13.72	1.06	-2.91		1260-1800	*

PHASE	T	CP	H	S	G	BT	P	
SOL	298	17.103	-73.0	23.35	-79.962	58.621		1
	300	17.126	-72.968	23.456	-80.005	58.292		
	400	18.020	-71.207	28.517	-82.614	45.144		
	500	18.604	-69.374	32.604	-85.676	37.454		
	600	19.065	-67.490	36.038	-89.113	32.464		
	700	19.469	-65.563	39.008	-92.869	28.999		
	800	19.842	-63.597	41.632	-96.903	26.476	.002	
	900	20.197	-61.595	43.990	-101.186	24.575	.115	
	1000	20.541	-59.558	46.136	-105.694	23.103	2.594	
	1100	20.878	-57.487	48.109	-110.408	21.939	32.326	
	1200	21.209	-55.383	49.940	-115.311	21.004	258.396	
	1260	21.407	-54.104	50.980	-118.339	20.529	760.	
			53.8	42.698				1
GAS	1260	14.872	-.304	93.678	-118.339	20.529		
	1300	14.926	.292	94.144	-122.095	20.529		
	1400	15.056	1.791	95.255	-131.566	20.541		
	1500	15.181	3.303	96.298	-141.144	20.567		
	1600	15.302	4.827	97.281	-150.824	20.604		
	1700	15.421	6.363	98.213	-160.599	20.649		
	1800	15.538	7.911	99.097	-170.465	20.700		

NICKEL BROMIDE $NiBr_2$

		A	B	C	D	RANGE	REF.
SOL	CP	16.5	4.7			298-1192	*
	** MPT= 1236, L= 13.						1
GAS	CP	14.	1.06	-2.91		1192-2000	*

PHASE	T	CP	H	S	G	BT	P	
SOL	298	17.901	-50.7	32.5	-60.390	44.273		19/*
	300	17.910	-50.667	32.611	-60.450	44.044		
	400	18.380	-48.852	37.828	-63.983	34.964		
	500	18.850	-46.991	41.979	-67.981	29.718		
	600	19.320	-45.082	45.458	-72.357	26.360		
	700	19.790	-43.127	48.471	-77.057	24.061		
	800	20.260	-41.124	51.144	-82.040	22.415	.008	
	900	20.730	-39.075	53.558	-87.277	21.197	.416	
	1000	21.200	-36.978	55.766	-92.745	20.272	9.216	
	1100	21.670	-34.835	57.809	-98.425	19.558	112.702	
	1192	22.102	-32.821	59.567	-103.825	19.039	760.	
			53.7	45.050				1
GAS	1192	15.059	20.879	104.617	-103.825	19.039		
	1200	15.070	20.999	104.718	-104.662	19.064		
	1300	15.206	22.513	105.929	-115.195	19.369		
	1400	15.336	24.040	107.061	-125.845	19.648		
	1500	15.461	25.580	108.123	-136.605	19.906		
	1600	15.582	27.132	109.125	-147.468	20.146		
	1700	15.701	28.696	110.073	-158.428	20.370		
	1800	15.818	30.272	110.974	-169.481	20.581		
	1900	15.933	31.860	111.832	-180.622	20.779		
	2000	16.047	33.459	112.653	-191.846	20.967		

NiJ_2 NICKEL IODIDE

		A	B	C	D	RANGE	REF.
SOL	CP	15.75	5.8			298-1070	*
	** MPT= 1070						1

PHASE	T	CP	H	S	G	BT	
SOL	298	17.479	-18.7	36.8	-29.672	21.753	19/*
	300	17.490	-18.668	36.908	-29.740	21.669	
	400	18.070	-16.890	42.019	-33.697	18.414	
	500	18.650	-15.054	46.114	-38.110	16.660	
	600	19.230	-13.160	49.565	-42.899	15.628	
	700	19.810	-11.208	52.573	-48.009	14.991	
	800	20.390	-9.198	55.256	-53.403	14.591	
	900	20.970	-7.130	57.691	-59.052	14.342	
	1000	21.550	-5.004	59.931	-64.934	14.193	
	1070	21.956	-3.481	61.402	-69.181	14.132	

NICKEL DISULFIDE Ni_3S_2

		A	B	C	D	RANGE	REF.
SOL-A	CP	23.298	16.17			298- 828	63
	** NI3S(2-X)						
SOL-B	CP	23.298	16.17			828-1063	63
	** MPT= 1063						1

PHASE	T	CP	H	S	G	BT	
SOL-A	298	28.119	-39.0	32.01	-48.544	35.588	63
	300	28.149	-38.948	32.184	-48.603	35.412	
	400	29.766	-36.052	40.503	-52.254	28.554	
	500	31.383	-32.995	47.319	-56.654	24.767	
	600	33.000	-29.776	53.184	-61.686	22.472	
	700	34.617	-26.395	58.392	-67.269	21.005	
	800	36.234	-22.852	63.120	-73.349	20.041	
	828	36.687	-21.831	64.375	-75.134	19.834	
SOL-B	828	36.687	-21.831	64.375	-75.134	19.834	
	900	37.851	-19.148	67.481	-79.881	19.400	
	1000	39.468	-15.282	71.553	-86.835	18.980	
	1063	40.487	-12.763	73.995	-91.420	18.798	

* * * * * * * * * *

NICKEL 0.84-SULFIDE $NiS_{0.84}$

		A	B	C	D	RANGE	REF.
SOL	CP	8.6	5.7			298- 833	63
	** NI7S6-NI6S5, DPT= 833						63

PHASE	T	CP	H	S	G	BT	
SOL	298	10.299	-19.7	11.7	-23.188	17.000	63
	300	10.310	-19.681	11.764	-23.210	16.911	
	400	10.880	-18.621	14.808	-24.545	13.412	
	500	11.450	-17.505	17.297	-26.153	11.433	
	600	12.020	-16.331	19.435	-27.992	10.198	
	700	12.590	-15.101	21.331	-30.032	9.378	
	800	13.160	-13.813	23.049	-32.253	8.812	
	833	13.348	-13.376	23.585	-33.022	8.665	

NiS

NICKEL MONOSULFIDE

		A	B	C	D	RANGE	REF.
SOL-A	CP	9.30	6.40			298- 670	63
	** NIS(1+X)						63
SOL-B	CP	9.305	6.394			670- 900	63
	** MPT= 1253, L= 7.4						63

PHASE	T	CP	H	S	G	BT	
SOL-A	298	11.208	-22.5	12.66	-26.275	19.262	63
	300	11.220	-22.479	12.729	-26.298	19.161	
	400	11.860	-21.325	16.045	-27.743	15.160	
	500	12.500	-20.107	18.760	-29.487	12.891	
	600	13.140	-18.825	21.096	-31.483	11.469	
	670	13.588	-17.890	22.570	-33.012	10.770	
			0.63	.940			63
SOL-B	670	13.589	-17.260	23.510	-33.012	10.770	
	700	13.781	-16.849	24.110	-33.726	10.531	
	800	14.420	-15.439	25.991	-36.232	9.900	
	900	15.060	-13.965	27.727	-38.919	9.452	

* * * * * * * * * *

$NiSO_4$

NICKEL SULFATE

		A	B	C	D	RANGE	REF.
SOL	CP	30.1	9.92			298-1200	58

PHASE	T	CP	H	S	G	BT	
SOL	298	33.058	-208.1	27.2	-216.210	158.507	66/*
	300	33.076	-208.039	27.405	-216.260	157.567	
	400	34.068	-204.682	37.056	-219.504	119.948	
	500	35.060	-201.225	44.764	-223.607	97.752	
	600	36.052	-197.670	51.244	-228.416	83.212	
	700	37.044	-194.015	56.876	-233.828	73.014	
	800	38.036	-190.261	61.888	-239.771	65.511	
	900	39.028	-186.408	66.425	-246.190	59.791	
	1000	40.020	-182.455	70.588	-253.043	55.310	
	1100	41.012	-178.404	74.449	-260.297	51.723	
	1200	42.004	-174.253	78.060	-267.925	48.802	

NICKEL 1.05-SELENIDE $NiSe_{1.05}$

		A	B	C	D	RANGE	REF.
SOL	CP	9.667	7.023	0.89		298-1100	63
	** NI7SE8, MPT= 1253						62

PHASE	T	CP	H	S	G	BT	
SOL	298	12.762	-17.9	17.97	-23.258	17.051	63
	300	12.763	-17.876	18.049	-23.291	16.970	
	400	13.032	-16.590	21.749	-25.289	13.819	
	500	13.534	-15.262	24.708	-27.617	12.073	
	600	14.128	-13.880	27.227	-30.216	11.008	
	700	14.765	-12.435	29.453	-33.052	10.321	
	800	15.424	-10.926	31.467	-36.100	9.863	
	900	16.098	-9.350	33.323	-39.340	9.554	
	1000	16.779	-7.706	35.054	-42.760	9.346	
	1100	17.466	-5.994	36.685	-46.348	9.210	

$NiSe_{1.143}$ NICKEL 1.143-SELENIDE

		A	B	C	D	RANGE	REF.
SOL	CP	9.558	12.088			298- 480	63
	** NI7SE8						
	CP	-155.572	356.087			480- 503	63
	CP	18.608	-6.596			503- 550	63
	CP	10.990	6.400			550-1000	63

PHASE	T	CP	H	S	G	BT	
SOL	298	13.162	-19.05	18.44	-24.548	17.997	63
	300	13.184	-19.026	18.521	-24.582	17.910	
	400	14.393	-17.647	22.480	-26.639	14.557	
	480	15.360	-16.457	25.190	-28.548	13.000	
	480	15.350	-16.457	25.190	-28.548	13.000	
	500	22.471	-16.078	25.961	-29.059	12.703	
	503	23.540	-16.009	26.098	-29.137	12.661	
	503	15.290	-16.009	26.098	-29.137	12.661	
	550	14.980	-15.298	27.450	-30.396	12.080	
	550	14.510	-15.298	27.450	-30.396	12.080	
	600	14.830	-14.565	28.727	-31.801	11.585	
	700	15.470	-13.050	31.061	-34.792	10.864	
	800	16.110	-11.471	33.168	-38.005	10.384	
	900	16.750	-9.828	35.103	-41.420	10.060	
	1000	17.390	-8.121	36.901	-45.021	9.841	

NICKEL 1.25-SELENIDE

$NiSe_{1.25}$

		A	B	C	D	RANGE	REF.
SOL-A	CP	12.764	5.287	-0.667		298- 500	63
	** NI4SE5						63
	CP	-8.568	47.416			500- 589	63
SOL-B	CP	0.396	16.708	20.772		589- 995	63
SOL-C	CP	11.122	7.636			995-1100	63

PHASE	T	CP	H	S	G	BT	
SOL-A	298	13.590	-19.85	19.14	-25.557	18.736	63
	300	13.609	-19.825	19.224	-25.592	18.646	
	400	14.462	-18.419	23.263	-27.724	15.150	
	500	15.141	-16.938	26.565	-30.220	13.211	
	500	15.140	-16.938	26.565	-30.220	13.211	
	589	19.360	-15.403	29.381	-32.708	12.138	
SOL-B	589	16.225	-15.403	29.381	-32.708	12.138	
	600	16.191	-15.224	29.681	-33.033	12.034	
	700	16.331	-13.604	32.178	-36.129	11.281	
	800	17.008	-11.941	34.399	-39.460	10.781	
	900	17.998	-10.192	36.457	-43.003	10.444	
	995	19.119	-8.430	38.317	-46.556	10.227	
SOL-C	995	18.720	-8.430	38.317	-46.556	10.227	
	1000	18.758	-8.337	38.411	-46.747	10.218	
	1100	19.522	-6.423	40.234	-50.681	10.071	

$NiSe_2$ NICKEL DISELENIDE

		A	B	C	D	RANGE	REF.
SOL	CP	17.393	4.943	-0.735		298- 900	63

** HOMOGENEITY RANGE....

PHASE	T	CP	H	S	G	BT	
SOL	298	18.040	-26.	24.74	-33.376	24.469	63
	300	18.059	-25.967	24.852	-33.422	24.351	
	400	18.911	-24.116	30.171	-36.184	19.773	
	500	19.570	-22.191	34.464	-39.422	17.234	
	600	20.155	-20.204	38.084	-43.054	15.685	
	700	20.703	-18.161	41.233	-47.024	14.683	
	800	21.233	-16.064	44.032	-51.289	14.013	
	900	21.751	-13.915	46.563	-55.821	13.557	

* * * * * * * * * *

$NiSeO_3$ NICKEL SELENITE

		A	B	C	D	RANGE	REF.
SOL	CP	19.0	14.3			298- 800	*

PHASE	T	CP	H	S	G	BT	
SOL	298	23.264	-135.6	24.7	-142.964	104.810	62
	300	23.290	-135.557	24.844	-143.010	104.197	
	400	24.720	-133.156	31.740	-145.852	79.701	
	500	26.150	-130.613	37.410	-149.318	65.276	
	600	27.580	-127.926	42.304	-153.309	55.850	
	700	29.010	-125.097	46.663	-157.761	49.262	
	800	30.440	-122.124	50.630	-162.628	44.434	

NICKEL TELLURIDE

NiTe

		A	B	C	D	RANGE	REF.
SOL	CP	11.57	3.30			298- 800	1

PHASE	T	CP	H	S	G	ΘT	
SOL	298	12.554	-8.52	19.13	-14.224	10.428	159/106
	300	12.560	-8.497	19.208	-14.259	10.389	
	400	12.890	-7.224	22.866	-16.371	8.946	
	500	13.220	-5.919	25.778	-18.808	8.222	
	600	13.550	-4.580	28.217	-21.511	7.836	
	700	13.880	-3.209	30.331	-24.440	7.632	
	800	14.210	-1.804	32.206	-27.569	7.533	

Ni_3P 3-NICKEL PHOSPHIDE

		A	B	C	D	RANGE	REF.
SOL	CP	18.6	8.0			298-1243	*
	** MPT=	1243					1

PHASE	T	CP	H	S	G	BT	
SOL	298	20.985	-52.57	25.4	-60.143	44.092	64/*
	300	21.000	-52.531	25.530	-60.190	43.854	
	400	21.800	-50.391	31.681	-63.063	34.461	
	500	22.600	-48.171	36.631	-66.487	29.065	
	600	23.400	-45.871	40.822	-70.365	25.634	
	700	24.200	-43.491	44.490	-74.634	23.305	
	800	25.000	-41.031	47.773	-79.250	21.653	
	900	25.800	-38.491	50.764	-84.179	20.444	
	1000	26.600	-35.871	53.524	-89.395	19.540	
	1100	27.400	-33.171	56.097	-94.877	18.853	
	1200	28.200	-30.391	58.515	-100.609	18.326	
	1243	28.544	-29.171	59.514	-103.147	18.138	

* * * * * * * * * *

Ni_5P_2 5-NICKEL 2-PHOSPHIDE

		A	B	C	D	RANGE	REF.
SOL	CP	32.3	13.5			298-1453	*
	** MPT=	1453					1

PHASE	T	CP	H	S	G	BT	
SOL	298	36.325	-104.24	44.2	-117.418	86.081	64/*
	300	36.350	-104.173	44.425	-117.500	85.610	
	400	37.700	-100.470	55.067	-122.497	66.938	
	500	39.050	-96.633	63.624	-128.445	56.151	
	600	40.400	-92.660	70.863	-135.178	49.245	
	700	41.750	-88.553	77.192	-142.588	44.524	
	800	43.100	-84.310	82.856	-150.595	41.146	
	900	44.450	-79.933	88.010	-159.142	38.650	
	1000	45.800	-75.420	92.763	-168.183	36.761	
	1100	47.150	-70.773	97.192	-177.684	35.307	
	1200	48.500	-65.990	101.352	-187.613	34.174	
	1300	49.850	-61.073	105.287	-197.946	33.282	
	1400	51.200	-56.020	109.031	-208.664	32.578	
	1453	51.915	-53.288	110.947	-214.493	32.267	

2-NICKEL PHOSPHIDE Ni_2P

		A	B	C	D	RANGE	REF.
SOL	CP	13.85	5.5			298-1383	*
	** MPT= 1383						1

PHASE	T	CP	H	S	G	BT	
SOL	298	15.490	-44.17	18.5	-49.686	36.426	64/*
	300	15.500	-44.141	18.596	-49.720	36.226	
	400	16.050	-42.564	23.130	-51.816	28.315	
	500	16.600	-40.931	26.771	-54.317	23.745	
	600	17.150	-39.244	29.846	-57.151	20.820	
	700	17.700	-37.501	32.531	-60.273	18.821	
	800	18.250	-35.704	34.930	-63.648	17.390	
	900	18.800	-33.851	37.112	-67.252	16.333	
	1000	19.350	-31.944	39.121	-71.065	15.533	
	1100	19.900	-29.981	40.991	-75.071	14.917	
	1200	20.450	-27.964	42.746	-79.259	14.437	
	1300	21.000	-25.891	44.405	-83.617	14.059	
	1383	21.457	-24.129	45.718	-87.358	13.807	

* * * * * * * * * *

6-NICKEL 5-PHOSPHIDE Ni_6P_5

		A	B	C	D	RANGE	REF.
SOL	CP	57.6	17.0			298-1200	*

PHASE	T	CP	H	S	G	BT	
SOL	298	62.669	-148.526	66.	-168.204	123.313	64/*
	300	62.700	-148.410	66.388	-168.326	122.642	
	400	64.400	-142.055	84.658	-175.918	96.130	
	500	66.100	-135.530	99.211	-185.136	80.934	
	600	67.800	-128.835	111.413	-195.683	71.287	
	700	69.500	-121.970	121.992	-207.365	64.751	
	800	71.200	-114.935	131.384	-220.042	60.121	
	900	72.900	-107.730	139.868	-233.611	56.736	
	1000	74.600	-100.355	147.637	-247.992	54.206	
	1100	76.300	-92.810	154.826	-263.119	52.284	
	1200	78.000	-85.095	161.538	-278.941	50.809	

NiP_2 NICKEL 2-PHOSPHIDE

		A	B	C	D	RANGE	REF.
SOL	CP	15.5	5.0			298-1100	*

PHASE	T	CP	H	S	G	BT	
SOL	298	16.991	-34.115	17.5	-39.333	28.835	228/*
	300	17.000	-34.084	17.605	-39.365	28.681	
	400	17.500	-32.359	22.564	-41.384	22.614	
	500	18.000	-30.584	26.523	-43.845	19.167	
	600	18.500	-28.759	29.849	-46.668	17.001	
	700	19.000	-26.884	32.738	-49.800	15.550	
	800	19.500	-24.959	35.308	-53.205	14.537	
	900	20.000	-22.984	37.634	-56.854	13.808	
	1000	20.500	-20.959	39.767	-60.725	13.273	
	1100	21.000	-18.884	41.744	-64.802	12.877	

* * * * * * * * * *

NiP_3 NICKEL 3-PHOSPHIDE

		A	B	C	D	RANGE	REF.
SOL	CP	22.4	7.0			298-1000	*

PHASE	T	CP	H	S	G	BT	
SOL	298	24.487	-39.318	23.5	-46.325	33.961	228/*
	300	24.500	-39.273	23.652	-46.368	33.784	
	400	25.200	-36.788	30.796	-49.106	26.834	
	500	25.900	-34.233	36.494	-52.480	22.942	
	600	26.600	-31.608	41.278	-56.374	20.537	
	700	27.300	-28.913	45.431	-60.714	18.958	
	800	28.000	-26.148	49.122	-65.445	17.881	
	900	28.700	-23.313	52.460	-70.527	17.129	
	1000	29.400	-20.408	55.521	-75.928	16.596	

NICKEL ARSENIDE

NiAs

		A	B	C	D	RANGE	REF.
SOL	CP	10.45	3.1			298-1237	*
	** MPT= 1237						62

PHASE	T	CP	H	S	G	BT	
SOL	298	11.374	-17.2	12.4	-20.897	15.320	159/*
	300	11.380	-17.179	12.470	-20.920	15.242	
	400	11.690	-16.025	15.787	-22.340	12.208	
	500	12.000	-14.841	18.429	-24.055	10.516	
	600	12.310	-13.625	20.644	-26.012	9.476	
	700	12.620	-12.379	22.565	-28.174	8.798	
	800	12.930	-11.101	24.270	-30.517	8.338	
	900	13.240	-9.793	25.811	-33.023	8.020	
	1000	13.550	-8.453	27.222	-35.675	7.798	
	1100	13.860	-7.083	28.528	-38.464	7.643	
	1200	14.170	-5.681	29.747	-41.378	7.537	
	1237	14.285	-5.155	30.179	-42.487	7.507	

* * * * * * * * * *

NICKEL ANTIMONIDE

NiSb

		A	B	C	D	RANGE	REF.
SOL	CP	11.05	2.78			298-1423	*
	** MPT= 1423						1

PHASE	T	CP	H	S	G	BT	
SOL	298	11.879	-20.	18.7	-25.575	18.750	159/*
	300	11.884	-19.978	18.773	-25.610	18.659	
	400	12.162	-18.776	22.230	-27.668	15.119	
	500	12.440	-17.546	24.974	-30.033	13.129	
	600	12.718	-16.288	27.267	-32.648	11.894	
	700	12.996	-15.002	29.248	-35.476	11.078	
	800	13.274	-13.689	31.002	-38.490	10.516	
	900	13.552	-12.347	32.581	-41.670	10.120	
	1000	13.830	-10.978	34.023	-45.002	9.836	
	1100	14.108	-9.581	35.355	-48.471	9.632	
	1200	14.386	-8.157	36.594	-52.069	9.484	
	1300	14.664	-6.704	37.757	-55.787	9.380	
	1400	14.942	-5.224	38.853	-59.618	9.308	
	1423	15.006	-4.879	39.097	-60.515	9.295	

NiBi

NICKEL BISMUTHIDE

		A	B	C	D	RANGE	REF.
SOL	CP	11.0	4.6			298- 928	*
	** DPT= 928						62

PHASE	T	CP	H	S	G	BT	
SOL	298	12.371	-1.85	21.1	-8.141	5.968	159/*
	300	12.380	-1.827	21.177	-8.180	5.960	
	400	12.840	-.566	24.801	-10.487	5.730	
	500	13.300	.741	27.716	-13.117	5.734	
	600	13.760	2.094	30.181	-16.015	5.834	
	700	14.220	3.493	32.337	-19.143	5.977	
	800	14.680	4.938	34.266	-22.475	6.141	
	900	15.140	6.429	36.021	-25.990	6.312	
	928	15.269	6.855	36.487	-27.005	6.361	

NICKEL CARBIDE Ni_3C

		A	B	C	D	RANGE	REF.
SOL	CP	24.0	5.0			298- 800	*

PHASE	T	CP	H	S	G	BT	
SOL	298	25.491	9.	25.4	1.427	-1.046	1/*
	300	25.500	9.047	25.558	1.380	-1.005	
	400	26.000	11.622	32.962	-1.563	.854	
	500	26.500	14.247	38.818	-5.162	2.256	
	600	27.000	16.922	43.693	-9.294	3.386	
	700	27.500	19.647	47.893	-13.878	4.333	
	800	28.000	22.422	51.598	-18.856	5.152	

* * * * * * * * * *

NICKEL CARBONATE $NiCO_3$

		A	B	C	D	RANGE	REF.
SOL	CP	22.0	9.3	-2.95		298- 700	*

PHASE	T	CP	H	S	G	BT	
SOL	298	21.454	-162.5	20.4	-168.582	123.591	166/130
	300	21.512	-162.460	20.533	-168.620	122.856	
	400	23.876	-160.181	27.075	-171.011	93.448	
	500	25.470	-157.710	32.582	-174.001	76.066	
	600	26.761	-155.096	37.343	-177.502	64.664	
	700	27.908	-152.362	41.556	-181.451	56.659	

NiSi

NICKEL SILICON

		A	B	C	D	RANGE	REF.
SOL	CP	11.654	1.463	-1.558		298-1265	221
LIQ	CP	19.0				1265-1800	221

PHASE	T	CP	H	S	G	BT	
SOL	298	10.338	-21.41	10.634	-24.581	18.020	221
	300	10.362	-21.391	10.698	-24.600	17.924	
	400	11.265	-20.304	13.818	-25.831	14.116	
	500	11.762	-19.151	16.390	-27.346	11.954	
	600	12.099	-17.957	18.566	-29.096	10.600	
	700	12.360	-16.733	20.451	-31.049	9.695	
	800	12.581	-15.486	22.116	-33.179	9.065	
	900	12.778	-14.218	23.610	-35.467	8.614	
	1000	12.961	-12.931	24.966	-37.896	8.283	
	1100	13.135	-11.626	26.209	-40.456	8.039	
	1200	13.301	-10.304	27.359	-43.135	7.857	
	1265	13.407	-9.436	28.064	-44.937	7.765	
			10.277	8.124			221
LIQ	1265	19.000	.841	36.188	-44.937	7.765	
	1300	19.000	1.506	36.706	-46.212	7.770	
	1400	19.000	3.406	38.114	-49.954	7.799	
	1500	19.000	5.306	39.425	-53.832	7.844	
	1600	19.000	7.206	40.651	-57.836	7.901	
	1700	19.000	9.106	41.803	-61.960	7.967	
	1800	19.000	11.006	42.889	-66.195	8.038	

7-NICKEL 13-SILICON Ni_7Si_{13}

		A	B	C	D	RANGE	REF.
SOL	CP	119.598	17.639	-17.256		298-1245	221
	** MPT= 1245						221

PHASE	T	CP	H	S	G	BT	
SOL	298	105.445	-150.1	105.108	-181.438	133.015	221
	300	105.716	-149.905	105.761	-181.633	132.337	
	400	115.869	-138.766	137.737	-193.860	105.935	
	500	121.515	-126.875	164.247	-208.998	91.365	
	600	125.388	-114.520	186.762	-226.577	82.542	
	700	128.424	-101.825	206.326	-246.253	76.894	
	800	131.013	-88.850	223.647	-267.768	73.161	
	900	133.343	-75.631	239.215	-290.924	70.655	
	1000	135.511	-62.187	253.377	-315.564	68.976	
	1100	137.575	-48.532	266.390	-341.561	67.871	
	1200	139.566	-34.674	278.447	-368.810	67.179	
	1245	140.445	-28.374	283.601	-381.457	66.971	

* * * * * * * * * *

NICKEL ORTHOSILICATE $2NiO \cdot SiO_2$

		A	B	C	D	RANGE	REF.
SOL	CP	39.0	4.72	-5.81	2.11	298-1818	5
	** MPT= 1818						5

PHASE	T	CP	H	S	G	BT	
SOL	298	34.059	-335.85	26.3	-343.691	251.966	5
	300	34.150	-335.787	26.511	-343.740	250.448	
	400	37.594	-332.180	36.864	-346.926	189.577	
	500	39.563	-328.315	45.480	-351.055	153.467	
	600	40.978	-324.285	52.824	-355.979	129.683	
	700	42.152	-320.127	59.231	-361.589	112.908	
	800	43.219	-315.858	64.930	-367.802	100.492	
	900	44.240	-311.485	70.079	-374.556	90.967	
	1000	45.249	-307.011	74.793	-381.803	83.454	
	1100	46.265	-302.435	79.153	-389.503	77.398	
	1200	47.299	-297.757	83.223	-397.624	72.427	
	1300	48.358	-292.974	87.050	-406.140	68.287	
	1400	49.447	-288.084	90.674	-415.027	64.797	
	1500	50.569	-283.084	94.123	-424.269	61.824	
	1600	51.727	-277.969	97.424	-433.847	59.269	
	1700	52.921	-272.737	100.595	-443.749	57.055	
	1800	54.153	-267.384	103.655	-453.962	55.126	
	1818	54.379	-266.407	104.195	-455.833	54.805	

Ni_2Ge 2-NICKEL GERMANIUM

		A	B	C	D	RANGE	REF.
SOL	CP	17.00	5.20	-1.64		298-1400	*

PHASE	T	CP	H	S	G	BT	
SOL	298	16.705	-26.3	21.7	-32.770	24.024	159/*
	300	16.738	-26.269	21.803	-32.810	23.905	
	400	18.055	-24.524	26.815	-35.250	19.262	
	500	18.944	-22.672	30.944	-38.144	16.675	
	600	19.664	-20.740	34.464	-41.419	15.089	
	700	20.305	-18.741	37.544	-45.022	14.058	
	800	20.904	-16.681	40.295	-48.916	13.365	
	900	21.478	-14.562	42.790	-53.072	12.889	
	1000	22.036	-12.386	45.082	-57.468	12.561	
	1100	22.584	-10.155	47.208	-62.083	12.336	
	1200	23.126	-7.869	49.196	-66.905	12.187	
	1300	23.663	-5.530	51.069	-71.919	12.092	
	1400	24.196	-3.137	52.842	-77.115	12.040	

3-NICKEL TITANIUM Ni_3Ti

		A	B	C	D	RANGE	REF.
SOL	CP	26.04	4.03	-4.35		298-1651	*
	** MPT= 1651						1

PHASE	T	CP	H	S	G	BT	
SOL	298	22.348	-33.5	25.	-40.954	30.024	1/*
	300	22.416	-33.459	25.138	-41.000	29.873	
	400	24.933	-31.076	31.975	-43.866	23.971	
	500	26.315	-28.508	37.700	-47.358	20.703	
	600	27.250	-25.828	42.585	-51.378	18.717	
	700	27.973	-23.065	46.841	-55.854	17.441	
	800	28.584	-20.237	50.617	-60.731	16.593	
	900	29.130	-17.350	54.016	-65.965	16.021	
	1000	29.635	-14.412	57.112	-71.524	15.634	
	1100	30.113	-11.424	59.959	-77.379	15.376	
	1200	30.574	-8.390	62.599	-83.509	15.211	
	1300	31.022	-5.310	65.064	-89.893	15.114	
	1400	31.460	-2.186	67.379	-96.516	15.069	
	1500	31.892	.982	69.564	-103.365	15.062	
	1600	32.318	4.192	71.636	-110.425	15.085	
	1651	32.534	5.846	72.654	-114.105	15.107	

NiTi

NICKEL TITANIUM

		A	B	C	D	RANGE	REF.
SOL	CP	12.67	2.30	-1.94		298-1513	*
	** HOMOGENEITY RANGE, MPT= 1513						1

PHASE	T	CP	H	S	G	BT	
SOL	298	11.173	-15.9	12.7	-19.687	14.433	1/*
	300	11.204	-15.879	12.769	-19.710	14.361	
	400	12.377	-14.693	16.173	-21.163	11.564	
	500	13.044	-13.420	19.012	-22.926	10.022	
	600	13.511	-12.091	21.433	-24.951	9.090	
	700	13.884	-10.721	23.545	-27.202	8.494	
	800	14.207	-9.316	25.420	-29.652	8.102	
	900	14.500	-7.880	27.111	-32.280	7.840	
	1000	14.776	-6.416	28.653	-35.069	7.665	
	1100	15.040	-4.926	30.074	-38.006	7.552	
	1200	15.295	-3.409	31.393	-41.081	7.483	
	1300	15.545	-1.867	32.627	-44.282	7.446	
	1400	15.791	-.300	33.788	-47.604	7.432	
	1500	16.034	1.291	34.886	-51.038	7.437	
	1513	16.065	1.500	35.025	-51.492	7.439	

* * * * * * * * * *

$NiTi_2$

NICKEL 2-TITANIUM

		A	B	C	D	RANGE	REF.
SOL	CP	16.25	5.6			298-1288	*
	** MPT= 1268 (PERITEC.)						1

PHASE	T	CP	H	S	G	BT	
SOL	298	17.920	-20.	20.	-25.963	19.034	1/*
	300	17.930	-19.967	20.111	-26.000	18.944	
	400	18.490	-18.146	25.346	-28.284	15.456	
	500	19.050	-16.269	29.532	-31.035	13.567	
	600	19.610	-14.336	33.055	-34.169	12.448	
	700	20.170	-12.347	36.119	-37.630	11.750	
	800	20.730	-10.302	38.849	-41.381	11.306	
	900	21.290	-8.201	41.323	-45.392	11.024	
	1000	21.850	-6.044	43.595	-49.639	10.850	
	1100	22.410	-3.831	45.704	-54.105	10.751	
	1200	22.970	-1.562	47.678	-58.776	10.706	
	1288	23.463	.481	49.321	-63.044	10.699	

NICKEL TITANATE

$NiO \cdot TiO_2$

		A	B	C	D	RANGE	REF.
SOL	CP	27.51	3.82	-4.38		298-1700	*

PHASE	T	CP	H	S	G	BT	
SOL	298	23.722	-287.35	23.733	-294.426	215.849	5
	300	23.789	-287.306	23.880	-294.470	214.550	
	400	26.300	-284.786	31.111	-297.231	162.421	
	500	27.668	-282.082	37.139	-300.652	131.433	
	600	28.585	-279.267	42.269	-304.629	110.976	
	700	29.290	-276.372	46.731	-309.084	96.513	
	800	29.882	-273.413	50.681	-313.958	85.781	
	900	30.407	-270.398	54.232	-319.207	77.524	
	1000	30.892	-267.333	57.461	-324.794	70.993	
	1100	31.350	-264.221	60.427	-330.690	65.711	
	1200	31.790	-261.064	63.174	-336.872	61.361	
	1300	32.217	-257.863	65.735	-343.319	57.725	
	1400	32.635	-254.620	68.138	-350.014	54.647	
	1500	33.045	-251.336	70.404	-356.942	52.013	
	1600	33.451	-248.012	72.549	-364.090	49.739	
	1700	33.852	-244.646	74.589	-371.448	47.759	

Ni_3Sn 3-NICKEL TIN

		A	B	C	D	RANGE	REF.
SOL	CP	20.78	10.20			298- 900	58
	** NISN(0.325-0.345) / TPT= 1194, MPT= 1447						1

PHASE	T	CP	H	S	G	BT	
SOL	298	23.821	-24.6	31.4	-33.962	24.898	*
	300	23.840	-24.556	31.547	-34.020	24.787	
	400	24.860	-22.121	38.545	-37.539	20.513	
	500	25.880	-19.584	44.202	-41.685	18.223	
	600	26.900	-16.945	49.011	-46.352	16.886	
	700	27.920	-14.204	53.234	-51.468	16.071	
	800	28.940	-11.361	57.029	-56.984	15.569	
	900	29.960	-8.416	60.497	-62.863	15.267	

3-NICKEL 2-TIN

Ni_3Sn_2

		A	B	C	D	RANGE	REF.
SOL	CP	23.2	9.3			298-1537	*
	** NISN(0.62-0.715) / TPT= 873, MFT= 1537						1

PHASE	T	CP	H	S	G	BT	
SOL	298	25.973	-46.	41.5	-58.373	42.794	159/*
	300	25.990	-45.952	41.661	-58.450	42.587	
	400	26.920	-43.306	49.265	-63.012	34.433	
	500	27.850	-40.568	55.372	-68.254	29.838	
	600	28.780	-37.736	60.532	-74.055	26.978	
	700	29.710	-34.812	65.038	-80.339	25.086	
	800	30.640	-31.794	69.066	-87.047	23.783	
	873	31.319	-29.533	71.771	-92.189	23.082	
	873	31.319	-29.533	71.771	-92.189	23.082	
	900	31.570	-28.684	72.729	-94.140	22.863	
	1000	32.500	-25.480	76.103	-101.583	22.204	
	1100	33.430	-22.184	79.244	-109.352	21.729	
	1200	34.360	-18.794	82.193	-117.426	21.389	
	1300	35.290	-15.312	84.980	-125.786	21.149	
	1400	36.220	-11.736	87.629	-134.417	20.986	
	1500	37.150	-8.068	90.160	-143.307	20.883	
	1537	37.494	-6.687	91.069	-146.660	20.857	

* * * * * * * * * *

3-NICKEL 4-TIN

Ni_3Sn_4

		A	B	C	D	RANGE	REF.
SOL	CP	42.65	3.1	-4.2		298-1000	*

PHASE	T	CP	H	S	G	BT	
SOL	298	38.850	-56.3	61.6	-74.666	54.739	84/*
	300	38.913	-56.228	61.841	-74.780	54.485	
	400	41.265	-52.205	73.399	-81.564	44.571	
	500	42.520	-48.010	82.754	-89.387	39.076	
	600	43.343	-43.715	90.583	-98.065	35.725	
	700	43.963	-39.348	97.313	-107.467	33.557	
	800	44.474	-34.926	103.218	-117.500	32.104	
	900	44.921	-30.455	108.482	-128.089	31.109	
	1000	45.330	-25.943	113.237	-139.179	30.422	

NiB

NICKEL MONOBORIDE

		A	B	C	D	RANGE	REF.
SOL	CP	10.26	3.5	-2.69		298-1300	*

PHASE	T	CP	H	S	G	BT	
SOL	298	8.277	-24.	7.2	-26.147	19.169	102/*
	300	8.321	-23.985	7.251	-26.160	19.060	
	400	9.979	-23.060	9.899	-27.020	14.765	
	500	10.934	-22.011	12.236	-28.129	12.297	
	600	11.613	-20.882	14.292	-29.458	10.731	
	700	12.161	-19.693	16.125	-30.980	9.674	
	800	12.640	-18.453	17.780	-32.677	8.928	
	900	13.078	-17.166	19.295	-34.532	8.387	
	1000	13.491	-15.838	20.694	-36.532	7.985	
	1100	13.888	-14.469	21.999	-38.667	7.684	
	1200	14.273	-13.061	23.224	-40.929	7.455	
	1300	14.651	-11.614	24.381	-43.310	7.282	

* * * * * * * * * *

Ni_4B_3

TETRANICKEL TRIBORIDE

		A	B	C	D	RANGE	REF.
SOL	CP	37.28	11.74	-9.03		298-1400	*
	** MPT= 1853						62

PHASE	T	CP	H	S	G	BT	
SOL	298	30.622	-74.5	27.4	-82.669	60.606	102/*
	300	30.769	-74.443	27.590	-82.720	60.270	
	400	36.332	-71.057	37.294	-85.974	46.981	
	500	39.538	-67.252	45.771	-90.137	39.404	
	600	41.816	-63.179	53.190	-95.093	34.642	
	700	43.655	-58.903	59.778	-100.748	31.459	
	800	45.261	-54.456	65.714	-107.027	29.242	
	900	46.731	-49.855	71.131	-113.873	27.656	
	1000	48.117	-45.113	76.127	-121.239	26.500	
	1100	49.448	-40.234	80.776	-129.087	25.651	
	1200	50.741	-35.224	85.134	-137.385	25.025	
	1300	52.008	-30.087	89.245	-146.106	24.566	
	1400	53.255	-24.823	93.145	-155.227	24.235	

3-NICKEL ALUMINIUM

Ni_3Al

		A	B	C	D	RANGE	REF.
SOL	CP	21.15	7.7			298-1668	*
	** NI3AL(0.90-1.14) / MPT= 1668 (FERITEC.)						21

PHASE	T	CP	H	S	G	BT	
SOL	298	23.446	-36.6	27.2	-44.710	32.777	1/*
	300	23.460	-36.557	27.345	-44.760	32.612	
	400	24.230	-34.172	34.200	-47.852	26.149	
	500	25.000	-31.711	39.689	-51.555	22.538	
	600	25.770	-29.172	44.315	-55.761	20.314	
	700	26.540	-26.557	48.345	-60.398	18.860	
	800	27.310	-23.864	51.940	-65.416	17.873	
	900	28.080	-21.095	55.201	-70.775	17.189	
	1000	28.850	-18.248	58.199	-76.447	16.710	
	1100	29.620	-15.325	60.985	-82.408	16.375	
	1200	30.390	-12.324	63.595	-88.638	16.145	
	1300	31.160	-9.247	66.058	-95.122	15.994	
	1400	31.930	-6.092	68.395	-101.846	15.901	
	1500	32.700	-2.861	70.625	-108.798	15.854	
	1600	33.470	.448	72.760	-115.968	15.843	
	1668	33.994	2.742	74.164	-120.963	15.851	

NiAl

NICKEL ALUMINIUM

		A	B	C	D	RANGE	REF.
SOL	CP	10.0	3.3			298-1912	167,21
	** HOMOGENEITY RANGE						
LIQ	CP	17.0				1912-2100	*

PHASE	T	CP	H	S	G	BT	
SOL	298	10.984	-28.3	12.93	-32.155	23.573	1/167
	300	10.990	-28.280	12.998	-32.179	23.446	
	400	11.320	-27.164	16.205	-33.646	18.386	
	500	11.650	-26.016	18.766	-35.399	15.475	
	600	11.980	-24.834	20.919	-37.386	13.620	
	700	12.310	-23.620	22.791	-39.573	12.357	
	800	12.640	-22.372	24.456	-41.937	11.458	
	900	12.970	-21.092	25.964	-44.459	10.798	
	1000	13.300	-19.778	27.348	-47.126	10.301	
	1100	13.630	-18.432	28.631	-49.926	9.921	
	1200	13.960	-17.052	29.831	-52.849	9.626	
	1300	14.290	-15.640	30.961	-55.889	9.397	
	1400	14.620	-14.194	32.032	-59.040	9.218	
	1500	14.950	-12.716	33.052	-62.294	9.077	
	1600	15.280	-11.204	34.028	-65.649	8.968	
	1700	15.610	-9.660	34.964	-69.098	8.884	
	1800	15.940	-8.082	35.866	-72.640	8.821	
	1900	16.270	-6.472	36.736	-76.271	8.774	
	1912	16.310	-6.276	36.839	-76.712	8.770	
			15.	7.845			*
LIQ	1912	17.000	8.724	44.684	-76.712	8.770	
	2000	17.000	10.220	45.449	-80.678	8.817	
	2100	17.000	11.920	46.278	-85.265	8.875	

2-NICKEL 3-ALUMINIUM — Ni_2Al_3

		A	B	C	D	RANGE	REF.
SOL	CP	25.35	8.2			298-1406	*
	** NI2AL(3-3.52), MPT= 1406 (PERITEC.)						21

PHASE	T	CP	H	S	G	BT	
SOL	298	27.795	-67.5	32.6	-77.220	56.611	60/*
	300	27.810	-67.449	32.772	-77.280	56.306	
	400	28.630	-64.627	40.885	-80.980	44.252	
	500	29.450	-61.723	47.361	-85.403	37.335	
	600	30.270	-58.737	52.803	-90.419	32.939	
	700	31.090	-55.669	57.531	-95.940	29.958	
	800	31.910	-52.519	61.736	-101.907	27.844	
	900	32.730	-49.287	65.542	-108.274	26.296	
	1000	33.550	-45.973	69.033	-115.005	25.138	
	1100	34.370	-42.577	72.269	-122.072	24.257	
	1200	35.190	-39.099	75.295	-129.452	23.580	
	1300	36.010	-35.539	78.144	-137.125	23.056	
	1400	36.830	-31.897	80.842	-145.076	22.650	
	1406	36.879	-31.675	81.000	-145.561	22.629	

* * * * * * * * * *

NICKEL 3-ALUMINIUM — $NiAl_3$

		A	B	C	D	RANGE	REF.
SOL	CP	20.1	8.4			298-1127	*
	** MPT= 1127 (PERITEC.)						21

PHASE	T	CP	H	S	G	BT	
SOL	298	22.604	-36.	26.45	-43.886	32.174	1/*
	300	22.620	-35.958	26.590	-43.935	32.011	
	400	23.460	-33.654	33.212	-46.939	25.650	
	500	24.300	-31.266	38.537	-50.535	22.092	
	600	25.140	-28.794	43.042	-54.619	19.898	
	700	25.980	-26.238	46.981	-59.125	18.462	
	800	26.820	-23.598	50.505	-64.002	17.487	
	900	27.660	-20.874	53.712	-69.215	16.810	
	1000	28.500	-18.066	56.670	-74.736	16.336	
	1100	29.340	-15.174	59.425	-80.542	16.004	
	1127	29.567	-14.379	60.140	-82.156	15.934	

$NiO \cdot Al_2O_3$ NICKEL ALUMINATE

		A	B	C	D	RANGE	REF.
SOL	CP	38.05	5.58	-7.35		298-2000	*
	** MPT= 2383						62

PHASE	T	CP	H	S	G	BT	
SOL	298	31.445	-459.25	23.5	-466.257	341.821	5,*
	300	31.557	-459.192	23.695	-466.300	339.745	
	400	35.688	-455.804	33.413	-469.169	256.377	
	500	37.900	-452.115	41.634	-472.933	206.746	
	600	39.356	-448.248	48.681	-477.457	173.937	
	700	40.456	-444.256	54.833	-482.639	150.707	
	800	41.366	-440.163	60.296	-488.401	133.443	
	900	42.165	-435.986	65.215	-494.680	120.141	
	1000	42.895	-431.733	69.696	-501.429	109.602	
	1100	43.581	-427.409	73.817	-508.607	101.065	
	1200	44.236	-423.018	77.637	-516.182	94.022	
	1300	44.869	-418.562	81.203	-524.126	88.125	
	1400	45.487	-414.044	84.551	-532.416	83.125	
	1500	46.093	-409.465	87.710	-541.030	78.839	
	1600	46.691	-404.826	90.704	-549.952	75.130	
	1700	47.282	-400.127	93.552	-559.166	71.895	
	1800	47.867	-395.370	96.271	-568.658	69.054	
	1900	48.448	-390.554	98.875	-578.417	66.542	
	2000	49.026	-385.680	101.375	-588.430	64.309	

NICKEL CHROMITE

		A	B	C	D	RANGE	REF.
SOL	CP	39.950	4.270	-5.030		298-1500	*

PHASE	T	CP	H	S	G	BT	
SOL	298	35.565	-328.416	31.01	-337.662	247.546	233,234,*
	300	35.642	-328.350	31.230	-337.719	246.061	
	400	38.514	-324.625	41.928	-341.396	186.555	
	500	40.073	-320.689	50.703	-346.041	151.275	
	600	41.115	-316.627	58.107	-351.491	128.048	
	700	41.912	-312.474	64.507	-357.629	111.672	
	800	42.580	-308.249	70.148	-364.367	99.554	
	900	43.172	-303.961	75.198	-371.639	90.258	
	1000	43.717	-299.616	79.775	-379.391	82.927	
	1100	44.231	-295.218	83.966	-387.581	77.016	
	1200	44.725	-290.770	87.836	-396.173	72.163	
	1300	45.203	-286.274	91.435	-405.139	68.119	
	1400	45.671	-281.730	94.802	-414.453	64.708	
	1500	46.131	-277.140	97.969	-424.093	61.799	

NpF_3 NEPTUNIUM TRIFLUORIDE

		A	B	C	D	RANGE	REF.
SOL	CP	24.0	6.5			298-1000	*

PHASE	T	CP	H	S	G	BT	
SOL	298	25.938	-360.	28.3	-368.438	270.108	19/*
	300	25.950	-359.952	28.460	-368.490	268.481	
	400	26.600	-357.325	36.015	-371.730	203.131	
	500	27.250	-354.632	42.020	-375.642	164.215	
	600	27.900	-351.875	47.046	-380.102	138.471	
	700	28.550	-349.052	51.396	-385.029	120.228	
	800	29.200	-346.165	55.250	-390.365	106.657	
	900	29.850	-343.212	58.727	-396.066	96.191	
	1000	30.500	-340.195	61.906	-402.100	87.891	

* * * * * * * * * *

$NpF_6(g)$ NEPTUNIUM HEXAFLUORIDE (GAS)

		A	B	C	D	RANGE	REF.
GAS	CP	37.6	0.	-6.0		298- 600	*

PHASE	T	CP	H	S	G	BT	
GAS	298	30.850	-410.	88.7	-436.446	319.967	*/57
	300	30.933	-409.943	88.891	-436.610	318.113	
	400	33.850	-406.683	98.250	-445.983	243.706	
	500	35.200	-403.223	105.965	-456.205	199.434	
	600	35.933	-399.663	112.453	-467.135	170.177	

NEPTUNIUM TRICHLORIDE

$NpCl_3$

		A	B	C	D	RANGE	REF.
SOL	CP	22.2	7.0			298-1000	*

PHASE	T	CP	H	S	G	BT	
SOL	298	24.287	-216.	38.3	-227.419	166.725	1/*
	300	24.300	-215.955	38.450	-227.490	165.749	
	400	25.000	-213.490	45.537	-231.705	126.615	
	500	25.700	-210.955	51.191	-236.550	103.410	
	600	26.400	-208.350	55.938	-241.913	88.129	
	700	27.100	-205.675	60.060	-247.717	77.351	
	800	27.800	-202.930	63.725	-253.910	69.374	
	900	28.500	-200.115	67.039	-260.451	63.255	
	1000	29.200	-197.230	70.078	-267.309	58.428	

* * * * * * * * * *

NEPTUNIUM TETRACHLORIDE

$NpCl_4$

		A	B	C	D	RANGE	REF.
SOL	CP	27.2	8.57	-0.79		298- 800	*

PHASE	T	CP	H	S	G	BT	
SOL	298	28.866	-236.	47.7	-250.222	183.442	99/*
	300	28.893	-235.947	47.879	-250.310	182.375	
	400	30.134	-232.992	56.369	-255.540	139.639	
	500	31.169	-229.926	63.206	-261.529	114.330	
	600	32.123	-226.761	68.974	-268.146	97.685	
	700	33.038	-223.503	73.995	-275.299	85.964	
	800	33.933	-220.154	78.465	-282.926	77.302	

$NpOCl_2$ NEPTUNIUM OXYDICHLORIDE

		A	B	C	D	RANGE	REF.
SOL	CP	23.2	5.1	-2.3		298- 800	*

PHASE	T	CP	H	S	G	BT	
SOL	298	22.133	-245.8	33.5	-255.788	187.523	99/*
	300	22.174	-245.759	33.637	-255.850	186.412	
	400	23.803	-243.452	40.262	-259.557	141.834	
	500	24.830	-241.018	45.690	-263.863	115.350	
	600	25.621	-238.494	50.290	-268.668	97.875	
	700	26.301	-235.897	54.291	-273.901	85.527	
	800	26.921	-233.236	57.844	-279.511	76.369	

NEPTUNIUM DIOXIDE NpO_2

		A	B	C	D	RANGE	REF.
SOL	CP	19.2	1.62	-3.96		298-1500	*

PHASE	T	CP	H	S	G	BT	
SOL	298	15.228	-246.	19.2	-251.724	184.544	1
	300	15.286	-245.972	19.294	-251.760	183.432	
	400	17.373	-244.325	24.017	-253.932	138.761	
	500	18.426	-242.530	28.018	-256.539	112.148	
	600	19.072	-240.653	31.439	-259.516	94.541	
	700	19.526	-238.722	34.415	-262.812	82.065	
	800	19.877	-236.751	37.046	-266.388	72.784	
	900	20.169	-234.749	39.404	-270.212	65.625	
	1000	20.424	-232.719	41.543	-274.261	59.948	
	1100	20.655	-230.665	43.500	-278.515	55.343	
	1200	20.869	-228.588	45.307	-282.956	51.540	
	1300	21.072	-226.491	46.985	-287.572	48.352	
	1400	21.266	-224.374	48.554	-292.350	45.644	
	1500	21.454	-222.238	50.028	-297.280	43.319	

* * * * * * * * * *

NEPTUNIUM TRIOXIDE MONOHYDRATE $NpO_3 \cdot H_2O$

		A	B	C	D	RANGE	REF.
SOL	CP	29.5	15.6	-5.3		298- 800	*

PHASE	T	CP	H	S	G	BT	
SOL	298	28.189	-332.4	32.8	-342.179	250.858	99/*
	300	28.291	-332.348	32.975	-342.240	249.355	
	400	32.427	-329.293	41.733	-345.987	189.064	
	500	35.180	-325.906	49.280	-350.546	153.244	
	600	37.388	-322.275	55.894	-355.812	129.622	
	700	39.338	-318.437	61.806	-361.702	112.944	
	800	41.152	-314.412	67.179	-368.155	100.589	

OsO_2 OSMIUM DIOXIDE

		A	B	C	D	RANGE	REF.
SOL	CP	16.72	2.48	-3.39		298-1200	96

PHASE	T	CP	H	S	G	BT	
SOL	298	13.646	-70.4	12.4	-74.097	54.322	96
	300	13.697	-70.375	12.485	-74.120	54.004	
	400	15.593	-68.898	16.719	-75.586	41.304	
	500	16.604	-67.284	20.316	-77.442	33.855	
	600	17.266	-65.589	23.405	-79.632	29.010	
	700	17.764	-63.836	26.106	-82.111	25.640	
	800	18.174	-62.039	28.506	-84.843	23.181	
	900	18.533	-60.203	30.667	-87.804	21.325	
	1000	18.861	-58.333	32.637	-90.970	19.884	
	1100	19.168	-56.432	34.449	-94.326	18.743	
	1200	19.461	-54.500	36.130	-97.856	17.824	

OSMIUM TETROXIDE (YELLOW)

OsO_4

		A	B	C	D	RANGE	REF.
SOL	CP	36.2				298- 314	*
LIQ	CP	37.7				314- 404	*
GAS	CP	20.55	4.88	-3.82		404-1000	58

PHASE	T	CP	H	S	G	ET	P	
SOL	298	36.200	-94.1	32.7	-103.850	76.134	6.939	62
	300	36.200	-94.033	32.924	-103.910	75.709	8.047	
	314	36.200	-93.526	34.575	-104.383	72.662	23.097	
			3.41	10.860				62
LIQ	314	37.700	-90.116	45.435	-104.383	72.662	23.097	
	400	37.700	-86.874	54.561	-108.698	59.398	679.773	
	404	37.700	-86.723	54.936	-108.917	58.928	760.	
			8.9	22.030				62
GAS	404	20.181	-77.823	76.966	-108.917	58.928		
	500	21.462	-75.820	81.409	-116.525	50.940		
	600	22.417	-73.624	85.410	-124.870	45.490		
	700	23.186	-71.343	88.925	-133.591	41.715		
	800	23.857	-68.990	92.066	-142.643	38.974		
	900	24.470	-66.573	94.912	-151.994	36.914		
	1000	25.048	-64.097	97.520	-161.618	35.326		

OsS_2

OSMIUM DISULFIDE

		A	B	C	D	RANGE	REF.
SOL	CP	16.38	2.83	-2.1		298-1400	63

PHASE	T	CP	H	S	G	BT	
SOL	298	14.861	-35.3	13.0	-39.176	28.721	63
	300	14.896	-35.272	13.092	-39.200	28.561	
	400	16.199	-33.710	17.577	-40.741	22.263	
	500	16.955	-32.050	21.279	-42.689	18.662	
	600	17.495	-30.326	24.420	-44.978	16.386	
	700	17.932	-28.554	27.150	-47.560	14.851	
	800	18.316	-26.742	29.570	-50.398	13.770	
	900	18.668	-24.892	31.748	-53.466	12.985	
	1000	19.000	-23.009	33.732	-56.741	12.402	
	1100	19.319	-21.093	35.558	-60.207	11.964	
	1200	19.630	-19.145	37.253	-63.849	11.630	
	1300	19.935	-17.167	38.836	-67.654	11.375	
	1400	20.235	-15.158	40.324	-71.613	11.181	

* * * * * * * * * *

$OsSe_2$

OSMIUM DISELENIDE

		A	B	C	D	RANGE	REF.
SOL	CP	17.60	-2.65	-1.		298-1200	63

PHASE	T	CP	H	S	G	BT	
SOL	298	15.685	-28.7	19.5	-34.514	25.303	63
	300	15.694	-28.671	19.597	-34.550	25.173	
	400	15.915	-27.087	24.152	-36.748	20.081	
	500	15.875	-25.496	27.702	-39.347	17.201	
	600	15.732	-23.915	30.585	-42.266	15.398	
	700	15.541	-22.351	32.996	-45.449	14.192	
	800	15.324	-20.808	35.057	-48.854	13.348	
	900	15.092	-19.287	36.849	-52.451	12.739	
	1000	14.850	-17.790	38.426	-56.216	12.288	
	1100	14.602	-16.317	39.830	-60.131	11.948	
	1200	14.351	-14.870	41.090	-64.178	11.690	

OSMIUM DIPHOSPHIDE

OsP_2

		A	B	C	D	RANGE	REF.
SOL	CP	15.9	3.7			298-1400	*

PHASE	T	CP	H	S	G	BT	
SOL	298	17.003	-31.5	19.6	-37.344	27.377	60,*
	300	17.010	-31.469	19.705	-37.380	27.235	
	400	17.380	-29.749	24.649	-39.609	21.644	
	500	17.750	-27.993	28.567	-42.276	18.481	
	600	18.120	-26.199	31.836	-45.301	16.503	
	700	18.490	-24.369	34.657	-48.629	15.185	
	800	18.860	-22.501	37.150	-52.221	14.268	
	900	19.230	-20.597	39.393	-56.050	13.613	
	1000	19.600	-18.655	41.438	-60.093	13.135	
	1100	19.970	-16.677	43.324	-64.333	12.783	
	1200	20.340	-14.661	45.077	-68.754	12.523	
	1300	20.710	-12.609	46.720	-73.344	12.332	
	1400	21.080	-10.519	48.268	-78.095	12.193	

P

PHOSPHORUS (RED)

		A	B	C	D	RANGE	REF.
SOL	CP	4.051	3.559			298- 870	3
	** MPT= 870, L= 4.5						3

PHASE	T	CP	H	S	G	BT	
SOL	298	5.112	-4.17	5.45	-5.795	4.248	3*
	300	5.119	-4.161	5.482	-5.805	4.230	
	400	5.475	-3.631	7.003	-6.432	3.515	
	500	5.830	-3.066	8.263	-7.197	3.146	
	550	6.008	-2.770	8.827	-7.624	3.030	
	550	6.008	-2.770	8.827	-7.624	3.030	
	600	6.186	-2.465	9.357	-8.079	2.943	
	700	6.542	-1.828	10.338	-9.065	2.831	
	800	6.898	-1.156	11.234	-10.144	2.772	
	870	7.147	-.665	11.823	-10.951	2.751	

PHOSPHORUS (WHITE)

		A	B	C	D	RANGE	REF.
SOL	CP	4.57	3.78			298- 317	3
	** MPT= 317.3						3
LIQ	CP	6.292				317- 550	3
GAS	CP	6.292				550- 870	3
	** METASTABLE, BPT= 550, GAS(P4)						3
	CP	6.292				870-2000	3

PHASE	T	CP	H	S	G	BT	P	
SOL	298	5.697	0	9.82	-2.928	2.146	.039	40
	300	5.704	.011	9.855	-2.946	2.146	.045	
	317	5.769	.110	10.177	-3.119	2.149	.163	
			0.157	.495				3,62
LIQ	317	6.292	.267	10.672	-3.119	2.149	.163	
	400	6.292	.787	12.129	-4.064	2.221	11.616	
	500	6.292	1.416	13.533	-5.350	2.339	257.979	
	550	6.292	1.731	14.133	-6.042	2.401	760.	
GAS	550	6.292	1.731	14.133	-6.042	2.401		
	600	6.292	2.046	14.680	-6.763	2.464		
	700	6.292	2.675	15.650	-8.280	2.586		
	800	6.292	3.304	16.490	-9.888	2.702		
	870	6.292	3.744	17.018	-11.061	2.779		
	870	6.292	3.744	17.018	-11.061	2.779		
	900	6.292	3.933	17.231	-11.575	2.811		
	1000	6.292	4.562	17.894	-13.332	2.914		
	1100	6.292	5.192	18.494	-15.152	3.011		
	1200	6.292	5.821	19.041	-17.029	3.102		
	1300	6.292	6.450	19.545	-18.959	3.188		
	1400	6.292	7.079	20.011	-20.937	3.269		
	1500	6.292	7.708	20.445	-22.960	3.346		
	1600	6.292	8.338	20.852	-25.025	3.419		
	1700	6.292	8.967	21.233	-27.129	3.488		
	1800	6.292	9.596	21.593	-29.271	3.554		
	1900	6.292	10.225	21.933	-31.447	3.618		
	2000	6.292	10.854	22.256	-33.657	3.678		

P(g)

PHOSPHORUS (MONATOMIC GAS)

		A	B	C	D	RANGE	REF.
GAS	CP	4.94	0.041			298-2000	3

PHASE	T	CP	H	S	G	BT	
GAS	298	4.952	75.63	38.98	64.008	-46.926	3*
	300	4.952	75.639	39.011	63.936	-46.584	
	400	4.956	76.135	40.436	59.960	-32.765	
	500	4.960	76.630	41.542	55.859	-24.419	
	550	4.963	76.879	42.015	53.770	-21.369	
	550	4.963	76.879	42.015	53.770	-21.369	
	600	4.965	77.127	42.447	51.658	-18.819	
	700	4.969	77.623	43.213	47.374	-14.793	
	800	4.973	78.120	43.876	43.019	-11.754	
	900	4.977	78.618	44.462	38.602	-9.375	
	1000	4.981	79.116	44.987	34.129	-7.460	
	1100	4.985	79.614	45.462	29.606	-5.883	
	1200	4.989	80.113	45.896	25.038	-4.561	
	1300	4.993	80.612	46.295	20.428	-3.435	
	1400	4.997	81.111	46.666	15.780	-2.464	
	1500	5.001	81.611	47.010	11.096	-1.617	
	1600	5.006	82.112	47.333	6.378	-.871	
	1700	5.010	82.613	47.637	1.630	-.210	
	1800	5.014	83.114	47.923	-3.148	.382	
	1900	5.018	83.615	48.195	-7.954	.915	
	2000	5.022	84.117	48.452	-12.787	1.397	

PHOSPHORUS (DIATOMIC GAS) $P_2(g)$

		A	B	C	D	RANGE	REF.
GAS	CP	8.675	0.191	-0.994		298-2000	3

PHASE	T	CP	H	S	G	BT	
GAS	298	7.614	34.34	52.1	18.806	-13.787	3*
	300	7.628	34.354	52.147	18.710	-13.632	
	400	8.130	35.145	54.420	13.377	-7.310	
	500	8.373	35.972	56.263	7.840	-3.427	
	550	8.451	36.393	57.065	5.007	-1.990	
	550	8.451	36.393	57.065	5.007	-1.990	
	600	8.513	36.817	57.803	2.135	-.778	
	700	8.606	37.673	59.123	-3.713	1.159	
	800	8.672	38.537	60.277	-9.684	2.646	
	900	8.724	39.407	61.301	-15.764	3.829	
	1000	8.767	40.282	62.223	-21.941	4.796	
	1100	8.803	41.160	63.060	-28.206	5.605	
	1200	8.835	42.042	63.827	-34.551	6.293	
	1300	8.864	42.927	64.536	-40.970	6.889	
	1400	8.892	43.815	65.194	-47.456	7.409	
	1500	8.917	44.705	65.808	-54.007	7.870	
	1600	8.942	45.598	66.384	-60.617	8.281	
	1700	8.965	46.494	66.927	-67.283	8.651	
	1800	8.988	47.391	67.440	-74.001	8.986	
	1900	9.010	48.291	67.927	-80.770	9.292	
	2000	9.032	49.193	68.390	-87.586	9.572	

$P_4(g)$ PHOSPHORUS (TETRATOMIC GAS)

		A	B	C	D	RANGE	REF.
GAS	CP	19.562	0.162	-3.213		298-2000	3

PHASE	T	CP	H	S	G	BT	
GAS	298	15.996	14.082	66.89	-5.861	4.297	3*
	300	16.041	14.112	66.989	-5.985	4.361	
	400	17.619	15.806	71.852	-12.935	7.068	
	500	18.358	17.609	75.872	-20.327	8.886	
	550	18.589	18.533	77.633	-24.166	9.604	
	550	18.589	18.533	77.633	-24.166	9.604	
	600	18.767	19.467	79.258	-28.088	10.233	
	700	19.020	21.357	82.172	-36.163	11.292	
	800	19.190	23.268	84.723	-44.511	12.161	
	900	19.311	25.193	86.991	-53.098	12.896	
	1000	19.403	27.129	89.030	-61.901	13.530	
	1100	19.475	29.073	90.883	-70.898	14.088	
	1200	19.533	31.024	92.580	-80.073	14.585	
	1300	19.582	32.979	94.146	-89.410	15.033	
	1400	19.625	34.940	95.598	-98.898	15.441	
	1500	19.662	36.904	96.954	-108.526	15.814	
	1600	19.696	38.872	98.224	-118.286	16.159	
	1700	19.726	40.843	99.419	-128.169	16.479	
	1800	19.754	42.817	100.547	-138.167	16.778	
	1900	19.781	44.794	101.616	-148.276	17.058	
	2000	19.806	46.773	102.631	-158.489	17.321	

PHOSPHINE (GAS) PH_3 (g)

		A	B	C	D	RANGE	REF.
GAS	CP	6.287	9.675	-0.273		298- 800	3
	CP	16.315	1.295	-21.577		800-2000	3

PHASE	T	CP	H	S	G	BT	
GAS	298	8.864	1.33	50.24	-13.649	10.006	3
	300	8.886	1.346	50.295	-13.742	10.012	
	400	9.986	2.291	53.005	-18.911	10.334	
	500	11.015	3.341	55.344	-24.331	10.636	
	600	12.016	4.493	57.441	-29.972	10.919	
	700	13.004	5.744	59.368	-35.813	11.183	
	800	13.984	7.094	61.169	-41.841	11.432	
	800	13.980	7.094	61.169	-41.841	11.432	
	900	14.817	8.536	62.866	-48.044	11.668	
	1000	15.452	10.050	64.461	-54.411	11.893	
	1100	15.956	11.622	65.959	-60.933	12.108	
	1200	16.371	13.239	67.365	-67.600	12.313	
	1300	16.722	14.894	68.690	-74.403	12.510	
	1400	17.027	16.581	69.940	-81.335	12.699	
	1500	17.299	18.298	71.125	-88.389	12.880	
	1600	17.544	20.040	72.249	-95.558	13.054	
	1700	17.770	21.806	73.319	-102.837	13.222	
	1800	17.980	23.594	74.341	-110.220	13.384	
	1900	18.178	25.402	75.319	-117.704	13.541	
	2000	18.366	27.229	76.256	-125.283	13.692	

$PF_3(g)$

PHOSPORUS TRIFLUORIDE (GAS)

		A	B	C	D	RANGE	REF.
GAS	CP	16.116	3.583	-2.812		298- 800	3
	CP	19.712	0.061	-7.808		800-2000	3

PHASE	T	CP	H	S	G	BT	
GAS	298	14.021	-229.07	65.23	-248.518	182.193	3
	300	14.066	-229.044	65.317	-248.639	181.158	
	400	15.792	-227.541	69.628	-255.393	139.559	
	500	16.783	-225.909	73.266	-262.542	114.773	
	600	17.485	-224.194	76.391	-270.029	98.371	
	700	18.050	-222.417	79.130	-277.808	86.747	
	800	18.543	-220.587	81.573	-285.845	78.100	
	800	18.541	-220.587	81.573	-285.845	78.100	
	900	18.803	-218.719	83.773	-294.114	71.430	
	1000	18.992	-216.828	85.764	-302.592	66.140	
	1100	19.134	-214.922	87.581	-311.261	61.850	
	1200	19.243	-213.003	89.251	-320.104	58.307	
	1300	19.329	-211.074	90.795	-329.107	55.335	
	1400	19.399	-209.137	92.230	-338.259	52.812	
	1500	19.456	-207.194	93.570	-347.550	50.645	
	1600	19.505	-205.246	94.827	-356.970	48.766	
	1700	19.546	-203.294	96.011	-366.513	47.125	
	1800	19.581	-201.337	97.129	-376.170	45.679	
	1900	19.612	-199.378	98.189	-385.937	44.399	
	2000	19.639	-197.415	99.196	-395.806	43.258	

PHOSPHORUS PENTAFLUORIDE (GAS)

PF_5 (g)

		A	B	C	D	RANGE	REF.
GAS	CP	24.282	7.123	-5.460		298- 800	3
	CP	31.332	0.165	-14.973		800-2000	3

PHASE	T	CP	H	S	G	BT	
GAS	298	20.264	-381.07	71.87	-402.498	295.079	3,104
	300	20.352	-381.032	71.996	-402.631	293.356	
	400	23.719	-378.810	78.366	-410.156	224.129	
	500	25.659	-376.334	83.883	-418.276	182.853	
	600	27.039	-373.696	88.689	-426.909	155.523	
	700	28.154	-370.935	92.943	-435.995	136.142	
	800	29.127	-368.070	96.767	-445.484	121.717	
	800	29.124	-368.070	96.767	-445.484	121.717	
	900	29.632	-365.131	100.228	-455.336	110.586	
	1000	30.000	-362.148	103.370	-465.519	101.753	
	1100	30.276	-359.134	106.243	-476.001	94.585	
	1200	30.490	-356.095	108.887	-486.760	88.663	
	1300	30.661	-353.037	111.334	-497.772	83.694	
	1400	30.799	-349.964	113.612	-509.021	79.472	
	1500	30.914	-346.878	115.741	-520.490	75.845	
	1600	31.011	-343.782	117.739	-532.165	72.700	
	1700	31.094	-340.677	119.622	-544.034	69.950	
	1800	31.167	-337.563	121.401	-556.086	67.527	
	1900	31.231	-334.443	123.088	-568.311	65.379	
	2000	31.288	-331.317	124.691	-580.700	63.465	

PCl_3

PHOSPHORUS TRICHLORIDE

		A	B	C	D	RANGE	REF.
LIQ	CP	31.4				298- 349	*
	** MPT= 182 / BPT= 348.5						3
GAS	CP	19.686	0.097	-2.252		348-1600	3

PHASE	T	CP	H	S	G	BT	P	
LIQ	298	31.400	-76.7	52.22	-92.269	67.644	117.336	3*
	300	31.400	-76.642	52.414	-92.366	67.298	127.505	
	348	31.400	-75.119	57.120	-95.025	59.600	760.	
			7.3	20.947				*
GAS	348	17.866	-67.819	78.067	-95.025	59.600		
	400	18.317	-66.887	80.561	-99.111	54.159		
	500	18.834	-65.026	84.711	-107.381	46.943		
	600	19.119	-63.127	88.172	-116.030	42.270		
	700	19.294	-61.206	91.133	-124.999	39.032		
	800	19.412	-59.270	93.718	-134.245	36.679		
	900	19.495	-57.325	96.009	-143.733	34.908		
	1000	19.558	-55.372	98.067	-153.439	33.538		
	1100	19.607	-53.414	99.933	-163.340	32.457		
	1200	19.646	-51.451	101.641	-173.420	31.588		
	1300	19.679	-49.485	103.215	-183.664	30.881		
	1400	19.707	-47.515	104.674	-194.059	30.298		
	1500	19.731	-45.543	106.035	-204.595	29.814		
	1600	19.753	-43.569	107.309	-215.263	29.408		

PHOSPHORUS PENTACHLORIDE (GAS) $PCl_5\,(g)$

		A	B	C	D	RANGE	REF.
GAS	CP	31.419	0.197	-4.267		298-2000	3

PHASE	T	CP	H	S	G	BT	
GAS	298	26.678	-86.07	87.0	-112.009	82.116	3*
	300	26.737	-86.021	87.165	-112.170	81.727	
	400	28.831	-83.227	95.186	-121.302	66.285	
	500	29.811	-80.290	101.737	-131.159	57.337	
	600	30.352	-77.279	107.224	-141.614	51.590	
	700	30.686	-74.226	111.930	-152.577	47.643	
	800	30.910	-71.146	116.043	-163.980	44.803	
	900	31.070	-68.046	119.694	-175.771	42.689	
	1000	31.189	-64.933	122.974	-187.907	41.073	
	1100	31.283	-61.809	125.951	-200.355	39.812	
	1200	31.359	-58.677	128.676	-213.089	38.814	
	1300	31.423	-55.538	131.189	-226.083	38.013	
	1400	31.477	-52.393	133.519	-239.320	37.365	
	1500	31.525	-49.243	135.693	-252.782	36.835	
	1600	31.568	-46.088	137.729	-266.454	36.401	
	1700	31.606	-42.930	139.644	-280.324	36.043	
	1800	31.642	-39.767	141.451	-294.379	35.747	
	1900	31.675	-36.601	143.163	-308.611	35.503	
	2000	31.706	-33.432	144.789	-323.009	35.302	

PBr_3 (g) PHOSPHORUS TRIBROMIDE (GAS)

		A	B	C	D	RANGE	REF.
GAS	CP	19.792	0.042	-1.467		298-2000	3

PHASE	T	CP	H	S	G	BT	
GAS	298	18.154	-34.87	83.206	-59.678	43.751	3
	300	18.175	-34.836	83.318	-59.832	43.593	
	400	18.892	-32.978	88.660	-68.442	37.400	
	500	19.226	-31.070	92.915	-77.528	33.892	
	600	19.410	-29.138	96.438	-87.001	31.694	
	700	19.522	-27.191	99.440	-96.798	30.226	
	800	19.596	-25.234	102.052	-106.876	29.201	
	900	19.649	-23.272	104.363	-117.199	28.464	
	1000	19.687	-21.305	106.435	-127.740	27.921	
	1100	19.717	-19.335	108.313	-138.479	27.517	
	1200	19.741	-17.362	110.030	-149.398	27.213	
	1300	19.760	-15.387	111.611	-160.481	26.983	
	1400	19.776	-13.410	113.075	-171.716	26.810	
	1500	19.790	-11.432	114.440	-183.092	26.680	
	1600	19.802	-9.452	115.718	-194.601	26.585	
	1700	19.813	-7.472	116.919	-206.233	26.517	
	1800	19.822	-5.490	118.052	-217.982	26.470	
	1900	19.831	-3.507	119.123	-229.842	26.441	
	2000	19.839	-1.524	120.141	-241.805	26.427	

* * * * * * * * * *

PJ_3 (g) PHOSPHORUS TRIIODIDE (GAS)

		A	B	C	D	RANGE	REF.
GAS	CP	19.8		-0.92		298- 600	19,*
	** MPT=	334.15					62

PHASE	T	CP	H	S	G	BT	
GAS	298	18.765	4.3	89.45	-22.370	16.400	1*/19
	300	18.778	4.335	89.566	-22.535	16.419	
	400	19.225	6.238	95.039	-31.777	17.365	
	500	19.432	8.172	99.353	-41.505	18.144	
	600	19.544	10.121	102.907	-51.623	18.806	

PHOSPHORUS OXYTRICHLORIDE $POCl_3$

		A	B	C	D	RANGE	REF.
LIQ	CP	33.17				298- 383	62
GAS	CP	24.594	0.623	-4.126		383-2000	3

PHASE	T	CP	H	S	G	BT	P	
LIQ	298	33.170	-142.8	53.17	-158.653	116.311	28.529	62
	300	33.170	-142.739	53.375	-158.751	115.666	31.411	
	383	33.170	-139.986	61.477	-163.531	93.328	760.	
			8.24	21.514				
GAS	383	22.020	-131.746	82.991	-163.531	93.328		62
	400	22.264	-131.369	83.953	-164.950	90.137		
	500	23.255	-129.088	89.039	-173.608	75.894		
	600	23.822	-126.732	93.333	-182.732	66.569		
	700	24.188	-124.330	97.035	-192.255	60.033		
	800	24.448	-121.898	100.283	-202.124	55.225		
	900	24.645	-119.443	103.174	-212.299	51.560		
	1000	24.804	-116.970	105.779	-222.749	48.688		
	1100	24.938	-114.483	108.150	-233.447	46.388		
	1200	25.055	-111.983	110.325	-244.372	44.512		
	1300	25.160	-109.472	112.334	-255.507	42.960		
	1400	25.256	-106.951	114.202	-266.835	41.660		
	1500	25.345	-104.421	115.948	-278.343	40.560		
	1600	25.430	-101.882	117.586	-290.021	39.620		
	1700	25.510	-99.335	119.131	-301.857	38.812		
	1800	25.588	-96.780	120.591	-313.844	38.111		
	1900	25.663	-94.218	121.976	-325.973	37.500		
	2000	25.737	-91.648	123.295	-338.237	36.966		

$POBr_3$ (g)   PHOSPHORUS OXYTRIBROMIDE (GAS)

		A	B	C	D	RANGE	REF.
GAS	CP	23.064	2.515	-2.079		298- 800	3
	CP	25.638	0.075	-6.069		800-2000	3

PHASE	T	CP	H	S	G	BT	
GAS	298	21.475	-97.17	85.978	-122.804	90.030	3*
	300	21.508	-97.130	86.111	-122.964	89.591	
	400	22.771	-94.909	92.492	-131.906	72.080	
	500	23.490	-92.593	97.656	-141.422	61.824	
	600	23.995	-90.218	101.986	-151.410	55.158	
	700	24.400	-87.798	105.716	-161.799	50.523	
	800	24.751	-85.340	108.998	-172.538	47.142	
	800	24.750	-85.340	108.998	-172.538	47.142	
	900	24.956	-82.854	111.925	-183.587	44.587	
	1000	25.106	-80.350	114.563	-194.913	42.604	
	1100	25.219	-77.834	116.961	-206.491	41.032	
	1200	25.307	-75.307	119.160	-218.299	39.763	
	1300	25.376	-72.773	121.188	-230.318	38.725	
	1400	25.433	-70.233	123.071	-242.532	37.866	
	1500	25.481	-67.687	124.827	-254.928	37.148	
	1600	25.521	-65.137	126.473	-267.493	36.543	
	1700	25.555	-62.583	128.021	-280.219	36.029	
	1800	25.586	-60.026	129.483	-293.095	35.591	
	1900	25.612	-57.466	130.867	-306.113	35.216	
	2000	25.636	-54.903	132.181	-319.266	34.892	

PHOSPHORUS MONOXIDE (GAS) PO (g)

		A	B	C	D	RANGE	REF.
GAS	CP	7.071	1.700			298- 800	3
	CP	8.790	0.111	-2.982		800-2000	3

PHASE	T	CP	H	S	G	BT	
GAS	298	7.578	-1.46	53.22	-17.328	12.703	3
	300	7.581	-1.446	53.267	-17.426	12.697	
	400	7.751	-.679	55.471	-22.868	12.496	
	500	7.921	.104	57.219	-28.505	12.461	
	600	8.091	.905	58.678	-34.302	12.496	
	700	8.261	1.722	59.938	-40.234	12.563	
	800	8.431	2.557	61.052	-46.285	12.646	
	800	8.413	2.557	61.052	-46.285	12.646	
	900	8.522	3.404	62.050	-52.441	12.736	
	1000	8.603	4.260	62.952	-58.692	12.829	
	1100	8.666	5.124	63.775	-65.029	12.922	
	1200	8.716	5.993	64.531	-71.444	13.014	
	1300	8.758	6.867	65.231	-77.933	13.103	
	1400	8.793	7.745	65.881	-84.489	13.191	
	1500	8.824	8.625	66.489	-91.108	13.276	
	1600	8.851	9.509	67.059	-97.785	13.359	
	1700	8.876	10.396	67.597	-104.519	13.439	
	1800	8.898	11.284	68.104	-111.304	13.516	
	1900	8.918	12.175	68.586	-118.139	13.591	
	2000	8.937	13.068	69.044	-125.020	13.663	

$P_4O_6(g)$ DIPHOSPHORUS TRIOXIDE (DIMERIC GAS)

		A	B	C	D	RANGE	REF.
GAS	CP	43.597	11.490	-11.241		298- 800	3
	CP	55.119	0.204	-27.215		800-2000	3

PHASE	T	CP	H	S	G	PT	
GAS	298	34.377	-529.18	82.6	-553.807	406.006	3*
	300	34.554	-529.116	82.813	-553.960	403.614	
	400	41.167	-525.291	93.772	-562.800	307.541	
	500	44.846	-520.976	103.385	-572.669	250.347	
	600	47.368	-516.359	111.796	-583.437	212.545	
	700	49.346	-511.521	119.251	-594.996	185.791	
	800	51.033	-506.500	125.953	-607.262	165.919	
	800	51.030	-506.500	125.953	-607.262	165.919	
	900	51.943	-501.349	132.019	-620.166	150.617	
	1000	52.601	-496.120	137.527	-633.647	138.502	
	1100	53.094	-490.834	142.565	-647.655	128.695	
	1200	53.474	-485.505	147.202	-662.147	120.610	
	1300	53.774	-480.142	151.494	-677.084	113.844	
	1400	54.016	-474.752	155.489	-692.436	108.109	
	1500	54.215	-469.340	159.222	-708.173	103.195	
	1600	54.382	-463.910	162.727	-724.273	98.944	
	1700	54.524	-458.464	166.028	-740.712	95.238	
	1800	54.646	-453.006	169.148	-757.472	91.982	
	1900	54.753	-447.536	172.106	-774.536	89.104	
	2000	54.847	-442.055	174.916	-791.888	86.545	

PHOSPHORUS DIOXIDE (GAS) PO_2 (g)

		A	B	C	D	RANGE	REF.
GAS	CP	10.517	3.173	-1.397		298- 800	3
	CP	13.747	0.060	-6.138		800-2000	3

PHASE	T	CP	H	S	G	BT	
GAS	298	9.891	-75.17	60.6	-93.238	68.354	3*
	300	9.917	-75.152	60.661	-93.350	68.015	
	400	10.913	-74.105	63.665	-99.571	54.410	
	500	11.545	-72.981	66.172	-106.066	46.368	
	600	12.033	-71.801	68.321	-112.794	41.091	
	700	12.453	-70.576	70.208	-119.722	37.384	
	800	12.837	-69.312	71.896	-126.829	34.653	
	800	12.836	-69.312	71.896	-126.829	34.653	
	900	13.043	-68.017	73.421	-134.096	32.567	
	1000	13.193	-66.705	74.803	-141.508	30.931	
	1100	13.306	-65.380	76.066	-149.052	29.618	
	1200	13.393	-64.045	77.228	-156.718	28.546	
	1300	13.462	-62.702	78.303	-164.495	27.658	
	1400	13.518	-61.353	79.302	-172.376	26.913	
	1500	13.564	-59.998	80.237	-180.353	26.281	
	1600	13.603	-58.640	81.113	-188.421	25.741	
	1700	13.637	-57.278	81.939	-196.574	25.275	
	1800	13.666	-55.913	82.719	-204.807	24.870	
	1900	13.691	-54.545	83.459	-213.117	24.517	
	2000	13.714	-53.175	84.162	-221.498	24.207	

P_4O_{10} DIPHOSPHORUS PENTOXIDE (DIMERIC)

		A	B	C	D	RANGE	REF.
SOL	CP	35.792	77.620	-7.424		298- 631	3
GAS	CP	76.616	1.199	-43.857		631-2000	3

PHASE	T	CP	H	S	G	BT	P	
SOL	298	50.583	-719.38	54.68	-735.683	539.343		3*
	300	50.829	-719.286	54.994	-735.784	536.091		
	400	62.200	-713.609	71.248	-742.108	405.524	.014	
	500	71.632	-706.908	86.161	-749.989	327.864	6.370	
	600	80.302	-699.307	99.995	-759.304	276.614	305.637	
	631	82.906	-696.777	104.106	-762.468	264.120	760.	
			21.844	34.618				3*
GAS	631	66.358	-674.933	138.724	-762.468	264.120		
	700	68.505	-670.277	145.725	-772.285	241.151		
	800	70.723	-663.309	155.027	-787.330	215.118		
	900	72.281	-656.154	163.452	-803.261	195.085		
	1000	73.429	-648.866	171.130	-819.996	179.234		
	1100	74.310	-641.477	178.171	-837.466	166.411		
	1200	75.009	-634.010	184.668	-855.612	155.849		
	1300	75.580	-626.480	190.695	-874.384	147.017		
	1400	76.057	-618.897	196.314	-893.737	139.537		
	1500	76.465	-611.271	201.576	-913.634	133.134		
	1600	76.821	-603.606	206.523	-934.042	127.601		
	1700	77.137	-595.908	211.189	-954.930	122.781		
	1800	77.421	-588.180	215.607	-976.271	118.551		
	1900	77.679	-580.424	219.800	-998.044	114.817		
	2000	77.918	-572.644	223.790	-1020.225	111.500		

4-PHOSPHORUS 3-SULFIDE

P_4S_3

		A	B	C	D	RANGE	REF.
SOL-A	CP	4.364	115.723			298- 314	63
	** TPT= 313.9						63
SOL-B	CP	38.93	13.6			314- 446	63
LIQ	CP	55.				446- 677	*
	** BPT= 677.551 GAS(P4S2 + P4S3 + ...)						63

PHASE	T	CP	H	S	G	BT	
SOL-A	298	38.867	-53.5	48.6	-67.990	49.845	63*
	300	39.081	-53.428	48.841	-68.080	49.603	
	314	40.689	-52.873	50.647	-68.772	47.888	
			2.464	7.850			63
SOL-B	314	43.199	-50.409	58.497	-68.772	47.888	
	400	44.370	-46.640	69.104	-74.281	40.591	
	446	44.996	-44.584	73.967	-77.574	38.018	
			4.82	10.807			63
LIQ	446	55.000	-39.764	84.775	-77.574	38.018	
	500	55.000	-36.794	91.061	-82.324	35.989	
	600	55.000	-31.294	101.088	-91.947	33.496	
	678	55.000	-27.029	107.774	-100.051	32.277	

PS (g)

PHOSPHORUS MONOSULFIDE (GAS)

		A	B	C	D	RANGE	REF.
GAS	CP	8.85	0.118	-0.43		298-2000	63

** H298 S. REF 63

PHASE	T	CP	H	S	G	BT	
GAS	298	8.401	53.	56.26	36.226	-26.558	63*
	300	8.408	53.016	56.312	36.122	-26.318	
	400	8.628	53.869	58.765	30.363	-16.592	
	500	8.737	54.738	60.704	24.386	-10.661	
	600	8.801	55.615	62.303	18.233	-6.642	
	700	8.845	56.497	63.663	11.933	-3.726	
	800	8.877	57.383	64.846	5.507	-1.505	
	900	8.903	58.272	65.893	-1.031	.250	
	1000	8.925	59.164	66.832	-7.668	1.676	
	1100	8.944	60.057	67.684	-14.395	2.860	
	1200	8.962	60.953	68.463	-21.203	3.862	
	1300	8.978	61.850	69.181	-28.085	4.722	
	1400	8.993	62.748	69.847	-35.037	5.470	
	1500	9.008	63.648	70.468	-42.053	6.128	
	1600	9.022	64.550	71.050	-49.130	6.712	
	1700	9.036	65.453	71.597	-56.262	7.234	
	1800	9.049	66.357	72.114	-63.448	7.705	
	1900	9.062	67.263	72.603	-70.684	8.132	
	2000	9.075	68.169	73.069	-77.968	8.521	

4-PHOSPHORUS 5-SULFIDE

P_4S_5

		A	B	C	D	RANGE	REF.
SOL	CP	41.87	28.8			298- 400	*

PHASE	T	CP	H	S	G	BT	
SOL	298	50.457	-72.88	60.4	-90.888	66.632	63*
	300	50.510	-72.787	60.712	-91.000	66.303	
	400	53.390	-67.592	75.638	-97.847	53.468	

* * * * * * * * * *

2-PHOSPHORUS 3-SULFIDE

P_2S_3

		A	B	C	D	RANGE	REF.
SOL	CP	18.98	25.9			298- 503	63
LIQ	CP	40.0				503- 700	63

PHASE	T	CP	H	S	G	BT	
SOL	298	26.702	-37.34	33.65	-47.373	34.730	63*/63
	300	26.750	-37.291	33.815	-47.435	34.561	
	400	29.340	-34.486	41.866	-51.232	27.996	
	500	31.930	-31.423	48.691	-55.768	24.379	
	503	32.008	-31.327	48.882	-55.914	24.298	
			3.525	7.008			63
LIQ	503	40.000	-27.802	55.890	-55.914	24.298	
	600	40.000	-23.922	62.944	-61.688	22.473	
	700	40.000	-19.922	69.110	-68.298	21.327	

P_4S_7 4-PHOSPHORUS 7-SULFIDE

		A	B	C	D	RANGE	REF.
SOL	CP	44.8	44.0			298- 581	63
LIQ	CP	88.0				581- 700	63

PHASE	T	CP	H	S	G	BT	
SOL	298	57.919	-77.28	73.5	-99.194	72.721	63*
	300	58.000	-77.173	73.859	-99.330	72.372	
	400	62.400	-71.153	91.147	-107.611	58.804	
	500	66.800	-64.693	105.544	-117.465	51.351	
	581	70.364	-59.138	115.834	-126.437	47.567	
			8.75	15.060			63
LIQ	581	88.000	-50.388	130.894	-126.437	47.567	
	600	88.000	-48.716	133.726	-128.951	46.977	
	700	88.000	-39.916	147.291	-143.019	44.659	

* * * * * * * * * *

P_2S_5 2-PHOSPHORUS 5-SULFIDE

		A	B	C	D	RANGE	REF.
SOL	CP	25.37	33.53			298- 560	63
LIQ	CP	70.2				560- 800	63

PHASE	T	CP	H	S	G	BT	
SOL	298	35.367	-44.5	45.62	-58.102	42.595	*/63
	300	35.429	-44.435	45.839	-58.186	42.394	
	400	38.782	-40.724	56.490	-63.320	34.601	
	500	42.135	-36.678	65.505	-69.430	30.352	
	560	44.147	-34.090	70.392	-73.509	28.692	
			4.91	8.768			63
LIQ	560	70.200	-29.180	79.159	-73.509	28.692	
	600	70.200	-26.372	84.003	-76.773	27.968	
	700	70.200	-19.352	94.824	-85.729	26.769	
	800	70.200	-12.332	104.198	-95.690	26.145	

PHOSPHORUS MONONITRIDE (GAS)

		A	B	C	D	RANGE	REF.
GAS	CP	6.554	2.135	-0.085		298- 800	3
	CP	8.724	0.129	-3.708		800-2000	3

PHASE	T	CP	H	S	G	BT	
GAS	298	7.095	20.83	50.437	5.792	-4.246	3*
	300	7.100	20.843	50.481	5.699	-4.152	
	400	7.355	21.566	52.559	.542	-.296	
	500	7.587	22.313	54.226	-4.799	2.098	
	600	7.811	23.083	55.629	-10.294	3.750	
	700	8.031	23.876	56.850	-15.919	4.971	
	800	8.249	24.690	57.936	-21.659	5.918	
	800	8.248	24.690	57.936	-21.659	5.918	
	900	8.382	25.521	58.916	-27.503	6.679	
	1000	8.482	26.365	59.804	-33.440	7.309	
	1100	8.559	27.217	60.617	-39.461	7.841	
	1200	8.621	28.076	61.364	-45.561	8.299	
	1300	8.672	28.941	62.056	-51.732	8.698	
	1400	8.715	29.810	62.701	-57.970	9.051	
	1500	8.753	30.684	63.303	-64.271	9.366	
	1600	8.786	31.561	63.869	-70.630	9.649	
	1700	8.815	32.441	64.403	-77.044	9.906	
	1800	8.842	33.324	64.907	-83.509	10.141	
	1900	8.866	34.209	65.386	-90.024	10.357	
	2000	8.889	35.097	65.841	-96.586	10.556	

Pb LEAD

		A	B	C	D	RANGE	REF.
SOL	CP	5.789	2.082			298- 600	3
	** MPT=	600.58					3
LIQ	CP	7.765	-0.738			600-1200	3
	CP	7.110	-0.195			1200-1400	3
	CP	6.491	0.246			1400-1700	3
	CP	6.143	0.453			1700-2026	3
GAS	CP	4.846	-0.846	2.945	0.666	2026-3000	3

PHASE	T	CP	H	S	G	BT	P	
SOL	298	6.410	0	15.484	-4.617	3.384		3
	300	6.414	.012	15.524	-4.645	3.385		
	400	6.622	.664	17.397	-6.295	3.440		
	500	6.830	1.336	18.897	-8.112	3.546		
	600	7.038	2.030	20.161	-10.067	3.667		
	601	7.039	2.034	20.168	-10.079	3.668		
			1.141	1.900				3
LIQ	601	7.322	3.175	22.068	-10.079	3.668		
	700	7.248	3.899	23.184	-12.330	3.850		
	800	7.175	4.620	24.147	-14.697	4.016		
	900	7.101	5.334	24.987	-17.155	4.166		
	1000	7.027	6.040	25.732	-19.692	4.304	.012	
	1100	6.953	6.739	26.398	-22.299	4.431	.088	
	1200	6.879	7.431	27.000	-24.969	4.548	.468	
	1200	6.876	7.431	27.000	-24.969	4.548	.468	
	1300	6.857	8.118	27.550	-27.697	4.657	1.919	
	1400	6.837	8.802	28.057	-30.477	4.758	6.394	
	1400	6.835	8.802	28.057	-30.477	4.758	6.394	
	1500	6.860	9.487	28.529	-33.307	4.853	18.076	
	1600	6.885	10.174	28.973	-36.182	4.943	44.717	
	1700	6.909	10.864	29.391	-39.101	5.027	99.154	
	1700	6.913	10.864	29.391	-39.101	5.027	99.154	
	1800	6.958	11.558	29.787	-42.060	5.107	200.753	
	1900	7.004	12.256	30.165	-45.058	5.184	376.599	
	2000	7.049	12.958	30.525	-48.092	5.256	662.277	
	2026	7.061	13.142	30.616	-48.887	5.274	760.	
			42.53	20.992				3
GAS	2026	5.937	55.672	51.608	-48.887	5.274		
	2100	6.073	56.116	51.824	-52.714	5.487		

PHASE	T	CP	H	S	G	BT	P
GAS	2200	6.269	56.733	52.111	-57.911	5.754	
	2300	6.479	57.370	52.394	-63.136	6.000	
	2400	6.703	58.029	52.675	-68.390	6.229	
	2500	6.941	58.711	52.953	-73.671	6.441	
	2600	7.192	59.418	53.230	-78.980	6.640	
	2700	7.457	60.150	53.506	-84.317	6.826	
	2800	7.736	60.910	53.783	-89.681	7.001	
	2900	8.029	61.698	54.059	-95.074	7.166	
	3000	8.335	62.516	54.336	-100.493	7.322	

Pb_3O_4 TRILEAD TETROXIDE

		A	B	C	D	RANGE	REF.
SOL	CP	44.740	3.463	-7.985	2.241	298-1500	3A

PHASE	T	CP	H	S	G	BT	
SOL	298	36.989	-171.77	50.66	-186.874	137.001	3A
	300	37.108	-171.701	50.889	-186.968	136.225	
	400	41.493	-167.744	62.244	-192.642	105.269	
	500	43.838	-163.468	71.776	-199.356	87.150	
	600	45.407	-159.002	79.915	-206.951	75.392	
	700	46.633	-154.398	87.009	-215.304	67.230	
	800	47.697	-149.680	93.307	-224.326	61.291	
	900	48.686	-144.861	98.982	-233.945	56.817	
	1000	49.645	-139.944	104.162	-244.106	53.356	
	1100	50.601	-134.932	108.938	-254.764	50.624	
	1200	51.568	-129.824	113.383	-265.883	48.430	
	1300	52.557	-124.618	117.549	-277.431	46.647	
	1400	53.573	-119.311	121.481	-289.385	45.181	
	1500	54.622	-113.902	125.213	-301.721	43.967	

* * * * * * * * * *

PbO_2 LEAD DIOXIDE

		A	B	C	D	RANGE	REF.
SOL	CP	15.109	7.413	-2.146	-3.344	298-1200	3A

PHASE	T	CP	H	S	G	BT	
SOL	298	14.608	-65.6	17.16	-70.716	51.843	3A
	300	14.647	-65.573	17.250	-70.748	51.547	
	400	16.198	-64.023	21.700	-72.703	39.728	
	500	17.121	-62.353	25.421	-75.064	32.815	
	600	17.757	-60.608	28.602	-77.769	28.331	
	700	18.222	-58.808	31.375	-80.771	25.221	
	800	18.564	-56.968	33.832	-84.033	22.960	
	900	18.807	-55.098	36.034	-87.529	21.258	
	1000	18.963	-53.209	38.024	-91.233	19.942	
	1100	19.040	-51.308	39.836	-95.127	18.903	
	1200	19.040	-49.404	41.493	-99.195	18.068	

LEAD SULFIDE

PbS

			A	B	C	D	RANGE	REF.
SOL	CP		11.17	2.2			298-1392	63*
LIQ	CP		14.8				1392-1609	*
	**	BPT=	1609.5					*
GAS	CP		8.920	-0.490			1609-2400	58,63

PHASE	T	CP	H	S	G	BT	P	
SOL	298	11.826	-23.5	21.8	-30.000	21.993		63
	300	11.830	-23.478	21.873	-30.040	21.887		
	400	12.050	-22.284	25.307	-32.407	17.709		
	500	12.270	-21.068	28.019	-35.078	15.334		
	600	12.490	-19.830	30.276	-37.995	13.842		
	700	12.710	-18.570	32.217	-41.122	12.841		
	800	12.930	-17.288	33.929	-44.431	12.140		
	900	13.150	-15.984	35.465	-47.902	11.634	.004	
	1000	13.370	-14.658	36.862	-51.520	11.261	.072	
	1100	13.590	-13.310	38.146	-55.271	10.983	.780	
	1200	13.810	-11.940	39.338	-59.146	10.773	5.548	
	1300	14.030	-10.548	40.452	-63.136	10.616	28.677	
	1392	14.232	-9.248	41.418	-66.902	10.505	104.043	
			8.7	6.250				78
LIQ	1392	14.800	-.548	47.668	-66.902	10.505	104.043	
	1400	14.800	-.430	47.753	-67.284	10.505	113.325	
	1500	14.800	1.050	48.774	-72.111	10.508	302.831	
	1600	14.800	2.530	49.729	-77.037	10.524	705.813	
	1609	14.800	2.671	49.817	-77.510	10.526	760.	
			39.947	24.820				*
GAS	1609	8.131	42.618	74.637	-77.510	10.526		
	1700	8.087	43.352	75.080	-84.284	10.837		
	1800	8.038	44.158	75.541	-91.816	11.149		
	1900	7.989	44.959	75.974	-99.392	11.434		
	2000	7.940	45.756	76.383	-107.010	11.695		
	2100	7.891	46.547	76.769	-114.668	11.935		
	2200	7.842	47.334	77.135	-122.363	12.157		
	2300	7.793	48.116	77.482	-130.094	12.363		
	2400	7.744	48.893	77.813	-137.859	12.555		

$PbO \cdot PbSO_4$ DILEAD OXIDE SULFATE

		A	B	C	D	RANGE	REF.
SOL	CP	21.	34.7	4.20		298-1000	*
	** MPT= 1243						2

PHASE	T	CP	H	S	G	BT	
SOL	298	36.071	-280.0	49.4	-294.729	216.071	19
	300	36.077	-279.933	49.623	-294.820	214.805	
	400	37.505	-276.269	60.155	-300.331	164.115	
	500	40.030	-272.397	68.784	-306.789	134.115	
	600	42.987	-268.249	76.339	-314.052	114.409	
	700	46.147	-263.793	83.201	-322.034	100.557	
	800	49.416	-259.016	89.576	-330.676	90.349	
	900	52.749	-253.908	95.588	-339.937	82.559	
	1000	56.120	-248.465	101.320	-349.785	76.456	

* * * * * * * * * *

$2\,PbO \cdot PbSO_4$ TRILEAD OXIDE SULFATE

		A	B	C	D	RANGE	REF.
SOL	CP	30.8	38.3	4.20		298-1000	*
	** MPT= 1223						2

PHASE	T	CP	H	S	G	BT	
SOL	298	46.944	-334.4	65.6	-353.959	259.494	19
	300	46.957	-334.313	65.890	-354.080	257.982	
	400	48.745	-329.543	79.602	-361.383	197.477	
	500	51.630	-324.529	90.777	-369.918	161.713	
	600	54.947	-319.203	100.479	-379.490	138.248	
	700	58.467	-313.533	109.212	-389.982	121.774	
	800	62.096	-307.506	117.255	-401.310	109.647	
	900	65.789	-301.112	124.782	-413.415	100.404	
	1000	69.520	-294.347	131.906	-426.253	93.170	

TETRALEAD OXIDE SULFATE $3\,PbO \cdot PbSO_4$

		A	B	C	D	RANGE	REF.
SOL	CP	40.7	42.	4.2		298- 800	*

PHASE	T	CP	H	S	G	BT	
SOL	298	57.947	-388.8	81.4	-413.069	302.829	19
	300	57.967	-388.693	81.759	-413.220	301.071	
	400	60.125	-382.803	98.688	-422.278	230.753	
	500	63.380	-376.633	112.442	-432.854	189.226	
	600	67.067	-370.113	124.320	-444.705	162.005	
	700	70.957	-363.213	134.948	-457.677	142.912	
	800	74.956	-355.918	144.683	-471.665	128.870	

PbSe

LEAD SELENIDE

		A	B	C	D	RANGE	REF.
SOL	CP	11.29	2.39			298-1350	63
LIQ	CP	15.0				1350-1607	*
GAS	CP	8.94		-0.25		1607-2000	63

PHASE	T	CP	H	S	G	BT	P	
SOL	298	12.003	-23.9	24.5	-31.205	22.877		63
	300	12.007	-23.878	24.574	-31.250	22.769		
	400	12.246	-22.665	28.061	-33.890	18.519		
	500	12.485	-21.429	30.819	-36.838	16.104		
	600	12.724	-20.168	33.117	-40.038	14.586		
	700	12.963	-18.884	35.096	-43.451	13.568		
	800	13.202	-17.576	36.843	-47.050	12.855		
	900	13.441	-16.243	38.412	-50.814	12.341	.006	
	1000	13.680	-14.887	39.840	-54.727	11.962	.115	
	1100	13.919	-13.507	41.155	-58.778	11.680	1.192	
	1200	14.158	-12.104	42.377	-62.955	11.467	8.196	
	1300	14.397	-10.676	43.519	-67.251	11.307	41.193	
	1350	14.516	-9.953	44.065	-69.440	11.243	83.925	
			11.8	8.741				78
LIQ	1350	15.000	1.847	52.806	-69.440	11.243	83.925	
	1400	15.000	2.597	53.351	-72.094	11.256	138.348	
	1500	15.000	4.097	54.386	-77.482	11.291	336.539	
	1600	15.000	5.597	55.354	-82.969	11.335	723.280	
	1607	15.000	5.702	55.420	-83.357	11.338	760.	
			36.131	22.484				63,78,*
GAS	1607	8.930	41.833	77.903	-83.357	11.338		
	1700	8.931	42.664	78.405	-90.626	11.652		
	1800	8.932	43.557	78.916	-98.492	11.960		
	1900	8.933	44.450	79.399	-106.408	12.241		
	2000	8.934	45.343	79.857	-114.371	12.500		

LEAD SELENITE

$PbSeO_3$

		A	B	C	D	RANGE	REF.
SOL	CP	20.5	11.0			298- 953	*
	** MPT= 953						62

PHASE	T	CP	H	S	G	BT	
SOL	298	23.780	-127.3	30.7	-136.453	100.036	62/*
	300	23.800	-127.256	30.847	-136.510	99.461	
	400	24.900	-124.821	37.845	-139.959	76.480	
	500	26.000	-122.276	43.519	-144.036	62.966	
	600	27.100	-119.621	48.357	-148.635	54.148	
	700	28.200	-116.856	52.617	-153.688	47.990	
	800	29.300	-113.981	56.454	-159.144	43.482	
	900	30.400	-110.996	59.969	-164.968	40.065	
	953	30.983	-109.369	61.725	-168.193	38.577	

PbTe

LEAD TELLURIDE

		A	B	C	D	RANGE	REF.
SOL	CP	11.28	2.69			298-1197	63
LIQ	CP	15.0				1197-1700	*
GAS	CP	8.94		-0.14		1700-2000	63

PHASE	T	CP	H	S	G	BT	P	
SOL	298	12.082	-16.4	26.3	-24.241	17.772		63
	300	12.087	-16.378	26.375	-24.290	17.698		
	400	12.356	-15.155	29.889	-27.111	14.815		
	500	12.625	-13.906	32.675	-30.244	13.221		
	600	12.894	-12.630	35.000	-33.631	12.252		
	700	13.163	-11.328	37.008	-37.233	11.626		
	800	13.432	-9.998	38.784	-41.025	11.209		
	900	13.701	-8.641	40.381	-44.984	10.925	.009	
	1000	13.970	-7.258	41.839	-49.096	10.731	.163	
	1100	14.239	-5.847	43.183	-53.348	10.601	1.627	
	1197	14.500	-4.453	44.397	-57.596	10.517	10.326	
			13.7	11.445				63
LIQ	1197	15.000	9.247	55.842	-57.596	10.517	10.326	
	1200	15.000	9.292	55.880	-57.764	10.522	10.722	
	1300	15.000	10.792	57.080	-63.413	10.662	33.753	
	1400	15.000	12.292	58.192	-69.177	10.800	88.695	
	1500	15.000	13.792	59.227	-75.049	10.936	201.940	
	1600	15.000	15.292	60.195	-81.020	11.068	409.589	
	1700	15.000	16.792	61.104	-87.086	11.197	760.	
			32.802	19.295				63*
GAS	1700	8.935	49.594	80.400	-87.086	11.197		
	1800	8.936	50.487	80.910	-95.151	11.555		
	1900	8.936	51.381	81.393	-103.267	11.880		
	2000	8.936	52.274	81.852	-111.429	12.178		

LEAD OXIDE CARBONATE $PbO \cdot PbCO_3$

		A	B	C	D	RANGE	REF.
SOL	CP	21.44	35.0			298- 700	*

PHASE	T	CP	H	S	G	BT	
SOL	298	31.875	-195.2	48.8	-209.750	153.771	66
	300	31.940	-195.141	48.997	-209.840	152.889	
	400	35.440	-191.772	58.665	-215.238	117.616	
	500	38.940	-188.053	66.949	-221.528	96.843	
	600	42.440	-183.984	74.358	-228.599	83.278	
	700	45.940	-179.565	81.163	-236.379	73.811	

* * * * * * * * * *

LEAD TITANATE $PbO \cdot TiO_2$

		A	B	C	D	RANGE	REF.
SOL-A	CP	28.57	4.28	-4.35		298- 763	*
SOL-B	CP	26.07	5.45	-3.19		763-1443	*
	** MPT= 1443						1

PHASE	T	CP	H	S	G	BT	
SOL-A	298	24.953	-286.5	26.75	-294.476	215.885	142
	300	25.021	-286.454	26.905	-294.525	214.590	
	400	27.563	-283.809	34.494	-297.607	162.627	
	500	28.970	-280.977	40.808	-301.381	131.751	
	600	29.930	-278.030	46.179	-305.738	111.380	
	700	30.678	-274.998	50.851	-310.594	96.985	
	763	31.088	-273.053	53.513	-313.883	89.919	
			1.15	1.507			1
SOL-B	763	29.680	-271.903	55.020	-313.883	89.919	
	800	29.932	-270.800	56.431	-315.945	86.324	
	900	30.581	-267.774	59.994	-321.769	78.147	
	1000	31.201	-264.684	63.249	-327.933	71.679	
	1100	31.801	-261.534	66.251	-334.410	66.450	
	1200	32.388	-258.325	69.043	-341.176	62.145	
	1300	32.966	-255.057	71.658	-348.213	58.548	
	1400	33.537	-251.732	74.122	-355.503	55.504	
	1443	33.781	-250.284	75.141	-358.712	54.336	

Pd

PALLADIUM

		A	B	C	D	RANGE	REF.
SOL	CP	5.683	1.767			298- 400	21
	CP	5.873	1.274			400-1400	21
	CP	0.841	3.632	34.042		1400-1825	21
LIQ	CP	8.3				1825-3200	21
	** BPT= 3237, L= 85.457						21

PHASE	T	CP	H	S	G	BT	P	
SOL	298	6.210	0	9.04	-2.695	1.976		21
	300	6.213	.011	9.078	-2.712	1.976		
	400	6.390	.642	10.890	-3.714	2.030		
	400	6.383	.642	10.890	-3.714	2.030		
	500	6.510	1.286	12.328	-4.878	2.132		
	600	6.637	1.944	13.526	-6.172	2.248		
	700	6.765	2.614	14.559	-7.577	2.366		
	800	6.892	3.297	15.470	-9.080	2.481		
	900	7.020	3.992	16.290	-10.668	2.591		
	1000	7.147	4.701	17.036	-12.335	2.696		
	1100	7.274	5.422	17.723	-14.074	2.797		
	1200	7.402	6.155	18.361	-15.878	2.892		
	1300	7.529	6.902	18.959	-17.745	2.984		
	1400	7.657	7.661	19.521	-19.669	3.071		
	1400	7.663	7.661	19.521	-19.669	3.071		
	1500	7.802	8.434	20.055	-21.648	3.155		
	1600	7.982	9.223	20.564	-23.679	3.235	.001	
	1700	8.193	10.032	21.054	-25.760	3.312	.005	
	1800	8.429	10.862	21.529	-27.889	3.387	.022	
	1825	8.491	11.074	21.645	-28.429	3.405	.031	
			4.197	2.300				21
LIQ	1825	8.300	15.271	23.945	-28.429	3.405	.031	
	1900	8.300	15.893	24.279	-30.237	3.479	.078	
	2000	8.300	16.723	24.705	-32.687	3.572	.235	
	2100	8.300	17.553	25.110	-35.178	3.661	.637	
	2200	8.300	18.383	25.496	-37.708	3.746	1.580	
	2300	8.300	19.213	25.865	-40.276	3.828	3.625	
	2400	8.300	20.043	26.218	-42.881	3.905	7.773	
	2500	8.300	20.873	26.557	-45.520	3.980	15.704	
	2600	8.300	21.703	26.883	-48.192	4.051	30.101	
	2700	8.300	22.533	27.196	-50.896	4.120	55.064	
	2800	8.300	23.363	27.498	-53.631	4.187	96.613	
	2900	8.300	24.193	27.789	-56.395	4.251	163.287	
	3000	8.300	25.023	28.071	-59.188	4.312	266.814	
	3100	8.300	25.853	28.343	-62.009	4.372	422.842	
	3200	8.300	26.683	28.606	-64.856	4.430	651.703	

PALLADIUM FLUORIDE

PdF_2

		A	B	C	D	RANGE	REF.
SOL	CP	14.3	4.9			298-1100	*

PHASE	T	CP	H	S	G	BT	
SOL	298	15.761	-112.	21.2	-118.321	86.743	1/*
	300	15.770	-111.971	21.298	-118.360	86.237	
	400	16.260	-110.369	25.901	-120.730	65.973	
	500	16.750	-108.719	29.582	-123.510	53.993	
	600	17.240	-107.019	32.680	-126.627	46.130	
	700	17.730	-105.271	35.374	-130.033	40.603	
	800	18.220	-103.473	37.773	-133.692	36.528	
	900	18.710	-101.627	39.948	-137.580	33.413	
	1000	19.200	-99.731	41.944	-141.676	30.967	
	1100	19.690	-97.787	43.797	-145.964	29.004	

$PdCl_2$ PALLADIUM CHLORIDE

		A	B	C	D	RANGE	REF.
SOL	CP	16.5	5.0			298- 952	*
LIQ	CP	22.5				952-1608	*
	** DEC./ BPT= 1608						1/*

PHASE	T	CP	H	S	G	BT	P	
SOL	298	17.991	-41.39	24.8	-48.784	35.765		74,*
	300	18.000	-41.357	24.911	-48.830	35.577		
	400	18.500	-39.532	30.158	-51.595	28.194		
	500	19.000	-37.657	34.340	-54.827	23.968		
	600	19.500	-35.732	37.848	-58.441	21.290		
	700	20.000	-33.757	40.892	-62.381	19.479		
	800	20.500	-31.732	43.595	-66.608	18.199		
	900	21.000	-29.657	46.038	-71.091	17.266		
	952	21.260	-28.558	47.225	-73.516	16.879	.002	
			4.4	4.622				74
LIQ	952	22.500	-24.158	51.847	-73.516	16.879	.002	
	1000	22.500	-23.078	52.954	-76.032	16.619	.008	
	1100	22.500	-20.828	55.098	-81.436	16.182	.132	
	1200	22.500	-18.578	57.056	-87.045	15.855	1.366	
	1300	22.500	-16.328	58.857	-92.842	15.610	9.601	
	1400	22.500	-14.078	60.524	-98.812	15.427	49.998	
	1500	22.500	-11.828	62.077	-104.943	15.292	205.136	
	1600	22.500	-9.578	63.529	-111.224	15.195	694.274	
	1608	22.500	-9.398	63.641	-111.733	15.188	760.	

PALLADIUM IODIDE

PdJ_2

		A	B	C	D	RANGE	REF.
SOL-A	CP	16.3	5.5			298- 833	*
	** TPT=	833					62
SOL-B	CP	16.3	5.5			833- 900	*

PHASE	T	CP	H	S	G	BT	
SOL-A	298	17.940	-15.2	36.	-25.933	19.012	19
	300	17.950	-15.167	36.111	-26.000	18.944	
	400	18.500	-13.344	41.350	-29.884	16.330	
	500	19.050	-11.467	45.537	-34.236	14.966	
	600	19.600	-9.534	49.059	-38.970	14.197	
	700	20.150	-7.547	52.122	-44.032	13.749	
	800	20.700	-5.504	54.849	-49.383	13.493	
	833	20.881	-4.818	55.689	-51.207	13.437	
SOL-B	833	20.881	-4.818	55.689	-51.207	13.437	
	900	21.250	-3.407	57.318	-54.993	13.356	

PdO

PALLADIUM OXIDE

		A	B	C	D	RANGE	REF.
SOL	CP	10.828	1.681	-0.304	0.09	298-1143	*
	** DEC. / MPT= 1143						1

PHASE	T	CP	H	S	G	BT	
SOL	298	10.995	-26.9	9.4	-29.703	21.776	62*
	300	11.003	-26.880	9.468	-29.720	21.654	
	400	11.325	-25.762	12.680	-30.834	16.849	
	500	11.569	-24.617	15.235	-32.234	14.092	
	600	11.785	-23.449	17.363	-33.867	12.338	
	700	11.987	-22.261	19.195	-35.697	11.147	
	800	12.183	-21.052	20.809	-37.699	10.300	
	900	12.376	-19.824	22.255	-39.853	9.679	
	1000	12.569	-18.577	23.569	-42.146	9.212	
	1100	12.761	-17.310	24.776	-44.564	8.855	
	1143	12.844	-16.760	25.266	-45.640	8.728	

TETRAPALLADIUM SULFIDE

Pd_4S

		A	B	C	D	RANGE	REF.
SOL	CP	24.0	11.66			298-1034	63
	** DPT= 1034						62

PHASE	T	CP	H	S	G	BT	
SOL	298	27.476	-16.5	43.18	-29.374	21.535	63
	300	27.498	-16.449	43.350	-29.454	21.460	
	400	28.664	-13.641	51.420	-34.209	18.694	
	500	29.830	-10.716	57.942	-39.687	17.350	
	600	30.996	-7.675	63.484	-45.765	16.672	
	700	32.162	-4.517	68.349	-52.362	16.350	
	800	33.328	-1.243	72.720	-59.419	16.235	
	900	34.494	2.148	76.713	-66.893	16.246	
	1000	35.660	5.656	80.407	-74.751	16.339	
	1034	36.056	6.875	81.606	-77.506	16.384	

* * * * * * * * * *

PALLADIUM MONOSULFICE

PdS

		A	B	C	D	RANGE	REF.
SOL	CP	9.97	4.11	-0.73		298-1243	63
	** MPT= 1243						62

PHASE	T	CP	H	S	G	BT	
SOL	298	10.374	-16.9	13.5	-20.925	15.341	63
	300	10.392	-16.881	13.564	-20.950	15.264	
	400	11.158	-15.801	16.666	-22.467	12.277	
	500	11.733	-14.655	19.220	-24.265	10.608	
	600	12.233	-13.457	21.404	-26.299	9.581	
	700	12.698	-12.210	23.325	-28.537	8.911	
	800	13.144	-10.918	25.050	-30.957	8.458	
	900	13.579	-9.581	26.623	-33.542	8.146	
	1000	14.007	-8.202	28.076	-36.278	7.930	
	1100	14.431	-6.780	29.431	-39.154	7.780	
	1200	14.851	-5.316	30.704	-42.161	7.680	
	1243	15.031	-4.674	31.230	-43.493	7.648	

PdS_2 PALLADIUM DISULFIDE

		A	B	C	D	RANGE	REF.
SOL	CP	16.39	3.77	-1.57		298-1245	63
	** MPT= 1245						63

PHASE	T	CP	H	S	G	BT	
SOL	298	15.748	-18.7	21.	-24.961	18.299	63
	300	15.777	-18.671	21.098	-25.000	18.215	
	400	16.917	-17.031	25.808	-27.354	14.948	
	500	17.647	-15.301	29.666	-30.133	13.173	
	600	18.216	-13.507	32.935	-33.268	12.119	
	700	18.709	-11.660	35.781	-36.706	11.462	
	800	19.161	-9.766	38.309	-40.413	11.042	
	900	19.589	-7.829	40.590	-44.360	10.774	
	1000	20.003	-5.849	42.676	-48.525	10.606	
	1100	20.407	-3.828	44.601	-52.890	10.510	
	1200	20.805	-1.768	46.394	-57.441	10.463	
	1245	20.982	-.827	47.163	-59.546	10.454	

* * * * * * * * * *

PdTe PALLADIUM TELLURIDE

		A	B	C	D	RANGE	REF.
SOL	CP	11.34	3.09			298- 993	63
	** HOMOGENEITY RANGE.../ MPT= 993						62

PHASE	T	CP	H	S	G	BT	
SOL	298	12.261	-9.	21.42	-15.386	11.280	63
	300	12.267	-8.977	21.496	-15.426	11.239	
	400	12.576	-7.735	25.067	-17.762	9.706	
	500	12.885	-6.462	27.907	-20.415	8.925	
	600	13.194	-5.158	30.283	-23.328	8.498	
	700	13.503	-3.823	32.340	-26.461	8.263	
	800	13.812	-2.458	34.163	-29.788	8.139	
	900	14.121	-1.061	35.808	-33.288	8.085	
	993	14.408	.266	37.211	-36.684	8.075	

PRASEODYMIUM

		A	B	C	D	RANGE	REF.
SOL-A	CP	4.405	4.941	0.612		298- 700	21
	CP	4.106	5.521			700-1068	21
SOL-B	CP	9.19				1068-1204	21
LIQ	CP	10.27				1204-3000	21
	** BPT= 3785, L= 70.931						21

PHASE	T	CP	H	S	G	BT	P	
SOL-A	298	6.567	0	17.67	-5.268	3.862		21
	300	6.567	.012	17.711	-5.301	3.862		
	400	6.764	.677	19.621	-7.172	3.919		
	500	7.120	1.370	21.167	-9.213	4.028		
	600	7.540	2.103	22.501	-11.398	4.152		
	700	7.989	2.879	23.697	-13.709	4.281		
	700	7.971	2.879	23.697	-13.709	4.281		
	800	8.523	3.704	24.797	-16.134	4.408		
	900	9.075	4.583	25.833	-18.666	4.533		
	1000	9.627	5.519	26.818	-21.299	4.656		
	1068	10.002	6.186	27.463	-23.145	4.737		
			0.757	.709				21
SOL-B	1068	9.190	6.943	28.172	-23.145	4.737		
	1100	9.190	7.237	28.443	-24.051	4.779		
	1200	9.190	8.156	29.243	-26.936	4.906		
	1204	9.190	8.193	29.274	-27.053	4.911		
			1.646	1.367				21
LIQ	1204	10.270	9.839	30.641	-27.053	4.911		
	1300	10.270	10.825	31.429	-30.032	5.050		
	1400	10.270	11.852	32.190	-33.214	5.186		
	1500	10.270	12.879	32.898	-36.469	5.314		
	1600	10.270	13.906	33.561	-39.792	5.436		
	1700	10.270	14.933	34.184	-43.179	5.552	.003	
	1800	10.270	15.960	34.771	-46.627	5.662	.011	
	1900	10.270	16.987	35.326	-50.133	5.767	.036	
	2000	10.270	18.014	35.853	-53.692	5.868	.101	
	2100	10.270	19.041	36.354	-57.302	5.964	.257	
	2200	10.270	20.068	36.832	-60.962	6.057	.597	
	2300	10.270	21.095	37.288	-64.668	6.146	1.284	
	2400	10.270	22.122	37.725	-68.419	6.231	2.581	
	2500	10.270	23.149	38.144	-72.212	6.314	4.894	
	2600	10.270	24.176	38.547	-76.047	6.393	8.806	
	2700	10.270	25.203	38.935	-79.921	6.470	15.128	
	2800	10.270	26.230	39.308	-83.833	6.544	24.941	
	2900	10.270	27.257	39.669	-87.782	6.616	39.628	
	3000	10.270	28.284	40.017	-91.767	6.686	60.911	

PrH_2 PRASEODYMIUM DIHYDRIDE

		A	B	C	D	RANGE	REF.
SOL	CP	8.48	4.5			298-1200	76,*

PHASE	T	CP	H	S	G	BT	
SOL	298	9.822	-47.4	13.57	-51.446	37.716	19/76
	300	9.830	-47.382	13.631	-51.471	37.502	
	400	10.280	-46.376	16.520	-52.984	28.953	
	500	10.730	-45.326	18.863	-54.757	23.938	
	600	11.180	-44.230	20.859	-56.746	20.672	
	700	11.630	-43.090	22.616	-58.921	18.398	
	800	12.080	-41.904	24.198	-61.263	16.738	
	900	12.530	-40.674	25.647	-63.756	15.484	
	1000	12.980	-39.398	26.990	-66.389	14.511	
	1100	13.430	-38.078	28.249	-69.151	13.741	
	1200	13.880	-36.712	29.437	-72.036	13.121	

* * * * * * * * * *

$PrCl_3$ PRASEODYMIUM CHLORIDE

		A	B	C	D	RANGE	REF.
SOL	CP	20.6	11.4			298-1059	89*
LIQ	CP	32.0				1059-1300	89*

PHASE	T	CP	H	S	G	BT	
SOL	298	23.999	-252.6	34.5	-262.886	192.727	19/*
	300	24.020	-252.556	34.649	-262.950	191.585	
	400	25.160	-250.097	41.715	-266.782	145.783	
	500	26.300	-247.524	47.452	-271.249	118.579	
	600	27.440	-244.837	52.347	-276.245	100.636	
	700	28.580	-242.036	56.663	-281.700	87.962	
	800	29.720	-239.121	60.554	-287.563	78.569	
	900	30.860	-236.092	64.120	-293.800	71.354	
	1000	32.000	-232.949	67.430	-300.379	65.657	
	1059	32.673	-231.041	69.284	-304.412	62.831	
			12.1	11.426			89
LIQ	1059	32.000	-218.941	80.710	-304.412	62.831	
	1100	32.000	-217.629	81.925	-307.747	61.152	
	1200	32.000	-214.429	84.710	-316.080	57.574	
	1300	32.000	-211.229	87.271	-324.681	54.591	

PRASEODYMIUM IODIDE PrJ_3

		A	B	C	D	RANGE	REF.
SOL	CP	21.3	9.7			298-1011	89*
LIQ	CP	34.2				1011-1300	89

PHASE	T	CP	H	S	G	BT	
SOL	298	24.192	-156.4	51.3	-171.695	125.873	19/*
	300	24.210	-156.355	51.450	-171.790	125.166	
	400	25.180	-153.886	58.547	-177.305	96.888	
	500	26.150	-151.319	64.270	-183.454	80.199	
	600	27.120	-148.656	69.124	-190.130	69.264	
	700	28.090	-145.895	73.377	-197.259	61.595	
	800	29.060	-143.038	77.191	-204.791	55.954	
	900	30.030	-140.083	80.670	-212.686	51.654	
	1000	31.000	-137.032	83.884	-220.916	48.288	
	1011	31.107	-136.690	84.224	-221.841	47.962	
			12.5	12.364			89
LIQ	1011	34.200	-124.190	96.588	-221.841	47.962	
	1100	34.200	-121.146	99.474	-230.567	45.816	
	1200	34.200	-117.726	102.449	-240.665	43.837	
	1300	34.200	-114.306	105.187	-251.049	42.211	

Pr_2O_3 PRASEODYMIUM OXIDE

		A	B	C	D	RANGE	REF.
SOL	CP	23.76	12.5			298-1100	133

PHASE	T	CP	H	S	G	BT	
SOL	298	27.487	-436.3	37.9	-447.600	328.144	1/191
	300	27.510	-436.249	38.070	-447.670	326.171	
	400	28.760	-433.436	46.155	-451.898	246.939	
	500	30.010	-430.497	52.707	-456.851	199.716	
	600	31.260	-427.434	58.289	-462.407	168.454	
	700	32.510	-424.245	63.202	-468.486	146.288	
	800	33.760	-420.932	67.625	-475.031	129.790	
	900	35.010	-417.493	71.673	-481.999	117.061	
	1000	36.260	-413.930	75.426	-489.356	106.963	
	1100	37.510	-410.241	78.941	-497.076	98.773	

* * * * * * * * * *

$PrO_{1.72}$ PRASEODYMIUM 1.72-OXIDE

		A	B	C	D	RANGE	REF.
SOL	CP	14.55	3.55	-1.04		298-1100	191

PHASE	T	CP	H	S	G	BT	
SOL	298	14.438	-224.6	19.2	-230.324	168.855	191
	300	14.459	-224.573	19.289	-230.360	167.840	
	400	15.320	-223.081	23.577	-232.512	127.056	
	500	15.909	-221.518	27.062	-235.049	102.754	
	600	16.391	-219.902	30.006	-237.906	86.669	
	700	16.823	-218.241	32.566	-241.037	75.265	
	800	17.227	-216.539	34.839	-244.410	66.779	
	900	17.617	-214.796	36.891	-247.998	60.230	
	1000	17.996	-213.016	38.766	-251.782	55.034	
	1100	18.369	-211.197	40.499	-255.746	50.819	

PRASEODYMIUM 1.83-OXIDE

$PrO_{1.83}$

		A	B	C	D	RANGE	REF.
SOL	CP	15.88	4.36	-1.55		298-1100	58

PHASE	T	CP	H	S	G	BT	
SOL	298	15.436	-228.	19.1	-233.695	171.326	191
	300	15.466	-227.971	19.196	-233.730	170.295	
	400	16.655	-226.360	23.823	-235.889	128.901	
	500	17.440	-224.653	27.628	-238.467	104.248	
	600	18.065	-222.877	30.865	-241.396	87.940	
	700	18.616	-221.043	33.692	-244.627	76.386	
	800	19.126	-219.155	36.211	-248.124	67.794	
	900	19.613	-217.218	38.492	-251.861	61.168	
	1000	20.085	-215.233	40.583	-255.816	55.916	
	1100	20.548	-213.202	42.519	-259.973	51.659	

* * * * * * * * * *

PRASEODYMIUM DIOXIDE

PrO_2

		A	B	C	D	RANGE	REF.
SOL	CP	18.54	5.98	-2.57		298-1100	*

PHASE	T	CP	H	S	G	BT	
SOL	298	17.432	-232.9	19.1	-238.595	174.918	191
	300	17.478	-232.868	19.208	-238.630	173.865	
	400	19.326	-231.019	24.515	-240.825	131.598	
	500	20.502	-229.024	28.961	-243.504	106.450	
	600	21.414	-226.927	32.782	-246.596	89.835	
	700	22.202	-224.745	36.143	-250.046	78.078	
	800	22.922	-222.489	39.156	-253.813	69.348	
	900	23.605	-220.162	41.895	-257.868	62.627	
	1000	24.263	-217.768	44.416	-262.185	57.308	
	1100	24.906	-215.310	46.759	-266.745	53.004	

PrS

PRASEODYMIUM MONOSULFIDE

		A	B	C	D	RANGE	REF.
SOL	CP	12.42	1.052			298-2000	63
	** MPT=	2500					63

PHASE	T	CP	H	S	G	BT	
SOL	298	12.734	-108.	18.6	-113.546	83.242	63
	300	12.736	-107.976	18.679	-113.580	82.754	
	400	12.841	-106.698	22.357	-115.640	63.191	
	500	12.946	-105.408	25.234	-118.025	51.596	
	600	13.051	-104.108	27.603	-120.670	43.960	
	700	13.156	-102.798	29.623	-123.534	38.574	
	800	13.262	-101.477	31.387	-126.586	34.586	
	900	13.367	-100.146	32.955	-129.805	31.525	
	1000	13.472	-98.804	34.369	-133.172	29.109	
	1100	13.577	-97.451	35.657	-136.675	27.158	
	1200	13.682	-96.088	36.843	-140.300	25.556	
	1300	13.788	-94.715	37.943	-144.040	24.219	
	1400	13.893	-93.331	38.968	-147.886	23.089	
	1500	13.998	-91.936	39.930	-151.832	22.125	
	1600	14.103	-90.531	40.837	-155.871	21.294	
	1700	14.208	-89.116	41.695	-159.998	20.572	
	1800	14.314	-87.690	42.510	-164.208	19.940	
	1900	14.419	-86.253	43.287	-168.498	19.384	
	2000	14.524	-84.806	44.029	-172.865	18.892	

TRIPRASEODYMIUM TETRASULFIDE

Pr_3S_4

		A	B	C	D	RANGE	REF.
SOL	CP	41.70	3.76			298-2000	63

PHASE	T	CP	H	S	G	BT	
SOL	298	42.821	-371.5	61.2	-389.747	285.731	63
	300	42.828	-371.421	61.465	-389.860	284.051	
	400	43.204	-367.119	73.837	-396.654	216.751	
	500	43.580	-362.780	83.518	-404.539	176.848	
	600	43.956	-358.403	91.497	-413.301	150.565	
	700	44.332	-353.989	98.301	-422.800	132.022	
	800	44.708	-349.537	104.245	-432.933	118.288	
	900	45.084	-345.047	109.533	-443.627	107.742	
	1000	45.460	-340.520	114.303	-454.823	99.415	
	1100	45.836	-335.955	118.653	-466.473	92.692	
	1200	46.212	-331.353	122.657	-478.542	87.166	
	1300	46.588	-326.713	126.371	-490.995	82.555	
	1400	46.964	-322.035	129.837	-503.808	78.658	
	1500	47.340	-317.320	133.090	-516.956	75.331	
	1600	47.716	-312.567	136.158	-530.420	72.462	
	1700	48.092	-307.777	139.062	-544.182	69.969	
	1800	48.468	-302.949	141.821	-558.227	67.787	
	1900	48.844	-298.083	144.452	-572.542	65.866	
	2000	49.220	-293.180	146.967	-587.114	64.165	

$PrAl_2$ PRASEODYMIUM 2-ALUMINIUM

		A	B	C	D	RANGE	REF.
SOL	CP	16.65	3.4			298-1200	*

PHASE	T	CP	H	S	G	BT	
SOL	298	17.664	-40.	27.42	-48.175	35.318	*/85
	300	17.670	-39.967	27.529	-48.226	35.137	
	400	18.010	-38.183	32.659	-51.247	28.004	
	500	18.350	-36.365	36.715	-54.723	23.922	
	600	18.690	-34.513	40.090	-58.567	21.336	
	700	19.030	-32.627	42.997	-62.725	19.586	
	800	19.370	-30.707	45.560	-67.155	18.348	
	900	19.710	-28.753	47.861	-71.828	17.445	
	1000	20.050	-26.765	49.955	-76.721	16.770	
	1100	20.390	-24.743	51.882	-81.814	16.257	
	1200	20.730	-22.687	53.671	-87.093	15.864	

PLATINUM DICHLORIDE

$PtCl_2$

		A	B	C	D	RANGE	REF.
SOL	CP	15.4	8.8			298- 700	*
	** DEC.						1

PHASE	T	CP	H	S	G	BT	
SOL	298	18.024	-26.5	31.	-35.743	26.204	19/*
	300	18.040	-26.467	31.112	-35.800	26.084	
	400	18.920	-24.619	36.422	-39.187	21.414	
	500	19.800	-22.683	40.738	-43.052	18.820	
	600	20.680	-20.659	44.426	-47.314	17.237	
	700	21.560	-18.547	47.680	-51.923	16.213	

* * * * * * * * * *

PLATINUM TRICHLORIDE

$PtCl_3$

		A	B	C	D	RANGE	REF.
SOL	CP	29.0				298- 500	*
	** DEC.						1

PHASE	T	CP	H	S	G	BT	
SOL	298	29.000	-41.6	35.5	-52.184	38.257	19/1
	300	29.000	-41.546	35.679	-52.250	38.069	
	400	29.000	-38.646	44.022	-56.255	30.741	
	500	29.000	-35.746	50.493	-60.993	26.664	

* * * * * * * * * *

PLATINUM TETRACHLORIDE

$PtCl_4$

		A	B	C	D	RANGE	REF.
SOL	CP	36.0				298- 400	*
	** DEC.						1

PHASE	T	CP	H	S	G	BT	
SOL	298	36.000	-56.6	49.	-71.209	52.205	19/*
	300	36.000	-56.533	49.223	-71.300	51.949	
	400	36.000	-52.933	59.579	-76.765	41.948	

$PtBr_2$ PLATINUM DIBROMIDE

		A	B	C	D	RANGE	REF.
SOL	CP	16.5	5.5			298- 889	*
	** DEC. / BPT= 889						1/

PHASE	T	CP	H	S	G	BT	
SOL	298	18.140	-19.2	37.	-30.232	22.163	1
	300	18.150	-19.166	37.112	-30.300	22.077	
	400	18.700	-17.324	42.409	-34.288	18.736	
	500	19.250	-15.426	46.641	-38.747	16.939	
	600	19.800	-13.474	50.199	-43.593	15.881	
	700	20.350	-11.466	53.293	-48.771	15.229	
	800	20.900	-9.404	56.046	-54.241	14.820	
	889	21.389	-7.522	58.276	-59.329	14.587	

* * * * * * * * * *

$PtBr_3$ PLATINUM TRIBROMIDE

		A	B	C	D	RANGE	REF.
SOL	CP	22.2	6.0			298-1000	*
	** DEC.						1

PHASE	T	CP	H	S	G	BT	
SOL	298	23.989	-30.5	43.	-43.320	31.759	1
	300	24.000	-30.456	43.148	-43.400	31.621	
	400	24.600	-28.026	50.135	-48.080	26.273	
	500	25.200	-25.536	55.689	-53.380	23.336	
	600	25.800	-22.986	60.336	-59.187	21.562	
	700	26.400	-20.376	64.358	-65.427	20.430	
	800	27.000	-17.706	67.923	-72.044	19.684	
	900	27.600	-14.976	71.138	-78.999	19.186	
	1000	28.200	-12.186	74.077	-86.262	18.855	

PLATINUM TETRABROMIDE $PtBr_4$

		A	B	C	D	RANGE	REF.
SOL	CP	35.0				298- 500	*
	** DEC.						1

PHASE	T	CP	H	S	G	BT	
SOL	298	35.000	-33.6	60.	-51.489	37.748	1
	300	35.000	-33.535	60.217	-51.600	37.596	
	400	35.000	-30.035	70.285	-58.149	31.776	
	500	35.000	-26.535	78.095	-65.583	28.670	

* * * * * * * * * *

PLATINUM TETRAIODICE PtJ_4

		A	B	C	D	RANGE	REF.
SOL	CP	35.5				298- 700	*
	** DEC.						1

PHASE	T	CP	H	S	G	BT	
SOL	298	35.500	-17.4	67.2	-37.436	27.445	19/*
	300	35.500	-17.334	67.420	-37.560	27.366	
	400	35.500	-13.784	77.632	-44.837	24.501	
	500	35.500	-10.234	85.554	-53.011	23.174	
	600	35.500	-6.684	92.026	-61.900	22.550	
	700	35.500	-3.134	97.499	-71.383	22.290	

PtO_2 (g) PLATINUM DIOXIDE (GAS)

		A	B	C	D	RANGE	REF.
GAS	CP	13.25	0.5	-2.75		298-2000	*

PHASE	T	CP	H	S	G	BT	
GAS	298	10.305	40.3	61.15	22.068	-16.179	218*
	300	10.344	40.319	61.214	21.955	-15.996	
	400	11.731	41.432	64.407	15.670	-8.563	
	500	12.400	42.642	67.105	9.090	-3.974	
	600	12.786	43.903	69.402	2.262	-.824	
	700	13.039	45.195	71.393	-4.780	1.493	
	800	13.220	46.509	73.147	-12.009	3.281	
	900	13.360	47.838	74.712	-19.403	4.712	
	1000	13.475	49.180	76.126	-26.946	5.890	
	1100	13.573	50.532	77.415	-34.624	6.880	
	1200	13.659	51.894	78.600	-42.426	7.728	
	1300	13.737	53.264	79.696	-50.341	8.464	
	1400	13.810	54.641	80.717	-58.363	9.112	
	1500	13.878	56.026	81.672	-66.483	9.688	
	1600	13.943	57.417	82.570	-74.695	10.204	
	1700	14.005	58.814	83.417	-82.995	10.671	
	1800	14.065	60.218	84.219	-91.377	11.096	
	1900	14.124	61.627	84.981	-99.837	11.485	
	2000	14.181	63.042	85.707	-108.372	11.844	

PLATINUM MONOSULFIDE

PtS

		A	B	C	D	RANGE	REF.
SOL	CP	9.97	4.11	-0.73		298-1500	63
	** DEC.						

PHASE	T	CP	H	S	G	BT	
SOL	298	10.374	-19.86	13.16	-23.784	17.436	62,63
	300	10.392	-19.841	13.224	-23.808	17.346	
	400	11.158	-18.761	16.326	-25.291	13.820	
	500	11.733	-17.615	18.880	-27.055	11.827	
	600	12.233	-16.417	21.064	-29.055	10.585	
	700	12.698	-15.170	22.985	-31.259	9.761	
	800	13.144	-13.878	24.710	-33.645	9.193	
	900	13.579	-12.541	26.283	-36.196	8.791	
	1000	14.007	-11.162	27.736	-38.898	8.502	
	1100	14.431	-9.740	29.091	-41.740	8.294	
	1200	14.851	-8.276	30.364	-44.713	8.144	
	1300	15.270	-6.770	31.570	-47.811	8.039	
	1400	15.687	-5.222	32.717	-51.025	7.966	
	1500	16.103	-3.633	33.813	-54.352	7.920	

* * * * * * * * * *

PLATINUM DISULFIDE

PtS_2

		A	B	C	D	RANGE	REF.
SOL	CP	16.39	3.77	-1.57		298-1500	63
	** DEC.						1

PHASE	T	CP	H	S	G	BT	
SOL	298	15.748	-26.4	17.85	-31.722	23.256	62,63
	300	15.777	-26.371	17.948	-31.755	23.137	
	400	16.917	-24.731	22.658	-33.794	18.467	
	500	17.647	-23.001	26.516	-36.258	15.851	
	600	18.216	-21.207	29.785	-39.078	14.236	
	700	18.709	-19.360	32.631	-42.201	13.178	
	800	19.161	-17.466	35.159	-45.593	12.457	
	900	19.589	-15.529	37.440	-49.225	11.955	
	1000	20.003	-13.549	39.526	-53.075	11.601	
	1100	20.407	-11.528	41.451	-57.125	11.351	
	1200	20.805	-9.468	43.244	-61.361	11.177	
	1300	21.198	-7.367	44.925	-65.770	11.058	
	1400	21.588	-5.228	46.510	-70.342	10.982	
	1500	21.975	-3.050	48.013	-75.069	10.939	

Pt_5Se_4 PENTAPLATINUM TETRASELENIDE

		A	B	C	D	RANGE	REF.
SOL	CP	45.0	18.75			298-1100	63

PHASE	T	CP	H	S	G	BT	
SOL	298	50.590	-58.	80.5	-82.001	60.116	63
	300	50.625	-57.906	80.813	-82.150	59.854	
	400	52.500	-52.750	95.634	-91.004	49.729	
	500	54.375	-47.406	107.550	-101.181	44.232	
	600	56.250	-41.875	117.630	-112.453	40.966	
	700	58.125	-36.156	126.441	-124.665	38.928	
	800	60.000	-30.250	134.325	-137.710	37.626	
	900	61.875	-24.156	141.501	-151.507	36.796	
	1000	63.750	-17.875	148.117	-165.992	36.282	
	1100	65.625	-11.406	154.281	-181.115	35.989	

		A	B	C	D	RANGE	REF.
SOL-A	CP	5.916	5.782			298- 395	21
SOL-B	CP	5.198	7.088			395- 480	21
SOL-C	CP	2.983	11.084			480- 588	21
SOL-D	CP	9.0				588- 730	21
S.-D1	CP	9.0				730- 753	21
SOL-E	CP	8.4				753- 913	21
LIQ	CP	10.				913-3503	21
GAS	CP	10.818				3503-4000	21

PHASE	T	CP	H	S	G	BT	P	
SOL-A	298	7.640	0	12.3	-3.667	2.689		21
	300	7.651	.014	12.347	-3.690	2.689		
	395	8.200	.767	14.524	-4.970	2.750		
			0.80	2.025				21
SOL-B	395	7.998	1.567	16.549	-4.970	2.750		
	400	8.033	1.607	16.650	-5.053	2.761		
	480	8.600	2.272	18.165	-6.447	2.936		
			0.14	.292				21
SOL-C	480	8.303	2.412	18.457	-6.447	2.936		
	500	8.525	2.581	18.800	-6.819	2.981		
	588	9.500	3.374	20.259	-8.538	3.174		
			0.13	.221				21
SOL-D	588	9.000	3.504	20.480	-8.538	3.174		
	600	9.000	3.612	20.662	-8.785	3.200		
	700	9.000	4.512	22.049	-10.923	3.411		
	730	9.000	4.782	22.427	-11.590	3.470		
			0.02	.027				21
S.-D1	730	9.000	4.802	22.454	-11.590	3.470		
	753	9.000	5.009	22.734	-12.110	3.515		
			0.44	.584				21
SOL-E	753	8.400	5.449	23.318	-12.110	3.515		
	800	8.400	5.844	23.827	-13.218	3.611		
	900	8.400	6.684	24.816	-15.651	3.801		
	913	8.400	6.793	24.936	-15.974	3.824		
			0.68	.745				21
LIQ	913	10.000	7.473	25.681	-15.974	3.824		

Pu

PLUTONIUM (CONTINUED)

PHASE	T	CP	H	S	G	BT	P	
LIQ	1000	10.000	8.343	26.591	-18.249	3.989		
	1100	10.000	9.343	27.544	-20.956	4.164		
	1200	10.000	10.343	28.415	-23.755	4.327		
	1300	10.000	11.343	29.215	-26.637	4.479		
	1400	10.000	12.343	29.956	-29.596	4.621		
	1500	10.000	13.343	30.646	-32.626	4.754		
	1600	10.000	14.343	31.291	-35.723	4.880		
	1700	10.000	15.343	31.898	-38.883	4.999	.003	
	1800	10.000	16.343	32.469	-42.102	5.113	.013	
	1900	10.000	17.343	33.010	-45.376	5.220	.042	
	2000	10.000	18.343	33.523	-48.703	5.323	.121	
	2100	10.000	19.343	34.011	-52.080	5.421	.316	
	2200	10.000	20.343	34.476	-55.504	5.515	.757	
	2300	10.000	21.343	34.920	-58.974	5.605	1.685	
	2400	10.000	22.343	35.346	-62.488	5.691	3.513	
	2500	10.000	23.343	35.754	-66.043	5.774	6.920	
	2600	10.000	24.343	36.146	-69.638	5.854	12.952	
	2700	10.000	25.343	36.524	-73.272	5.932	23.171	
	2800	10.000	26.343	36.888	-76.942	6.006	39.805	
	2900	10.000	27.343	37.238	-80.649	6.079	65.935	
	3000	10.000	28.343	37.577	-84.390	6.149	105.687	
	3100	10.000	29.343	37.905	-88.164	6.216	164.438	
	3200	10.000	30.343	38.223	-91.970	6.282	249.024	
	3300	10.000	31.343	38.531	-95.808	6.346	367.944	
	3400	10.000	32.343	38.829	-99.676	6.408	531.549	
	3500	10.000	33.343	39.119	-103.574	6.468	752.217	
	3503	10.000	33.373	39.128	-103.691	6.470	760.	
			82.145	23.450				21
GAS	3503	10.818	115.518	62.577	-103.691	6.470		
	3600	10.818	116.567	62.873	-109.775	6.665		
	3700	10.818	117.649	63.169	-116.078	6.857		
	3800	10.818	118.731	63.458	-122.409	7.041		
	3900	10.818	119.813	63.739	-128.769	7.217		
	4000	10.818	120.894	64.013	-135.157	7.386		

PLUTONIUM DIHYDRIDE PuH_2

		A	B	C	D	RANGE	REF.
SOL	CP	9.46	3.85	-1.14		298-1000	152

PHASE	T	CP	H	S	G	BT	
SOL	298	9.325	-33.3	14.3	-37.564	27.539	152
	300	9.348	-33.283	14.358	-37.590	27.388	
	400	10.287	-32.297	17.187	-39.172	21.405	
	500	10.929	-31.235	19.555	-41.012	17.929	
	600	11.453	-30.115	21.595	-43.072	15.691	
	700	11.922	-28.946	23.396	-45.323	14.152	
	800	12.362	-27.731	25.017	-47.745	13.045	
	900	12.784	-26.474	26.498	-50.322	12.221	
	1000	13.196	-25.175	27.866	-53.041	11.594	

* * * * * * * * * *

PLUTONIUM TRIHYDRICE PuH_3

		A	B	C	D	RANGE	REF.
SOL	CP	7.95	8.0			298- 500	152

PHASE	T	CP	H	S	G	BT	
SOL	298	10.335	-33.	15.5	-37.621	27.581	152
	300	10.350	-32.981	15.564	-37.650	27.432	
	400	11.150	-31.906	18.651	-39.366	21.512	
	500	11.950	-30.751	21.225	-41.363	18.082	

PuF_3 PLUTONIUM TRIFLUORIDE

		A	B	C	D	RANGE	REF.
SOL	CP	20.7	8.2			298-1699	152
LIQ	CP	35.0				1699-2493	152
	** BPT=	2493					1

PHASE	T	CP	H	S	G	BT	
SOL	298	23.145	-371.	27.	-379.050	277.889	152
	300	23.160	-370.957	27.143	-379.100	276.211	
	400	23.980	-368.600	33.918	-382.167	208.835	
	500	24.800	-366.161	39.357	-385.840	168.673	
	600	25.620	-363.640	43.951	-390.011	142.081	
	700	26.440	-361.037	47.962	-394.611	123.220	
	800	27.260	-358.352	51.546	-399.589	109.177	
	900	28.080	-355.585	54.804	-404.909	98.339	
	1000	28.900	-352.736	57.805	-410.542	89.736	
	1100	29.720	-349.805	60.598	-416.463	82.755	
	1200	30.540	-346.792	63.220	-422.656	76.986	
	1300	31.360	-343.697	65.696	-429.102	72.148	
	1400	32.180	-340.520	68.050	-435.791	68.039	
	1500	33.000	-337.261	70.299	-442.709	64.511	
	1600	33.820	-333.920	72.455	-449.847	61.455	
	1699	34.632	-330.532	74.509	-457.123	58.810	
			13.	7.652			152
LIQ	1699	35.000	-317.532	82.161	-457.123	58.810	
	1700	35.000	-317.497	82.181	-457.205	58.786	
	1800	35.000	-313.997	84.182	-465.524	56.530	
	1900	35.000	-310.497	86.074	-474.038	54.534	
	2000	35.000	-306.997	87.869	-482.736	52.758	
	2100	35.000	-303.497	89.577	-491.609	51.169	
	2200	35.000	-299.997	91.205	-500.648	49.742	
	2300	35.000	-296.497	92.761	-509.847	48.453	
	2400	35.000	-292.997	94.251	-519.198	47.286	
	2493	35.000	-289.742	95.581	-528.026	46.296	

PLUTONIUM TETRAFLUORIDE

PuF_4

		A	B	C	D	RANGE	REF.
SOL	CP	26.83	6.9			298-1310	152
LIQ	CP	41.0				1310-1500	152

PHASE	T	CP	H	S	G	BT	
SOL	298	28.887	-414.4	38.7	-425.938	312.263	152
	300	28.900	-414.347	38.879	-426.010	310.390	
	400	29.590	-411.422	47.287	-430.337	235.157	
	500	30.280	-408.429	53.964	-435.411	190.343	
	600	30.970	-405.366	59.546	-441.094	160.690	
	700	31.660	-402.235	64.372	-447.295	139.670	
	800	32.350	-399.034	68.644	-453.950	124.030	
	900	33.040	-395.765	72.494	-461.010	111.963	
	1000	33.730	-392.426	76.011	-468.437	102.391	
	1100	34.420	-389.019	79.258	-476.203	94.626	
	1200	35.110	-385.542	82.283	-484.282	88.212	
	1300	35.800	-381.997	85.121	-492.653	82.834	
	1310	35.869	-381.638	85.395	-493.506	82.344	
			10.2	7.786			152
LIQ	1310	41.000	-371.438	93.181	-493.506	82.344	
	1400	41.000	-367.748	95.906	-502.016	78.379	
	1500	41.000	-363.648	98.734	-511.750	74.572	

PuF_6 PLUTONIUM HEXAFLUORIDE

		A	B	C	D	RANGE	REF.
SOL	CP	41.3				298- 325	*
LIQ	CP	45.0				325- 335	*
GAS	CP	37.18	0.29	-5.65		335-2000	65,152

PHASE	T	CP	H	S	G	BT	P	
SOL	298	41.300	-424.6	52.058	-440.121	322.661	103.023	65,152,*
	300	41.300	-424.524	52.313	-440.218	320.741	116.779	
	325	41.300	-423.491	55.619	-441.567	296.977	542.976	
			4.455	13.708				65,152,*
LIQ	325	45.000	-419.036	69.327	-441.567	296.977	542.976	
	335	45.000	-418.586	70.691	-442.267	288.569	760.	
			7.2	21.493				1
GAS	335	32.243	-411.386	92.183	-442.267	288.569		
	400	33.765	-409.237	98.044	-448.454	245.057		
	500	35.065	-405.788	105.734	-458.655	200.505		
	600	35.785	-402.242	112.196	-469.560	171.060		
	700	36.230	-398.640	117.748	-481.064	150.215		
	800	36.529	-395.001	122.607	-493.087	134.723		
	900	36.743	-391.337	126.922	-505.567	122.785		
	1000	36.905	-387.654	130.802	-518.456	113.324		
	1100	37.032	-383.957	134.326	-531.716	105.656		
	1200	37.136	-380.249	137.553	-545.312	99.328		
	1300	37.223	-376.531	140.529	-559.218	94.026		
	1400	37.298	-372.804	143.290	-573.410	89.525		
	1500	37.364	-369.071	145.866	-587.870	85.664		
	1600	37.423	-365.332	148.279	-602.578	82.319		
	1700	37.477	-361.587	150.549	-617.521	79.398		
	1800	37.528	-357.837	152.693	-632.684	76.829		
	1900	37.574	-354.081	154.723	-648.055	74.553		
	2000	37.619	-350.322	156.652	-663.625	72.527		

PLUTONIUM TRICHLORICE

$PuCl_3$

		A	B	C	D	RANGE	REF.
SOL	CP	22.2	7.0			298-1040	*
LIQ	CP	31.5				1040-2063	*
	** BPT= 2063, L= 44.4						1

PHASE	T	CP	H	S	G	BT	
SOL	298	24.287	-229.8	38.	-241.130	176.777	152
	300	24.300	-229.755	38.150	-241.200	175.738	
	400	25.000	-227.290	45.237	-245.385	134.090	
	500	25.700	-224.755	50.891	-250.200	109.377	
	600	26.400	-222.150	55.638	-255.533	93.090	
	700	27.100	-219.475	59.760	-261.307	81.595	
	800	27.800	-216.730	63.425	-267.470	73.079	
	900	28.500	-213.915	66.739	-273.981	66.541	
	1000	29.200	-211.030	69.778	-280.809	61.379	
	1040	29.480	-209.856	70.929	-283.623	59.610	
			15.2	14.615			152
LIQ	1040	31.500	-194.656	85.545	-283.623	59.610	
	1100	31.500	-192.766	87.311	-288.809	57.389	
	1200	31.500	-189.616	90.052	-297.679	54.222	
	1300	31.500	-186.466	92.574	-306.812	51.587	
	1400	31.500	-183.316	94.908	-316.188	49.366	
	1500	31.500	-180.166	97.081	-325.788	47.474	
	1600	31.500	-177.016	99.114	-335.599	45.847	
	1700	31.500	-173.866	101.024	-345.607	44.437	
	1800	31.500	-170.716	102.824	-355.800	43.206	
	1900	31.500	-167.566	104.527	-366.169	42.125	
	2000	31.500	-164.416	106.143	-376.703	41.170	
	2063	31.500	-162.432	107.120	-383.421	40.624	

$PuBr_3$ PLUTONIUM TRIBROMICE

		A	B	C	D	RANGE	REF.
SOL	CP	22.8	7.5			298- 954	*
LIQ	CP	32.0				954-1748	*
	** BPT= 1748, L= 46.2						1

PHASE	T	CP	H	S	G	BT	
SOL	298	25.036	-198.8	45.7	-212.425	155.733	152/*
	300	25.050	-198.754	45.855	-212.510	154.834	
	400	25.800	-196.211	53.164	-217.477	118.840	
	500	26.550	-193.594	59.002	-223.095	97.528	
	600	27.300	-190.901	63.909	-229.246	83.514	
	700	28.050	-188.134	68.173	-235.855	73.647	
	800	28.800	-185.291	71.968	-242.865	66.357	
	900	29.550	-182.374	75.403	-250.237	60.774	
	954	29.955	-180.767	77.137	-254.356	58.278	
			14.	14.675			1
LIQ	954	32.000	-166.767	91.812	-254.356	58.278	
	1000	32.000	-165.295	93.319	-258.614	56.528	
	1100	32.000	-162.095	96.369	-268.101	53.274	
	1200	32.000	-158.895	99.153	-277.879	50.615	
	1300	32.000	-155.695	101.714	-287.924	48.411	
	1400	32.000	-152.495	104.086	-298.215	46.560	
	1500	32.000	-149.295	106.294	-308.736	44.989	
	1600	32.000	-146.095	108.359	-319.469	43.643	
	1700	32.000	-142.895	110.299	-330.403	42.482	
	1748	32.000	-141.359	111.190	-335.719	41.980	

PLUTONIUM TRIIODIDE

PuJ_3

		A	B	C	D	RANGE	REF.
SOL	CP	25.27	4.9			298-1050	152
LIQ	CP	34.0				1050-1500	152

PHASE	T	CP	H	S	G	BT	
SOL	298	26.731	-155.	53.5	-170.951	125.327	152
	300	26.740	-154.951	53.665	-171.050	124.627	
	400	27.230	-152.252	61.425	-176.822	96.624	
	500	27.720	-149.505	67.554	-183.282	80.123	
	600	28.210	-146.708	72.651	-190.299	69.326	
	700	28.700	-143.863	77.037	-197.788	61.761	
	800	29.190	-140.968	80.901	-205.689	56.199	
	900	29.680	-138.025	84.367	-213.955	51.962	
	1000	30.170	-135.032	87.520	-222.552	48.645	
	1050	30.415	-133.517	88.998	-226.965	47.247	
			12.	11.429			152
LIQ	1050	34.000	-121.517	100.426	-226.965	47.247	
	1100	34.000	-119.817	102.008	-232.026	46.106	
	1200	34.000	-116.417	104.966	-242.377	44.149	
	1300	34.000	-113.017	107.688	-253.012	42.541	
	1400	34.000	-109.617	110.207	-263.908	41.203	
	1500	34.000	-106.217	112.553	-275.047	40.080	

PuOF

PLUTONIUM OXYFLUORIDE

		A	B	C	D	RANGE	REF.
SOL	CP	17.29	5.71			298-1500	152
	** MPT>1500						

PHASE	T	CP	H	S	G	BT	
SOL	298	18.992	-269.8	21.9	-276.329	202.582	152
	300	19.003	-269.765	22.018	-276.370	201.363	
	400	19.574	-267.836	27.563	-278.861	152.383	
	500	20.145	-265.850	31.992	-281.846	123.211	
	600	20.716	-263.807	35.715	-285.236	103.911	
	700	21.287	-261.707	38.951	-288.973	90.233	
	800	21.858	-259.550	41.831	-293.014	80.059	
	900	22.429	-257.335	44.439	-297.330	72.211	
	1000	23.000	-255.064	46.831	-301.895	65.988	
	1100	23.571	-252.735	49.050	-306.690	60.942	
	1200	24.142	-250.350	51.126	-311.700	56.776	
	1300	24.713	-247.907	53.080	-316.911	53.285	
	1400	25.284	-245.407	54.933	-322.313	50.322	
	1500	25.855	-242.850	56.697	-327.895	47.781	

* * * * * * * * * *

PuOCl

PLUTONIUM OXYCHLORIDE

		A	B	C	D	RANGE	REF.
SOL	CP	18.29	5.71			298-1500	152

PHASE	T	CP	H	S	G	BT	
SOL	298	19.992	-222.7	26.	-230.452	168.949	152
	300	20.003	-222.663	26.124	-230.500	167.942	
	400	20.574	-220.634	31.956	-233.417	127.550	
	500	21.145	-218.548	36.609	-236.853	103.542	
	600	21.716	-216.405	40.514	-240.714	87.692	
	700	22.287	-214.205	43.905	-244.938	76.483	
	800	22.858	-211.948	46.918	-249.482	68.165	
	900	23.429	-209.633	49.643	-254.312	61.764	
	1000	24.000	-207.262	52.141	-259.403	56.700	
	1100	24.571	-204.833	54.456	-264.735	52.605	
	1200	25.142	-202.348	56.618	-270.289	49.233	
	1300	25.713	-199.805	58.653	-276.054	46.415	
	1400	26.284	-197.205	60.579	-282.016	44.031	
	1500	26.855	-194.548	62.412	-288.167	41.992	

PLUTONIUM OXYBROMICE

PuOBr

		A	B	C	D	RANGE	REF.
SOL	CP	19.28	5.71			298-1500	152

PHASE	T	CP	H	S	G	BT	
SOL	298	20.982	-212.4	28.5	-220.897	161.944	152
	300	20.993	-212.361	28.630	-220.950	160.984	
	400	21.564	-210.233	34.747	-224.132	122.477	
	500	22.135	-208.048	39.621	-227.859	99.610	
	600	22.706	-205.806	43.707	-232.030	84.528	
	700	23.277	-203.507	47.250	-236.582	73.874	
	800	23.848	-201.151	50.395	-241.467	65.975	
	900	24.419	-198.738	53.237	-246.651	59.903	
	1000	24.990	-196.267	55.839	-252.107	55.105	
	1100	25.561	-193.740	58.248	-257.812	51.229	
	1200	26.132	-191.155	60.497	-263.751	48.042	
	1300	26.703	-188.513	62.611	-269.907	45.382	
	1400	27.274	-185.814	64.611	-276.269	43.133	
	1500	27.845	-183.058	66.512	-282.826	41.213	

* * * * * * * * * *

PLUTONIUM OXYIODIDE

PuOJ

		A	B	C	D	RANGE	REF.
SOL	CP	20.29	5.71			298-1500	152

PHASE	T	CP	H	S	G	BT	
SOL	298	21.992	-197.8	30.2	-206.804	151.612	152
	300	22.003	-197.759	30.336	-206.860	150.718	
	400	22.574	-195.530	36.744	-210.228	114.879	
	500	23.145	-193.245	41.843	-214.166	93.624	
	600	23.716	-190.901	46.113	-218.569	79.625	
	700	24.287	-188.501	49.812	-223.370	69.748	
	800	24.858	-186.044	53.092	-228.518	62.437	
	900	25.429	-183.530	56.053	-233.977	56.825	
	1000	26.000	-180.958	58.762	-239.720	52.398	
	1100	26.571	-178.330	61.267	-245.723	48.827	
	1200	27.142	-175.644	63.603	-251.968	45.896	
	1300	27.713	-172.901	65.798	-258.439	43.453	
	1400	28.284	-170.101	67.873	-265.123	41.393	
	1500	28.855	-167.245	69.844	-272.010	39.637	

PuO_2 PLUTONIUM DIOXIDE

		A	B	C	D	RANGE	REF.
SOL	CP	22.18	0.21	-4.94		298-2663	134
	** MPT= 2663						134

PHASE	T	CP	H	S	G	BT	
SOL	298	16.685	-252.4	16.35	-257.275	188.613	202/134
	300	16.754	-252.369	16.453	-257.305	187.472	
	400	19.176	-250.555	21.655	-259.217	141.649	
	500	20.309	-248.575	26.069	-261.609	114.365	
	600	20.934	-246.510	29.832	-264.409	96.324	
	700	21.319	-244.396	33.090	-267.559	83.547	
	800	21.576	-242.250	35.955	-271.014	74.048	
	900	21.759	-240.083	38.507	-274.740	66.725	
	1000	21.896	-237.900	40.807	-278.707	60.920	
	1100	22.003	-235.705	42.899	-282.894	56.213	
	1200	22.089	-233.500	44.818	-287.281	52.328	
	1300	22.161	-231.288	46.588	-291.853	49.072	
	1400	22.222	-229.069	48.233	-296.595	46.307	
	1500	22.275	-226.844	49.768	-301.496	43.934	
	1600	22.323	-224.614	51.207	-306.545	41.878	
	1700	22.366	-222.379	52.562	-311.734	40.082	
	1800	22.406	-220.141	53.841	-317.055	38.501	
	1900	22.442	-217.898	55.054	-322.500	37.101	
	2000	22.476	-215.652	56.206	-328.064	35.854	
	2100	22.509	-213.403	57.303	-333.740	34.737	
	2200	22.540	-211.150	58.351	-339.523	33.733	
	2300	22.570	-208.895	59.354	-345.408	32.826	
	2400	22.598	-206.637	60.315	-351.392	32.003	
	2500	22.626	-204.375	61.238	-357.470	31.254	
	2600	22.653	-202.111	62.126	-363.639	30.571	
	2663	22.670	-200.684	62.668	-367.570	30.170	

PLUTONIUM MONOSULFICE

PuS

		A	B	C	D	RANGE	REF.
SOL	CP	12.7	1.6	-0.9		298-1800	152
	** MPT=	2620					

PHASE	T	CP	H	S	G	BT	
SOL	298	12.165	-104.3	18.7	-109.875	80.552	152*
	300	12.180	-104.277	18.775	-109.910	80.080	
	400	12.777	-103.026	22.370	-111.975	61.188	
	500	13.140	-101.729	25.263	-114.361	49.994	
	600	13.410	-100.401	27.683	-117.011	42.627	
	700	13.636	-99.049	29.768	-119.886	37.435	
	800	13.839	-97.675	31.602	-122.957	33.595	
	900	14.029	-96.281	33.243	-126.200	30.650	
	1000	14.210	-94.869	34.731	-129.600	28.328	
	1100	14.386	-93.440	36.093	-133.142	26.457	
	1200	14.557	-91.992	37.352	-136.815	24.921	
	1300	14.727	-90.528	38.524	-140.610	23.642	
	1400	14.894	-89.047	39.622	-144.518	22.563	
	1500	15.060	-87.549	40.655	-148.532	21.644	
	1600	15.225	-86.035	41.632	-152.647	20.853	
	1700	15.389	-84.505	42.560	-156.857	20.168	
	1800	15.552	-82.957	43.445	-161.158	19.570	

Pu_2S_3 DIPLUTONIUM TRISULFIDE

		A	B	C	D	RANGE	REF.
SOL	CP	30.5	4.0	-0.9		298-1500	152
	** MPT= 2000						63

PHASE	T	CP	H	S	G	BT	
SOL	298	30.680	-235.	46.	-248.715	182.337	152*
	300	30.700	-234.943	46.190	-248.800	181.275	
	400	31.537	-231.828	55.145	-253.886	138.736	
	500	32.140	-228.643	62.250	-259.768	113.560	
	600	32.650	-225.403	68.156	-266.297	97.012	
	700	33.116	-222.115	73.224	-273.372	85.362	
	800	33.559	-218.781	77.675	-280.921	76.754	
	900	33.989	-215.403	81.653	-288.891	70.162	
	1000	34.410	-211.983	85.256	-297.239	64.970	
	1100	34.826	-208.521	88.555	-305.932	60.791	
	1200	35.237	-205.018	91.603	-314.942	57.366	
	1300	35.647	-201.474	94.440	-324.246	54.518	
	1400	36.054	-197.889	97.096	-333.824	52.119	
	1500	36.460	-194.263	99.598	-343.660	50.078	

* * * * * * * * * *

$Pu(SO_4)_2$ PLUTONIUM DISULFATE

		A	B	C	D	RANGE	REF.
SOL	CP	28.1	51.5			298-1500	152

PHASE	T	CP	H	S	G	BT	
SOL	298	43.455	-526.	39.	-537.628	394.145	152
	300	43.550	-525.920	39.269	-537.700	391.767	
	400	48.700	-521.307	52.503	-542.308	296.343	
	500	53.850	-516.180	63.923	-548.141	239.625	
	600	59.000	-510.537	74.197	-555.055	202.206	
	700	64.150	-504.380	83.678	-562.954	175.786	
	800	69.300	-497.707	92.580	-571.771	156.222	
	900	74.450	-490.520	101.040	-581.456	141.216	
	1000	79.600	-482.817	109.151	-591.968	129.392	
	1100	84.750	-474.600	116.979	-603.276	119.876	
	1200	89.900	-465.867	124.574	-615.356	112.087	
	1300	95.050	-456.620	131.973	-628.185	105.622	
	1400	100.200	-446.857	139.206	-641.745	100.194	
	1500	105.350	-436.580	146.294	-656.021	95.595	

PLUTONIUM NITRIDE

PuN

		A	B	C	D	RANGE	REF.
SOL	CP	12.10	2.55	-1.25		298-2200	152
	** MPT= 2843						152

PHASE	T	CP	H	S	G	BT	
SOL	298	11.454	-71.5	17.3	-76.658	56.199	152*
	300	11.476	-71.479	17.371	-76.690	55.876	
	400	12.339	-70.284	20.803	-78.605	42.954	
	500	12.875	-69.021	23.617	-80.830	35.336	
	600	13.283	-67.713	26.002	-83.314	30.351	
	700	13.630	-66.367	28.076	-86.020	26.860	
	800	13.945	-64.988	29.917	-88.922	24.296	
	900	14.241	-63.579	31.577	-91.998	22.343	
	1000	14.525	-62.140	33.092	-95.232	20.816	
	1100	14.802	-60.674	34.489	-98.612	19.595	
	1200	15.073	-59.180	35.789	-102.127	18.602	
	1300	15.341	-57.659	37.006	-105.767	17.783	
	1400	15.606	-56.112	38.153	-109.526	17.100	
	1500	15.869	-54.538	39.238	-113.396	16.524	
	1600	16.131	-52.938	40.271	-117.372	16.034	
	1700	16.392	-51.312	41.257	-121.448	15.615	
	1800	16.651	-49.660	42.201	-125.622	15.255	
	1900	16.910	-47.982	43.108	-129.887	14.942	
	2000	17.169	-46.278	43.982	-134.242	14.671	
	2100	17.427	-44.548	44.826	-138.683	14.435	
	2200	17.684	-42.792	45.643	-143.207	14.228	

$PuC_{0.88}$ PLUTONIUM 0.88-CARBIDE

		A	B	C	D	RANGE	REF.
SOL	CP	13.08	1.14	-3.23		298-1400	65
	**	H298 AFTER REF. 214, 215, 216					

PHASE	T	CP	H	S	G	BT	
SOL	298	9.786	-11.1	17.3	-16.258	11.919	214
	300	9.833	-11.082	17.361	-16.290	11.869	
	400	11.517	-10.003	20.452	-18.184	9.937	
	500	12.358	-8.805	23.122	-20.366	8.903	
	600	12.867	-7.542	25.423	-22.796	8.305	
	700	13.219	-6.237	27.434	-25.441	7.944	
	800	13.487	-4.901	29.218	-28.276	7.726	
	900	13.707	-3.541	30.819	-31.279	7.597	
	1000	13.897	-2.161	32.274	-34.435	7.527	
	1100	14.067	-.762	33.606	-37.729	7.497	
	1200	14.224	.652	34.837	-41.152	7.496	
	1300	14.371	2.082	35.981	-44.694	7.515	
	1400	14.511	3.526	37.052	-48.346	7.548	

PLUTONIUM 1.5-CARBICE

$PuC_{1.5}$

		A	B	C	D	RANGE	REF.
SOL	CP	8.99	9.23	-0.01	-1.59	298-2200	214,*

PHASE	T	CP	H	S	G	BT	
SOL	298	11.589	-13.2	20.7	-19.372	14.202	214,*
	300	11.605	-13.179	20.772	-19.410	14.142	
	400	12.421	-11.977	24.223	-21.666	11.839	
	500	13.203	-10.695	27.079	-24.235	10.595	
	600	13.953	-9.337	29.553	-27.069	9.861	
	700	14.670	-7.906	31.758	-30.137	9.410	
	800	15.355	-6.404	33.762	-33.414	9.130	
	900	16.008	-4.836	35.609	-36.884	8.958	
	1000	16.629	-3.204	37.328	-40.532	8.859	
	1100	17.218	-1.511	38.941	-44.346	8.812	
	1200	17.776	.239	40.463	-48.317	8.801	
	1300	18.301	2.043	41.907	-52.436	8.816	
	1400	18.795	3.898	43.281	-56.696	8.852	
	1500	19.257	5.801	44.594	-61.090	8.902	
	1600	19.687	7.748	45.851	-65.613	8.964	
	1700	20.086	9.737	47.056	-70.259	9.034	
	1800	20.452	11.764	48.215	-75.023	9.110	
	1900	20.787	13.826	49.330	-79.900	9.192	
	2000	21.090	15.921	50.404	-84.887	9.277	
	2100	21.361	18.043	51.440	-89.980	9.366	
	2200	21.600	20.192	52.439	-95.174	9.456	

PuC_2 PLUTONIUM DICARBIDE

		A	B	C	D	RANGE	REF.
SOL	CP	14.38	6.46	-2.75	-1.34	298-2525	152
	** MPT= 2525						

PHASE	T	CP	H	S	G	BT	
SOL	298	13.093	-8.25	24.1	-15.435	11.316	*/214
	300	13.142	-8.226	24.181	-15.480	11.279	
	400	15.031	-6.807	28.249	-18.107	9.894	
	500	16.175	-5.243	31.734	-21.110	9.229	
	600	17.010	-3.582	34.760	-24.438	8.903	
	700	17.684	-1.847	37.434	-28.051	8.759	
	800	18.261	-.049	39.834	-31.916	8.720	
	900	18.769	1.803	42.015	-36.010	8.746	
	1000	19.225	3.703	44.016	-40.313	8.812	
	1100	19.637	5.647	45.868	-44.808	8.904	
	1200	20.011	7.629	47.593	-49.482	9.013	
	1300	20.351	9.648	49.209	-54.323	9.134	
	1400	20.657	11.699	50.728	-59.321	9.262	
	1500	20.933	13.778	52.163	-64.466	9.394	
	1600	21.178	15.884	53.522	-69.751	9.529	
	1700	21.394	18.013	54.813	-75.168	9.665	
	1800	21.582	20.162	56.041	-80.711	9.801	
	1900	21.740	22.328	57.212	-86.375	9.937	
	2000	21.871	24.509	58.331	-92.152	10.071	
	2100	21.974	26.702	59.400	-98.039	10.204	
	2200	22.050	28.903	60.424	-104.031	10.336	
	2300	22.097	31.111	61.406	-110.123	10.465	
	2400	22.118	33.322	62.347	-116.310	10.593	
	2500	22.111	35.533	63.250	-122.591	10.718	
	2525	22.105	36.086	63.470	-124.175	10.749	

RUBIDIUM

		A	B	C	D	RANGE	REF.
SOL	CP	0.840	22.084			298- 313	21
	** MPT=	312.64					21
LIQ	CP	9.768	-6.265	0.081	3.382	313- 961	21
GAS	CP	4.968				961-2000	21

PHASE	T	CP	H	S	G	BT	P	
SOL	298	7.424	0	18.35	-5.471	4.011		21
	300	7.465	.014	18.396	-5.505	4.011		
	313	7.744	.110	18.710	-5.740	4.013		
			0.524	1.676				21
LIQ	313	8.223	.634	20.386	-5.740	4.013		
	400	7.854	1.336	22.367	-7.611	4.159	.001	
	500	7.513	2.103	24.081	-9.937	4.344	.139	
	600	7.249	2.841	25.427	-12.415	4.523	2.903	
	700	7.056	3.555	26.529	-15.015	4.688	24.749	
	800	6.933	4.254	27.462	-17.715	4.840	121.177	
	900	6.879	4.944	28.275	-20.503	4.980	411.127	
	961	6.879	5.364	28.726	-22.242	5.059	760.	
			17.26	17.960				21
GAS	961	4.968	22.624	46.687	-22.242	5.059		
	1000	4.968	22.818	46.884	-24.067	5.260		
	1100	4.968	23.314	47.358	-28.779	5.719		
	1200	4.968	23.811	47.790	-33.537	6.109		
	1300	4.968	24.308	48.188	-38.336	6.446		
	1400	4.968	24.805	48.556	-43.173	6.741		
	1500	4.968	25.302	48.899	-48.046	7.001		
	1600	4.968	25.798	49.219	-52.952	7.234		
	1700	4.968	26.295	49.520	-57.889	7.443		
	1800	4.968	26.792	49.804	-62.856	7.633		
	1900	4.968	27.289	50.073	-67.850	7.806		
	2000	4.968	27.786	50.328	-72.870	7.964		

Rb (g)

RUBIDIUM (MONATOMIC GAS)

		A	B	C	D	RANGE	REF.
GAS	CP	4.968				298-2000	21*

PHASE	T	CP	H	S	G	BT	
GAS	298	4.968	19.331	40.873	7.145	-5.238	21,*
	300	4.968	19.340	40.904	7.069	-5.151	
	400	4.968	19.837	42.333	2.904	-1.587	
	500	4.968	20.334	43.442	-1.387	.606	
	600	4.968	20.831	44.347	-5.778	2.105	
	700	4.968	21.327	45.113	-10.252	3.201	
	800	4.968	21.824	45.776	-14.797	4.043	
	900	4.968	22.321	46.362	-19.404	4.713	
	961	4.968	22.624	46.687	-22.243	5.059	
	961	4.968	22.624	46.687	-22.243	5.059	
	1000	4.968	22.818	46.885	-24.067	5.261	
	1100	4.968	23.315	47.359	-28.780	5.719	
	1200	4.968	23.811	47.791	-33.538	6.109	
	1300	4.968	24.308	48.188	-38.337	6.446	
	1400	4.968	24.805	48.557	-43.174	6.741	
	1500	4.968	25.302	48.899	-48.047	7.001	
	1600	4.968	25.799	49.220	-52.953	7.234	
	1700	4.968	26.295	49.521	-57.891	7.443	
	1800	4.968	26.792	49.805	-62.857	7.633	
	1900	4.968	27.289	50.074	-67.851	7.806	
	2000	4.968	27.786	50.329	-72.871	7.964	

RUBIDIUM (DIATOMIC GAS)

		A	B	C	D	RANGE	REF.
GAS	CP	8.911	0.477			298-2000	21

PHASE	T	CP	H	S	G	BT	
GAS	298	9.053	13.77	32.35	4.125	-3.024	21
	300	9.054	13.787	32.406	4.065	-2.962	
	400	9.102	14.695	35.017	.688	-.376	
	500	9.149	15.607	37.053	-2.920	1.276	
	600	9.197	16.524	38.726	-6.711	2.445	
	700	9.245	17.447	40.147	-10.656	3.328	
	800	9.293	18.373	41.385	-14.734	4.026	
	900	9.340	19.305	42.482	-18.929	4.597	
	961	9.369	19.876	43.095	-21.539	4.899	
	961	9.369	19.876	43.095	-21.539	4.899	
	1000	9.388	20.241	43.469	-23.227	5.077	
	1100	9.436	21.183	44.366	-27.619	5.488	
	1200	9.483	22.129	45.189	-32.098	5.847	
	1300	9.531	23.079	45.950	-36.655	6.163	
	1400	9.579	24.035	46.658	-41.286	6.446	
	1500	9.626	24.995	47.320	-45.985	6.701	
	1600	9.674	25.960	47.943	-50.749	6.933	
	1700	9.722	26.930	48.531	-55.572	7.145	
	1800	9.770	27.905	49.088	-60.454	7.341	
	1900	9.817	28.884	49.617	-65.389	7.522	
	2000	9.865	29.868	50.122	-70.376	7.691	

RbF

RUBIDIUM FLUORIDE

		A	B	C	D	RANGE	REF.
SOL	CP	7.966	9.21	1.20		298-1048	121
LIQ	CP	-11.303	0.883	350.67		1048-1200	121
	CP	14.1				1200-1663	*
	** BPT= 1663, L= 42.5						1

PHASE	T	CP	H	S	G	BT	
SOL	298	12.062	-132.2	17.6	-137.447	100.765	19,*
	300	12.062	-132.178	17.675	-137.480	100.168	
	400	12.400	-130.959	21.179	-139.430	76.191	
	500	13.051	-129.688	24.013	-141.694	61.943	
	600	13.825	-128.345	26.459	-144.220	52.539	
	700	14.658	-126.921	28.652	-146.977	45.895	
	800	15.521	-125.412	30.666	-149.945	40.968	
	900	16.403	-123.816	32.545	-153.106	37.184	
	1000	17.296	-122.131	34.319	-156.450	34.197	
	1048	17.727	-121.290	35.140	-158.117	32.978	
			5.49	5.239			121
LIQ	1048	21.551	-115.800	40.379	-158.117	32.978	
	1100	18.649	-114.757	41.351	-160.243	31.842	
	1200	14.109	-113.129	42.770	-164.453	29.955	
	1200	14.100	-113.129	42.770	-164.453	29.955	
	1300	14.100	-111.719	43.899	-168.787	28.380	
	1400	14.100	-110.309	44.944	-173.230	27.046	
	1500	14.100	-108.899	45.916	-177.774	25.905	
	1600	14.100	-107.489	46.826	-182.411	24.920	
	1663	14.100	-106.601	47.371	-185.379	24.366	

RUBIDIUM OXIDE Rb_2O

		A	B	C	D	RANGE	REF.
SOL	CP	16.1	8.0			298- 900	*
	** DEC.						1

PHASE	T	CP	H	S	G	BT	
SOL	298	18.485	-78.9	27.4	-87.069	63.832	19A/*
	300	18.500	-78.866	27.514	-87.120	63.475	
	400	19.300	-76.976	32.946	-90.154	49.265	
	500	20.100	-75.006	37.339	-93.675	40.951	
	600	20.900	-72.956	41.074	-97.600	35.556	
	700	21.700	-70.826	44.356	-101.875	31.811	
	800	22.500	-68.616	47.306	-106.460	29.088	
	900	23.300	-66.326	50.002	-111.328	27.038	

* * * * * * * * * *

RUBIDIUM CARBONATE Rb_2CO_3

		A	B	C	D	RANGE	REF.
SOL	CP	25.5	19.3	-2.5		298-1113	*
	** MPT= 1113						1

PHASE	T	CP	H	S	G	BT	
SOL	298	28.442	-269.6	41.8	-282.063	206.785	66
	300	28.512	-269.547	41.976	-282.140	205.567	
	400	31.657	-266.530	50.634	-286.784	156.713	
	500	34.150	-263.237	57.973	-292.223	127.748	
	600	36.386	-259.708	64.400	-298.348	108.688	
	700	38.500	-255.964	70.168	-305.081	95.264	
	800	40.549	-252.011	75.444	-312.366	85.346	
	900	42.561	-247.855	80.336	-320.157	77.755	
	1000	44.550	-243.499	84.924	-328.423	71.786	
	1100	46.523	-238.945	89.262	-337.134	66.991	
	1113	46.779	-238.339	89.810	-338.298	66.437	

$Rb_2O \cdot SiO_2$ RUBIDIUM METASILICATE

		A	B	C	D	RANGE	REF.
SOL	CP	29.56	11.76	-4.44		298-1143	5
LIQ	CP	42.5				1143-1800	5

PHASE	T	CP	H	S	G	BT	
SOL	298	28.071	-365.1	38.5	-376.579	276.077	5
	300	28.155	-365.048	38.674	-376.650	274.426	
	400	31.489	-362.050	47.275	-380.960	208.175	
	500	33.664	-358.787	54.547	-386.061	168.770	
	600	35.383	-355.332	60.841	-391.837	142.746	
	700	36.886	-351.718	66.410	-398.205	124.342	
	800	38.274	-347.959	71.427	-405.101	110.683	
	900	39.596	-344.065	76.012	-412.476	100.176	
	1000	40.876	-340.041	80.251	-420.292	91.867	
	1100	42.129	-335.891	84.206	-428.517	85.150	
	1143	42.662	-334.068	85.831	-432.173	82.646	
			10.	8.749			5
LIQ	1143	42.500	-324.068	94.580	-432.173	82.646	
	1200	42.500	-321.645	96.648	-437.623	79.713	
	1300	42.500	-317.395	100.050	-447.460	75.235	
	1400	42.500	-313.145	103.200	-457.625	71.448	
	1500	42.500	-308.895	106.132	-468.093	68.210	
	1600	42.500	-304.645	108.875	-478.845	65.416	
	1700	42.500	-300.395	111.451	-489.863	62.985	
	1800	42.500	-296.145	113.881	-501.130	60.854	

RUBIDIUM DISILICATE $Rb_2O \cdot 2SiO_2$

		A	B	C	D	RANGE	REF.
SOL	CP	46.19	12.04	-7.98		298-1363	5
LIQ	CP	62.0				1363-1800	5

PHASE	T	CP	H	S	G	BT	
SOL	298	40.803	-589.3	46.5	-603.164	442.191	5
	300	40.935	-589.224	46.753	-603.250	439.527	
	400	46.018	-584.849	59.305	-608.571	332.553	
	500	49.018	-580.087	69.919	-615.046	268.873	
	600	51.197	-575.072	79.056	-622.506	226.778	
	700	52.989	-569.860	87.086	-630.821	196.978	
	800	54.575	-564.481	94.267	-639.895	174.835	
	900	56.041	-558.949	100.781	-649.652	157.778	
	1000	57.432	-553.275	106.758	-660.033	144.270	
	1100	58.774	-547.465	112.295	-670.989	133.331	
	1200	60.084	-541.521	117.465	-682.480	124.313	
	1300	61.370	-535.449	122.326	-694.472	116.767	
	1363	62.171	-531.557	125.249	-702.271	112.621	
			13.	9.538			5
LIQ	1363	62.000	-518.557	134.787	-702.271	112.621	
	1400	62.000	-516.263	136.447	-707.289	110.428	
	1500	62.000	-510.063	140.725	-721.150	105.086	
	1600	62.000	-503.863	144.726	-735.425	100.468	
	1700	62.000	-497.663	148.485	-750.087	96.443	
	1800	62.000	-491.463	152.029	-765.115	92.910	

$Rb_2O \cdot 4SiO_2$ RUBIDIUM TETRASILICATE

		A	B	C	D	RANGE	REF.
SOL	CP	61.03	29.77	-4.59		298-1173	5
LIQ	CP	95.0				1173-1800	5

PHASE	T	CP	H	S	G	BT	
SOL	298	64.742	-1030.	66.5	-1049.827	769.647	5
	300	64.861	-1029.880	66.901	-1049.950	764.991	
	400	70.069	-1023.118	86.319	-1057.645	577.948	
	500	74.079	-1015.905	102.399	-1067.104	466.493	
	600	77.617	-1008.317	116.222	-1078.050	392.732	
	700	80.932	-1000.388	128.438	-1090.295	340.451	
	800	84.129	-992.135	139.454	-1103.698	301.557	
	900	87.256	-983.565	149.544	-1118.155	271.562	
	1000	90.341	-974.685	158.898	-1133.583	247.778	
	1100	93.398	-965.498	167.652	-1149.915	228.498	
	1173	95.617	-958.599	173.724	-1162.376	216.600	
			11.	9.378			5
LIQ	1173	95.000	-947.599	183.101	-1162.376	216.600	
	1200	95.000	-945.034	185.263	-1167.349	212.632	
	1300	95.000	-935.534	192.867	-1186.261	199.455	
	1400	95.000	-926.034	199.907	-1205.904	188.275	
	1500	95.000	-916.534	206.462	-1226.226	178.685	
	1600	95.000	-907.034	212.593	-1247.182	170.380	
	1700	95.000	-897.534	218.352	-1268.732	163.129	
	1800	95.000	-888.034	223.782	-1290.842	156.751	

RHENIUM

		A	B	C	D	RANGE	REF.
SOL	CP	5.662	1.302			298-3453	21
LIQ	CP	10.				3453-4200	21
	** BPT= 5869, L= 170.85						21

PHASE	T	CP	H	S	G	BT	P	
SOL	298	6.050	0	8.73	-2.603	1.908		21
	300	6.053	.011	8.767	-2.619	1.908		
	400	6.183	.623	10.526	-3.588	1.960		
	500	6.313	1.248	11.920	-4.712	2.060		
	600	6.443	1.886	13.083	-5.964	2.173		
	700	6.573	2.536	14.086	-7.324	2.287		
	800	6.704	3.200	14.972	-8.777	2.398		
	900	6.834	3.877	15.769	-10.315	2.505		
	1000	6.964	4.567	16.496	-11.929	2.607		
	1100	7.094	5.270	17.166	-13.612	2.705		
	1200	7.224	5.986	17.788	-15.360	2.798		
	1300	7.355	6.715	18.372	-17.169	2.887		
	1400	7.485	7.457	18.922	-19.034	2.972		
	1500	7.615	8.212	19.442	-20.952	3.053		
	1600	7.745	8.980	19.938	-22.921	3.131		
	1700	7.875	9.761	20.412	-24.939	3.207		
	1800	8.006	10.555	20.865	-27.003	3.279		
	1900	8.136	11.362	21.302	-29.111	3.349		
	2000	8.266	12.182	21.722	-31.263	3.417		
	2100	8.396	13.015	22.129	-33.455	3.482		
	2200	8.526	13.861	22.522	-35.688	3.546		
	2300	8.657	14.720	22.904	-37.959	3.607		
	2400	8.787	15.593	23.275	-40.268	3.667		
	2500	8.917	16.478	23.637	-42.614	3.726		
	2600	9.047	17.376	23.989	-44.996	3.783		
	2700	9.177	18.287	24.333	-47.412	3.838		
	2800	9.308	19.211	24.669	-49.862	3.892		
	2900	9.438	20.149	24.998	-52.345	3.945		
	3000	9.568	21.099	25.320	-54.861	3.997		
	3100	9.698	22.062	25.636	-57.409	4.048	.002	
	3200	9.828	23.039	25.946	-59.988	4.098	.004	
	3300	9.959	24.028	26.250	-62.598	4.146	.009	
	3400	10.089	25.030	26.550	-65.238	4.194	.020	
	3453	10.158	25.567	26.706	-66.649	4.219	.031	
			7.942	2.300				21
LIQ	3453	10.000	33.509	29.006	-66.649	4.219	.031	
	3500	10.000	33.979	29.141	-68.016	4.248	.043	
	3600	10.000	34.979	29.423	-70.944	4.307	.083	
	3700	10.000	35.979	29.697	-73.900	4.366	.158	
	3800	10.000	36.979	29.964	-76.883	4.422	.288	
	3900	10.000	37.979	30.223	-79.893	4.478	.510	
	4000	10.000	38.979	30.477	-82.928	4.532	.877	
	4100	10.000	39.979	30.724	-85.988	4.584	1.469	
	4200	10.000	40.979	30.965	-89.072	4.636	2.401	

$ReF_6(g)$ RHENIUM HEXAFLUORIDE (GAS)

		A	B	C	D	RANGE	REF.
GAS	CP	37.1	0.31	-7.65		298-2000	*

PHASE	T	CP	H	S	G	BT	
GAS	298	28.587	-271.1	84.8	-296.383	217.284	1/*
	300	28.693	-271.047	84.977	-296.540	216.058	
	400	32.443	-267.964	93.822	-305.492	166.936	
	500	34.195	-264.622	101.271	-315.258	137.818	
	600	35.161	-261.150	107.598	-325.709	118.655	
	700	35.756	-257.602	113.067	-336.749	105.152	
	800	36.153	-254.006	117.869	-348.300	95.164	
	900	36.435	-250.375	122.144	-360.305	87.506	
	1000	36.645	-246.721	125.994	-372.715	81.468	
	1100	36.809	-243.048	129.495	-385.492	76.601	
	1200	36.941	-239.360	132.703	-398.604	72.606	
	1300	37.050	-235.661	135.665	-412.025	69.277	
	1400	37.144	-231.951	138.414	-425.730	66.468	
	1500	37.225	-228.232	140.979	-439.701	64.073	
	1600	37.297	-224.506	143.384	-453.921	62.011	
	1700	37.362	-220.773	145.647	-468.373	60.222	
	1800	37.422	-217.034	147.784	-483.046	58.658	
	1900	37.477	-213.289	149.809	-497.926	57.282	
	2000	37.529	-209.538	151.733	-513.004	56.066	

* * * * * * * * * *

$ReCl_3$ RHENIUM TRICHLORIDE

		A	B	C	D	RANGE	REF.
SOL	CP	25.21	6.6	-4.56		298- 700	19,*

PHASE	T	CP	H	S	G	BT	
SOL	298	22.048	-63.	29.6	-71.825	52.656	19
	300	22.123	-62.959	29.737	-71.880	52.372	
	400	25.000	-60.587	36.541	-75.203	41.095	
	500	26.686	-57.997	42.313	-79.154	34.603	
	600	27.903	-55.265	47.291	-83.640	30.470	
	700	28.899	-52.424	51.669	-88.592	27.663	

RHENIUM TRIBROMIDE $ReBr_3$

		A	B	C	D	RANGE	REF.
SOL	CP	19.2	16.3			298- 773	*
	** SPT= 773						46

PHASE	T	CP	H	S	G	BT	
SOL	298	24.060	-39.3	35.	-49.735	36.462	1
	300	24.090	-39.255	35.149	-49.800	36.284	
	400	25.720	-36.765	42.302	-53.686	29.337	
	500	27.350	-34.111	48.217	-58.220	25.451	
	600	28.980	-31.295	53.347	-63.303	23.061	
	700	30.610	-28.315	57.937	-68.871	21.506	
	773	31.800	-26.038	61.032	-73.215	20.703	

* * * * * * * * * *

RHENIUM DIOXIDE ReO_2

		A	B	C	D	RANGE	REF.
SOL	CP	16.1	3.03	-3.09		298-1300	96

PHASE	T	CP	H	S	G	BT	
SOL	298	13.527	-103.4	15.	-107.872	79.083	96,*
	300	13.576	-103.375	15.084	-107.900	78.616	
	400	15.381	-101.916	19.267	-109.623	59.903	
	500	16.379	-100.325	22.815	-111.732	48.845	
	600	17.060	-98.651	25.865	-114.170	41.592	
	700	17.590	-96.918	28.536	-116.893	36.500	
	800	18.041	-95.135	30.915	-119.867	32.751	
	900	18.446	-93.311	33.064	-123.068	29.889	
	1000	18.821	-91.447	35.027	-126.474	27.645	
	1100	19.178	-89.547	36.837	-130.068	25.846	
	1200	19.521	-87.612	38.521	-133.837	24.378	
	1300	19.856	-85.643	40.097	-137.769	23.164	

ReO_3 RHENIUM TRIOXIDE

		A	B	C	D	RANGE	REF.
SOL	CP	26.0				298- 433	*
	** DEC./ MPT= 433						1

PHASE	T	CP	H	S	G	BT	
SOL	298	26.000	-146.	19.3	-151.754	111.254	96
	300	26.000	-145.952	19.461	-151.790	110.594	
	400	26.000	-143.352	26.941	-154.128	84.223	
	433	26.000	-142.494	29.002	-155.052	78.270	

* * * * * * * * * *

Re_2O_7 DIRHENIUM HEPTOXIDE

		A	B	C	D	RANGE	REF.
SOL	CP	29.15	44.0	-2.25		298- 570	10
LIQ	CP	71.1				570- 600	10

PHASE	T	CP	H	S	G	BT	
SOL	298	39.737	-298.4	49.54	-313.170	229.591	96,*
	300	39.850	-298.326	49.786	-313.262	228.242	
	400	45.344	-294.059	62.025	-318.869	174.245	
	500	50.250	-289.276	72.677	-325.615	142.345	
	570	53.537	-285.643	79.472	-330.943	126.907	
			15.	26.316			10
LIQ	570	71.100	-270.643	105.788	-330.943	126.907	
	600	71.100	-268.510	109.435	-334.171	121.738	

RHENIUM DISULFIDE ReS_2

		A	B	C	D	RANGE	REF.
SOL	CP	16.39	3.77	-1.57		298-1500	63

PHASE	T	CP	H	S	G	BT	
SOL	298	15.748	-42.7	14.5	-47.023	34.474	63
	300	15.777	-42.671	14.598	-47.050	34.281	
	400	16.917	-41.031	19.308	-48.754	26.641	
	500	17.647	-39.301	23.166	-50.883	22.244	
	600	18.216	-37.507	26.435	-53.368	19.442	
	700	18.709	-35.660	29.281	-56.156	17.535	
	800	19.161	-33.766	31.809	-59.213	16.178	
	900	19.589	-31.829	34.090	-62.510	15.182	
	1000	20.003	-29.849	36.176	-66.025	14.432	
	1100	20.407	-27.828	38.101	-69.740	13.858	
	1200	20.805	-25.768	39.894	-73.641	13.414	
	1300	21.198	-23.667	41.575	-77.715	13.067	
	1400	21.588	-21.528	43.160	-81.952	12.795	
	1500	21.975	-19.350	44.663	-86.344	12.582	

* * * * * * * * * *

DIRHENIUM HEPTASULFIDE Re_2S_7

		A	B	C	D	RANGE	REF.
SOL	CP	44.4	12.0			298- 700	63

PHASE	T	CP	H	S	G	BT	
SOL	298	47.978	-107.9	40.	-119.826	87.847	63
	300	48.000	-107.811	40.297	-119.900	87.359	
	400	49.200	-102.951	54.270	-124.659	68.120	
	500	50.400	-97.971	65.378	-130.660	57.119	
	600	51.600	-92.871	74.673	-137.675	50.155	
	700	52.800	-87.651	82.717	-145.553	45.450	

Re_2Te_5 DIRHENIUM PENTATELLURIDE

		A	B	C	D	RANGE	REF.
SOL	CP	37.5	15.0			298-1200	*

PHASE	T	CP	H	S	G	BT	
SOL	298	41.972	-29.3	60.2	-47.249	34.639	154*
	300	42.000	-29.222	60.460	-47.360	34.507	
	400	43.500	-24.947	72.748	-54.046	29.534	
	500	45.000	-20.522	82.616	-61.830	27.030	
	600	46.500	-15.947	90.953	-70.519	25.690	
	700	48.000	-11.222	98.233	-79.986	24.976	
	800	49.500	-6.347	104.741	-90.140	24.628	
	900	51.000	-1.322	110.658	-100.914	24.509	
	1000	52.500	3.853	116.109	-112.256	24.537	
	1100	54.000	9.178	121.183	-124.123	24.664	
	1200	55.500	14.653	125.946	-136.482	24.860	

* * * * * * * * * *

Re_3As_7 TRIRHENIUM HEPTARSENIDE

		A	B	C	D	RANGE	REF.
SOL	CP	56.0	13.0	-0.40		298-1100	*

PHASE	T	CP	H	S	G	BT	
SOL	298	59.426	-22.8	80.5	-46.801	34.311	64,*
	300	59.456	-22.690	80.868	-46.950	34.208	
	400	60.950	-16.668	98.181	-55.941	30.569	
	500	62.340	-10.503	111.932	-66.469	29.058	
	600	63.689	-4.202	123.417	-78.252	28.507	
	700	65.018	2.234	133.335	-91.101	28.447	
	800	66.337	8.802	142.103	-104.881	28.656	
	900	67.651	15.501	149.992	-119.492	29.021	
	1000	68.960	22.332	157.188	-134.856	29.477	
	1100	70.267	29.293	163.822	-150.911	29.987	

5-RHENIUM 3-SILICON

Re_5Si_3

		A	B	C	D	RANGE	REF.
SOL	CP	45.60	10.80	-3.36		298-1500	*

PHASE	T	CP	H	S	G	BT	
SOL	298	45.040	-37.6	61.15	-55.832	40.931	221
	300	45.107	-37.517	61.429	-55.945	40.762	
	400	47.820	-32.859	74.810	-62.783	34.308	
	500	49.656	-27.981	85.688	-70.825	30.962	
	600	51.147	-22.939	94.876	-79.864	29.095	
	700	52.474	-17.757	102.862	-89.760	28.028	
	800	53.715	-12.447	109.950	-100.407	27.434	
	900	54.905	-7.015	116.346	-111.727	27.135	
	1000	56.064	-1.467	122.191	-123.658	27.029	
	1100	57.202	4.197	127.588	-136.150	27.054	
	1200	58.327	9.973	132.614	-149.163	27.170	
	1300	59.441	15.862	137.327	-162.663	27.350	
	1400	60.549	21.861	141.772	-176.620	27.575	
	1500	61.651	27.971	145.987	-191.009	27.834	

ReSi

RHENIUM SILICON

		A	B	C	D	RANGE	REF.
SOL	CP	12.57	2.30	-0.90		298-2000	*

PHASE	T	CP	H	S	G	BT	
SOL	298	12.243	-12.6	13.23	-16.545	12.129	221
	300	12.260	-12.577	13.306	-16.569	12.072	
	400	12.927	-11.315	16.933	-18.088	9.884	
	500	13.360	-9.999	19.867	-19.933	8.714	
	600	13.700	-8.646	22.334	-22.046	8.031	
	700	13.996	-7.261	24.468	-24.388	7.615	
	800	14.269	-5.847	26.355	-26.931	7.358	
	900	14.529	-4.407	28.051	-29.653	7.202	
	1000	14.780	-2.942	29.595	-32.537	7.112	
	1100	15.026	-1.452	31.015	-35.568	7.068	
	1200	15.267	.063	32.333	-38.736	7.056	
	1300	15.507	1.602	33.564	-42.032	7.067	
	1400	15.744	3.164	34.722	-45.447	7.095	
	1500	15.980	4.751	35.816	-48.974	7.136	
	1600	16.215	6.360	36.855	-52.608	7.187	
	1700	16.449	7.994	37.845	-56.343	7.244	
	1800	16.682	9.650	38.792	-60.176	7.307	
	1900	16.915	11.330	39.700	-64.101	7.374	
	2000	17.147	13.033	40.574	-68.115	7.444	

RHENIUM 2-SILICON

$ReSi_2$

		A	B	C	D	RANGE	REF.
SOL	CP	16.2	2.64	-1.46		298-2000	*

PHASE	T	CP	H	S	G	BT	
SOL	298	15.345	-21.6	17.7	-26.877	19.704	221
	300	15.370	-21.572	17.795	-26.910	19.607	
	400	16.343	-19.981	22.365	-28.927	15.807	
	500	16.936	-18.315	26.079	-31.355	13.707	
	600	17.378	-16.599	29.208	-34.123	12.431	
	700	17.750	-14.842	31.915	-37.182	11.610	
	800	18.084	-13.050	34.307	-40.496	11.064	
	900	18.396	-11.226	36.456	-44.036	10.695	
	1000	18.694	-9.371	38.409	-47.780	10.444	
	1100	18.983	-7.487	40.205	-51.712	10.276	
	1200	19.267	-5.575	41.869	-55.817	10.167	
	1300	19.546	-3.634	43.422	-60.082	10.102	
	1400	19.822	-1.666	44.880	-64.498	10.070	
	1500	20.095	.330	46.257	-69.056	10.063	
	1600	20.367	2.353	47.563	-73.747	10.075	
	1700	20.637	4.404	48.806	-78.566	10.102	
	1800	20.907	6.481	49.993	-83.506	10.140	
	1900	21.176	8.585	51.130	-88.563	10.188	
	2000	21.443	10.716	52.223	-93.731	10.244	

Rh

RHODIUM

		A	B	C	D	RANGE	REF.
SOL	CP	4.974	3.214	0.081	-0.542	298-2233	21
LIQ	CP	10.				2233-3970	21
	** BPT= 3970, L= 117.893						21

PHASE	T	CP	H	S	G	BT	P	
SOL	298	5.975	0	7.53	-2.245	1.646		21
	300	5.979	.011	7.567	-2.259	1.646		
	400	6.224	.621	9.320	-3.107	1.698		
	500	6.478	1.256	10.736	-4.112	1.798		
	600	6.730	1.917	11.939	-5.247	1.912		
	700	6.975	2.602	12.995	-6.495	2.028		
	800	7.211	3.311	13.942	-7.843	2.143		
	900	7.438	4.044	14.805	-9.281	2.254		
	1000	7.654	4.798	15.600	-10.801	2.361		
	1100	7.860	5.574	16.339	-12.399	2.464		
	1200	8.056	6.370	17.031	-14.068	2.562		
	1300	8.241	7.185	17.684	-15.804	2.657		
	1400	8.415	8.018	18.301	-17.603	2.748		
	1500	8.579	8.868	18.887	-19.463	2.836		
	1600	8.732	9.733	19.446	-21.380	2.921		
	1700	8.874	10.614	19.979	-23.351	3.002		
	1800	9.006	11.508	20.490	-25.375	3.081		
	1900	9.126	12.415	20.980	-27.448	3.158		
	2000	9.236	13.333	21.451	-29.570	3.232		
	2100	9.335	14.261	21.904	-31.738	3.303		
	2200	9.423	15.199	22.341	-33.950	3.373	.003	
	2233	9.450	15.511	22.481	-34.690	3.396	.005	
			5.136	2.300				21
LIQ	2233	10.000	20.647	24.781	-34.690	3.396	.005	
	2300	10.000	21.317	25.077	-36.360	3.455	.011	
	2400	10.000	22.317	25.503	-38.889	3.542	.034	
	2500	10.000	23.317	25.911	-41.460	3.625	.094	
	2600	10.000	24.317	26.303	-44.071	3.705	.243	
	2700	10.000	25.317	26.680	-46.720	3.782	.584	
	2800	10.000	26.317	27.044	-49.407	3.857	1.315	
	2900	10.000	27.317	27.395	-52.129	3.929	2.792	
	3000	10.000	28.317	27.734	-54.885	3.999	5.628	
	3100	10.000	29.317	28.062	-57.675	4.067	10.822	
	3200	10.000	30.317	28.379	-60.497	4.132	19.942	
	3300	10.000	31.317	28.687	-63.351	4.196	35.355	
	3400	10.000	32.317	28.986	-66.235	4.258	60.515	
	3500	10.000	33.317	29.276	-69.148	4.318	100.305	
	3600	10.000	34.317	29.557	-72.089	4.377	161.443	
	3700	10.000	35.317	29.831	-75.059	4.434	252.932	
	3800	10.000	36.317	30.098	-78.055	4.490	386.560	
	3900	10.000	37.317	30.358	-81.078	4.544	577.433	
	3970	10.000	38.017	30.536	-83.209	4.581	760.	

RHODIUM TRICHLORIDE $RhCl_3$

		A	B	C	D	RANGE	REF.
SOL	CP	25.2	6.6	-4.56		298-1200	*
	** DEC.						1

PHASE	T	CP	H	S	G	BT	
SOL	298	22.038	-65.8	30.3	-74.834	54.862	74,*
	300	22.113	-65.759	30.437	-74.890	54.565	
	400	24.990	-63.388	37.238	-78.283	42.778	
	500	26.676	-60.799	43.008	-82.303	35.980	
	600	27.893	-58.068	47.984	-86.858	31.642	
	700	28.889	-55.228	52.360	-91.880	28.690	
	800	29.767	-52.294	56.276	-97.315	26.589	
	900	30.577	-49.276	59.830	-103.123	25.045	
	1000	31.344	-46.180	63.091	-109.271	23.884	
	1100	32.083	-43.009	66.114	-115.734	22.997	
	1200	32.803	-39.764	68.936	-122.488	22.311	

* * * * * * * * * *

RHODIUM TRIBROMIDE $RhBr_3$

		A	B	C	D	RANGE	REF.
SOL	CP	22.3	6.3			298-1000	*
	** DEC.						1

PHASE	T	CP	H	S	G	BT	
SOL	298	24.178	-50.4	45.	-63.817	46.785	1/*
	300	24.190	-50.355	45.150	-63.900	46.557	
	400	24.820	-47.905	52.195	-68.783	37.586	
	500	25.450	-45.391	57.801	-74.292	32.477	
	600	26.080	-42.815	62.497	-80.313	29.258	
	700	26.710	-40.175	66.564	-86.770	27.095	
	800	27.340	-37.473	70.172	-93.610	25.577	
	900	27.970	-34.707	73.429	-100.793	24.479	
	1000	28.600	-31.879	76.408	-108.287	23.669	

Rh_2O_3 DIRHODIUM TRIOXIDE

		A	B	C	D	RANGE	REF.
SOL	CP	20.74	13.8			298-1300	58
	** DEC.						1

PHASE	T	CP	H	S	G	BT	
SOL	298	24.854	-91.5	22.	-98.059	71.889	125/*
	300	24.880	-91.454	22.154	-98.100	71.476	
	400	26.260	-88.897	29.500	-100.697	55.026	
	500	27.640	-86.202	35.508	-103.956	45.445	
	600	29.020	-83.369	40.670	-107.771	39.261	
	700	30.400	-80.398	45.247	-112.071	34.995	
	800	31.780	-77.289	49.396	-116.806	31.914	
	900	33.160	-74.042	53.219	-121.939	29.615	
	1000	34.540	-70.657	56.784	-127.441	27.856	
	1100	35.920	-67.134	60.141	-133.289	26.486	
	1200	37.300	-63.473	63.326	-139.464	25.403	
	1300	38.680	-59.674	66.366	-145.949	24.540	

* * * * * * * * * *

$RhO_2(g)$ RHODIUM DIOXIDE (GAS)

		A	B	C	D	RANGE	REF.
GAS	CP	14.0	0.	-3.3		298-1500	*

PHASE	T	CP	H	S	G	BT	
GAS	298	10.288	-46.8	63.	-65.583	48.080	217,*
	300	10.333	-46.781	63.064	-65.700	47.869	
	400	11.937	-45.656	66.289	-72.172	39.438	
	500	12.680	-44.421	69.042	-78.942	34.510	
	600	13.083	-43.131	71.393	-85.967	31.318	
	700	13.327	-41.809	73.429	-93.210	29.105	
	800	13.484	-40.468	75.220	-100.644	27.498	
	900	13.593	-39.114	76.815	-108.248	26.290	
	1000	13.670	-37.751	78.251	-116.002	25.356	
	1100	13.727	-36.381	79.557	-123.893	24.619	
	1200	13.771	-35.006	80.753	-131.910	24.027	
	1300	13.805	-33.627	81.857	-140.041	23.546	
	1400	13.832	-32.245	82.881	-148.278	23.150	
	1500	13.853	-30.861	83.836	-156.615	22.822	

RUTHENIUM

		A	B	C	D	RANGE	REF.
SOL	CP	5.249	1.178	0.109	0.266	298-1600	21
	CP	3.719	2.575			1600-2523	21
LIQ	CP	10.				2523-3700	21
	** BPT= 4423, L= 142.337						21

PHASE	T	CP	H	S	G	BT	P	
SOL	298	5.746	0	6.82	-2.033	1.491		21
	300	5.747	.011	6.856	-2.046	1.491		
	400	5.831	.589	8.519	-2.819	1.540		
	500	5.948	1.178	9.833	-3.738	1.634		
	600	6.082	1.779	10.929	-4.778	1.741		
	700	6.226	2.395	11.877	-5.919	1.848		
	800	6.379	3.025	12.718	-7.150	1.953		
	900	6.538	3.671	13.479	-8.460	2.055		
	1000	6.704	4.333	14.176	-9.843	2.152		
	1100	6.876	5.012	14.823	-11.294	2.244		
	1200	7.053	5.708	15.429	-12.807	2.333		
	1300	7.236	6.422	16.000	-14.378	2.418		
	1400	7.425	7.155	16.544	-16.006	2.499		
	1500	7.619	7.908	17.063	-17.686	2.577		
	1600	7.819	8.679	17.561	-19.417	2.653		
	1600	7.839	8.679	17.561	-19.417	2.653		
	1700	8.096	9.476	18.044	-21.198	2.726		
	1800	8.354	10.299	18.514	-23.026	2.796		
	1900	8.611	11.147	18.972	-24.900	2.865		
	2000	8.869	12.021	19.420	-26.820	2.931		
	2100	9.126	12.921	19.859	-28.784	2.996		
	2200	9.384	13.846	20.290	-30.791	3.059		
	2300	9.641	14.798	20.713	-32.842	3.121		
	2400	9.899	15.775	21.129	-34.934	3.182		
	2500	10.156	16.777	21.538	-37.067	3.241	.002	
	2523	10.216	17.012	21.631	-37.564	3.254	.003	
			5.803	2.300				21
LIQ	2523	10.000	22.815	23.931	-37.564	3.254	.003	
	2600	10.000	23.585	24.232	-39.418	3.314	.007	
	2700	10.000	24.585	24.609	-41.860	3.389	.021	
	2800	10.000	25.585	24.973	-44.339	3.461	.056	
	2900	10.000	26.585	25.324	-46.854	3.532	.138	
	3000	10.000	27.585	25.663	-49.404	3.600	.319	
	3100	10.000	28.585	25.991	-51.986	3.666	.702	
	3200	10.000	29.585	26.308	-54.601	3.730	1.465	
	3300	10.000	30.585	26.616	-57.248	3.792	2.923	
	3400	10.000	31.585	26.914	-59.924	3.852	5.592	
	3500	10.000	32.585	27.204	-62.630	3.911	10.301	
	3600	10.000	33.585	27.486	-65.365	3.969	18.325	
	3700	10.000	34.585	27.760	-68.127	4.025	31.575	

RuF_5 RUTHENIUM PENTAFLUORIDE

		A	B	C	D	RANGE	REF.
SOL	CP	39.0				298- 374	*
LIQ	CP	43.5				374- 545	*
	** BPT= 545, L= 12.						22

PHASE	T	CP	H	S	G	BT	
SOL	298	39.000	-213.41	38.5	-224.889	164.870	203/*
	300	39.000	-213.338	38.741	-224.960	163.905	
	374	39.000	-210.452	47.340	-228.157	133.343	
			4.5	12.032			22
LIQ	374	43.500	-205.952	59.372	-228.157	133.343	
	400	43.500	-204.821	62.295	-229.739	125.540	
	500	43.500	-200.471	72.002	-236.472	103.376	
	545	43.500	-198.513	75.751	-239.798	96.174	

RUTHENIUM TRICHLORICE $RuCl_3$

		A	B	C	D	RANGE	REF.
SOL	CP	27.5				298-1000	*
	** DEC./ TPT= 723; SPT= 1000, L= 67.4						1/62

PHASE	T	CP	H	S	G	BT	P	
SOL	298	27.500	-60.5	30.5	-69.594	51.020		1
	300	27.500	-60.449	30.670	-69.650	50.747		
	400	27.500	-57.699	38.581	-73.132	39.963		
	500	27.500	-54.949	44.718	-77.308	33.796		
	600	27.500	-52.199	49.732	-82.038	29.886		
	700	27.500	-49.449	53.971	-87.229	27.238		
	800	27.500	-46.699	57.643	-92.813	25.359		
	900	27.500	-43.949	60.882	-98.743	23.981	.006	
	1000	27.500	-41.199	63.779	-104.978	22.946	.217	

* * * * * * * * * *

RUTHENIUM TRICHLORIDE (GAS) $RuCl_3(g)$

		A	B	C	D	RANGE	REF.
GAS	CP	13.60	1.83	-0.72		298-1500	*

PHASE	T	CP	H	S	G	BT	
GAS	298	13.336	13.4	95.	-14.924	10.941	73
	300	13.349	13.425	95.083	-15.100	11.002	
	400	13.882	14.789	99.003	-24.812	13.559	
	500	14.227	16.195	102.140	-34.875	15.246	
	600	14.498	17.632	104.758	-45.223	16.475	
	700	14.734	19.094	107.011	-55.814	17.428	
	800	14.952	20.578	108.993	-66.617	18.201	
	900	15.158	22.083	110.766	-77.606	18.848	
	1000	15.358	23.609	112.374	-88.764	19.402	
	1000	15.358	23.609	112.374	-88.764	19.402	
	1100	15.553	25.155	113.847	-100.076	19.886	
	1200	15.746	26.720	115.208	-111.530	20.315	
	1300	15.936	28.304	116.476	-123.115	20.700	
	1400	16.125	29.907	117.664	-134.822	21.050	
	1500	16.313	31.529	118.783	-146.645	21.369	

$RuCl_4(g)$ RUTHENIUM TETRACHLORIDE (GAS)

		A	B	C	D	RANGE	REF.
GAS	CP	22.9	0.	-2.5		298-1500	*

PHASE	T	CP	H	S	G	BT	
GAS	298	20.088	-22.3	89.5	-48.984	35.911	1,*
	300	20.122	-22.263	89.624	-49.150	35.811	
	400	21.337	-20.181	95.605	-58.423	31.925	
	500	21.900	-18.016	100.433	-68.233	29.829	
	600	22.206	-15.809	104.456	-78.483	28.591	
	700	22.390	-13.579	107.894	-89.105	27.823	
	800	22.509	-11.334	110.892	-100.047	27.335	
	900	22.591	-9.078	113.548	-111.272	27.024	
	1000	22.650	-6.816	115.931	-122.748	26.830	
	1100	22.693	-4.549	118.092	-134.450	26.716	
	1200	22.726	-2.278	120.068	-146.360	26.659	
	1300	22.752	-.004	121.889	-158.459	26.643	
	1400	22.772	2.272	123.575	-170.733	26.656	
	1500	22.789	4.551	125.147	-183.170	26.691	

RUTHENIUM DIOXIDE RuO_2

		A	B	C	D	RANGE	REF.
SOL	CP	16.7	2.5	-3.55		298-1300	*

PHASE	T	CP	H	S	G	BT	
SOL	298	13.452	-72.8	14.5	-77.123	56.540	62,82
	300	13.506	-72.775	14.583	-77.150	56.211	
	400	15.481	-71.313	18.775	-78.823	43.073	
	500	16.530	-69.708	22.352	-80.884	35.359	
	600	17.214	-68.019	25.430	-83.277	30.338	
	700	17.726	-66.271	28.123	-85.958	26.841	
	800	18.145	-64.477	30.518	-88.892	24.287	
	900	18.512	-62.644	32.677	-92.053	22.357	
	1000	18.845	-60.776	34.645	-95.421	20.857	
	1100	19.157	-58.876	36.456	-98.977	19.668	
	1200	19.453	-56.945	38.136	-102.708	18.708	
	1300	19.740	-54.985	39.704	-106.601	17.924	

* * * * * * * * * *

RUTHENIUM TRIOXIDE (GAS) RuO_3 (g)

		A	B	C	D	RANGE	REF.
GAS	CP	18.8	0.6	-4.25		298-1900	*

PHASE	T	CP	H	S	G	BT	
GAS	298	14.198	-18.7	66.	-38.378	28.136	62
	300	14.258	-18.674	66.088	-38.500	28.051	
	400	16.384	-17.127	70.523	-45.336	24.774	
	500	17.400	-15.432	74.300	-52.583	22.987	
	600	17.979	-13.661	77.528	-60.178	21.923	
	700	18.353	-11.843	80.330	-68.074	21.257	
	800	18.616	-9.994	82.799	-76.233	20.829	
	900	18.815	-8.122	85.003	-84.625	20.553	
	1000	18.975	-6.232	86.994	-93.226	20.377	
	1100	19.109	-4.328	88.809	-102.018	20.272	
	1200	19.225	-2.411	90.477	-110.983	20.216	
	1300	19.329	-.483	92.020	-120.109	20.195	
	1400	19.423	1.454	93.456	-129.384	20.200	
	1500	19.511	3.401	94.799	-138.797	20.225	
	1600	19.594	5.356	96.061	-148.341	20.265	
	1700	19.673	7.320	97.251	-158.007	20.316	
	1800	19.749	9.291	98.378	-167.789	20.375	
	1900	19.822	11.269	99.447	-177.681	20.441	

$RuO_4(g)$ RUTHENIUM TETROXIDE (GAS)

		A	B	C	D	RANGE	REF.
GAS	CP	24.33	0.73	-5.76		298-2000	62,*

PHASE	T	CP	H	S	G	BT	
GAS	298	18.068	-44.	69.47	-64.712	47.442	62
	300	18.149	-43.966	69.582	-64.841	47.243	
	400	21.022	-41.988	75.254	-72.090	39.393	
	500	22.391	-39.810	80.108	-79.864	34.913	
	600	23.168	-37.529	84.265	-88.088	32.090	
	700	23.665	-35.186	87.877	-96.699	30.195	
	800	24.014	-32.801	91.061	-105.649	28.866	
	900	24.276	-30.386	93.905	-114.900	27.905	
	1000	24.484	-27.947	96.474	-124.421	27.196	
	1100	24.657	-25.490	98.816	-134.187	26.664	
	1200	24.806	-23.017	100.968	-144.178	26.262	
	1300	24.938	-20.529	102.958	-154.375	25.956	
	1400	25.058	-18.030	104.811	-164.765	25.724	
	1500	25.169	-15.518	106.544	-175.334	25.550	
	1600	25.273	-12.996	108.171	-186.070	25.419	
	1700	25.372	-10.464	109.707	-196.965	25.325	
	1800	25.466	-7.922	111.159	-208.009	25.259	
	1900	25.557	-5.371	112.539	-219.194	25.216	
	2000	25.646	-2.810	113.852	-230.514	25.193	

RUTHENIUM DISULFIDE RuS_2

		A	B	C	D	RANGE	REF.
SOL	CP	16.38	2.83	-2.10		298-1500	63

PHASE	T	CP	H	S	G	BT	
SOL	298	14.861	-49.2	13.	-53.076	38.911	63
	300	14.896	-49.172	13.092	-53.100	38.689	
	400	16.199	-47.610	17.577	-54.641	29.859	
	500	16.955	-45.950	21.279	-56.589	24.739	
	600	17.495	-44.226	24.420	-58.878	21.449	
	700	17.932	-42.454	27.150	-61.460	19.191	
	800	18.316	-40.642	29.570	-64.298	17.568	
	900	18.668	-38.792	31.748	-67.366	16.361	
	1000	19.000	-36.909	33.732	-70.641	15.441	
	1100	19.319	-34.993	35.558	-74.107	14.726	
	1200	19.630	-33.045	37.253	-77.749	14.162	
	1300	19.935	-31.067	38.836	-81.554	13.712	
	1400	20.235	-29.058	40.324	-85.513	13.351	
	1500	20.532	-27.020	41.731	-89.616	13.059	

* * * * * * * * * *

RUTHENIUM DISELENIDE $RuSe_2$

		A	B	C	D	RANGE	REF.
SOL	CP	17.60	2.65	-1.0		298-1500	63

PHASE	T	CP	H	S	G	BT	
SOL	298	17.265	-38.6	19.5	-44.414	32.561	63
	300	17.284	-38.568	19.607	-44.450	32.386	
	400	18.035	-36.799	24.692	-46.675	25.506	
	500	18.525	-34.969	28.772	-49.355	21.576	
	600	18.912	-33.097	32.185	-52.408	19.092	
	700	19.251	-31.189	35.126	-55.777	17.417	
	800	19.564	-29.248	37.717	-59.421	16.235	
	900	19.862	-27.276	40.039	-63.311	15.376	
	1000	20.150	-25.276	42.146	-67.422	14.737	
	1100	20.432	-23.246	44.080	-71.734	14.254	
	1200	20.711	-21.189	45.870	-76.233	13.886	
	1300	20.986	-19.104	47.538	-80.904	13.603	
	1400	21.259	-16.992	49.104	-85.737	13.386	
	1500	21.531	-14.853	50.580	-90.722	13.220	

S

SULFUR (RHOMBIC,MONOCLINIC)

		A	B	C	D	RANGE	REF.
S.-RH	CP	3.54	5.75	0.174		298- 368	63
	** TPT=368.5						63
S.-MO	CP	16.337	-28.333			368- 374	63
	CP	3.270	7.162			374- 388	63
	** MPT=388.3						
LIQ	CP	-493.393	828.729	270.387		388- 440	63
	CP	-6.109	13.809	21.182		440- 718	63
	** BPT=717.75, L=2.30 GAS(S8 + S7 + S6 + ...)						3/*

PHASE	T	CP	H	S	G	BT	P	
S.-RH	298	5.450	0	7.62	-2.272	1.666		63
	300	5.458	.010	7.654	-2.286	1.666		
	368	5.787	.395	8.808	-2.851	1.691	.012	
			0.096	.261				63
S.-MO	368	5.896	.491	9.069	-2.851	1.691	.012	
	374	5.740	.523	9.155	-2.901	1.695	.018	
	374	5.949	.523	9.155	-2.901	1.695	.018	
	388	6.051	.609	9.380	-3.033	1.708	.055	
			0.41	1.056				63
LIQ	388	7.732	1.019	10.436	-3.033	1.708	.055	
	400	7.090	1.105	10.654	-3.157	1.725	.114	
	440	10.910	1.437	11.442	-3.598	1.787	.950	
	440	10.908	1.437	11.442	-3.598	1.787	.950	
	500	9.268	2.037	12.724	-4.325	1.891	9.782	
	600	8.060	2.892	14.285	-5.679	2.069	120.886	
	700	7.880	3.683	15.505	-7.171	2.239	604.076	
	718	7.914	3.823	15.703	-7.448	2.268	760.	

SULFUR (MONATOMIC GAS) S(g)

		A	B	C	D	RANGE	REF.
GAS	CP	5.26	-0.10	0.36		298-2000	63

PHASE	T	CP	H	S	G	BT	
GAS	298	5.635	66.78	40.09	54.827	-40.195	63
	300	5.630	66.790	40.125	54.753	-39.893	
	400	5.445	67.343	41.716	50.657	-27.681	
	500	5.354	67.882	42.920	46.423	-20.294	
	600	5.300	68.415	43.891	42.080	-15.330	
	700	5.263	68.943	44.705	37.650	-11.756	
	718	5.258	69.036	44.837	36.855	-11.224	
	718	5.258	69.036	44.837	36.855	-11.224	
	800	5.236	69.468	45.406	33.143	-9.056	
	900	5.214	69.990	46.021	28.571	-6.939	
	1000	5.196	70.511	46.570	23.941	-5.233	
	1100	5.180	71.030	47.064	19.259	-3.827	
	1200	5.165	71.547	47.514	14.530	-2.647	
	1300	5.151	72.063	47.927	9.757	-1.641	
	1400	5.138	72.577	48.308	4.945	-.772	
	1500	5.126	73.090	48.662	.097	-.014	
	1600	5.114	73.602	48.993	-4.786	.654	
	1700	5.102	74.113	49.303	-9.701	1.247	
	1800	5.091	74.623	49.594	-14.646	1.779	
	1900	5.080	75.131	49.869	-19.619	2.257	
	2000	5.069	75.639	50.129	-24.619	2.691	

$S_2(g)$

SULFUR (DIATOMIC GAS)

		A	B	C	D	RANGE	REF.
GAS	CP	8.72	0.16	-0.90		298-2000	63

PHASE	T	CP	H	S	G	BT	
GAS	298	7.755	30.75	54.50	14.501	-10.631	63
	300	7.768	30.764	54.548	14.400	-10.492	
	400	8.221	31.567	56.854	8.825	-4.823	
	500	8.440	32.401	58.714	3.044	-1.331	
	600	8.566	33.252	60.265	-2.907	1.059	
	700	8.648	34.113	61.592	-9.002	2.811	
	718	8.660	34.267	61.809	-10.097	3.075	
	718	8.660	34.267	61.809	-10.097	3.075	
	800	8.707	34.981	62.751	-15.220	4.158	
	900	8.753	35.854	63.779	-21.548	5.233	
	1000	8.790	36.731	64.704	-27.972	6.114	
	1100	8.822	37.612	65.543	-34.485	6.853	
	1200	8.849	38.495	66.312	-41.079	7.482	
	1300	8.875	39.382	67.021	-47.746	8.028	
	1400	8.898	40.270	67.680	-54.481	8.506	
	1500	8.920	41.161	68.294	-61.280	8.930	
	1600	8.941	42.054	68.871	-68.139	9.309	
	1700	8.961	42.949	69.413	-75.053	9.650	
	1800	8.980	43.846	69.926	-82.021	9.960	
	1900	8.999	44.745	70.412	-89.038	10.243	
	2000	9.017	45.646	70.874	-96.102	10.503	

SULFUR (3-ATOMIC GAS) S_3 (g)

		A	B	C	D	RANGE	REF.
GAS	CP	13.766	0.076	-1.512		298-2000	63

PHASE	T	CP	H	S	G	BT	
GAS	298	12.088	35.0	65.954	15.336	-11.243	63
	300	12.109	35.022	66.029	15.214	-11.085	
	400	12.851	36.276	69.629	8.424	-4.603	
	500	13.199	37.580	72.538	1.311	-.573	
	600	13.392	38.910	74.964	-6.068	2.210	
	700	13.511	40.256	77.037	-13.670	4.269	
	718	13.527	40.496	77.376	-15.041	4.580	
	718	13.527	40.496	77.376	-15.041	4.580	
	800	13.591	41.611	78.847	-21.466	5.865	
	900	13.648	42.973	80.451	-29.433	7.148	
	1000	13.691	44.340	81.892	-37.551	8.208	
	1100	13.725	45.711	83.198	-45.807	9.102	
	1200	13.752	47.085	84.393	-54.187	9.870	
	1300	13.775	48.461	85.495	-62.682	10.539	
	1400	13.795	49.840	86.517	-71.283	11.129	
	1500	13.813	51.220	87.469	-79.983	11.655	
	1600	13.829	52.603	88.361	-88.775	12.128	
	1700	13.843	53.986	89.200	-97.654	12.556	
	1800	13.856	55.371	89.992	-106.614	12.946	
	1900	13.869	56.757	90.741	-115.651	13.305	
	2000	13.880	58.145	91.453	-124.761	13.635	

S_4 (g)

SULFUR (4-ATOMIC GAS)

		A	B	C	D	RANGE	REF.
GAS	CP	19.73	0.75	-1.70		298-2000	63

PHASE	T	CP	H	S	G	BT	
GAS	298	18.041	45.	77.9	21.774	-15.963	63
	300	18.066	45.033	78.012	21.630	-15.759	
	400	18.967	46.891	83.349	13.551	-7.405	
	500	19.425	48.813	87.636	4.995	-2.184	
	600	19.708	50.770	91.204	-3.952	1.440	
	700	19.908	52.752	94.258	-13.229	4.131	
	718	19.938	53.105	94.757	-14.906	4.540	
	718	19.938	53.105	94.757	-14.906	4.540	
	800	20.064	54.750	96.927	-22.791	6.227	
	900	20.195	56.764	99.298	-32.604	7.918	
	1000	20.310	58.789	101.432	-42.643	9.321	
	1100	20.415	60.825	103.372	-52.884	10.509	
	1200	20.512	62.872	105.153	-63.312	11.532	
	1300	20.604	64.928	106.798	-73.910	12.427	
	1400	20.693	66.992	108.329	-84.668	13.219	
	1500	20.779	69.066	109.759	-95.573	13.927	
	1600	20.864	71.148	111.103	-106.617	14.565	
	1700	20.946	73.239	112.370	-117.791	15.145	
	1800	21.028	75.337	113.570	-129.088	15.676	
	1900	21.108	77.444	114.709	-140.503	16.164	
	2000	21.187	79.559	115.794	-152.028	16.615	

SULFUR (5-ATOMIC GAS) S_5 (g)

		A	B	C	D	RANGE	REF.
GAS	CP	25.45	0.203	-3.931		298-2000	63

PHASE	T	CP	H	S	G	BT	
GAS	298	21.088	26.53	76.70	3.662	-2.685	63*
	300	21.143	26.569	76.831	3.520	-2.565	
	400	23.074	28.794	83.217	-4.493	2.455	
	500	23.979	31.151	88.474	-13.086	5.721	
	600	24.480	33.576	92.894	-22.160	8.073	
	700	24.790	36.041	96.693	-31.644	9.881	
	718	24.833	36.481	97.314	-33.366	10.161	
	718	24.833	36.481	97.314	-33.366	10.161	
	800	24.998	38.531	100.017	-41.483	11.334	
	900	25.147	41.039	102.971	-51.635	12.540	
	1000	25.260	43.559	105.626	-62.067	13.567	
	1100	25.348	46.090	108.038	-72.752	14.457	
	1200	25.421	48.628	110.247	-83.668	15.240	
	1300	25.481	51.174	112.284	-94.796	15.939	
	1400	25.534	53.724	114.175	-106.120	16.568	
	1500	25.580	56.280	115.938	-117.627	17.141	
	1600	25.621	58.840	117.590	-129.304	17.664	
	1700	25.659	61.404	119.145	-141.142	18.147	
	1800	25.694	63.972	120.612	-153.130	18.595	
	1900	25.727	66.543	122.002	-165.261	19.012	
	2000	25.758	69.117	123.323	-177.528	19.402	

$S_6(g)$ SULFUR (6-ATOMIC GAS)

		A	B	C	D	RANGE	REF.
GAS	CP	31.49	0.16	-4.11		298-2000	63

PHASE	T	CP	H	S	G	BT	
GAS	298	26.914	23.6	84.58	-1.618	1.186	63*
	300	26.971	23.650	84.747	-1.774	1.293	
	400	28.985	26.462	92.823	-10.667	5.829	
	500	29.926	29.413	99.403	-20.289	8.869	
	600	30.444	32.433	104.909	-30.512	11.116	
	700	30.763	35.495	109.628	-41.245	12.879	
	718	30.807	36.041	110.399	-43.197	13.155	
	718	30.807	36.041	110.399	-43.197	13.155	
	800	30.976	38.583	113.751	-52.418	14.322	
	900	31.127	41.688	117.408	-63.979	15.538	
	1000	31.239	44.807	120.694	-75.887	16.587	
	1100	31.326	47.935	123.676	-88.108	17.508	
	1200	31.397	51.071	126.404	-100.614	18.327	
	1300	31.455	54.214	128.920	-113.382	19.064	
	1400	31.504	57.362	131.253	-126.392	19.733	
	1500	31.547	60.515	133.428	-139.627	20.346	
	1600	31.585	63.671	135.465	-153.073	20.912	
	1700	31.620	66.832	137.381	-166.716	21.436	
	1800	31.651	69.995	139.189	-180.545	21.924	
	1900	31.680	73.162	140.901	-194.551	22.381	
	2000	31.707	76.331	142.527	-208.723	22.811	

SULFUR (7-ATOMIC GAS) S_7 (g)

		A	B	C	D	RANGE	REF.
GAS	CP	37.263	0.261	-5.578		298-2000	63

PHASE	T	CP	H	S	G	BT	
GAS	298	31.066	24.6	94.192	-3.483	2.554	63*
	300	31.144	24.658	94.384	-3.658	2.665	
	400	33.881	27.928	103.775	-13.582	7.422	
	500	35.162	31.387	111.488	-24.357	10.648	
	600	35.870	34.942	117.967	-35.838	13.056	
	700	36.307	38.552	123.532	-47.920	14.963	
	718	36.368	39.197	124.442	-50.121	15.263	
	718	36.368	39.197	124.442	-50.121	15.263	
	800	36.600	42.199	128.400	-60.522	16.536	
	900	36.809	45.870	132.724	-73.582	17.871	
	1000	36.966	49.559	136.611	-87.052	19.028	
	1100	37.089	53.262	140.140	-100.892	20.048	
	1200	37.189	56.976	143.372	-115.070	20.960	
	1300	37.272	60.699	146.352	-129.558	21.784	
	1400	37.344	64.430	149.117	-144.333	22.534	
	1500	37.407	68.168	151.695	-159.375	23.224	
	1600	37.463	71.911	154.111	-174.667	23.862	
	1700	37.514	75.660	156.384	-190.193	24.454	
	1800	37.561	79.414	158.529	-205.939	25.008	
	1900	37.604	83.172	160.561	-221.895	25.527	
	2000	37.646	86.934	162.491	-238.048	26.016	

$S_8(g)$ SULFUR (OCTATOMIC GAS)

		A	B	C	D	RANGE	REF.
GAS	CP	43.302	0.221	-5.446		298-2000	63

PHASE	T	CP	H	S	G	BT	
GAS	298	37.241	23.145	101.13	-7.007	5.137	63*
	300	37.317	23.214	101.361	-7.194	5.242	
	400	39.987	27.098	112.516	-17.908	9.786	
	500	41.234	31.166	121.588	-29.628	12.952	
	600	41.922	35.327	129.172	-42.177	15.365	
	700	42.345	39.542	135.669	-55.427	17.307	
	718	42.403	40.294	136.730	-57.844	17.616	
	718	42.403	40.294	136.730	-57.844	17.616	
	800	42.628	43.791	141.343	-69.283	18.930	
	900	42.829	48.065	146.376	-83.674	20.322	
	1000	42.978	52.355	150.896	-98.541	21.539	
	1100	43.095	56.659	154.998	-113.839	22.621	
	1200	43.189	60.973	158.752	-129.529	23.594	
	1300	43.267	65.296	162.212	-145.580	24.477	
	1400	43.334	69.626	165.421	-161.964	25.287	
	1500	43.391	73.963	168.413	-178.657	26.034	
	1600	43.443	78.305	171.215	-195.640	26.727	
	1700	43.489	82.651	173.850	-212.894	27.373	
	1800	43.532	87.002	176.337	-230.405	27.979	
	1900	43.571	91.357	178.692	-248.157	28.548	
	2000	43.608	95.716	180.928	-266.139	29.086	

SULFUR MONOCHLORIDE (GAS) SCl (g)

		A	B	C	D	RANGE	REF.
GAS	CP	8.86	0.24	-0.66		298-2000	63

PHASE	T	CP	H	S	G	BT	
GAS	298	8.189	35.	57.25	17.931	-13.145	63
	300	8.199	35.015	57.301	17.825	-12.987	
	400	8.543	35.855	59.713	11.969	-6.541	
	500	8.716	36.718	61.640	5.898	-2.579	
	600	8.821	37.596	63.239	-.348	.127	
	700	8.893	38.481	64.604	-6.742	2.105	
	800	8.949	39.374	65.796	-13.263	3.624	
	900	8.995	40.271	66.852	-19.896	4.832	
	1000	9.034	41.172	67.802	-26.630	5.821	
	1100	9.069	42.078	68.665	-33.454	6.648	
	1200	9.102	42.986	69.456	-40.360	7.352	
	1300	9.133	43.898	70.185	-47.343	7.960	
	1400	9.162	44.813	70.863	-54.396	8.493	
	1500	9.191	45.730	71.496	-61.514	8.964	
	1600	9.218	46.651	72.090	-68.694	9.384	
	1700	9.245	47.574	72.650	-75.931	9.763	
	1800	9.272	48.500	73.179	-83.223	10.106	
	1900	9.298	49.428	73.681	-90.566	10.419	
	2000	9.323	50.359	74.159	-97.958	10.706	

S_2Cl_2 SULFUR MONOCHLORIDE (DIMERIC)

		A	B	C	D	RANGE	REF.
LIQ	CP	30.0				298- 411	63,*
	** MPT=	193					63
GAS	CP	19.309	0.727	-1.855		411-1000	63

PHASE	T	CP	H	S	G	BT	P	
LIQ	298	30.000	-13.9	53.	-29.702	21.775	10.375	63
	300	30.000	-13.844	53.186	-29.800	21.712	11.497	
	400	30.000	-10.844	61.816	-35.571	19.438	568.519	
	411	30.000	-10.514	62.630	-36.255	19.281	760.	
			8.525	20.742				62,*
GAS	411	18.510	-1.989	83.372	-36.255	19.281		
	500	18.930	-.322	87.043	-43.844	19.167		
	600	19.230	1.587	90.523	-52.727	19.208		
	700	19.439	3.521	93.504	-61.932	19.339		
	800	19.601	5.473	96.111	-71.415	19.512		
	900	19.734	7.440	98.427	-81.144	19.707		
	1000	19.850	9.420	100.513	-91.093	19.911		

SULFUR DICHLORIDE (GAS)

SCl_2 (g)

		A	B	C	D	RANGE	REF.
GAS	CP	13.75	0.085	-1.435		298-2000	63

PHASE	T	CP	H	S	G	BT	
GAS	298	12.161	-5.4	67.2	-25.436	18.647	63
	300	12.181	-5.377	67.275	-25.560	18.623	
	400	12.887	-4.119	70.891	-32.475	17.746	
	500	13.218	-2.812	73.806	-39.715	17.362	
	600	13.402	-1.480	76.234	-47.220	17.202	
	700	13.517	-.134	78.309	-54.950	17.158	
	800	13.594	1.222	80.119	-62.873	17.179	
	900	13.649	2.584	81.724	-70.967	17.235	
	1000	13.691	3.951	83.164	-79.213	17.314	
	1100	13.725	5.322	84.471	-87.595	17.406	
	1200	13.752	6.696	85.666	-96.103	17.505	
	1300	13.776	8.073	86.768	-104.725	17.608	
	1400	13.796	9.451	87.789	-113.454	17.713	
	1500	13.814	10.832	88.742	-122.281	17.819	
	1600	13.830	12.214	89.634	-131.200	17.924	
	1700	13.845	13.598	90.473	-140.206	18.027	
	1800	13.859	14.983	91.264	-149.293	18.129	
	1900	13.872	16.369	92.014	-158.457	18.229	
	2000	13.884	17.757	92.726	-167.695	18.327	

S_2Br_2 (g) SULFUR MONOBROMIDE (DIMERIC GAS)

		A	B	C	D	RANGE	REF.
GAS	CP	19.445	0.647	-1.469		298-2000	63
	** MPT= 227, BPT= 360						63

PHASE	T	CP	H	S	G	BT	
GAS	298	17.985	7.4	83.75	-17.570	12.881	63
	300	18.007	7.433	83.861	-17.725	12.914	
	400	18.786	9.278	89.163	-26.387	14.419	
	500	19.181	11.178	93.401	-35.523	15.529	
	600	19.425	13.109	96.922	-45.044	16.409	
	700	19.598	15.061	99.930	-54.890	17.140	
	800	19.733	17.028	102.556	-65.017	17.764	
	900	19.846	19.007	104.887	-75.391	18.310	
	1000	19.945	20.996	106.983	-85.986	18.795	
	1100	20.035	22.995	108.888	-96.781	19.231	
	1200	20.119	25.003	110.635	-107.759	19.628	
	1300	20.199	27.019	112.249	-118.904	19.992	
	1400	20.276	29.043	113.748	-130.205	20.329	
	1500	20.350	31.074	115.150	-141.650	20.641	
	1600	20.423	33.113	116.465	-153.232	20.933	
	1700	20.494	35.159	117.706	-164.941	21.207	
	1800	20.564	37.212	118.879	-176.771	21.466	
	1900	20.634	39.272	119.993	-188.715	21.710	
	2000	20.702	41.338	121.053	-200.767	21.942	

SULFUR DIBROMIDE (GAS)

SBr_2 (g)

		A	B	C	D	RANGE	REF.
GAS	CP	13.833	0.041	-1.069		298-2000	63

PHASE	T	CP	H	S	G	BT	
GAS	298	12.643	-3.	72.07	-24.488	17.952	63
	300	12.658	-2.977	72.148	-24.621	17.939	
	400	13.181	-1.681	75.872	-32.030	17.503	
	500	13.426	-.349	78.843	-39.771	17.386	
	600	13.561	1.001	81.303	-47.781	17.407	
	700	13.644	2.361	83.401	-56.019	17.492	
	800	13.699	3.728	85.226	-64.452	17.610	
	900	13.738	5.100	86.842	-73.057	17.743	
	1000	13.767	6.476	88.291	-81.815	17.883	
	1100	13.790	7.854	89.604	-90.711	18.025	
	1200	13.808	9.234	90.805	-99.732	18.166	
	1300	13.823	10.615	91.911	-108.869	18.305	
	1400	13.836	11.998	92.936	-118.112	18.441	
	1500	13.847	13.382	93.891	-127.454	18.573	
	1600	13.857	14.767	94.785	-136.888	18.701	
	1700	13.866	16.154	95.625	-146.409	18.825	
	1800	13.874	17.541	96.418	-156.011	18.945	
	1900	13.881	18.928	97.168	-165.691	19.061	
	2000	13.888	20.317	97.880	-175.444	19.174	

$SOCl_2$ (g) SULFUR OXYDICHLORIDE (GAS)

		A	B	C	D	RANGE	REF.
GAS	CP	17.75	1.72	-2.09		298-1100	2

PHASE	T	CP	H	S	G	BT	
GAS	298	15.912	-50.75	73.6	-72.694	53.293	62
	300	15.944	-50.721	73.699	-72.830	53.064	
	400	17.132	-49.060	78.469	-80.447	43.960	
	500	17.774	-47.312	82.367	-88.495	38.686	
	600	18.201	-45.512	85.647	-96.900	35.301	
	700	18.527	-43.675	88.478	-105.608	32.977	
	800	18.799	-41.808	90.970	-114.584	31.307	
	900	19.040	-39.916	93.199	-123.795	30.065	
	1000	19.261	-38.001	95.216	-133.217	29.118	
	1100	19.469	-36.064	97.062	-142.832	28.382	

* * * * * * * * * *

SO_2Cl_2 (g) SULFUR DIOXYDICHLORIDE (GAS)

		A	B	C	D	RANGE	REF.
GAS	CP	12.84	19.0			298- 500	58
	** MPT= 219, BPT= 342.65						62

PHASE	T	CP	H	S	G	BT	
GAS	298	18.505	-86.2	74.4	-108.382	79.457	62
	300	18.540	-86.166	74.515	-108.520	79.067	
	400	20.440	-84.217	80.108	-116.260	63.530	
	500	22.340	-82.078	84.874	-124.515	54.433	

DISULFUR MONOXIDE (GAS) S_2O (g)

		A	B	C	D	RANGE	REF.
GAS	CP	12.90	0.518	-2.268		298-2000	63

PHASE	T	CP	H	S	G	BT	
GAS	298	10.503	-9.7	63.8	-28.722	21.057	63
	300	10.535	-9.681	63.865	-28.840	21.013	
	400	11.690	-8.561	67.077	-35.392	19.340	
	500	12.252	-7.361	69.752	-42.237	18.464	
	600	12.581	-6.119	72.017	-49.329	17.970	
	700	12.800	-4.849	73.974	-56.631	17.683	
	800	12.960	-3.561	75.694	-64.116	17.518	
	900	13.086	-2.258	77.228	-71.763	17.429	
	1000	13.191	-.944	78.612	-79.556	17.389	
	1100	13.282	.380	79.874	-87.482	17.383	
	1200	13.364	1.712	81.033	-95.528	17.400	
	1300	13.439	3.052	82.106	-103.685	17.433	
	1400	13.509	4.400	83.104	-111.946	17.478	
	1500	13.576	5.754	84.039	-120.304	17.531	
	1600	13.640	7.115	84.917	-128.752	17.589	
	1700	13.702	8.482	85.746	-137.286	17.652	
	1800	13.762	9.855	86.531	-145.900	17.717	
	1900	13.821	11.235	87.276	-154.591	17.784	
	2000	13.879	12.620	87.987	-163.354	17.853	

SO (g)

SULFUR MONOXIDE (GAS)

		A	B	C	D	RANGE	REF.
GAS	CP	8.26	0.32	-1.		298-2000	63

PHASE	T	CP	H	S	G	BT	
GAS	298	7.230	1.6	53.	-14.202	10.412	3,63
	300	7.245	1.613	53.045	-14.300	10.419	
	400	7.763	2.367	55.210	-19.717	10.774	
	500	8.020	3.158	56.973	-25.329	11.073	
	600	8.174	3.968	58.449	-31.102	11.330	
	700	8.280	4.791	59.718	-37.012	11.557	
	800	8.360	5.623	60.829	-43.040	11.760	
	900	8.425	6.462	61.817	-49.173	11.943	
	1000	8.480	7.308	62.708	-55.400	12.109	
	1100	8.529	8.158	63.519	-61.712	12.263	
	1200	8.575	9.013	64.263	-68.102	12.405	
	1300	8.617	9.873	64.951	-74.563	12.537	
	1400	8.657	10.737	65.591	-81.090	12.660	
	1500	8.696	11.604	66.189	-87.680	12.777	
	1600	8.733	12.476	66.752	-94.327	12.886	
	1700	8.769	13.351	67.282	-101.029	12.990	
	1800	8.805	14.230	67.785	-107.783	13.088	
	1900	8.840	15.112	68.262	-114.585	13.182	
	2000	8.875	15.998	68.716	-121.434	13.271	

ANTIMONY

		A	B	C	D	RANGE	REF.
SOL	CP	7.283	-3.677	-0.478	4.287	298- 904	21
LIQ	CP	7.5				904-1860	21
	** BPT= 1860 GAS(83 SB4 + 13 SB2 + 4 SB)						21

PHASE	T	CP	H	S	G	BT	P	
SOL	298	6.030	0	10.88	-3.244	2.378		21
	300	6.035	.011	10.917	-3.264	2.378		
	400	6.199	.624	12.679	-4.448	2.430		
	500	6.325	1.250	14.075	-5.788	2.530		
	600	6.487	1.890	15.242	-7.255	2.643		
	700	6.712	2.549	16.258	-8.831	2.758		
	800	7.010	3.235	17.173	-10.503	2.870	.006	
	900	7.387	3.954	18.020	-12.263	2.978	.140	
	904	7.404	3.984	18.052	-12.336	2.983	.156	
			4.75	5.254				21
LIQ	904	7.500	8.734	23.307	-12.336	2.983	.156	
	1000	7.500	9.454	24.064	-14.610	3.193	.654	
	1100	7.500	10.204	24.779	-17.053	3.389	2.235	
	1200	7.500	10.954	25.431	-19.564	3.564	6.407	
	1300	7.500	11.704	26.031	-22.137	3.722	16.300	
	1400	7.500	12.454	26.587	-24.768	3.867	37.937	
	1500	7.500	13.204	27.105	-27.453	4.000	81.870	
	1600	7.500	13.954	27.589	-30.188	4.124	164.856	
	1700	7.500	14.704	28.043	-32.970	4.239	311.173	
	1800	7.500	15.454	28.472	-35.796	4.347	553.301	
	1860	7.500	15.904	28.718	-37.512	4.408	760.	

Sb (g)

ANTIMONY (1-ATOMIC GAS)

		A	B	C	D	RANGE	REF.
GAS	CP	4.968				298- 800	21
	CP	4.261	0.456	2.313		800-1300	21
	CP	4.261	0.456	2.313		1300-2000	21

PHASE	T	CP	H	S	G	BT	
GAS	298	4.968	63.23	43.058	50.392	-36.943	21
	300	4.968	63.239	43.089	50.313	-36.658	
	400	4.968	63.736	44.518	45.929	-25.098	
	500	4.968	64.233	45.627	41.420	-18.107	
	600	4.968	64.730	46.532	36.810	-13.410	
	700	4.968	65.226	47.298	32.118	-10.029	
	800	4.968	65.723	47.961	27.354	-7.474	
	800	4.987	65.723	47.961	27.354	-7.474	
	900	4.957	66.220	48.547	22.528	-5.471	
	1000	4.948	66.715	49.069	17.647	-3.857	
	1100	4.954	67.210	49.540	12.716	-2.527	
	1200	4.969	67.706	49.972	7.740	-1.410	
	1300	4.991	68.204	50.371	2.723	-.458	
	1300	4.991	68.204	50.371	2.723	-.458	
	1400	5.017	68.705	50.741	-2.333	.364	
	1500	5.048	69.208	51.088	-7.425	1.082	
	1600	5.081	69.714	51.415	-12.550	1.715	
	1700	5.116	70.224	51.724	-17.707	2.277	
	1800	5.153	70.738	52.018	-22.895	2.780	
	1900	5.191	71.255	52.298	-28.110	3.234	
	2000	5.231	71.776	52.565	-33.354	3.645	

ANTIMONY (2-ATOMIC GAS) $Sb_2(g)$

		A	B	C	D	RANGE	REF.
GAS	CP	8.929	0.009	-0.217		298-1300	21
	CP	8.909	0.014			1300-1860	21
	CP	8.936				1860-2000	21

PHASE	T	CP	H	S	G	BT	
GAS	298	8.688	55.26	60.90	37.103	-27.201	21
	300	8.691	55.276	60.954	36.990	-26.951	
	400	8.797	56.151	63.471	30.763	-16.810	
	500	8.847	57.034	65.440	24.314	-10.629	
	600	8.874	57.920	67.055	17.687	-6.443	
	700	8.891	58.808	68.424	10.911	-3.407	
	800	8.902	59.698	69.612	4.008	-1.095	
	900	8.910	60.588	70.661	-3.007	.730	
	1000	8.916	61.480	71.601	-10.121	2.212	
	1100	8.921	62.372	72.451	-17.324	3.442	
	1200	8.925	63.264	73.227	-24.608	4.482	
	1300	8.928	64.157	73.942	-31.967	5.375	
	1300	8.927	64.157	73.942	-31.967	5.375	
	1400	8.929	65.049	74.603	-39.395	6.151	
	1500	8.930	65.942	75.219	-46.886	6.832	
	1600	8.931	66.835	75.796	-54.438	7.437	
	1700	8.933	67.729	76.337	-62.044	7.977	
	1800	8.934	68.622	76.848	-69.704	8.464	
	1860	8.935	69.158	77.141	-74.324	8.734	
	1860	8.936	69.158	77.141	-74.324	8.734	
	1900	8.936	69.516	77.331	-77.413	8.906	
	2000	8.936	70.409	77.789	-85.169	9.308	

Sb_4 (g)

ANTIMONY (4-ATOMIC GAS)

		A	B	C	D	RANGE	REF.
GAS	CP	19.817	0.040	-0.430		298-1300	21
	CP	19.793	0.034			1300-1860	21
	CP	14.852				1860-2000	21

PHASE	T	CP	H	S	G	BT	
GAS	298	19.345	49.36	83.652	24.419	-17.902	21
	300	19.351	49.396	83.772	24.264	-17.679	
	400	19.564	51.343	89.372	15.594	-8.521	
	500	19.665	53.305	93.750	6.430	-2.811	
	600	19.722	55.275	97.341	-3.130	1.140	
	700	19.757	57.249	100.384	-13.020	4.066	
	800	19.782	59.226	103.023	-23.193	6.337	
	900	19.800	61.205	105.355	-33.614	8.164	
	1000	19.814	63.186	107.441	-44.256	9.673	
	1100	19.825	65.168	109.330	-55.096	10.948	
	1200	19.835	67.151	111.056	-66.117	12.043	
	1300	19.844	69.135	112.644	-77.303	12.997	
	1300	19.837	69.135	112.644	-77.303	12.997	
	1400	19.841	71.118	114.114	-88.641	13.839	
	1500	19.844	73.103	115.483	-100.122	14.590	
	1600	19.847	75.087	116.764	-111.735	15.264	
	1700	19.851	77.072	117.967	-123.472	15.876	
	1800	19.854	79.057	119.102	-135.326	16.433	
	1860	19.856	80.249	119.753	-142.492	16.745	
	1860	14.852	80.249	119.753	-142.492	16.745	
	1900	14.852	80.843	120.069	-147.288	16.944	
	2000	14.852	82.328	120.831	-159.334	17.414	

ANTIMONY HYDRIDE (GAS) $SbH_3(g)$

		A	B	C	D	RANGE	REF.
GAS	CP	12.07	4.47	-3.15		298-1600	*

PHASE	T	CP	H	S	G	BT	
GAS	298	9.859	34.68	55.68	18.079	-13.254	62
	300	9.911	34.698	55.741	17.976	-13.097	
	400	11.889	35.799	58.895	12.241	-6.689	
	500	13.045	37.050	61.681	6.209	-2.715	
	600	13.877	38.398	64.136	-.084	.031	
	700	14.556	39.820	66.327	-6.609	2.064	
	800	15.154	41.306	68.311	-13.342	3.645	
	900	15.704	42.849	70.128	-20.266	4.922	
	1000	16.225	44.446	71.810	-27.363	5.981	
	1100	16.727	46.094	73.380	-34.624	6.880	
	1200	17.215	47.791	74.856	-42.036	7.657	
	1300	17.695	49.537	76.253	-49.592	8.338	
	1400	18.167	51.330	77.582	-57.285	8.944	
	1500	18.635	53.170	78.851	-65.107	9.487	
	1600	19.099	55.057	80.069	-73.053	9.980	

* * * * * * * * * *

ANTIMONY FLUORIDE SbF_3

		A	B	C	D	RANGE	REF.
SOL	CP	25.6				298- 564	*
LIQ	CP	30.5				564- 592	*
	** BPT=	592					62

PHASE	T	CP	H	S	G	BT	
SOL	298	25.600	-218.8	30.4	-227.864	167.051	19
	300	25.600	-218.753	30.558	-227.920	166.062	
	400	25.600	-216.193	37.923	-231.362	126.427	
	500	25.600	-213.633	43.635	-235.450	102.929	
	564	25.600	-211.994	46.719	-238.344	92.371	
			5.1	9.043			*
LIQ	564	30.500	-206.894	55.761	-238.344	92.371	
	592	30.500	-206.040	57.239	-239.926	88.586	

$SbCl_5$ (g) ANTIMONY PENTACHLORIDE (GAS)

		A	B	C	D	RANGE	REF.
GAS	CP	31.6	0.	-2.34		298- 800	40,*

PHASE	T	CP	H	S	G	BT	
GAS	298	28.968	-94.25	96.04	-122.884	90.089	40
	300	29.000	-94.196	96.219	-123.062	89.663	
	400	30.137	-91.231	104.741	-133.128	72.747	
	500	30.664	-88.188	111.529	-143.953	62.930	
	600	30.950	-85.106	117.148	-155.395	56.610	
	700	31.122	-82.002	121.933	-167.355	52.258	
	800	31.234	-78.884	126.096	-179.761	49.115	

* * * * * * * * * *

SbOCl ANTIMONY OXIDE CHLORIDE

		A	B	C	D	RANGE	REF.
SOL	CP	16.25	5.25			298- 800	*
	** DEC.						1

PHASE	T	CP	H	S	G	BT	
SOL	298	17.815	-91.	25.7	-98.662	72.331	62/*
	300	17.825	-90.967	25.810	-98.710	71.920	
	400	18.350	-89.158	31.010	-101.562	55.499	
	500	18.875	-87.297	35.161	-104.878	45.848	
	600	19.400	-85.383	38.649	-108.573	39.553	
	700	19.925	-83.417	41.679	-112.592	35.158	
	800	20.450	-81.398	44.374	-116.897	31.939	

ANTIMONY OXIDE

Sb_2O_3

		A	B	C	D	RANGE	REF.
SOL-A	CP	22.0	15.8			298- 845	*
SOL-B	CP	22.0	15.8			845- 929	*
LIQ	CP	37.5				929-1729	*
	** BPT= 1729 GAS(SB406)						62

PHASE	T	CP	H	S	G	BT	
SOL-A	298	26.711	-171.	31.7	-180.451	132.292	62
	300	26.740	-170.951	31.865	-180.510	131.519	
	400	28.320	-168.198	39.774	-184.107	100.605	
	500	29.900	-165.287	46.263	-188.418	82.369	
	600	31.480	-162.218	51.855	-193.330	70.430	
	700	33.060	-158.991	56.826	-198.769	62.067	
	800	34.640	-155.606	61.344	-204.680	55.924	
	845	35.351	-154.031	63.259	-207.484	53.671	
			1.6	1.893			62
SOL-B	845	35.351	-152.431	65.152	-207.484	53.671	
	900	36.220	-150.463	67.408	-211.130	51.276	
	929	36.678	-149.406	68.564	-213.102	50.140	
			13.15	14.155			62
LIQ	929	37.500	-136.256	82.719	-213.102	50.140	
	1000	37.500	-133.593	85.481	-219.074	47.885	
	1100	37.500	-129.843	89.055	-227.804	45.266	
	1200	37.500	-126.093	92.318	-236.875	43.147	
	1300	37.500	-122.343	95.320	-246.259	41.405	
	1400	37.500	-118.593	98.099	-255.931	39.958	
	1500	37.500	-114.843	100.686	-265.872	38.743	
	1600	37.500	-111.093	103.106	-276.063	37.713	
	1700	37.500	-107.343	105.379	-286.488	36.836	
	1729	37.500	-106.256	106.014	-289.553	36.605	

SbO_2 ANTIMONY DIOXIDE

		A	B	C	D	RANGE	REF.
SOL	CP	11.3	8.2			298-1200	*
	** DEC.						1

PHASE	T	CP	H	S	G	BT	
SOL	298	13.745	-108.45	15.2	-112.982	82.829	62
	300	13.760	-108.425	15.285	-113.010	82.339	
	400	14.580	-107.008	19.356	-114.750	62.705	
	500	15.400	-105.509	22.697	-116.857	51.085	
	600	16.220	-103.928	25.578	-119.274	43.451	
	700	17.040	-102.265	28.140	-121.962	38.083	
	800	17.860	-100.520	30.468	-124.894	34.124	
	900	18.680	-98.693	32.619	-128.050	31.099	
	1000	19.500	-96.784	34.630	-131.414	28.724	
	1100	20.320	-94.793	36.527	-134.972	26.820	
	1200	21.140	-92.720	38.330	-138.716	25.267	

* * * * * * * * * *

Sb_2O_5 DIANTIMONY PENTOXIDE

		A	B	C	D	RANGE	REF.
SOL	CP	21.5	37.			298- 500	*
	** DEC.						

PHASE	T	CP	H	S	G	BT	
SOL	298	32.532	-240.8	29.9	-249.715	183.070	62
	300	32.600	-240.740	30.101	-249.770	181.982	
	400	36.300	-237.295	39.987	-253.289	138.410	
	500	40.000	-233.480	48.484	-257.722	112.665	

ANTIMONY SULFIDE (BLACK) Sb_2S_3

		A	B	C	D	RANGE	REF.
SOL	CP	24.34	14.47			298- 823	63
LIQ	CP	40.				823-1100	63

PHASE	T	CP	H	S	G	BT	
SOL	298	28.654	-49.	43.5	-61.970	45.431	63
	300	28.681	-48.947	43.677	-62.050	45.210	
	400	30.128	-46.007	52.127	-66.857	36.534	
	500	31.575	-42.921	59.005	-72.424	31.661	
	600	33.022	-39.692	64.890	-78.625	28.643	
	700	34.469	-36.317	70.089	-85.379	26.660	
	800	35.916	-32.798	74.786	-92.626	25.308	
	823	36.249	-31.968	75.808	-94.358	25.060	
			11.45	13.913			63
LIQ	823	40.000	-20.518	89.721	-94.358	25.060	
	900	40.000	-17.438	93.298	-101.406	24.628	
	1000	40.000	-13.438	97.513	-110.951	24.252	
	1100	40.000	-9.438	101.325	-120.896	24.023	

* * * * * * * * * *

ANTIMONY SULFATE $Sb_2(SO_4)_3$

		A	B	C	D	RANGE	REF.
SOL	CP	52.6	44.4			298-1000	*

PHASE	T	CP	H	S	G	BT	
SOL	298	65.838	-574.2	69.6	-594.951	436.170	19
	300	65.920	-574.078	70.008	-595.080	433.574	
	400	70.360	-567.264	89.580	-603.096	329.561	
	500	74.800	-560.006	105.757	-612.885	267.928	
	600	79.240	-552.304	119.787	-624.176	227.387	
	700	83.680	-544.158	132.335	-636.793	198.842	
	800	88.120	-535.568	143.799	-650.607	177.762	
	900	92.560	-526.534	154.435	-665.525	161.633	
	1000	97.000	-517.056	164.416	-681.473	148.956	

Sb_2Se_3 ANTIMONY SELENIDE

		A	B	C	D	RANGE	REF.
SOL	CP	28.38	5.0			298- 888	63,193
LIQ	CP	41.				888-1000	63

PHASE	T	CP	H	S	G	BT	
SOL	298	29.871	-30.5	50.5	-45.557	33.398	1/193
	300	29.880	-30.445	50.685	-45.650	33.261	
	400	30.380	-27.432	59.349	-51.171	27.963	
	500	30.880	-24.369	66.182	-57.460	25.119	
	600	31.380	-21.256	71.856	-64.370	23.450	
	700	31.880	-18.093	76.731	-71.805	22.421	
	800	32.380	-14.880	81.021	-79.696	21.775	
	888	32.820	-12.011	84.422	-86.978	21.409	
			12.85	14.471			63,79
LIQ	888	41.000	.839	98.893	-86.978	21.409	
	900	41.000	1.331	99.444	-88.168	21.413	
	1000	41.000	5.431	103.763	-98.332	21.493	

* * * * * * * * * *

Sb_2Te_3 ANTIMONY TELLURIDE

		A	B	C	D	RANGE	REF.
SOL	CP	26.98	12.7			298- 892	63
LIQ	CP	49.5				892-1000	115*

PHASE	T	CP	H	S	G	BT	
SOL	298	30.767	-13.5	58.4	-30.912	22.662	115/193
	300	30.790	-13.443	58.590	-31.020	22.601	
	400	32.060	-10.301	67.622	-37.349	20.409	
	500	33.330	-7.031	74.912	-44.487	19.448	
	600	34.600	-3.635	81.101	-52.295	19.051	
	700	35.870	-.111	86.530	-60.682	18.948	
	800	37.140	3.539	91.403	-69.583	19.012	
	892	38.308	7.010	95.508	-78.183	19.158	
			23.65	26.513			115
LIQ	892	49.500	30.660	122.022	-78.183	19.158	
	900	49.500	31.056	122.464	-79.161	19.226	
	1000	49.500	36.006	127.679	-91.673	20.038	

SCANDIUM

		A	B	C	D	RANGE	REF.
SOL-A	CP	5.653	1.537			298- 500	21
	CP	5.909	0.328	0.086	1.214	500-1608	21
SOL-B	CP	10.57				1608-1812	21
LIQ	CP	10.57				1812-3000	21
	** BPT= 3104, L= 75.086						21

PHASE	T	CP	H	S	G	BT	P	
SOL-A	298	6.111	0	8.28	-2.469	1.810		21
	300	6.114	.011	8.318	-2.484	1.810		
	400	6.268	.630	10.098	-3.409	1.863		
	500	6.421	1.265	11.513	-4.492	1.964		
	500	6.411	1.265	11.513	-4.492	1.964		
	600	6.567	1.913	12.695	-5.704	2.078		
	700	6.751	2.579	13.721	-7.025	2.194		
	800	6.962	3.265	14.636	-8.444	2.307		
	900	7.198	3.972	15.469	-9.950	2.416		
	1000	7.460	4.705	16.241	-11.536	2.521		
	1100	7.746	5.465	16.965	-13.196	2.622		
	1200	8.057	6.255	17.652	-14.928	2.719		
	1300	8.392	7.077	18.310	-16.726	2.812		
	1400	8.752	7.934	18.945	-18.589	2.902		
	1500	9.136	8.829	19.562	-20.514	2.989		
	1600	9.545	9.762	20.164	-22.501	3.074	.004	
	1608	9.579	9.839	20.212	-22.662	3.081	.004	
			0.958	.596				21
SOL-B	1608	10.570	10.797	20.808	-22.662	3.081	.004	
	1700	10.570	11.769	21.396	-24.604	3.163	.018	
	1800	10.570	12.826	22.000	-26.774	3.251	.072	
	1812	10.570	12.953	22.070	-27.038	3.262	.084	
			3.369	1.859				21
LIQ	1812	10.570	16.322	23.930	-27.038	3.262	.084	
	1900	10.570	17.252	24.431	-29.166	3.355	.240	
	2000	10.570	18.309	24.973	-31.637	3.458	.699	
	2100	10.570	19.366	25.489	-34.160	3.556	1.830	
	2200	10.570	20.423	25.981	-36.734	3.650	4.362	
	2300	10.570	21.480	26.450	-39.356	3.740	9.592	
	2400	10.570	22.537	26.900	-42.023	3.827	19.657	
	2500	10.570	23.594	27.332	-44.735	3.911	37.870	
	2600	10.570	24.651	27.746	-47.489	3.992	69.097	
	2700	10.570	25.708	28.145	-50.284	4.071	120.149	
	2800	10.570	26.765	28.530	-53.118	4.147	200.180	
	2900	10.570	27.822	28.901	-55.989	4.220	321.043	
	3000	10.570	28.879	29.259	-58.897	4.291	497.609	

ScF_3 SCANDIUM FLUORIDE

		A	B	C	D	RANGE	REF.
SOL	CP	23.75	1.77	-2.25		298-1300	*

PHASE	T	CP	H	S	G	BT	
SOL	298	21.747	-396.	21.5	-402.410	295.014	93/*
	300	21.781	-395.960	21.635	-402.450	293.224	
	400	23.052	-393.710	28.097	-404.949	221.284	
	500	23.735	-391.368	33.321	-408.029	178.373	
	600	24.187	-388.971	37.690	-411.585	149.940	
	700	24.530	-386.534	41.446	-415.546	129.757	
	800	24.814	-384.067	44.740	-419.859	114.716	
	900	25.065	-381.573	47.678	-424.482	103.092	
	1000	25.295	-379.054	50.330	-429.385	93.855	
	1100	25.511	-376.514	52.752	-434.541	86.347	
	1200	25.718	-373.952	54.980	-439.929	80.133	
	1300	25.918	-371.371	57.047	-445.531	74.911	

SCANDIUM CHLORIDE $ScCl_3$

		A	B	C	D	RANGE	REF.
SOL	CP	22.864	3.68	-1.743		298-1240	89
LIQ	CP	34.284				1240-1300	89

PHASE	T	CP	H	S	G	BT	
SOL	298	22.000	-215.	29.	-223.646	163.959	1/*
	300	22.031	-214.959	29.136	-223.700	162.987	
	400	23.247	-212.689	35.658	-226.953	124.018	
	500	24.007	-210.324	40.932	-230.790	100.892	
	600	24.588	-207.894	45.362	-235.111	85.651	
	700	25.084	-205.410	49.190	-239.843	74.892	
	800	25.536	-202.878	52.570	-244.934	66.922	
	900	25.961	-200.303	55.602	-250.345	60.800	
	1000	26.370	-197.687	58.359	-256.045	55.966	
	1100	26.768	-195.030	60.891	-262.010	52.064	
	1200	27.159	-192.333	63.237	-268.217	48.856	
	1240	27.314	-191.244	64.130	-270.765	47.729	
			16.1	12.984			89
LIQ	1240	34.284	-175.144	77.114	-270.765	47.729	
	1300	34.284	-173.087	78.734	-275.441	46.312	

* * * * * * * * * *

SCANDIUM BROMIDE $ScBr_3$

		A	B	C	D	RANGE	REF.
SOL	CP	21.0	6.0			298-1202	*
	** SPT= 1202						1

PHASE	T	CP	H	S	G	BT	
SOL	298	22.789	-170.	40.	-181.926	133.373	1/*
	300	22.800	-169.958	40.141	-182.000	132.605	
	400	23.400	-167.648	46.782	-186.361	101.836	
	500	24.000	-165.278	52.068	-191.312	83.634	
	600	24.600	-162.848	56.497	-196.746	71.674	
	700	25.200	-160.358	60.334	-202.592	63.261	
	800	25.800	-157.808	63.738	-208.799	57.049	
	900	26.400	-155.198	66.812	-215.329	52.296	
	1000	27.000	-152.528	69.624	-222.152	48.558	
	1100	27.600	-149.798	72.226	-229.246	45.553	
	1200	28.200	-147.008	74.653	-236.592	43.095	
	1202	28.212	-146.951	74.700	-236.741	43.050	

ScN

SCANDIUM NITRIDE

		A	B	C	D	RANGE	REF.
SOL	CP	10.95	1.3	-2.2		298-2000	*
	** MPT=	2923					1

PHASE	T	CP	H	S	G	BT	
SOL	298	8.863	-75.	7.1	-77.117	56.536	116/*
	300	8.896	-74.984	7.155	-77.130	56.197	
	400	10.095	-74.026	9.900	-77.987	42.616	
	500	10.720	-72.983	12.226	-79.096	34.577	
	600	11.119	-71.890	14.218	-80.421	29.297	
	700	11.411	-70.763	15.955	-81.931	25.584	
	800	11.646	-69.609	17.495	-83.605	22.843	
	900	11.848	-68.434	18.878	-85.425	20.747	
	1000	12.030	-67.240	20.136	-87.377	19.099	
	1100	12.198	-66.029	21.291	-89.449	17.774	
	1200	12.357	-64.801	22.359	-91.632	16.691	
	1300	12.510	-63.558	23.354	-93.918	15.791	
	1400	12.658	-62.299	24.287	-96.301	15.035	
	1500	12.802	-61.026	25.165	-98.774	14.393	
	1600	12.944	-59.739	25.996	-101.332	13.843	
	1700	13.084	-58.437	26.785	-103.971	13.368	
	1800	13.222	-57.122	27.536	-106.688	12.955	
	1900	13.359	-55.793	28.255	-109.478	12.594	
	2000	13.495	-54.450	28.944	-112.338	12.277	

SELENIUM

		A	B	C	D	RANGE	REF.
SOL-G	CP	4.276	6.			298- 493	63
	** HEXAGONAL, GREY SE						
LIQ	CP	8.4				493- 958	63
	** BPT= 958, GAS(SE2 + SE3 + SE4 + SE5 + SE6)						1/

PHASE	T	CP	H	S	G	BT	P	
SOL-G	298	6.065	0	10.1	-3.011	2.208		63
	300	6.076	.011	10.138	-3.030	2.208		
	400	6.676	.649	11.968	-4.138	2.261		
	493	7.234	1.296	13.420	-5.320	2.359	.008	
			1.4	2.840				63
LIQ	493	8.400	2.696	16.259	-5.320	2.359	.008	
	500	8.400	2.754	16.378	-5.434	2.376	.011	
	600	8.400	3.594	17.909	-7.151	2.605	.480	
	700	8.400	4.434	19.204	-9.008	2.813	7.175	
	800	8.400	5.274	20.326	-10.986	3.002	59.038	
	900	8.400	6.114	21.315	-13.069	3.174	325.944	
	958	8.400	6.602	21.840	-14.321	3.267	760.	

Se(g) SELENIUM (1-ATOMIC GAS)

		A	B	C	D	RANGE	REF.
GAS	CP	5.13	0.36	-0.22		298-2000	63

PHASE	T	CP	H	S	G	BT	
GAS	298	4.990	56.25	42.21	43.665	-32.012	63
	300	4.994	56.259	42.241	43.587	-31.757	
	400	5.136	56.767	43.699	39.287	-21.468	
	500	5.222	57.285	44.855	34.857	-15.238	
	600	5.285	57.810	45.813	30.322	-11.046	
	700	5.337	58.341	46.632	25.699	-8.025	
	800	5.384	58.877	47.347	20.999	-5.738	
	900	5.427	59.418	47.984	16.232	-3.942	
	958	5.451	59.733	48.324	13.439	-3.066	
	958	5.451	59.733	48.324	13.439	-3.066	
	1000	5.468	59.963	48.558	11.405	-2.493	
	1100	5.508	60.512	49.081	6.522	-1.296	
	1200	5.547	61.064	49.562	1.590	-.290	
	1300	5.585	61.621	50.007	-3.389	.570	
	1400	5.623	62.181	50.423	-8.411	1.313	
	1500	5.660	62.745	50.812	-13.473	1.963	
	1600	5.697	63.313	51.178	-18.572	2.537	
	1700	5.734	63.885	51.525	-23.708	3.048	
	1800	5.771	64.460	51.854	-28.877	3.507	
	1900	5.808	65.039	52.167	-34.078	3.920	
	2000	5.844	65.622	52.466	-39.310	4.296	

SELENIUM (2-ATOMIC GAS)

		A	B	C	D	RANGE	REF.
GAS	CP	10.66	-0.635	-0.598		298- 958	63
GAS	CP	10.66	-0.635	-0.593		958-2000	63

PHASE	T	CP	H	S	G	BT	
GAS	298	9.798	33.027	59.	15.436	-11.317	63*
	300	9.805	33.045	59.061	15.327	-11.167	
	400	10.032	34.039	61.918	9.272	-5.066	
	500	10.103	35.047	64.166	2.963	-1.295	
	600	10.113	36.058	66.010	-3.548	1.293	
	700	10.093	37.068	67.568	-10.229	3.194	
	800	10.059	38.076	68.913	-17.055	4.660	
	900	10.015	39.080	70.096	-24.006	5.830	
	958	9.987	39.660	70.720	-28.090	6.409	
	958	9.987	39.660	70.720	-28.090	6.409	
	1000	9.966	40.079	71.148	-31.069	6.791	
	1100	9.912	41.073	72.096	-38.232	7.597	
	1200	9.857	42.061	72.956	-45.486	8.285	
	1300	9.799	43.044	73.742	-52.821	8.881	
	1400	9.741	44.021	74.466	-60.232	9.404	
	1500	9.681	44.992	75.136	-67.713	9.867	
	1600	9.621	45.957	75.759	-75.258	10.281	
	1700	9.560	46.916	76.341	-82.863	10.654	
	1800	9.499	47.869	76.885	-90.525	10.993	
	1900	9.437	48.816	77.397	-98.239	11.302	
	2000	9.375	49.757	77.880	-106.003	11.585	

Se_3 (g)

SELENIUM (3-ATOMIC GAS)

		A	B	C	D	RANGE	REF.
GAS	CP	13.897	0.726	-0.528		298-2000	63

PHASE	T	CP	H	S	G	BT	
GAS	298	13.519	42.1	75.27	19.658	-14.412	63
	300	13.528	42.125	75.354	19.519	-14.221	
	400	13.857	43.496	79.296	11.778	-6.436	
	500	14.049	44.892	82.410	3.687	-1.612	
	600	14.186	46.304	84.984	-4.686	1.707	
	700	14.297	47.728	87.179	-13.297	4.152	
	800	14.395	49.163	89.095	-22.113	6.042	
	900	14.485	50.607	90.796	-31.109	7.555	
	958	14.535	51.449	91.702	-36.402	8.306	
	958	14.535	51.449	91.702	-36.402	8.306	
	1000	14.570	52.060	92.327	-40.266	8.801	
	1100	14.652	53.521	93.719	-49.570	9.850	
	1200	14.732	54.990	94.997	-59.006	10.748	
	1300	14.810	56.467	96.180	-68.566	11.529	
	1400	14.886	57.952	97.280	-78.240	12.215	
	1500	14.963	59.445	98.310	-88.020	12.826	
	1600	15.038	60.945	99.278	-97.900	13.374	
	1700	15.113	62.452	100.192	-107.873	13.870	
	1800	15.188	63.967	101.058	-117.936	14.321	
	1900	15.262	65.490	101.881	-128.084	14.735	
	2000	15.336	67.020	102.665	-138.311	15.116	

SELENIUM (4-ATOMIC GAS) Se_4 (g)

		A	B	C	D	RANGE	REF.
GAS	CP	19.857	0.008	-0.599		298-2000	63

PHASE	T	CP	H	S	G	BT	
GAS	298	19.186	43.8	90.606	16.786	-12.306	63
	300	19.194	43.836	90.725	16.618	-12.108	
	400	19.486	45.772	96.292	7.255	-3.964	
	500	19.621	47.728	100.657	-2.601	1.137	
	600	19.695	49.694	104.241	-12.851	4.682	
	700	19.740	51.666	107.281	-23.431	7.316	
	800	19.770	53.641	109.919	-34.294	9.370	
	900	19.790	55.619	112.249	-45.404	11.027	
	958	19.799	56.768	113.485	-51.951	11.853	
	958	19.799	56.768	113.485	-51.951	11.853	
	1000	19.805	57.599	114.335	-56.735	12.401	
	1100	19.816	59.580	116.223	-68.265	13.565	
	1200	19.825	61.562	117.948	-79.975	14.567	
	1300	19.832	63.545	119.535	-91.850	15.443	
	1400	19.838	65.529	121.005	-103.878	16.218	
	1500	19.842	67.513	122.373	-116.047	16.910	
	1600	19.846	69.497	123.654	-128.349	17.534	
	1700	19.850	71.482	124.857	-140.776	18.100	
	1800	19.853	73.467	125.992	-153.319	18.618	
	1900	19.856	75.453	127.066	-165.972	19.094	
	2000	19.858	77.438	128.084	-178.730	19.533	

Se_5 (g)

SELENIUM (5-ATOMIC GAS)

		A	B	C	D	RANGE	REF.
GAS	CP	25.795	0.021	-1.413		298-2000	63

PHASE	T	CP	H	S	G	BT	
GAS	298	24.212	33.	92.077	5.547	-4.067	63
	300	24.231	33.045	92.227	5.377	-3.917	
	400	24.920	35.507	99.306	-4.215	2.303	
	500	25.240	38.017	104.905	-14.436	6.311	
	600	25.415	40.551	109.524	-25.164	9.167	
	700	25.521	43.098	113.450	-36.317	11.340	
	800	25.591	45.654	116.863	-47.837	13.070	
	900	25.639	48.215	119.880	-59.677	14.493	
	958	25.661	49.703	121.482	-66.677	15.213	
	958	25.661	49.703	121.482	-66.677	15.213	
	1000	25.675	50.781	122.584	-71.802	15.695	
	1100	25.701	53.350	125.032	-84.185	16.728	
	1200	25.722	55.921	127.269	-96.802	17.632	
	1300	25.739	58.494	129.329	-109.633	18.433	
	1400	25.752	61.069	131.237	-122.663	19.151	
	1500	25.764	63.645	133.014	-135.876	19.800	
	1600	25.773	66.222	134.677	-149.262	20.391	
	1700	25.782	68.799	136.240	-162.808	20.933	
	1800	25.789	71.378	137.714	-176.507	21.434	
	1900	25.796	73.957	139.108	-190.348	21.898	
	2000	25.802	76.537	140.431	-204.326	22.331	

SELENIUM (6-ATOMIC GAS) Se_6 (g)

		A	B	C	D	RANGE	REF.
GAS	CP	31.765	0.016	-1.416		298-2000	63

PHASE	T	CP	H	S	G	BT	
GAS	298	30.177	32.3	103.61	1.409	-1.033	63
	300	30.196	32.356	103.797	1.217	-.887	
	400	30.886	35.415	112.592	-9.622	5.258	
	500	31.207	38.521	119.523	-21.240	9.285	
	600	31.381	41.652	125.229	-33.486	12.199	
	700	31.487	44.795	130.075	-46.257	14.444	
	800	31.557	47.948	134.285	-59.480	16.251	
	900	31.605	51.106	138.004	-73.098	17.753	
	958	31.626	52.940	139.979	-81.160	18.518	
	958	31.626	52.940	139.979	-81.160	18.518	
	1000	31.639	54.268	141.336	-87.068	19.031	
	1100	31.666	57.434	144.353	-101.355	20.140	
	1200	31.686	60.601	147.109	-115.930	21.117	
	1300	31.702	63.771	149.646	-130.769	21.987	
	1400	31.715	66.941	151.996	-145.853	22.772	
	1500	31.726	70.114	154.185	-161.163	23.485	
	1600	31.735	73.287	156.232	-176.685	24.137	
	1700	31.743	76.461	158.157	-192.406	24.739	
	1800	31.750	79.635	159.971	-208.313	25.296	
	1900	31.756	82.811	161.688	-224.397	25.815	
	2000	31.762	85.986	163.317	-240.648	26.300	

SeF_4 (g)   SELENIUM TETRAFLUORIDE (GAS)

		A	B	C	D	RANGE	REF.
GAS	CP	24.15	0.88	-6.58		298-2000	63

PHASE	T	CP	H	S	G	BT	
GAS	298	17.010	-194.	70.85	-215.124	157.711	63
	300	17.103	-193.968	70.956	-215.255	156.834	
	400	20.389	-192.071	76.392	-222.628	121.654	
	500	21.958	-189.945	81.128	-230.510	100.769	
	600	22.850	-187.701	85.217	-238.832	87.006	
	700	23.423	-185.386	88.786	-247.536	77.295	
	800	23.826	-183.022	91.941	-256.575	70.102	
	900	24.130	-180.624	94.766	-265.913	64.581	
	1000	24.372	-178.198	97.321	-275.519	60.223	
	1100	24.574	-175.751	99.654	-285.370	56.705	
	1200	24.749	-173.284	101.799	-295.444	53.815	
	1300	24.905	-170.802	103.787	-305.724	51.404	
	1400	25.046	-168.304	105.638	-316.197	49.367	
	1500	25.178	-165.793	107.370	-326.848	47.628	
	1600	25.301	-163.269	108.999	-337.667	46.129	
	1700	25.418	-160.733	110.536	-348.645	44.827	
	1800	25.531	-158.185	111.992	-359.772	43.688	
	1900	25.640	-155.627	113.376	-371.041	42.685	
	2000	25.745	-153.057	114.694	-382.445	41.797	

SELENIUM MONOCHLORIDE (DIMERIC) Se_2Cl_2

		A	B	C	D	RANGE	REF.
LIQ	CP	30.4				298- 647	*
	** DEC., BPT= 646.2						63
GAS	CP	19.690	0.367	-1.082		646-1000	63

PHASE	T	CP	H	S	G	BT	P	
LIQ	298	30.400	-20.	59.	-37.591	27.559	.004	63
	300	30.400	-19.944	59.188	-37.700	27.468	.005	
	400	30.400	-16.904	67.934	-44.077	24.086	1.909	
	500	30.400	-13.864	74.717	-51.222	22.392	51.770	
	600	30.400	-10.824	80.260	-58.980	21.486	388.225	
	646	30.400	-9.419	82.515	-62.740	21.222	760.	
			10.937	16.925				63*
GAS	646	19.668	1.518	99.440	-62.740	21.222		
	700	19.726	2.577	101.015	-68.133	21.275		
	800	19.815	4.555	103.655	-78.369	21.412		
	900	19.887	6.540	105.993	-88.854	21.580		
	1000	19.949	8.532	108.092	-99.560	21.762		

$SeCl_2(g)$ SELENIUM DICHLORIDE (GAS)

		A	B	C	D	RANGE	REF.
GAS	CP	13.85	0.032	-0.944		298-2000	63

PHASE	T	CP	H	S	G	BT	
GAS	298	12.798	-8.	70.65	-29.064	21.308	63
	300	12.811	-7.976	70.729	-29.195	21.271	
	400	13.273	-6.669	74.487	-36.464	19.926	
	500	13.488	-5.330	77.475	-44.067	19.264	
	600	13.607	-3.974	79.946	-51.942	18.922	
	700	13.680	-2.610	82.049	-60.044	18.749	
	800	13.728	-1.239	83.879	-68.342	18.673	
	900	13.762	.135	85.498	-76.813	18.655	
	1000	13.788	1.513	86.949	-85.436	18.675	
	1100	13.807	2.893	88.264	-94.198	18.718	
	1200	13.823	4.274	89.467	-103.086	18.777	
	1300	13.836	5.657	90.573	-112.088	18.846	
	1400	13.847	7.041	91.599	-121.198	18.922	
	1500	13.856	8.427	92.555	-130.406	19.003	
	1600	13.864	9.813	93.449	-139.706	19.086	
	1700	13.872	11.199	94.290	-149.094	19.170	
	1800	13.878	12.587	95.083	-158.563	19.255	
	1900	13.885	13.975	95.834	-168.109	19.340	
	2000	13.890	15.364	96.546	-177.728	19.424	

* * * * * * * * * *

$SeCl_4$ SELENIUM TETRACHLORIDE

		A	B	C	D	RANGE	REF.
SOL	CP	32.0				298- 578	63
	** DEC., MPT= 578						1

PHASE	T	CP	H	S	G	BT	
SOL	298	32.000	-45.1	46.5	-58.964	43.228	63
	300	32.000	-45.041	46.698	-59.050	43.024	
	400	32.00u	-41.841	55.904	-64.202	35.083	
	500	32.000	-38.641	63.044	-70.163	30.672	
	578	32.000	-36.145	67.683	-75.266	28.463	

SELENIUM MONOBROMIDE (GAS) Se_2Br_2 (g)

		A	B	C	D	RANGE	REF.
GAS	CP	19.70	0.437	-0.703		298-2000	63

PHASE	T	CP	H	S	G	BT	
GAS	298	19.039	7.	90.35	-19.938	14.617	63
	300	19.050	7.035	90.468	-20.105	14.649	
	400	19.435	8.962	96.008	-29.441	16.088	
	500	19.637	10.916	100.369	-39.268	17.166	
	600	19.767	12.887	103.961	-49.490	18.029	
	700	19.862	14.869	107.016	-60.042	18.749	
	800	19.940	16.859	109.673	-70.879	19.366	
	900	20.007	18.856	112.026	-81.967	19.907	
	1000	20.067	20.860	114.137	-93.277	20.388	
	1100	20.123	22.870	116.052	-104.787	20.822	
	1200	20.176	24.884	117.805	-116.481	21.217	
	1300	20.227	26.905	119.422	-128.344	21.579	
	1400	20.276	28.930	120.923	-140.362	21.914	
	1500	20.324	30.960	122.323	-152.525	22.226	
	1600	20.372	32.995	123.636	-164.824	22.517	
	1700	20.419	35.034	124.873	-177.250	22.790	
	1800	20.465	37.078	126.041	-189.796	23.047	
	1900	20.511	39.127	127.149	-202.456	23.291	
	2000	20.556	41.180	128.202	-215.224	23.522	

$SeBr_2$ (g)   SELENIUM DIBROMIDE (GAS)

		A	B	C	D	RANGE	REF.
GAS	CP	13.89	0.011	-0.576		298-2000	63

PHASE	T	CP	H	S	G	BT	
GAS	298	13.245	-5.	75.76	-27.588	20.225	63
	300	13.253	-4.975	75.842	-27.728	20.203	
	400	13.534	-3.634	79.699	-35.514	19.406	
	500	13.665	-2.273	82.735	-43.641	19.078	
	600	13.737	-.903	85.233	-52.043	18.959	
	700	13.780	.473	87.354	-60.675	18.946	
	800	13.809	1.853	89.196	-69.504	18.990	
	900	13.829	3.234	90.824	-78.507	19.067	
	1000	13.843	4.618	92.282	-87.664	19.161	
	1100	13.854	6.003	93.602	-96.959	19.267	
	1200	13.863	7.389	94.807	-106.380	19.377	
	1300	13.870	8.776	95.917	-115.917	19.490	
	1400	13.876	10.163	96.946	-125.561	19.604	
	1500	13.881	11.551	97.903	-135.304	19.716	
	1600	13.885	12.939	98.799	-145.139	19.828	
	1700	13.889	14.328	99.641	-155.062	19.937	
	1800	13.892	15.717	100.435	-165.066	20.044	
	1900	13.895	17.106	101.186	-175.147	20.149	
	2000	13.898	18.496	101.899	-185.302	20.252	

SELENIUM MONOXIDE (GAS) SeO (g)

		A	B	C	D	RANGE	REF.
GAS	CP	8.35	0.36	-0.88		298-2000	63

PHASE	T	CP	H	S	G	BT	
GAS	298	7.467	14.9	55.9	-1.767	1.295	63
	300	7.480	14.914	55.946	-1.870	1.363	
	400	7.944	15.688	58.170	-7.580	4.142	
	500	8.178	16.495	59.971	-13.490	5.897	
	600	8.322	17.321	61.475	-19.564	7.127	
	700	8.422	18.158	62.766	-25.778	8.049	
	800	8.500	19.004	63.896	-32.112	8.774	
	900	8.565	19.858	64.901	-38.553	9.363	
	1000	8.622	20.717	65.807	-45.089	9.856	
	1100	8.673	21.582	66.631	-51.712	10.276	
	1200	8.721	22.452	67.387	-58.413	10.640	
	1300	8.766	23.326	68.087	-65.187	10.960	
	1400	8.809	24.205	68.739	-72.029	11.246	
	1500	8.851	25.088	69.348	-78.934	11.502	
	1600	8.892	25.975	69.920	-85.897	11.735	
	1700	8.932	26.866	70.460	-92.917	11.947	
	1800	8.971	27.761	70.972	-99.988	12.142	
	1900	9.010	28.660	71.458	-107.110	12.322	
	2000	9.048	29.563	71.921	-114.279	12.490	

SeO_2 SELENIUM DIOXIDE

		A	B	C	D	RANGE	REF.
SOL	CP	16.63	0.93	-2.64		298- 601	63
	** SPT= 600.761, MPT= 613						*/63
GAS	CP	12.63	0.738	-2.365		601-2000	63

PHASE	T	CP	H	S	G	BT	P	
SOL	298	13.937	-53.8	15.94	-58.553	42.926		63
	300	13.976	-53.774	16.026	-58.582	42.683		
	400	15.352	-52.299	20.262	-60.403	33.007	.008	
	500	16.039	-50.726	23.769	-62.610	27.371	7.929	
	600	16.455	-49.100	26.732	-65.139	23.730	738.472	
	601	16.457	-49.087	26.753	-65.159	23.707	760.	
			26.81	44.627				63,*
GAS	601	12.418	-22.277	71.380	-65.159	23.707		
	700	12.664	-21.032	73.298	-72.340	22.589		
	800	12.851	-19.756	75.002	-79.757	21.792		
	900	13.002	-18.463	76.524	-87.335	21.211		
	1000	13.131	-17.156	77.901	-95.057	20.777		
	1100	13.246	-15.837	79.158	-102.911	20.449		
	1200	13.351	-14.507	80.315	-110.885	20.198		
	1300	13.449	-13.167	81.388	-118.971	20.004		
	1400	13.543	-11.817	82.388	-127.160	19.853		
	1500	13.632	-10.459	83.325	-135.447	19.737		
	1600	13.718	-9.091	84.208	-143.824	19.648		
	1700	13.803	-7.715	85.042	-152.286	19.580		
	1800	13.885	-6.331	85.833	-160.831	19.530		
	1900	13.967	-4.938	86.586	-169.452	19.494		
	2000	14.047	-3.537	87.305	-178.147	19.470		

SILICON MONOHYDRIDE (GAS) SiH (g)

		A	B	C	D	RANGE	REF.
GAS	CP	6.770	1.157			298- 700	3
	CP	7.242	0.873			700-2000	3

PHASE	T	CP	H	S	G	BT	
GAS	298	7.115	90.	47.306	75.896	-55.641	3
	300	7.117	90.013	47.350	75.808	-55.234	
	400	7.233	90.731	49.413	70.965	-38.779	
	500	7.349	91.460	51.040	65.940	-28.826	
	600	7.464	92.200	52.390	60.767	-22.137	
	700	7.580	92.953	53.549	55.468	-17.320	
	700	7.853	92.953	53.549	55.468	-17.320	
	800	7.940	93.742	54.603	50.060	-13.677	
	900	8.028	94.541	55.544	44.551	-10.820	
	1000	8.115	95.348	56.394	38.954	-8.514	
	1100	8.202	96.164	57.172	33.275	-6.612	
	1200	8.290	96.988	57.889	27.522	-5.013	
	1300	8.377	97.822	58.556	21.699	-3.648	
	1400	8.464	98.664	59.180	15.812	-2.469	
	1500	8.551	99.514	59.767	9.864	-1.437	
	1600	8.639	100.374	60.322	3.859	-.527	
	1700	8.726	101.242	60.848	-2.199	.283	
	1800	8.813	102.119	61.349	-8.309	1.009	
	1900	8.901	103.005	61.828	-14.468	1.664	
	2000	8.988	103.899	62.287	-20.674	2.259	

Si_2H_6 DISILANE

		A	B	C	D	RANGE	REF.
SOL	CP	12.89	32.6	-2.47	-10.7	298-1300	157

PHASE	T	CP	H	S	G	BT	
SOL	298	18.880	19.1	65.55	-.444	.325	62/157
	300	18.963	19.135	65.667	-.565	.412	
	400	22.674	21.227	71.660	-7.437	4.064	
	500	25.527	23.642	77.037	-14.877	6.503	
	600	27.912	26.317	81.908	-22.828	8.316	
	700	29.963	29.213	86.369	-31.245	9.756	
	800	31.736	32.301	90.488	-40.090	10.954	
	900	33.258	35.552	94.316	-49.332	11.981	
	1000	34.543	38.944	97.889	-58.945	12.884	
	1100	35.599	42.453	101.233	-68.903	13.692	
	1200	36.430	46.057	104.367	-79.184	14.423	
	1300	37.041	49.732	107.309	-89.770	15.094	

SILICON DICHLORIDE (GAS) $SiCl_2(g)$

		A	B	C	D	RANGE	REF.
GAS	CP	13.76	0.09	-1.35		298-2200	3

PHASE	T	CP	H	S	G	BT	
GAS	298	12.268	-39.3	67.36	-59.383	43.535	3
	300	12.287	-39.277	67.436	-59.508	43.357	
	400	12.952	-38.011	71.075	-66.441	36.306	
	500	13.265	-36.698	74.003	-73.700	32.218	
	600	13.439	-35.362	76.438	-81.225	29.590	
	700	13.547	-34.012	78.519	-88.975	27.783	
	800	13.621	-32.654	80.333	-96.920	26.481	
	900	13.674	-31.289	81.940	-105.035	25.509	
	1000	13.715	-29.919	83.383	-113.302	24.766	
	1100	13.747	-28.546	84.692	-121.707	24.184	
	1200	13.774	-27.170	85.889	-130.237	23.723	
	1300	13.797	-25.791	86.993	-138.882	23.351	
	1400	13.817	-24.411	88.016	-147.633	23.050	
	1500	13.835	-23.028	88.970	-156.483	22.803	
	1600	13.851	-21.644	89.863	-165.425	22.599	
	1700	13.866	-20.258	90.703	-174.454	22.431	
	1800	13.880	-18.871	91.496	-183.564	22.291	
	1900	13.894	-17.482	92.247	-192.752	22.174	
	2000	13.906	-16.092	92.960	-202.012	22.078	
	2100	13.918	-14.701	93.639	-211.342	21.998	
	2200	13.930	-13.308	94.287	-220.739	21.931	

$SiCl_3$ (g)   SILICON TRICHLORIDE (GAS)

		A	B	C	D	RANGE	REF.
GAS	CP	15.620	5.468			298- 700	3
	CP	19.056	0.408			700-2000	3

PHASE	T	CP	H	S	G	BT	
GAS	298	17.250	-96.	76.17	-118.710	87.029	3
	300	17.260	-95.968	76.277	-118.851	86.595	
	400	17.807	-94.215	81.317	-126.742	69.258	
	500	18.354	-92.407	85.349	-135.081	59.052	
	600	18.901	-90.544	88.744	-143.790	52.383	
	700	19.448	-88.626	91.699	-152.816	47.718	
	700	19.342	-88.626	91.699	-152.816	47.718	
	800	19.382	-86.690	94.284	-162.118	44.294	
	900	19.423	-84.750	96.569	-171.662	41.691	
	1000	19.464	-82.806	98.618	-181.424	39.655	
	1100	19.505	-80.857	100.475	-191.380	38.029	
	1200	19.546	-78.905	102.174	-201.513	36.706	
	1300	19.586	-76.948	103.740	-211.810	35.613	
	1400	19.627	-74.987	105.193	-222.258	34.701	
	1500	19.668	-73.023	106.548	-232.845	33.930	
	1600	19.709	-71.054	107.819	-243.564	33.274	
	1700	19.750	-69.081	109.015	-254.407	32.711	
	1800	19.790	-67.104	110.145	-265.365	32.224	
	1900	19.831	-65.123	111.216	-276.434	31.801	
	2000	19.872	-63.138	112.235	-287.607	31.432	

SILICON TETRACHLORIDE

$SiCl_4$

		A	B	C	D	RANGE	REF.
LIQ	CP	35.0				298- 335	*
	** MPT= 203 / BPT= 334.29						1/3*
GAS	CP	24.25	1.64	-2.75		334-2000	3

PHASE	T	CP	H	S	G	BT	P	
LIQ	298	35.000	-164.35	57.3	-181.434	133.013	210.842	105*/211
	300	35.000	-164.285	57.517	-181.540	132.270	227.341	
	334	35.000	-163.085	61.304	-183.579	120.035	760.	
			6.780	20.282				*
GAS	334	22.337	-156.305	81.586	-183.579	120.035		
	400	23.187	-154.807	85.675	-189.077	103.321		
	500	23.970	-152.446	90.941	-197.916	86.521		
	600	24.470	-150.022	95.358	-207.237	75.496		
	700	24.837	-147.556	99.159	-216.967	67.749		
	800	25.132	-145.057	102.495	-227.053	62.036		
	900	25.386	-142.531	105.470	-237.454	57.670		
	1000	25.615	-139.981	108.157	-248.138	54.238		
	1100	25.827	-137.409	110.608	-259.078	51.481		
	1200	26.027	-134.816	112.864	-270.253	49.226		
	1300	26.219	-132.204	114.955	-281.645	47.355		
	1400	26.406	-129.572	116.905	-293.239	45.783		
	1500	26.588	-126.923	118.733	-305.022	44.448		
	1600	26.767	-124.255	120.455	-316.982	43.304		
	1700	26.943	-121.569	122.083	-329.110	42.316		
	1800	27.117	-118.866	123.628	-341.396	41.457		
	1900	27.290	-116.146	125.098	-353.833	40.706		
	2000	27.461	-113.408	126.503	-366.414	40.045		

$SiBr_2$(g) SILICON DIBROMIDE (GAS)

		A	B	C	D	RANGE	REF.
GAS	CP	13.81	0.	-1.11		298-1000	*

PHASE	T	CP	H	S	G	BT	
GAS	298	12.561	-10.2	72.5	-31.816	23.325	62
	300	12.577	-10.177	72.578	-31.950	23.279	
	400	13.116	-8.888	76.281	-39.401	21.530	
	500	13.366	-7.563	79.238	-47.182	20.626	
	600	13.502	-6.219	81.688	-55.231	20.121	
	700	13.583	-4.864	83.776	-63.507	19.830	
	800	13.637	-3.503	85.593	-71.977	19.666	
	900	13.673	-2.137	87.201	-80.619	19.580	
	1000	13.699	-.769	88.643	-89.412	19.544	

* * * * * * * * * *

$SiBr_4$ SILICON TETRABROMIDE

		A	B	C	D	RANGE	REF.
LIQ	CP	35.50				298- 426	*
	** MPT=	278.65					62
GAS	CP	29.19	0.64	-1.94		426-1200	62

PHASE	T	CP	H	S	G	BT	P	
LIQ	298	35.500	-110.3	66.464	-130.116	95.391	7.397	62/*
	300	35.500	-110.234	66.684	-130.239	94.892	8.192	
	400	35.500	-106.684	76.896	-137.443	75.105	398.344	
	426	35.500	-105.770	79.111	-139.452	71.594	760.	
			8.34	19.589				62
GAS	426	28.392	-97.430	98.700	-139.452	71.594		
	500	28.734	-95.309	103.293	-146.955	64.243		
	600	29.035	-92.419	108.560	-157.555	57.397		
	700	29.242	-89.505	113.052	-168.641	52.659		
	800	29.399	-86.572	116.968	-180.146	49.220		
	900	29.526	-83.626	120.438	-192.020	46.635		
	1000	29.636	-80.668	123.555	-204.222	44.639		
	1100	29.734	-77.699	126.384	-216.721	43.064		
	1200	29.823	-74.721	128.975	-229.491	41.802		

SILICON DIIODIDE (GAS) $SiJ_2(g)$

		A	B	C	D	RANGE	REF.
GAS	CP	13.82	0.	-0.83		298-1800	*

PHASE	T	CP	H	S	G	BT	
GAS	298	12.886	19.6	75.8	-3.000	2.199	62,168,*
	300	12.898	19.624	75.880	-3.140	2.288	
	400	13.301	20.937	79.654	-10.925	5.970	
	500	13.488	22.277	82.644	-19.045	8.326	
	600	13.589	23.632	85.113	-27.436	9.995	
	700	13.651	24.994	87.213	-36.055	11.258	
	800	13.690	26.361	89.039	-44.870	12.260	
	900	13.718	27.731	90.653	-53.856	13.080	
	1000	13.737	29.104	92.099	-62.995	13.769	
	1100	13.751	30.479	93.409	-72.271	14.361	
	1200	13.762	31.854	94.606	-81.673	14.877	
	1300	13.771	33.231	95.708	-91.189	15.332	
	1400	13.778	34.608	96.729	-100.812	15.740	
	1500	13.783	35.987	97.680	-110.533	16.107	
	1600	13.788	37.365	98.569	-120.346	16.441	
	1700	13.791	38.744	99.405	-130.245	16.746	
	1800	13.794	40.123	100.194	-140.225	17.028	

SiJ_4 SILICON TETRAIODIDE

		A	B	C	D	RANGE	REF.
SOL	CP	34.0				298- 395	*
LIQ	CP	36.0				395- 574	*
	** BPT=	574.1					
GAS	CP	25.570	0.240	-1.440		574-1000	58

PHASE	T	CP	H	S	G	BT	P	
SOL	298	34.000	-47.6	63.44	-66.515	48.763	.004	168/*
	300	34.000	-47.537	63.650	-66.632	48.548	.005	
	395	34.000	-44.307	73.004	-73.144	40.475	5.484	
			3.5	8.861				168
LIQ	395	36.000	-40.807	81.865	-73.144	40.475	5.484	
	400	36.000	-40.627	82.317	-73.554	40.193	6.772	
	500	36.000	-37.027	90.351	-82.202	35.935	165.590	
	574	36.000	-34.360	95.326	-89.086	33.918	760.	
			11.312	19.704				168,*
GAS	574	25.271	-23.047	115.030	-89.086	33.918		
	600	25.314	-22.392	116.146	-92.080	33.545		
	700	25.444	-19.854	120.058	-103.895	32.442		
	800	25.537	-17.305	123.462	-116.075	31.714		
	900	25.608	-14.747	126.474	-128.574	31.226		
	1000	25.666	-12.184	129.175	-141.359	30.898		

SILICON MONOSELENIDE (GAS) $SiSe(g)$

		A	B	C	D	RANGE	REF.
GAS	CP	8.78	0.10	-0.69		298-2000	63

PHASE	T	CP	H	S	G	BT	
GAS	298	8.034	48.5	56.2	31.744	-23.272	63
	300	8.043	48.515	56.250	31.640	-23.053	
	400	8.389	49.339	58.618	25.892	-14.148	
	500	8.554	50.187	60.509	19.932	-8.714	
	600	8.648	51.047	62.078	13.801	-5.028	
	700	8.709	51.915	63.416	7.524	-2.349	
	800	8.752	52.789	64.582	1.123	-.307	
	900	8.785	53.666	65.615	-5.388	1.309	
	1000	8.811	54.545	66.542	-11.996	2.622	
	1100	8.833	55.428	67.383	-18.693	3.715	
	1200	8.852	56.312	68.152	-25.471	4.639	
	1300	8.869	57.198	68.861	-32.322	5.434	
	1400	8.885	58.086	69.519	-39.241	6.127	
	1500	8.899	58.975	70.133	-46.224	6.736	
	1600	8.913	59.865	70.707	-53.266	7.277	
	1700	8.926	60.757	71.248	-60.364	7.761	
	1800	8.939	61.651	71.759	-67.515	8.199	
	1900	8.951	62.545	72.242	-74.715	8.595	
	2000	8.963	63.441	72.702	-81.963	8.958	

SiP SILICON MONOPHOSPHICE

		A	B	C	D	RANGE	REF.
SOL	CP	10.25	2.6	-1.35		298-1413	*
	** DPT= 1413						194

PHASE	T	CP	H	S	G	BT	
SOL	298	9.507	-14.8	7.8	-17.126	12.555	64,*
	300	9.530	-14.782	7.859	-17.140	12.488	
	400	10.446	-13.779	10.739	-18.075	9.877	
	500	11.010	-12.704	13.135	-19.272	8.425	
	600	11.435	-11.581	15.181	-20.690	7.537	
	700	11.794	-10.420	16.971	-22.300	6.963	
	800	12.119	-9.224	18.568	-24.078	6.579	
	900	12.423	-7.996	20.013	-26.008	6.316	
	1000	12.715	-6.739	21.337	-28.076	6.137	
	1100	12.998	-5.454	22.562	-30.272	6.015	
	1200	13.276	-4.140	23.705	-32.586	5.936	
	1300	13.550	-2.799	24.779	-35.011	5.887	
	1400	13.821	-1.430	25.793	-37.540	5.861	
	1413	13.856	-1.250	25.921	-37.876	5.859	

		A	B	C	D	RANGE	REF.
SOL-A	CP	13.687	-1.570	-16.271	3.22	298- 800	21
	CP	2.704	6.416	17.225		800-1190	21
SOL-B	CP	11.22				1190-1345	21
LIQ	CP	12.				1345-2064	21
GAS	CP	6.281				2064-2300	21

PHASE	T	CP	H	S	G	BT	P	
SOL-A	298	7.064	0	16.61	-4.952	3.631		21
	300	7.063	.013	16.654	-4.983	3.631		
	400	7.921	.754	18.778	-6.758	3.693		
	500	8.970	1.600	20.663	-8.731	3.817		
	600	9.716	2.537	22.369	-10.884	3.965		
	700	10.206	3.535	23.906	-13.199	4.122		
	800	10.518	4.572	25.291	-15.660	4.279		
	800	10.528	4.572	25.291	-15.660	4.279		
	900	10.605	5.627	26.533	-18.253	4.433		
	1000	10.842	6.699	27.662	-20.963	4.582	.007	
	1100	11.185	7.799	28.711	-23.782	4.726	.065	
	1190	11.555	8.822	29.604	-26.407	4.850	.336	
			0.744	.625				21
SOL-B	1190	11.220	9.566	30.229	-26.407	4.850	.336	
	1200	11.220	9.678	30.323	-26.709	4.865	.396	
	1300	11.220	10.800	31.221	-29.787	5.008	1.753	
	1345	11.220	11.305	31.603	-31.201	5.071	3.173	
			2.06	1.532				21
LIQ	1345	12.000	13.365	33.135	-31.201	5.071	3.173	
	1400	12.000	14.025	33.616	-33.037	5.158	6.017	
	1500	12.000	15.225	34.444	-36.440	5.310	16.913	
	1600	12.000	16.425	35.218	-39.924	5.454	41.330	
	1700	12.000	17.625	35.946	-43.482	5.591	90.029	
	1800	12.000	18.825	36.632	-47.111	5.721	178.265	
	1900	12.000	20.025	37.280	-50.807	5.845	325.841	
	2000	12.000	21.225	37.896	-54.566	5.964	556.590	
	2064	12.000	21.993	38.274	-57.004	6.037	760.	
			39.773	19.270				21
GAS	2064	6.281	61.766	57.544	-57.004	6.037		
	2100	6.281	61.993	57.652	-59.077	6.149		
	2200	6.281	62.621	57.944	-64.857	6.444		
	2300	6.281	63.249	58.224	-70.666	6.716		

SOL-A CP(T) = A + B*T + C/(T**2) + E/(T**3)

$SmCl_2$ SAMARIUM DICHLORIDE

		A	B	C	D	RANGE	REF.
SOL	CP	18.5	4.0			298-1013	*
	** MPT= 1013						1

PHASE	T	CP	H	S	G	BT	
SOL	298	19.693	-194.9	30.5	-203.994	149.551	19/*
	300	19.700	-194.864	30.622	-204.050	148.670	
	400	20.100	-192.874	36.344	-207.411	113.339	
	500	20.500	-190.844	40.872	-211.280	92.363	
	600	20.900	-188.774	44.645	-215.561	78.528	
	700	21.300	-186.664	47.897	-220.191	68.756	
	800	21.700	-184.514	50.767	-225.127	61.510	
	900	22.100	-182.324	53.346	-230.335	55.941	
	1000	22.500	-180.094	55.695	-235.789	51.539	
	1013	22.552	-179.801	55.986	-236.515	51.034	

* * * * * * * * * *

$SmCl_3$ SAMARIUM CHLORIDE

		A	B	C	D	RANGE	REF.
SOL	CP	19.66	11.4	0.18		298- 951	*
LIQ	CP	34.3				951-1200	*

PHASE	T	CP	H	S	G	BT	
SOL	298	23.261	-245.2	34.8	-255.576	187.367	19/*
	300	23.280	-245.157	34.944	-255.640	186.259	
	400	24.332	-242.777	41.784	-259.490	141.798	
	500	25.432	-240.289	47.331	-263.954	115.390	
	600	26.550	-237.690	52.066	-268.930	97.971	
	700	27.677	-234.979	56.243	-274.349	85.667	
	800	28.808	-232.154	60.013	-280.165	76.548	
	900	29.942	-229.217	63.472	-286.341	69.543	
	951	30.521	-227.675	65.138	-289.621	66.567	
			10.6	11.146			*
LIQ	951	34.300	-217.075	76.284	-289.621	66.567	
	1000	34.300	-215.394	78.007	-293.402	64.132	
	1100	34.300	-211.964	81.276	-301.368	59.884	
	1200	34.300	-208.534	84.261	-309.647	56.402	

SAMARIUM OXYFLUORICE

		A	B	C	D	RANGE	REF.
SOL-A	CP	18.355	2.32	-2.15		298- 797	*
SOL-B	CP	18.355	2.32	-2.15		797-1195	*
SOL-C	CP	21.43				1195-2000	*

PHASE	T	CP	H	S	G	BT	
SOL-A	298	16.628	-274.6	22.5	-281.308	206.232	210
	300	16.662	-274.569	22.603	-281.350	204.991	
	400	17.939	-272.832	27.593	-283.869	155.120	
	500	18.655	-270.999	31.679	-286.839	125.394	
	600	19.150	-269.108	35.126	-290.183	105.713	
	700	19.540	-267.173	38.108	-293.848	91.756	
	797	19.866	-265.261	40.665	-297.671	81.637	
			1.25	1.568			*
SOL-B	797	19.866	-264.011	42.233	-297.671	81.637	
	800	19.875	-263.952	42.308	-297.798	81.366	
	900	20.178	-261.949	44.667	-302.149	73.382	
	1000	20.460	-259.917	46.807	-306.724	67.044	
	1100	20.729	-257.857	48.770	-311.504	61.899	
	1195	20.977	-255.876	50.497	-316.220	57.840	
							*
SOL-C	1195	21.430	-255.876	50.497	-316.220	57.840	
	1200	21.430	-255.769	50.587	-316.473	57.645	
	1300	21.430	-253.626	52.302	-321.619	54.076	
	1400	21.430	-251.483	53.890	-326.929	51.043	
	1500	21.430	-249.340	55.369	-332.393	48.436	
	1600	21.430	-247.197	56.752	-338.000	46.175	
	1700	21.430	-245.054	58.051	-343.741	44.197	
	1800	21.430	-242.911	59.276	-349.608	42.454	
	1900	21.430	-240.768	60.435	-355.594	40.908	
	2000	21.430	-238.625	61.534	-361.693	39.529	

Sm_2O_3 SAMARIUM OXIDE

		A	B	C	D	RANGE	REF.
SOL-A	CP	30.75	4.64	-4.3		298-1195	1
	** MONOCLINIC						
SOL-B	CP	36.9				1195-2000	*
	** MPT= 2623						1

PHASE	T	CP	H	S	G	BT	
SOL-A	298	27.296	-437.9	36.1	-448.663	328.923	1
	300	27.364	-437.849	36.269	-448.730	326.944	
	400	29.918	-434.970	44.534	-452.784	247.423	
	500	31.350	-431.902	51.376	-457.590	200.039	
	600	32.340	-428.715	57.184	-463.025	168.679	
	700	33.120	-425.440	62.229	-469.001	146.448	
	800	33.790	-422.094	66.697	-475.452	129.905	
	900	34.395	-418.685	70.712	-482.325	117.140	
	1000	34.960	-415.217	74.365	-489.582	107.012	
	1100	35.499	-411.693	77.723	-497.189	98.796	
	1195	35.994	-408.297	80.684	-504.714	92.318	
			0.25	.209			1
SOL-B	1195	36.900	-408.047	80.893	-504.714	92.318	
	1200	36.900	-407.863	81.047	-505.119	92.007	
	1300	36.900	-404.173	84.001	-513.374	86.318	
	1400	36.900	-400.483	86.735	-521.912	81.485	
	1500	36.900	-396.793	89.281	-530.714	77.335	
	1600	36.900	-393.103	91.662	-539.763	73.738	
	1700	36.900	-389.413	93.899	-549.042	70.594	
	1800	36.900	-385.723	96.009	-558.538	67.825	
	1900	36.900	-382.033	98.004	-568.240	65.371	
	2000	36.900	-378.343	99.896	-578.136	63.184	

SAMARIUM CARBIDE SmC_2

		A	B	C	D	RANGE	REF.
SOL	CP	16.4	2.7	-1.9		298-1200	*

PHASE	T	CP	H	S	G	BT	
SOL	298	15.068	-23.4	18.6	-28.946	21.221	-
	300	15.099	-23.372	18.693	-28.980	21.115	
	400	16.292	-21.796	23.219	-31.084	16.986	
	500	16.990	-20.129	26.935	-33.597	14.687	
	600	17.492	-18.404	30.079	-36.452	13.279	
	700	17.902	-16.634	32.807	-39.599	12.365	
	800	18.263	-14.825	35.222	-43.003	11.749	
	900	18.595	-12.982	37.392	-46.635	11.326	
	1000	18.910	-11.107	39.368	-50.475	11.033	
	1100	19.213	-9.201	41.184	-54.504	10.830	
	1200	19.508	-7.265	42.869	-58.707	10.694	

* * * * * * * * * *

SAMARIUM DIZIRCONATE $Sm_2O_3 \cdot 2ZrO_2$

		A	B	C	D	RANGE	REF.
SOL	CP	64.0	8.2	-11.0		298-1200	*

PHASE	T	CP	H	S	G	BT	
SOL	298	54.070	-973.	60.2	-990.949	726.483	129/*
	300	54.238	-972.900	60.535	-991.060	722.084	
	400	60.405	-967.129	77.093	-997.967	545.337	
	500	63.700	-960.910	90.957	-1006.389	439.951	
	600	65.864	-954.426	102.773	-1016.090	370.160	
	700	67.495	-947.755	113.053	-1026.892	320.653	
	800	68.841	-940.936	122.156	-1038.662	283.787	
	900	70.022	-933.992	130.334	-1051.293	255.323	
	1000	71.100	-926.935	137.768	-1064.704	232.722	
	1100	72.111	-919.774	144.593	-1078.826	214.372	
	1200	73.076	-912.515	150.909	-1093.605	199.199	

Sn

TIN (WHITE)

		A	B	C	D	RANGE	REF.
SOL-B	CP	5.161	4.325			298- 505	21
	** GREY= WHITE TPT= 286.2, MPT= 505.06						21
LIQ	CP	5.148	1.469	3.079		505- 800	21
	CP	6.8				800-2876	21
GAS	CP	6.314				2876-2900	21

PHASE	T	CP	H	S	G	BT	P	
SOL-B	298	6.450	0	12.236	-3.648	2.675		21
	300	6.458	.012	12.276	-3.671	2.675		
	400	6.891	.679	14.193	-4.998	2.731		
	500	7.323	1.390	15.777	-6.499	2.841		
	505	7.345	1.427	15.851	-6.579	2.847		
			1.68	3.326				21
LIQ	505	7.097	3.107	19.177	-6.579	2.847		
	600	6.885	3.770	20.380	-8.458	3.081		
	700	6.805	4.453	21.434	-10.550	3.294		
	800	6.804	5.133	22.341	-12.740	3.481		
	800	6.800	5.133	22.341	-12.740	3.481		
	900	6.800	5.813	23.142	-15.015	3.647		
	1000	6.800	6.493	23.859	-17.366	3.796		
	1100	6.800	7.173	24.507	-19.785	3.931		
	1200	6.800	7.853	25.099	-22.265	4.056		
	1300	6.800	8.533	25.643	-24.803	4.170		
	1400	6.800	9.213	26.147	-27.393	4.277	.002	
	1500	6.800	9.893	26.616	-30.031	4.376	.009	
	1600	6.800	10.573	27.055	-32.715	4.469	.039	
	1700	6.800	11.253	27.467	-35.441	4.557	.143	
	1800	6.800	11.933	27.856	-38.207	4.640	.459	
	1900	6.800	12.613	28.223	-41.011	4.718	1.303	
	2000	6.800	13.293	28.572	-43.851	4.793	3.331	
	2100	6.800	13.973	28.904	-46.725	4.863	7.789	
	2200	6.800	14.653	29.220	-49.632	4.931	16.862	
	2300	6.800	15.333	29.523	-52.569	4.996	34.128	
	2400	6.800	16.013	29.812	-55.536	5.058	65.124	
	2500	6.800	16.693	30.090	-58.531	5.117	117.989	
	2600	6.800	17.373	30.356	-61.553	5.175	204.172	
	2700	6.800	18.053	30.613	-64.602	5.230	339.157	
	2800	6.800	18.733	30.860	-67.676	5.283	543.186	
	2876	6.800	19.250	31.042	-70.028	5.322	760.	
			70.689	24.579				21
GAS	2876	6.314	89.939	55.621	-70.028	5.322		
	2900	6.314	90.090	55.674	-71.364	5.379		

TIN TETRAHYDRIDE (GAS) SnH_4 (g)

		A	B	C	D	RANGE	REF.
GAS	CP	12.38	9.0	-2.7		298- 800	62,*

PHASE	T	CP	H	S	G	BT	
GAS	298	12.026	38.9	54.65	22.606	-16.573	62
	300	12.080	38.922	54.725	22.505	-16.397	
	400	14.292	40.250	58.530	16.838	-9.201	
	500	15.800	41.758	61.889	10.814	-4.727	
	600	17.030	43.401	64.881	4.473	-1.629	
	700	18.129	45.160	67.590	-2.153	.672	
	800	19.158	47.025	70.078	-9.038	2.469	

$SnCl_2$ TIN DICHLORIDE

		A	B	C	D	RANGE	REF.
SOL	CP	12.1	20.0			298- 520	*
LIQ	CP	23.0				520- 925	*
GAS	CP	13.55	0.23			925-1500	*

PHASE	T	CP	H	S	G	BT	
SOL	298	18.063	-79.1	31.5	-88.492	64.875	62/*
	300	18.100	-79.067	31.612	-88.550	64.517	
	400	20.100	-77.157	37.093	-91.994	50.270	
	500	22.100	-75.047	41.793	-95.943	41.942	
	520	22.500	-74.601	42.667	-96.788	40.684	
			3.47	6.673			170
LIQ	520	23.000	-71.131	49.340	-96.788	40.684	
	600	23.000	-69.291	52.632	-100.870	36.747	
	700	23.000	-66.991	56.177	-106.315	33.197	
	800	23.000	-64.691	59.248	-112.089	30.626	
	900	23.000	-62.391	61.958	-118.152	28.695	
	925	23.000	-61.816	62.588	-119.709	28.287	
			19.5	21.081			22
GAS	925	13.763	-42.316	83.669	-119.709	28.287	
	1000	13.780	-41.283	84.742	-126.025	27.546	
	1100	13.803	-39.904	86.057	-134.566	26.739	
	1200	13.826	-38.522	87.259	-143.233	26.090	
	1300	13.849	-37.138	88.366	-152.015	25.559	
	1400	13.872	-35.752	89.394	-160.903	25.122	
	1500	13.895	-34.364	90.351	-169.891	24.756	

TIN TETRACHLORIDE

$SnCl_4$

		A	B	C	D	RANGE	REF.
LIQ	CP	39.50				298- 385	62
	** MPT=	240					1
GAS	CP	25.57	0.20	-1.87		385-1000	62

PHASE	T	CP	H	S	G	BT	P	
LIQ	298	39.500	-126.4	61.9	-144.855	106.196	21.803	62
	300	39.500	-126.327	62.144	-144.970	105.625	24.177	
	385	39.500	-122.969	71.998	-150.689	85.552	760.	
			8.60	22.338				62,*
GAS	385	24.385	-114.369	94.336	-150.689	85.552		
	400	24.481	-114.003	95.270	-152.111	83.121		
	500	24.922	-111.530	100.785	-161.923	70.786		
	600	25.171	-109.025	105.353	-172.236	62.745		
	700	25.328	-106.499	109.245	-182.971	57.134		
	800	25.438	-103.961	112.635	-194.069	53.024		
	900	25.519	-101.413	115.636	-205.485	49.905		
	1000	25.583	-98.857	118.328	-217.186	47.472		

$SnBr_2$ TIN DIBROMIDE

		A	B	C	D	RANGE	REF.
SOL	CP	27.4	7.9			298- 505	*
LIQ	CP	23.8				505- 914	*
GAS	CP	14.0	0.22			914-1500	*

PHASE	T	CP	H	S	G	BT	
SOL	298	29.755	-62.14	35.8	-72.814	53.381	62/*
	300	29.770	-62.085	35.984	-72.880	53.100	
	400	30.560	-59.068	44.657	-76.931	42.039	
	500	31.350	-55.973	51.561	-81.753	35.739	
	505	31.389	-55.816	51.873	-82.012	35.497	
			4.3	8.515			170
LIQ	505	23.800	-51.516	60.388	-82.012	35.497	
	600	23.800	-49.255	64.490	-87.949	32.040	
	700	23.800	-46.875	68.159	-94.586	29.535	
	800	23.800	-44.495	71.337	-101.565	27.750	
	900	23.800	-42.115	74.140	-108.841	26.434	
	914	23.800	-41.782	74.508	-109.882	26.278	
			23.3	25.492			62
GAS	914	14.201	-18.482	100.000	-109.882	26.278	
	1000	14.220	-17.260	101.278	-118.538	25.910	
	1100	14.242	-15.837	102.634	-128.734	25.581	
	1200	14.264	-14.411	103.874	-139.061	25.330	
	1300	14.286	-12.984	105.017	-149.506	25.138	
	1400	14.308	-11.554	106.076	-160.061	24.990	
	1500	14.330	-10.122	107.064	-170.719	24.877	

TIN DIIODIDE

		A	B	C	D	RANGE	REF.
SOL	CP	16.8	7.0			298- 593	*
LIQ	CP	22.6				593- 991	*
GAS	CP	14.50	0.21			991-1500	*

PHASE	T	CP	H	S	G	BT	
SOL	298	18.887	-34.7	40.1	-46.656	34.204	62/*
	300	18.900	-34.665	40.217	-46.730	34.047	
	400	19.600	-32.740	45.750	-51.040	27.891	
	500	20.300	-30.745	50.199	-55.844	24.413	
	593	20.951	-28.827	53.716	-60.680	22.367	
			4.5	7.589			170
LIQ	593	22.600	-24.327	61.304	-60.680	22.367	
	600	22.600	-24.169	61.569	-61.110	22.262	
	700	22.600	-21.909	65.053	-67.446	21.060	
	800	22.600	-19.649	68.071	-74.105	20.247	
	900	22.600	-17.389	70.733	-81.048	19.684	
	991	22.600	-15.332	72.910	-87.586	19.318	
			24.0	24.218			62
GAS	991	14.708	8.668	97.128	-87.586	19.318	
	1000	14.710	8.800	97.261	-88.460	19.336	
	1100	14.731	10.272	98.664	-98.258	19.525	
	1200	14.752	11.747	99.946	-108.189	19.707	
	1300	14.773	13.223	101.128	-118.244	19.881	
	1400	14.794	14.701	102.223	-128.412	20.049	
	1500	14.815	16.182	103.245	-138.686	20.209	

SnO (g)

TIN MONOXIDE (GAS)

		A	B	C	D	RANGE	REF.
GAS	CP	8.42	0.32	-0.84		298-2000	58

PHASE	T	CP	H	S	G	BT	
GAS	298	7.570	4.985	55.45	-11.547	8.466	62
	300	7.583	4.999	55.497	-11.650	8.488	
	400	8.023	5.782	57.747	-17.317	9.463	
	500	8.244	6.597	59.563	-23.185	10.136	
	600	8.379	7.428	61.079	-29.219	10.645	
	700	8.473	8.271	62.378	-35.394	11.052	
	800	8.545	9.122	63.514	-41.690	11.391	
	900	8.604	9.980	64.524	-48.092	11.680	
	1000	8.656	10.843	65.434	-54.591	11.932	
	1100	8.703	11.711	66.261	-61.176	12.156	
	1200	8.746	12.583	67.020	-67.841	12.357	
	1300	8.786	13.460	67.722	-74.578	12.539	
	1400	8.825	14.340	68.374	-81.384	12.706	
	1500	8.863	15.225	68.984	-88.252	12.860	
	1600	8.899	16.113	69.557	-95.179	13.003	
	1700	8.935	17.004	70.098	-102.162	13.136	
	1800	8.970	17.900	70.610	-109.198	13.260	
	1900	9.005	18.798	71.096	-116.283	13.377	
	2000	9.039	19.701	71.558	-123.416	13.488	

TIN MONOSULFIDE (GAS)

SnS(g)

		A	B	C	D	RANGE	REF.
GAS	CP	8.83	0.08	-0.55		298-2000	58

PHASE	T	CP	H	S	G	BT	
GAS	298	8.235	25.112	57.876	7.856	-5.760	62*
	300	8.243	25.127	57.927	7.749	-5.646	
	400	8.518	25.967	60.342	1.831	-1.000	
	500	8.650	26.826	62.258	-4.303	1.881	
	600	8.725	27.695	63.842	-10.610	3.865	
	700	8.774	28.570	65.191	-17.063	5.328	
	800	8.808	29.450	66.365	-23.642	6.460	
	900	8.834	30.332	67.404	-30.332	7.367	
	1000	8.855	31.216	68.336	-37.120	8.114	
	1100	8.873	32.103	69.181	-43.996	8.742	
	1200	8.888	32.991	69.953	-50.953	9.281	
	1300	8.901	33.880	70.665	-57.985	9.749	
	1400	8.914	34.771	71.326	-65.085	10.162	
	1483	8.924	35.511	71.839	-71.026	10.469	
	1483	8.924	35.511	71.839	-71.026	10.469	
	1500	8.926	35.663	71.941	-72.248	10.528	
	1600	8.937	36.556	72.517	-79.472	10.857	
	1700	8.947	37.450	73.059	-86.751	11.154	
	1800	8.957	38.345	73.571	-94.083	11.425	
	1900	8.967	39.242	74.056	-101.464	11.673	
	2000	8.976	40.139	74.516	-108.893	11.901	

Sn_3S_4 3-TIN 4-SULFIDE

		A	B	C	D	RANGE	REF.
SOL	CP	36.08	14.9			298- 983	63
	** DEC.,MPT= 983						63

PHASE	T	CP	H	S	G	BT	
SOL	298	40.522	-88.5	58.2	-105.852	77.602	63
	300	40.550	-88.425	58.451	-105.960	77.202	
	400	42.040	-84.296	70.320	-112.424	61.434	
	500	43.530	-80.017	79.861	-119.948	52.436	
	600	45.020	-75.590	87.929	-128.347	46.757	
	700	46.510	-71.013	94.981	-137.500	42.935	
	800	48.000	-66.288	101.289	-147.319	40.251	
	900	49.490	-61.413	107.029	-157.739	38.309	
	983	50.727	-57.254	111.448	-166.808	37.091	

* * * * * * * * * *

Sn_2S_3 2-TIN 3-SULFIDE

		A	B	C	D	RANGE	REF.
SOL	CP	25.58	10.50			298-1000	63
	** DEC., 1/2SN2S3 = SN3S4 + 1/4S2						63

PHASE	T	CP	H	S	G	BT	
SOL	298	28.711	-63.	39.3	-74.717	54.777	63
	300	28.730	-62.947	39.478	-74.790	54.492	
	400	29.780	-60.021	47.887	-79.176	43.266	
	500	30.830	-56.991	54.645	-84.313	36.858	
	600	31.880	-53.855	60.358	-90.070	32.813	
	700	32.930	-50.615	65.352	-96.361	30.089	
	800	33.980	-47.269	69.817	-103.123	28.176	
	900	35.030	-43.819	73.880	-110.311	26.791	
	1000	36.080	-40.263	77.625	-117.889	25.768	

TIN DISULFIDE SnS_2

		A	B	C	D	RANGE	REF.
SOL	CP	15.51	4.20			298-1038	1
	** DEC., MPT= 1038						63

PHASE	T	CP	H	S	G	BT	
SOL	298	16.762	-36.7	20.9	-42.931	31.474	63/1
	300	16.770	-36.669	21.004	-42.970	31.308	
	400	17.190	-34.971	25.886	-45.325	24.768	
	500	17.610	-33.231	29.767	-48.114	21.034	
	600	18.030	-31.449	33.014	-51.258	18.673	
	700	18.450	-29.625	35.825	-54.703	17.081	
	800	18.870	-27.759	38.316	-58.412	15.960	
	900	19.290	-25.851	40.563	-62.358	15.145	
	1000	19.710	-23.901	42.617	-66.518	14.540	
	1038	19.870	-23.149	43.355	-68.152	14.351	

$SnSO_4$ TIN MONOSULFATE

		A	B	C	D	RANGE	REF.
SOL	CP	28.25	26.0			298- 900	*

PHASE	T	CP	H	S	G	BT	
SOL	298	36.002	-242.5	33.12	-252.375	185.021	19*/176
	300	36.050	-242.433	33.343	-252.436	183.924	
	400	38.650	-238.698	44.070	-256.326	140.069	
	500	41.250	-234.703	52.974	-261.190	114.182	
	600	43.850	-230.448	60.724	-266.883	97.225	
	700	46.450	-225.933	67.679	-273.309	85.342	
	800	49.050	-221.158	74.051	-280.399	76.612	
	900	51.650	-216.123	79.979	-288.104	69.971	

* * * * * * * * * *

$Sn(SO_4)_2$ TIN DISULFATE

		A	B	C	D	RANGE	REF.
SOL	CP	68.0				298- 700	*

PHASE	T	CP	H	S	G	BT	
SOL	298	68.000	-389.4	35.8	-400.074	293.301	19/*
	300	68.000	-389.274	36.221	-400.140	291.541	
	400	68.000	-382.474	55.783	-404.787	221.195	
	500	68.000	-375.674	70.957	-411.153	179.739	
	600	68.000	-368.874	83.355	-418.887	152.600	
	700	68.000	-362.074	93.837	-427.760	133.571	

TIN MONOSELENIDE

SnSe

		A	B	C	D	RANGE	REF.
SOL-B	CP	11.15	4.77			298- 813	78
	** TPT= 813, MPT= 1153						62,63

PHASE	T	CP	H	S	G	BT	P	
SOL-B	298	12.572	-22.6	23.45	-29.592	21.694		143
	300	12.581	-22.577	23.528	-29.635	21.592		
	400	13.058	-21.295	27.212	-32.180	17.585		
	500	13.535	-19.965	30.178	-35.054	15.324		
	600	14.012	-18.588	32.687	-38.200	13.916		
	700	14.489	-17.163	34.883	-41.581	12.984		
	800	14.966	-15.690	36.849	-45.169	12.341	.001	
	813	15.028	-15.495	37.091	-45.650	12.273	.002	

* * * * * * * * * *

TIN MONOSELENIDE (GAS)

SnSe (g)

		A	B	C	D	RANGE	REF.
GAS	CP	8.93		-0.31		298-2000	63

PHASE	T	CP	H	S	G	BT	
GAS	298	8.581	28.9	62.85	10.161	-7.449	63*
	300	8.586	28.916	62.903	10.045	-7.319	
	400	8.736	29.783	65.397	3.624	-1.981	
	500	8.806	30.661	67.355	-3.017	1.319	
	600	8.844	31.543	68.964	-9.835	3.583	
	700	8.867	32.429	70.329	-16.801	5.246	
	800	8.882	33.316	71.514	-23.895	6.529	
	900	8.892	34.205	72.561	-31.100	7.553	
	1000	8.899	35.095	73.498	-38.403	8.394	
	1100	8.904	35.985	74.346	-45.796	9.100	
	1200	8.908	36.875	75.121	-53.270	9.703	
	1300	8.912	37.766	75.834	-60.818	10.226	
	1400	8.914	38.658	76.495	-68.435	10.685	
	1500	8.916	39.549	77.110	-76.116	11.092	
	1600	8.918	40.441	77.686	-83.856	11.456	
	1700	8.919	41.333	78.226	-91.652	11.784	
	1800	8.920	42.225	78.736	-99.500	12.083	
	1900	8.921	43.117	79.218	-107.398	12.355	
	2000	8.922	44.009	79.676	-115.343	12.606	

$SnSe_2$ TIN DISELENIDE

		A	B	C	D	RANGE	REF.
SOL	CP	14.83	7.56			298- 948	*
	** MPT= 948						62,63

PHASE	T	CP	H	S	G	BT	
SOL	298	17.084	-29.8	28.2	-38.208	28.011	143*
	300	17.098	-29.768	28.306	-38.260	27.876	
	400	17.854	-28.021	33.328	-41.352	22.597	
	500	18.610	-26.198	37.393	-44.894	19.626	
	600	19.366	-24.299	40.853	-48.811	17.782	
	700	20.122	-22.324	43.895	-53.051	16.565	
	800	20.878	-20.274	46.631	-57.580	15.732	
	900	21.634	-18.149	49.134	-62.370	15.147	
	948	21.997	-17.102	50.268	-64.755	14.931	

TIN MONOTELLURIDE

SnTe

		A	B	C	D	RANGE	REF.
SOL	CP	11.7	2.42			298-1079	78
LIQ	CP	15.2				1079-1645	*
GAS	CP	8.94		-0.22		1645-2000	21

PHASE	T	CP	H	S	G	BT	P	
SOL	298	12.422	-14.5	23.6	-21.536	15.789		78/*
	300	12.426	-14.477	23.677	-21.580	15.723		
	400	12.668	-13.222	27.285	-24.136	13.189		
	500	12.910	-11.943	30.138	-27.012	11.809		
	600	13.152	-10.640	32.513	-30.148	10.983		
	700	13.394	-9.313	34.558	-33.504	10.462		
	800	13.636	-7.962	36.363	-37.052	10.123		
	900	13.878	-6.586	37.983	-40.770	9.902	.026	
	1000	14.120	-5.186	39.457	-44.643	9.758	.401	
	1079	14.311	-4.063	40.538	-47.804	9.684	2.391	
			10.8	10.009				78
LIQ	1079	15.200	6.737	50.547	-47.804	9.684	2.391	
	1100	15.200	7.056	50.840	-48.868	9.711	3.335	
	1200	15.200	8.576	52.163	-54.019	9.840	13.665	
	1300	15.200	10.096	53.380	-59.297	9.970	44.164	
	1400	15.200	11.616	54.506	-64.692	10.100	118.632	
	1500	15.200	13.136	55.555	-70.196	10.229	275.157	
	1600	15.200	14.656	56.536	-75.801	10.355	566.987	
	1645	15.200	15.340	56.957	-78.354	10.411	760.	
			33.874	20.592				*
GAS	1645	8.932	49.214	77.549	-78.354	10.411		
	1700	8.932	49.706	77.843	-82.628	10.624		
	1800	8.933	50.599	78.354	-90.438	10.982		
	1900	8.934	51.492	78.837	-98.298	11.308		
	2000	8.934	52.386	79.295	-106.204	11.607		

Sr

STRONTIUM

		A	B	C	D	RANGE	REF.
SOL-A	CP	5.31	3.32			298- 830	21
SOL-C	CP	3.03	6.40			830-1041	21
LIQ	CP	7.4				1041-1650	21
GAS	CP	4.974				1650-2000	21

PHASE	T	CP	H	S	G	BT	P	
SOL-A	298	6.300	0	12.5	-3.727	2.732		21
	300	6.306	.012	12.539	-3.750	2.732		
	400	6.638	.659	14.399	-5.101	2.787		
	500	6.970	1.339	15.915	-6.618	2.893		
	600	7.302	2.053	17.216	-8.276	3.015		
	700	7.634	2.800	18.366	-10.057	3.140		
	800	7.966	3.580	19.407	-11.946	3.264	.005	
	830	8.066	3.820	19.702	-12.533	3.300	.012	
			0.191	.230				21
SOL-C	830	8.342	4.011	19.932	-12.533	3.300	.012	
	900	8.790	4.611	20.626	-13.952	3.389	.076	
	1000	9.430	5.522	21.585	-16.063	3.511	.651	
	1041	9.692	5.914	21.969	-16.956	3.560	1.384	
			1.978	1.900				21
LIQ	1041	7.400	7.892	23.869	-16.956	3.560	1.384	
	1100	7.400	8.328	24.277	-18.377	3.652	3.512	
	1200	7.400	9.068	24.921	-20.837	3.795	13.708	
	1300	7.400	9.808	25.513	-23.359	3.928	43.055	
	1400	7.400	10.548	26.062	-25.938	4.050	114.064	
	1500	7.400	11.288	26.572	-28.570	4.163	263.829	
	1600	7.400	12.028	27.050	-31.251	4.269	546.716	
	1650	7.400	12.398	27.278	-32.610	4.320	760.	
			34.5	20.909				21
GAS	1650	4.974	46.898	48.187	-32.610	4.320		
	1700	4.974	47.147	48.335	-35.023	4.503		
	1800	4.974	47.644	48.619	-39.871	4.842		
	1900	4.974	48.142	48.888	-44.746	5.148		
	2000	4.974	48.639	49.144	-49.648	5.426		

STRONTIUM HYDRIDE

		A	B	C	D	RANGE	REF.
SOL	CP	8.0	5.4			298-1000	*

PHASE	T	CP	H	S	G	BT	
SOL	298	9.610	-43.	11.9	-46.548	34.125	1/*
	300	9.620	-42.982	11.959	-46.570	33.931	
	400	10.160	-41.993	14.801	-47.914	26.182	
	500	10.700	-40.950	17.126	-49.513	21.645	
	600	11.240	-39.853	19.125	-51.328	18.699	
	700	11.780	-38.702	20.898	-53.331	16.653	
	800	12.320	-37.497	22.506	-55.502	15.165	
	900	12.860	-36.238	23.988	-57.828	14.044	
	1000	13.400	-34.925	25.371	-60.296	13.180	

SrO

STRONTIUM OXIDE

		A	B	C	D	RANGE	REF.
SOL	CP	12.13	1.26	-1.55		298-1800	5
	** MPT= 2693						5

PHASE	T	CP	H	S	G	BT	
SOL	298	10.762	-141.5	13.	-145.376	106.578	19
	300	10.786	-141.480	13.067	-145.400	105.938	
	400	11.665	-140.352	16.305	-146.874	80.259	
	500	12.140	-139.160	18.964	-148.642	64.980	
	600	12.455	-137.929	21.207	-150.653	54.883	
	700	12.696	-136.671	23.145	-152.873	47.736	
	800	12.896	-135.391	24.854	-155.275	42.425	
	900	13.073	-134.093	26.383	-157.838	38.333	
	1000	13.235	-132.777	27.769	-160.547	35.092	
	1100	13.388	-131.446	29.038	-163.388	32.467	
	1200	13.534	-130.100	30.209	-166.351	30.301	
	1300	13.676	-128.740	31.298	-169.427	28.487	
	1400	13.815	-127.365	32.317	-172.608	26.949	
	1500	13.951	-125.977	33.274	-175.888	25.630	
	1600	14.085	-124.575	34.179	-179.261	24.489	
	1700	14.218	-123.160	35.037	-182.723	23.494	
	1800	14.350	-121.731	35.853	-186.267	22.619	

* * * * * * * * * *

SrO_2

STRONTIUM PEROXIDE

		A	B	C	D	RANGE	REF.
SOL	CP	17.68	4.4			298- 600	13,*
	** DPT= 427, SRO2 = SRO + 1/2O2						

PHASE	T	CP	H	S	G	BT	
SOL	298	18.992	-151.4	14.1	-155.604	114.076	19/*
	300	19.000	-151.365	14.218	-155.630	113.392	
	400	19.440	-149.443	19.744	-157.340	85.978	
	500	19.880	-147.477	24.129	-159.541	69.745	
	600	20.320	-145.467	27.792	-162.142	59.068	

STRONTIUM SULFIDE

		A	B	C	D	RANGE	REF.
SOL	CP	12.98	1.26	-1.55		298-2275	63
	** MPT= 2275						63

PHASE	T	CP	H	S	G	BT	
SOL	298	11.612	-108.2	16.3	-113.060	82.886	19,63
	300	11.636	-108.178	16.372	-113.090	82.397	
	400	12.515	-106.966	19.855	-114.908	62.791	
	500	12.990	-105.688	22.703	-117.040	51.165	
	600	13.305	-104.373	25.101	-119.433	43.509	
	700	13.546	-103.030	27.171	-122.049	38.111	
	800	13.746	-101.665	28.993	-124.859	34.115	
	900	13.923	-100.281	30.622	-127.842	31.048	
	1000	14.085	-98.881	32.098	-130.979	28.629	
	1100	14.238	-97.465	33.448	-134.257	26.678	
	1200	14.384	-96.033	34.693	-137.665	25.076	
	1300	14.526	-94.588	35.850	-141.193	23.740	
	1400	14.665	-93.128	36.931	-144.832	22.612	
	1500	14.801	-91.655	37.948	-148.577	21.651	
	1600	14.935	-90.168	38.907	-152.420	20.822	
	1700	15.068	-88.668	39.817	-156.356	20.104	
	1800	15.200	-87.155	40.682	-160.382	19.476	
	1900	15.331	-85.628	41.507	-164.491	18.923	
	2000	15.461	-84.088	42.297	-168.682	18.435	
	2100	15.591	-82.536	43.054	-172.950	18.002	
	2200	15.720	-80.970	43.783	-177.292	17.615	
	2275	15.817	-79.788	44.311	-180.595	17.351	

Sr_3N_2 STRONTIUM NITRIDE

		A	B	C	D	RANGE	REF.
SOL	CP	29.05	6.5			298-1303	*
	** MPT=	1303					1

PHASE	T	CP	H	S	G	BT	
SOL	298	30.988	-93.5	32.2	-103.100	75.585	19/*
	300	31.000	-93.443	32.392	-103.160	75.162	
	400	31.650	-90.310	41.399	-106.870	58.399	
	500	32.300	-87.113	48.531	-111.378	48.690	
	600	32.950	-83.850	54.478	-116.537	42.454	
	700	33.600	-80.523	59.606	-122.247	38.172	
	800	34.250	-77.130	64.135	-128.438	35.092	
	900	34.900	-73.673	68.206	-135.058	32.801	
	1000	35.550	-70.150	71.917	-142.067	31.053	
	1100	36.200	-66.563	75.336	-149.432	29.693	
	1200	36.850	-62.910	78.514	-157.126	28.620	
	1300	37.500	-59.193	81.489	-165.128	27.764	
	1303	37.519	-59.080	81.575	-165.373	27.741	

STRONTIUM CARBIDE

SrC_2

		A	B	C	D	RANGE	REF.
SOL	CP	16.4	2.7	-1.9		298-1000	*

PHASE	T	CP	H	S	G	BT	
SOL	298	15.068	-20.2	17.	-25.269	18.525	95
	300	15.099	-20.172	17.093	-25.300	18.434	
	400	16.292	-18.596	21.619	-27.244	14.887	
	500	16.990	-16.929	25.335	-29.597	12.939	
	600	17.492	-15.204	28.479	-32.292	11.764	
	700	17.902	-13.434	31.207	-35.279	11.016	
	800	18.263	-11.625	33.622	-38.523	10.525	
	900	18.595	-9.782	35.792	-41.995	10.199	
	1000	18.910	-7.907	37.768	-45.675	9.984	

* * * * * * * * * *

STRONTIUM CARBONATE

$SrCO_3$

		A	B	C	D	RANGE	REF.
SOL-A	CP	21.42	8.56	-3.396		298-1197	66
SOL-B	CP	34.0				1197-1445	*

** DPT= 1445, SRCO3 = SRO + CO2

PHASE	T	CP	H	S	G	BT	
SOL-A	298	20.152	-291.55	23.2	-298.467	218.812	66
	300	20.215	-291.513	23.325	-298.510	217.494	
	400	22.721	-289.354	29.518	-301.161	164.569	
	500	24.342	-286.997	34.771	-304.382	133.063	
	600	25.613	-284.497	39.325	-308.092	112.238	
	700	26.719	-281.880	43.358	-312.230	97.496	
	800	27.737	-279.156	46.993	-316.750	86.544	
	900	28.705	-276.334	50.316	-321.618	78.110	
	1000	29.640	-273.416	53.389	-326.805	71.433	
	1100	30.555	-270.406	56.257	-332.289	66.029	
	1197	31.429	-267.400	58.876	-337.874	61.698	
			4.7	3.926			66
SOL-B	1197	34.000	-262.700	62.802	-337.874	61.698	
	1200	34.000	-262.598	62.887	-338.063	61.578	
	1300	34.000	-259.198	65.609	-344.490	57.922	
	1400	34.000	-255.798	68.129	-351.178	54.829	
	1445	34.000	-254.268	69.204	-354.268	53.589	

$SrO \cdot SiO_2$ STRONTIUM METASILICATE

		A	B	C	D	RANGE	REF.
SOL	CP	27.90	2.65	-7.0		298-1300	5
	** MFT= 1853						5

PHASE	T	CP	H	S	G	BT	
SOL	298	20.815	-390.4	23.	-397.257	291.237	5
	300	20.917	-390.361	23.129	-397.300	289.472	
	400	24.585	-388.062	29.719	-399.950	218.552	
	500	26.425	-385.503	35.422	-403.214	176.268	
	600	27.546	-382.800	40.346	-407.008	148.273	
	700	28.326	-380.005	44.654	-411.263	128.419	
	800	28.926	-377.141	48.477	-415.923	113.640	
	900	29.421	-374.223	51.914	-420.945	102.233	
	1000	29.850	-371.259	55.036	-426.295	93.179	
	1100	30.236	-368.254	57.899	-431.944	85.831	
	1200	30.594	-365.213	60.546	-437.868	79.757	
	1300	30.931	-362.136	63.008	-444.047	74.661	

* * * * * * * * * *

$2SrO \cdot SiO_2$ STRONTIUM ORTHOSILICATE

		A	B	C	D	RANGE	REF.
SOL	CP	36.9	7.5	-7.0		298-1200	5

PHASE	T	CP	H	S	G	BT	
SOL	298	31.262	-550.7	35.8	-561.374	411.553	5
	300	31.372	-550.642	35.994	-561.440	409.064	
	400	35.525	-547.273	45.658	-565.536	309.036	
	500	37.850	-543.595	53.854	-570.523	249.409	
	600	39.456	-539.726	60.904	-576.269	209.934	
	700	40.721	-535.715	67.084	-582.674	181.944	
	800	41.806	-531.588	72.594	-589.663	161.110	
	900	42.786	-527.358	77.576	-597.176	145.034	
	1000	43.700	-523.033	82.131	-605.164	132.276	
	1100	44.571	-518.619	86.338	-613.590	121.926	
	1200	45.414	-514.120	90.252	-622.422	113.374	

STRONTIUM ALUMINATE $SrO \cdot Al_2O_3$

		A	B	C	D	RANGE	REF.
SOL-A	CP	42.35	1.18	-12.67		298- 932	80
SOL-B	CP	34.92	7.0			932-1600	80

PHASE	T	CP	H	S	G	BT	
SOL-A	298	28.449	-559.4	26.	-567.152	415.789	169,*
	300	28.626	-559.347	26.177	-567.200	413.261	
	400	34.903	-556.127	35.398	-570.286	311.632	
	500	37.872	-552.472	43.541	-574.243	251.035	
	600	39.539	-548.595	50.606	-578.958	210.914	
	700	40.590	-544.585	56.786	-584.334	182.462	
	800	41.314	-540.487	62.256	-590.292	161.282	
	900	41.848	-536.328	67.154	-596.767	144.934	
	932	41.991	-534.987	68.619	-598.939	140.467	
			0.46	.494			80
SOL-B	932	41.444	-534.527	69.112	-598.939	140.467	
	1000	41.920	-531.692	72.047	-603.739	131.965	
	1100	42.620	-527.465	76.076	-611.148	121.440	
	1200	43.320	-523.168	79.814	-618.945	112.740	
	1300	44.020	-518.801	83.309	-627.103	105.440	
	1400	44.720	-514.364	86.597	-635.600	99.235	
	1500	45.420	-509.857	89.706	-644.416	93.904	
	1600	46.120	-505.280	92.660	-653.536	89.281	

* * * * * * * * * *

STRONTIUM MOLYBDATE $SrO \cdot MoO_3$

		A	B	C	D	RANGE	REF.
SOL	CP	32.06	7.02	-5.486		298-1000	5

PHASE	T	CP	H	S	G	BT	
SOL	298	27.982	-370.3	30.8	-379.483	278.206	5
	300	28.070	-370.248	30.973	-379.540	276.532	
	400	31.439	-367.254	39.565	-383.080	209.333	
	500	33.376	-364.006	46.804	-387.408	169.359	
	600	34.748	-360.597	53.016	-392.406	142.953	
	700	35.854	-357.065	58.458	-397.986	124.273	
	800	36.819	-353.431	63.310	-404.078	110.404	
	900	37.701	-349.704	67.698	-410.632	99.728	
	1000	38.531	-345.892	71.713	-417.605	91.280	

$SrO \cdot WO_3$ STRONTIUM TUNGSTATE

		A	B	C	D	RANGE	REF.
SOL	CP	28.84	8.63			298-1100	163
	** MPT= 1808						163

PHASE	T	CP	H	S	G	BT	
SOL	298	31.413	-390.78	32.87	-400.580	293.673	5*
	300	31.429	-390.722	33.064	-400.641	291.906	
	400	32.292	-387.536	42.224	-404.425	220.998	
	500	33.155	-384.263	49.523	-409.025	178.809	
	600	34.018	-380.905	55.644	-414.291	150.926	
	700	34.881	-377.460	60.952	-420.127	131.187	
	800	35.744	-373.929	65.666	-426.462	116.520	
	900	36.607	-370.311	69.926	-433.245	105.220	
	1000	37.470	-366.607	73.828	-440.435	96.270	
	1100	38.333	-362.817	77.440	-448.001	89.022	

* * * * * * * * * *

$SrO \cdot HfO_2$ STRONTIUM HAFNATE

		A	B	C	D	RANGE	REF.
SOL	CP	29.2	3.3	-4.75		298-1900	*

PHASE	T	CP	H	S	G	BT	
SOL	298	24.840	-426.26	27.	-434.310	318.401	137/*
	300	24.912	-426.214	27.154	-434.360	316.474	
	400	27.551	-423.574	34.730	-437.466	239.053	
	500	28.950	-420.743	41.041	-441.264	192.902	
	600	29.861	-417.800	46.405	-445.643	162.347	
	700	30.541	-414.779	51.061	-450.521	140.678	
	800	31.098	-411.696	55.176	-455.837	124.546	
	900	31.584	-408.562	58.868	-461.542	112.093	
	1000	32.025	-405.381	62.218	-467.599	102.207	
	1100	32.437	-402.157	65.290	-473.977	94.183	
	1200	32.830	-398.894	68.130	-480.650	87.550	
	1300	33.209	-395.592	70.773	-487.596	81.983	
	1400	33.578	-392.253	73.247	-494.799	77.252	
	1500	33.939	-388.877	75.576	-502.241	73.186	
	1600	34.294	-385.465	77.778	-509.910	69.660	
	1700	34.646	-382.018	79.868	-517.793	66.576	
	1800	34.993	-378.536	81.858	-525.880	63.859	
	1900	35.338	-375.019	83.759	-534.161	61.451	

TANTALUM

		A	B	C	D	RANGE	REF.
SOL	CP	6.537	-0.222	-0.365		298- 500	21
	CP	5.980	0.590			500-1200	21
	CP	5.446	0.951	1.616		1200-1700	21
	CP	2.545	2.051	31.779		1700-2600	21
	CP	-15.572	6.748	431.684		2600-3000	21
	CP	-156.712	37.821	4744.833		3000-3287	21
LIQ	CP	10.				3287-4400	21
	** BPT= 5731, L= 177.609						21

PHASE	T	CP	H	S	G	BT	P	
SOL	298	6.060	0	9.92	-2.958	2.168		21
	300	6.065	.011	9.958	-2.976	2.168		
	400	6.220	.627	11.727	-4.064	2.221		
	500	6.280	1.252	13.123	-5.309	2.321		
	500	6.275	1.252	13.123	-5.309	2.321		
	600	6.334	1.883	14.272	-6.680	2.434		
	700	6.393	2.519	15.253	-8.158	2.547		
	800	6.452	3.161	16.110	-9.727	2.658		
	900	6.511	3.809	16.874	-11.377	2.763		
	1000	6.570	4.463	17.563	-13.099	2.863		
	1100	6.629	5.123	18.192	-14.887	2.958		
	1200	6.688	5.789	18.771	-16.736	3.048		
	1200	6.699	5.789	18.771	-16.736	3.048		
	1300	6.778	6.463	19.310	-18.640	3.134		
	1400	6.860	7.145	19.815	-20.597	3.216		
	1500	6.944	7.835	20.292	-22.602	3.294		
	1600	7.031	8.534	20.743	-24.654	3.368		
	1700	7.119	9.241	21.171	-26.750	3.439		
	1700	7.131	9.241	21.171	-26.750	3.439		
	1800	7.218	9.959	21.581	-28.888	3.508		
	1900	7.322	10.685	21.974	-31.066	3.574		
	2000	7.441	11.424	22.353	-33.282	3.637		
	2100	7.573	12.174	22.719	-35.536	3.699		
	2200	7.714	12.938	23.075	-37.826	3.758		
	2300	7.863	13.717	23.421	-40.151	3.816		
	2400	8.019	14.511	23.759	-42.510	3.872		

Ta

TANTALUM (CONTINUED)

PHASE	T	CP	H	S	G	BT	P	
SOL	2500	8.181	15.321	24.089	-44.902	3.926		
	2600	8.348	16.148	24.413	-47.327	3.979		
	2600	8.359	16.148	24.413	-47.327	3.979		
	2700	8.569	16.994	24.733	-49.785	4.030		
	2800	8.829	17.863	25.049	-52.274	4.081		
	2900	9.130	18.761	25.364	-54.794	4.130		
	3000	9.468	19.690	25.679	-57.346	4.178		
	3000	9.471	19.690	25.679	-57.346	4.178		
	3100	9.907	20.656	25.996	-59.930	4.226	.001	
	3200	10.651	21.682	26.321	-62.546	4.272	.003	
	3287	11.522	22.645	26.618	-64.849	4.312	.007	
			7.56	2.300				21
LIQ	3287	10.000	30.205	28.918	-64.849	4.312	.007	
	3300	10.000	30.335	28.958	-65.225	4.320	.007	
	3400	10.000	31.335	29.256	-68.136	4.380	.017	
	3500	10.000	32.335	29.546	-71.076	4.439	.035	
	3600	10.000	33.335	29.828	-74.045	4.496	.072	
	3700	10.000	34.335	30.102	-77.041	4.551	.142	
	3800	10.000	35.335	30.368	-80.065	4.605	.269	
	3900	10.000	36.335	30.628	-83.114	4.658	.493	
	4000	10.000	37.335	30.881	-86.190	4.710	.876	
	4100	10.000	38.335	31.128	-89.291	4.760	1.513	
	4200	10.000	39.335	31.369	-92.415	4.810	2.545	
	4300	10.000	40.335	31.604	-95.564	4.858	4.179	
	4400	10.000	41.335	31.834	-98.736	4.905	6.708	

TANTALUM PENTAFLUORIDE

TaF_5

		A	B	C	D	RANGE	REF.
SOL	CP	38.5				298- 368	32*
LIQ	CP	42.5				368- 502	*
GAS	CP	31.23	0.26	-5.900		502-2000	*

PHASE	T	CP	H	S	G	BT	P	
SOL	298	38.500	-454.9	41.	-467.124	342.457	.014	65/*
	300	38.500	-454.829	41.238	-467.200	340.401	.018	
	368	38.500	-452.211	49.104	-470.281	279.331	6.282	
			4.5	12.228				32
LIQ	368	42.500	-447.711	61.332	-470.281	279.331	6.282	
	400	42.500	-446.351	64.876	-472.301	258.088	28.385	
	500	42.500	-442.101	74.359	-479.280	209.521	724.377	
	502	42.500	-442.016	74.529	-479.429	208.752	760.	
			12.1	24.104				*
GAS	502	29.019	-429.916	98.633	-479.429	208.752		
	600	29.747	-427.033	103.876	-489.359	178.273		
	700	30.208	-424.034	108.499	-499.983	156.123		
	800	30.516	-420.997	112.554	-511.040	139.628		
	900	30.736	-417.933	116.162	-522.479	126.892		
	1000	30.900	-414.851	119.409	-534.260	116.778		
	1100	31.028	-411.755	122.360	-546.351	108.564		
	1200	31.132	-408.646	125.064	-558.724	101.771		
	1300	31.219	-405.529	127.560	-571.357	96.067		
	1400	31.293	-402.403	129.876	-584.230	91.215		
	1500	31.358	-399.270	132.038	-597.327	87.042		
	1600	31.416	-396.132	134.063	-610.633	83.420		
	1700	31.468	-392.988	135.969	-624.135	80.249		
	1800	31.516	-389.838	137.769	-637.823	77.453		
	1900	31.561	-386.684	139.475	-651.686	74.971		
	2000	31.602	-383.526	141.094	-665.715	72.756		

$TaCl_5$ TANTALUM PENTACHLORIDE

		A	B	C	D	RANGE	REF.
SOL	CP	33.0	0.	-3.0		298- 490	33,*
	** MPT=	489.65					32
LIQ	CP	44.0				490- 506	*
GAS	CP	31.6		-3.7		506-2000	33,*

PHASE	T	CP	H	S	G	BT	P	
SOL	298	29.625	-205.3	55.909	-221.969	162.730		33,*
	300	29.667	-205.245	56.092	-222.073	161.802		
	400	31.125	-202.195	64.857	-228.138	124.665	2.863	
	490	31.749	-199.374	71.218	-234.246	104.567	490.108	
			8.906	18.189				32,*
LIQ	490	44.000	-190.468	89.407	-234.246	104.567	490.108	
	500	44.000	-190.013	90.327	-235.176	102.809	649.895	
	506	44.000	-189.749	90.852	-235.720	101.825	760.	
			13.107	25.903				32,33,*
GAS	506	30.155	-176.642	116.755	-235.720	101.825		
	600	30.572	-173.786	121.931	-246.944	89.962		
	700	30.845	-170.714	126.666	-259.380	80.993		
	800	31.022	-167.620	130.797	-272.257	74.387		
	900	31.143	-164.511	134.458	-285.524	69.344		
	1000	31.230	-161.392	137.744	-299.137	65.385		
	1100	31.294	-158.266	140.724	-313.062	62.208		
	1200	31.343	-155.134	143.449	-327.273	59.613		
	1300	31.381	-151.998	145.959	-341.745	57.460		
	1400	31.411	-148.858	148.286	-356.459	55.653		
	1500	31.436	-145.716	150.454	-371.397	54.120		
	1600	31.455	-142.571	152.484	-386.545	52.807		
	1700	31.472	-139.425	154.391	-401.890	51.673		
	1800	31.486	-136.277	156.190	-417.419	50.688		
	1900	31.498	-133.128	157.893	-433.124	49.827		
	2000	31.507	-129.977	159.509	-448.995	49.071		

TANTALUM PENTABROMIDE

		A	B	C	D	RANGE	REF.
SOL	CP	28.48	29.3			298- 513	32
LIQ	CP	44.				513- 595	*
	** BPT= 595, L= 14.0						32

PHASE	T	CP	H	S	G	BT	
SOL	298	37.216	-143.	73.	-164.765	120.792	65/32
	300	37.270	-142.931	73.230	-164.900	120.146	
	400	40.200	-139.058	84.354	-172.799	94.426	
	500	43.130	-134.891	93.639	-181.710	79.436	
	513	43.511	-134.328	94.751	-182.935	77.945	
			9.0	17.544			32
LIQ	513	44.000	-125.328	112.294	-182.935	77.945	
	595	44.000	-121.720	118.819	-192.417	70.686	

TaJ_5 TANTALUM PENTAIODIDE

		A	B	C	D	RANGE	REF.
SOL	CP	30.98	20.87			298- 640	32
LIQ	CP	44.				640- 670	*
GAS	CP	28.05	0.78			670-2000	*

PHASE	T	CP	H	S	G	BT	
SOL	298	37.202	-79.70	82.	-104.148	76.353	32
	300	37.241	-79.631	82.230	-104.300	75.993	
	400	39.328	-75.803	93.230	-113.095	61.800	
	500	41.415	-71.766	102.230	-122.880	53.718	
	600	43.502	-67.520	109.965	-133.499	48.633	
	640	44.337	-65.763	112.799	-137.954	47.116	
			10.0	15.625			32
LIQ	640	44.000	-55.763	128.424	-137.954	47.116	
	670	44.000	-54.443	130.440	-141.838	46.273	
			15.5	23.134			32
GAS	670	28.573	-38.943	153.574	-141.838	46.273	
	700	28.596	-38.085	154.826	-146.464	45.734	
	800	28.674	-35.222	158.650	-162.142	44.301	
	900	28.752	-32.351	162.032	-178.179	43.274	
	1000	28.830	-29.471	165.065	-194.536	42.522	
	1100	28.908	-26.585	167.816	-211.183	41.964	
	1200	28.986	-23.690	170.335	-228.092	41.547	
	1300	29.064	-20.787	172.658	-245.243	41.235	
	1400	29.142	-17.877	174.815	-262.618	41.002	
	1500	29.220	-14.959	176.828	-280.201	40.831	
	1600	29.298	-12.033	178.716	-297.979	40.708	
	1700	29.376	-9.099	180.495	-315.941	40.622	
	1800	29.454	-6.158	182.176	-334.075	40.568	
	1900	29.532	-3.209	183.771	-352.373	40.538	
	2000	29.610	-.251	185.288	-370.827	40.528	

TANTALUM OXYTRICHLORIDE $TaOCl_3$

		A	B	C	D	RANGE	REF.
SOL	CP	31.9	0.	-2.9		298- 600	*

PHASE	T	CP	H	S	G	BT	P	
SOL	298	28.638	-213.3	42.4	-225.942	165.642		65
	300	28.678	-213.247	42.577	-226.020	164.678		
	400	30.087	-210.299	51.049	-230.718	126.076	.016	
	500	30.740	-207.254	57.841	-236.174	103.246	9.024	
	600	31.094	-204.160	63.480	-242.249	88.251	570.745	

* * * * * * * * * *

TANTALUM DISULFIDE TaS_2

		A	B	C	D	RANGE	REF.
SOL	CP	16.7				298- 600	63,*

PHASE	T	CP	H	S	G	BT	
SOL	298	16.700	-84.6	18.0	-89.967	65.956	63
	300	16.700	-84.569	18.103	-90.000	65.574	
	400	16.700	-82.899	22.908	-92.062	50.307	
	500	16.700	-81.229	26.634	-94.546	41.332	
	600	16.700	-79.559	29.679	-97.366	35.470	

Ta_2N DITANTALUM NITRIDE

		A	B	C	D	RANGE	REF.
SOL	CP	16.845	4.22	-1.69		298-3000	65
LIQ	CP	22.5				3000-4000	10

PHASE	T	CP	H	S	G	BT	
SOL	298	16.202	-65.2	17.8	-70.507	51.690	65/*
	300	16.233	-65.170	17.900	-70.540	51.395	
	400	17.477	-63.479	22.758	-72.582	39.662	
	500	18.279	-61.689	26.748	-75.063	32.814	
	600	18.908	-59.828	30.138	-77.911	28.383	
	700	19.454	-57.910	33.095	-81.076	25.317	
	800	19.957	-55.939	35.726	-84.520	23.093	
	900	20.434	-53.919	38.104	-88.213	21.424	
	1000	20.896	-51.853	40.281	-92.134	20.138	
	1100	21.347	-49.740	42.294	-96.264	19.128	
	1200	21.792	-47.583	44.170	-100.588	18.322	
	1300	22.231	-45.382	45.932	-105.094	17.670	
	1400	22.667	-43.137	47.595	-109.771	17.138	
	1500	23.100	-40.849	49.174	-114.610	16.701	
	1600	23.531	-38.518	50.679	-119.603	16.339	
	1700	23.961	-36.143	52.118	-124.744	16.039	
	1800	24.389	-33.725	53.500	-130.025	15.789	
	1900	24.816	-31.265	54.830	-135.442	15.581	
	2000	25.243	-28.762	56.114	-140.989	15.409	
	2100	25.669	-26.217	57.355	-146.663	15.265	
	2200	26.094	-23.629	58.559	-152.459	15.147	
	2300	26.519	-20.998	59.729	-158.374	15.051	
	2400	26.944	-18.325	60.866	-164.404	14.973	
	2500	27.368	-15.609	61.975	-170.546	14.911	
	2600	27.792	-12.851	63.056	-176.798	14.863	
	2700	28.216	-10.051	64.113	-183.157	14.827	
	2800	28.639	-7.208	65.147	-189.620	14.802	
	2900	29.063	-4.323	66.159	-196.185	14.787	
	3000	29.486	-1.395	67.152	-202.851	14.780	
			22.	7.333			10
LIQ	3000	22.500	20.605	74.485	-202.851	14.780	
	3100	22.500	22.855	75.223	-210.337	14.831	
	3200	22.500	25.105	75.937	-217.895	14.884	
	3300	22.500	27.355	76.630	-225.523	14.938	
	3400	22.500	29.605	77.301	-233.220	14.993	
	3500	22.500	31.855	77.954	-240.983	15.050	
	3600	22.500	34.105	78.587	-248.810	15.107	
	3700	22.500	36.355	79.204	-256.700	15.165	
	3800	22.500	38.605	79.804	-264.650	15.223	
	3900	22.500	40.855	80.388	-272.660	15.281	
	4000	22.500	43.105	80.958	-280.728	15.340	

TANTALUM MONONITRIDE

TaN

		A	B	C	D	RANGE	REF.
SOL	CP	13.21	0.65	-3.02		298-3363	10,*
LIQ	CP	15.0				3363-3500	10

PHASE	T	CP	H	S	G	BT	
SOL	298	10.006	-60.3	10.2	-63.341	46.437	65,*
	300	10.049	-60.281	10.262	-63.360	46.164	
	400	11.582	-59.189	13.393	-64.547	35.271	
	500	12.327	-57.990	16.066	-66.023	28.863	
	600	12.761	-56.734	18.355	-67.747	24.680	
	700	13.049	-55.443	20.345	-69.684	21.759	
	800	13.258	-54.127	22.102	-71.808	19.620	
	900	13.422	-52.793	23.673	-74.099	17.996	
	1000	13.558	-51.443	25.095	-76.538	16.730	
	1100	13.675	-50.082	26.393	-79.113	15.720	
	1200	13.780	-48.709	27.587	-81.813	14.902	
	1300	13.876	-47.326	28.694	-84.628	14.229	
	1400	13.966	-45.934	29.726	-87.549	13.669	
	1500	14.051	-44.533	30.692	-90.571	13.198	
	1600	14.132	-43.124	31.601	-93.686	12.799	
	1700	14.211	-41.706	32.461	-96.889	12.458	
	1800	14.287	-40.282	33.275	-100.177	12.165	
	1900	14.361	-38.849	34.049	-103.543	11.912	
	2000	14.434	-37.409	34.788	-106.985	11.692	
	2100	14.507	-35.962	35.494	-110.500	11.501	
	2200	14.578	-34.508	36.170	-114.083	11.335	
	2300	14.648	-33.047	36.820	-117.733	11.189	
	2400	14.718	-31.579	37.445	-121.446	11.061	
	2500	14.787	-30.103	38.047	-125.221	10.948	
	2600	14.855	-28.621	38.628	-129.055	10.850	
	2700	14.924	-27.132	39.190	-132.946	10.763	
	2800	14.991	-25.637	39.734	-136.892	10.686	
	2900	15.059	-24.134	40.262	-140.892	10.619	
	3000	15.126	-22.625	40.773	-144.944	10.561	
	3100	15.194	-21.109	41.270	-149.047	10.509	
	3200	15.261	-19.586	41.754	-153.198	10.464	
	3300	15.327	-18.057	42.224	-157.397	10.425	
	3363	15.369	-17.090	42.515	-160.066	10.404	
			16.	4.758			10
LIQ	3363	15.000	-1.090	47.272	-160.066	10.404	
	3400	15.000	-.535	47.436	-161.818	10.403	
	3500	15.000	.965	47.871	-166.584	10.403	

Ta_2C DITANTALIUM CARBIDE

		A	B	C	D	RANGE	REF.
SOL	CP	15.88	3.33	-2.05		298-3773	10
	** MPT= 3773 (PERITEC.)						10

PHASE	T	CP	H	S	G	ΘT	
SOL	298	14.567	-48.5	20.	-54.463	39.928	65/10
	300	14.601	-48.473	20.090	-54.500	39.709	
	400	15.931	-46.939	24.493	-56.737	31.004	
	500	16.725	-45.304	28.139	-59.374	25.956	
	600	17.309	-43.601	31.242	-62.346	22.713	
	700	17.793	-41.845	33.948	-65.609	20.487	
	800	18.224	-40.044	36.352	-69.126	18.887	
	900	18.624	-38.202	38.522	-72.871	17.698	
	1000	19.005	-36.320	40.504	-76.824	16.792	
	1100	19.374	-34.401	42.333	-80.967	16.089	
	1200	19.734	-32.446	44.034	-85.286	15.535	
	1300	20.088	-30.455	45.627	-89.770	15.094	
	1400	20.437	-28.428	47.129	-94.409	14.740	
	1500	20.784	-26.367	48.551	-99.193	14.454	
	1600	21.128	-24.272	49.903	-104.117	14.224	
	1700	21.470	-22.142	51.194	-109.172	14.037	
	1800	21.811	-19.978	52.431	-114.354	13.886	
	1900	22.150	-17.780	53.619	-119.657	13.765	
	2000	22.489	-15.548	54.764	-125.076	13.670	
	2100	22.827	-13.282	55.870	-130.608	13.594	
	2200	23.164	-10.982	56.939	-136.249	13.537	
	2300	23.500	-8.649	57.976	-141.995	13.494	
	2400	23.836	-6.282	58.984	-147.843	13.465	
	2500	24.172	-3.882	59.964	-153.791	13.446	
	2600	24.508	-1.448	60.918	-159.835	13.437	
	2700	24.843	1.020	61.849	-165.974	13.436	
	2800	25.178	3.521	62.759	-172.204	13.443	
	2900	25.513	6.055	63.648	-178.525	13.456	
	3000	25.847	8.623	64.519	-184.933	13.474	
	3100	26.182	11.225	65.372	-191.428	13.497	
	3200	26.516	13.859	66.208	-198.007	13.525	
	3300	26.850	16.528	67.029	-204.669	13.556	
	3400	27.184	19.229	67.836	-211.412	13.591	
	3500	27.518	21.965	68.629	-218.236	13.629	
	3600	27.852	24.733	69.409	-225.138	13.670	
	3700	28.186	27.535	70.176	-232.117	13.712	
	3773	28.430	29.602	70.729	-237.260	13.745	

TaC

TANTALUM MONOCARBICE

		A	B	C	D	RANGE	REF.
SOL	CP	10.347	2.713	-2.103		298-4000	10,65
	** TAC(0.85-1.0) / MPT= 4273, L= 25.						1/10

PHASE	T	CP	H	S	G	BT	
SOL	298	8.790	-34.2	10.1	-37.211	27.280	65
	300	8.824	-34.184	10.154	-37.230	27.126	
	400	10.118	-33.229	12.891	-38.386	20.976	
	500	10.862	-32.178	15.235	-39.795	17.397	
	600	11.391	-31.064	17.264	-41.422	15.090	
	700	11.817	-29.903	19.053	-43.240	13.502	
	800	12.189	-28.702	20.656	-45.227	12.357	
	900	12.529	-27.466	22.111	-47.366	11.504	
	1000	12.850	-26.197	23.448	-49.645	10.851	
	1100	13.157	-24.897	24.687	-52.052	10.343	
	1200	13.457	-23.566	25.845	-54.580	9.942	
	1300	13.749	-22.206	26.934	-57.219	9.621	
	1400	14.038	-20.816	27.963	-59.964	9.362	
	1500	14.323	-19.398	28.941	-62.810	9.153	
	1600	14.606	-17.952	29.875	-65.751	8.982	
	1700	14.886	-16.477	30.769	-68.784	8.844	
	1800	15.165	-14.974	31.627	-71.904	8.731	
	1900	15.443	-13.444	32.455	-75.108	8.641	
	2000	15.720	-11.886	33.254	-78.394	8.568	
	2100	15.997	-10.300	34.028	-81.758	8.510	
	2200	16.272	-8.686	34.778	-85.199	8.465	
	2300	16.547	-7.045	35.508	-88.713	8.431	
	2400	16.822	-5.377	36.218	-92.299	8.406	
	2500	17.096	-3.681	36.910	-95.956	8.390	
	2600	17.370	-1.958	37.586	-99.681	8.380	
	2700	17.643	-.207	38.246	-103.473	8.377	
	2800	17.917	1.571	38.893	-107.330	8.379	
	2900	18.190	3.376	39.527	-111.251	8.385	
	3000	18.463	5.209	40.148	-115.235	8.396	
	3100	18.735	7.069	40.758	-119.280	8.410	
	3200	19.008	8.956	41.357	-123.386	8.428	
	3300	19.281	10.870	41.946	-127.551	8.448	
	3400	19.553	12.812	42.525	-131.775	8.472	
	3500	19.825	14.781	43.096	-136.056	8.497	
	3600	20.098	16.777	43.658	-140.394	8.524	
	3700	20.370	18.800	44.213	-144.787	8.553	
	3800	20.642	20.851	44.760	-149.236	8.584	
	3900	20.914	22.929	45.299	-153.739	8.616	
	4000	21.186	25.034	45.832	-158.296	8.650	

Ta_2Si 2-TANTALIUM SILICON

		A	B	C	D	RANGE	REF.
SOL	CP	17.3	2.6	-1.4		298-2000	*

PHASE	T	CP	H	S	G	BT	
SOL	298	16.500	-30.	25.2	-37.513	27.502	221
	300	16.524	-29.969	25.302	-37.560	27.366	
	400	17.465	-28.265	30.199	-40.345	22.046	
	500	18.040	-26.488	34.162	-43.569	19.047	
	600	18.471	-24.662	37.490	-47.156	17.179	
	700	18.834	-22.796	40.365	-51.052	15.941	
	800	19.161	-20.896	42.902	-55.218	15.087	
	900	19.467	-18.965	45.177	-59.624	14.481	
	1000	19.760	-17.003	47.243	-64.246	14.043	
	1100	20.044	-15.013	49.140	-69.067	13.724	
	1200	20.323	-12.994	50.896	-74.069	13.492	
	1300	20.597	-10.948	52.533	-79.242	13.324	
	1400	20.869	-8.875	54.070	-84.573	13.204	
	1500	21.138	-6.775	55.519	-90.053	13.122	
	1600	21.405	-4.648	56.891	-95.674	13.070	
	1700	21.672	-2.494	58.197	-101.429	13.041	
	1800	21.937	-.313	59.443	-107.311	13.031	
	1900	22.201	1.894	60.637	-113.316	13.036	
	2000	22.465	4.127	61.782	-119.437	13.053	

5-TANTALUM 3-SILICON

Ta_5Si_3

		A	B	C	D	RANGE	REF.
SOL	CP	42.95	9.35	-2.13		298-2000	*
	** MPT= 2773						1

PHASE	T	CP	H	S	G	BT	
SOL	298	43.342	-80.	67.1	-100.006	73.316	221
	300	43.388	-79.920	67.368	-100.130	72.955	
	400	45.359	-75.475	80.141	-107.532	58.760	
	500	46.773	-70.866	90.421	-116.076	50.744	
	600	47.968	-66.128	99.056	-125.561	45.742	
	700	49.060	-61.275	106.534	-135.849	42.420	
	800	50.097	-56.317	113.153	-146.840	40.120	
	900	51.102	-51.257	119.112	-158.458	38.484	
	1000	52.087	-46.098	124.547	-170.645	37.299	
	1100	53.059	-40.840	129.557	-183.353	36.434	
	1200	54.022	-35.486	134.215	-196.544	35.800	
	1300	54.979	-30.036	138.577	-210.186	35.340	
	1400	55.931	-24.490	142.686	-224.251	35.012	
	1500	56.880	-18.850	146.578	-238.716	34.786	
	1600	57.827	-13.114	150.279	-253.560	34.639	
	1700	58.771	-7.284	153.813	-268.766	34.557	
	1800	59.714	-1.360	157.199	-284.318	34.526	
	1900	60.656	4.658	160.453	-300.202	34.536	
	2000	61.597	10.771	163.588	-316.405	34.580	

$TaSi_2$ TANTALUM 2-SILICON

		A	B	C	D	RANGE	REF.
SOL	CP	17.51	1.843	-2.165		298-2200	221
	** MPT= 2473						221

PHASE	T	CP	H	S	G	BT	
SOL	298	15.624	-28.466	13.47	-32.482	23.813	221
	300	15.657	-28.437	13.567	-32.507	23.685	
	400	16.894	-26.802	18.262	-34.107	18.638	
	500	17.565	-25.076	22.110	-36.131	15.795	
	600	18.014	-23.296	25.355	-38.509	14.029	
	700	18.358	-21.477	28.158	-41.188	12.861	
	800	18.646	-19.626	30.629	-44.129	12.057	
	900	18.901	-17.749	32.840	-47.305	11.489	
	1000	19.136	-15.847	34.844	-50.691	11.080	
	1100	19.358	-13.922	36.678	-54.268	10.783	
	1200	19.571	-11.975	38.372	-58.022	10.569	
	1300	19.778	-10.008	39.947	-61.938	10.414	
	1400	19.980	-8.020	41.420	-66.007	10.306	
	1500	20.178	-6.012	42.805	-70.219	10.232	
	1600	20.374	-3.984	44.113	-74.566	10.187	
	1700	20.568	-1.937	45.354	-79.040	10.163	
	1800	20.761	.129	46.536	-83.635	10.156	
	1900	20.952	2.215	47.663	-88.345	10.163	
	2000	21.142	4.320	48.743	-93.166	10.182	
	2100	21.331	6.443	49.779	-98.092	10.210	
	2200	21.520	8.586	50.775	-103.120	10.245	

TANTALUM 2-CHROMIUM

$TaCr_2$

		A	B	C	D	RANGE	REF.
SOL	CP	17.65	5.45	-1.72		298-1700	141
	** MPT= 2300 (APPROX.)						21

PHASE	T	CP	H	S	G	BT	
SOL	298	17.340	-6.45	21.05	-12.726	9.330	141
	300	17.374	-6.418	21.157	-12.765	9.301	
	400	18.755	-4.605	26.362	-15.150	8.279	
	500	19.687	-2.681	30.652	-18.007	7.872	
	600	20.442	-.674	34.310	-21.260	7.745	
	700	21.114	1.404	37.512	-24.854	7.761	
	800	21.741	3.548	40.373	-28.751	7.855	
	900	22.343	5.752	42.968	-32.920	7.995	
	1000	22.928	8.016	45.353	-37.337	8.161	
	1100	23.503	10.337	47.565	-41.985	8.343	
	1200	24.071	12.716	49.635	-46.846	8.533	
	1300	24.633	15.151	51.584	-51.908	8.728	
	1400	25.192	17.642	53.430	-57.159	8.924	
	1500	25.749	20.189	55.187	-62.590	9.121	
	1600	26.303	22.792	56.866	-68.194	9.316	
	1700	26.855	25.450	58.477	-73.961	9.510	

$TaFe_2$ TANTALUM 2-IRON

		A	B	C	D	RANGE	REF.
SOL	CP	16.0	6.3			298-1300	65

PHASE	T	CP	H	S	G	BT	
SOL	298	17.878	-13.8	25.5	-21.403	15.691	65
	300	17.890	-13.767	25.611	-21.450	15.628	
	400	18.520	-11.946	30.844	-24.284	13.270	
	500	19.150	-10.063	35.044	-27.585	12.059	
	600	19.780	-8.116	38.591	-31.271	11.392	
	700	20.410	-6.107	41.687	-35.288	11.019	
	800	21.040	-4.034	44.454	-39.598	10.819	
	900	21.670	-1.899	46.968	-44.170	10.728	
	1000	22.300	.300	49.284	-48.985	10.707	
	1100	22.930	2.561	51.439	-54.022	10.735	
	1200	23.560	4.886	53.461	-59.268	10.796	
	1300	24.190	7.273	55.372	-64.711	10.880	

TERBIUM

		A	B	C	D	RANGE	REF.
SOL-A	CP	7.127	-0.744			298- 500	21
	CP	6.872	0.223	-1.576	1.631	500-1300	21
	CP	3.083	5.184			1300-1560	21
SOL-B	CP	6.63				1560-1630	21
LIQ	CP	11.11				1630-2500	21
	** BPT= 3496, L= 79.080						21

PHASE	T	CP	H	S	G	BT	P	
SOL-A	298	6.905	0	17.52	-5.224	3.830		21
	300	6.904	.013	17.563	-5.256	3.830		
	400	6.829	.699	19.539	-7.116	3.889		
	500	6.755	1.379	21.055	-9.149	3.999		
	500	6.761	1.379	21.055	-9.149	3.999		
	600	7.155	2.075	22.323	-11.319	4.123		
	700	7.506	2.808	23.453	-13.609	4.249		
	800	7.848	3.576	24.477	-16.006	4.373		
	900	8.199	4.378	25.422	-18.501	4.493		
	1000	8.568	5.216	26.305	-21.088	4.609		
	1100	8.961	6.093	27.139	-23.761	4.721		
	1200	9.379	7.009	27.937	-26.515	4.830		
	1300	9.825	7.969	28.705	-29.347	4.934		
	1300	9.822	7.969	28.705	-29.347	4.934		
	1400	10.341	8.977	29.452	-32.255	5.036		
	1500	10.859	10.037	30.183	-35.237	5.135		
	1560	11.170	10.698	30.615	-37.061	5.193		
			1.2	.769				21
SOL-B	1560	6.630	11.898	31.384	-37.061	5.193		
	1600	6.630	12.164	31.552	-38.320	5.235		
	1630	6.630	12.362	31.675	-39.268	5.266		
			2.58	1.583				21
LIQ	1630	11.110	14.942	33.258	-39.268	5.266		
	1700	11.110	15.720	33.725	-41.613	5.350	.002	
	1800	11.110	16.831	34.360	-45.017	5.467	.010	
	1900	11.110	17.942	34.961	-48.484	5.578	.035	
	2000	11.110	19.053	35.531	-52.008	5.684	.107	
	2100	11.110	20.164	36.073	-55.589	5.786	.295	
	2200	11.110	21.275	36.590	-59.222	5.884	.736	
	2300	11.110	22.386	37.083	-62.906	5.978	1.692	
	2400	11.110	23.497	37.556	-66.638	6.069	3.617	
	2500	11.110	24.608	38.010	-70.417	6.157	7.253	

$TbCl_3$ TERBIUM TRICHLORIDE

		A	B	C	D	RANGE	REF.
SOL-A	CP	22.494	6.155	-0.737		298- 783	89
SOL-B	CP	29.62				783- 855	89
LIQ	CP	34.53				855-1000	89

PHASE	T	CP	H	S	G	BT	
SOL-A	298	23.500	-238.3	35.3	-248.825	182.418	19/*
	300	23.522	-238.257	35.445	-248.890	181.341	
	400	24.495	-235.853	42.353	-252.794	138.139	
	500	25.277	-233.364	47.905	-257.316	112.488	
	600	25.982	-230.800	52.577	-262.346	95.572	
	700	26.652	-228.168	56.632	-267.811	83.626	
	783	27.193	-225.934	59.649	-272.638	76.109	
			3.4	4.342			89
SOL-B	783	29.620	-222.534	63.991	-272.638	76.109	
	800	29.620	-222.030	64.627	-273.732	74.790	
	855	29.620	-220.401	66.596	-277.341	70.902	
			4.65	5.439			89
LIQ	855	34.530	-215.751	72.035	-277.341	70.902	
	900	34.530	-214.197	73.806	-280.623	68.154	
	1000	34.530	-210.744	77.444	-288.188	62.992	

DITERBIUM TRIOXIDE Tb_2O_3

		A	B	C	D	RANGE	REF.
SOL	CP	25.77	9.9	-2.2		298-1200	*

PHASE	T	CP	H	S	G	BT	
SOL	298	26.247	-445.5	37.5	-456.681	334.801	94/191
	300	26.296	-445.451	37.663	-456.750	332.787	
	400	28.355	-442.711	45.531	-460.924	251.871	
	500	29.840	-439.799	52.024	-465.811	203.633	
	600	31.099	-436.751	57.578	-471.298	171.693	
	700	32.251	-433.582	62.460	-477.304	149.041	
	800	33.346	-430.302	66.838	-483.773	132.178	
	900	34.408	-426.914	70.827	-490.659	119.164	
	1000	35.450	-423.421	74.507	-497.928	108.837	
	1100	36.478	-419.825	77.934	-505.552	100.457	
	1200	37.497	-416.126	81.151	-513.508	93.535	

* * * * * * * * * *

TERBIUM 1.72-OXIDE $TbO_{1.72}$

		A	B	C	D	RANGE	REF.
SOL	CP	13.95	4.75	-1.8		298-1200	*

PHASE	T	CP	H	S	G	BT	
SOL	298	13.341	-227.8	19.3	-233.554	171.223	94/191
	300	13.375	-227.775	19.383	-233.590	170.193	
	400	14.725	-226.364	23.433	-235.737	128.818	
	500	15.605	-224.845	26.819	-238.255	104.155	
	600	16.300	-223.249	29.727	-241.085	87.827	
	700	16.908	-221.588	32.286	-244.188	76.249	
	800	17.469	-219.869	34.581	-247.534	67.632	
	900	18.003	-218.095	36.669	-251.098	60.983	
	1000	18.520	-216.269	38.593	-254.862	55.708	
	1100	19.026	-214.392	40.382	-258.812	51.428	
	1200	19.525	-212.464	42.059	-262.935	47.893	

$TbO_{1.83}$ TERBIUM 1.83-OXIDE

		A	B	C	D	RANGE	REF.
SOL	CP	14.52	4.52	-1.52		298-1200	*

PHASE	T	CP	H	S	G	BT	
SOL	298	14.158	-230.2	19.4	-235.984	173.004	94/191
	300	14.187	-230.174	19.488	-236.020	171.964	
	400	15.378	-228.690	23.747	-238.189	130.158	
	500	16.172	-227.111	27.268	-240.745	105.244	
	600	16.810	-225.461	30.275	-243.626	88.753	
	700	17.374	-223.751	32.909	-246.788	77.061	
	800	17.898	-221.987	35.264	-250.198	68.360	
	900	18.400	-220.172	37.401	-253.833	61.647	
	1000	18.888	-218.308	39.365	-257.673	56.322	
	1100	19.366	-216.395	41.188	-261.701	52.002	
	1200	19.838	-214.435	42.893	-265.906	48.435	

* * * * * * * * * *

TbO_2 TERBIUM DIOXIDE

		A	B	C	D	RANGE	REF.
SOL	CP	15.49	4.23	-1.815		298-1300	*

PHASE	T	CP	H	S	G	BT	
SOL	298	14.709	-232.2	19.8	-238.103	174.558	94/191
	300	14.742	-232.173	19.891	-238.140	173.508	
	400	16.048	-230.627	24.329	-240.359	131.344	
	500	16.879	-228.978	28.004	-242.981	106.221	
	600	17.524	-227.257	31.141	-245.942	89.596	
	700	18.081	-225.476	33.885	-249.196	77.813	
	800	18.590	-223.643	36.333	-252.709	69.046	
	900	19.073	-221.759	38.550	-256.455	62.284	
	1000	19.538	-219.829	40.584	-260.413	56.921	
	1100	19.993	-217.852	42.468	-264.566	52.572	
	1200	20.440	-215.830	44.226	-268.902	48.980	
	1300	20.882	-213.764	45.880	-273.408	45.970	

TECHNETIUM DIOXIDE

TcO_2

		A	B	C	D	RANGE	REF.
SOL	CP	16.4	2.75	-3.48		298-1200	*

PHASE	T	CP	H	S	G	BT	
SOL	298	13.305	-103.5	14.	-107.674	78.938	1/*
	300	13.358	-103.475	14.082	-107.700	78.470	
	400	15.325	-102.029	18.230	-109.321	59.738	
	500	16.383	-100.439	21.773	-111.326	48.667	
	600	17.083	-98.764	24.825	-113.659	41.406	
	700	17.615	-97.028	27.500	-116.278	36.309	
	800	18.056	-95.244	29.882	-119.149	32.554	
	900	18.445	-93.419	32.031	-122.247	29.690	
	1000	18.802	-91.556	33.993	-125.549	27.442	
	1100	19.137	-89.659	35.801	-129.040	25.641	
	1200	19.458	-87.729	37.480	-132.705	24.172	

* * * * * * * * * *

TECHNETIUM TRIOXIDE

TcO_3

		A	B	C	D	RANGE	REF.
SOL	CP	25.8				298- 400	*
	** DISP.						1

PHASE	T	CP	H	S	G	BT	
SOL	298	25.800	-129.	17.	-134.069	98.288	1
	300	25.800	-128.952	17.160	-134.100	97.705	
	400	25.800	-126.372	24.582	-136.205	74.429	

Tc_2O_7 DITECHNETIUM HEPTOXIDE

		A	B	C	D	RANGE	REF.
SOL	CP	57.0				298- 392	*
LIQ	CP	60.0				392- 585	*
	** BPT= 585, L= 14.						1

PHASE	T	CP	H	S	G	BT	
SOL	298	57.000	-266.	44.	-279.119	204.627	1/*
	300	57.000	-265.895	44.353	-279.200	203.425	
	392	57.000	-260.651	59.599	-284.013	158.366	
			11.5	29.337			1
LIQ	392	60.000	-249.151	88.936	-284.013	158.366	
	400	60.000	-248.671	90.148	-284.730	155.590	
	500	60.000	-242.671	103.536	-294.439	128.716	
	585	60.000	-237.571	112.957	-303.650	113.456	

		A	B	C	D	RANGE	REF.
SOL	CP	4.57	5.28			298- 723	63
	** MPT= 722.65						63
LIQ	CP	9.0				723-1282	63
	** BPT=1282.4 GAS (TE+TE2)						15

PHASE	T	CP	H	S	G	BT	P	
SOL	298	6.144	0	11.83	-3.527	2.586		63
	300	6.154	.011	11.868	-3.549	2.586		
	400	6.682	.653	13.711	-4.831	2.640		
	500	7.210	1.348	15.259	-6.281	2.746		
	600	7.738	2.095	16.620	-7.877	2.869	.002	
	700	8.266	2.895	17.852	-9.601	2.998	.147	
	723	8.387	3.087	18.121	-10.015	3.028	.329	
			4.18	5.784				63
LIQ	723	9.000	7.267	23.906	-10.009	3.027	.329	
	800	9.000	7.963	24.821	-11.894	3.250	2.029	
	900	9.000	8.863	25.881	-14.430	3.504	12.742	
	1000	9.000	9.763	26.829	-17.066	3.730	52.897	
	1100	9.000	10.663	27.687	-19.793	3.933	163.354	
	1200	9.000	11.563	28.470	-22.601	4.117	405.662	
	1282	9.000	12.305	29.068	-24.972	4.256	760.	

Te (g)

TELLURIUM (MONATOMIC GAS)

		A	B	C	D	RANGE	REF.
GAS	CP	4.64	0.44	0.18		298-2000	63

PHASE	T	CP	H	S	G	BT	
GAS	298	4.974	50.6	43.64	37.589	-27.557	63
	300	4.972	50.609	43.671	37.508	-27.328	
	400	4.928	51.104	45.093	33.066	-18.069	
	500	4.932	51.596	46.193	28.500	-12.459	
	600	4.954	52.091	47.094	23.834	-8.683	
	700	4.985	52.587	47.860	19.086	-5.960	
	800	5.020	53.088	48.528	14.266	-3.898	
	900	5.058	53.592	49.121	9.383	-2.279	
	1000	5.098	54.099	49.656	4.443	-.971	
	1100	5.139	54.611	50.144	-.547	.109	
	1200	5.180	55.127	50.593	-5.584	1.017	
	1282	5.215	55.556	50.938	-9.768	1.665	
	1282	5.215	55.556	50.938	-9.768	1.665	
	1300	5.223	55.647	51.009	-10.665	1.793	
	1400	5.265	56.172	51.398	-15.785	2.465	
	1500	5.308	56.700	51.763	-20.943	3.052	
	1600	5.351	57.233	52.106	-26.137	3.571	
	1700	5.394	57.771	52.432	-31.364	4.033	
	1800	5.438	58.312	52.742	-36.623	4.447	
	1900	5.481	58.858	53.037	-41.912	4.822	
	2000	5.524	59.408	53.319	-47.230	5.162	

TELLURIUM (DIATOMIC GAS) Te_2 (g)

		A	B	C	D	RANGE	REF.
GAS	CP	8.28	1.581	-0.061		298-1283	63
GAS	CP	8.28	1.591	-0.061		1282-2000	63

PHASE	T	CP	H	S	G	BT	
GAS	298	8.683	38.03	62.283	19.460	-14.267	63*
	300	8.687	38.046	62.337	19.345	-14.095	
	400	8.874	38.924	64.862	12.980	-7.093	
	500	9.046	39.820	66.861	6.390	-2.793	
	600	9.212	40.733	68.525	-.382	.139	
	700	9.374	41.663	69.957	-7.307	2.282	
	800	9.535	42.608	71.219	-14.367	3.926	
	900	9.695	43.570	72.352	-21.547	5.233	
	1000	9.855	44.547	73.381	-28.834	6.303	
	1100	10.014	45.541	74.328	-36.220	7.197	
	1200	10.173	46.550	75.206	-43.698	7.959	
	1282	10.304	47.394	75.886	-49.923	8.509	
	1282	10.317	47.394	75.886	-49.923	8.509	
	1300	10.345	47.575	76.027	-51.260	8.619	
	1400	10.504	48.618	76.800	-58.901	9.196	
	1500	10.664	49.676	77.530	-66.618	9.708	
	1600	10.823	50.751	78.223	-74.406	10.165	
	1700	10.983	51.841	78.884	-82.262	10.577	
	1800	11.142	52.947	79.516	-90.182	10.951	
	1900	11.301	54.069	80.123	-98.164	11.293	
	2000	11.460	55.207	80.707	-106.206	11.607	

TeF (g)

TELLURIUM MONOFLUORIDE (GAS)

		A	B	C	D	RANGE	REF.
GAS	CP	8.777	0.09	-7.759		298-2000	63

PHASE	T	CP	H	S	G	BT	
GAS	298	.075	-20.8	57.66	-37.991	27.852	63
	300	.183	-20.800	57.661	-38.098	27.758	
	400	3.964	-20.565	58.309	-43.889	23.983	
	500	5.718	-20.072	59.404	-49.773	21.759	
	600	6.676	-19.448	60.539	-55.771	20.317	
	700	7.257	-18.749	61.615	-61.879	19.322	
	800	7.637	-18.003	62.610	-68.091	18.604	
	900	7.900	-17.225	63.526	-74.399	18.069	
	1000	8.091	-16.425	64.368	-80.794	17.660	
	1100	8.235	-15.609	65.147	-87.270	17.341	
	1200	8.346	-14.779	65.868	-93.821	17.089	
	1300	8.435	-13.940	66.540	-100.442	16.888	
	1400	8.507	-13.093	67.168	-107.128	16.726	
	1500	8.567	-12.239	67.757	-113.874	16.594	
	1600	8.618	-11.380	68.311	-120.678	16.486	
	1700	8.662	-10.516	68.835	-127.536	16.398	
	1800	8.700	-9.648	69.331	-134.444	16.326	
	1900	8.733	-8.776	69.803	-141.401	16.267	
	2000	8.763	-7.901	70.251	-148.404	16.219	

TELLURIUM DIFLUORIDE (GAS) TeF_2 (g)

		A	B	C	D	RANGE	REF.
GAS	CP	13.424	0.255	-2.247		298-2000	63

PHASE	T	CP	H	S	G	BT	
GAS	298	10.972	-92.0	65.75	-111.603	81.818	63
	300	11.004	-91.980	65.818	-111.725	81.403	
	400	12.122	-90.816	69.159	-118.479	64.743	
	500	12.653	-89.574	71.927	-125.538	54.880	
	600	12.953	-88.293	74.263	-132.850	48.397	
	700	13.144	-86.987	76.275	-140.380	43.834	
	800	13.277	-85.666	78.039	-148.097	40.464	
	900	13.376	-84.333	79.609	-155.981	37.882	
	1000	13.454	-82.991	81.023	-164.014	35.850	
	1100	13.519	-81.642	82.308	-172.181	34.214	
	1200	13.574	-80.288	83.487	-180.472	32.873	
	1300	13.623	-78.928	84.575	-188.876	31.757	
	1400	13.666	-77.563	85.586	-197.384	30.817	
	1500	13.707	-76.195	86.531	-205.991	30.017	
	1600	13.744	-74.822	87.416	-214.688	29.329	
	1700	13.780	-73.446	88.251	-223.472	28.733	
	1800	13.814	-72.066	89.039	-232.337	28.213	
	1900	13.846	-70.683	89.787	-241.279	27.757	
	2000	13.878	-69.297	90.498	-250.293	27.354	

$TeF_4(g)$ TELLURIUM TETRAFLUORIDE (GAS)

		A	B	C	D	RANGE	REF.
GAS	CP	24.89	0.496	-4.184		298-2000	63
	** MPT= 403, L= 6.35						63

PHASE	T	CP	H	S	G	BT	
GAS	298	20.331	-226.6	77.46	-249.695	183.056	63
	300	20.390	-226.562	77.586	-249.838	182.031	
	400	22.473	-224.405	83.779	-257.916	140.938	
	500	23.464	-222.103	88.912	-266.558	116.528	
	600	24.025	-219.726	93.244	-275.672	100.427	
	700	24.383	-217.304	96.976	-285.187	89.051	
	800	24.633	-214.853	100.249	-295.052	80.615	
	900	24.820	-212.380	103.162	-305.225	74.129	
	1000	24.968	-209.890	105.785	-315.675	69.000	
	1100	25.090	-207.387	108.170	-326.374	64.853	
	1200	25.195	-204.873	110.358	-337.302	61.439	
	1300	25.287	-202.348	112.378	-348.440	58.586	
	1400	25.371	-199.815	114.256	-359.773	56.171	
	1500	25.448	-197.274	116.009	-371.287	54.104	
	1600	25.520	-194.726	117.653	-382.971	52.318	
	1700	25.588	-192.170	119.203	-394.815	50.764	
	1800	25.654	-189.608	120.667	-406.809	49.400	
	1900	25.716	-187.040	122.056	-418.946	48.196	
	2000	25.777	-184.465	123.376	-431.218	47.128	

TELLURIUM PENTAFLUORIDE (GAS) $TeF_5 (g)$

		A	B	C	D	RANGE	REF.
GAS	CP	30.67	0.591	-6.05		298-2000	63
	** TE2F10: MPT= 240, BPT= 326						63

PHASE	T	CP	H	S	G	BT	
GAS	298	24.040	-277.2	81.45	-301.484	221.024	63
	300	24.125	-277.155	81.599	-301.635	219.771	
	400	27.125	-274.572	89.011	-310.176	169.495	
	500	28.545	-271.781	95.233	-319.397	139.627	
	600	29.344	-268.883	100.514	-329.192	119.924	
	700	29.849	-265.922	105.078	-339.476	106.004	
	800	30.197	-262.918	109.088	-350.189	95.680	
	900	30.455	-259.885	112.660	-361.279	87.742	
	1000	30.656	-256.829	115.880	-372.709	81.466	
	1100	30.820	-253.755	118.810	-384.446	76.393	
	1200	30.959	-250.666	121.497	-396.463	72.215	
	1300	31.080	-247.564	123.980	-408.738	68.724	
	1400	31.189	-244.450	126.288	-421.253	65.769	
	1500	31.288	-241.327	128.443	-433.991	63.241	
	1600	31.379	-238.193	130.465	-446.937	61.057	
	1700	31.465	-235.051	132.370	-460.080	59.155	
	1800	31.547	-231.900	134.171	-473.408	57.487	
	1900	31.625	-228.742	135.879	-486.911	56.015	
	2000	31.701	-225.575	137.503	-500.581	54.708	

TeF_6 (g)

TELLURIUM HEXAFLUORIDE (GAS)

		A	B	C	D	RANGE	REF.
GAS	CP	36.347	0.741	-7.58		298-2000	63
	** MPT= 235, L= 1.90						63

PHASE	T	CP	H	S	G	BT	
GAS	298	28.041	-327.2	80.28	-351.135	257.424	63
	300	28.147	-327.148	80.454	-351.284	255.945	
	400	31.906	-324.119	89.142	-359.776	196.599	
	500	33.685	-320.830	96.474	-369.067	161.341	
	600	34.686	-317.407	102.712	-379.034	138.082	
	700	35.319	-313.905	108.109	-389.581	121.649	
	800	35.755	-310.350	112.856	-400.634	109.463	
	900	36.078	-306.758	117.086	-412.135	100.094	
	1000	36.330	-303.137	120.901	-424.038	92.686	
	1100	36.536	-299.493	124.374	-436.304	86.697	
	1200	36.710	-295.831	127.560	-448.903	81.767	
	1300	36.862	-292.152	130.505	-461.808	77.647	
	1400	36.998	-288.459	133.242	-474.997	74.160	
	1500	37.122	-284.753	135.799	-488.451	71.177	
	1600	37.237	-281.035	138.198	-502.152	68.600	
	1700	37.344	-277.306	140.459	-516.085	66.356	
	1800	37.447	-273.566	142.596	-530.239	64.388	
	1900	37.545	-269.816	144.624	-544.601	62.652	
	2000	37.639	-266.057	146.552	-559.161	61.110	

TELLURIUM DICHLORIDE (GAS) $TeCl_2$ (g)

		A	B	C	D	RANGE	REF.
GAS	CP	13.869	0.022	-0.788		298-2000	63
	** MPT= 488, L= 2.706						63

PHASE	T	CP	H	S	G	BT	
GAS	298	12.989	-27.	73.029	-48.774	35.757	63
	300	13.000	-26.976	73.109	-48.909	35.635	
	400	13.385	-25.654	76.910	-56.418	30.829	
	500	13.565	-24.305	79.918	-64.265	28.094	
	600	13.663	-22.944	82.401	-72.384	26.369	
	700	13.724	-21.574	84.512	-80.732	25.209	
	800	13.763	-20.200	86.347	-89.277	24.393	
	900	13.792	-18.822	87.970	-97.995	23.800	
	1000	13.812	-17.442	89.424	-106.866	23.359	
	1100	13.828	-16.059	90.742	-115.875	23.025	
	1200	13.841	-14.676	91.945	-125.010	22.771	
	1300	13.851	-13.291	93.054	-134.261	22.574	
	1400	13.860	-11.906	94.080	-143.618	22.423	
	1500	13.867	-10.520	95.037	-153.075	22.306	
	1600	13.873	-9.133	95.932	-162.624	22.216	
	1700	13.879	-7.745	96.773	-172.259	22.148	
	1800	13.884	-6.357	97.567	-181.977	22.098	
	1900	13.889	-4.968	98.317	-191.771	22.062	
	2000	13.893	-3.579	99.030	-201.639	22.037	

$TeCl_4$ TELLURIUM TETRACHLORIDE

		A	B	C	D	RANGE	REF.
SOL	CP	33.1				298- 497	63
LIQ	CP	55.				497- 700	63
GAS	CP	23.115	0.037	-1.313		700-2000	63

PHASE	T	CP	H	S	G	BT	P	
SOL	298	33.100	-77.4	48.	-91.711	67.235		63
	300	33.100	-77.339	48.205	-91.800	66.885		
	400	33.100	-74.029	57.727	-97.120	53.071	.005	
	497	33.100	-70.818	64.914	-103.080	45.334	3.478	
			4.51	9.074				63
LIQ	497	55.000	-66.308	73.988	-103.080	45.334	3.478	
	500	55.000	-66.143	74.319	-103.303	45.160	3.956	
	600	55.000	-60.643	84.347	-111.251	40.529	106.537	
	700	55.000	-55.143	92.825	-120.121	37.508	760.	
			14.70	21.000				*
GAS	700	22.873	-40.443	113.825	-120.121	37.508		
	800	22.939	-38.152	116.884	-131.660	35.973		
	900	22.986	-35.856	119.589	-143.486	34.848		
	1000	23.021	-33.555	122.013	-155.568	34.004		
	1100	23.047	-31.252	124.208	-167.881	33.359		
	1200	23.068	-28.946	126.214	-180.403	32.860		
	1300	23.085	-26.638	128.061	-193.118	32.471		
	1400	23.100	-24.329	129.773	-206.011	32.164		
	1500	23.112	-22.019	131.367	-219.069	31.923		
	1600	23.123	-19.707	132.859	-232.281	31.732		
	1700	23.132	-17.394	134.261	-245.638	31.583		
	1800	23.141	-15.080	135.584	-259.131	31.467		
	1900	23.149	-12.766	136.835	-272.752	31.378		
	2000	23.156	-10.451	138.022	-286.495	31.311		

TELLURIUM TETRABROMIDE

$TeBr_4$

		A	B	C	D	RANGE	REF.
SOL	CP	26.7	14.3			298- 388	*
LIQ	CP	40.5				388- 700	*
	** DEC., TEBR4(S) = TEBR2(G) + BR2(G)						63

PHASE	T	CP	H	S	G	BT	
SOL	298	30.964	-45.5	58.2	-62.852	46.078	63/*
	300	30.990	-45.443	58.392	-62.960	45.873	
	388	32.248	-42.660	66.518	-68.469	38.572	
			5.9	15.206			*
LIQ	388	40.500	-36.760	81.724	-68.469	38.572	
	400	40.500	-36.274	82.958	-69.457	37.955	
	500	40.500	-32.224	91.995	-78.222	34.195	
	600	40.500	-28.174	99.379	-87.802	31.986	
	700	40.500	-24.124	105.622	-98.060	30.620	

TeO (g) TELLURIUM MONOXIDE (GAS)

		A	B	C	D	RANGE	REF.
GAS	CP	8.44	0.32	-0.83		298-2000	63

PHASE	T	CP	H	S	G	BT	
GAS	298	7.602	17.8	57.5	.656	-.481	63
	300	7.614	17.814	57.547	.550	-.401	
	400	8.049	18.600	59.805	-5.322	2.908	
	500	8.268	19.417	61.627	-11.397	4.982	
	600	8.401	20.251	63.147	-17.637	6.425	
	700	8.495	21.096	64.450	-24.019	7.500	
	800	8.566	21.949	65.589	-30.522	8.339	
	900	8.626	22.809	66.601	-37.132	9.018	
	1000	8.677	23.674	67.513	-43.839	9.582	
	1100	8.723	24.544	68.342	-50.632	10.061	
	1200	8.766	25.419	69.103	-57.505	10.475	
	1300	8.807	26.297	69.806	-64.451	10.837	
	1400	8.846	27.180	70.460	-71.465	11.158	
	1500	8.883	28.066	71.072	-78.542	11.445	
	1600	8.920	28.956	71.647	-85.678	11.705	
	1700	8.955	29.850	72.188	-92.870	11.941	
	1800	8.990	30.748	72.701	-100.115	12.157	
	1900	9.025	31.648	73.188	-107.409	12.357	
	2000	9.059	32.553	73.652	-114.752	12.541	

TELLURIUM MONOXIDE (DIMERIC GAS) Te_2O_2 (g)

		A	B	C	D	RANGE	REF.
GAS	CP	19.625	1.333	-2.821		298-2000	63

PHASE	T	CP	H	S	G	BT	
GAS	298	16.849	-26.	78.2	-49.315	36.154	63
	300	16.890	-25.969	78.304	-49.460	36.037	
	400	18.395	-24.195	83.398	-57.554	31.450	
	500	19.163	-22.313	87.593	-66.110	28.900	
	600	19.641	-20.372	91.132	-75.051	27.341	
	700	19.982	-18.390	94.186	-84.320	26.329	
	800	20.251	-16.377	96.873	-93.876	25.649	
	900	20.476	-14.341	99.271	-103.685	25.182	
	1000	20.676	-12.283	101.439	-113.722	24.857	
	1100	20.858	-10.206	103.419	-123.967	24.633	
	1200	21.029	-8.112	105.241	-134.401	24.481	
	1300	21.191	-6.001	106.930	-145.010	24.382	
	1400	21.347	-3.874	108.507	-155.783	24.322	
	1500	21.499	-1.731	109.985	-166.708	24.293	
	1600	21.648	.426	111.377	-177.777	24.286	
	1700	21.793	2.598	112.694	-188.981	24.298	
	1800	21.937	4.785	113.943	-200.314	24.325	
	1900	22.080	6.985	115.133	-211.768	24.362	
	2000	22.220	9.200	116.269	-223.339	24.409	

TeO_2 TELLURIUM DIOXIDE

		A	B	C	D	RANGE	REF.
SOL	CP	15.58	3.48	-1.2		298-1006	63
LIQ	CP	26.92	0.52			1006-1200	63

PHASE	T	CP	H	S	G	BT	P	
SOL	298	15.268	-77.3	17.7	-82.577	60.539		63
	300	15.291	-77.272	17.795	-82.610	60.189		
	400	16.222	-75.692	22.333	-84.625	46.243		
	500	16.840	-74.037	26.023	-87.049	38.054		
	600	17.335	-72.328	29.138	-89.811	32.718		
	700	17.771	-70.572	31.843	-92.863	28.997		
	800	18.176	-68.775	34.243	-96.169	26.276		
	900	18.564	-66.938	36.406	-99.703	24.215	.004	
	1000	18.940	-65.062	38.382	-103.444	22.611	.112	
	1006	18.962	-64.949	38.495	-103.675	22.526	.134	
			6.95	6.909				63
LIQ	1006	27.443	-57.999	45.404	-103.675	22.526	.134	
	1100	27.492	-55.417	47.857	-108.060	21.472	1.237	
	1200	27.544	-52.665	50.252	-112.967	20.577	8.540	

THORIUM

		A	B	C	D	RANGE	REF.
SOL-A	CP	5.63	3.04			298-1636	21
SOL-B	CP	11.				1636-2028	21
LIQ	CP	11.				2028-3500	21
	** BPT= 5061, L= 122.882						21

PHASE	T	CP	H	S	G	BT	P	
SOL-A	298	6.536	0	12.76	-3.804	2.789		21,40
	300	6.542	.012	12.800	-3.828	2.789		
	400	6.846	.681	14.724	-5.208	2.846		
	500	7.150	1.381	16.284	-6.761	2.956		
	600	7.454	2.111	17.615	-8.457	3.081		
	700	7.758	2.872	18.787	-10.279	3.210		
	800	8.062	3.663	19.843	-12.211	3.336		
	900	8.366	4.484	20.810	-14.244	3.459		
	1000	8.670	5.336	21.707	-16.371	3.578		
	1100	8.974	6.218	22.547	-18.584	3.693		
	1200	9.278	7.131	23.341	-20.878	3.803		
	1300	9.582	8.074	24.096	-23.251	3.909		
	1400	9.886	9.047	24.817	-25.697	4.012		
	1500	10.190	10.051	25.510	-28.213	4.111		
	1600	10.494	11.085	26.177	-30.798	4.207		
	1636	10.603	11.465	26.412	-31.744	4.241		
			0.654	.400				21
SOL-B	1636	11.000	12.119	26.811	-31.744	4.241		
	1700	11.000	12.823	27.234	-33.474	4.304		
	1800	11.000	13.923	27.862	-36.229	4.399		
	1900	11.000	15.023	28.457	-39.045	4.492		
	2000	11.000	16.123	29.021	-41.919	4.581		
	2028	11.000	16.431	29.174	-42.734	4.606		
			3.853	1.900				21
LIQ	2028	11.000	20.284	31.074	-42.734	4.606		
	2100	11.000	21.076	31.458	-44.985	4.682		
	2200	11.000	22.176	31.970	-48.157	4.785		
	2300	11.000	23.276	32.459	-51.378	4.883		
	2400	11.000	24.376	32.927	-54.648	4.977		
	2500	11.000	25.476	33.376	-57.963	5.068	.002	
	2600	11.000	26.576	33.807	-61.322	5.155	.005	
	2700	11.000	27.676	34.222	-64.724	5.240	.014	
	2800	11.000	28.776	34.622	-68.166	5.321	.032	
	2900	11.000	29.876	35.008	-71.648	5.400	.070	
	3000	11.000	30.976	35.381	-75.167	5.477	.146	
	3100	11.000	32.076	35.742	-78.724	5.551	.290	
	3200	11.000	33.176	36.091	-82.315	5.623	.550	
	3300	11.000	34.276	36.430	-85.942	5.692	1.004	
	3400	11.000	35.376	36.758	-89.601	5.760	1.768	
	3500	11.000	36.476	37.077	-93.293	5.826	3.010	

ThH_2 THORIUM DIHYDRIDE

		A	B	C	D	RANGE	REF.
SOL	CP	12.61	2.65			298-1000	*

PHASE	T	CP	H	S	G	BT	
SOL	298	13.400	-34.3	14.2	-38.534	28.250	65
	300	13.405	-34.275	14.283	-38.560	28.095	
	400	13.670	-32.921	18.176	-40.192	21.963	
	500	13.935	-31.541	21.254	-42.168	18.434	
	600	1 .200	-30.134	23.818	-44.426	16.184	
	700	14.465	-28.701	26.027	-46.920	14.651	
	800	14.730	-27.241	27.976	-49.622	13.558	
	900	14.995	-25.755	29.726	-52.509	12.753	
	1000	15.260	-24.242	31.320	-55.562	12.145	

THORIUM TETRAFLUORIDE

		A	B	C	D	RANGE	REF.
SOL	CP	26.75	5.854	-1.805		298-1383	65
LIQ	CP	36.5				1383-2055	65
GAS	CP	25.5	0.2	-3.3		2055-2200	*

PHASE	T	CP	H	S	G	BT	P	
SOL	298	26.465	-504.6	33.95	-514.722	377.352		65
	300	26.501	-504.551	34.114	-514.785	375.071		
	400	27.963	-501.822	41.956	-518.604	283.390		
	500	28.955	-498.973	48.307	-523.127	228.689		
	600	29.761	-496.037	53.660	-528.232	192.434		
	700	30.479	-493.024	58.302	-533.835	166.693		
	800	31.151	-489.942	62.416	-539.875	147.507		
	900	31.796	-486.795	66.123	-546.305	132.679		
	1000	32.423	-483.584	69.505	-553.089	120.894		
	1100	33.040	-480.310	72.625	-560.197	111.316		
	1200	33.649	-476.976	75.526	-567.607	103.389	.010	
	1300	34.253	-473.581	78.243	-575.296	96.729	.114	
	1383	34.752	-470.717	80.378	-581.880	91.964	.630	
			10.51	7.599				65
LIQ	1383	36.500	-460.207	87.978	-581.880	91.964	.630	
	1400	36.500	-459.586	88.424	-583.379	91.082	.831	
	1500	36.500	-455.936	90.942	-592.349	86.317	3.677	
	1600	36.500	-452.286	93.297	-601.562	82.181	13.210	
	1700	36.500	-448.636	95.510	-611.004	78.560	40.017	
	1800	36.500	-444.986	97.597	-620.660	75.369	105.294	
	1900	36.500	-441.336	99.570	-630.519	72.536	246.301	
	2000	36.500	-437.686	101.442	-640.571	70.008	521.744	
	2055	36.500	-435.679	102.432	-646.177	68.730	760.	
			55.7	27.105				65
GAS	2055	25.833	-379.979	129.537	-646.177	68.730		
	2100	25.845	-378.816	130.097	-652.019	67.866		
	2200	25.872	-376.230	131.300	-665.090	66.079		

$ThBr_4$ THORIUM TETRABROMIDE

		A	B	C	D	RANGE	REF.
SOL-A	CP	30.5	3.6	-1.47		298- 693	65
SOL-B	CP	30.5	3.6	-1.47		693- 952	65
LIQ	CP	41.0				952-1178	65
	** BPT= 1178, L= 26.5						65

PHASE	T	CP	H	S	G	BT	
SOL-A	298	29.920	-230.8	54.5	-247.049	181.116	65
	300	29.947	-230.745	54.685	-247.150	180.073	
	400	31.021	-227.691	63.462	-253.076	138.293	
	500	31.712	-224.553	70.463	-259.784	113.567	
	600	32.252	-221.354	76.294	-267.130	97.315	
	693	32.689	-218.334	80.972	-274.447	86.564	
			1.	1.443			65
SOL-B	693	32.689	-217.334	82.415	-274.447	86.564	
	700	32.720	-217.105	82.744	-275.025	85.878	
	800	33.150	-213.811	87.142	-283.524	77.466	
	900	33.559	-210.475	91.070	-292.438	71.023	
	952	33.765	-208.725	92.961	-297.223	68.243	
			13.	13.655			65
LIQ	952	41.000	-195.725	106.616	-297.223	68.243	
	1000	41.000	-193.757	108.633	-302.390	66.096	
	1100	41.000	-189.657	112.541	-313.452	62.285	
	1178	41.000	-186.459	115.349	-322.341	59.811	

THORIUM TETRAIODIDE ThJ_4

		A	B	C	D	RANGE	REF.
SOL	CP	31.0	3.1	-1.47		298- 843	65
LIQ	CP	42.0				843-1200	65
	** DEC.						219

PHASE	T	CP	H	S	G	BT	
SOL	298	30.271	-158.8	63.5	-177.733	130.299	65
	300	30.297	-158.744	63.687	-177.850	129.581	
	400	31.321	-155.658	72.558	-184.681	100.919	
	500	31.962	-152.492	79.620	-192.302	84.066	
	600	32.452	-149.270	85.492	-200.566	73.066	
	700	32.870	-146.004	90.527	-209.373	65.378	
	800	33.250	-142.698	94.941	-218.651	59.741	
	843	33.406	-141.265	96.686	-222.771	57.762	
			11.5	13.642			65
LIQ	843	42.000	-129.765	110.328	-222.771	57.762	
	900	42.000	-127.371	113.076	-229.139	55.650	
	1000	42.000	-123.171	117.501	-240.672	52.606	
	1100	42.000	-118.971	121.504	-252.625	50.199	
	1200	42.000	-114.771	125.158	-264.961	48.262	

$ThOF_2$ THORIUM OXYFLUORIDE

		A	B	C	D	RANGE	REF.
SOL	CP	21.66	4.04	-2.0		298-1500	65

PHASE	T	CP	H	S	G	BT	
SOL	298	20.615	-401.	24.2	-408.215	299.270	65
	300	20.650	-400.962	24.328	-408.260	297.457	
	400	22.026	-398.821	30.477	-411.012	224.597	
	500	22.880	-396.573	35.489	-414.318	181.123	
	600	23.528	-394.252	39.720	-418.084	152.307	
	700	24.080	-391.871	43.389	-422.243	131.848	
	800	24.579	-389.437	46.638	-426.748	116.598	
	900	25.049	-386.956	49.560	-431.560	104.811	
	1000	25.500	-384.428	52.223	-436.651	95.443	
	1100	25.939	-381.856	54.674	-441.997	87.829	
	1200	26.369	-379.241	56.949	-447.580	81.526	
	1300	26.794	-376.583	59.077	-453.382	76.231	
	1400	27.214	-373.882	61.078	-459.391	71.724	
	1500	27.631	-371.140	62.969	-465.594	67.846	

* * * * * * * * * *

$ThOBr_2$ THORIUMOXYBROMIDE

		A	B	C	D	RANGE	REF.
SOL	CP	23.53	2.916	-1.833		298-1500	65

PHASE	T	CP	H	S	G	BT	
SOL	298	22.337	-270.	31.3	-279.332	204.783	65
	300	22.368	-269.959	31.438	-279.390	203.563	
	400	23.551	-267.656	38.054	-282.878	154.578	
	500	24.255	-265.264	43.389	-286.958	125.446	
	600	24.770	-262.811	47.859	-291.527	106.203	
	700	25.197	-260.313	51.710	-296.510	92.587	
	800	25.576	-257.774	55.100	-301.854	82.474	
	900	25.928	-255.198	58.133	-307.518	74.686	
	1000	26.263	-252.589	60.882	-313.471	68.518	
	1100	26.586	-249.946	63.401	-319.687	63.524	
	1200	26.902	-247.272	65.727	-326.145	59.407	
	1300	27.212	-244.566	67.893	-332.827	55.961	
	1400	27.519	-241.829	69.921	-339.719	53.040	
	1500	27.823	-239.062	71.830	-346.807	50.537	

THORIUM OXYIODIDE

$ThOJ_2$

		A	B	C	D	RANGE	REF.
SOL	CP	23.78	2.666	-1.833		298-1500	65

PHASE	T	CP	H	S	G	BT	
S L	298	22.513	-237.2	34.6	-247.516	181.459	65
	300	22.543	-237.158	34.739	-247.580	180.386	
	400	23.701	-234.840	41.402	-251.400	137.377	
	500	24.380	-232.433	46.768	-255.818	111.833	
	600	24.870	-229.970	51.258	-260.725	94.982	
	700	25.272	-227.462	55.123	-266.049	83.075	
	800	25.626	-224.917	58.521	-271.734	74.244	
	900	25.953	-222.338	61.559	-277.741	67.454	
	1000	26.263	-219.727	64.309	-284.036	62.084	
	1100	26.561	-217.086	66.827	-290.595	57.744	
	1200	26.852	-214.415	69.150	-297.395	54.170	
	1300	27.137	-211.716	71.311	-304.419	51.184	
	1400	27.419	-208.988	73.332	-311.653	48.658	
	1500	27.698	-206.232	75.233	-319.082	46.496	

ThO (g)

THORIUM MONOXIDE (GAS)

		A	B	C	D	RANGE	REF.
GAS	CP	6.64	2.5			298-2700	65

PHASE	T	CP	H	S	G	BT	
GAS	298	7.385	-6.8	57.37	-23.905	17.525	65
	300	7.390	-6.786	57.416	-24.011	17.494	
	400	7.640	-6.035	59.576	-29.865	16.320	
	500	7.890	-5.258	61.308	-35.912	15.699	
	600	8.140	-4.457	62.768	-42.118	15.343	
	700	8.390	-3.630	64.042	-48.460	15.132	
	800	8.640	-2.779	65.178	-54.922	15.006	
	900	8.890	-1.902	66.210	-61.492	14.934	
	1000	9.140	-1.001	67.160	-68.161	14.899	
	1100	9.390	-.074	68.043	-74.922	14.888	
	1200	9.640	.877	68.871	-81.768	14.894	
	1300	9.890	1.854	69.652	-88.694	14.913	
	1400	10.140	2.855	70.394	-95.697	14.941	
	1500	10.390	3.882	71.102	-102.772	14.976	
	1600	10.640	4.933	71.781	-109.916	15.016	
	1700	10.890	6.010	72.433	-117.127	15.060	
	1800	11.140	7.111	73.063	-124.402	15.107	
	1900	11.390	8.238	73.672	-131.739	15.155	
	2000	11.640	9.389	74.263	-139.136	15.206	
	2100	11.890	10.566	74.837	-146.591	15.258	
	2200	12.140	11.767	75.395	-154.103	15.311	
	2300	12.390	12.994	75.941	-161.670	15.364	
	2400	12.640	14.245	76.473	-169.290	15.418	
	2500	12.890	15.522	76.994	-176.964	15.472	
	2600	13.140	16.823	77.505	-184.689	15.527	
	2700	13.390	18.150	78.005	-192.465	15.581	

THORIUM DIOXIDE

		A	B	C	D	RANGE	REF.
SOL	CP	16.56	2.232	-2.195		298-3493	65
LIQ	CP	23.0				3493-4000	*

PHASE	T	CP	H	S	G	BT	P	
SOL	298	14.756	-293.12	15.59	-297.768	218.299		40
	300	14.791	-293.093	15.681	-297.797	216.974		
	400	16.081	-291.541	20.135	-299.596	163.713		
	500	16.798	-289.895	23.807	-301.798	131.934		
	600	17.289	-288.189	26.915	-304.338	110.870		
	700	17.674	-286.440	29.610	-307.167	95.915		
	800	18.003	-284.656	31.992	-310.250	84.768		
	900	18.298	-282.841	34.130	-313.558	76.152		
	1000	18.572	-280.997	36.072	-317.069	69.305		
	1100	18.834	-279.127	37.854	-320.767	63.739		
	1200	19.086	-277.231	39.504	-324.636	59.132		
	1300	19.332	-275.310	41.041	-328.664	55.261		
	1400	19.573	-273.365	42.483	-332.841	51.966		
	1500	19.810	-271.395	43.841	-337.158	49.130		
	1600	20.045	-269.403	45.127	-341.607	46.668		
	1700	20.278	-267.386	46.350	-346.181	44.511		
	1800	20.510	-265.347	47.515	-350.875	42.608		
	1900	20.740	-263.284	48.630	-355.682	40.918		
	2000	20.969	-261.199	49.700	-360.599	39.410		
	2100	21.197	-259.091	50.729	-365.621	38.056		
	2200	21.425	-256.960	51.720	-370.744	36.835		
	2300	21.652	-254.806	52.677	-375.964	35.730		
	2400	21.879	-252.629	53.604	-381.278	34.725		
	2500	22.105	-250.430	54.501	-386.684	33.808		
	2600	22.331	-248.208	55.373	-392.178	32.970	.003	
	2700	22.556	-245.964	56.220	-397.757	32.201	.008	
	2800	22.782	-243.697	57.044	-403.421	31.493	.024	
	2900	23.007	-241.408	57.848	-409.166	30.840	.066	
	3000	23.232	-239.096	58.631	-414.990	30.236	.164	
	3100	23.456	-236.761	59.397	-420.891	29.677	.385	
	3200	23.681	-234.404	60.145	-426.868	29.158	.853	
	3300	23.905	-232.025	60.877	-432.920	28.675	1.795	
	3400	24.130	-229.623	61.594	-439.043	28.225	3.598	
	3493	24.338	-227.369	62.248	-444.802	27.834	6.607	
			25.15	7.200				*
LIQ	3493	23.000	-202.219	69.448	-444.802	27.834	6.607	
	3500	23.000	-202.058	69.494	-445.288	27.809	6.857	
	3600	23.000	-199.758	70.142	-452.270	27.460	11.440	
	3700	23.000	-197.458	70.772	-459.316	27.134	18.508	
	3800	23.000	-195.158	71.386	-466.424	26.829	29.109	
	3900	23.000	-192.858	71.983	-473.593	26.543	44.609	
	4000	23.000	-190.558	72.565	-480.820	26.274	66.743	

ThS

THORIUM MONOSULFIDE

		A	B	C	D	RANGE	REF.
SOL	CP	11.98	1.305	-0.857		298-2000	65
	** MPT= 2600 (APPROX.)						65

PHASE	T	CP	H	S	G	BT	
SOL	298	11.405	-95.5	16.68	-100.473	73.659	65
	300	11.419	-95.479	16.751	-100.504	73.227	
	400	11.966	-94.307	20.119	-102.354	55.931	
	500	12.290	-93.093	22.827	-104.506	45.686	
	600	12.525	-91.852	25.089	-106.905	38.945	
	700	12.719	-90.589	27.035	-109.513	34.196	
	800	12.890	-89.309	28.744	-112.304	30.684	
	900	13.049	-88.012	30.272	-115.256	27.992	
	1000	13.199	-86.699	31.654	-118.354	25.870	
	1100	13.345	-85.372	32.919	-121.583	24.160	
	1200	13.486	-84.030	34.087	-124.934	22.757	
	1300	13.626	-82.675	35.172	-128.398	21.589	
	1400	13.763	-81.305	36.186	-131.966	20.604	
	1500	13.899	-79.922	37.141	-135.633	19.764	
	1600	14.035	-78.525	38.042	-139.392	19.043	
	1700	14.169	-77.115	38.897	-143.240	18.417	
	1800	14.303	-75.692	39.710	-147.170	17.871	
	1900	14.436	-74.255	40.487	-151.181	17.392	
	2000	14.569	-72.804	41.231	-155.267	16.969	

DITHORIUM TRISULFIDE

Th_2S_3

		A	B	C	D	RANGE	REF.
SOL	CP	29.15	3.6	-0.9		298-2270	*
	** MPT= 2560 (APPROX.)						65

PHASE	T	CP	H	S	G	BT	
SOL	298	29.211	-258.9	42.	-271.422	198.985	65
	300	29.230	-258.846	42.181	-271.500	197.814	
	400	30.027	-255.880	50.708	-276.163	150.909	
	500	30.590	-252.848	57.471	-281.584	123.097	
	600	31.060	-249.765	63.091	-287.620	104.779	
	700	31.486	-246.637	67.911	-294.175	91.858	
	800	31.889	-243.468	72.142	-301.182	82.290	
	900	32.279	-240.260	75.921	-308.589	74.946	
	1000	32.660	-237.013	79.342	-316.355	69.149	
	1100	33.036	-233.728	82.472	-324.447	64.470	
	1200	33.407	-230.406	85.362	-332.841	60.627	
	1300	33.777	-227.047	88.051	-341.513	57.421	
	1400	34.144	-223.651	90.568	-350.445	54.714	
	1500	34.510	-220.218	92.936	-359.622	52.404	
	1600	34.875	-216.749	95.175	-369.028	50.414	
	1700	35.239	-213.243	97.300	-378.653	48.686	
	1800	35.602	-209.701	99.324	-388.485	47.175	
	1900	35.965	-206.123	101.259	-398.515	45.846	
	2000	36.327	-202.508	103.113	-408.734	44.670	
	2100	36.690	-198.857	104.894	-419.135	43.626	
	2200	37.051	-195.170	106.609	-429.711	42.694	
	2270	37.305	-192.568	107.774	-437.214	42.100	

ThS_2 THORIUM DISULFIDE

		A	B	C	D	RANGE	REF.
SOL	CP	17.16	2.3			298-2188	*
	** MPT= 2188						65

PHASE	T	CP	H	S	G	BT	
SOL	298	17.846	-150.	23.	-156.857	114.995	65
	300	17.850	-149.967	23.110	-156.900	114.317	
	400	18.080	-148.170	28.277	-159.481	87.148	
	500	18.310	-146.351	32.336	-162.519	71.047	
	600	18.540	-144.508	35.695	-165.925	60.446	
	700	18.770	-142.643	38.570	-169.642	52.972	
	800	19.000	-140.754	41.091	-173.628	47.439	
	900	19.230	-138.843	43.343	-177.851	43.194	
	1000	19.460	-136.908	45.381	-182.289	39.845	
	1100	19.690	-134.951	47.246	-186.922	37.143	
	1200	19.920	-132.970	48.969	-191.734	34.924	
	1300	20.150	-130.967	50.573	-196.712	33.075	
	1400	20.380	-128.940	52.074	-201.845	31.514	
	1500	20.610	-126.891	53.488	-207.124	30.182	
	1600	20.840	-124.818	54.826	-212.540	29.035	
	1700	21.070	-122.723	56.096	-218.086	28.041	
	1800	21.300	-120.604	57.307	-223.757	27.171	
	1900	21.530	-118.463	58.465	-229.546	26.407	
	2000	21.760	-116.298	59.575	-235.448	25.732	
	2100	21.990	-114.111	60.642	-241.460	25.132	
	2188	22.192	-112.167	61.549	-246.836	24.659	

THORIUM MONONITRIDE

ThN

		A	B	C	D	RANGE	REF.
SOL	CP	11.34	2.276	-1.144		298-2000	65
	** MPT=	3080					65

PHASE	T	CP	H	S	G	BT	
SOL	298	10.732	-90.5	13.7	-94.585	69.342	65
	300	10.752	-90.480	13.766	-94.610	68.933	
	400	11.535	-89.362	16.978	-96.153	52.543	
	500	12.020	-88.183	19.608	-97.986	42.836	
	600	12.388	-86.962	21.833	-100.061	36.452	
	700	12.700	-85.707	23.766	-102.343	31.957	
	800	12.982	-84.423	25.481	-104.807	28.636	
	900	13.247	-83.111	27.025	-107.434	26.092	
	1000	13.502	-81.773	28.434	-110.208	24.089	
	1100	13.749	-80.411	29.733	-113.117	22.477	
	1200	13.992	-79.024	30.940	-116.151	21.157	
	1300	14.231	-77.613	32.069	-119.302	20.059	
	1400	14.468	-76.178	33.132	-122.563	19.136	
	1500	14.703	-74.719	34.139	-125.927	18.350	
	1600	14.937	-73.237	35.095	-129.389	17.676	
	1700	15.170	-71.732	36.007	-132.944	17.093	
	1800	15.401	-70.203	36.881	-136.589	16.586	
	1900	15.633	-68.651	37.720	-140.320	16.143	
	2000	15.863	-67.077	38.528	-144.132	15.752	

Th_3N_4 TRITHORIUM TETRANITRIDE

		A	B	C	D	RANGE	REF.
SOL	CP	39.33	6.24	-5.33		298-2000	65

PHASE	T	CP	H	S	G	BT	
SOL	298	35.195	-312.	43.7	-325.029	238.285	65
	300	35.280	-311.935	43.918	-325.110	236.874	
	400	38.495	-308.228	54.561	-330.052	180.356	
	500	40.318	-304.280	63.362	-335.961	146.868	
	600	41.593	-300.182	70.831	-342.680	124.838	
	700	42.610	-295.970	77.321	-350.095	109.319	
	800	43.489	-291.664	83.069	-358.120	97.847	
	900	44.288	-287.275	88.238	-366.689	89.056	
	1000	45.037	-282.808	92.944	-375.752	82.132	
	1100	45.754	-278.269	97.270	-385.265	76.555	
	1200	46.448	-273.658	101.281	-395.195	71.985	
	1300	47.127	-268.979	105.026	-405.513	68.182	
	1400	47.794	-264.233	108.543	-416.193	64.979	
	1500	48.453	-259.421	111.862	-427.215	62.254	
	1600	49.106	-254.543	115.010	-438.560	59.913	
	1700	49.754	-249.600	118.007	-450.212	57.886	
	1800	50.397	-244.592	120.869	-462.157	56.121	
	1900	51.038	-239.521	123.611	-474.382	54.574	
	2000	51.677	-234.385	126.245	-486.875	53.210	

DITHORIUM DINITRIDE MONOXIDE Th_2N_2O

		A	B	C	D	RANGE	REF.
SOL	CP	27.94	4.48	-3.76		298-2000	65

PHASE	T	CP	H	S	G	BT	
SOL	298	25.046	-308.	29.7	-316.855	232.292	65
	300	25.106	-307.954	29.855	-316.910	230.900	
	400	27.382	-305.316	37.427	-320.287	175.020	
	500	28.676	-302.509	43.687	-324.352	141.793	
	600	29.584	-299.593	48.999	-328.993	119.852	
	700	30.309	-296.598	53.615	-334.129	104.334	
	800	30.936	-293.535	57.704	-339.698	92.814	
	900	31.508	-290.412	61.382	-345.656	83.948	
	1000	32.044	-287.235	64.729	-351.964	76.932	
	1100	32.557	-284.004	67.808	-358.593	71.255	
	1200	33.055	-280.724	70.662	-365.518	66.579	
	1300	33.542	-277.394	73.327	-372.719	62.668	
	1400	34.020	-274.016	75.830	-380.178	59.356	
	1500	34.493	-270.590	78.193	-387.880	56.522	
	1600	34.961	-267.117	80.435	-395.812	54.073	
	1700	35.426	-263.598	82.568	-403.963	51.940	
	1800	35.888	-260.032	84.606	-412.323	50.070	
	1900	36.348	-256.420	86.559	-420.882	48.419	
	2000	36.806	-252.763	88.435	-429.632	46.954	

ThP

THORIUM MONOPHOSPHIDE

		A	B	C	D	RANGE	REF.
SOL	CP	12.93	1.3	-2.06		298-3223	*
	** MPT= 3223						65

PHASE	T	CP	H	S	G	BT	
SOL	298	11.000	-83.	16.8	-88.009	64.521	65
	300	11.031	-82.980	16.868	-88.040	64.146	
	400	12.162	-81.813	20.217	-89.900	49.125	
	500	12.756	-80.564	23.001	-92.065	40.247	
	600	13.138	-79.268	25.362	-94.486	34.421	
	700	13.420	-77.940	27.409	-97.127	30.328	
	800	13.648	-76.586	29.217	-99.960	27.311	
	900	13.846	-75.211	30.836	-102.964	25.006	
	1000	14.024	-73.818	32.304	-106.122	23.196	
	1100	14.190	-72.407	33.649	-109.420	21.743	
	1200	14.347	-70.980	34.890	-112.848	20.555	
	1300	14.498	-69.538	36.044	-116.396	19.570	
	1400	14.645	-68.081	37.124	-120.055	18.744	
	1500	14.788	-66.609	38.140	-123.818	18.043	
	1600	14.930	-65.123	39.098	-127.681	17.443	
	1700	15.069	-63.623	40.008	-131.636	16.925	
	1800	15.206	-62.109	40.873	-135.681	16.476	
	1900	15.343	-60.582	41.699	-139.810	16.084	
	2000	15.478	-59.041	42.489	-144.019	15.740	
	2100	15.613	-57.486	43.248	-148.306	15.437	
	2200	15.747	-55.918	43.977	-152.668	15.168	
	2300	15.881	-54.337	44.680	-157.101	14.930	
	2400	16.014	-52.742	45.359	-161.603	14.718	
	2500	16.147	-51.134	46.015	-166.172	14.529	
	2600	16.280	-49.513	46.651	-170.805	14.359	
	2700	16.412	-47.878	47.268	-175.501	14.208	
	2800	16.544	-46.230	47.867	-180.258	14.072	
	2900	16.676	-44.569	48.450	-185.074	13.949	
	3000	16.807	-42.895	49.018	-189.948	13.840	
	3100	16.939	-41.208	49.571	-194.877	13.741	
	3200	17.070	-39.507	50.111	-199.862	13.652	
	3223	17.100	-39.114	50.233	-201.015	13.633	

TRITHORIUM TETRAPHOSPHIDE Th_3P_4

		A	B	C	D	RANGE	REF.
SOL	CP	45.7	3.54	-6.9		298-3500	*

PHASE	T	CP	H	S	G	BT	
SOL	298	38.993	-274.	59.	-291.591	213.771	65
	300	39.095	-273.928	59.242	-291.700	212.532	
	400	42.803	-269.809	71.066	-298.235	162.970	
	500	44.710	-265.425	80.841	-305.845	133.703	
	600	45.907	-260.890	89.105	-314.353	114.518	
	700	46.770	-256.254	96.250	-323.629	101.055	
	800	47.454	-251.542	102.541	-333.575	91.141	
	900	48.034	-246.767	108.165	-344.115	83.574	
	1000	48.550	-241.937	113.253	-355.190	77.637	
	1100	49.024	-237.058	117.903	-366.751	72.876	
	1200	49.469	-232.133	122.187	-378.758	68.991	
	1300	49.894	-227.165	126.164	-391.178	65.772	
	1400	50.304	-222.155	129.877	-403.982	63.073	
	1500	50.703	-217.105	133.361	-417.146	60.786	
	1600	51.094	-212.015	136.646	-430.648	58.832	
	1700	51.479	-206.886	139.755	-444.469	57.148	
	1800	51.859	-201.719	142.708	-458.593	55.688	
	1900	52.235	-196.514	145.522	-473.006	54.415	
	2000	52.607	-191.272	148.211	-487.694	53.300	
	2100	52.978	-185.993	150.787	-502.644	52.318	
	2200	53.345	-180.677	153.260	-517.848	51.450	
	2300	53.712	-175.324	155.639	-533.293	50.681	
	2400	54.076	-169.934	157.933	-548.973	49.997	
	2500	54.440	-164.509	160.147	-564.877	49.388	
	2600	54.802	-159.046	162.290	-581.000	48.844	
	2700	55.163	-153.548	164.365	-597.333	48.357	
	2800	55.524	-148.014	166.377	-613.870	47.921	
	2900	55.884	-142.443	168.332	-630.606	47.530	
	3000	56.243	-136.837	170.233	-647.535	47.179	
	3100	56.602	-131.195	172.083	-664.651	46.864	
	3200	56.961	-125.517	173.885	-681.950	46.581	
	3300	57.319	-119.803	175.644	-699.427	46.327	
	3400	57.676	-114.053	177.360	-717.077	46.099	
	3500	58.034	-108.267	179.037	-734.898	45.895	

Th_3Si_2

3-THORIUM 2-SILICIUM

		A	B	C	D	RANGE	REF.
SOL	CP	29.7	6.8	-2.34		298-1900	*

PHASE	T	CP	H	S	G	BT	
SOL	298	29.095	-68.1	39.	-79.728	58.450	65
	300	29.140	-68.046	39.180	-79.800	58.142	
	400	30.957	-65.033	47.836	-84.167	45.993	
	500	32.164	-61.874	54.880	-89.314	39.044	
	600	33.130	-58.608	60.832	-95.107	34.647	
	700	33.982	-55.252	66.004	-101.454	31.680	
	800	34.774	-51.814	70.594	-108.288	29.587	
	900	35.531	-48.298	74.733	-115.558	28.065	
	1000	36.266	-44.708	78.515	-123.223	26.934	
	1100	36.987	-41.045	82.006	-131.251	26.081	
	1200	37.697	-37.311	85.254	-139.616	25.431	
	1300	38.402	-33.506	88.300	-148.296	24.934	
	1400	39.101	-29.631	91.171	-157.270	24.554	
	1500	39.796	-25.686	93.892	-166.525	24.266	
	1600	40.489	-21.672	96.483	-176.045	24.050	
	1700	41.179	-17.588	98.958	-185.818	23.892	
	1800	41.868	-13.436	101.331	-195.833	23.781	
	1900	42.555	-9.215	103.614	-206.081	23.708	

THORIUM SILICON

ThSi

		A	B	C	D	RANGE	REF.
SOL	CP	11.81	2.46	-1.11		298-1900	*

PHASE	T	CP	H	S	G	BT	
SOL	298	11.295	-30.6	13.9	-34.744	25.472	65
	300	11.315	-30.579	13.970	-34.770	25.333	
	400	12.100	-29.404	17.344	-36.342	19.859	
	500	12.596	-28.168	20.100	-38.218	16.707	
	600	12.978	-26.889	22.431	-40.348	14.699	
	700	13.305	-25.575	24.457	-42.694	13.332	
	800	13.605	-24.229	26.254	-45.232	12.358	
	900	13.887	-22.854	27.872	-47.939	11.643	
	1000	14.159	-21.452	29.350	-50.801	11.104	
	1100	14.424	-20.023	30.712	-53.805	10.692	
	1200	14.685	-18.567	31.978	-56.941	10.372	
	1300	14.942	-17.086	33.164	-60.198	10.122	
	1400	15.197	-15.579	34.280	-63.571	9.925	
	1500	15.451	-14.046	35.337	-67.052	9.771	
	1600	15.703	-12.489	36.343	-70.637	9.650	
	1700	15.954	-10.906	37.302	-74.319	9.556	
	1800	16.204	-9.298	38.221	-78.096	9.483	
	1900	16.453	-7.665	39.104	-81.962	9.429	

Th_3Si_5 3-THORIUM 2-SILICON

		A	B	C	D	RANGE	REF.
SOL	CP	46.87	8.56	-5.32		298-1900	*

PHASE	T	CP	H	S	G	BT	
SOL	298	43.437	-116.2	51.	-131.406	96.336	65
	300	43.527	-116.120	51.269	-131.500	95.811	
	400	46.969	-111.576	64.316	-137.303	75.029	
	500	49.022	-106.770	75.032	-144.286	63.076	
	600	50.528	-101.790	84.108	-152.254	55.466	
	700	51.776	-96.673	91.993	-161.068	50.294	
	800	52.887	-91.439	98.981	-170.623	46.618	
	900	53.917	-86.098	105.270	-180.841	43.920	
	1000	54.898	-80.657	111.002	-191.659	41.893	
	1100	55.846	-75.120	116.279	-203.026	40.343	
	1200	56.773	-69.489	121.178	-214.902	39.144	
	1300	57.683	-63.766	125.758	-227.251	38.210	
	1400	58.583	-57.952	130.066	-240.044	37.478	
	1500	59.474	-52.049	134.138	-253.256	36.904	
	1600	60.358	-46.058	138.005	-266.865	36.457	
	1700	61.238	-39.978	141.690	-280.851	36.111	
	1800	62.114	-33.810	145.215	-295.198	35.847	
	1900	62.987	-27.555	148.597	-309.890	35.650	

THORIUM 2-SILICON

$ThSi_2$

		A	B	C	D	RANGE	REF.
SOL	CP	17.53	3.05	-2.1		298-1900	*

PHASE	T	CP	H	S	G	BT	
SOL	298	16.077	-41.6	19.6	-47.444	34.782	65
	300	16.112	-41.570	19.700	-47.480	34.594	
	400	17.437	-39.885	24.537	-49.700	27.159	
	500	18.215	-38.100	28.518	-52.359	22.889	
	600	18.777	-36.249	31.890	-55.384	20.176	
	700	19.236	-34.348	34.820	-58.722	18.336	
	800	19.642	-32.404	37.416	-62.337	17.032	
	900	20.016	-30.421	39.751	-66.197	16.077	
	1000	20.370	-28.401	41.879	-70.280	15.362	
	1100	20.711	-26.347	43.836	-74.567	14.817	
	1200	21.044	-24.259	45.653	-79.043	14.398	
	1300	21.371	-22.139	47.350	-83.694	14.072	
	1400	21.693	-19.985	48.945	-88.509	13.819	
	1500	22.012	-17.800	50.453	-93.480	13.622	
	1600	22.328	-15.583	51.884	-98.597	13.470	
	1700	22.642	-13.335	53.247	-103.854	13.353	
	1800	22.955	-11.055	54.550	-109.245	13.266	
	1900	23.267	-8.744	55.799	-114.762	13.202	

$ThAl_3$ THORIUM 3-ALUMINIUM

		A	B	C	D	RANGE	REF.
SOL	CP	24.0	3.2			298-1400	*

PHASE	T	CP	H	S	G	BT	
SOL	298	24.954	-26.7	33.2	-36.599	26.831	65,*
	300	24.960	-26.654	33.354	-36.660	26.710	
	400	25.280	-24.142	40.579	-40.373	22.062	
	500	25.600	-21.598	46.254	-44.725	19.552	
	600	25.920	-19.022	50.950	-49.592	18.066	
	700	26.240	-16.414	54.970	-54.892	17.141	
	800	26.560	-13.774	58.494	-60.569	16.549	
	900	26.880	-11.102	61.641	-66.579	16.170	
	1000	27.200	-8.398	64.490	-72.888	15.932	
	1100	27.520	-5.662	67.097	-79.469	15.791	
	1200	27.840	-2.894	69.505	-86.300	15.720	
	1300	28.160	-.094	71.746	-93.364	15.698	
	1400	28.480	2.738	73.845	-100.645	15.713	

* * * * * * * * * *

$ThMg_2$ THORIUM 2-MAGNESIUM

		A	B	C	D	RANGE	REF.
SOL	CP	16.75	6.77	-0.33		298-1100	*

PHASE	T	CP	H	S	G	BT	
SOL	298	18.397	-7.5	22.1	-14.089	10.329	151*
	300	18.414	-7.466	22.214	-14.130	10.295	
	400	19.252	-5.581	27.629	-16.633	9.089	
	500	20.003	-3.618	32.007	-19.622	8.578	
	600	20.720	-1.582	35.718	-23.013	8.383	
	700	21.422	.525	38.964	-26.750	8.353	
	800	22.114	2.702	41.870	-30.794	8.414	
	900	22.802	4.948	44.515	-35.115	8.528	
	1000	23.487	7.262	46.953	-39.690	8.675	
	1100	24.170	9.645	49.223	-44.500	8.843	

THORIUM 2-RHENIUM

$ThRe_2$

		A	B	C	D	RANGE	REF.
SOL	CP	16.96	5.63			298-1500	*

PHASE	T	CP	H	S	G	BT	
SOL	298	18.639	-41.6	29.55	-50.410	36.957	164
	300	18.649	-41.566	29.665	-50.465	36.769	
	400	19.212	-39.672	35.107	-53.715	29.353	
	500	19.775	-37.723	39.455	-57.451	25.115	
	600	20.338	-35.717	43.110	-61.584	22.435	
	700	20.901	-33.656	46.287	-66.057	20.627	
	800	21.464	-31.537	49.115	-70.829	19.352	
	900	22.027	-29.363	51.676	-75.871	18.426	
	1000	22.590	-27.132	54.026	-81.158	17.739	
	1100	23.153	-24.845	56.205	-86.670	17.222	
	1200	23.716	-22.501	58.244	-92.394	16.829	
	1300	24.279	-20.102	60.164	-98.315	16.531	
	1400	24.842	-17.645	61.984	-104.423	16.303	
	1500	25.405	-15.133	63.717	-110.709	16.132	

TiF_2 (g)

TITANIUM DIFLUORIDE (GAS)

		A	B	C	D	RANGE	REF.
GAS	CP	14.213	0.612	-1.550		298-2000	3

PHASE	T	CP	H	S	G	BT	
GAS	298	12.652	-164.5	61.1	-182.717	133.953	3
	300	12.674	-164.477	61.178	-182.830	133.210	
	400	13.489	-163.163	64.952	-189.144	103.357	
	500	13.899	-161.792	68.010	-195.797	85.594	
	600	14.150	-160.388	70.568	-202.729	73.854	
	700	14.325	-158.964	72.763	-209.898	65.542	
	800	14.460	-157.525	74.685	-217.273	59.364	
	900	14.572	-156.073	76.395	-224.828	54.603	
	1000	14.670	-154.611	77.935	-232.546	50.830	
	1100	14.758	-153.139	79.338	-240.411	47.772	
	1200	14.840	-151.659	80.625	-248.410	45.248	
	1300	14.917	-150.171	81.816	-256.532	43.133	
	1400	14.991	-148.676	82.924	-264.770	41.338	
	1500	15.062	-147.173	83.961	-273.115	39.798	
	1600	15.132	-145.664	84.935	-281.560	38.465	
	1700	15.200	-144.147	85.855	-290.100	37.300	
	1800	15.267	-142.624	86.725	-298.729	36.276	
	1900	15.333	-141.094	87.553	-307.444	35.369	
	2000	15.398	-139.557	88.341	-316.239	34.562	

TITANIUM DICHLORIDE

$TiCl_2$

		A	B	C	D	RANGE	REF.
SOL	CP	16.339	4.308	-0.826		298-1582	3
	** SPT=	1581.5					
GAS	CP	14.371	.530	-.662		1581-2000	3

PHASE	T	CP	H	S	G	BT	P	
SOL	298	16.694	-123.2	20.88	-129.425	94.884		3
	300	16.714	-123.169	20.983	-129.464	94.327		
	400	17.546	-121.453	25.914	-131.819	72.032		
	500	18.163	-119.667	29.898	-134.616	58.848		
	600	18.694	-117.823	33.257	-137.778	50.192		
	700	19.186	-115.929	36.176	-141.252	44.107		
	800	19.656	-113.987	38.769	-145.002	39.618		
	900	20.114	-111.998	41.110	-148.998	36.186		
	1000	20.564	-109.964	43.253	-153.217	33.490	.008	
	1100	21.010	-107.886	45.234	-157.643	31.325	.147	
	1200	21.451	-105.763	47.081	-162.260	29.556	1.595	
	1300	21.891	-103.595	48.815	-167.055	28.088	11.722	
	1400	22.328	-101.385	50.454	-172.020	26.857	63.578	
	1500	22.764	-99.130	52.009	-177.143	25.813	270.503	
	1581	23.119	-97.260	53.223	-181.432	25.076	760.	
			59.4	37.559				3
GAS	1581	15.183	-37.860	90.782	-181.432	25.076		
	1600	15.193	-37.579	90.959	-183.113	25.015		
	1700	15.249	-36.057	91.881	-192.255	24.719		
	1800	15.305	-34.529	92.755	-201.488	24.467		
	1900	15.360	-32.996	93.584	-210.805	24.251		
	2000	15.414	-31.457	94.373	-220.203	24.066		

Ti_2Cl_6 (g) TITANIUM TRICHLORIDE (DIMERIC GAS)

		A	B	C	D	RANGE	REF.
GAS	CP	32.174	1.06	1.324		298-2000	*

PHASE	T	CP	H	S	G	BT	
GAS	298	33.979	-298.4	115.	-332.687	243.899	227,*
	300	33.963	-298.337	115.210	-332.900	242.550	
	400	33.425	-294.972	124.894	-344.930	188.486	
	500	33.234	-291.641	132.328	-357.805	156.418	
	600	33.178	-288.321	138.381	-371.350	135.282	
	700	33.186	-285.003	143.496	-385.450	120.359	
	800	33.229	-281.683	147.929	-400.026	109.297	
	900	33.291	-278.357	151.847	-415.019	100.794	
	1000	33.366	-275.024	155.358	-430.382	94.073	
	1100	33.449	-271.683	158.542	-446.080	88.640	
	1200	33.538	-268.334	161.456	-462.082	84.168	
	1300	33.630	-264.976	164.144	-478.363	80.431	
	1400	33.726	-261.608	166.640	-494.904	77.268	
	1500	33.823	-258.230	168.970	-511.686	74.563	
	1600	33.922	-254.843	171.156	-528.693	72.226	
	1700	34.022	-251.446	173.216	-545.913	70.191	
	1800	34.123	-248.039	175.163	-563.333	68.407	
	1900	34.225	-244.622	177.011	-580.942	66.833	
	2000	34.327	-241.194	178.769	-598.732	65.435	

TITANIUM DIIODIDE

TiJ_2

		A	B	C	D	RANGE	REF.
SOL	CP	20.09	1.74	0.001		298-1359	3
	** SPT=	1358.2					3
GAS	CP	14.386	0.525			1358-2000	3

PHASE	T	CP	H	S	G	BT	P	
SOL	298	20.610	-63.6	29.3	-72.336	53.031		3
	300	20.613	-63.562	29.427	-72.390	52.743		
	400	20.787	-61.492	35.381	-75.644	41.336		
	500	20.960	-59.405	40.038	-79.424	34.721		
	600	21.134	-57.300	43.875	-83.625	30.464		
	700	21.308	-55.178	47.146	-88.180	27.535		
	800	21.482	-53.038	50.003	-93.040	25.421		
	900	21.656	-50.881	52.543	-98.170	23.842	.031	
	1000	21.830	-48.707	54.834	-103.541	22.632	.650	
	1100	22.004	-46.515	56.923	-109.130	21.685	7.700	
	1200	22.178	-44.306	58.845	-114.920	20.933	58.790	
	1300	22.352	-42.080	60.627	-120.894	20.327	320.828	
	1358	22.453	-40.776	61.608	-124.452	20.028	760.	
			51.80	38.139				3
GAS	1358	15.099	11.024	99.747	-124.452	20.028		
	1400	15.121	11.656	100.205	-128.631	20.083		
	1500	15.173	13.171	101.250	-138.704	20.212		
	1600	15.226	14.691	102.231	-148.878	20.339		
	1700	15.279	16.216	103.155	-159.148	20.463		
	1800	15.331	17.746	104.030	-169.508	20.584		
	1900	15.383	19.282	104.860	-179.953	20.702		
	2000	15.436	20.823	105.651	-190.479	20.817		

TiJ_3 TITANIUM TRIIODIDE

		A	B	C	D	RANGE	REF.
SOL	CP	27.390	1.740			298-1000	3
GAS	CP	20.738	-0.115			1000-2000	3

PHASE	T	CP	H	S	G	BT	P	
SOL	298	27.909	-77.	46.0	-90.715	66.505		3
	300	27.912	-76.948	46.173	-90.800	66.157		
	400	28.086	-74.148	54.226	-95.839	52.371		
	500	28.260	-71.331	60.512	-101.587	44.410		
	600	28.434	-68.496	65.680	-107.904	39.309	.003	
	700	28.608	-65.644	70.076	-114.698	35.815	.264	
	800	28.782	-62.775	73.908	-121.901	33.306	7.763	
	900	28.956	-59.888	77.308	-129.465	31.443	101.676	
	1000	29.130	-56.984	80.367	-137.351	30.022	760.	
			35.531	35.531				3*
GAS	1000	20.623	-21.453	115.898	-137.351	30.022		
	1100	20.611	-19.391	117.864	-149.041	29.616		
	1200	20.600	-17.330	119.656	-160.918	29.311		
	1300	20.588	-15.271	121.305	-172.967	29.082		
	1400	20.577	-13.213	122.830	-185.175	28.911		
	1500	20.566	-11.156	124.249	-197.530	28.784		
	1600	20.554	-9.100	125.576	-210.022	28.692		
	1700	20.543	-7.045	126.822	-222.642	28.626		
	1800	20.531	-4.991	127.996	-235.384	28.583		
	1900	20.519	-2.939	129.106	-248.239	28.558		
	2000	20.508	-.887	130.158	-261.203	28.547		

TITANIUM TRIIODIDE (GAS) $TiJ_3(g)$

		A	B	C	D	RANGE	REF.
GAS	CP	21.396	-0.607	-1.694		298- 800	3
GAS	CP	20.738	-0.115			800-2000	3

PHASE	T	CP	H	S	G	BT	
GAS	298	19.309	-35.794	91.3	-63.015	46.198	3*
	300	19.332	-35.758	91.420	-63.184	46.036	
	400	20.094	-33.781	97.102	-72.622	39.684	
	500	20.415	-31.753	101.625	-82.566	36.095	
	600	20.561	-29.704	105.362	-92.921	33.851	
	700	20.625	-27.644	108.537	-103.620	32.356	
	800	20.646	-25.580	111.293	-114.615	31.315	
	800	20.646	-25.580	111.293	-114.615	31.315	
	900	20.635	-23.516	113.724	-125.868	30.569	
	1000	20.623	-21.453	115.898	-137.351	30.022	
	1100	20.611	-19.391	117.863	-149.040	29.616	
	1200	20.600	-17.331	119.656	-160.918	29.311	
	1300	20.588	-15.271	121.304	-172.967	29.082	
	1400	20.577	-13.213	122.829	-185.174	28.911	
	1500	20.566	-11.156	124.249	-197.529	28.784	
	1600	20.554	-9.100	125.576	-210.021	28.691	
	1700	20.543	-7.045	126.821	-222.641	28.626	
	1800	20.531	-4.992	127.995	-235.383	28.583	
	1900	20.519	-2.939	129.105	-248.238	28.558	
	2000	20.508	-.888	130.157	-261.202	28.547	

TiJ_4 TITANIUM TETRAIODIDE

		A	B	C	D	RANGE	REF.
SOL-A	CP	18.702	38.000			298- 379	3
SOL-B	CP	35.4				379- 428	3
LIQ	CP	37.4				428- 653	3
	** BPT=	652.6					3
GAS	CP	25.816	.010	-.803		653-2000	3

PHASE	T	CP	H	S	G	BT	P	
SOL-A	298	30.032	-89.8	58.8	-107.331	78.686		3
	300	30.102	-89.744	58.986	-107.440	78.281		
	379	33.104	-87.248	66.360	-112.398	64.823	.070	
			2.37	6.253				3
SOL-B	379	35.400	-84.878	72.613	-112.398	64.823	.070	
	400	35.400	-84.134	74.522	-113.943	62.264	.308	
	428	35.400	-83.143	76.917	-116.064	59.274	1.745	
			4.74	11.075				3
LIQ	428	37.400	-78.403	87.992	-116.064	59.274	1.745	
	500	37.400	-75.710	93.807	-122.614	53.602	25.145	
	600	37.400	-71.970	100.626	-132.346	48.213	298.489	
	653	37.400	-70.003	103.769	-137.723	46.128	760.	
			13.50	20.686				3
GAS	653	25.634	-56.503	124.455	-137.723	46.128		
	700	25.659	-55.287	126.253	-143.665	44.860		
	800	25.699	-52.719	129.683	-156.465	42.750		
	900	25.726	-50.148	132.711	-169.588	41.187		
	1000	25.746	-47.574	135.423	-182.997	39.999		
	1100	25.761	-44.999	137.877	-196.664	39.079		
	1200	25.772	-42.422	140.119	-210.565	38.354		
	1300	25.781	-39.845	142.182	-224.682	37.778		
	1400	25.789	-37.266	144.093	-238.997	37.314		
	1500	25.795	-34.687	145.873	-253.496	36.939		
	1600	25.801	-32.107	147.538	-268.168	36.635		
	1700	25.805	-29.527	149.102	-283.000	36.387		
	1800	25.809	-26.946	150.577	-297.985	36.185		
	1900	25.813	-24.365	151.973	-313.113	36.021		
	2000	25.816	-21.784	153.297	-328.377	35.888		

TITANIUM MONOSULFIDE

TiS

		A	B	C	D	RANGE	REF.
SOL	CP	10.97	1.76			298-2200	63
	** MPT= 2200						63

PHASE	T	CP	H	S	G	BT	P	
SOL	298	11.495	-65.	13.5	-69.025	50.604		63
	300	11.498	-64.979	13.571	-69.050	50.310		
	400	11.674	-63.820	16.903	-70.581	38.569		
	500	11.850	-62.644	19.527	-72.407	31.653		
	600	12.026	-61.450	21.703	-74.472	27.130		
	700	12.202	-60.239	23.570	-76.738	23.962		
	800	12.378	-59.010	25.211	-79.178	21.633		
	900	12.554	-57.763	26.679	-81.774	19.860		
	1000	12.730	-56.499	28.011	-84.510	18.472		
	1100	12.906	-55.217	29.232	-87.373	17.362		
	1200	13.082	-53.918	30.363	-90.353	16.458		
	1300	13.258	-52.601	31.417	-93.443	15.711		
	1400	13.434	-51.266	32.406	-96.634	15.087		
	1500	13.610	-49.914	33.339	-99.922	14.561		
	1600	13.786	-48.544	34.223	-103.300	14.112		
	1700	13.962	-47.157	35.064	-106.765	13.727		
	1800	14.138	-45.752	35.867	-110.312	13.395		
	1900	14.314	-44.329	36.636	-113.937	13.108		
	2000	14.490	-42.889	37.375	-117.638	12.857		
	2100	14.666	-41.431	38.086	-121.411	12.637		
	2200	14.842	-39.956	38.772	-125.254	12.445	.004	

TiS (g)

TITANIUM MONOSULFIDE (GAS)

		A	B	C	D	RANGE	REF.
GAS	CP	8.842	0.053	-0.706		298-2200	63

PHASE	T	CP	H	S	G	BT	
GAS	298	8.064	79.	58.89	61.442	-45.044	63
	300	8.073	79.015	58.940	61.333	-44.687	
	400	8.422	79.842	61.317	55.315	-30.227	
	500	8.586	80.693	63.216	49.085	-21.458	
	600	8.678	81.557	64.790	42.683	-15.549	
	700	8.735	82.428	66.133	36.135	-11.283	
	800	8.774	83.303	67.302	29.462	-8.050	
	900	8.803	84.182	68.337	22.679	-5.508	
	1000	8.824	85.064	69.266	15.798	-3.453	
	1100	8.842	85.947	70.108	8.829	-1.754	
	1200	8.857	86.832	70.878	1.779	-.324	
	1300	8.869	87.718	71.587	-5.345	.899	
	1400	8.880	88.606	72.245	-12.537	1.957	
	1500	8.890	89.494	72.858	-19.792	2.884	
	1600	8.899	90.384	73.432	-27.107	3.703	
	1700	8.908	91.274	73.971	-34.477	4.433	
	1800	8.916	92.165	74.481	-41.900	5.088	
	1900	8.923	93.057	74.963	-49.373	5.680	
	2000	8.930	93.950	75.421	-56.892	6.218	
	2100	8.937	94.843	75.857	-64.456	6.709	
	2200	8.944	95.737	76.273	-72.063	7.160	

TITANIUM DISULFIDE

TiS_2

		A	B	C	D	RANGE	REF.
SOL-A	CP	8.08	27.34			298- 420	1
	** TIS(1.5-2.0)						1
SOL-B	CP	14.99	5.14			420-1000	1

PHASE	T	CP	H	S	G	BT	
SOL-A	298	16.231	-97.3	18.73	-102.884	75.426	63
	300	16.282	-97.270	18.831	-102.919	74.987	
	400	19.016	-95.505	23.889	-105.061	57.410	
	420	19.563	-95.119	24.830	-105.548	54.930	
							1
SOL-B	420	17.149	-95.119	24.830	-105.548	54.930	
	500	17.560	-93.731	27.855	-107.658	47.064	
	600	18.074	-91.949	31.102	-110.610	40.295	
	700	18.588	-90.116	33.927	-113.865	35.555	
	800	19.102	-88.232	36.442	-117.385	32.072	
	900	19.616	-86.296	38.722	-121.145	29.422	
	1000	20.130	-84.308	40.815	-125.123	27.349	

Ti_5Si_3 5-TITANIUM 3-SILICON

		A	B	C	D	RANGE	REF.
SOL	CP	46.95	10.70	-4.8		298-2300	*

PHASE	T	CP	H	S	G	BT	
SOL	298	44.740	-138.4	52.1	-153.934	112.852	21/*
	300	44.827	-138.317	52.377	-154.030	112.226	
	400	48.230	-133.648	65.787	-159.962	87.411	
	500	50.380	-128.711	76.794	-167.108	73.053	
	600	52.037	-123.588	86.130	-175.266	63.849	
	700	53.460	-118.311	94.261	-184.294	57.547	
	800	54.760	-112.900	101.485	-194.088	53.029	
	900	55.987	-107.362	108.006	-204.568	49.682	
	1000	57.170	-101.704	113.967	-215.671	47.141	
	1100	58.323	-95.929	119.470	-227.346	45.176	
	1200	59.457	-90.040	124.594	-239.552	43.634	
	1300	60.576	-84.038	129.397	-252.254	42.413	
	1400	61.685	-77.925	133.927	-265.422	41.440	
	1500	62.787	-71.701	138.220	-279.031	40.660	
	1600	63.882	-65.368	142.307	-293.059	40.035	
	1700	64.974	-58.925	146.213	-307.487	39.535	
	1800	66.062	-52.373	149.958	-322.297	39.137	
	1900	67.147	-45.713	153.558	-337.473	38.824	
	2000	68.230	-38.944	157.030	-353.004	38.580	
	2100	69.311	-32.067	160.385	-368.876	38.395	
	2200	70.391	-25.081	163.635	-385.077	38.259	
	2300	71.469	-17.988	166.787	-401.599	38.166	

TITANIUM SILICON

		A	B	C	D	RANGE	REF.
SOL	CP	11.50	2.73	-1.30		298-2000	*

PHASE	T	CP	H	S	G	BT	
SOL	298	10.852	-31.	11.7	-34.488	25.284	21/*
	300	10.875	-30.980	11.767	-34.510	25.144	
	400	11.779	-29.843	15.033	-35.856	19.593	
	500	12.345	-28.635	17.725	-37.498	16.392	
	600	12.777	-27.378	20.016	-39.387	14.349	
	700	13.146	-26.082	22.014	-41.491	12.956	
	800	13.481	-24.750	23.791	-43.783	11.963	
	900	13.797	-23.386	25.397	-46.244	11.231	
	1000	14.100	-21.991	26.867	-48.858	10.679	
	1100	14.396	-20.566	28.224	-51.613	10.256	
	1200	14.686	-19.112	29.489	-54.500	9.927	
	1300	14.972	-17.629	30.676	-57.508	9.669	
	1400	15.256	-16.118	31.796	-60.633	9.466	
	1500	15.537	-14.578	32.858	-63.866	9.306	
	1600	15.817	-13.010	33.870	-67.203	9.181	
	1700	16.096	-11.415	34.837	-70.638	9.082	
	1800	16.374	-9.791	35.765	-74.169	9.007	
	1900	16.651	-8.140	36.658	-77.790	8.949	
	2000	16.927	-6.461	37.519	-81.499	8.907	

$TiSi_2$ TITANIUM 2-SILICON

		A	B	C	D	RANGE	REF.
SOL	CP	16.83	4.20	-2.16		298-1800	*

PHASE	T	CP	H	S	G	BT	
SOL	298	15.652	-32.1	14.6	-36.453	26.724	21/*
	300	15.690	-32.071	14.697	-36.480	26.579	
	400	17.160	-30.421	19.434	-38.194	20.871	
	500	18.066	-28.657	23.366	-40.340	17.635	
	600	18.750	-26.815	26.723	-42.849	15.610	
	700	19.329	-24.910	29.657	-45.671	14.261	
	800	19.852	-22.951	32.273	-48.769	13.325	
	900	20.343	-20.941	34.640	-52.117	12.657	
	1000	20.814	-18.883	36.808	-55.691	12.173	
	1100	21.271	-16.779	38.813	-59.473	11.818	
	1200	21.720	-14.629	40.683	-63.449	11.557	
	1300	22.162	-12.435	42.439	-67.606	11.367	
	1400	22.600	-10.197	44.098	-71.934	11.231	
	1500	23.034	-7.915	45.672	-76.423	11.136	
	1600	23.466	-5.590	47.172	-81.065	11.075	
	1700	23.895	-3.222	48.608	-85.855	11.039	
	1800	24.323	-.811	49.986	-90.785	11.024	

TITANIUM DIBORIDE

TiB_2

		A	B	C	D	RANGE	REF.
SOL	CP	13.475	6.18	-4.174	-0.80	298-3193	3
LIQ	CP	26.0				3193-3500	3
	** BPT=	4250					3

PHASE	T	CP	H	S	G	BT	
SOL	298	10.551	-77.4	6.81	-79.430	58.232	1/3
	300	10.619	-77.380	6.875	-79.443	57.882	
	400	13.210	-76.174	10.327	-80.305	43.883	
	500	14.695	-74.774	13.447	-81.497	35.627	
	600	15.736	-73.250	16.222	-82.983	30.231	
	700	16.557	-71.634	18.712	-84.732	26.458	
	800	17.255	-69.942	20.969	-86.718	23.693	
	900	17.874	-68.185	23.038	-88.920	21.596	
	1000	18.438	-66.369	24.951	-91.320	19.961	
	1100	18.960	-64.499	26.733	-93.905	18.660	
	1200	19.449	-62.579	28.404	-96.663	17.607	
	1300	19.910	-60.610	29.979	-99.583	16.744	
	1400	20.346	-58.597	31.471	-102.656	16.027	
	1500	20.759	-56.542	32.889	-105.875	15.428	
	1600	21.152	-54.446	34.241	-109.232	14.922	
	1700	21.525	-52.312	35.535	-112.721	14.493	
	1800	21.878	-50.142	36.775	-116.337	14.127	
	1900	22.213	-47.937	37.967	-120.074	13.814	
	2000	22.531	-45.700	39.114	-123.929	13.544	
	2100	22.830	-43.432	40.221	-127.896	13.312	
	2200	23.113	-41.134	41.290	-131.972	13.112	
	2300	23.378	-38.810	42.323	-136.153	12.939	
	2400	23.627	-36.459	43.323	-140.435	12.790	
	2500	23.858	-34.085	44.293	-144.816	12.662	
	2600	24.073	-31.688	45.233	-149.293	12.551	
	2700	24.272	-29.271	46.145	-153.862	12.456	
	2800	24.454	-26.834	47.031	-158.521	12.375	
	2900	24.619	-24.381	47.892	-163.267	12.306	
	3000	24.769	-21.911	48.729	-168.098	12.248	
	3100	24.902	-19.427	49.543	-173.012	12.199	
	3193	25.011	-17.106	50.281	-177.654	12.161	
			24.	7.516			3
LIQ	3193	26.000	6.894	57.798	-177.654	12.161	
	3200	26.000	7.076	57.855	-178.059	12.162	
	3300	26.000	9.676	58.655	-183.885	12.180	
	3400	26.000	12.276	59.431	-189.789	12.201	
	3500	26.000	14.876	60.184	-195.770	12.226	

TiAl TITANIUM ALUMINIUM

		A	B	C	D	RANGE	REF.
SOL	CP	13.37	1.42	-1.8		298-1733	*

** HOMOGENEITY RANGE

PHASE	T	CP	H	S	G	BT	
SOL	298	11.768	-17.4	12.5	-21.127	15.488	64/220*
	300	11.796	-17.378	12.573	-21.150	15.410	
	400	12.813	-16.142	16.124	-22.591	12.345	
	500	13.360	-14.831	19.047	-24.354	10.647	
	600	13.722	-13.476	21.516	-26.385	9.612	
	700	13.997	-12.089	23.653	-28.646	8.945	
	800	14.225	-10.678	25.537	-31.107	8.499	
	900	14.426	-9.245	27.224	-33.747	8.196	
	1000	14.610	-7.793	28.754	-36.547	7.988	
	1100	14.783	-6.323	30.155	-39.494	7.848	
	1200	14.949	-4.837	31.448	-42.574	7.755	
	1300	15.109	-3.334	32.651	-45.780	7.697	
	1400	15.266	-1.815	33.777	-49.102	7.666	
	1500	15.420	-.281	34.835	-52.533	7.655	
	1600	15.572	1.269	35.835	-56.067	7.659	
	1700	15.722	2.834	36.784	-59.699	7.676	
	1733	15.771	3.353	37.086	-60.917	7.683	

TITANIUM 3-ALUMINIUM

$TiAl_3$

		A	B	C	D	RANGE	REF.
SOL	CP	24.74	4.0	-2.15		298-1613	*

** HOMOGENEITY RANGE

PHASE	T	CP	H	S	G	BT	
SOL	298	23.514	-34.	22.6	-40.738	29.866	64/220,*
	300	23.551	-33.956	22.746	-40.780	29.712	
	400	24.996	-31.522	29.740	-43.418	23.726	
	500	25.880	-28.975	35.419	-46.685	20.409	
	600	26.543	-26.353	40.198	-50.472	18.387	
	700	27.101	-23.670	44.333	-54.703	17.081	
	800	27.604	-20.934	47.985	-59.322	16.208	
	900	28.075	-18.150	51.264	-64.287	15.613	
	1000	28.525	-15.320	54.245	-69.565	15.205	
	1100	28.962	-12.446	56.984	-75.128	14.929	
	1200	29.391	-9.528	59.523	-80.955	14.746	
	1300	29.813	-6.568	61.892	-87.027	14.633	
	1400	30.230	-3.566	64.117	-93.329	14.571	
	1500	30.644	-.522	66.216	-99.846	14.550	
	1600	31.056	2.563	68.207	-106.568	14.559	
	1613	31.109	2.967	68.459	-107.457	14.562	

Tl

THALLIUM

		A	B	C	D	RANGE	REF.
SOL-A	CP	3.372	6.680	0.823		298- 507	21
SOL-B	CP	6.490	2.289			507- 577	21
LIQ	CP	7.10				577-1746	21
GAS	CP	5.234				1746-2000	21

PHASE	T	CP	H	S	G	BT	P	
SOL-A	298	6.289	0	15.34	-4.574	3.353		21
	300	6.290	.012	15.379	-4.602	3.353		
	400	6.558	.651	17.217	-6.236	3.407		
	500	7.041	1.330	18.730	-8.035	3.513		
	507	7.079	1.380	18.828	-8.166	3.521		
			0.09	.178				21
SOL-B	507	7.651	1.470	19.006	-8.166	3.521		
	577	7.811	2.011	20.005	-9.532	3.611		
			0.99	1.716				21
LIQ	577	7.100	3.001	21.721	-9.532	3.611		
	600	7.100	3.164	21.999	-10.035	3.656		
	700	7.100	3.874	23.093	-12.291	3.838		
	800	7.100	4.584	24.041	-14.649	4.002		
	900	7.100	5.294	24.877	-17.096	4.152	.014	
	1000	7.100	6.004	25.625	-19.621	4.289	.137	
	1100	7.100	6.714	26.302	-22.218	4.415	.877	
	1200	7.100	7.424	26.920	-24.880	4.532	4.095	
	1300	7.100	8.134	27.488	-27.601	4.641	14.989	
	1400	7.100	8.844	28.014	-30.376	4.743	45.318	
	1500	7.100	9.554	28.504	-33.202	4.838	117.640	
	1600	7.100	10.264	28.962	-36.076	4.928	269.902	
	1700	7.100	10.974	29.393	-38.994	5.014	559.549	
	1746	7.100	11.301	29.582	-40.350	5.051	760.	
			39.215	22.460				21
GAS	1746	5.234	50.516	52.042	-40.350	5.051		
	1800	5.234	50.798	52.202	-43.165	5.242		
	1900	5.234	51.322	52.485	-48.399	5.568		
	2000	5.234	51.845	52.753	-53.661	5.865		

THALLIUM FLUORIDE

		A	B	C	D	RANGE	REF.
SOL-A	CP	13.0				298- 356	140
SOL-B	CP	13.25				356- 595	140
LIQ	CP	16.08				595- 973	1
	** BPT= 973, L= 27.7						1

PHASE	T	CP	H	S	G	BT	
SOL-A	298	13.000	-77.8	22.87	-84.619	62.036	140
	300	13.000	-77.776	22.950	-84.661	61.684	
	356	13.000	-77.048	25.175	-86.010	52.809	
			0.08	.225			140
SOL-B	356	13.250	-76.968	25.400	-86.010	52.809	
	400	13.250	-76.385	26.944	-87.163	47.630	
	500	13.250	-75.060	29.901	-90.010	39.349	
	595	13.250	-73.801	32.206	-92.964	34.151	
			3.32	5.580			1
LIQ	595	16.080	-70.481	37.785	-92.964	34.151	
	600	16.080	-70.401	37.920	-93.153	33.935	
	700	16.080	-68.793	40.399	-97.072	30.311	
	800	16.080	-67.185	42.546	-101.222	27.656	
	900	16.080	-65.577	44.440	-105.573	25.640	
	973	16.080	-64.403	45.694	-108.863	24.456	

$TlCl_3$ THALLIUM TRICHLORIDE

		A	B	C	D	RANGE	REF.
SOL	CP	16.45	32.0			298- 500	*

PHASE	T	CP	H	S	G	BT	
SOL	298	25.991	-75.3	36.4	-86.153	63.160	19/*
	300	26.050	-75.252	36.561	-86.220	62.820	
	400	29.250	-72.487	44.493	-90.284	49.336	
	500	32.450	-69.402	51.364	-95.084	41.567	

* * * * * * * * * *

TlBr THALLIUM BROMIDE

		A	B	C	D	RANGE	REF.
SOL	CP	11.07	4.95			298- 733	1*
LIQ	CP	25.25	-9.04			733-1098	1
	** BPT= 1098, L= 24.7						1

PHASE	T	CP	H	S	G	BT	
SOL	298	12.546	-41.27	29.3	-50.006	36.660	62
	300	12.555	-41.247	29.378	-50.060	36.474	
	400	13.050	-39.967	33.057	-53.189	29.065	
	500	13.545	-38.637	36.022	-56.648	24.764	
	600	14.040	-37.258	38.536	-60.379	21.996	
	700	14.535	-35.829	40.737	-64.345	20.092	
	733	14.698	-35.346	41.411	-65.700	19.592	
			3.92	5.348			1
LIQ	733	18.624	-31.426	46.758	-65.700	19.592	
	800	18.018	-30.199	48.361	-68.888	18.822	
	900	17.114	-28.442	50.431	-73.830	17.931	
	1000	16.210	-26.776	52.188	-78.964	17.260	
	1098	15.324	-25.231	53.662	-84.152	16.752	

THALLIUM IODIDE

		A	B	C	D	RANGE	REF.
SOL-A	CP	11.590	3.240			298- 451	62
SOL-B	CP	7.720	11.260			451- 715	1
LIQ	CP	17.200				715-1082	1
GAS	CP	9.23		-0.3		1082-1600	62*
	** BPT= 1082.34						

PHASE	T	CP	H	S	G	BT	P	
SOL-A	298	12.556	-29.56	30.52	-38.660	28.342		62
	300	12.562	-29.537	30.598	-38.716	28.208		
	400	12.886	-28.264	34.256	-41.967	22.933		
	451	13.051	-27.603	35.812	-43.754	21.206		
			0.218	.483				62*
SOL-B	451	12.798	-27.385	36.295	-43.754	21.206		
	500	13.350	-26.744	37.643	-45.566	19.920		
	600	14.476	-25.353	40.177	-49.459	18.018	.019	
	700	15.602	-23.849	42.493	-53.594	16.735	.850	
	715	15.771	-23.614	42.825	-54.234	16.580	1.362	
			3.52	4.923				62
LIQ	715	17.200	-20.094	47.749	-54.234	16.580	1.362	
	800	17.200	-18.632	49.681	-58.376	15.950	10.586	
	900	17.200	-16.912	51.707	-63.448	15.409	68.319	
	1000	17.200	-15.192	53.519	-68.711	15.019	290.332	
	1082	17.200	-13.776	54.880	-73.174	14.778	760.	
			24.8	22.913				1
GAS	1082	9.204	11.024	77.793	-73.174	14.778		
	1100	9.205	11.187	77.942	-74.549	14.814		
	1200	9.209	12.108	78.743	-82.384	15.006		
	1300	9.212	13.029	79.480	-90.296	15.182		
	1400	9.215	13.950	80.163	-98.278	15.344		
	1500	9.217	14.872	80.799	-106.327	15.494		
	1600	9.218	15.793	81.394	-114.437	15.633		

Tl_2O THALLIUM OXIDE

		A	B	C	D	RANGE	REF.
SOL	CP	13.4	10.0			298- 852	*
LIQ	CP	22.7				852-1000	*

PHASE	T	CP	H	S	G	BT	
SOL	298	16.381	-40.	32.1	-49.571	36.341	62/*
	300	16.400	-39.970	32.201	-49.630	36.160	
	400	17.400	-38.280	37.056	-53.102	29.018	
	500	18.400	-36.490	41.046	-57.013	24.924	
	600	19.400	-34.600	44.490	-61.293	22.329	
	700	20.400	-32.610	47.555	-65.898	20.577	
	800	21.400	-30.520	50.345	-70.795	19.343	
	852	21.920	-29.393	51.708	-73.449	18.843	
			7.24	8.498			62
LIQ	852	22.700	-22.153	60.206	-73.449	18.843	
	900	22.700	-21.064	61.450	-76.369	18.547	
	1000	22.700	-18.794	63.842	-82.636	18.062	

* * * * * * * * * *

Tl_2O_3 DITHALLIUM TRIOXIDE

		A	B	C	D	RANGE	REF.
SOL	CP	31.52	0.85	-5.32		298-1107	145
	** MPT= 1107						62

PHASE	T	CP	H	S	G	BT	
SOL	298	25.789	-93.3	32.8	-103.079	75.569	62/*
	300	25.864	-93.252	32.960	-103.140	75.148	
	400	28.535	-90.514	40.819	-106.842	58.383	
	500	29.817	-87.590	47.339	-111.259	48.638	
	600	30.552	-84.568	52.846	-116.276	42.359	
	700	31.029	-81.488	57.594	-121.803	38.034	
	800	31.369	-78.367	61.761	-127.775	34.911	
	900	31.628	-75.216	65.471	-134.140	32.578	
	1000	31.838	-72.043	68.814	-140.857	30.788	
	1100	32.015	-68.850	71.857	-147.893	29.388	
	1107	32.027	-68.626	72.061	-148.397	29.301	

THALLIUM SULFIDE Tl_2S

		A	B	C	D	RANGE	REF.
SOL	CP	17.1	7.0			298- 730	*
LIQ	CP	23.8				730-1400	*

PHASE	T	CP	H	S	G	BT	
SOL	298	19.187	-22.7	38.	-34.030	24.948	63
	300	19.200	-22.664	38.119	-34.100	24.845	
	400	19.900	-20.709	43.738	-38.205	20.877	
	500	20.600	-18.684	48.254	-42.811	18.715	
	600	21.300	-16.589	52.072	-47.832	17.425	
	700	22.000	-14.424	55.408	-53.210	16.615	
	730	22.210	-13.761	56.335	-54.886	16.434	
			5.5	7.534			63
LIQ	730	23.800	-8.261	63.869	-54.886	16.434	
	800	23.800	-6.595	66.049	-59.434	16.239	
	900	23.800	-4.215	68.852	-66.182	16.073	
	1000	23.800	-1.835	71.359	-73.195	15.999	
	1100	23.800	.545	73.628	-80.446	15.985	
	1200	23.800	2.925	75.699	-87.914	16.013	
	1300	23.800	5.305	77.604	-95.580	16.071	
	1400	23.800	7.685	79.368	-103.430	16.148	

Tl_2SO_4 THALLIUM SULFATE

		A	B	C	D	RANGE	REF.
SOL-A	CP	24.0	30.0			298- 773	*
SOL-B	CP	24.0	30.0			773- 905	*
LIQ	CP	49.				905-1200	*

PHASE	T	CP	H	S	G	BT	
SOL-A	298	32.944	-223.15	48.	-237.461	174.087	62/*
	300	33.000	-223.089	48.204	-237.550	173.078	
	400	36.000	-219.639	58.108	-242.882	132.723	
	500	39.000	-215.889	66.464	-249.121	108.905	
	600	42.000	-211.839	73.839	-256.143	93.312	
	700	45.000	-207.489	80.539	-263.866	82.394	
	773	47.190	-204.124	85.110	-269.914	76.323	
							62
SOL-B	773	47.190	-204.124	85.110	-269.914	76.323	
	800	48.000	-202.839	86.744	-272.234	74.381	
	900	51.000	-197.889	92.571	-281.203	68.294	
	905	51.150	-197.634	92.854	-281.666	68.029	
			5.7	6.298			62
LIQ	905	49.000	-191.934	99.152	-281.666	68.029	
	1000	49.000	-187.279	104.043	-291.322	63.677	
	1100	49.000	-182.379	108.713	-301.963	60.003	
	1200	49.000	-177.479	112.977	-313.051	57.022	

DITHALLIUM SELENIDE

Tl_2Se

		A	B	C	D	RANGE	REF.
SOL	CP	16.67	7.8			298- 663	*
	** MPT= 663						63

PHASE	T	CP	H	S	G	BT	
SOL	298	18.996	-22.12	43.05	-34.955	25.626	185
	300	19.010	-22.085	43.168	-35.035	25.526	
	400	19.790	-20.145	48.743	-39.642	21.662	
	500	20.570	-18.127	53.243	-44.748	19.562	
	600	21.350	-16.031	57.062	-50.268	18.313	
	663	21.841	-14.670	59.218	-53.932	17.780	

* * * * * * * * * *

THALLIUM MONOSELENIDE

$TlSe$

		A	B	C	D	RANGE	REF.
SOL	CP	9.4	8.4			298- 607	185,*
	** MPT= 607						63

PHASE	T	CP	H	S	G	BT	
SOL	298	11.904	-14.57	24.6	-21.904	16.059	185
	300	11.920	-14.548	24.674	-21.950	15.993	
	400	12.760	-13.314	28.218	-24.601	13.443	
	500	13.600	-11.996	31.155	-27.574	12.054	
	600	14.440	-10.594	33.709	-30.820	11.228	
	607	14.499	-10.493	33.877	-31.056	11.183	

Tl_2Te DITHALLIUM TELLURICE

		A	B	C	D	RANGE	REF.
SOL-C	CP	16.25	6.5			298- 726	63
	** HOMOGENEITY RANGE / MPT= 726, L= 5.60						63

PHASE	T	CP	H	S	G	BT	
SOL-C	298	18.188	-19.2	41.6	-31.603	23.169	63
	300	18.200	-19.166	41.713	-31.680	23.082	
	400	18.850	-17.314	47.037	-36.129	19.743	
	500	19.500	-15.396	51.313	-41.053	17.947	
	600	20.150	-13.414	54.926	-46.370	16.892	
	700	20.800	-11.366	58.081	-52.023	16.245	
	726	20.969	-10.823	58.843	-53.543	16.120	

THULIUM

		A	B	C	D	RANGE	REF.
SOL	CP	8.759	-7.401	-0.619	6.772	298- 800	21
	CP	6.128	1.591	-1.992		800-1818	21
LIQ	CP	9.89				1818-2220	21
GAS	CP	5.153				2220-2400	21

PHASE	T	CP	H	S	G	BT	P	
SOL	298	6.458	0	17.69	-5.274	3.867		21
	300	6.460	.012	17.730	-5.307	3.867		
	400	6.495	.661	19.596	-7.178	3.922		
	500	6.504	1.310	21.046	-9.213	4.027		
	600	6.584	1.964	22.237	-11.378	4.145		
	700	6.770	2.631	23.265	-13.655	4.264		
	800	7.076	3.322	24.187	-16.028	4.379		
	800	7.090	3.322	24.187	-16.028	4.379		
	900	7.314	4.042	25.036	-18.490	4.490		
	1000	7.520	4.784	25.817	-21.033	4.597		
	1100	7.713	5.546	26.543	-23.651	4.700	.005	
	1200	7.899	6.327	27.222	-26.340	4.798	.040	
	1300	8.078	7.126	27.861	-29.094	4.892	.223	
	1400	8.254	7.942	28.466	-31.911	4.982	.969	
	1500	8.426	8.776	29.042	-34.786	5.069	3.441	
	1600	8.596	9.627	29.591	-37.718	5.153	10.353	
	1700	8.764	10.495	30.117	-40.704	5.234	27.182	
	1800	8.930	11.380	30.623	-43.741	5.312	63.717	
	1818	8.960	11.541	30.712	-44.293	5.325	73.494	
			4.025	2.214				21
LIQ	1818	9.890	15.566	32.926	-44.293	5.325	73.494	
	1900	9.890	16.377	33.362	-47.011	5.408	129.340	
	2000	9.890	17.366	33.869	-50.373	5.505	240.628	
	2100	9.890	18.355	34.352	-53.784	5.598	419.549	
	2200	9.890	19.344	34.812	-57.242	5.687	691.845	
	2220	9.890	19.542	34.902	-57.940	5.705	760.	
			45.577	20.530				21
GAS	2220	5.153	65.119	55.432	-57.940	5.705		
	2300	5.153	65.531	55.614	-62.381	5.928		
	2400	5.153	66.046	55.833	-67.954	6.189		

$TmCl_3$ THULIUM CHLORIDE

		A	B	C	D	RANGE	REF.
SOL	CP	22.85	2.8	-0.3		298-1118	*
LIQ	CP	35.5				1118-1300	*

PHASE	T	CP	H	S	G	BT	
SOL	298	23.347	-235.8	35.1	-246.265	180.541	19/*
	300	23.357	-235.757	35.244	-246.330	179.476	
	400	23.782	-233.399	42.025	-250.209	136.726	
	500	24.130	-231.003	47.370	-254.688	111.339	
	600	24.447	-228.574	51.798	-259.653	94.591	
	700	24.749	-226.114	55.589	-265.026	82.756	
	800	25.043	-223.624	58.913	-270.755	73.977	
	900	25.333	-221.105	61.880	-276.797	67.225	
	1000	25.620	-218.558	64.564	-283.121	61.884	
	1100	25.905	-215.982	67.019	-289.702	57.566	
	1118	25.956	-215.515	67.440	-290.912	56.876	
			8.35	7.469			*
LIQ	1118	35.500	-207.165	74.908	-290.912	56.876	
	1200	35.500	-204.254	77.421	-297.159	54.127	
	1300	35.500	-200.704	80.263	-305.045	51.290	

THULIUM OXIDE

Tm_2O_3

		A	B	C	D	RANGE	REF.
SOL	CP	30.75	4.65	-3.8		298-1800	*

PHASE	T	CP	H	S	G	BT	
SOL	298	27.862	-451.4	33.4	-461.358	338.230	1
	300	27.923	-451.348	33.573	-461.420	336.190	
	400	30.235	-448.427	41.960	-465.211	254.214	
	500	31.555	-445.333	48.859	-469.763	205.361	
	600	32.484	-442.129	54.698	-474.948	173.023	
	700	33.229	-438.842	59.764	-480.677	150.094	
	800	33.876	-435.486	64.244	-486.881	133.028	
	900	34.466	-432.069	68.268	-493.510	119.857	
	1000	35.020	-428.594	71.929	-500.523	109.404	
	1100	35.551	-425.066	75.291	-507.886	100.921	
	1200	36.066	-421.485	78.407	-515.573	93.911	
	1300	36.570	-417.853	81.314	-523.561	88.030	
	1400	37.066	-414.171	84.042	-531.830	83.034	
	1500	37.556	-410.440	86.616	-540.364	78.742	
	1600	38.042	-406.660	89.055	-549.148	75.020	
	1700	38.524	-402.832	91.376	-558.171	71.767	
	1800	39.003	-398.955	93.592	-567.420	68.903	

U URANIUM

		A	B	C	D	RANGE	REF.
SOL-A	CP	6.547	-0.870	-0.229	6.518	298- 941	21
SOL-B	CP	10.26				941-1048	21
SOL-C	CP	9.15				1048-1405	21
LIQ	CP	11.45				1405-4407	21
GAS	CP	10.572				4407-5000	21

PHASE	T	CP	H	S	G	BT	P	
SOL-A	298	6.609	0	12.02	-3.584	2.627		21
	300	6.618	.012	12.061	-3.606	2.627		
	400	7.099	.698	14.030	-4.914	2.685		
	500	7.650	1.434	15.671	-6.401	2.798		
	600	8.308	2.231	17.122	-8.042	2.930		
	700	9.085	3.100	18.460	-9.822	3.067		
	800	9.987	4.053	19.731	-11.732	3.205		
	900	11.015	5.102	20.965	-13.767	3.344		
	941	11.474	5.563	21.466	-14.637	3.400		
			0.667	.709				21
SOL-B	941	10.260	6.230	22.175	-14.637	3.400		
	1000	10.260	6.835	22.798	-15.964	3.489		
	1048	10.260	7.327	23.279	-17.070	3.560		
			1.137	1.085				21
SOL-C	1048	9.150	8.464	24.364	-17.070	3.560		
	1100	9.150	8.940	24.808	-18.348	3.646		
	1200	9.150	9.855	25.604	-20.869	3.801		
	1300	9.150	10.770	26.336	-23.467	3.946		
	1400	9.150	11.685	27.014	-26.135	4.080		
	1405	9.150	11.731	27.047	-26.270	4.087		
			2.036	1.449				21
LIQ	1405	11.450	13.767	28.496	-26.270	4.087		
	1500	11.450	14.855	29.245	-29.013	4.228		
	1600	11.450	16.000	29.984	-31.975	4.368		
	1700	11.450	17.145	30.678	-35.008	4.501		
	1800	11.450	18.290	31.333	-38.109	4.628		
	1900	11.450	19.435	31.952	-41.274	4.748		
	2000	11.450	20.580	32.539	-44.498	4.863		
	2100	11.450	21.725	33.098	-47.780	4.973		
	2200	11.450	22.870	33.630	-51.117	5.079	.002	
	2300	11.450	24.015	34.139	-54.506	5.180	.006	
	2400	11.450	25.160	34.627	-57.944	5.277	.016	
	2500	11.450	26.305	35.094	-61.430	5.371	.042	
	2600	11.450	27.450	35.543	-64.962	5.461	.102	
	2700	11.450	28.595	35.975	-68.538	5.549	.231	

URANIUM (CONTINUED)

PHASE	T	CP	H	S	G	BT	P	
LIQ	2800	11.450	29.740	36.392	-72.157	5.633	.490	
	2900	11.450	30.885	36.793	-75.816	5.714	.989	
	3000	11.450	32.030	37.182	-79.515	5.793	1.899	
	3100	11.450	33.175	37.557	-83.252	5.870	3.496	
	3200	11.450	34.320	37.921	-87.026	5.944	6.187	
	3300	11.450	35.465	38.273	-90.836	6.017	10.568	
	3400	11.450	36.610	38.615	-94.680	6.087	17.482	
	3500	11.450	37.755	38.947	-98.558	6.155	28.079	
	3600	11.450	38.900	39.269	-102.469	6.222	43.903	
	3700	11.450	40.045	39.583	-106.412	6.286	66.971	
	3800	11.450	41.190	39.888	-110.386	6.349	99.867	
	3900	11.450	42.335	40.186	-114.389	6.411	145.841	
	4000	11.450	43.480	40.476	-118.422	6.471	208.906	
	4100	11.450	44.625	40.758	-122.484	6.530	293.940	
	4200	11.450	45.770	41.034	-126.574	6.587	406.790	
	4300	11.450	46.915	41.304	-130.691	6.643	554.368	
	4400	11.450	48.060	41.567	-134.834	6.698	744.743	
	4407	11.450	48.140	41.585	-135.125	6.702	760.	
			110.924	25.170				21
GAS	4407	10.572	159.064	66.755	-135.125	6.702		
	4500	10.572	160.047	66.976	-141.344	6.866		
	4600	10.572	161.104	67.208	-148.053	7.035		
	4700	10.572	162.161	67.435	-154.785	7.198		
	4800	10.572	163.219	67.658	-161.540	7.356		
	4900	10.572	164.276	67.876	-168.317	7.508		
	5000	10.572	165.333	68.090	-175.115	7.655		

UBr_3 URANIUM TRIBROMIDE

		A	B	C	D	RANGE	REF.
SOL	CP	22.9	7.0			298-1000	*
LIQ	CP	31.7				1000-1800	*

** DEC.

PHASE	T	CP	H	S	G	BT	
SOL	298	24.987	-172.3	45.	-185.717	136.152	27
	300	25.000	-172.254	45.155	-185.800	135.374	
	400	25.700	-169.719	52.443	-190.696	104.205	
	500	26.400	-167.114	58.253	-196.240	85.788	
	600	27.100	-164.439	63.128	-202.315	73.703	
	700	27.800	-161.694	67.358	-208.844	65.213	
	800	28.500	-158.879	71.116	-215.771	58.954	
	900	29.200	-155.994	74.513	-223.055	54.173	
	1000	29.900	-153.039	77.626	-230.664	50.418	
			10.5	10.500			27
LIQ	1000	31.700	-142.539	88.126	-230.664	50.418	
	1100	31.700	-139.369	91.147	-239.630	47.617	
	1200	31.700	-136.199	93.905	-248.885	45.334	
	1300	31.700	-133.029	96.443	-258.404	43.448	
	1400	31.700	-129.859	98.792	-268.167	41.868	
	1500	31.700	-126.689	100.979	-278.157	40.533	
	1600	31.700	-123.519	103.025	-288.358	39.393	
	1700	31.700	-120.349	104.946	-298.758	38.413	
	1800	31.700	-117.179	106.758	-309.344	37.565	

URANIUM TRIIODIDE

UJ_3

		A	B	C	D	RANGE	REF.
SOL	CP	23.3	7.3			298- 900	*

PHASE	T	CP	H	S	G	BT	
SOL	298	25.476	-90.	53.5	-105.951	77.675	182*/*
	300	25.490	-89.953	53.658	-106.050	77.268	
	400	26.220	-87.367	61.091	-111.804	61.095	
	500	26.950	-84.709	67.020	-118.219	51.680	
	600	27.680	-81.977	71.998	-125.176	45.602	
	700	28.410	-79.173	76.320	-132.597	41.404	
	800	29.140	-76.295	80.161	-140.424	38.367	
	900	29.870	-73.345	83.635	-148.617	36.094	

* * * * * * * * * *

URANIUM TETRAIODIDE

UJ_4

		A	B	C	D	RANGE	REF.
SOL	CP	34.8	2.38	-4.72		298- 779	27
LIQ	CP	39.6				779-1030	27
	** BPT= 1030, GAS(UJ4 + J2 + J)						27

PHASE	T	CP	H	S	G	BT	
SOL	298	30.200	-102.2	63.5	-121.133	88.804	182*/*
	300	30.270	-102.144	63.687	-121.250	88.343	
	400	32.802	-98.974	72.789	-128.090	69.994	
	500	34.102	-95.623	80.262	-135.754	59.346	
	600	34.917	-92.169	86.556	-144.103	52.497	
	700	35.503	-88.647	91.984	-153.036	47.786	
	779	35.876	-85.827	95.801	-160.456	45.022	
			16.9	21.694			27
LIQ	779	39.600	-68.927	117.495	-160.456	45.022	
	800	39.600	-68.096	118.549	-162.935	44.518	
	900	39.600	-64.136	123.213	-175.027	42.508	
	1000	39.600	-60.176	127.385	-187.561	40.997	
	1030	39.600	-58.988	128.556	-191.400	40.618	

UO_2Br_2 URANYL BROMIDE

		A	B	C	D	RANGE	REF.
SOL	CP	28.19	4.19	-2.56		298-1000	*

PHASE	T	CP	H	S	G	BT	
SOL	298	26.559	-276.6	40.5	-288.675	211.633	27
	300	26.603	-276.551	40.664	-288.750	210.383	
	400	28.266	-273.799	48.571	-293.227	160.233	
	500	29.261	-270.919	54.992	-298.415	130.455	
	600	29.993	-267.955	60.395	-304.192	110.817	
	700	30.601	-264.924	65.065	-310.470	96.946	
	800	31.142	-261.837	69.187	-317.186	86.663	
	900	31.645	-258.697	72.884	-324.293	78.760	
	1000	32.124	-255.509	76.243	-331.752	72.514	

* * * * * * * * * *

$UOBr_3$ URANIUM OXYTRIBROMIDE

		A	B	C	D	RANGE.	REF.
SOL	CP	31.2	4.9	-3.3		298-1100	*

PHASE	T	CP	H	S	G	BT	
SOL	298	28.949	-236.	49.	-250.609	183.726	27/*
	300	29.003	-235.946	49.179	-250.700	182.660	
	400	31.097	-232.930	57.843	-256.067	139.927	
	500	32.330	-229.754	64.924	-262.216	114.630	
	600	33.223	-226.475	70.900	-269.015	98.002	
	700	33.957	-223.115	76.078	-276.370	86.298	
	800	34.604	-219.686	80.656	-284.211	77.653	
	900	35.203	-216.196	84.766	-292.485	71.035	
	1000	35.770	-212.647	88.505	-301.152	65.826	
	1100	36.317	-209.042	91.940	-310.176	61.635	

URANIUM TRIOXIDE MONOHYDRATE $UO_3 \cdot H_2O$

		A	B	C	D	RANGE	REF.
SOL	CP	29.5	15.5	-5.4		298- 800	*

PHASE	T	CP	H	S	G	BT	
SOL	298	28.047	-367.5	32.1	-377.071	276.437	27/*
	300	28.150	-367.448	32.274	-377.130	274.776	
	400	32.325	-364.406	40.998	-380.805	208.090	
	500	35.090	-361.028	48.523	-385.290	168.433	
	600	37.300	-357.406	55.122	-390.479	142.251	
	700	39.248	-353.577	61.020	-396.291	123.744	
	800	41.056	-349.561	66.380	-402.665	110.018	

* * * * * * * * * *

URANIUM MONOSULFIDE US

		A	B	C	D	RANGE	REF.
SOL	CP	12.63	1.56	-0.905		298-2000	135
	** MPT=	2735					63

PHASE	T	CP	H	S	G	BT	
SOL	298	12.077	-76.5	18.64	-82.058	60.158	135*/27
	300	12.092	-76.478	18.715	-82.092	59.812	
	400	12.688	-75.235	22.284	-84.149	45.983	
	500	13.048	-73.948	25.157	-86.526	37.826	
	600	13.315	-72.629	27.560	-89.165	32.483	
	700	13.537	-71.286	29.630	-92.027	28.736	
	800	13.737	-69.922	31.451	-95.083	25.979	
	900	13.922	-68.539	33.079	-98.311	23.876	
	1000	14.099	-67.138	34.555	-101.693	22.228	
	1100	14.271	-65.719	35.907	-105.217	20.908	
	1200	14.439	-64.284	37.156	-108.871	19.831	
	1300	14.604	-62.832	38.319	-112.646	18.940	
	1400	14.768	-61.363	39.407	-116.533	18.194	
	1500	14.930	-59.878	40.431	-120.525	17.563	
	1600	15.091	-58.377	41.400	-124.617	17.024	
	1700	15.251	-56.860	42.320	-128.803	16.561	
	1800	15.410	-55.327	43.196	-133.080	16.160	
	1900	15.569	-53.778	44.033	-137.441	15.811	
	2000	15.727	-52.213	44.836	-141.885	15.507	

U_2S_3 DIURANIUM TRISULFICE

		A	B	C	D	RANGE	REF.
SOL	CP	29.8	3.85	-0.9		298-2000	27
	** MPT= 2300						63

PHASE	T	CP	H	S	G	BT	
SOL	298	29.935	-196.	45.	-209.417	153.527	27,*
	300	29.955	-195.945	45.185	-209.500	152.641	
	400	30.777	-192.905	53.924	-214.475	117.199	
	500	31.365	-189.797	60.858	-220.226	96.273	
	600	31.860	-186.635	66.621	-226.607	82.553	
	700	32.311	-183.426	71.567	-233.523	72.919	
	800	32.739	-180.173	75.909	-240.901	65.820	
	900	33.154	-176.879	79.789	-248.689	60.398	
	1000	33.560	-173.543	83.304	-256.846	56.141	
	1100	33.961	-170.167	86.521	-265.340	52.725	
	1200	34.357	-166.751	89.493	-274.143	49.935	
	1300	34.752	-163.295	92.259	-283.232	47.622	
	1400	35.144	-159.801	94.848	-292.588	45.681	
	1500	35.535	-156.267	97.286	-302.196	44.036	
	1600	35.925	-152.694	99.592	-312.041	42.629	
	1700	36.314	-149.082	101.782	-322.111	41.416	
	1800	36.702	-145.431	103.869	-332.394	40.364	
	1900	37.090	-141.741	105.863	-342.882	39.446	
	2000	37.477	-138.013	107.776	-353.564	38.641	

URANIUM DISULFIDE

US_2

		A	B	C	D	RANGE	REF.
SOL-B	CP	17.16	2.30			298-2000	63
	** DEC. US2 = US(2-X) + X/2 S2						63

PHASE	T	CP	H	S	G	BT	
SOL-B	298	17.846	-110.	26.4	-117.871	86.414	27,135,*
	300	17.850	-109.967	26.510	-117.920	85.916	
	400	18.080	-108.170	31.677	-120.841	66.033	
	500	18.310	-106.351	35.736	-124.219	54.303	
	600	18.540	-104.508	39.095	-127.965	46.618	
	700	18.770	-102.643	41.970	-132.022	41.225	
	800	19.000	-100.754	44.491	-136.348	37.253	
	900	19.230	-98.843	46.743	-140.911	34.223	
	1000	19.460	-96.908	48.781	-145.689	31.845	
	1100	19.690	-94.951	50.646	-150.662	29.938	
	1200	19.920	-92.970	52.369	-155.814	28.381	
	1300	20.150	-90.967	53.973	-161.132	27.092	
	1400	20.380	-88.940	55.474	-166.605	26.012	
	1500	20.610	-86.891	56.888	-172.224	25.096	
	1600	20.840	-84.818	58.226	-177.980	24.314	
	1700	21.070	-82.723	59.496	-183.866	23.641	
	1800	21.300	-80.604	60.707	-189.877	23.057	
	1900	21.530	-78.463	61.865	-196.006	22.549	
	2000	21.760	-76.298	62.975	-202.248	22.104	

* * * * * * * * * *

URANIUM DISULFATE

$U(SO_4)_2$

		A	B	C	D	RANGE	REF.
SOL	CP	25.0	55.2			298-1000	*

PHASE	T	CP	H	S	G	BT	
SOL	298	41.458	-542.	38.5	-553.479	405.765	27
	300	41.560	-541.923	38.757	-553.550	403.315	
	400	47.080	-537.491	51.469	-558.079	304.961	
	500	52.600	-532.507	62.567	-563.791	246.466	
	600	58.120	-526.971	72.645	-570.558	207.854	
	700	63.640	-520.883	82.019	-578.297	180.577	
	800	69.160	-514.243	90.877	-586.945	160.368	
	900	74.680	-507.051	99.342	-596.459	144.860	
	1000	80.200	-499.307	107.496	-606.803	132.635	

USe

URANIUM MONOSELENIDE

		A	B	C	D	RANGE	REF.
SOL	CP	12.644	1.53			298-1600	*

PHASE	T	CP	H	S	G	BT	
SOL	298	13.100	-65.9	23.07	-72.778	53.355	63/192
	300	13.103	-65.876	23.151	-72.821	53.057	
	400	13.256	-64.558	26.941	-75.334	41.166	
	500	13.409	-63.225	29.916	-78.183	34.178	
	600	13.562	-61.876	32.374	-81.301	29.618	
	700	13.715	-60.512	34.476	-84.646	26.431	
	800	13.868	-59.133	36.318	-88.187	24.095	
	900	14.021	-57.739	37.960	-91.902	22.320	
	1000	14.174	-56.329	39.445	-95.774	20.934	
	1100	14.327	-54.904	40.803	-99.787	19.829	
	1200	14.480	-53.463	42.056	-103.931	18.931	
	1300	14.633	-52.008	43.221	-108.196	18.192	
	1400	14.786	-50.537	44.311	-112.573	17.576	
	1500	14.939	-49.051	45.337	-117.056	17.057	
	1600	15.092	-47.549	46.306	-121.638	16.617	

URANIUM BISMUTHIDE

		A	B	C	D	RANGE	REF.
SOL	CP	11.5	3.5			298-1723	*
	** MPT= 1723 (PERITEC.)						1

PHASE	T	CP	H	S	G	BT	
SOL	298	12.544	-17.6	23.2	-24.517	17.974	21*/*
	300	12.550	-17.577	23.278	-24.560	17.894	
	400	12.900	-16.304	26.936	-27.079	14.797	
	500	13.250	-14.997	29.852	-29.923	13.081	
	600	13.600	-13.654	32.299	-33.034	12.034	
	700	13.950	-12.277	34.422	-36.372	11.357	
	800	14.300	-10.864	36.307	-39.910	10.904	
	900	14.650	-9.417	38.012	-43.627	10.596	
	1000	15.000	-7.934	39.573	-47.508	10.384	
	1100	15.350	-6.417	41.019	-51.538	10.241	
	1200	15.700	-4.864	42.370	-55.708	10.147	
	1300	16.050	-3.277	43.640	-60.009	10.090	
	1400	16.400	-1.654	44.843	-64.434	10.060	
	1500	16.750	.003	45.986	-68.976	10.051	
	1600	17.100	1.696	47.078	-73.630	10.059	
	1700	17.450	3.423	48.126	-78.390	10.079	
	1723	17.530	3.825	48.361	-79.500	10.085	

U_3Bi_4 3-URANIUM 4-BISMUTHIDE

		A	B	C	D	RANGE	REF.
SOL	CP	40.0	13.3			298-1423	*
	** MPT= 1423 (PERITEC.)						21

PHASE	T	CP	H	S	G	BT	
SOL	298	43.965	-63.7	81.9	-88.118	64.601	21*/*
	300	43.990	-63.619	82.172	-88.270	64.313	
	400	45.320	-59.153	95.009	-97.157	53.091	
	500	46.650	-54.555	105.265	-107.187	46.858	
	600	47.980	-49.823	113.888	-118.156	43.044	
	700	49.310	-44.959	121.384	-129.927	40.571	
	800	50.640	-39.961	128.055	-142.405	38.909	
	900	51.970	-34.831	134.097	-155.518	37.770	
	1000	53.300	-29.567	139.641	-169.208	36.985	
	1100	54.630	-24.171	144.783	-183.432	36.450	
	1200	55.960	-18.641	149.594	-198.154	36.094	
	1300	57.290	-12.979	154.126	-213.342	35.871	
	1400	58.620	-7.183	158.420	-228.971	35.749	
	1423	58.926	-5.831	159.378	-232.626	35.732	

* * * * * * * * * *

UBi_2 URANIUM 2-BISMUTHIDE

		A	B	C	D	RANGE	REF.
SOL	CP	16.9	7.0			298-1283	*
	** MPT= 1283 (PERITEC.)						21

PHASE	T	CP	H	S	G	BT	
SOL	298	18.987	-24.5	36.3	-35.323	25.896	21*/*
	300	19.000	-24.465	36.417	-35.390	25.785	
	400	19.700	-22.530	41.979	-39.322	21.487	
	500	20.400	-20.525	46.450	-43.750	19.126	
	600	21.100	-18.450	50.232	-48.589	17.701	
	700	21.800	-16.305	53.537	-53.781	16.793	
	800	22.500	-14.090	56.494	-59.285	16.198	
	900	23.200	-11.805	59.184	-65.070	15.803	
	1000	23.900	-9.450	61.665	-71.114	15.544	
	1100	24.600	-7.025	63.975	-77.398	15.380	
	1200	25.300	-4.530	66.146	-83.905	15.283	
	1283	25.881	-2.406	67.857	-89.467	15.242	

2-URANIUM 3-CARBIDE

U_2C_3

		A	B	C	D	RANGE	REF.
SOL	CP	29.9	3.06	-3.71		298-2000	65

PHASE	T	CP	H	S	G	BT	
SOL	298	26.639	-43.4	33.07	-53.260	39.046	208/1
	300	26.696	-43.351	33.235	-53.321	38.850	
	400	28.805	-40.563	41.241	-57.059	31.180	
	500	29.946	-37.621	47.802	-61.521	26.895	
	600	30.705	-34.586	53.332	-66.585	24.257	
	700	31.285	-31.485	58.111	-72.163	22.533	
	800	31.768	-28.332	62.320	-78.188	21.363	
	900	32.196	-25.134	66.087	-84.612	20.549	
	1000	32.589	-21.894	69.500	-91.394	19.977	
	1100	32.959	-18.616	72.624	-98.503	19.573	
	1200	33.314	-15.303	75.507	-105.911	19.292	
	1300	33.658	-11.954	78.187	-113.597	19.100	
	1400	33.995	-8.571	80.694	-121.543	18.976	
	1500	34.325	-5.155	83.050	-129.731	18.904	
	1600	34.651	-1.706	85.276	-138.148	18.873	
	1700	34.974	1.775	87.387	-146.782	18.873	
	1800	35.293	5.288	89.395	-155.622	18.898	
	1900	35.611	8.834	91.311	-164.658	18.943	
	2000	35.927	12.410	93.146	-173.882	19.003	

$UC_{1.93}$ URANIUM 1.93-CARBIDE

		A	B	C	D	RANGE	REF.
SOL-A	CP	22.352	-8.946	-5.138	6.917	298-2038	9
SOL-B	CP	29.44				2038-2800	9

PHASE	T	CP	H	S	G	BT	
SOL-A	298	14.520	-20.5	16.33	-25.369	18.598	9
	300	14.582	-20.473	16.420	-25.399	18.506	
	400	16.669	-18.894	20.949	-27.273	14.904	
	500	17.553	-17.177	24.775	-29.565	12.925	
	600	18.047	-15.396	28.022	-32.209	11.734	
	700	18.431	-13.572	30.834	-35.155	10.977	
	800	18.819	-11.709	33.320	-38.365	10.482	
	900	19.269	-9.806	35.561	-41.811	10.154	
	1000	19.809	-7.853	37.619	-45.471	9.939	
	1100	20.456	-5.840	39.536	-49.330	9.802	
	1200	21.220	-3.757	41.348	-53.375	9.722	
	1300	22.108	-1.592	43.081	-57.597	9.684	
	1400	23.123	.668	44.755	-61.989	9.678	
	1500	24.268	3.037	46.389	-66.547	9.697	
	1600	25.545	5.526	47.995	-71.266	9.736	
	1700	26.956	8.150	49.586	-76.145	9.790	
	1800	28.502	10.922	51.170	-81.183	9.858	
	1900	30.183	13.855	52.755	-86.379	9.937	
	2000	32.000	16.963	54.349	-91.734	10.026	
	2038	32.726	18.193	54.958	-93.811	10.061	
			2.57	1.261			9
SOL-B	2038	29.440	20.763	56.219	-93.811	10.061	
	2100	29.440	22.588	57.101	-97.324	10.130	
	2200	29.440	25.532	58.471	-103.103	10.244	
	2300	29.440	28.476	59.779	-109.016	10.360	
	2400	29.440	31.420	61.032	-115.057	10.479	
	2500	29.440	34.364	62.234	-121.221	10.599	
	2600	29.440	37.308	63.389	-127.503	10.719	
	2700	29.440	40.252	64.500	-133.897	10.840	
	2800	29.440	43.196	65.570	-140.401	10.960	

3-URANIUM SILICON U_3Si

		A	B	C	D	RANGE	REF.
SOL	CP	34.67	0.0	-7.5		298-1200	*

PHASE	T	CP	H	S	G	BT	
SOL	298	26.233	-22.	40.	-33.926	24.872	27
	300	26.337	-21.951	40.163	-34.000	24.772	
	400	29.983	-19.109	48.314	-38.435	21.003	
	500	31.670	-16.017	55.206	-43.620	19.069	
	600	32.587	-12.800	61.069	-49.442	18.012	
	700	33.139	-9.512	66.137	-55.808	17.426	
	800	33.498	-6.179	70.587	-62.649	17.117	
	900	33.744	-2.816	74.548	-69.909	16.979	
	1000	33.920	.568	78.113	-77.545	16.950	
	1100	34.050	3.966	81.352	-85.521	16.994	
	1200	34.149	7.377	84.319	-93.806	17.087	

* * * * * * * * * *

3-URANIUM 2-SILICON U_3Si_2

		A	B	C	D	RANGE	REF.
SOL	CP	40.48	0.58	-8.41		298-1800	27,*

PHASE	T	CP	H	S	G	BT	
SOL	298	31.192	-40.8	47.2	-54.873	40.228	27
	300	31.310	-40.742	47.393	-54.960	40.044	
	400	35.456	-37.375	57.053	-60.196	32.894	
	500	37.406	-33.721	65.197	-66.320	28.992	
	600	38.492	-29.922	72.122	-73.195	26.665	
	700	39.170	-26.036	78.110	-80.713	25.203	
	800	39.630	-22.095	83.372	-88.792	24.260	
	900	39.964	-18.114	88.060	-97.368	23.647	
	1000	40.219	-14.105	92.284	-106.389	23.254	
	1100	40.423	-10.072	96.128	-115.813	23.013	
	1200	40.592	-6.021	99.652	-125.604	22.879	
	1300	40.736	-1.955	102.907	-135.734	22.822	
	1400	40.863	2.125	105.931	-146.178	22.822	
	1500	40.976	6.218	108.754	-156.913	22.865	
	1600	41.079	10.320	111.402	-167.923	22.940	
	1700	41.175	14.433	113.895	-179.189	23.039	
	1800	41.264	18.555	116.251	-190.697	23.157	

USi

URANIUM SILICON

		A	B	C	D	RANGE	REF.
SOL	CP	15.43	0.39	-3.11		298-1800	27,*

PHASE	T	CP	H	S	G	BT	
SOL	298	12.048	-20.2	15.9	-24.941	18.284	27
	300	12.091	-20.178	15.975	-24.970	18.193	
	400	13.642	-18.880	19.697	-26.759	14.622	
	500	14.381	-17.475	22.829	-28.890	12.629	
	600	14.800	-16.014	25.491	-31.309	11.406	
	700	15.068	-14.520	27.794	-33.976	10.609	
	800	15.256	-13.003	29.819	-36.859	10.071	
	900	15.397	-11.470	31.624	-39.932	9.698	
	1000	15.509	-9.925	33.253	-43.178	9.438	
	1100	15.602	-8.369	34.735	-46.578	9.255	
	1200	15.682	-6.805	36.096	-50.121	9.129	
	1300	15.753	-5.233	37.354	-53.794	9.045	
	1400	15.817	-3.655	38.524	-57.589	8.991	
	1500	15.877	-2.070	39.618	-61.496	8.961	
	1600	15.933	-.479	40.644	-65.510	8.949	
	1700	15.985	1.117	41.612	-69.623	8.952	
	1800	16.036	2.718	42.527	-73.830	8.965	

3-URANIUM 5-SILICON U_3Si_5

		A	B	C	D	RANGE	REF.
SOL	CP	57.91	2.32	-11.16		298-1800	27,*

PHASE	T	CP	H	S	G	BT	
SOL	298	46.047	-84.7	55.3	-101.188	74.183	27
	300	46.206	-84.615	55.585	-101.290	73.800	
	400	51.863	-79.672	69.764	-107.578	58.786	
	500	54.606	-74.335	81.663	-115.167	50.346	
	600	56.202	-68.788	91.771	-123.851	45.119	
	700	57.256	-63.112	100.519	-133.476	41.679	
	800	58.022	-57.347	108.217	-143.920	39.322	
	900	58.620	-51.513	115.087	-155.092	37.666	
	1000	59.114	-45.626	121.289	-166.915	36.484	
	1100	59.540	-39.693	126.944	-179.331	35.635	
	1200	59.919	-33.720	132.141	-192.289	35.025	
	1300	60.266	-27.710	136.951	-205.747	34.594	
	1400	60.589	-21.667	141.429	-219.668	34.296	
	1500	60.894	-15.593	145.620	-234.023	34.102	
	1600	61.186	-9.489	149.559	-248.784	33.987	
	1700	61.468	-3.356	153.277	-263.927	33.935	
	1800	61.742	2.804	156.798	-279.433	33.932	

USi_2 URANIUM 2-SILICON

		A	B	C	D	RANGE	REF.
SOL	CP	21.24	0.97	-4.02		298-1800	27,*

PHASE	T	CP	H	S	G	BT	
SOL	298	17.007	-31.	19.6	-36.844	27.011	27
	300	17.064	-30.968	19.705	-36.880	26.871	
	400	19.115	-29.146	24.936	-39.120	21.377	
	500	20.117	-27.179	29.320	-41.839	18.290	
	600	20.705	-25.136	33.044	-44.962	16.380	
	700	21.099	-23.044	36.267	-48.431	15.123	
	800	21.388	-20.919	39.104	-52.202	14.263	
	900	21.617	-18.769	41.637	-56.242	13.659	
	1000	21.808	-16.597	43.924	-60.522	13.229	
	1100	21.975	-14.408	46.011	-65.020	12.920	
	1200	22.125	-12.203	47.930	-69.718	12.699	
	1300	22.263	-9.983	49.706	-74.601	12.543	
	1400	22.393	-7.750	51.361	-79.655	12.436	
	1500	22.516	-5.505	52.910	-84.870	12.367	
	1600	22.635	-3.247	54.367	-90.234	12.327	
	1700	22.750	-.978	55.743	-95.740	12.310	
	1800	22.862	1.303	57.046	-101.380	12.311	

URANIUM 3-SILICON

		A	B	C	D	RANGE	REF.
SOL	CP	27.05	1.54	-4.94		298-1800	27,*

PHASE	T	CP	H	S	G	BT	
SOL	298	21.952	-31.2	25.4	-38.773	28.425	27
	300	22.023	-31.159	25.536	-38.820	28.284	
	400	24.578	-28.812	32.271	-41.721	22.798	
	500	25.844	-26.285	37.905	-45.237	19.776	
	600	26.602	-23.660	42.689	-49.273	17.950	
	700	27.120	-20.972	46.831	-53.754	16.785	
	800	27.510	-18.240	50.479	-58.623	16.017	
	900	27.826	-15.473	53.738	-63.837	15.504	
	1000	28.096	-12.676	56.684	-69.360	15.161	
	1100	28.336	-9.854	59.373	-75.165	14.936	
	1200	28.555	-7.010	61.848	-81.228	14.796	
	1300	28.760	-4.144	64.142	-87.529	14.717	
	1400	28.954	-1.258	66.281	-94.051	14.684	
	1500	29.140	1.647	68.285	-100.780	14.686	
	1600	29.321	4.570	70.171	-107.704	14.714	
	1700	29.497	7.511	71.954	-114.811	14.762	
	1800	29.670	10.469	73.645	-122.092	14.826	

U_5Ge_3 5-URANIUM 3-GERMANIUM

		A	B	C	D	RANGE	REF.
SOL	CP	50.21	16.38	-4.74		298-1400	27,*

PHASE	T	CP	H	S	G	BT	
SOL	298	49.761	-56.2	89.6	-82.914	60.786	27
	300	49.857	-56.108	89.908	-83.080	60.532	
	400	53.799	-50.909	104.839	-92.844	50.734	
	500	56.504	-45.387	117.147	-103.961	45.447	
	600	58.721	-39.624	127.650	-116.214	42.336	
	700	60.709	-33.651	136.853	-129.448	40.421	
	800	62.573	-27.486	145.083	-143.552	39.222	
	900	64.367	-21.138	152.557	-158.439	38.480	
	1000	66.116	-14.614	159.429	-174.043	38.042	
	1100	67.836	-7.916	165.812	-190.309	37.816	
	1200	69.537	-1.047	171.787	-207.192	37.740	
	1300	71.224	5.991	177.420	-224.655	37.773	
	1400	72.900	13.197	182.759	-242.666	37.887	

* * * * * * * * * *

UGe URANIUM GERMANIUM

		A	B	C	D	RANGE	REF.
SOL	CP	12.26	3.64	-0.94		298-1400	27,*

PHASE	T	CP	H	S	G	BT	
SOL	298	12.288	-14.7	21.6	-21.140	15.498	27
	300	12.308	-14.677	21.676	-21.180	15.432	
	400	13.128	-13.402	25.339	-23.538	12.862	
	500	13.704	-12.059	28.333	-26.226	11.465	
	600	14.183	-10.665	30.874	-29.189	10.634	
	700	14.616	-9.224	33.094	-32.390	10.114	
	800	15.025	-7.742	35.072	-35.800	9.781	
	900	15.420	-6.220	36.865	-39.398	9.568	
	1000	15.806	-4.658	38.510	-43.168	9.436	
	1100	16.186	-3.059	40.034	-47.096	9.358	
	1200	16.563	-1.421	41.458	-51.171	9.321	
	1300	16.936	.254	42.799	-55.385	9.312	
	1400	17.308	1.966	44.068	-59.729	9.325	

3-URANIUM 5-GERMANIUM U_3Ge_5

		A	B	C	D	RANGE	REF.
SOL	CP	47.87	12.78	-2.78		298-1400	27,*

PHASE	T	CP	H	S	G	BT	
SOL	298	48.553	-57.3	84.1	-82.374	60.390	27
	300	48.615	-57.210	84.401	-82.530	60.131	
	400	51.244	-52.207	98.774	-91.717	50.119	
	500	53.148	-46.984	110.421	-102.195	44.675	
	600	54.766	-41.587	120.257	-113.741	41.436	
	700	56.249	-36.036	128.812	-126.204	39.408	
	800	57.660	-30.340	136.416	-139.472	38.107	
	900	59.029	-24.505	143.286	-153.463	37.271	
	1000	60.372	-18.535	149.575	-168.110	36.745	
	1100	61.698	-12.431	155.392	-183.362	36.436	
	1200	63.013	-6.196	160.817	-199.175	36.280	
	1300	64.320	.171	165.912	-215.514	36.236	
	1400	65.620	6.668	170.726	-232.348	36.276	

* * * * * * * * * *

URANIUM 2-GERMANIUM UGe_2

		A	B	C	D	RANGE	REF.
SOL	CP	17.80	4.57	-0.92		298-1400	27,*

PHASE	T	CP	H	S	G	BT	
SOL	298	18.128	-20.9	31.2	-30.202	22.142	27
	300	18.149	-20.866	31.312	-30.260	22.047	
	400	19.053	-19.003	36.666	-33.670	18.399	
	500	19.717	-17.064	40.992	-37.559	16.419	
	600	20.286	-15.063	44.638	-41.846	15.244	
	700	20.811	-13.008	47.805	-46.471	14.511	
	800	21.312	-10.901	50.617	-51.395	14.042	
	900	21.799	-8.746	53.155	-56.585	13.743	
	1000	22.278	-6.542	55.477	-62.019	13.556	
	1100	22.751	-4.290	57.622	-67.675	13.448	
	1200	23.220	-1.992	59.622	-73.538	13.395	
	1300	23.687	.354	61.499	-79.595	13.383	
	1400	24.151	2.746	63.271	-85.835	13.401	

UGe_3

URANIUM 3-GERMANIUM

		A	B	C	D	RANGE	REF.
SOL	CP	23.35	5.5	-0.9		298-1400	27,*

PHASE	T	CP	H	S	G	BT	
SOL	298	23.977	-25.5	40.8	-37.665	27.613	27
	300	24.000	-25.456	40.948	-37.740	27.497	
	400	24.987	-23.003	47.997	-42.202	23.061	
	500	25.740	-20.466	53.656	-47.294	20.675	
	600	26.400	-17.858	58.408	-52.903	19.273	
	700	27.016	-15.187	62.525	-58.954	18.409	
	800	27.609	-12.456	66.171	-65.392	17.867	
	900	28.189	-9.666	69.457	-72.177	17.529	
	1000	28.760	-6.818	72.456	-79.274	17.328	
	1100	29.326	-3.914	75.224	-86.660	17.220	
	1200	29.887	-.953	77.800	-94.313	17.179	
	1300	30.447	2.064	80.214	-102.215	17.186	
	1400	31.004	5.136	82.491	-110.351	17.229	

URANIUM 2-BORIDE

UB_2

		A	B	C	D	RANGE	REF.
SOL	CP	25.58	-9.62	-8.48	5.86	298-2300	98
	** MPT= 2703						1

PHASE	T	CP	H	S	G	BT	
SOL	298	13.693	-39.3	13.17	-43.227	31.690	196
	300	13.799	-39.275	13.255	-43.251	31.513	
	400	17.370	-37.688	17.796	-44.806	24.484	
	500	18.843	-35.867	21.852	-46.793	20.456	
	600	19.562	-33.943	25.358	-49.158	17.908	
	700	19.987	-31.965	28.407	-51.850	16.190	
	800	20.309	-29.949	31.098	-54.827	14.980	
	900	20.622	-27.903	33.507	-58.060	14.101	
	1000	20.972	-25.824	35.698	-61.522	13.447	
	1100	21.388	-23.706	37.716	-65.194	12.955	
	1200	21.886	-21.543	39.597	-69.060	12.579	
	1300	22.476	-19.326	41.372	-73.109	12.292	
	1400	23.165	-17.045	43.062	-77.332	12.074	
	1500	23.958	-14.690	44.687	-81.720	11.908	
	1600	24.858	-12.250	46.261	-86.267	11.785	
	1700	25.868	-9.714	47.798	-90.971	11.697	
	1800	26.989	-7.073	49.307	-95.826	11.636	
	1900	28.222	-4.313	50.799	-100.831	11.600	
	2000	29.568	-1.424	52.281	-105.986	11.583	
	2100	31.028	1.604	53.758	-111.287	11.583	
	2200	32.603	4.785	55.237	-116.737	11.598	
	2300	34.293	8.129	56.724	-122.335	11.626	

UB_4 URANIUM 4-BORIDE

		A	B	C	D	RANGE	REF.
SOL	CP	26.8	6.95	-8.9		298-1200	*
	** MPT= 2803						1

PHASE	T	CP	H	S	G	BT	
SOL	298	18.860	-58.7	17.	-63.769	46.750	27
	300	18.996	-58.665	17.117	-63.800	46.485	
	400	24.017	-56.483	23.359	-65.827	35.971	
	500	26.715	-53.936	29.033	-68.452	29.924	
	600	28.498	-51.170	34.070	-71.612	26.088	
	700	29.849	-48.250	38.568	-75.248	23.497	
	800	30.969	-45.208	42.629	-79.311	21.670	
	900	31.956	-42.061	46.335	-83.762	20.343	
	1000	32.860	-38.819	49.749	-88.569	19.359	
	1100	33.709	-35.491	52.921	-93.704	18.620	
	1200	34.522	-32.079	55.889	-99.146	18.059	

* * * * * * * * * *

UB_{12} URANIUM 12-BORIDE

		A	B	C	D	RANGE	REF.
SOL	CP	73.75	3.66	-30.5		298-1200	*
	** MPT= 2533						1

PHASE	T	CP	H	S	G	BT	
SOL	298	40.530	-103.5	33.4	-113.458	83.178	27
	300	40.959	-103.425	33.652	-113.520	82.711	
	400	56.151	-98.463	47.821	-117.592	64.258	
	500	63.380	-92.448	61.213	-123.055	53.795	
	600	67.474	-85.889	73.161	-129.786	47.281	
	700	70.088	-79.002	83.772	-137.643	42.980	
	800	71.912	-71.897	93.257	-146.503	40.028	
	900	73.279	-64.635	101.809	-156.263	37.951	
	1000	74.360	-57.251	109.588	-166.839	36.467	
	1100	75.255	-49.769	116.718	-178.159	35.402	
	1200	76.024	-42.204	123.300	-190.164	34.638	

URANIUM 2-ALUMINIUM

UAl_2

		A	B	C	D	RANGE	REF.
SOL	CP	18.0	2.5			298-1863	*
	** MPT=	1863					1

PHASE	T	CP	H	S	G	BT	
SOL	298	18.745	-23.6	25.5	-31.203	22.875	84/27
	300	18.750	-23.565	25.616	-31.250	22.769	
	400	19.000	-21.678	31.044	-34.096	18.631	
	500	19.250	-19.765	35.311	-37.421	16.359	
	600	19.500	-17.828	38.843	-41.133	14.985	
	700	19.750	-15.865	41.867	-45.172	14.105	
	800	20.000	-13.878	44.521	-49.495	13.523	
	900	20.250	-11.865	46.891	-54.067	13.131	
	1000	20.500	-9.828	49.037	-58.865	12.867	
	1100	20.750	-7.765	51.003	-63.869	12.691	
	1200	21.000	-5.678	52.819	-69.061	12.579	
	1300	21.250	-3.565	54.510	-74.428	12.514	
	1400	21.500	-1.428	56.094	-79.959	12.484	
	1500	21.750	.735	57.586	-85.644	12.480	
	1600	22.000	2.922	58.998	-91.474	12.496	
	1700	22.250	5.135	60.339	-97.441	12.529	
	1800	22.500	7.372	61.618	-103.540	12.573	
	1863	22.657	8.795	62.394	-107.446	12.606	

UAl_3 URANIUM 3-ALUMINIUM

		A	B	C	D	RANGE	REF.
SOL	CP	24.0	3.2			298-1623	*
	** MPT= 1623						1

PHASE	T	CP	H	S	G	BT	
SOL	298	24.954	-27.3	32.5	-36.990	27.118	84/27
	300	24.960	-27.254	32.654	-37.050	26.995	
	400	25.280	-24.742	39.879	-40.693	22.237	
	500	25.600	-22.198	45.554	-44.975	19.661	
	600	25.920	-19.622	50.250	-49.772	18.132	
	700	26.240	-17.014	54.270	-55.002	17.175	
	800	26.560	-14.374	57.794	-60.609	16.560	
	900	26.880	-11.702	60.941	-66.549	16.162	
	1000	27.200	-8.998	63.790	-72.788	15.910	
	1100	27.520	-6.262	66.397	-79.299	15.757	
	1200	27.840	-3.494	68.805	-86.060	15.676	
	1300	28.160	-.694	71.046	-93.054	15.646	
	1400	28.480	2.138	73.145	-100.265	15.654	
	1500	28.800	5.002	75.121	-107.679	15.691	
	1600	29.120	7.898	76.990	-115.286	15.749	
	1623	29.194	8.569	77.406	-117.061	15.765	

* * * * * * * * * *

UAl_4 URANIUM 4-ALUMINIUM

		A	B	C	D	RANGE	REF.
SOL	CP	28.5	8.0			298-1003	*
	** MPT= 1003						1

PHASE	T	CP	H	S	G	BT	
SOL	298	30.885	-31.	39.	-42.628	31.251	84/27
	300	30.900	-30.943	39.191	-42.700	31.111	
	400	31.700	-27.813	48.190	-47.089	25.732	
	500	32.500	-24.603	55.350	-52.278	22.854	
	600	33.300	-21.313	61.346	-58.120	21.173	
	700	34.100	-17.943	66.539	-64.520	20.147	
	800	34.900	-14.493	71.145	-71.409	19.511	
	900	35.700	-10.963	75.302	-78.734	19.122	
	1000	36.500	-7.353	79.104	-86.457	18.898	
	1003	36.524	-7.243	79.214	-86.695	18.893	

URANIUM 11-CADMIUM

UCd_{11}

		A	B	C	D	RANGE	REF.
SOL	CP	78.35	12.85	-6.35		298- 900	*

PHASE	T	CP	H	S	G	BT	
SOL	298	75.038	-10.9	139.4	-52.462	38.461	119
	300	75.149	-10.761	139.865	-52.720	38.412	
	400	79.521	-3.005	162.146	-67.864	37.084	
	500	82.235	5.090	180.200	-85.010	37.163	
	600	84.296	13.420	195.382	-103.809	37.817	
	700	86.049	21.939	208.510	-124.018	38.725	
	800	87.638	30.625	220.106	-145.460	39.743	
	900	89.131	39.464	230.515	-168.000	40.801	

* * * * * * * * * *

URANIUM 2-IRON

UFe_2

		A	B	C	D	RANGE	REF.
SOL	CP	16.7	7.0			298-1473	*
LIQ	CP	33.0				1473-1600	*

PHASE	T	CP	H	S	G	BT	
SOL	298	18.787	-7.7	25.	-15.154	11.109	69/27
	300	18.800	-7.665	25.116	-15.200	11.075	
	400	19.500	-5.750	30.621	-17.998	9.835	
	500	20.200	-3.765	35.047	-21.289	9.307	
	600	20.900	-1.710	38.792	-24.985	9.102	
	700	21.600	.415	42.066	-29.032	9.065	
	800	22.300	2.610	44.996	-33.387	9.122	
	900	23.000	4.875	47.663	-38.022	9.234	
	1000	23.700	7.210	50.123	-42.913	9.380	
	1100	24.400	9.615	52.414	-48.041	9.546	
	1200	25.100	12.090	54.567	-53.391	9.725	
	1300	25.800	14.635	56.604	-58.951	9.912	
	1400	26.500	17.250	58.542	-64.709	10.103	
	1473	27.011	19.203	59.902	-69.032	10.244	
			16.18	10.984			*
LIQ	1473	33.000	35.383	70.886	-69.032	10.244	
	1500	33.000	36.274	71.485	-70.954	10.339	
	1600	33.000	39.574	73.615	-78.210	10.684	

URh_3 URANIUM 3-RHODIUM

		A	B	C	D	RANGE	REF.
SOL	CP	24.70	7.30	-2.09		298-1200	*

PHASE	T	CP	H	S	G	BT	
SOL	298	24.525	-62.	35.4	-72.555	53.191	113,*
	300	24.568	-61.955	35.552	-72.620	52.911	
	400	26.314	-59.403	42.880	-76.555	41.833	
	500	27.514	-56.709	48.886	-81.152	35.476	
	600	28.499	-53.907	53.992	-86.302	31.440	
	700	29.383	-51.013	58.452	-91.929	28.705	
	800	30.213	-48.033	62.430	-97.977	26.770	
	900	31.012	-44.971	66.035	-104.403	25.356	
	1000	31.791	-41.831	69.343	-111.174	24.300	
	1100	32.557	-38.613	72.409	-118.264	23.500	
	1200	33.315	-35.320	75.275	-125.649	22.887	

* * * * * * * * * *

URu_3 URANIUM 3-RUTHENIUM

		A	B	C	D	RANGE	REF.
SOL	CP	23.20	5.96	-1.0		298-1850	*

** MPT= 1850 (PERITEC.)

PHASE	T	CP	H	S	G	BT	
SOL	298	23.852	-52.	25.9	-59.722	43.783	113,*
	300	23.877	-51.956	26.048	-59.770	43.548	
	400	24.959	-49.511	33.075	-62.740	34.284	
	500	25.780	-46.972	38.735	-66.340	29.001	
	600	26.498	-44.358	43.500	-70.458	25.668	
	700	27.168	-41.674	47.635	-75.019	23.425	
	800	27.812	-38.925	51.305	-79.970	21.850	
	900	28.441	-36.112	54.618	-85.268	20.709	
	1000	29.060	-33.237	57.646	-90.884	19.865	
	1100	29.673	-30.301	60.445	-96.790	19.233	
	1200	30.283	-27.303	63.053	-102.966	18.755	
	1300	30.889	-24.244	65.501	-109.395	18.393	
	1400	31.493	-21.125	67.812	-116.062	18.121	
	1500	32.096	-17.946	70.005	-122.954	17.917	
	1600	32.697	-14.706	72.096	-130.059	17.768	
	1700	33.297	-11.406	74.096	-137.370	17.662	
	1800	33.897	-8.047	76.016	-144.876	17.593	
	1850	34.197	-6.344	76.949	-148.700	17.569	

VANADIUM

		A	B	C	D	RANGE	REF.
SOL	CP	6.331	0.629	-0.505		298- 600	21
	CP	3.994	3.028	2.732		600-1400	21
	CP	22.782	-12.060	-86.732	3.511	1400-2175	21
LIQ	CP	10.				2175-3200	21
	** BPT= 3682, L= 107.993						21

PHASE	T	CP	H	S	G	BT	P	
SOL	298	5.950	0	6.91	-2.060	1.510		21
	300	5.959	.011	6.947	-2.073	1.510		
	400	6.267	.624	8.708	-2.859	1.562		
	500	6.443	1.260	10.127	-3.803	1.663		
	600	6.568	1.911	11.313	-4.877	1.777		
	600	6.570	1.911	11.313	-4.877	1.777		
	700	6.671	2.572	12.333	-6.060	1.892		
	800	6.843	3.248	13.234	-7.340	2.005		
	900	7.056	3.942	14.052	-8.705	2.114		
	1000	7.295	4.660	14.808	-10.148	2.218		
	1100	7.551	5.402	15.515	-11.664	2.318		
	1200	7.817	6.170	16.183	-13.250	2.413		
	1300	8.092	6.966	16.820	-14.900	2.505		
	1400	8.373	7.789	17.430	-16.613	2.594		
	1400	8.354	7.789	17.430	-16.613	2.594		
	1500	8.737	8.644	18.019	-18.385	2.679		
	1600	9.086	9.535	18.595	-20.216	2.762		
	1700	9.426	10.461	19.156	-22.104	2.842		
	1800	9.773	11.421	19.704	-24.047	2.920		
	1900	10.140	12.416	20.242	-26.044	2.996		
	2000	10.538	13.450	20.772	-28.095	3.070	.002	
	2100	10.973	14.525	21.297	-30.198	3.143	.008	
	2175	11.327	15.361	21.688	-31.810	3.197	.020	
			5.002	2.300				21
LIQ	2175	10.000	20.363	23.988	-31.810	3.197	.020	
	2200	10.000	20.613	24.102	-32.412	3.220	.027	
	2300	10.000	21.613	24.547	-34.844	3.311	.085	
	2400	10.000	22.613	24.972	-37.320	3.399	.237	
	2500	10.000	23.613	25.380	-39.838	3.483	.609	
	2600	10.000	24.613	25.773	-42.396	3.564	1.450	
	2700	10.000	25.613	26.150	-44.992	3.642	3.229	
	2800	10.000	26.613	26.514	-47.625	3.718	6.775	
	2900	10.000	27.613	26.865	-50.294	3.791	13.472	
	3000	10.000	28.613	27.204	-52.998	3.861	25.533	
	3100	10.000	29.613	27.531	-55.735	3.930	46.342	
	3200	10.000	30.613	27.849	-58.504	3.996	80.885	

VF_3 VANADIUM TRIFLUORICE

		A	B	C	D	RANGE	REF.
SOL	CP	19.70	6.42			298-1000	19,*

PHASE	T	CP	H	S	G	BT	
SOL	298	21.614	-275.	23.18	-281.911	206.674	*/19
	300	21.626	-274.960	23.314	-281.954	205.431	
	400	22.268	-272.765	29.623	-284.615	155.527	
	500	22.910	-270.506	34.661	-287.837	125.830	
	600	23.552	-268.183	38.895	-291.520	106.200	
	700	24.194	-265.796	42.574	-295.597	92.302	
	800	24.836	-263.345	45.846	-300.021	81.973	
	900	25.478	-260.829	48.808	-304.756	74.015	
	1000	26.120	-258.249	51.526	-309.775	67.710	

* * * * * * * * * *

VF_4 VANADIUM TETRAFLUORIDE

		A	B	C	D	RANGE	REF.
SOL	CP	22.75	9.5			298-1000	*

PHASE	T	CP	H	S	G	BT	
SOL	298	25.582	-335.4	29.	-344.046	252.227	19/*
	300	25.600	-335.353	29.158	-344.100	250.710	
	400	26.550	-332.745	36.653	-347.406	189.840	
	500	27.500	-330.043	42.680	-351.382	153.610	
	600	28.450	-327.245	47.777	-355.912	129.658	
	700	29.400	-324.353	52.234	-360.917	112.698	
	800	30.350	-321.365	56.222	-366.343	100.094	
	900	31.300	-318.283	59.852	-372.149	90.382	
	1000	32.250	-315.105	63.199	-378.304	82.689	

VANADIUM PENTAFLUORIDE (GAS) $VF_5(g)$

		A	B	C	D	RANGE	REF.
GAS	CP	31.18	0.15	-6.87		298-2000	19,*

PHASE	T	CP	H	S	G	BT	
GAS	298	23.496	-342.7	76.67	-365.559	267.998	19
	300	23.592	-342.656	76.816	-365.701	266.449	
	400	26.946	-340.106	84.131	-373.758	204.239	
	500	28.507	-337.324	90.331	-382.490	167.209	
	600	29.362	-334.427	95.610	-391.793	142.730	
	700	29.883	-331.463	100.179	-401.588	125.398	
	800	30.227	-328.456	104.193	-411.811	112.517	
	900	30.467	-325.421	107.768	-422.412	102.589	
	1000	30.643	-322.365	110.987	-433.353	94.722	
	1100	30.777	-319.294	113.914	-444.600	88.346	
	1200	30.883	-316.211	116.597	-456.127	83.083	
	1300	30.968	-313.118	119.073	-467.912	78.674	
	1400	31.039	-310.017	121.370	-479.936	74.931	
	1500	31.100	-306.910	123.514	-492.181	71.720	
	1600	31.152	-303.798	125.523	-504.634	68.939	
	1700	31.197	-300.680	127.413	-517.282	66.510	
	1800	31.238	-297.559	129.197	-530.113	64.373	
	1900	31.275	-294.433	130.887	-543.118	62.481	
	2000	31.308	-291.304	132.492	-556.288	60.796	

VBr_2 VANADIUM DIBROMIDE

		A	B	C	D	RANGE	REF.
SOL	CP	17.6	3.0			298- 900	*

PHASE	T	CP	H	S	G	BT	
SOL	298	18.494	-83.	30.	-91.944	67.406	1
	300	18.500	-82.966	30.114	-92.000	67.031	
	400	18.800	-81.101	35.478	-95.292	52.072	
	500	19.100	-79.206	39.705	-99.058	43.304	
	600	19.400	-77.281	43.214	-103.209	37.599	
	700	19.700	-75.326	46.227	-107.685	33.625	
	800	20.000	-73.341	48.877	-112.442	30.722	
	900	20.300	-71.326	51.250	-117.451	28.525	

* * * * * * * * * *

VBr_3 VANADIUM TRIBROMIDE

		A	B	C	D	RANGE	REF.
SOL	CP	22.0	7.7			298-1000	*

PHASE	T	CP	H	S	G	BT	
SOL	298	24.296	-107.	34.	-117.137	85.875	1
	300	24.310	-106.955	34.150	-117.200	85.392	
	400	25.080	-104.486	41.249	-120.985	66.112	
	500	25.850	-101.939	46.928	-125.403	54.821	
	600	26.620	-99.316	51.710	-130.341	47.483	
	700	27.390	-96.615	55.871	-135.725	42.381	
	800	28.160	-93.838	59.579	-141.500	38.661	
	900	28.930	-90.983	62.940	-147.629	35.854	
	1000	29.700	-88.052	66.028	-154.079	33.679	

VANADIUM TETRABROMIDE (GAS)

$VBr_4(g)$

		A	B	C	D	RANGE	REF.
GAS	CP	25.75	0.2	-1.75		298-1000	*

PHASE	T	CP	H	S	G	BT	
GAS	298	23.841	-94.	80.	-117.852	86.399	1
	300	23.866	-93.956	80.148	-118.000	85.975	
	400	24.736	-91.520	87.150	-126.380	69.060	
	500	25.150	-89.023	92.719	-135.383	59.184	
	600	25.384	-86.496	97.327	-144.892	52.784	
	700	25.533	-83.949	101.252	-154.825	48.345	
	800	25.637	-81.390	104.668	-165.125	45.116	
	900	25.714	-78.823	107.693	-175.746	42.683	
	1000	25.775	-76.248	110.405	-186.653	40.799	

$VN_{0.465}$ VANADIUM 0.465-NITRIDE

		A	B	C	D	RANGE	REF.
SOL	CP	7.535	2.724	-1.291		298-2000	3A
	** DPT= 2875						3A

PHASE	T	CP	H	S	G	BT	
SOL	298	6.895	-31.6	6.384	-33.503	24.562	3A
	300	6.918	-31.587	6.427	-33.515	24.419	
	400	7.818	-30.846	8.553	-34.267	18.725	
	500	8.381	-30.034	10.362	-35.215	15.395	
	600	8.811	-29.174	11.929	-36.331	13.236	
	700	9.178	-28.274	13.315	-37.595	11.739	
	800	9.512	-27.340	14.563	-38.990	10.653	
	900	9.827	-26.372	15.702	-40.504	9.837	
	1000	10.130	-25.375	16.753	-42.127	9.208	
	1100	10.425	-24.347	17.732	-43.852	8.714	
	1200	10.714	-23.290	18.652	-45.672	8.319	
	1300	11.000	-22.204	19.521	-47.581	8.000	
	1400	11.283	-21.090	20.346	-49.574	7.740	
	1500	11.564	-19.948	21.134	-51.649	7.526	
	1600	11.843	-18.777	21.889	-53.800	7.350	
	1700	12.121	-17.579	22.616	-56.026	7.204	
	1800	12.398	-16.353	23.316	-58.322	7.082	
	1900	12.675	-15.099	23.994	-60.688	6.982	
	2000	12.951	-13.818	24.651	-63.121	6.898	

VANADIUM MONOCARBICE

		A	B	C	D	RANGE	REF.
SOL	CP	8.7	3.2	-1.7		298-2000	61
	** HYPOTHETICAL						

PHASE	T	CP	H	S	G	BT	
SOL	298	7.742	-24.1	6.6	-26.068	19.111	61
	300	7.771	-24.086	6.648	-26.080	19.002	
	400	8.917	-23.245	9.058	-26.868	14.682	
	500	9.620	-22.316	11.128	-27.880	12.188	
	600	10.148	-21.327	12.930	-29.085	10.596	
	700	10.593	-20.289	14.528	-30.459	9.511	
	800	10.994	-19.210	15.970	-31.985	8.739	
	900	11.370	-18.091	17.286	-33.649	8.172	
	1000	11.730	-16.936	18.503	-35.439	7.746	
	1100	12.080	-15.746	19.638	-37.347	7.421	
	1200	12.422	-14.521	20.703	-39.365	7.170	
	1300	12.759	-13.262	21.711	-41.486	6.975	
	1400	13.093	-11.969	22.669	-43.705	6.824	
	1500	13.424	-10.643	23.583	-46.018	6.706	
	1600	13.754	-9.284	24.460	-48.421	6.615	
	1700	14.081	-7.892	25.304	-50.909	6.546	
	1800	14.408	-6.468	26.118	-53.480	6.494	
	1900	14.733	-5.011	26.906	-56.132	6.457	
	2000	15.057	-3.521	27.670	-58.861	6.433	

V_3Si 3-VANADIUM SILICON

		A	B	C	D	RANGE	REF.
SOL	CP	22.408	4.368	-1.661		298-1400	5

PHASE	T	CP	H	S	G	BT	
SOL	298	21.842	-36.	24.25	-43.230	31.693	221/*
	300	21.873	-35.960	24.385	-43.275	31.530	
	400	23.117	-33.704	30.865	-46.050	25.164	
	500	23.928	-31.350	36.115	-49.407	21.599	
	600	24.567	-28.924	40.536	-53.246	19.397	
	700	25.127	-26.439	44.365	-57.495	17.953	
	800	25.643	-23.900	47.755	-62.104	16.968	
	900	26.134	-21.311	50.803	-67.035	16.280	
	1000	26.610	-18.674	53.582	-72.256	15.794	
	1100	27.076	-15.990	56.140	-77.744	15.448	
	1200	27.534	-13.259	58.515	-83.478	15.205	
	1300	27.988	-10.483	60.737	-89.441	15.039	
	1400	28.438	-7.662	62.828	-95.621	14.929	

* * * * * * * * * *

V_5Si_3 5-VANADIUM 3-SILICON

		A	B	C	D	RANGE	REF.
SOL	CP	45.04	28.4	-4.13		298-1800	*

PHASE	T	CP	H	S	G	BT	
SOL	298	48.861	-110.4	49.9	-125.278	91.843	221/*
	300	48.971	-110.310	50.203	-125.370	91.344	
	400	53.819	-105.156	64.996	-131.154	71.669	
	500	57.588	-99.580	77.422	-138.291	60.455	
	600	60.933	-93.652	88.221	-146.584	53.401	
	700	64.077	-87.400	97.852	-155.896	48.680	
	800	67.115	-80.840	106.607	-166.126	45.390	
	900	70.090	-73.979	114.685	-177.195	43.035	
	1000	73.027	-66.823	122.222	-189.045	41.321	
	1100	75.939	-59.375	129.319	-201.625	40.065	
	1200	78.833	-51.636	136.050	-214.896	39.143	
	1300	81.716	-43.608	142.474	-228.825	38.474	
	1400	84.589	-35.293	148.635	-243.382	37.999	
	1500	87.456	-26.691	154.569	-258.544	37.675	
	1600	90.319	-17.802	160.305	-274.290	37.471	
	1700	93.177	-8.627	165.866	-290.599	37.364	
	1800	96.033	.833	171.273	-307.458	37.335	

VANADIUM 2-SILICON

VSi_2

		A	B	C	D	RANGE	REF.
SOL	CP	17.079	2.786	-2.25		298-1950	5
LIQ	CP	28.50				1950-2100	5

PHASE	T	CP	H	S	G	BT	
SOL	298	15.379	-30.	19.183	-35.719	26.187	221
	300	15.415	-29.972	19.278	-35.755	26.051	
	400	16.787	-28.354	23.923	-37.923	20.723	
	500	17.572	-26.633	27.760	-40.513	17.710	
	600	18.126	-24.847	31.015	-43.456	15.831	
	700	18.570	-23.011	33.843	-46.702	14.583	
	800	18.956	-21.135	36.349	-50.213	13.720	
	900	19.309	-19.221	38.602	-53.963	13.106	
	1000	19.640	-17.274	40.654	-57.927	12.662	
	1100	19.958	-15.294	42.540	-62.088	12.337	
	1200	20.266	-13.282	44.290	-66.431	12.100	
	1300	20.568	-11.241	45.924	-70.942	11.928	
	1400	20.865	-9.169	47.459	-75.612	11.805	
	1500	21.158	-7.068	48.909	-80.431	11.720	
	1600	21.449	-4.937	50.284	-85.392	11.666	
	1700	21.737	-2.778	51.593	-90.486	11.634	
	1800	22.024	-.590	52.843	-95.708	11.622	
	1900	22.310	1.627	54.042	-101.053	11.625	
	1950	22.453	2.746	54.623	-103.770	11.632	
			37.83	19.400			5
LIQ	1950	28.500	40.576	74.023	-103.770	11.632	
	2000	28.500	42.001	74.745	-107.489	11.747	
	2100	28.500	44.851	76.135	-115.033	11.973	

VB

VANADIUM MONOBORIDE

		A	B	C	D	RANGE	REF.
SOL	CP	9.055	5.405	-1.753	-0.964	298-1200	197
	CP	10.657	1.625	-0.267	1.023	1200-2500	197

PHASE	T	CP	H	S	G	BT	
SOL	298	8.609	-33.1	7.	-35.187	25.796	175
	300	8.642	-33.084	7.053	-35.200	25.647	
	400	9.967	-32.147	9.739	-36.043	19.696	
	500	10.815	-31.106	12.059	-37.136	16.234	
	600	11.464	-29.991	14.091	-38.445	14.006	
	700	12.008	-28.816	15.900	-39.946	12.473	
	800	12.488	-27.591	17.535	-41.619	11.371	
	900	12.922	-26.320	19.032	-43.449	10.552	
	1000	13.321	-25.008	20.414	-45.422	9.928	
	1100	13.689	-23.657	21.701	-47.528	9.444	
	1200	14.031	-22.271	22.907	-49.759	9.064	
	1200	14.062	-22.271	22.907	-49.759	9.064	
	1300	14.483	-20.844	24.049	-52.108	8.761	
	1400	14.923	-19.374	25.138	-54.567	8.520	
	1500	15.384	-17.859	26.184	-57.134	8.326	
	1600	15.865	-16.296	27.192	-59.803	8.170	
	1700	16.367	-14.685	28.168	-62.571	8.045	
	1800	16.888	-13.022	29.119	-65.436	7.946	
	1900	17.430	-11.307	30.046	-68.394	7.868	
	2000	17.992	-9.536	30.954	-71.444	7.808	
	2100	18.575	-7.707	31.846	-74.585	7.763	
	2200	19.178	-5.820	32.724	-77.813	7.731	
	2300	19.801	-3.871	33.590	-81.129	7.710	
	2400	20.445	-1.859	34.447	-84.531	7.699	
	2500	21.109	.219	35.295	-88.018	7.696	

VANADIUM 2-BORIDE VB_2

		A	B	C	D	RANGE	REF.
SOL	CP	11.963	10.635	-3.238	-2.951	298-1200	197
	CP	15.167	3.075	-0.267	1.023	1200-2500	197

PHASE	T	CP	H	S	G	BT	
SOL	298	11.229	-48.7	7.2	-50.847	37.277	175
	300	11.290	-48.679	7.270	-50.860	37.057	
	400	13.721	-47.417	10.884	-51.771	28.290	
	500	15.248	-45.964	14.120	-53.024	23.180	
	600	16.382	-44.380	17.005	-54.583	19.885	
	700	17.301	-42.695	19.601	-56.415	17.616	
	800	18.076	-40.925	21.963	-58.495	15.982	
	900	18.744	-39.083	24.132	-60.802	14.767	
	1000	19.323	-37.179	26.138	-63.316	13.840	
	1100	19.823	-35.221	28.003	-66.024	13.120	
	1200	20.251	-33.217	29.747	-68.913	12.552	
	1200	20.312	-33.217	29.747	-68.913	12.552	
	1300	20.878	-31.157	31.395	-71.971	12.101	
	1400	21.463	-29.040	32.964	-75.189	11.739	
	1500	22.069	-26.864	34.465	-78.561	11.448	
	1600	22.695	-24.626	35.909	-82.080	11.213	
	1700	23.342	-22.324	37.304	-85.741	11.024	
	1800	24.008	-19.957	38.657	-89.540	10.873	
	1900	24.695	-17.522	39.974	-93.472	10.753	
	2000	25.402	-15.017	41.258	-97.534	10.659	
	2100	26.130	-12.441	42.515	-101.722	10.588	
	2200	26.878	-9.790	43.748	-106.036	10.535	
	2300	27.646	-7.064	44.959	-110.471	10.499	
	2400	28.435	-4.261	46.153	-115.027	10.476	
	2500	29.244	-1.377	47.330	-119.701	10.466	

V_3B_2 3-VANADIUM 2-BORIDE

		A	B	C	D	RANGE	REF.
SOL	CP	24.257	10.986	-3.774	-0.905	298-1200	197
	CP	27.461	3.426	-0.803	3.069	1200-2500	197

PHASE	T	CP	H	S	G	BT	
SOL	298	23.206	-72.6	20.78	-78.796	57.766	175
	300	23.278	-72.557	20.924	-78.834	57.438	
	400	26.148	-70.072	28.052	-81.293	44.422	
	500	28.014	-67.359	34.098	-84.408	36.900	
	600	29.474	-64.483	39.339	-88.086	32.090	
	700	30.734	-61.471	43.979	-92.256	28.808	
	800	31.877	-58.340	48.158	-96.866	26.466	
	900	32.945	-55.098	51.975	-101.876	24.742	
	1000	33.961	-51.753	55.499	-107.252	23.443	
	1100	34.935	-48.308	58.782	-112.967	22.448	
	1200	35.875	-44.767	61.862	-119.001	21.676	
	1200	35.936	-44.767	61.862	-119.001	21.676	
	1300	37.054	-41.118	64.782	-125.335	21.073	
	1400	38.232	-37.354	67.571	-131.953	20.602	
	1500	39.470	-33.469	70.250	-138.845	20.232	
	1600	40.768	-29.458	72.839	-146.000	19.945	
	1700	42.127	-25.314	75.351	-153.410	19.725	
	1800	43.547	-21.031	77.799	-161.068	19.559	
	1900	45.027	-16.603	80.193	-168.968	19.438	
	2000	46.569	-12.023	82.541	-177.105	19.356	
	2100	48.172	-7.287	84.852	-185.475	19.305	
	2200	49.836	-2.387	87.131	-194.075	19.282	
	2300	51.561	2.682	89.384	-202.901	19.283	
	2400	53.347	7.927	91.616	-211.951	19.303	
	2500	55.194	13.354	93.831	-221.223	19.342	

2-VANADIUM 3-BORIDE V_2B_3

		A	B	C	D	RANGE	REF.
SOL	CP	21.018	16.041	-4.992	-3.915	298-1200	197
	CP	25.824	4.701	-0.535	2.046	1200-2500	197

PHASE	T	CP	H	S	G	BT	
SOL	298	19.837	-82.5	14.2	-86.734	63.586	175
	300	19.931	-82.463	14.323	-86.760	63.213	
	400	23.688	-80.264	20.623	-88.514	48.368	
	500	26.063	-77.770	26.180	-90.860	39.720	
	600	27.847	-75.071	31.095	-93.728	34.145	
	700	29.310	-72.211	35.501	-97.062	30.308	
	800	30.565	-69.216	39.499	-100.815	27.545	
	900	31.667	-66.103	43.164	-104.950	25.489	
	1000	32.645	-62.887	46.552	-109.438	23.921	
	1100	33.513	-59.578	49.705	-114.253	22.703	
	1200	34.283	-56.187	52.654	-119.372	21.744	
	1200	34.374	-56.187	52.654	-119.372	21.744	
	1300	35.361	-52.701	55.445	-124.779	20.980	
	1400	36.388	-49.114	58.102	-130.457	20.368	
	1500	37.455	-45.422	60.649	-136.395	19.875	
	1600	38.562	-41.621	63.102	-142.584	19.479	
	1700	39.710	-37.708	65.474	-149.013	19.159	
	1800	40.898	-33.678	67.777	-155.676	18.904	
	1900	42.127	-29.527	70.021	-162.566	18.702	
	2000	43.397	-25.251	72.214	-169.679	18.544	
	2100	44.707	-20.846	74.363	-177.008	18.424	
	2200	46.058	-16.308	76.473	-184.550	18.336	
	2300	47.450	-11.633	78.551	-192.301	18.275	
	2400	48.882	-6.817	80.601	-200.259	18.239	
	2500	50.355	-1.855	82.626	-208.421	18.223	

V_3B_4 3-VANADIUM 4-BORIDE

		A	B	C	D	RANGE	REF.
SOL	CP	30.073	21.447	-6.746	-4.879	298-1200	197
	CP	36.481	6.326	-0.803	3.069	1200-2500	197

PHASE	T	CP	H	S	G	BT	
SOL	298	28.445	-116.3	21.19	-122.618	89.893	175
	300	28.572	-116.247	21.366	-122.657	89.368	
	400	33.655	-113.112	30.352	-125.252	68.444	
	500	36.878	-109.576	38.229	-128.690	56.258	
	600	39.311	-105.762	45.176	-132.867	48.403	
	700	41.318	-101.728	51.391	-137.701	42.998	
	800	43.054	-97.507	57.024	-143.126	39.105	
	900	44.590	-93.123	62.185	-149.090	36.209	
	1000	45.966	-88.594	66.956	-155.550	34.000	
	1100	47.204	-83.935	71.396	-162.470	32.284	
	1200	48.315	-79.158	75.552	-169.820	30.933	
	1200	48.436	-79.158	75.552	-169.820	30.933	
	1300	49.844	-74.244	79.484	-177.573	29.857	
	1400	51.312	-69.187	83.231	-185.711	28.995	
	1500	52.840	-63.980	86.823	-194.214	28.301	
	1600	54.428	-58.617	90.284	-203.071	27.742	
	1700	56.077	-53.092	93.632	-212.267	27.293	
	1800	57.787	-47.400	96.886	-221.794	26.933	
	1900	59.557	-41.533	100.057	-231.642	26.648	
	2000	61.389	-35.486	103.159	-241.803	26.427	
	2100	63.282	-29.253	106.199	-252.272	26.258	
	2200	65.236	-22.828	109.188	-263.041	26.134	
	2300	67.251	-16.204	112.132	-274.108	26.050	
	2400	69.327	-9.376	115.038	-285.466	25.999	
	2500	71.464	-2.337	117.911	-297.114	25.977	

5-VANADIUM 6-BORIDE

V_5B_6

		A	B	C	D	RANGE	REF.
SOL	CP	48.183	32.258	-10.252	-6.807	298-1200	197
	CP	57.795	9.578	-1.339	5.115	1200-2500	197

PHASE	T	CP	H	S	G	BT	
SOL	298	45.663	-182.5	18.2	-187.926	137.772	175
	300	45.857	-182.415	18.483	-187.960	136.947	
	400	53.590	-177.406	32.840	-190.542	104.122	
	500	58.509	-171.787	45.358	-194.466	85.013	
	600	62.240	-165.743	56.368	-199.564	72.701	
	700	65.336	-159.360	66.201	-205.701	64.231	
	800	68.031	-152.689	75.105	-212.773	58.135	
	900	70.436	-145.764	83.259	-220.697	53.600	
	1000	72.609	-138.610	90.795	-229.404	50.143	
	1100	74.583	-131.249	97.809	-238.838	47.459	
	1200	76.379	-123.699	104.377	-248.951	45.346	
	1200	76.561	-123.699	104.377	-248.951	45.346	
	1300	78.812	-115.931	110.593	-259.702	43.666	
	1400	81.161	-107.933	116.519	-271.060	42.320	
	1500	83.611	-99.696	122.202	-282.998	41.238	
	1600	86.162	-91.208	127.679	-295.494	40.368	
	1700	88.814	-82.460	132.981	-308.528	39.669	
	1800	91.567	-73.442	138.135	-322.085	39.112	
	1900	94.421	-64.143	143.162	-336.151	38.671	
	2000	97.378	-54.554	148.080	-350.714	38.329	
	2100	100.436	-44.664	152.904	-365.764	38.071	
	2200	103.596	-34.464	157.649	-381.292	37.883	
	2300	106.857	-23.942	162.326	-397.291	37.756	
	2400	110.221	-13.089	166.944	-413.755	37.683	
	2500	113.687	-1.894	171.514	-430.678	37.655	

$WO_2J_2(g)$ TUNGSTEN OXYDIIODIDE (GAS)

		A	B	C	D	RANGE	REF.
GAS	CP	24.51	1.1	-2.53		298-1000	107

PHASE	T	CP	H	S	G	BT	
GAS	298	21.992	-102.8	90.1	-129.663	95.059	107,*
	300	22.029	-102.759	90.236	-129.830	94.594	
	400	23.369	-100.481	96.782	-139.194	76.062	
	500	24.048	-98.107	102.077	-149.145	65.200	
	600	24.467	-95.679	106.501	-159.580	58.135	
	700	24.764	-93.217	110.296	-170.424	53.216	
	800	24.995	-90.729	113.618	-181.624	49.624	
	900	25.188	-88.220	116.574	-193.136	46.906	
	1000	25.357	-85.692	119.236	-204.929	44.793	

* * * * * * * * * *

WS_2 TUNGSTEN DISULFIDE

		A	B	C	D	RANGE	REF.
SOL	CP	16.403	3.73	-2.069		298-1500	63
	** DEC.						

PHASE	T	CP	H	S	G	BT	
SOL	298	15.188	-62.	15.5	-66.621	48.841	63
	300	15.223	-61.972	15.594	-66.650	48.561	
	400	16.602	-60.373	20.183	-68.447	37.403	
	500	17.440	-58.669	23.983	-70.660	30.890	
	600	18.066	-56.892	27.221	-73.225	26.676	
	700	18.592	-55.059	30.046	-76.091	23.760	
	800	19.064	-53.176	32.560	-79.223	21.646	
	900	19.505	-51.247	34.831	-82.595	20.059	
	1000	19.926	-49.275	36.908	-86.183	18.838	
	1100	20.335	-47.262	38.826	-89.971	17.878	
	1200	20.735	-45.209	40.613	-93.944	17.112	
	1300	21.130	-43.115	42.288	-98.090	16.493	
	1400	21.519	-40.983	43.868	-102.399	15.987	
	1500	21.906	-38.812	45.366	-106.861	15.572	

DITUNGSTEN CARBIDE

W_2C

		A	B	C	D	RANGE	REF.
SOL	CP	21.45	2.60	-3.48		298-3068	10
	** MPT=	3068					21

PHASE	T	CP	H	S	G	BT	
SOL	298	18.310	-6.3	19.113	-11.999	8.796	21
	300	18.363	-6.266	19.226	-12.034	8.768	
	400	20.315	-4.320	24.811	-14.245	7.784	
	500	21.358	-2.232	29.466	-16.965	7.416	
	600	22.043	-.060	33.424	-20.115	7.328	
	700	22.560	2.171	36.863	-23.633	7.379	
	800	22.986	4.449	39.904	-27.474	7.507	
	900	23.360	6.767	42.633	-31.603	7.675	
	1000	23.702	9.120	45.112	-35.992	7.867	
	1100	24.022	11.506	47.387	-40.619	8.071	
	1200	24.328	13.924	49.490	-45.464	8.281	
	1300	24.624	16.372	51.449	-50.512	8.493	
	1400	24.912	18.848	53.284	-55.750	8.704	
	1500	25.195	21.354	55.013	-61.165	8.913	
	1600	25.474	23.887	56.648	-66.749	9.119	
	1700	25.750	26.449	58.200	-72.492	9.321	
	1800	26.023	29.037	59.680	-78.387	9.519	
	1900	26.294	31.653	61.094	-84.426	9.713	
	2000	26.563	34.296	62.450	-90.604	9.902	
	2100	26.831	36.966	63.752	-96.914	10.087	
	2200	27.098	39.662	65.007	-103.353	10.269	
	2300	27.364	42.385	66.217	-109.914	10.446	
	2400	27.630	45.135	67.387	-116.595	10.619	
	2500	27.894	47.911	68.521	-123.390	10.788	
	2600	28.159	50.714	69.620	-130.298	10.954	
	2700	28.422	53.543	70.687	-137.313	11.116	
	2800	28.686	56.398	71.726	-144.434	11.275	
	2900	28.949	59.280	72.737	-151.658	11.431	
	3000	29.211	62.188	73.723	-158.981	11.583	
	3068	29.390	64.180	74.380	-164.016	11.685	

W_5Si_3 5-TUNGSTEN 3-SILICCN

		A	B	C	D	RANGE	REF.
SOL	CP	42.944	9.369	-2.122		298-2200	221
	** MPT= 2593						221

PHASE	T	CP	H	S	G	BT	
SOL	298	43.350	-32.16	59.10	-49.781	36.495	221
	300	43.397	-32.080	59.368	-49.890	36.350	
	400	45.365	-27.634	72.144	-56.492	30.870	
	500	46.780	-23.024	82.425	-64.237	28.082	
	600	47.976	-18.285	91.061	-72.922	26.565	
	700	49.069	-13.433	98.540	-82.410	25.733	
	800	50.108	-8.473	105.160	-92.602	25.301	
	900	51.114	-3.412	111.121	-103.421	25.117	
	1000	52.101	1.749	116.557	-114.808	25.095	
	1100	53.075	7.008	121.569	-126.718	25.180	
	1200	54.039	12.363	126.228	-139.110	25.339	
	1300	54.998	17.815	130.592	-151.954	25.549	
	1400	55.952	23.363	134.702	-165.220	25.796	
	1500	56.903	29.006	138.595	-178.887	26.067	
	1600	57.852	34.743	142.298	-192.933	26.357	
	1700	58.798	40.576	145.833	-207.341	26.659	
	1800	59.743	46.503	149.221	-222.095	26.970	
	1900	60.686	52.524	152.476	-237.181	27.286	
	2000	61.629	58.640	155.613	-252.586	27.605	
	2100	62.571	64.850	158.643	-268.300	27.926	
	2200	63.512	71.154	161.575	-284.311	28.248	

TUNGSTEN 2-SILICON

WSi_2

		A	B	C	D	RANGE	REF.
SOL	CP	16.212	2.639	-1.456		298-2200	221
	** MPT= 2433						221

PHASE	T	CP	H	S	G	BT	
SOL	298	15.361	-22.168	15.30	-26.730	19.596	221
	300	15.386	-22.140	15.395	-26.758	19.496	
	400	16.358	-20.547	19.969	-28.535	15.593	
	500	16.949	-18.880	23.687	-30.724	13.431	
	600	17.391	-17.162	26.817	-33.253	12.114	
	700	17.762	-15.404	29.527	-36.073	11.264	
	800	18.096	-13.611	31.921	-39.148	10.696	
	900	18.407	-11.786	34.070	-42.449	10.309	
	1000	18.705	-9.930	36.025	-45.955	10.045	
	1100	18.995	-8.045	37.822	-49.649	9.866	
	1200	19.278	-6.131	39.486	-53.515	9.748	
	1300	19.557	-4.190	41.041	-57.542	9.675	
	1400	19.832	-2.220	42.500	-61.720	9.636	
	1500	20.106	-.223	43.878	-66.040	9.623	
	1600	20.378	1.801	45.184	-70.493	9.630	
	1700	20.648	3.852	46.427	-75.074	9.653	
	1800	20.917	5.930	47.615	-79.777	9.688	
	1900	21.186	8.036	48.753	-84.596	9.732	
	2000	21.454	10.168	49.847	-89.526	9.784	
	2100	21.721	12.326	50.900	-94.564	9.843	
	2200	21.988	14.512	51.917	-99.705	9.906	

Y

YTTRIUM

		A	B	C	D	RANGE	REF.
SOL-A	CP	5.880	1.535			298- 500	21
	CP	5.727	1.816			500-1752	21
SOL-B	CP	8.37				1752-1799	21
LIQ	CP	9.51				1799-3000	223
	** BPT= 3611						21

PHASE	T	CP	H	S	G	BT	P	
SOL-A	298	6.338	0	10.62	-3.166	2.321		21
	300	6.340	.012	10.659	-3.186	2.321		
	400	6.494	.653	12.504	-4.348	2.376		
	500	6.647	1.311	13.970	-5.674	2.481		
	500	6.635	1.311	13.970	-5.674	2.481		
	600	6.817	1.983	15.196	-7.134	2.599		
	700	6.998	2.674	16.260	-8.708	2.719		
	800	7.180	3.383	17.206	-10.382	2.837		
	900	7.361	4.110	18.063	-12.146	2.950		
	1000	7.543	4.855	18.848	-13.992	3.058		
	1100	7.725	5.618	19.575	-15.914	3.162		
	1200	7.906	6.400	20.255	-17.906	3.262		
	1300	8.088	7.200	20.895	-19.964	3.357		
	1400	8.269	8.018	21.501	-22.084	3.448		
	1500	8.451	8.854	22.078	-24.263	3.536		
	1600	8.633	9.708	22.629	-26.498	3.620		
	1700	8.814	10.580	23.158	-28.788	3.701		
	1752	8.909	11.041	23.425	-29.999	3.743	.001	
			1.193	.681				21
SOL-B	1752	8.370	12.234	24.106	-29.999	3.743	.001	
	1799	8.370	12.627	24.327	-31.137	3.783	.002	
			2.724	1.514				21
LIQ	1799	9.510	15.351	25.841	-31.137	3.783	.002	
	1800	9.510	15.361	25.847	-31.163	3.784	.002	
	1900	9.510	16.312	26.361	-33.774	3.885	.009	
	2000	9.510	17.263	26.849	-36.434	3.982	.030	
	2100	9.510	18.214	27.313	-39.143	4.074	.092	
	2200	9.510	19.165	27.755	-41.896	4.163	.252	
	2300	9.510	20.116	28.178	-44.693	4.247	.631	
	2400	9.510	21.067	28.582	-47.531	4.329	1.458	
	2500	9.510	22.018	28.971	-50.409	4.407	3.140	
	2600	9.510	22.969	29.344	-53.325	4.483	6.359	
	2700	9.510	23.920	29.703	-56.277	4.556	12.190	
	2800	9.510	24.871	30.048	-59.265	4.626	22.254	
	2900	9.510	25.822	30.382	-62.286	4.695	38.894	
	3000	9.510	26.773	30.705	-65.341	4.761	65.376	

YTTRIUM FLUORIDE

		A	B	C	D	RANGE	REF.
SOL-A	CP	23.76	1.775	-1.36		298-1350	174
	** ORTHOROMBIC						
SOL-B	CP	-76.35	50.86	673.5		1350-1428	174
	** HEXAGONAL						
LIQ	CP	31.96	-0.006	-0.115		1428-1900	174

PHASE	T	CP	H	S	G	BT	
SOL-A	298	22.759	-410.7	26.2	-418.512	306.818	1
	300	22.781	-410.658	26.341	-418.560	304.962	
	400	23.620	-408.333	33.023	-421.542	230.351	
	500	24.103	-405.945	38.350	-425.120	185.845	
	600	24.447	-403.517	42.776	-429.182	156.351	
	700	24.725	-401.058	46.566	-433.654	135.411	
	800	24.967	-398.573	49.884	-438.480	119.803	
	900	25.190	-396.065	52.837	-443.619	107.740	
	1000	25.399	-393.536	55.502	-449.038	98.150	
	1100	25.600	-390.986	57.932	-454.711	90.355	
	1200	25.796	-388.416	60.168	-460.618	83.901	
	1300	25.987	-385.827	62.241	-466.740	78.477	
	1350	26.082	-384.525	63.223	-469.876	76.078	
			7.75	5.741			174
SOL-B	1350	29.266	-376.775	68.964	-469.876	76.078	
	1400	29.216	-375.314	70.027	-473.351	73.903	
	1428	29.306	-374.495	70.606	-475.320	72.756	
			6.685	4.681			174
LIQ	1428	31.946	-367.810	75.287	-475.320	72.756	
	1500	31.946	-365.510	76.859	-480.798	70.062	
	1600	31.946	-362.315	78.920	-488.588	66.747	
	1700	31.946	-359.121	80.857	-496.578	63.848	
	1800	31.946	-355.926	82.683	-504.756	61.294	
	1900	31.945	-352.731	84.410	-513.111	59.029	

YCl_3 YTTRIUM CHLORIDE

		A	B	C	D	RANGE	REF.
SOL	CP	25.027	0.77	-2.895		298- 994	89
LIQ	CP	32.436				994-1783	89
	** BPT= 1783						1

PHASE	T	CP	H	S	G	BT	
SOL	298	22.000	-232.7	32.9	-242.509	177.788	1/*
	300	22.041	-232.659	33.036	-242.570	176.736	
	400	23.526	-230.371	39.609	-246.215	134.544	
	500	24.254	-227.978	44.945	-250.451	109.487	
	600	24.685	-225.530	49.408	-255.175	92.960	
	700	24.975	-223.046	53.237	-260.312	81.284	
	800	25.191	-220.537	56.586	-265.806	72.625	
	900	25.363	-218.009	59.564	-271.616	65.966	
	994	25.499	-215.619	62.090	-277.336	60.986	
			7.5	7.545			89,*
LIQ	994	32.436	-208.119	69.635	-277.336	60.986	
	1000	32.436	-207.924	69.830	-277.754	60.711	
	1100	32.436	-204.680	72.922	-284.894	56.611	
	1200	32.436	-201.437	75.744	-292.330	53.248	
	1300	32.436	-198.193	78.340	-300.036	50.447	
	1400	32.436	-194.950	80.744	-307.991	48.086	
	1500	32.436	-191.706	82.982	-316.179	46.073	
	1600	32.436	-188.462	85.075	-324.583	44.342	
	1700	32.436	-185.219	87.042	-333.190	42.840	
	1783	32.436	-182.527	88.588	-340.479	41.740	

YTTRIUM IODIDE

YJ_3

			A	B	C	D	RANGE	REF.
SOL	CP		24.12	2.75	-1.77		298-1273	*
	** MPT= 1273							1

PHASE	T	CP	H	S	G	BT	
SOL	298	22.949	-147.4	49.5	-162.158	118.881	19/*
	300	22.978	-147.358	49.642	-162.250	118.215	
	400	24.114	-144.997	56.426	-167.567	91.567	
	500	24.787	-142.550	61.884	-173.491	75.843	
	600	25.278	-140.045	66.448	-179.914	65.543	
	700	25.684	-137.497	70.376	-186.760	58.317	
	800	26.043	-134.910	73.830	-193.974	52.998	
	900	26.376	-132.289	76.917	-201.514	48.941	
	1000	26.693	-129.635	79.712	-209.347	45.759	
	1100	26.999	-126.951	82.271	-217.448	43.209	
	1200	27.297	-124.236	84.633	-225.795	41.128	
	1273	27.512	-122.235	86.251	-232.033	39.841	

YN

YTTRIUM NITRIDE

		A	B	C	D	RANGE	REF.
SOL	CP	10.9	1.55	-1.75		298-2200	*

PHASE	T	CP	H	S	G	BT	
SOL	298	9.393	-71.5	9.	-74.183	54.385	1/*
	300	9.421	-71.483	9.058	-74.200	54.062	
	400	10.426	-70.484	11.924	-75.254	41.122	
	500	10.975	-69.412	14.314	-76.569	33.473	
	600	11.344	-68.295	16.349	-78.105	28.453	
	700	11.628	-67.146	18.120	-79.830	24.927	
	800	11.867	-65.971	19.689	-81.722	22.328	
	900	12.079	-64.773	21.099	-83.762	20.343	
	1000	12.275	-63.556	22.382	-85.937	18.784	
	1100	12.460	-62.319	23.560	-88.235	17.533	
	1200	12.638	-61.064	24.652	-90.647	16.511	
	1300	12.811	-59.791	25.671	-93.163	15.664	
	1400	12.981	-58.502	26.626	-95.779	14.954	
	1500	13.147	-57.195	27.528	-98.487	14.351	
	1600	13.312	-55.872	28.381	-101.283	13.836	
	1700	13.474	-54.533	29.193	-104.162	13.393	
	1800	13.636	-53.177	29.968	-107.120	13.008	
	1900	13.797	-51.806	30.710	-110.154	12.672	
	2000	13.956	-50.418	31.421	-113.261	12.378	
	2100	14.115	-49.015	32.106	-116.438	12.119	
	2200	14.274	-47.595	32.767	-119.682	11.891	

YTTRIUM 2-RHENIUM YRe_2

		A	B	C	D	RANGE	REF.
SOL	CP	17.0	4.44	-0.1		298-1500	*

PHASE	T	CP	H	S	G	BT	
SOL	298	18.211	-32.45	26.3	-40.291	29.538	164
	300	18.221	-32.416	26.413	-40.340	29.392	
	400	18.713	-30.569	31.723	-43.258	23.638	
	500	19.180	-28.674	35.949	-46.649	20.393	
	600	19.636	-26.734	39.487	-50.425	18.370	
	700	20.088	-24.747	42.547	-54.531	17.027	
	800	20.536	-22.716	45.259	-58.923	16.099	
	900	20.984	-20.640	47.704	-63.573	15.440	
	1000	21.430	-18.519	49.938	-68.457	14.963	
	1100	21.876	-16.354	52.001	-73.555	14.616	
	1200	22.321	-14.144	53.924	-78.853	14.363	
	1300	22.766	-11.890	55.728	-84.336	14.180	
	1400	23.211	-9.591	57.431	-89.995	14.051	
	1500	23.656	-7.248	59.048	-95.819	13.963	

* * * * * * * * * *

YTTRIUM DIZIRCONATE $Y_2O_3 \cdot 2ZrO_2$

		A	B	C	D	RANGE	REF.
SOL	CP	62.9	4.8	-11.5		298-1500	*

PHASE	T	CP	H	S	G	BT	
SOL	298	51.394	-985.18	47.8	-999.432	732.702	129/*
	300	51.562	-985.085	48.118	-999.520	728.248	
	400	57.632	-979.585	63.898	-1005.144	549.259	
	500	60.700	-973.654	77.120	-1012.214	442.498	
	600	62.586	-967.483	88.366	-1020.503	371.768	
	700	63.913	-961.155	98.118	-1029.838	321.573	
	800	64.943	-954.711	106.722	-1040.088	284.177	
	900	65.800	-948.172	114.422	-1051.152	255.289	
	1000	66.550	-941.554	121.394	-1062.949	232.338	
	1100	67.230	-934.865	127.770	-1075.411	213.693	
	1200	67.861	-928.110	133.647	-1088.486	198.267	
	1300	68.460	-921.293	139.102	-1102.127	185.309	
	1400	69.033	-914.419	144.197	-1116.294	174.285	
	1500	69.589	-907.487	148.979	-1130.956	164.802	

Yb

YTTERBIUM

		A	B	C	D	RANGE	REF.
SOL-A	CP	-0.104	13.210	2.271		298- 553	21
	** TPT= 553 (HYPOTHETIC.)						
	CP	6.375	1.251			553-1033	21
SOL-B	CP	8.63				1033-1097	21
LIQ	CP	8.79				1097-1467	21
GAS	CP	4.968				1467-2000	21

PHASE	T	CP	H	S	G	BT	P	
SOL-A	298	6.389	0	14.30	-4.264	3.126		21
	300	6.382	.012	14.340	-4.290	3.126		
	400	6.599	.653	16.183	-5.820	3.180		
	500	7.409	1.351	17.736	-7.517	3.286		
	553	7.944	1.757	18.508	-8.478	3.351		
	553	7.067	1.757	18.508	-8.478	3.351		
	600	7.126	2.091	19.087	-9.362	3.410		
	700	7.251	2.810	20.195	-11.327	3.537	.002	
	800	7.376	3.541	21.171	-13.396	3.660	.052	
	900	7.501	4.285	22.047	-15.558	3.778	.606	
	1000	7.626	5.041	22.844	-17.803	3.891	4.267	
	1033	7.667	5.293	23.092	-18.561	3.927	7.459	
			0.418	.405				21
SOL-B	1033	8.630	5.711	23.497	-18.561	3.927	7.459	
	1097	8.630	6.264	24.016	-20.082	4.001	19.694	
			1.83	1.668				21
LIQ	1097	8.790	8.094	25.684	-20.082	4.001	19.694	
	1100	8.790	8.120	25.708	-20.159	4.006	20.503	
	1200	8.790	8.999	26.473	-22.768	4.147	69.481	
	1300	8.790	9.878	27.176	-25.451	4.279	192.751	
	1400	8.790	10.757	27.828	-28.202	4.403	457.328	
	1467	8.790	11.346	28.239	-30.080	4.482	760.	
			30.792	20.990				21
GAS	1467	4.968	42.138	49.228	-30.080	4.482		
	1500	4.968	42.302	49.339	-31.706	4.620		
	1600	4.968	42.799	49.660	-36.657	5.008		
	1700	4.968	43.296	49.961	-41.638	5.354		
	1800	4.968	43.792	50.245	-46.648	5.665		
	1900	4.968	44.289	50.513	-51.686	5.946		
	2000	4.968	44.786	50.768	-56.750	6.202		

YTTERBIUM DICHLORIDE

$YbCl_2$

		A	B	C	D	RANGE	REF.
SOL	CP	18.6	4.1			298-1000	*

PHASE	T	CP	H	S	G	BT	
SOL	298	19.822	-191.1	31.2	-200.402	146.919	19/*
	300	19.830	-191.063	31.323	-200.460	146.055	
	400	20.240	-189.060	37.084	-203.893	111.417	
	500	20.650	-187.015	41.644	-207.837	90.858	
	600	21.060	-184.930	45.445	-212.197	77.303	
	700	21.470	-182.803	48.722	-216.909	67.731	
	800	21.880	-180.636	51.616	-221.929	60.636	
	900	22.290	-178.427	54.217	-227.222	55.185	
	1000	22.700	-176.178	56.587	-232.764	50.877	

* * * * * * * * * *

YTTERBIUM CHLORIDE

$YbCl_3$

		A	B	C	D	RANGE	REF.
SOL	CP	22.63	2.23	-0.45		298-1127	*
LIQ	CP	29.0				1127-1500	*

PHASE	T	CP	H	S	G	BT	
SOL	298	22.789	-229.4	35.3	-239.925	175.893	19/*
	300	22.799	-229.358	35.441	-239.990	174.856	
	400	23.241	-227.054	42.065	-243.880	133.268	
	500	23.565	-224.713	47.287	-248.357	108.571	
	600	23.843	-222.343	51.608	-253.308	92.280	
	700	24.099	-219.946	55.303	-258.658	80.767	
	800	24.344	-217.523	58.537	-264.353	72.228	
	900	24.581	-215.077	61.418	-270.354	65.660	
	1000	24.815	-212.607	64.020	-276.628	60.465	
	1100	25.046	-210.114	66.396	-283.150	56.264	
	1127	25.108	-209.437	67.004	-284.951	55.266	
			8.45	7.498			*
LIQ	1127	29.000	-200.987	74.502	-284.951	55.266	
	1200	29.000	-198.870	76.322	-290.457	52.907	
	1300	29.000	-195.970	78.644	-298.207	50.140	
	1400	29.000	-193.070	80.793	-306.180	47.803	
	1500	29.000	-190.170	82.794	-314.360	45.808	

Yb_2O_3 YTTERBIUM OXIDE (CUBIC)

		A	B	C	D	RANGE	REF.
SOL	CP	30.75	4.65	-4.1		298-1800	*

PHASE	T	CP	H	S	G	BT	
SOL	298	27.524	-433.7	31.8	-443.181	324.904	1
	300	27.589	-433.649	31.970	-443.240	322.944	
	400	30.047	-430.753	40.285	-446.867	244.190	
	500	31.435	-427.674	47.151	-451.249	197.267	
	600	32.401	-424.480	52.971	-456.262	166.216	
	700	33.168	-421.200	58.025	-461.818	144.205	
	800	33.829	-417.849	62.498	-467.848	127.827	
	900	34.429	-414.436	66.518	-474.302	115.192	
	1000	34.990	-410.965	70.175	-481.140	105.167	
	1100	35.526	-407.439	73.535	-488.328	97.035	
	1200	36.045	-403.860	76.649	-495.839	90.317	
	1300	36.552	-400.230	79.554	-503.650	84.683	
	1400	37.051	-396.550	82.281	-511.743	79.897	
	1500	37.543	-392.820	84.854	-520.101	75.789	
	1600	38.030	-389.042	87.293	-528.710	72.228	
	1700	38.513	-385.215	89.613	-537.556	69.117	
	1800	38.993	-381.339	91.828	-546.629	66.379	

ZINC

		A	B	C	D	RANGE	REF.
SOL	CP	4.956	2.990	0.199		298- 693	21
	** MPT=	692.655					21
LIQ	CP	7.5				693-1180	21
GAS	CP	4.968				1180-2000	21

PHASE	T	CP	H	S	G	BT	P	
SOL	298	6.071	0	9.95	-2.967	2.175		21
	300	6.074	.011	9.988	-2.985	2.175		
	400	6.276	.628	11.761	-4.076	2.227		
	500	6.531	1.268	13.188	-5.326	2.328		
	600	6.805	1.935	14.403	-6.707	2.443	.005	
	693	7.069	2.578	15.398	-8.088	2.552	.156	
			1.75	2.527				21
LIQ	693	7.500	4.328	17.925	-8.088	2.552	.156	
	700	7.500	4.383	18.004	-8.220	2.567	.195	
	800	7.500	5.133	19.005	-10.072	2.752	2.557	
	900	7.500	5.883	19.889	-12.017	2.919	18.624	
	1000	7.500	6.633	20.679	-14.046	3.070	89.907	
	1100	7.500	7.383	21.394	-16.151	3.209	322.225	
	1180	7.500	7.983	21.920	-17.883	3.313	760.	
			27.565	23.360				21
GAS	1180	4.968	35.548	45.281	-17.883	3.313		
	1200	4.968	35.647	45.364	-18.790	3.423		
	1300	4.968	36.144	45.762	-23.346	3.925		
	1400	4.968	36.641	46.130	-27.941	4.362		
	1500	4.968	37.137	46.473	-32.572	4.746		
	1600	4.968	37.634	46.793	-37.235	5.087		
	1700	4.968	38.131	47.094	-41.930	5.391		
	1800	4.968	38.628	47.378	-46.653	5.665		
	1900	4.968	39.125	47.647	-51.405	5.914		
	2000	4.968	39.621	47.902	-56.182	6.140		

ZnF_2 ZINC FLUORIDE

		A	B	C	D	RANGE	REF.
SOL-A	CP	14.89	2.715			298-1148	2,*
LIQ	CP	22.5				1148-1778	*
	** BPT= 1778, L= 44.2						62

PHASE	T	CP	H	S	G	BT	
SOL-A	298	15.699	-182.7	17.61	-187.950	137.790	62
	300	15.704	-182.671	17.707	-187.983	136.964	
	400	15.976	-181.087	22.262	-189.992	103.821	
	500	16.247	-179.476	25.856	-192.404	84.111	
	600	16.519	-177.837	28.843	-195.143	71.090	
	700	16.790	-176.172	31.409	-198.159	61.876	
	800	17.062	-174.479	33.669	-201.415	55.031	
	900	17.333	-172.760	35.694	-204.885	49.759	
	1000	17.605	-171.013	37.535	-208.547	45.584	
	1100	17.876	-169.239	39.225	-212.387	42.203	
	1148	18.007	-168.377	39.992	-214.288	40.800	
			10.	8.711			62
LIQ	1148	22.500	-158.377	48.703	-214.288	40.800	
	1200	22.500	-157.207	49.699	-216.847	39.498	
	1300	22.500	-154.957	51.500	-221.908	37.311	
	1400	22.500	-152.707	53.168	-227.142	35.463	
	1500	22.500	-150.457	54.720	-232.537	33.885	
	1600	22.500	-148.207	56.172	-238.083	32.525	
	1700	22.500	-145.957	57.536	-243.769	31.343	
	1778	22.500	-144.202	58.546	-248.296	30.524	

ZINC SULFIDE (SPHALERITE) ZnS

		A	B	C	D	RANGE	REF.
SOL-S	CP	11.77	1.26	-1.16		298-1293	63
	** TPT= 1293 (SPHALERITE → WURTZITE)						63

PHASE	T	CP	H	S	G	BT	P	
SOL-S	298	10.841	-49.04	13.78	-53.149	38.964		63*
	300	10.859	-49.020	13.847	-53.174	38.742		
	400	11.549	-47.895	17.077	-54.726	29.905		
	500	11.936	-46.720	19.699	-56.569	24.730		
	600	12.204	-45.512	21.900	-58.652	21.367		
	700	12.415	-44.281	23.798	-60.939	19.029		
	800	12.597	-43.030	25.468	-63.404	17.324		
	900	12.761	-41.762	26.961	-66.027	16.036		
	1000	12.914	-40.478	28.313	-68.792	15.036		
	1100	13.060	-39.180	29.551	-71.686	14.245		
	1200	13.201	-37.866	30.694	-74.699	13.606		
	1293	13.330	-36.633	31.684	-77.600	13.118		

ZnS (W)

ZINC SULFIDE (WURTZITE)

		A	B	C	D	RANGE	REF.
SOL-W	CP	11.82	1.06	-1.04		298-1400	63

PHASE	T	CP	H	S	G	BT	P	
SOL-W	298	10.966	-45.85	16.25	-50.695	37.165		63
	300	10.982	-45.830	16.318	-50.725	36.958		
	400	11.594	-44.697	19.572	-52.526	28.703		
	500	11.934	-43.520	22.198	-54.619	23.877		
	600	12.167	-42.314	24.396	-56.951	20.747		
	700	12.350	-41.088	26.285	-59.487	18.575		
	800	12.505	-39.845	27.945	-62.201	16.995		
	900	12.646	-38.587	29.426	-65.071	15.803		
	1000	12.776	-37.316	30.765	-68.081	14.881		
	1100	12.900	-36.032	31.989	-71.220	14.152		
	1200	13.020	-34.736	33.116	-74.476	13.566		
	1293	13.128	-33.520	34.092	-77.601	13.118		
	1293	13.128	-33.520	34.092	-77.601	13.118		
	1300	13.136	-33.428	34.163	-77.840	13.088		
	1400	13.251	-32.109	35.141	-81.306	12.694	.001	

ZINC SULFIDE (GAS) ZnS(g)

		A	B	C	D	RANGE	REF.
GAS	CP	8.91	0.017	-0.456		298-2000	63

PHASE	T	CP	H	S	G	BT	
GAS	298	8.402	48.3	59.2	30.650	-22.470	63
	300	8.408	48.316	59.252	30.540	-22.251	
	400	8.632	49.169	61.706	24.487	-13.381	
	500	8.736	50.038	63.645	18.216	-7.963	
	600	8.794	50.915	65.243	11.769	-4.287	
	700	8.829	51.796	66.601	5.175	-1.616	
	800	8.852	52.680	67.782	-1.545	.422	
	900	8.869	53.566	68.826	-8.377	2.034	
	1000	8.881	54.454	69.761	-15.307	3.346	
	1100	8.891	55.343	70.608	-22.326	4.436	
	1200	8.899	56.232	71.382	-29.426	5.360	
	1300	8.905	57.122	72.094	-36.600	6.154	
	1400	8.911	58.013	72.754	-43.843	6.845	
	1500	8.915	58.904	73.369	-51.150	7.453	
	1600	8.919	59.796	73.945	-58.516	7.994	
	1700	8.923	60.688	74.486	-65.937	8.478	
	1800	8.927	61.581	74.996	-73.412	8.915	
	1900	8.930	62.473	75.478	-80.936	9.311	
	2000	8.933	63.367	75.937	-88.507	9.673	

$ZnSO_4$ ZINC SULFATE

		A	B	C	D	RANGE	REF.
SOL-2	CP	18.25	18.2			298-1027	62*
SOL-1	CP	18.25	18.2			1027-1500	62*
	** DEC.						

PHASE	T	CP	H	S	G	BT	
SOL-2	298	23.676	-234.55	26.42	-242.427	177.728	62
	300	23.710	-234.506	26.567	-242.476	176.667	
	400	25.530	-232.044	33.637	-245.499	134.152	
	500	27.350	-229.400	39.529	-249.165	108.924	
	600	29.170	-226.574	44.676	-253.380	92.306	
	700	30.990	-223.566	49.310	-258.083	80.588	
	800	32.810	-220.376	53.567	-263.230	71.921	
	900	34.630	-217.004	57.536	-268.787	65.279	
	1000	36.450	-213.450	61.279	-274.729	60.050	
	1027	36.941	-212.459	62.257	-276.397	58.826	
			4.7	4.576			62
SOL-1	1027	36.941	-207.759	66.833	-276.397	58.826	
	1100	38.270	-205.014	69.415	-281.371	55.911	
	1200	40.090	-201.096	72.823	-288.484	52.547	
	1300	41.910	-196.996	76.104	-295.931	49.757	
	1400	43.730	-192.714	79.276	-303.701	47.416	
	1500	45.550	-188.250	82.355	-311.783	45.433	

* * * * * * * * * *

$ZnSO_4 \cdot H_2O$ ZINC SULFATE MONOHYDRATE

		A	B	C	D	RANGE	REF.
SOL	CP	30.5	20.8			298- 800	62*

PHASE	T	CP	H	S	G	BT	
SOL	298	36.702	-311.07	34.779	-321.439	235.653	62
	300	36.740	-311.002	35.006	-321.504	234.247	
	400	38.820	-307.224	45.860	-325.568	177.906	
	500	40.900	-303.238	54.746	-330.611	144.529	
	600	42.980	-299.044	62.387	-336.476	122.578	
	700	45.060	-294.642	69.169	-343.060	107.123	
	800	47.140	-290.032	75.321	-350.289	95.707	

ZINC SULFATE DIHYDRATE $ZnSO_4 \cdot 2\,H_2O$

		A	B	C	D	RANGE	REF.
SOL	CP	40.18	24.55			298- 365	62*
	** MPT= 365						62

PHASE	T	CP	H	S	G	BT	
SOL	298	47.500	-381.461	46.0	-395.176	289.711	62
	300	47.545	-381.373	46.294	-395.261	287.986	
	365	49.141	-378.231	55.770	-398.587	238.693	

* * * * * * * * * *

ZINC SULFATE HEXAHYDRATE $ZnSO_4 \cdot 6\,H_2O$

		A	B	C	D	RANGE	REF.
SOL	CP	75.55	33.56			298- 800	62*
	** DPT= 344						62

PHASE	T	CP	H	S	G	BT	
SOL	298	85.556	-664.19	85.06	-689.551	505.522	62
	300	85.618	-664.032	85.589	-689.708	502.520	
	344	87.095	-660.232	97.406	-693.740	440.805	
	344	87.095	-660.232	97.406	-693.740	440.805	
	400	88.974	-655.302	110.680	-699.574	382.281	
	500	92.330	-646.237	130.894	-711.684	311.119	
	600	95.686	-636.836	148.025	-725.651	264.354	
	700	99.042	-627.100	163.027	-741.218	231.450	
	800	102.398	-617.028	176.471	-758.205	207.160	

$ZnSO_4 \cdot 7\ H_2O$ ZINC SULFATE HEPTAHYDRATE

		A	B	C	D	RANGE	REF.
SOL	CP	79.85	36.12			298- 800	62*
	** DISPROPORT. PT= 321						62

PHASE	T	CP	H	S	G	BT	
SOL	298	90.619	-735.788	92.9	-763.486	559.726	62
	300	90.686	-735.620	93.461	-763.659	556.400	
	321	91.445	-733.708	99.622	-765.687	521.381	
	321	91.445	-733.708	99.622	-765.687	521.381	
	400	94.298	-726.371	120.044	-774.389	423.163	
	500	97.910	-716.761	141.474	-787.498	344.261	
	600	101.522	-706.789	159.645	-802.576	292.377	
	700	105.134	-696.456	175.565	-819.352	255.848	
	800	108.746	-685.762	189.840	-837.634	228.862	

* * * * * * * * * *

$ZnO \cdot ZnSO_4$ ZINC OXIDE SULFATE

		A	B	C	D	RANGE	REF.
SOL	CP	48.21	37.62	-2.18		298- 800	*

PHASE	T	CP	H	S	G	BT	
SOL	298	56.974	-554.6	63.2	-573.443	420.402	62/*
	300	57.074	-554.495	63.553	-573.560	417.895	
	400	61.895	-548.538	80.654	-580.800	317.377	
	500	66.148	-542.134	94.929	-589.598	257.748	
	600	70.176	-535.316	107.347	-599.724	218.479	
	700	74.099	-528.102	118.460	-611.024	190.796	
	800	77.965	-520.498	128.608	-623.384	170.324	

ZINC SELENIDE

ZnSe

		A	B	C	D	RANGE	REF.
SOL	CP	11.99	1.38			298-1300	63
	** MPT= 1799						63

PHASE	T	CP	H	S	G	BT	P	
SOL	298	12.401	-38.	16.8	-43.009	31.531		63
	300	12.404	-37.977	16.877	-43.040	31.359		
	400	12.542	-36.730	20.464	-44.915	24.544		
	500	12.680	-35.469	23.278	-47.107	20.593		
	600	12.818	-34.194	25.602	-49.555	18.053		
	700	12.956	-32.905	27.588	-52.217	16.305		
	800	13.094	-31.603	29.327	-55.064	15.045		
	900	13.232	-30.286	30.877	-58.076	14.105		
	1000	13.370	-28.956	32.278	-61.235	13.385		
	1100	13.508	-27.612	33.559	-64.527	12.822		
	1200	13.646	-26.255	34.740	-67.943	12.376		
	1300	13.784	-24.883	35.838	-71.473	12.017		

ZnSe (g)

ZINC SELENIDE (GAS)

		A	B	C	D	RANGE	REF.
GAS	CP	8.934	0.004	-0.257		298-2000	63

PHASE	T	CP	H	S	G	BT	
GAS	298	8.646	56.7	59.71	38.897	-28.516	63
	300	8.650	56.716	59.763	38.787	-28.260	
	400	8.775	57.588	62.272	32.679	-17.858	
	500	8.833	58.469	64.237	26.351	-11.519	
	600	8.865	59.354	65.850	19.844	-7.229	
	700	8.884	60.241	67.218	13.189	-4.118	
	800	8.897	61.131	68.406	6.406	-1.750	
	900	8.906	62.021	69.454	-.488	.118	
	1000	8.912	62.912	70.393	-7.481	1.635	
	1100	8.917	63.803	71.242	-14.563	2.894	
	1200	8.921	64.695	72.018	-21.727	3.958	
	1300	8.924	65.587	72.733	-28.965	4.870	
	1400	8.926	66.480	73.394	-36.272	5.663	
	1500	8.929	67.373	74.010	-43.642	6.360	
	1600	8.930	68.266	74.586	-51.072	6.977	
	1700	8.932	69.159	75.128	-58.558	7.529	
	1800	8.933	70.052	75.638	-66.097	8.026	
	1900	8.934	70.945	76.121	-73.685	8.477	
	2000	8.936	71.839	76.580	-81.320	8.887	

ZINC SELENITE $ZnSeO_3$

		A	B	C	D	RANGE	REF.
SOL	CP	18.45	13.2			298- 894	*
LIQ	CP	33.5				894-1000	*

PHASE	T	CP	H	S	G	BT	
SOL	298	22.386	-155.9	23.5	-162.907	119.430	62
	300	22.410	-155.859	23.639	-162.950	118.725	
	400	23.730	-153.552	30.266	-165.658	90.524	
	500	25.050	-151.113	35.703	-168.964	73.864	
	600	26.370	-148.542	40.387	-172.774	62.941	
	700	27.690	-145.839	44.551	-177.024	55.277	
	800	29.010	-143.004	48.335	-181.671	49.637	
	894	30.251	-140.218	51.625	-186.371	45.567	
			11.1	12.416			62
LIQ	894	33.500	-129.118	64.041	-186.371	45.567	
	900	33.500	-128.917	64.266	-186.756	45.357	
	1000	33.500	-125.567	67.795	-193.362	42.265	

ZnTe

ZINC TELLURIDE

		A	B	C	D	RANGE	REF.
SOL	CP	10.54	4.48			298-1300	63
	** MPT= 1570						63

PHASE	T	CP	H	S	G	BT	P	
SOL	298	11.876	-28.5	18.6	-34.046	24.959		63
	300	11.884	-28.478	18.673	-34.080	24.831		
	400	12.332	-27.267	22.154	-36.129	19.742		
	500	12.780	-26.012	24.954	-38.488	16.826		
	600	13.228	-24.711	27.323	-41.105	14.975		
	700	13.676	-23.366	29.396	-43.943	13.722		
	800	14.124	-21.976	31.251	-46.977	12.835		
	900	14.572	-20.541	32.941	-50.188	12.189		
	1000	15.020	-19.062	34.499	-53.561	11.707		
	1100	15.468	-17.537	35.952	-57.084	11.343		
	1200	15.916	-15.968	37.317	-60.748	11.065		
	1300	16.364	-14.354	38.609	-64.545	10.853		

ZINC TELLURIDE (GAS) — ZnTe (g)

		A	B	C	D	RANGE	REF.
GAS	CP	8.937	0.003	-0.189		298-2000	63

PHASE	T	CP	H	S	G	BT	
GAS	298	8.725	61.	61.60	42.634	-31.256	63
	300	8.728	61.016	61.654	42.520	-30.980	
	400	8.820	61.894	64.179	36.222	-19.794	
	500	8.863	62.779	66.153	29.702	-12.985	
	600	8.886	63.666	67.771	23.004	-8.380	
	700	8.901	64.556	69.142	16.156	-5.045	
	800	8.910	65.446	70.331	9.181	-2.509	
	900	8.916	66.337	71.381	2.095	-.509	
	1000	8.921	67.229	72.320	-5.091	1.113	
	1100	8.925	68.122	73.171	-12.366	2.457	
	1200	8.927	69.014	73.948	-19.723	3.593	
	1300	8.930	69.907	74.662	-27.154	4.566	
	1400	8.932	70.800	75.324	-34.654	5.410	
	1500	8.933	71.693	75.940	-42.217	6.152	
	1600	8.934	72.587	76.517	-49.840	6.809	
	1700	8.936	73.480	77.059	-57.519	7.396	
	1800	8.937	74.374	77.569	-65.251	7.924	
	1900	8.937	75.268	78.053	-73.032	8.402	
	2000	8.938	76.161	78.511	-80.861	8.837	

* * * * * * * * * *

ZINC NITRIDE — Zn_3N_2

		A	B	C	D	RANGE	REF.
SOL	CP	19.0	22.5			298- 700	1
	** MPT= 838						1

PHASE	T	CP	H	S	G	BT	
SOL	298	25.708	-5.3	33.5	-15.288	11.208	1/*
	300	25.750	-5.252	33.659	-15.350	11.184	
	400	28.000	-2.565	41.375	-19.115	10.445	
	500	30.250	.348	47.865	-23.585	10.310	
	600	32.500	3.485	53.579	-28.662	10.442	
	700	34.750	6.848	58.758	-34.283	10.705	

Zn_3P_2 TRIZINC DIPHOSPHIDE

		A	B	C	D	RANGE	REF.
SOL	CP	30.17	6.23	-3.64		298-1513	*
LIQ	CP	37.2				1513-1600	*

PHASE	T	CP	H	S	G	BT	
SOL	298	27.933	-38.0	36.	-48.733	35.727	208
	300	27.995	-37.948	36.173	-48.800	35.556	
	400	30.387	-35.017	44.591	-52.853	28.881	
	500	31.829	-31.901	51.536	-57.669	25.211	
	600	32.897	-28.663	57.438	-63.125	22.997	
	700	33.788	-25.328	62.577	-69.132	21.587	
	800	34.585	-21.908	67.142	-75.622	20.662	
	900	35.328	-18.412	71.259	-82.545	20.047	
	1000	36.036	-14.844	75.018	-89.862	19.642	
	1100	36.722	-11.206	78.485	-97.539	19.382	
	1200	37.393	-7.500	81.709	-105.550	19.226	
	1300	38.054	-3.728	84.728	-113.874	19.146	
	1400	38.706	.110	87.572	-122.490	19.124	
	1500	39.353	4.013	90.264	-131.383	19.145	
	1513	39.437	4.526	90.604	-132.559	19.150	
			40.0	26.438			208
LIQ	1513	37.200	44.526	117.042	-132.559	19.150	
	1600	37.200	47.762	119.122	-142.833	19.513	

ZINC DIPHOSPHIDE ZnP_2

		A	B	C	D	RANGE	REF.
SOL	CP	17.03	4.0	-0.65		298-1253	*
LIQ	CP	21.8				1253-1400	*

PHASE	T	CP	H	S	G	BT	
SOL	298	17.491	-24.3	14.4	-28.593	20.962	208
	300	17.508	-24.268	14.508	-28.620	20.853	
	400	18.224	-22.479	19.649	-30.339	16.578	
	500	18.770	-20.628	23.776	-32.517	14.215	
	600	19.249	-18.727	27.242	-35.072	12.777	
	700	19.697	-16.779	30.243	-37.949	11.850	
	800	20.128	-14.788	32.901	-41.109	11.232	
	900	20.550	-12.754	35.297	-44.521	10.813	
	1000	20.965	-10.678	37.483	-48.162	10.527	
	1100	21.376	-8.561	39.501	-52.012	10.335	
	1200	21.785	-6.403	41.378	-56.057	10.211	
	1253	22.001	-5.243	42.324	-58.275	10.166	
			22.2	17.717			208
LIQ	1253	21.800	16.957	60.042	-58.275	10.166	
	1300	21.800	17.982	60.845	-61.116	10.276	
	1400	21.800	20.162	62.460	-67.283	10.505	

Zn_3As_2 ZINC ARSENIDE

			A	B	C	D	RANGE	REF.
SOL-3	CP		26.95	10.0			298- 463	*
	**	TPT=	463					62
SOL-2	CP		26.95	10.0			463- 965	*
	**	TPT=	965					
SOL-1	CP		26.95	10.0			965-1288	*
LIQ	CP		37.0				1288-1500	*

PHASE	T	CP	H	S	G	BT	
SOL-3	298	29.932	-32.	40.162	-43.974	32.238	62/173
	300	29.950	-31.945	40.347	-44.049	32.094	
	400	30.950	-28.900	49.100	-48.540	26.524	
	463	31.580	-26.930	53.672	-51.780	24.445	
SOL-2	463	31.580	-26.930	53.672	-51.780	24.445	
	500	31.950	-25.755	56.114	-53.812	23.524	
	600	32.950	-22.510	62.028	-59.726	21.758	
	700	33.950	-19.165	67.182	-66.192	20.669	
	800	34.950	-15.720	71.781	-73.144	19.985	
	900	35.950	-12.175	75.955	-80.534	19.559	
	965	36.600	-9.817	78.484	-85.554	19.379	
SOL-1	965	36.600	-9.817	78.484	-85.554	19.379	
	1000	36.950	-8.530	79.794	-88.324	19.306	
	1100	37.950	-4.785	83.363	-96.484	19.172	
	1200	38.950	-.940	86.708	-104.989	19.124	
	1288	39.830	2.527	89.495	-112.743	19.133	
			37.	28.727			62
LIQ	1288	37.000	39.527	118.222	-112.743	19.133	
	1300	37.000	39.971	118.565	-114.164	19.195	
	1400	37.000	43.671	121.307	-126.159	19.697	
	1500	37.000	47.371	123.860	-138.419	20.170	

ZINC MONOANTIMONIDE

ZnSb

		A	B	C	D	RANGE	REF.
SOL	CP	10.68	4.14			298- 822	*
LIQ	CP	15.				822-1100	*

PHASE	T	CP	H	S	G	BT	
SOL	298	11.914	-4.552	19.756	-10.442	7.655	21
	300	11.922	-4.530	19.830	-10.479	7.635	
	400	12.336	-3.317	23.316	-12.644	6.909	
	500	12.750	-2.063	26.113	-15.119	6.610	
	600	13.164	-.767	28.475	-17.852	6.503	
	700	13.578	.570	30.535	-20.804	6.496	
	800	13.992	1.949	32.375	-23.951	6.544	
	822	14.083	2.257	32.756	-24.668	6.559	
			7.35	8.942			21*
LIQ	822	15.000	9.607	41.697	-24.668	6.559	
	900	15.000	10.777	43.057	-27.974	6.794	
	1000	15.000	12.277	44.638	-32.360	7.073	
	1100	15.000	13.777	46.067	-36.897	7.332	

$2ZnO \cdot SiO_2$ ZINC ORTHOSILICATE (WILLEMITE)

		A	B	C	D	RANGE	REF.
SOL	CP	34.63	8.83	-7.24		298-1785	5
	** MPT= 1785						5

PHASE	T	CP	H	S	G	BT	
SOL	298	29.118	-391.9	31.4	-401.262	294.172	5
	300	29.235	-391.846	31.580	-401.320	292.401	
	400	33.637	-388.677	40.666	-404.944	221.281	
	500	36.149	-385.179	48.462	-409.410	178.977	
	600	37.917	-381.472	55.216	-414.602	151.039	
	700	39.333	-377.607	61.171	-420.427	131.281	
	800	40.563	-373.611	66.505	-426.815	116.616	
	900	41.683	-369.498	71.348	-433.711	105.334	
	1000	42.736	-365.277	75.795	-441.072	96.409	
	1100	43.745	-360.952	79.916	-448.860	89.192	
	1200	44.723	-356.529	83.764	-457.046	83.251	
	1300	45.681	-352.008	87.382	-465.605	78.286	
	1400	46.623	-347.393	90.802	-474.515	74.085	
	1500	47.553	-342.684	94.050	-483.759	70.493	
	1600	48.475	-337.883	97.148	-493.320	67.393	
	1700	49.390	-332.989	100.115	-503.185	64.697	
	1785	50.164	-328.758	102.543	-511.798	62.671	

2-ZINC 2-TITANIUM 4-OXIDE

$2ZnO \cdot TiO_2$

		A	B	C	D	RANGE	REF.
SOL	CP	39.82	5.54	-7.69		298-2000	1

PHASE	T	CP	H	S	G	BT	
SOL	298	32.821	-393.1	34.6	-403.416	295.752	5/232
	300	32.938	-393.039	34.803	-403.480	293.975	
	400	37.230	-389.504	44.944	-407.482	222.668	
	500	39.514	-385.657	53.518	-412.416	180.291	
	600	41.008	-381.627	60.862	-418.144	152.329	
	700	42.129	-377.468	67.271	-424.558	132.571	
	800	43.050	-373.208	72.959	-431.575	117.917	
	900	43.857	-368.862	78.077	-439.131	106.650	
	1000	44.591	-364.439	82.736	-447.175	97.743	
	1100	45.278	-359.945	87.018	-455.665	90.545	
	1200	45.934	-355.384	90.986	-464.568	84.621	
	1300	46.567	-350.759	94.688	-473.854	79.673	
	1400	47.184	-346.071	98.162	-483.498	75.488	
	1500	47.788	-341.323	101.438	-493.479	71.910	
	1600	48.384	-336.514	104.541	-503.780	68.822	
	1700	48.972	-331.646	107.492	-514.383	66.137	
	1800	49.555	-326.720	110.308	-525.274	63.786	
	1900	50.133	-321.735	113.002	-536.440	61.713	
	2000	50.708	-316.693	115.589	-547.871	59.877	

$ZnO \cdot Fe_2O_3$ ZINC 2-IRON 4-OXIDE

		A	B	C	D	RANGE	REF.
SOL	CP	37.15	10.65	-3.82		298-1000	*

PHASE	T	CP	H	S	G	BT	
SOL	298	36.028	-281.7	32.2	-291.300	213.558	5
	300	36.101	-281.633	32.423	-291.360	212.284	
	400	39.022	-277.864	43.247	-295.163	161.291	
	500	40.947	-273.861	52.172	-299.947	131.124	
	600	42.479	-269.687	59.777	-305.553	111.313	
	700	43.825	-265.371	66.428	-311.870	97.383	
	800	45.073	-260.925	72.362	-318.815	87.108	
	900	46.263	-256.358	77.740	-326.324	79.253	
	1000	47.418	-251.674	82.674	-334.348	73.082	

ZIRCONIUM MONOFLUORIDE (GAS) ZrF (g)

		A	B	C	D	RANGE	REF.
GAS	CP	7.380	2.298			298- 800	3
	CP	8.357	1.175			800-2000	3

PHASE	T	CP	H	S	G	BT	
GAS	298	8.065	19.8	58.	2.507	-1.838	3
	300	8.069	19.815	58.050	2.400	-1.749	
	400	8.299	20.633	60.403	-3.528	1.928	
	500	8.529	21.475	62.279	-9.665	4.225	
	600	8.759	22.339	63.855	-15.974	5.819	
	700	8.989	23.227	65.222	-22.429	7.004	
	800	9.218	24.137	66.437	-29.013	7.927	
	800	9.297	24.137	66.437	-29.013	7.927	
	900	9.414	25.072	67.539	-35.713	8.673	
	1000	9.532	26.020	68.537	-42.517	9.293	
	1100	9.649	26.979	69.451	-49.418	9.820	
	1200	9.767	27.950	70.296	-56.405	10.274	
	1300	9.885	28.932	71.082	-63.475	10.673	
	1400	10.002	29.927	71.819	-70.620	11.026	
	1500	10.120	30.933	72.513	-77.837	11.342	
	1600	10.237	31.950	73.170	-85.122	11.629	
	1700	10.354	32.980	73.794	-92.470	11.889	
	1800	10.472	34.021	74.389	-99.879	12.129	
	1900	10.589	35.074	74.959	-107.347	12.349	
	2000	10.707	36.139	75.505	-114.870	12.554	

ZrF_2 ZIRCONIUM DIFLUORICE

		A	B	C	D	RANGE	REF.
SOL	CP	16.548	3.513	-1.640		298-1175	3
LIQ	CP	24.0				1175-2529	3
GAS	CP	12.078	1.140	4.779		2529-3000	3

PHASE	T	CP	H	S	G	BT	P	
SOL	298	15.750	-230.	18.	-235.367	172.552		3
	300	15.780	-229.971	18.098	-235.400	171.512		
	400	16.928	-228.330	22.811	-237.454	129.756		
	500	17.648	-226.599	26.670	-239.934	104.889		
	600	18.200	-224.806	29.938	-242.768	88.440		
	700	18.672	-222.961	32.780	-245.907	76.786		
	800	19.102	-221.072	35.302	-249.314	68.119		
	900	19.507	-219.142	37.575	-252.960	61.435		
	1000	19.897	-217.171	39.651	-256.822	56.136		
	1100	20.277	-215.163	41.565	-260.884	51.840		
	1175	20.557	-213.631	42.912	-264.053	49.120		
			9.	7.660				3
LIQ	1175	24.000	-204.631	50.571	-264.053	49.120		
	1200	24.000	-204.031	51.077	-265.323	48.328		
	1300	24.000	-201.631	52.998	-270.528	45.486		
	1400	24.000	-199.231	54.776	-275.918	43.079	.004	
	1500	24.000	-196.831	56.432	-281.479	41.017	.029	
	1600	24.000	-194.431	57.981	-287.201	39.235	.147	
	1700	24.000	-192.031	59.436	-293.072	37.682	.615	
	1800	24.000	-189.631	60.808	-299.085	36.319	2.159	
	1900	24.000	-187.231	62.105	-305.232	35.114	6.543	
	2000	24.000	-184.831	63.336	-311.504	34.044	17.523	
	2100	24.000	-182.431	64.507	-317.897	33.088	42.241	
	2200	24.000	-180.031	65.624	-324.404	32.231	93.037	
	2300	24.000	-177.631	66.691	-331.020	31.458	189.557	
	2400	24.000	-175.231	67.712	-337.740	30.760	360.924	
	2500	24.000	-172.831	68.692	-344.561	30.126	647.739	
	2529	24.000	-172.135	68.969	-346.557	29.953	760.	
			69.0	27.284				3
GAS	2529	15.036	-103.135	96.252	-346.557	29.953		
	2600	15.113	-102.065	96.669	-353.406	29.710		
	2700	15.222	-100.548	97.242	-363.102	29.395		
	2800	15.331	-99.021	97.797	-372.854	29.106		
	2900	15.441	-97.482	98.337	-382.660	28.842		
	3000	15.551	-95.933	98.863	-392.521	28.599		

ZIRCONIUM TRIFLUORIDE

ZrF_3

		A	B	C	D	RANGE	REF.
SOL	CP	22.879	3.437	-3.385		298- 800	3
	** DEC.	4ZRF3 = 3ZRF4 + ZR					3
	CP	27.932	-0.733	-14.390		800-1468	3
	** SPT=	1468.35					
GAS	CP	20.071	0.458			1468-2000	3

PHASE	T	CP	H	S	G	BT	P	
SOL	298	20.096	-335.	21.	-341.261	250.185		3
	300	20.149	-334.963	21.124	-341.300	248.670		
	400	22.138	-332.837	27.227	-343.728	187.829		
	500	23.243	-330.563	32.295	-346.711	151.568		
	600	24.001	-328.199	36.604	-350.161	127.563		
	700	24.594	-325.769	40.349	-354.013	110.543		
	800	25.100	-323.283	43.667	-358.217	97.874		
	800	25.097	-323.283	43.667	-358.217	97.874		
	900	25.496	-320.752	46.648	-362.735	88.096		
	1000	25.760	-318.189	49.349	-367.537	80.336	.018	
	1100	25.936	-315.603	51.813	-372.597	74.038	.393	
	1200	26.053	-313.003	54.075	-377.893	68.833	4.978	
	1300	26.128	-310.394	56.163	-383.406	64.465	41.885	
	1400	26.172	-307.779	58.102	-389.121	60.753	256.078	
	1468	26.188	-305.989	59.349	-393.135	58.522	760.	
			64.820	44.145				*
GAS	1468	20.744	-241.169	103.494	-393.135	58.522		
	1500	20.758	-240.513	103.937	-396.418	57.766		
	1600	20.804	-238.434	105.278	-406.879	55.585		
	1700	20.850	-236.352	106.541	-417.471	53.677		
	1800	20.895	-234.265	107.734	-428.185	51.996		
	1900	20.941	-232.173	108.865	-439.015	50.505		
	2000	20.987	-230.076	109.940	-449.956	49.176		

ZIRCONIUM TETRAFLUORIDE

		A	B	C	D	RANGE	REF.
SOL-A	CP	28.028	4.433	-4.116		298- 723	3
SOL-B	CP	28.028	4.433	-4.116		723-1179	3
	** MPT=	1205, L= 15.35					3
GAS	CP	25.244	0.285	-3.855		1179-2000	3

PHASE	T	CP	H	S	G	BT	P	
SOL-A	298	24.719	-456.80	25.024	-464.261	340.358		3
	300	24.785	-456.754	25.177	-464.307	338.293		
	400	27.229	-454.139	32.683	-467.213	255.307		
	500	28.598	-451.343	38.918	-470.802	205.815		
	600	29.544	-448.433	44.220	-474.965	173.029		
	700	30.291	-445.440	48.832	-479.623	149.765		
	723	30.446	-444.742	49.813	-480.757	145.344		
SOL-B	723	30.446	-444.742	49.813	-480.757	145.344		
	800	30.931	-442.379	52.919	-484.714	132.436	.016	
	900	31.510	-439.256	56.596	-490.193	119.051	.708	
	1000	32.049	-436.078	59.944	-496.022	108.420	13.935	
	1100	32.564	-432.847	63.023	-502.173	99.786	154.648	
	1179	32.958	-430.259	65.295	-507.242	94.040	760.	
			51.64	43.800				3
GAS	1179	25.303	-378.619	109.095	-507.242	94.040		
	1200	25.318	-378.087	109.542	-509.538	92.812		
	1300	25.386	-375.552	111.571	-520.595	87.532		
	1400	25.446	-373.010	113.455	-531.847	83.036		
	1500	25.500	-370.463	115.212	-543.282	79.167		
	1600	25.549	-367.911	116.860	-554.886	75.804		
	1700	25.595	-365.353	118.410	-566.650	72.858		
	1800	25.638	-362.792	119.874	-578.565	70.257		
	1900	25.679	-360.226	121.261	-590.623	67.946		
	2000	25.718	-357.656	122.580	-602.815	65.881		

ZIRCONIUM MONOCHLORIDE (GAS) $ZrCl(g)$

		A	B	C	D	RANGE	REF.
GAS	CP	8.669	0.833	-0.492		298-2000	3

PHASE	T	CP	H	S	G	BT	
GAS	298	8.364	49.1	60.8	30.972	-22.706	3
	300	8.372	49.115	60.852	30.860	-22.484	
	400	8.695	49.971	63.309	24.647	-13.468	
	500	8.889	50.850	65.272	18.214	-7.963	
	600	9.032	51.747	66.906	11.603	-4.227	
	700	9.152	52.656	68.307	4.841	-1.512	
	800	9.259	53.577	69.536	-2.052	.561	
	900	9.358	54.507	70.632	-9.062	2.201	
	1000	9.453	55.448	71.623	-16.175	3.536	
	1100	9.545	56.398	72.529	-23.384	4.647	
	1200	9.634	57.357	73.363	-30.679	5.588	
	1300	9.723	58.325	74.138	-38.054	6.398	
	1400	9.810	59.301	74.861	-45.505	7.105	
	1500	9.897	60.287	75.541	-53.025	7.727	
	1600	9.983	61.281	76.183	-60.612	8.280	
	1700	10.068	62.283	76.790	-68.260	8.777	
	1800	10.153	63.294	77.368	-75.969	9.225	
	1900	10.238	64.314	77.920	-83.733	9.633	
	2000	10.323	65.342	78.447	-91.552	10.006	

$ZrCl_2$ ZIRCONIUM DICHLORIDE

		A	B	C	D	RANGE	REF.
SOL	CP	16.295	4.388			298-1000	3
LIQ	CP	21.750				1000-1565	3
	** BPT=	1565.3					3
GAS	CP	14.493	0.405	-0.736		1565-2000	3

PHASE	T	CP	H	S	G	BT	P	
SOL	298	17.603	-103.0	26.3	-110.841	81.260		3
	300	17.611	-102.967	26.409	-110.890	80.794		
	400	18.050	-101.184	31.535	-113.799	62.185		
	500	18.489	-99.357	35.610	-117.163	51.219		
	600	18.928	-97.487	39.020	-120.899	44.043		
	700	19.367	-95.572	41.971	-124.951	39.017		
	800	19.805	-93.613	44.586	-129.282	35.323		
	900	20.244	-91.611	46.944	-133.860	32.510	.006	
	1000	20.683	-89.564	49.099	-138.664	30.309	.143	
			6.4	6.400				3
LIQ	1000	21.750	-83.164	55.499	-138.664	30.309	.143	
	1100	21.750	-80.989	57.572	-144.319	28.677	1.316	
	1200	21.750	-78.814	59.465	-150.172	27.354	8.149	
	1300	21.750	-76.639	61.206	-156.207	26.264	37.277	
	1400	21.750	-74.464	62.818	-162.409	25.357	134.684	
	1500	21.750	-72.289	64.318	-168.767	24.593	403.461	
	1565	21.750	-70.869	65.245	-172.997	24.157	760.	
			45.0	28.748				3
GAS	1565	15.097	-25.869	93.993	-172.997	24.157		
	1600	15.112	-25.345	94.325	-176.264	24.080		
	1700	15.156	-23.832	95.242	-185.743	23.882		
	1800	15.199	-22.314	96.110	-195.311	23.717		
	1900	15.242	-20.792	96.933	-204.964	23.579		
	2000	15.285	-19.265	97.715	-214.696	23.464		

ZIRCONIUM TRICHLORICE

$ZrCl_3$

		A	B	C	D	RANGE	REF.
SOL	CP	23.863	3.016	-1.574		298-1046	3
	** SPT=	1045.9					3
GAS	CP	19.743	0.990	-1.666		1046-2000	3

PHASE	T	CP	H	S	G	BT	P	
SOL	298	22.992	-170.7	34.8	-181.076	132.750		3
	300	23.019	-170.657	34.942	-181.140	131.978		
	400	24.086	-168.297	41.726	-184.987	101.086		
	500	24.741	-165.853	47.176	-189.441	82.816		
	600	25.235	-163.354	51.732	-194.393	70.817		
	700	25.653	-160.809	55.654	-199.767	62.378	.035	
	800	26.030	-158.224	59.104	-205.508	56.150	1.602	
	900	26.383	-155.604	62.191	-211.575	51.384	30.266	
	1000	26.722	-152.948	64.988	-217.936	47.636	307.208	
	1046	26.874	-151.718	66.191	-220.947	46.175	760.	
			40.842	39.050				3*
GAS	1046	20.626	-110.876	105.240	-220.947	46.175		
	1100	20.694	-109.759	106.282	-226.669	45.041		
	1200	20.815	-107.683	108.088	-237.389	43.240		
	1300	20.931	-105.596	109.759	-248.282	41.746		
	1400	21.044	-103.497	111.314	-259.337	40.490		
	1500	21.154	-101.387	112.770	-270.542	39.423		
	1600	21.262	-99.266	114.139	-281.888	38.509		
	1700	21.368	-97.135	115.431	-293.367	37.720		
	1800	21.474	-94.993	116.655	-304.972	37.034		
	1900	21.578	-92.840	117.819	-316.696	36.433		
	2000	21.681	-90.677	118.928	-328.534	35.905		

$ZrCl_4$ ZIRCONIUM TETRACHLORIDE

		A	B	C	D	RANGE	REF.
SOL	CP	29.868	3.380	-2.0		298- 607	3
	** SPT= 606.8, MPT= 710, L= 12.						3
GAS	CP	25.683	0.069	-1.974		607-2000	3

PHASE	T	CP	H	S	G	BT	P	
SOL	298	28.626	-234.17	43.36	-247.098	181.152		3
	300	28.660	-234.117	43.537	-247.178	180.093		
	400	29.970	-231.179	51.982	-251.971	137.689	.014	
	500	30.758	-228.140	58.759	-257.519	112.577	9.051	
	600	31.340	-225.034	64.421	-263.686	96.061	602.627	
	607	31.376	-224.820	64.774	-264.125	95.142	760.	
			24.662	40.643				3*
GAS	607	25.189	-200.158	105.417	-264.125	95.142		
	700	25.328	-197.804	109.026	-274.122	85.596		
	800	25.430	-195.266	112.415	-285.198	77.923		
	900	25.501	-192.719	115.415	-296.592	72.032		
	1000	25.555	-190.166	118.105	-308.271	67.382		
	1100	25.596	-187.608	120.542	-320.205	63.627		
	1200	25.629	-185.047	122.771	-332.372	60.541		
	1300	25.656	-182.483	124.823	-344.753	57.966		
	1400	25.679	-179.916	126.726	-357.332	55.790		
	1500	25.699	-177.347	128.498	-370.094	53.930		
	1600	25.716	-174.776	130.157	-383.028	52.326		
	1700	25.732	-172.204	131.717	-396.122	50.932		
	1800	25.746	-169.630	133.188	-409.368	49.711		
	1900	25.759	-167.055	134.580	-422.757	48.635		
	2000	25.772	-164.478	135.902	-436.282	47.681		

ZIRCONIUM MONOBROMIDE (GAS) $ZrBr(g)$

		A	B	C	D	RANGE	REF.
GAS	CP	8.373	1.064			298-2000	3

PHASE	T	CP	H	S	G	BT	
GAS	298	8.690	72.	63.4	53.097	-38.927	3
	300	8.692	72.016	63.454	52.980	-38.601	
	400	8.799	72.891	65.969	46.503	-25.412	
	500	8.905	73.776	67.944	39.804	-17.401	
	600	9.011	74.672	69.577	32.926	-11.995	
	700	9.118	75.578	70.974	25.896	-8.086	
	800	9.224	76.495	72.198	18.737	-5.119	
	900	9.331	77.423	73.291	11.461	-2.784	
	1000	9.437	78.361	74.279	4.082	-.892	
	1100	9.543	79.310	75.184	-3.392	.674	
	1200	9.650	80.270	76.019	-10.953	1.995	
	1300	9.756	81.240	76.795	-18.594	3.126	
	1400	9.863	82.221	77.522	-26.310	4.108	
	1500	9.969	83.213	78.206	-34.097	4.969	
	1600	10.075	84.215	78.853	-41.950	5.731	
	1700	10.182	85.228	79.467	-49.866	6.412	
	1800	10.288	86.251	80.052	-57.843	7.024	
	1900	10.395	87.286	80.611	-65.876	7.578	
	2000	10.501	88.330	81.147	-73.964	8.083	

$ZrBr_2$ ZIRCONIUM DIBROMIDE

		A	B	C	D	RANGE	REF.
SOL	CP	19.862	2.797	-.160		298- 900	3
LIQ	CP	21.750				900-1555	3
	** BPT=	1555.1					3
GAS	CP	14.402	0.508	0.057		1555-2000	3

PHASE	T	CP	H	S	G	BT	P	
SOL	298	20.516	-96.7	27.7	-104.959	76.947		3
	300	20.523	-96.662	27.827	-105.010	76.510		
	400	20.881	-94.591	33.782	-108.104	59.073		
	500	21.196	-92.487	38.475	-111.725	48.842		
	600	21.496	-90.353	42.367	-115.773	42.176		
	700	21.787	-88.188	45.702	-120.180	37.527		
	800	22.075	-85.995	48.630	-124.899	34.126	.007	
	900	22.360	-83.773	51.247	-129.896	31.547	.249	
			15.0	16.667				3
LIQ	900	21.750	-68.773	67.913	-129.896	31.547	.249	
	1000	21.750	-66.598	70.205	-136.803	29.902	1.819	
	1100	21.750	-64.423	72.278	-143.929	28.600	8.958	
	1200	21.750	-62.248	74.171	-151.253	27.551	32.970	
	1300	21.750	-60.073	75.911	-158.758	26.693	97.184	
	1400	21.750	-57.898	77.523	-166.431	25.985	241.025	
	1500	21.750	-55.723	79.024	-174.259	25.393	521.326	
	1555	21.750	-54.525	79.809	-178.635	25.108	760.	
			31.5	20.256				3
GAS	1555	15.194	-23.025	100.064	-178.635	25.108		
	1600	15.217	-22.342	100.497	-183.138	25.019		
	1700	15.268	-20.818	101.421	-193.234	24.845		
	1800	15.318	-19.289	102.295	-203.421	24.702		
	1900	15.369	-17.754	103.125	-213.692	24.583		
	2000	15.419	-16.215	103.915	-224.044	24.486		

ZIRCONIUM MONOSULFIDE (GAS)

		A	B	C	D	RANGE	REF.
GAS	CP	7.701	1.423			298- 800	63
	CP	8.669	0.136			800-2000	63

PHASE	T	CP	H	S	G	BT	
GAS	298	8.125	74.	59.	56.409	-41.355	63
	300	8.128	74.015	59.050	56.300	-41.020	
	400	8.270	74.835	61.408	50.272	-27.471	
	500	8.412	75.669	63.269	44.035	-19.250	
	600	8.555	76.517	64.815	37.628	-13.708	
	700	8.697	77.380	66.145	31.079	-9.705	
	800	8.839	78.257	67.315	24.405	-6.668	
	800	8.778	78.257	67.315	24.405	-6.668	
	900	8.791	79.135	68.350	17.621	-4.279	
	1000	8.805	80.015	69.277	10.738	-2.347	
	1100	8.819	80.896	70.117	3.768	-.749	
	1200	8.832	81.779	70.885	-3.283	.598	
	1300	8.846	82.663	71.592	-10.407	1.750	
	1400	8.859	83.548	72.248	-17.599	2.748	
	1500	8.873	84.435	72.860	-24.855	3.622	
	1600	8.887	85.323	73.433	-32.170	4.395	
	1700	8.900	86.212	73.972	-39.540	5.084	
	1800	8.914	87.103	74.481	-46.963	5.703	
	1900	8.927	87.995	74.963	-54.436	6.262	
	2000	8.941	88.888	75.422	-61.955	6.771	

ZrS_2 ZIRCONIUM DISULFIDE

		A	B	C	D	RANGE	REF.
SOL	CP	15.36	3.6			298-1823	*
	** HOMOGENEITY RANGE, MPT= 1823						63

PHASE	T	CP	H	S	G	BT	
SOL	298	16.433	-138.	18.7	-143.575	105.258	63
	300	16.440	-137.970	18.802	-143.610	104.634	
	400	16.800	-136.308	23.580	-145.740	79.639	
	500	17.160	-134.610	27.368	-148.294	64.828	
	600	17.520	-132.876	30.528	-151.193	55.079	
	700	17.880	-131.106	33.256	-154.385	48.208	
	800	18.240	-129.300	35.667	-157.833	43.124	
	900	18.600	-127.458	37.836	-161.510	39.225	
	1000	18.960	-125.580	39.815	-165.394	36.152	
	1100	19.320	-123.666	41.639	-169.468	33.675	
	1200	19.680	-121.716	43.335	-173.718	31.643	
	1300	20.040	-119.730	44.925	-178.132	29.951	
	1400	20.400	-117.708	46.423	-182.700	28.525	
	1500	20.760	-115.650	47.843	-187.414	27.310	
	1600	21.120	-113.556	49.194	-192.266	26.266	
	1700	21.480	-111.426	50.485	-197.250	25.362	
	1800	21.840	-109.260	51.723	-202.361	24.573	
	1823	21.923	-108.756	52.001	-203.554	24.406	

2-ZIRCONIUM SILICON

		A	B	C	D	RANGE	REF.
SOL	CP	17.3	2.63	-1.46		298-2000	*
	** MPT= 2380 (PERITEC.)						21

PHASE	T	CP	H	S	G	BT	
SOL	298	16.442	-49.8	24.	-56.956	41.755	21/*
	300	16.467	-49.770	24.102	-57.000	41.530	
	400	17.439	-48.069	28.987	-59.664	32.603	
	500	18.031	-46.294	32.946	-62.767	27.439	
	600	18.472	-44.468	36.274	-66.232	24.128	
	700	18.843	-42.602	39.150	-70.007	21.860	
	800	19.176	-40.700	41.688	-74.051	20.232	
	900	19.487	-38.767	43.965	-78.336	19.025	
	1000	19.784	-36.804	46.033	-82.837	18.106	
	1100	20.072	-34.811	47.933	-87.537	17.394	
	1200	20.355	-32.789	49.691	-92.419	16.834	
	1300	20.633	-30.740	51.331	-97.471	16.389	
	1400	20.908	-28.663	52.871	-102.682	16.031	
	1500	21.180	-26.558	54.322	-108.042	15.744	
	1600	21.451	-24.427	55.698	-113.544	15.511	
	1700	21.720	-22.268	57.007	-119.179	15.324	
	1800	21.989	-20.083	58.256	-124.943	15.172	
	1900	22.257	-17.871	59.452	-130.829	15.051	
	2000	22.523	-15.632	60.600	-136.832	14.954	

Zr_5Si_3 5-ZIRCON 3-SILICON

		A	B	C	D	RANGE	REF.
SOL	CP	45.2	7.35	-3.6		298-2000	*

PHASE	T	CP	H	S	G	BT	
SOL	298	43.342	-137.6	62.9	-156.354	114.626	21/*
	300	43.405	-137.520	63.168	-156.470	114.004	
	400	45.890	-133.043	76.032	-163.455	89.320	
	500	47.435	-128.372	86.448	-171.596	75.014	
	600	48.610	-123.568	95.204	-180.690	65.825	
	700	49.610	-118.655	102.774	-190.597	59.515	
	800	50.517	-113.649	109.458	-201.215	54.977	
	900	51.371	-108.554	115.458	-212.466	51.601	
	1000	52.190	-103.376	120.913	-224.288	49.025	
	1100	52.987	-98.116	125.925	-236.634	47.021	
	1200	53.770	-92.779	130.569	-249.461	45.439	
	1300	54.542	-87.363	134.903	-262.737	44.176	
	1400	55.306	-81.870	138.973	-276.433	43.159	
	1500	56.065	-76.302	142.815	-290.524	42.335	
	1600	56.819	-70.658	146.457	-304.989	41.665	
	1700	57.570	-64.938	149.925	-319.810	41.120	
	1800	58.319	-59.144	153.236	-334.969	40.676	
	1900	59.065	-53.274	156.410	-350.452	40.317	
	2000	59.810	-47.331	159.458	-366.247	40.027	

ZIRCON MONOSILICON

ZrSi

		A	B	C	D	RANGE	REF.
SOL	CP	10.8	2.1	-0.71		298-2000	*
	** MPT= 2380 (PERITEC.)						21

PHASE	T	CP	H	S	G	BT	
SOL	298	10.627	-37.	13.9	-41.144	30.164	21/*
	300	10.641	-36.980	13.966	-41.170	29.996	
	400	11.196	-35.886	17.110	-42.730	23.350	
	500	11.566	-34.747	19.650	-44.572	19.485	
	600	11.863	-33.575	21.786	-46.647	16.993	
	700	12.125	-32.376	23.635	-48.920	15.275	
	800	12.369	-31.151	25.270	-51.367	14.035	
	900	12.602	-29.902	26.740	-53.968	13.107	
	1000	12.829	-28.630	28.080	-56.710	12.396	
	1100	13.051	-27.336	29.313	-59.581	11.839	
	1200	13.271	-26.020	30.458	-62.570	11.397	
	1300	13.488	-24.682	31.529	-65.670	11.042	
	1400	13.704	-23.323	32.536	-68.874	10.753	
	1500	13.918	-21.942	33.489	-72.175	10.517	
	1600	14.132	-20.539	34.394	-75.570	10.324	
	1700	14.345	-19.115	35.257	-79.053	10.164	
	1800	14.558	-17.670	36.083	-82.620	10.033	
	1900	14.770	-16.204	36.876	-86.268	9.924	
	2000	14.982	-14.716	37.639	-89.994	9.835	

$ZrSi_2$ ZIRCONIUM 2-SILICON

		A	B	C	D	RANGE	REF.
SOL	CP	15.1	3.67	-0.668		298-1790	*
	** MPT= 1790 (PERITEC.)						21

PHASE	T	CP	H	S	G	BT	
SOL	298	15.443	-38.1	17.1	-43.198	31.670	21/*
	300	15.459	-38.071	17.196	-43.230	31.497	
	400	16.150	-36.489	21.744	-45.186	24.692	
	500	16.668	-34.847	25.406	-47.550	20.787	
	600	17.116	-33.157	28.485	-50.248	18.305	
	700	17.533	-31.425	31.155	-53.233	16.622	
	800	17.932	-29.651	33.522	-56.469	15.429	
	900	18.321	-27.839	35.657	-59.930	14.555	
	1000	18.703	-25.987	37.607	-63.594	13.900	
	1100	19.082	-24.098	39.407	-67.446	13.402	
	1200	19.458	-22.171	41.084	-71.472	13.019	
	1300	19.831	-20.207	42.656	-75.659	12.721	
	1400	20.204	-18.205	44.139	-80.000	12.490	
	1500	20.575	-16.166	45.546	-84.485	12.311	
	1600	20.946	-14.090	46.886	-89.107	12.173	
	1700	21.316	-11.977	48.166	-93.860	12.068	
	1790	21.648	-10.043	49.275	-98.245	11.997	

ZIRCONIUM ORTHOSILICATE $ZrO_2 \cdot SiO_2$

		A	B	C	D	RANGE	REF.
SOL	CP	28.353	11.450	-7.115	-3.998	298-1400	3
	CP	36.0				1400-1980	3
	** DPT= 1980 (APPROX.)						*

PHASE	T	CP	H	S	G	BT	
SOL	298	23.407	-485.47	20.1	-491.463	360.300	3,5
	300	23.523	-485.427	20.245	-491.500	358.106	
	400	27.846	-482.833	27.678	-493.904	269.893	
	500	30.232	-479.919	34.169	-497.004	217.269	
	600	31.807	-476.813	39.829	-500.710	182.408	
	700	32.957	-473.572	44.822	-504.947	157.673	
	800	33.843	-470.230	49.283	-509.657	139.250	
	900	34.541	-466.809	53.311	-514.790	125.025	
	1000	35.093	-463.327	56.980	-520.307	113.728	
	1100	35.522	-459.795	60.346	-526.176	104.555	
	1200	35.842	-456.226	63.451	-532.367	96.970	
	1300	36.060	-452.630	66.330	-538.858	90.602	
	1400	36.184	-449.017	69.007	-545.627	85.188	
	1400	36.000	-449.017	69.007	-545.627	85.188	
	1500	36.000	-445.417	71.491	-552.653	80.532	
	1600	36.000	-441.817	73.814	-559.919	76.492	
	1700	36.000	-438.217	75.997	-567.411	72.955	
	1800	36.000	-434.617	78.054	-575.115	69.838	
	1900	36.000	-431.017	80.001	-583.018	67.071	
	1980	36.000	-428.137	81.485	-589.478	65.075	

MIX
Papier aus verantwortungsvollen Quellen
Paper from responsible sources
FSC® C105338

If you have any concerns about our products,
you can contact us on
ProductSafety@springernature.com

In case Publisher is established outside the EU,
the EU authorized representative is:
Springer Nature Customer Service Center GmbH
Europaplatz 3, 69115 Heidelberg, Germany

Printed by Libri Plureos GmbH
in Hamburg, Germany